D0538470

Physical Geography Today

A Portrait of a Planet

Cover

The varied and complex interactions of all the earth's systems—weather, climate, vegetation, soil, oceans, rivers, and many more—combine to produce the physical features of the world we see around us. *Physical Geography Today: A Portrait of a Planet* examines each of these systems and their interrelationships to present an integrated picture of our planet.

Cotopaxi by Frederic Edwin Church. (On loan to the Metropolitan Museum of Art from John Astor)

Principal Academic Advisor

Robert A. Muller, Ph.D.,
Louisiana State University

Principal Contributor

Robert J. Kolenkow, Ph.D.,
Del Mar, California

Contributors

Reid A. Bryson, Ph.D.,
University of Wisconsin, Madison

Douglas B. Carter, Ph.D.,
Southern Illinois University

R. Keith Julian, Ph.D. Candidate,
University of California, Los Angeles

Theodore M. Oberlander, Ph.D.,
University of California, Berkeley

Robert P. Sharp, Ph.D.,
California Institute of Technology

M. Gordon Wolman, Ph.D.,
The Johns Hopkins University

Biographical sketches of the contributors and attributions of their work in this book are found on page 480.

Physical Geography Today

A Portrait of a Planet

CRM BOOKS
Del Mar, California

Published by CRM Books, a division of Ziff-Davis Publishing Company, Del Mar, California 92014.

Library of Congress Catalog Card Number: 73-86479
Manufactured in the United States of America.

9 8 7 6 5 4 3 2

Preface

There are many ways to perceive the world—the painter has a point of view, as does the poet, the architect, the physicist, and the geographer. What is the perspective of the physical geographer? The group of people who came together to plan and write this book attempted to answer that question.

At first glance physical geography may appear to be merely a composite view of several disciplines—of geology, meteorology, climatology, biology, pedology, oceanography. An introductory course may touch on each of these fields, but physical geography is more than a synthesis of other sciences. And it is more than an inventory of global distributions of physical features. This book presents the view that physical geography is both a description of *what is* at the surface of the earth and an explanation of *how* and *why* physical processes have acted to shape the earth.

The scope of physical geography stretches from the atmosphere through the lithosphere and the hydrosphere. In order to stress the global and regional relationships of physical systems, this text focuses on that thin surface layer where the land, air, and water meet and interact and where life is sustained. The book takes the point of view that basic principles and processes govern the operation of the world. The organizing thread that ties this concept to physical reality is that all systems on the earth—weather, climate, vegetation, soil, oceans, rivers, and so on—share energy and materials. Energy continually flows through the systems. Materials such as water also flow through systems in never-ending cycles that involve inputs and outputs, storage and equilibrium. And the systems interact and respond to one another: when one system gains energy or moisture, others must compensate.

The book begins with a discussion of the earth's evolution, which places the planet in a perspective of time and space. The text then goes on to emphasize that systems continue to evolve and change in response to inputs of energy from the sun and from the earth's interior. The principles of energy and moisture exchange that are important in precipitation processes, in soil formation, and in the growth of plants are also powerful activants in land-shaping processes of weathering, mass wasting, and erosion. The interdependent operation of all these systems occurs in the arena where man lives. Man is seen as a part of these environmental systems, as both an agent of environmental change and a recipient of change.

Although the book takes a somewhat novel approach to the field, it does not ignore the descriptions and distributions that make up the core of a physical geography course. After the first two introductory chapters, succeeding chapters discuss the interaction of radiant energy with temperature, with precipitation processes, and with the general atmospheric circulation that delivers moisture to the surface. The local water budget is introduced early in the text as a useful tool in hydrology and as an index of the effect of climate on vegetation and, indirectly, on soils. Classification systems are described with a discussion of their uses and limitations; the regional distributions of climate, soils, and vegetation are given on maps as well as in the text. The chapter on ecologic energetics traces the energy flow through ecosystems and discusses the complex interactions that affect food chains and world food productivity.

The second half of the text treats landforms and their response to forces that involve the earth's internal energy and the shaping effects associated with the sun's energy. The processes involved in the formation of landforms are presented along with descriptions and examples of the features themselves. The applicability of principles to the physical environment, which is emphasized throughout the text, is summarized in the Epilogue with regional examples from the United States.

Appendix II provides an extensive discussion of the tools of geography, including maps and map projections, map-reading, field techniques, and remote sensing. This material is a reference that can be adapted to the needs of a particular course. The glossary, which clarifies technical terms, also serves as a reference.

The illustrations are closely correlated with the text and, together with the captions, form an integral part of the material presented. Technical data are presented graphically, for the most part, and schematic diagrams are used to clarify and summarize important concepts. A large number of maps especially prepared for this book present regional and global distributions. The metric system with English equivalents is used to express numerical data in the text and in the graphs.

Man's response to the environment and his interaction with it are brought out throughout the text. Examples of the challenges and problems in physical geography are given in the timely case studies and the graphic essays.

Physical Geography Today: A Portrait of a Planet is designed to portray the earth as it is and as it is becoming. Throughout the book the reader is invited to consider several perspectives for viewing the world. The climatologists, geomorphologists, and environmental scientists who came together to write this book represented a number of points of view. Although they may not have always seen details in the same way, they shared a commitment to present a dynamic picture of an evolving earth. The result is this book: an integrated effort by scholars who are, as well, scientists, educators, and friends.

Overview

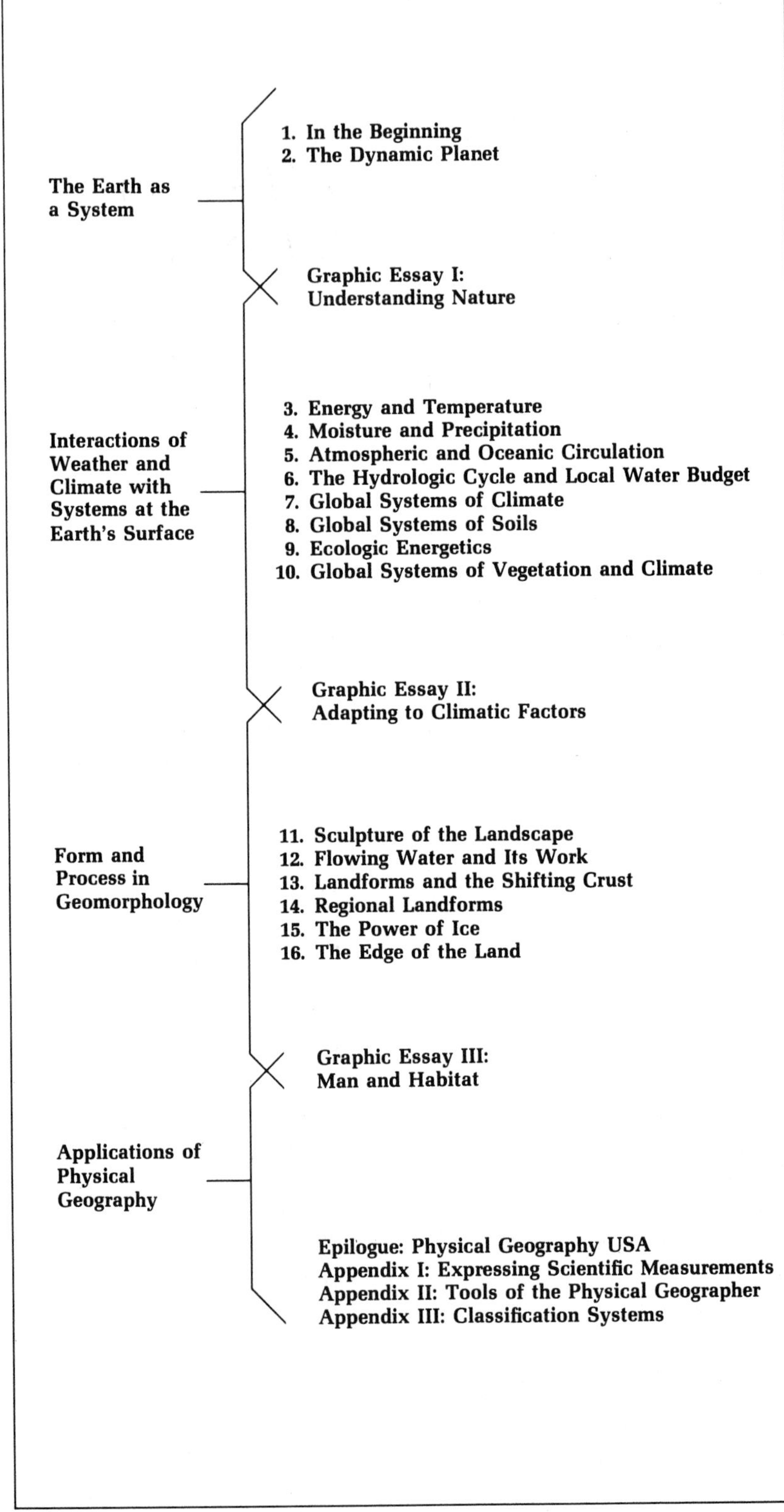
1. In the Beginning
2. The Dynamic Planet
The Earth as a System
Graphic Essay I: Understanding Nature
3. Energy and Temperature
4. Moisture and Precipitation
5. Atmospheric and Oceanic Circulation
6. The Hydrologic Cycle and Local Water Budget
7. Global Systems of Climate
8. Global Systems of Soils
9. Ecologic Energetics
10. Global Systems of Vegetation and Climate
Interactions of Weather and Climate with Systems at the Earth's Surface
Graphic Essay II: Adapting to Climatic Factors
11. Sculpture of the Landscape
12. Flowing Water and Its Work
13. Landforms and the Shifting Crust
14. Regional Landforms
15. The Power of Ice
16. The Edge of the Land
Form and Process in Geomorphology
Graphic Essay III: Man and Habitat
Applications of Physical Geography
Epilogue: Physical Geography USA
Appendix I: Expressing Scientific Measurements
Appendix II: Tools of the Physical Geographer
Appendix III: Classification Systems

Contents

1 In the Beginning | 1

The Changing Earth and the Time Scale of Change 2
Formation of the Universe 4
Formation of the Solar System 5
Structure of the Earth 6
Evolution of the Earth's Atmosphere and Landforms 10
The Early Atmosphere 10
The Early Landforms 11
Appearance and Development of Life 12
The Fossil Record 13
Organic Evolution 14
Interaction of Life With the Environment 15
Life and the Atmosphere 18
Life and the Landscape 18
Man's Place on Earth 19
Continental Drift and Sea Floor Spreading 19
The Scale of Time 22
Radioactivity and Time 23
Dating the Earth 23
Carbon 14 Dating 24
Other Dating Methods 24
Summary 24

2 The Dynamic Planet | 27

Units of Measure: The Metric System 29
Forms of Energy 30
Energy and Work 30
Heat Energy 31
Units of Energy 34
Water: An Energy Converter 34
Changes of Phase and Changes of Energy 35
Physical and Chemical Properties 36
The Atmosphere: An Energy Transporter 37
Systems: Energy Exchange 38
Budgets of Energy and Matter 39
Cycles of Materials: Synergy 40
The Hydrologic Cycle 40
Man's Input to Cycles 41
The Classification of Systems 41
Summary 43
Graphic Essay I: Understanding Nature 44

3 Energy and Temperature | 47

The Earth-Sun Relationship 48
The Earth's Tilted Axis 48
Solar Altitude and Azimuth 52
Solar Energy Input to the Earth 53
Total Daily Solar Radiation at the Top of the Atmosphere 55
The Electromagnetic Spectrum and the Quality of Sunlight 55
Interaction of Radiation With the Atmosphere 57
Energy Balance in the Earth-Atmosphere System 59
Reflection and Albedo 59
Energy Loss and the Atmosphere 62
Radiant Energy Balance 63
Conduction, Turbulent Transfer, and Evapotranspiration 64
Energy Balance at the Earth's Surface 65
Heating and Cooling of Land and Water 66
Temperature Regimes 67
Modification of the Energy Balance 70
Summary 74

4 Moisture and Precipitation | 77

Transfer of Water to the Atmosphere 80
Saturation Vapor Pressure and Evaporation Rate 81
Evapotranspiration 82
Moisture in the Atmosphere 83
Absolute, Relative, and Specific Humidity 83
Humidity and Temperature 84
Formation of Dew 85
Vertical Distribution of Water Vapor 86
Condensation Processes 87
The Coalescence Model of Rainfall 88
The Bergeron Ice Crystal Model of Rainfall 90
Formation of Hail 91
Formation of Fog 93
Formation of Clouds 93
Adiabatic Processes 96
Rainmaking: Artificial Condensation Processes 99
Modification of Harmful Weather Conditions 101
Summary 102

5 Atmospheric and Oceanic Circulation | 105

Forces and Atmospheric Motion 107
The Force of Friction 107
The Pressure Gradient Force 107
The Coriolis Force 108
The Geostrophic Wind 108
General Circulation of the Atmosphere 109
Winds and Pressure on a Uniform Earth 109
Observed Pressure and Wind Patterns 111
Circulation in the Upper Atmosphere 114
Midlatitude Secondary Circulations 114
Forces in Cyclones and Anticyclones 115
Air Masses 115
Weather Fronts 117
Jet Streams—The Atmospheric Vacuum Cleaner 120
Tornadoes 120
Tropical Secondary Circulations 120
Monsoons 122
Easterly Waves 124
Tropical Cyclones 124
General Circulation of the Ocean 126
The Gulf Stream 127
Case Study: Expanding Horizons in Atmospheric and Oceanic Research 128
Upwelling 130
The Deep Circulation 130
Summary 130

6 The Hydrologic Cycle and Local Water Budget | 133

Water Use and Water Management 134
Water on the Earth 135
Principles of Hydrology 138
Interception, Throughfall, and Stemflow 138
Infiltration and Soil Moisture 140
Groundwater and Aquifers 141
Runoff and Streamflow 144
The Local Water Budget: An Accounting Scheme for Water 145
Case Study: The California Water Plan 146
Potential and Actual Evapotranspiration 148
Measuring Evapotranspiration and Potential Evapotranspiration 149
The Water Balance Equation 150
A Local Water Budget: Baton Rouge 150
Improving the Water Budget Model 155
Summary 156

7
Global Systems of Climate | 159

Classification of Climate 160
Measures of Climate 161
Time and Space Scales of Climate 161
One View of Climate: Dynamic Global Delivery System 163
Distribution of Climatic Regions on a Hypothetical Continent 163
Distribution of Climatic Regions on the Earth 166
Another View of Climate: System Interactions at the Surface 166
Climate and the Water Budget 167
Distribution of Energy and Moisture 170
Three Functional Realms of Climate 172
The Köppen System of Climate Classification 172
The Thornthwaite System of Climate Classification 177
Summary 178

8
Global Systems of Soils | 181

Properties of a Soil 182
Texture and Structure 183
Chemistry of Solutions 184
Chemistry of Soils 185
Horizons and Profiles 185
Weathering: The Disintegration of Rocks 186
Physical Weathering 186
Chemical Weathering 187
Factors Affecting Soil Development 187
Inorganic Material 187
Topography 188
Climate 189
Organisms 189
Rate of Formation 189
Formation of Soil 190
The Leaching Process 190
Climate and Soils 190
Classification of Soil 191
Examples of Principal Soil Types 194
Tundra Soils 195
Podzols 196
Gray-Brown Podzolic Soils 196
Red-Yellow Podzolic Soils 196
Latosols 197
Chernozems 198
Prairie Soils 198
Chestnut-Brown Soils 198
Desert Soils 199
Bog and Meadow Soils 202
Soil Management 202
Summary 203

9
Ecologic Energetics | 205

Energy Use by Green Plants: Photosynthesis and Food Production 207
Plant Respiration and Net Production 210
Productivity Measures 210
Climate and Crop Growth 212
Potential Photosynthesis and Crop Productivity 213
Energy Flow Through Ecosystems 215
Food Chains and Food Webs 215
Laboratory Ecosystems 217
Energy Transfer Along Food Chains 217
Studies of Natural Ecosystems 218
Agricultural Ecosystems 220
Problems of World Agriculture 221
Ecosystems and Man 223
Case Study: The Tragedy of the Whale 224
Summary 227

10
Global Systems of Vegetation and Climate | 229

Vegetation and Climate 230
Plant Communities 231
Plant Succession and Climax 232
Natural Vegetation 233
Principal Vegetation Regions and Their Climates 233
Equatorial Rainforests 235
Tropical Seasonal Forests 237
Tropical Savannas 237
Tropical Thorn and Scrub Forests 238
Subtropical Semidesert and Desert Regions 239
Midlatitude Shrublands 240
Midlatitude Deciduous Forests 241
Midlatitude Grasslands 242
Subarctic Coniferous Forests 244
Tundra Vegetation 245
Changes in Climate 247
Case Study: Slash-and-Burn Agriculture: An Efficient or Destructive Land Use? 248
Climate of the Early Holocene 250
Climate in Historic Times 251
Theories of Climatic Change 252
Climates of the Future 253
Summary 253
Graphic Essay II: Adapting to Climatic Factors 254

11
Sculpture of the Landscape | 259

Orders of Relief 261
First Order of Relief: Continents and Ocean Basins 261
Second Order of Relief: Mountains, Oceanic Ridges, and Plateaus 264
Third Order of Relief: Hills and Valleys 267
The Differentiation of Landforms 268
Geologic Movement 268
Erosion and Deposition 269
Diversity of Slopes 272
Evolution of Landforms 274
The Cycle of Erosion 275
Relict Landforms 278
The Tempo of Geomorphic Change 279
Summary 280

12
Flowing Water and Its Work | 283

Energy of Moving Water 284
Principles of Water Flow 285
Turbulence 287
Transport of Material by Streams 287
Scour and Fill 289
The Geometry of Stream Drainage 291
Anatomy of a River 292
Stream Channels: Form and Process 295
Vertical Adjustments 297
Slopes in Stream Beds 298
Graded Streams 298
Knickpoints 298
Lateral Migration of Stream Channels 299
Landforms Associated With Rivers 302
Floodplains 302
Terraces 304
Underfit Streams 304
Deltas 305
Alluvial Fans 305
Summary 306
Case Study: The Flood of '73 308

13
Landforms and the Shifting Crust | 311

The Structural Foundation 314
Igneous Rock 315
Sedimentary Rock 315
Metamorphic Rock 318
Differentially Eroded Rock Structures 318
Volcanism: The Production of New Igneous Rock 325
Geothermal Energy 328
The Source of Volcanic Energy 330
Ejected Materials and Volcanic Hazards 330
Landforms Associated With Crustal Motion 336
Faults and Folds 337
Landforms and Faulted Structures 339
Landforms and Folded Sedimentary Rocks 343
Summary 345

14
Regional Landforms | 347

Climate and Landforms 348
Climatic Regions and Weathering Processes 349
Vegetation and Erosion 351
Slopes and Their Formation 352
Forces That Hold Slopes Together 352
Creep: Slippage of Slope Material 354
Soil Wash and Rainsplash 356
Local and Regional Climatic Differentiation 357
Hoodoo Rocks 357
Asymmetry of Slopes 357
Morphogenetic Regions 358
Slope Processes in Different Climates 360
Slopes in Periglacial Climates 360
Slopes in Moist Midlatitude Regions 360
Slopes in Arid and Semiarid Regions 360
Slopes in Hot, Moist Climates 361
A Gallery of Landscapes 362
Summary 373

15
The Power of Ice | 375

Glaciation: Present and Past 378
Glaciation of the Pleistocene 378
Indirect Effects of Pleistocene Glaciation 379
Anatomy and Dynamics of Glaciers 380
The Making of Glacial Ice 381
The Glacier Budget 382
Movement of Glaciers 384
The Work of Glaciers on the Landscape 385
Mechanisms of Glacial Erosion 386
Glacial Erosion and Landforms 387
Glacial Deposition and Landforms 391
Glaciation and Man 392
Summary 392

16
The Edge of the Land | 395

The Continental Shelves 397
The Geomorphic Significance of Tides 399
Cause of the Tides 399
Estuaries: Gift From the Sea 401
Wave Motion 401
Wave Action Against the Land 403
The Work of Waves 404
Tsunamis 405
Beaches 406
Sediment Transport and Deposition 406
Beach Erosion and Man 408
Relief Forms of the Ocean Floor 409
Summary 414
Graphic Essay III: Man and Habitat 416

Epilogue: Physical Geography USA | 421

The Eastern Seaboard: Boston, Massachusetts 422
Land's Edge: Cape Cod 423
First Stop: Niagara Falls 425
Rolling Farmland: The Midwest 430
The Land of the Big Sky: The Great Plains 432
Mudstone Mountains: The Badlands of South Dakota 433
Devil's Tower: A Wyoming Monument 435
The Continental Divide: The Rocky Mountains 436
Arches, Pillars, Balancing Rocks: The Southwest 439
A Man-made Garden: The Imperial Valley 441
Journey's End: California 443

Appendix I: Expressing Scientific Measurements | 444

Scientific Notation: "Powers of Ten" 444
Significant Figures 444
Metric to English Conversions 445

Appendix II: Tools of the Physical Geographer | 446

Maps: A Representation of the Earth's Surface 446
Scale and Distance 447
Location 449
Direction 451
The Shape of the Earth 452
Map Projections 452
Projections and Distortions 453
Azimuthal Projections 454
Cylindric Projections 454
Conic Projections 454
Pseudocylindric Global Projections 454
Interruption and Condensation of Projections 459
Simplification in Maps 459
Symbolization: The Representation of Relief 459
Mapping in the United States 462
USGS Topographic Maps 464
Using USGS Topographic Maps 465
Field Work and Mapping 468
An Approach to Field Work 468
Mapping Techniques 468
Remote Sensing 470

Appendix III: Classification Systems | 474

Köppen System of Climate Classification 474
Principal Subdivisions of A Regions 475
Principal Subdivisions of BS and BW Regions 475
Principal Subdivisions of C and D Regions 475
Thornthwaite's Formula for Potential Evapotranspiration 476
Soil Classification Systems 477
The 1938 Soil Classification System 477
The 7th Approximation 478
Morphogenetic Regions 479

Contributors 480
Bibliography and Suggested Readings 482
Glossary—Index 486
Credits and Acknowledgments 514

1 In the Beginning

The Changing Earth and the Time Scale of Change

Formation of the Universe
Formation of the Solar System
Structure of the Earth

Evolution of the Earth's Atmosphere and Landforms
The Early Atmosphere
The Early Landforms

Appearance and Development of Life
The Fossil Record
Organic Evolution

Interaction of Life With the Environment
Life and the Atmosphere
Life and the Landscape
Man's Place on Earth

Continental Drift and Sea Floor Spreading

The Scale of Time
Radioactivity and Time
Dating the Earth
Carbon 14 Dating
Other Dating Methods

A central "eye" evokes looking back and looking forward. From a vortex of gases, the earth evolved into the planet we know today. And the earth continues to evolve because of the dynamic interplay of energy with the air, land, and water. Physical geography is the study of the changes and interactions at the earth's surface.

The Psychology of Children's Art by Rhoda Kellogg and Scott O'Dell. (Copyright © 1967 CRM, Inc.)

In the Beginning

There is such a reassuring familiarity in the natural features of the earth—clouds and trees, hills and seacoasts—that you may take them for granted. But the earth's surface is the scene of ceaseless change. You can easily observe some of the changes—the sun warming the earth, clear skies coming after a rainstorm, plant growth slowing as winter approaches. Less obvious changes are the gradual eroding of a hillside by wind and water and the wearing away of a seaside cliff by pounding waves. Change is the essence of nature.

The thin layer of rock, water, and air on the surface of the earth is man's natural home. It is also the dynamic arena in which atmosphere, rock, soil, water, and vegetation interact under the driving energy of the sun and of the earth's internal heat engine. The primary concern of physical geography is with man's environment at the surface of the earth and with the forces that shape that environment. The discipline of physical geography embraces a wide range of natural phenomena, from the circulation of the upper atmosphere to the chemical processes that help form soils. In addition, physical geography focuses on interactions—the interaction of vegetation and climate, the interaction of radiant energy with the atmosphere, the interaction of water with rock and soil. The emphasis on physical processes and principles provides a unity to the study of the earth's surface that will give you a greater understanding of the land, water, and air around you.

Physical geography is also concerned with the interactions between man and the earth. Man depends on the earth for food and materials, and in turn, his agriculture and industry cause changes on the earth. The scale and tempo of human activities are steadily increasing in response to increasing needs for food, water, and power, and as a result, man's impact on the environment is becoming greater. Some of these activities have produced unexpected and occasionally undesirable changes. A knowledge of the interrelatedness of physical processes can supply a useful viewpoint for understanding how man's activities affect the earth.

The Changing Earth and the Time Scale of Change

The earth is a dynamic planet, and its present state represents the results of changes that have accumulated over long periods of time. This chapter places the earth in the time scale of geologic and evolutionary processes. Although most of this book discusses the forces that shape man's environment today, a look at the earth's history places the present in a framework of ongoing change.

Physical, chemical, and biological processes are steadily at work transforming the earth's surface. Some of the transformations are visible on the human time scale of a few years or decades: new volcanoes form, beaches come and go, shallow ponds become filled with vegetation. Other processes work so slowly that their consequences become apparent only over immense spans of time. There

Figure 1.2 (opposite) To describe the surface of the earth, physical geography draws upon the specialized knowledge of many disciplines. The unique contribution of physical geography is its concern with the interactions among different parts of the earth—how the earth's relation to the sun influences climate, for example, and how climate in turn affects the development of systems such as vegetation and soils.

Climatology
Meteorology
Hydrology
Pedology
Plant
Geography
Ecology
Astronomy
Geology
Geodesy
Geomorphology
Cartography
Physical Geography
Physical
Oceanography

is evidence, for instance, that the mountain ranges along the western coast of the United States have been lifted up from sea level during the last 1 or 2 million years and that they continue to rise today. Even the familiar patterns of the continents and oceans change with time. Recent research indicates that all of the earth's landmasses were grouped together in one or two supercontinents 250 million years ago. The gigantic landmasses fragmented into separate continents, and the fragments have since been drifting imperceptibly across the earth's surface. According to the theory of continental drift, the Atlantic Ocean basin is a relatively new feature that is opening up as North and South America gradually drift apart from Europe and Africa.

The time scale of geologic change goes back billions of years to the time when the earth was formed. The time scale of biologic evolution occupies millions and hundreds of millions of years. The life span of an individual plant or animal represents a fraction of an instant in such vast reaches of time. About a billion years ago, the only life forms on the earth were simple plants in the sea. More complex forms of life evolved in response to the stresses of particular environments. And life, in turn, acted on the environment, introducing new processes of change that redirected the course of the earth's development.

Figure 1.3 This is an artist's conception of the formation of the universe. According to the "big bang" hypothesis, all of the matter and energy in the universe were compressed together in a "cosmic egg." Heated by energy, the egg expanded rapidly outward, and in the process the chemical elements were formed from undifferentiated matter. The expansion of the universe is continuing today, but perhaps far in the future the universe will contract to form a cosmic egg once more.

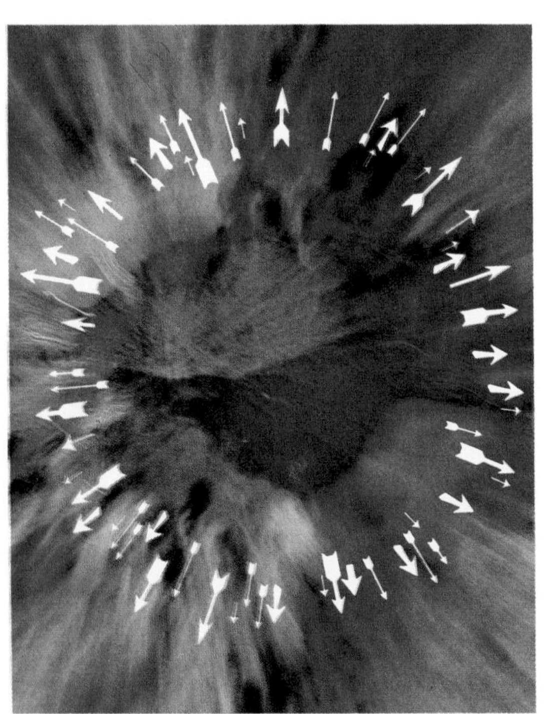

Because the written historical record extends only a few thousand years into the past, knowledge of most of the earth's long history has come by inference. Although it is true that large portions of the earth's history still remain subjects for conjecture because only a few remnants of early events are available for study, the scientific process permits us a degree of certainty that some speculations are correct. The underlying assumption for the reconstruction of the earth's history is the belief that processes of change in the past operated according to the fundamental principles of physical and biological science known today.

The great changes that have occurred in the earth's life forms and landforms through the course of time have left a variety of traces, including sediment layers, many kinds of rocks, fossilized plants and animals, radioactive decay products, and ancient magnetism in rocks. With careful observation, experimentation, and scientific reasoning, it has been possible to use such clues to reconstruct a plausible account of some of the major events in the history of the earth.

Formation of the Universe

Throughout recorded history, men have wondered about the origin of the universe. Many peoples in many lands have developed creation stories that assert that the universe did not always exist but was created at a definite point in time. Present scientific knowledge leads to a similar conclusion: there once was a time when all the matter and energy now in the universe was compressed into a single "cosmic egg." The energy soon heated the egg to a temperature of millions of degrees, turning it into a fireball hot enough to convert the primeval matter into various chemical elements by a series of nuclear reactions. The sudden generation of intense heat during this "big bang" caused the primordial fireball to begin expanding and cooling. With the expansion and the lower temperature, the initial period of element formation came to an end. Because of the many reaction steps required to form the heavier elements, about 99 percent of the matter was left in the form of hydrogen or helium, the two lightest elements.

The expansion of the universe appears to be continuing today, according to observations of light the earth receives from distant galaxies. By extrapolating the measured rate of expansion backward in time, scientists have estimated that all the matter in the universe was gathered together about 10 billion years ago, or 10×10^9 years ago using "powers-of-ten" notation. (This system of notation, commonly used in science to express very large and very small numbers, is illustrated in Figure 1.4 and explained in Appendix I.)

The glowing fireball would have given off copious amounts of energy in the form of light waves and radio waves. If the big bang hypothesis is correct, the degraded remnants of this radiation should still be present everywhere in the universe. The recent detection of radiation that has the expected properties gives powerful support to the big bang hypothesis.

It is difficult to imagine the passage of 10×10^9 years since the universe was formed. To put the time span into perspective, use the analogy of a 24-hour day and imagine that the beginning of the universe occurred at midnight 24 hours ago. On this scale, each second of the 24-hour day corresponds to about 100,000 years of actual time.

For a long period of time after the big bang occurred, there were no stars or planets; only clouds of separate atoms whirled through space. Urged by mutual gravitational attraction, the atoms slowly began to amass into dense clouds. These centers of aggregation tended to increase in size by sweeping up atoms from space. Atoms gained speed as the gravitational pull of a massive cloud drew them in, and when atoms collided with one another, their energy was converted to heat. The largest clouds reached temperatures high enough to initiate self-sustaining nuclear reactions, and these clouds began to shine as stars. Our sun was among them. Studies of the sun's energy-generating processes indicate that its active life probably began about 6×10^9 years ago, or about 10 hours after the formation of the universe in our analogy.

$$100 = 10 \times 10 = 10^2$$
$$1{,}000 = 10 \times 10 \times 10 = 10^3$$
$$10{,}000 = 10 \times 10 \times 10 \times 10 = 10^4$$
$$100{,}000 = 10 \times 10 \times 10 \times 10 \times 10 = 10^5$$
$$1{,}000{,}000 = 10 \times 10 \times 10 \times 10 \times 10 \times 10 = 10^6$$
$$1{,}000{,}000{,}000 = 10^9$$
$$2{,}000 = 2 \times 1{,}000 = 2 \times 10^3$$
$$2{,}000{,}000{,}000 = 2 \times 10^9$$
$$0.01 = \frac{1}{100} = \frac{1}{10^2} = 10^{-2}$$
$$0.001 = \frac{1}{1{,}000} = \frac{1}{10^3} = 10^{-3}$$
$$0.0000001 = 10^{-7}$$
$$0.005 = \frac{5}{1{,}000} = 5 \times 10^{-3}$$

Figure 1.4 In science, numbers larger than 1,000 or so and numbers smaller than 1/10 or so are usually expressed in "powers of ten," or what is called the *exponential notation.* When two powers of ten are multiplied, the exponents are added to find the exponent of the product: $10^2 \times 10^4 = 10^6$. When two powers of ten are divided, the exponent of the divisor is subtracted from the exponent of the dividend: $10^8/10^2 = 10^6$.

Formation of the Solar System

Numerous models have been devised to account for the origin of the planets, but none agrees fully with all known facts and principles. The chemist Harold Urey, an American Nobel Prize winner, presented a picture in which a massive swirling cloud of matter condensed at its center to form the sun. The outer portions of the cloud then broke up into eddies, which contracted into swarms of bodies called *planetesimals.* The planetesimals were smaller than the present planets, but some may have been as large as the moon. Later, swarms of planetesimals came together under gravitational

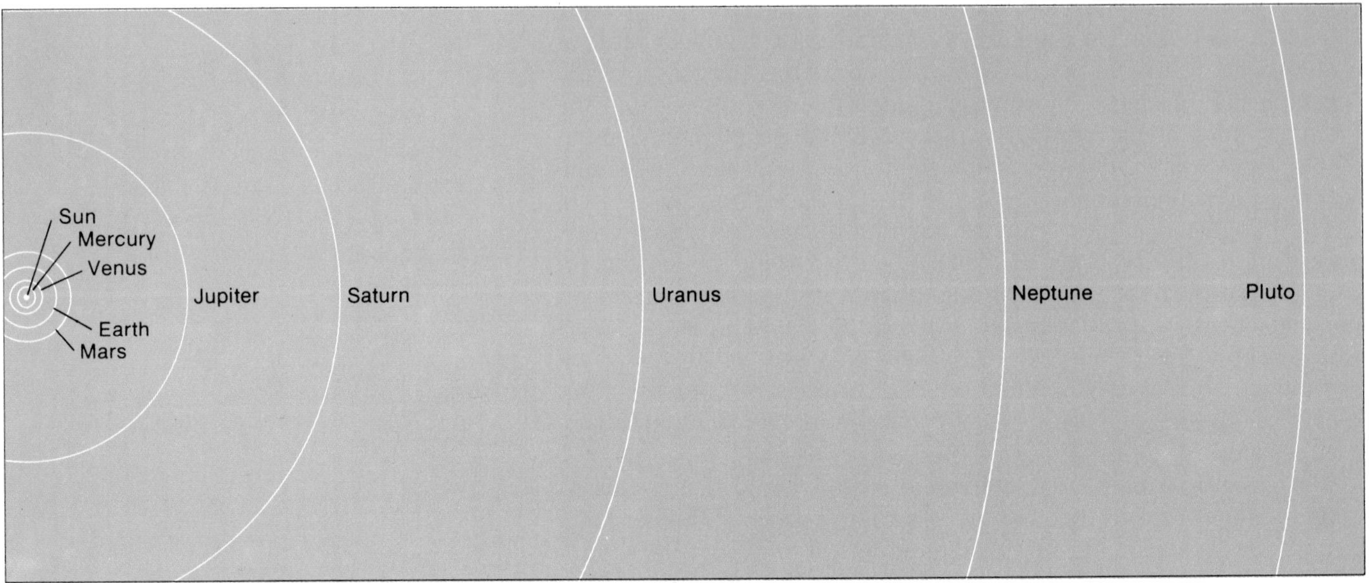
Sun
Mercury
Venus
Earth
Mars
Jupiter
Saturn
Uranus
Neptune
Pluto

Figure 1.5 This diagram shows the distances of the planets from the sun in correct scale. The planets themselves are not shown because their size would be disproportionate to the scale of their orbits. Between the orbits of Mars and Jupiter are swarms of asteroids, or minor planets, which may be fragments of earlier planets. The chunks of matter that form the asteroid belt range from dust particles to one lump that is as large as the British Isles.

The orbits of the planets lie nearly in the same plane, and every planet revolves about the sun in the same direction, which is consistent with the hypothesis that planetary bodies originally condensed from a whirling disk of gas. The inner planets Mercury, Venus, Earth, and Mars, together with Pluto, are denser than the other planets because they are too small to have retained much of the hydrogen that was present when the solar system was formed.

Figure 1.6 (opposite) These twenty-five images represent an interpretation of the earth's 5-billion-year history as it might have been seen by extraterrestrial observers who visited the earth every 200 million years. Significant events in the earth's history include the appearance of life and the subsequent formation of an oxygenated atmosphere. The evolution of higher forms of life occupies only the time period spanned by the last three images.

attraction and formed planets. Urey has suggested that the moon might be a planetesimal trapped in the gravitational pull of the earth. Evidence from radioactive dating, which is discussed later in this chapter, places the time of the earth's formation at about 5×10^9 years ago, which would be about noon, halfway through our hypothetical day.

The original cloud of matter that formed the solar system had a swirling, or rotational, motion that was imparted to the planets. Two kinds of rotational motion occur in the solar system: the planets revolve about the sun in approximately circular paths, and many of them, including the earth, rotate about their own axes. The rotation imparted to the earth at its formation has important consequences for the distribution of incoming solar energy over the earth's surface and for the development of weather systems. As a matter of chance, the earth's rotational axis was initially set at a considerable angle to its orbital motion; how this tilt influences solar energy distribution to the earth and how it determines the seasons is discussed in Chapter 3.

Structure of the Earth

It is believed that earth materials accumulated at comparatively low temperatures and that the earth was never fully molten. In Urey's model, the final slow build-up of the earth from colliding planetesimals would not have heated the whole planet significantly. The earth's interior slowly became hotter after its formation because of the energy released from radioactive elements that were incorporated into minerals. *Radioactivity* arises from instability in the central nucleus of an atom, where most of the atom's mass is concentrated. The instability causes the nucleus to change form spontaneously, releasing energy. When the interior of the earth became hot, materials flowed and diffused from place to place. Lighter materials rose toward the earth's surface and heavier substances sank toward

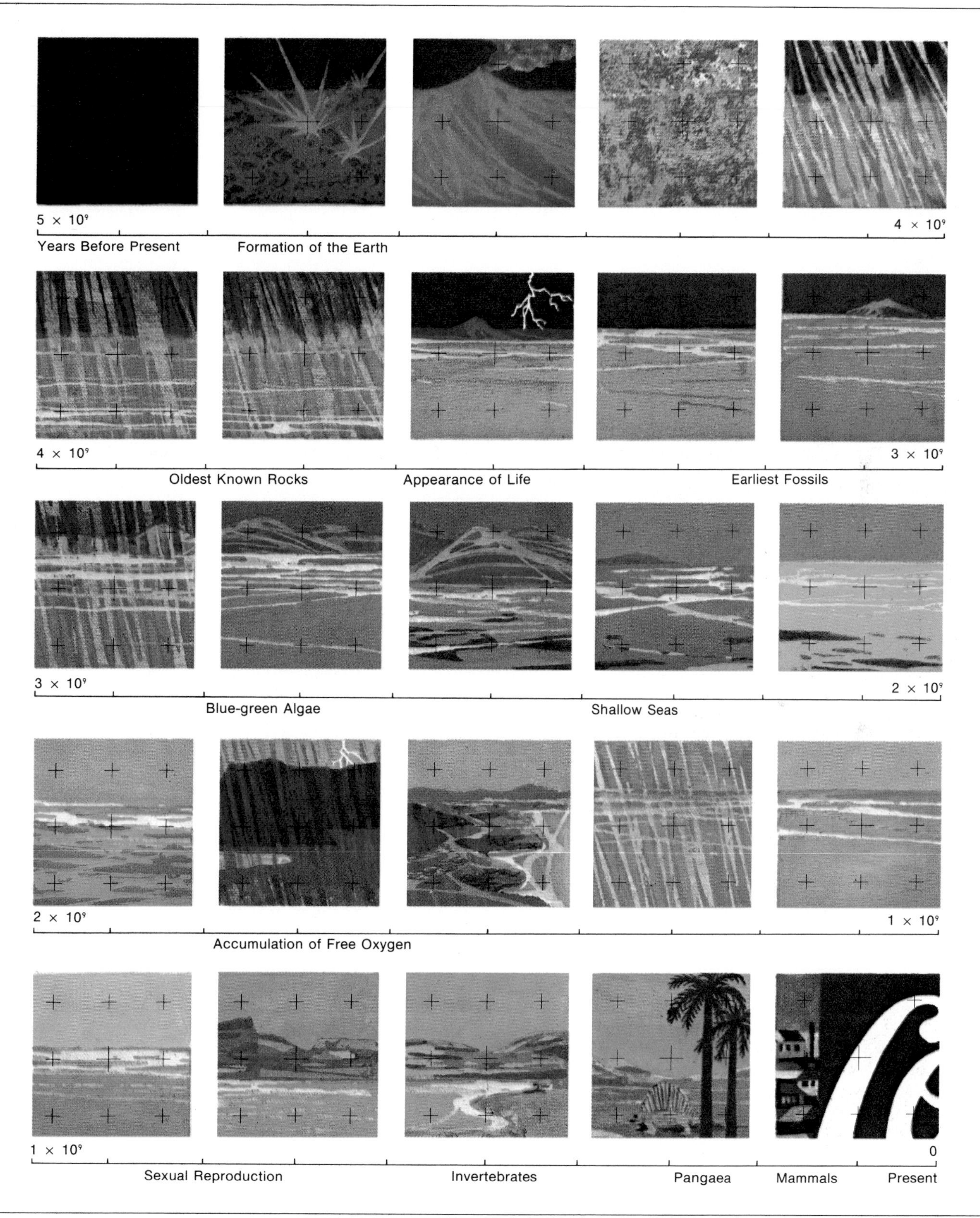
5 × 10⁹
4 × 10⁹
Years Before Present
Formation of the Earth
4 × 10⁹
3 × 10⁹
Oldest Known Rocks
Appearance of Life
Earliest Fossils
3 × 10⁹
2 × 10⁹
Blue-green Algae
Shallow Seas
2 × 10⁹
1 × 10⁹
Accumulation of Free Oxygen
1 × 10⁹
0
Sexual Reproduction
Invertebrates
Pangaea
Mammals
Present

Mid-Atlantic Ridge
Atlantic Ocean
Atlantic Ocean
South America
57°W 26°S
Nazca Ridge
Easter Island
Pacific Ocean
175 Km
620
2890
2000°
3000°
3500°
4000°
1
2
3
5
6
7
8
9
10
P
PcP
PKP
PKi KP

(a)

Africa

(b)

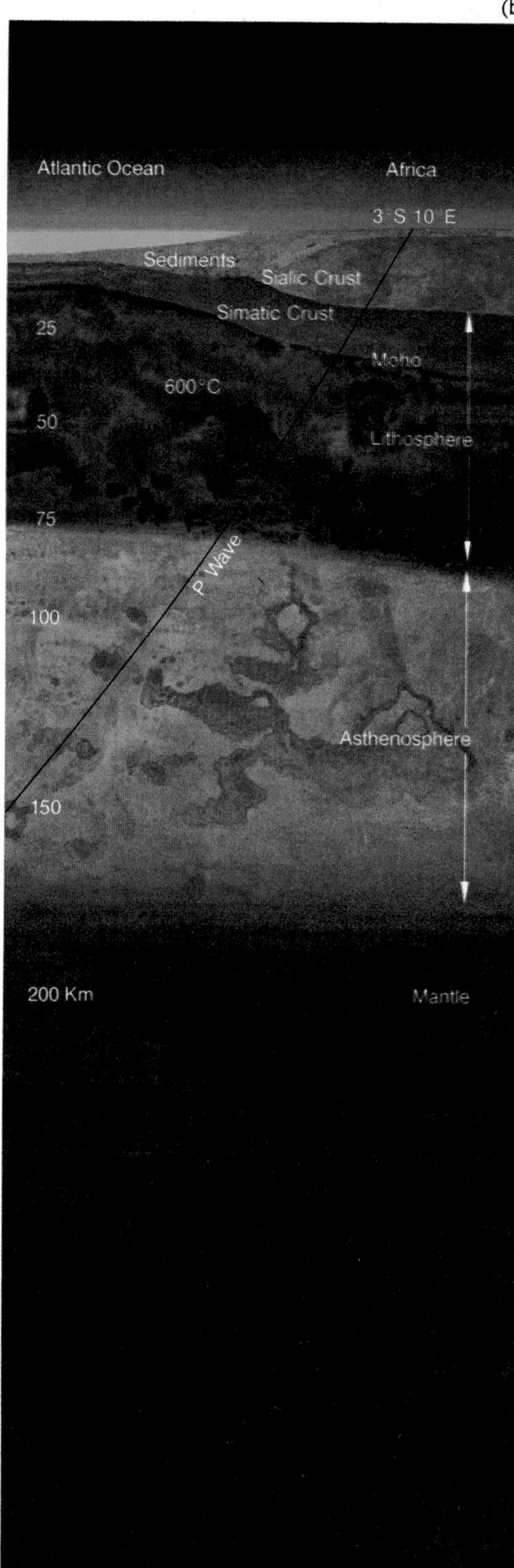

Atlantic Ocean
Africa
3°S 10°E
Sediments
Sialic Crust
Simatic Crust
25
Moho
600°C
50
Lithosphere
75
P Wave
100
Asthenosphere
150
200 Km
Mantle

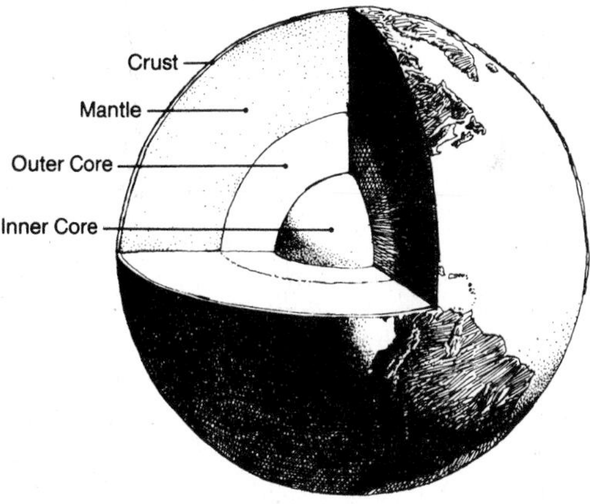

Crust
Mantle
Outer Core
Inner Core

Figure 1.7 (far left) Earth in cross section. The earth may be visualized as a series of concentric shells having different properties. The *crust* varies in thickness from about 10 or 12 kilometers under the oceans to 30 or more kilometers under the continents. The boundary between crust and *mantle* is recognized by an abrupt change in the velocities of compressional seismic waves. At depths of about 150 to 200 kilometers, there is a *low-velocity zone* in which the seismic velocities drop. Another marked change in seismic properties occurs at about 2,900 kilometers; here the *outer core*, which has the physical properties of a liquid, is recognized. The *inner core*, which is below 5,000 kilometers, behaves like a solid; it is believed to be an alloy of iron and nickel (that possibly includes some silicon or sulfur).

Seismology, the study of earthquake waves, is our most powerful tool for understanding interior composition and has provided information on the strength and compressibility of the earth. In this cross section, *P* waves of various types are illustrated. The numbers plotted on the path of the *P* wave represent the time (in minutes) it takes the wave to travel from Nazca Ridge to a point 3°S 10°E. *PKP* denotes the *P* waves that pass through the mantle and core (*K* represents *P* waves during their passage through the outer core). *PKiKP* are waves reflected from the earth's inner core; *PcP* are waves reflected from the boundary between the earth's mantle and the outer core (in this notation, *i* represents *P* waves reflected from the inner core and *c* represents *P* waves reflected from the outer core).

(Near left) The crust and upper mantle vary in composition. The chemical composition of the mantle varies, probably differing beneath ocean basins and continents; the chemistry of the mantles beneath different ocean basins (and different continents) probably varies as well. The outer 200 kilometers of the earth may be separated into two classes of material: The zone of low strength and low seismic velocity is called the *asthenosphere*; the more rigid layers that overlie it, which include both the crust and the upper mantle, are called the *lithosphere. Lithospheric plates* may be visualized as rigid slabs that are free to move on the low-strength asthenosphere. It is important to note that lithospheric plates include both crust and mantle and that a single plate may include both oceanic and continental crust.

the center. Ores and minerals became concentrated under the action of heat and pressure. The earth differentiated into three principal layers: the *crust*, the *mantle*, and the *core* (see Figure 1.7).

The thin outer layer, the crust, forms the continents and underlies the oceans. The second layer, the mantle, is much thicker than the crust and contains most of the earth's mass. Knowledge of the inner constitution of the earth has come primarily from a study of the waves generated by earthquakes. These waves, called *seismic waves*, travel through the earth at speeds that vary depending on the properties of the material through which they are traveling. Study of seismic data leads to the conclusion that the earth's core has a temperature of several thousand degrees. The inner core seems to be solid, whereas the outer core acts something like a liquid.

The flow of molten rock from an erupting volcano reveals that the interior of the earth is hot. The disintegration of radioactive elements generates heat that keeps the upper part of the mantle hot enough to be plastic; the consistency of the upper mantle gives the crust the freedom of movement important in geologic processes. The amount of heat that radioactivity generates in the earth decreases with time as radioactive elements disintegrate into stable forms; the amount of heat supplied by radioactivity was 15 times greater in the early days of the earth than it is today.

Figure 1.8 Every form of matter consists of an enormous number of tiny particles called atoms. The atoms making up a solid—a rock, a table, a book—are so firmly bound together that they are effectively locked in position. For this reason, solids have definite shape and volume. Two or more atoms can bond together strongly in definite combinations to form molecules. Molecules cannot be broken apart easily, and they retain their identity in liquids and gases. In liquids, the molecules are fairly close together and exert moderate forces of attraction on one another. The molecules in a liquid are able to move past one another, so that liquids have a definite volume but not a definite shape. The molecules in gases are far apart, move freely at high speeds, and interact with one another only when they collide. The density of air at normal sea level conditions is approximately 1/1,000 of the density of water.

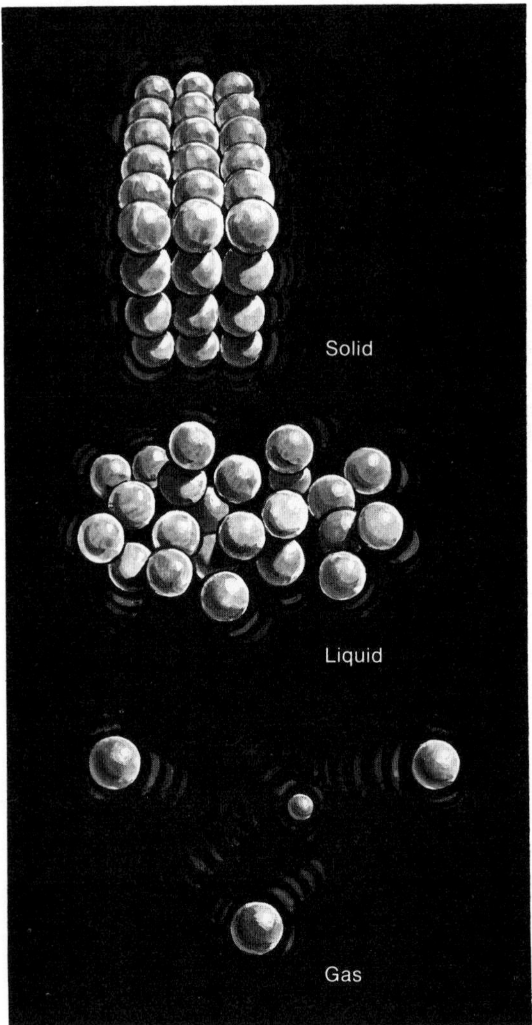
Solid
Liquid
Gas

Evolution of the Earth's Atmosphere and Landforms

The air you depend on for life may seem too commonplace to deserve special attention, but stop for a moment to consider its origin. After the earth formed, it was a barren, lifeless planet, probably without an atmosphere. From the beginning, however, the earth was carrying its full stock of chemical elements. The oxygen atoms you are breathing now were present on the earth 5×10^9 years ago, not in air but chemically bound in minerals in the earth. Nature has been recycling its materials for a long time.

There is strong evidence that any atmosphere the earth possessed at the time of its formation was not retained long. Compared to many other elements, the element neon is practically nonexistent on the earth, but it is relatively abundant on the sun. The difference is puzzling, because the earth and the sun are thought to have been formed from the same cloud of atoms, which would make their compositions similar initially. Neon is an inert gas—that is, it forms no chemical compounds—so it would be found primarily in the atmosphere. The relative lack of neon in the earth's atmosphere indicates that most of the original atmosphere was lost. Urey has suggested that when atoms from the original swirling cloud condensed to form solid planetesimals, enough heat was generated to drive away most of the atmospheric gases even before the planets were formed.

The Early Atmosphere

The earth lost its original atmosphere, but a new atmosphere began to form as gases released from molten rock escaped to the earth's surface from its interior. The composition of this second atmosphere—a witches' brew of strange vapors—may have been similar to the gases present-day volcanoes release: water vapor, carbon dioxide, sulfurous gases, and nitrogen. Eventually much of the water vapor condensed and fell to the ground as rain. The early atmosphere probably contained mostly carbon dioxide, with some water vapor and nitrogen and only traces of oxygen gas, which would make it similar to the present atmosphere of Venus.

The Early Landforms

According to reasonable speculation, the landscape 3 or 4×10^9 years ago must have presented a forbidding appearance. Volcanoes erupted, spreading molten rock over the earth's surface. The ground was raw and barren, and the earth was devoid of life. Water released from the earth's interior collected in the atmosphere and then fell as rain, forming the seas. The presence of moisture made possible chemical reactions that weakened the structure of rocks at the earth's surface, so that fragments and grains of rock littered the ground. There was no organic plant matter to help form true soils, and as water from rainstorms flowed down hillsides, it carried rock

Figure 1.9 The appearance of the earth's surface before the advent of life on the land may have been similar to this raw volcanic wasteland on the slopes of Haleakala Crater on the island of Maui, Hawaiian Islands.

debris and dissolved minerals toward the seas. Windstorms transported rock grains and dust across the landscape. As wind and water cleared debris from the ground, fresh surfaces of rock were exposed and the weathering process continued.

The early landscape probably had the steep slopes, exposed rock faces, and piles of debris characteristic of present-day deserts or volcanic regions, but the complete lack of vegetation and soil and the abundance of rainfall and flowing water gave the early landscape an appearance that does not exist anywhere on the earth today. The rock that originally covered the earth's surface was solidified molten rock, or *igneous* rock, such as basalt and granite. Most of the rock exposed at the earth's surface today is *sedimentary* rock, such as sandstone, which forms when rock grains derived from older rocks are cemented together. *Metamorphic* rock results when igneous or sedimentary rock is further transformed by heat and pressure.

If wind and water had been the only agents at work on the landscape, the land areas of the earth would have been worn down to a low, featureless plain in a few tens of millions of years. But the earth's internal energy has also shaped the surface. Mountain chains are formed when the internal energy lifts up the earth's crust more rapidly than the external energy can wear it down. The earth's internal energy probably will be exhausted eventually. Then the crust will no longer be active and in time the land will lose its features of relief.

Appearance and Development of Life

About 3×10^9 years ago—at 5:00 P.M. on our 24-hour clock—the history of the earth took a unique turn. Life appeared. The origin of life has not yet been fully explained in terms of physical laws. However, in 1953 Stanley Miller, working with Harold Urey at the University of Chicago, carried out an experiment that indicates how it might have been possible for life to have arisen from inanimate matter. Miller took mixtures of simple gases such as ammonia, methane, hydrogen, and water vapor and allowed them to react together for several days under the stimulation of an electric discharge. He discovered that the reacting gases produced complex molecules. Among the reaction products, he was able to identify amino acids, the building blocks of proteins. Miller's work showed that given sufficient energy to break chemical bonds, relatively simple chemical substances can recombine to form the complex molecules required in life processes.

Conditions 3×10^9 years ago could have been roughly analogous to the conditions in Miller's experiments. The earth's early atmosphere contained a variety of gases, and the energy to disrupt the molecules could have come from lightning and from intense, energetic ultraviolet light from the sun. Rain washed many of the reaction products out of the atmosphere into the oceans, and in time the oceans became an "organic soup" rich with synthesized complex molecules. No living organisms existed to digest the molecules and break them down. Perhaps by chance a collection of molecules complex enough to "live" came together in one of the collisions between molecules that occurred many times a second in the early oceans. The aggregation would have to be able to do two things to qualify as "living": it would have to be able to absorb energy from its surroundings and to reproduce. For energy, the earliest organisms probably depended on the supply of complex molecules present in the oceans. Yeast cells, for example, can generate energy by fermenting sugar.

The Fossil Record

The origin of life on the earth remains a matter for speculation, because the earliest life forms left no known traces. The oldest evidence of life on earth consists of fossil plant remains about 3×10^9 years old. The early plants closely resembled the blue-green algae that you can see today floating on ponds in the summer.

Most of our knowledge of the history of life on the earth comes from the *fossil record*. Wind and flowing water carried fine rock debris from the early landforms and deposited it in layers of sediment on the land and on the ocean floor. Later, when compression

Figure 1.10 (left) Layers of deposited sedimentary rock are clearly visible in this road cut through shale near Olathe, Kansas. By its position, the layer of dark shale is inferred to be more recent in geologic time than the lighter layer below and less recent than the layer above.

(right) The layers in this rock in England have been plastically deformed by slow shifts in the earth's crust, and some of the originally horizontal strata are now nearly vertical.

caused the sediment to consolidate, the remains of the plants and animals that were buried in the sediment were sealed in stone and protected for long ages. Often, additional sediment layers were deposited subsequently; the many layers of sedimentary rock exposed at the Grand Canyon of the Colorado River are examples of successive deposits of sediment. It can be inferred that in undisturbed sedimentary rock formations, the earliest fossils are found in the deepest layers. Layers of sedimentary rock therefore contain a partial history of life on the earth.

The fossil record is not a complete catalog of all the life forms that have appeared on the earth, however. Because soft tissues are not preserved, only hard remains such as bones, shells, and woody stems can leave a fossil record. Furthermore, because the process of sedimentation is variable, there may be long periods in which no deposition at all occurs in a particular location. And the later wearing away of sedimentary rock may destroy a portion of the local fossil record.

Organic Evolution

The primordial forms of life on the earth probably depended on the organic material in the ocean for food. The earliest known fossils are plants, which are complicated, specialized organisms capable of producing food energy using only carbon dioxide, water, and the sun's radiant energy. How did complex plants arise from the simple early forms of life? And what was the origin of higher forms of animal life, such as dinosaurs and human beings?

In 1859, the British naturalists Charles Darwin and Alfred Russel Wallace independently set forth the idea of *organic evolution*, which provides insight into the processes of biologic change. According to Darwin and later researchers, organisms do not duplicate themselves in their offspring because of inherent variability in the mechanism of heredity and because of accidental alterations in the controls of heredity. The individual members of a plant or animal species therefore differ in some ways. In a given environment, the individuals who are most efficient at such things as gathering food and escaping predators are most likely to survive, reproduce, and pass along some of their characteristics to their offspring. If a species is successful, and if the environment does not change, individuals with hereditary make-up like that of most other members of their species will be favored. According to the fossil record, opossums, for example, have changed little in the past 60 million years.

If the physical environment changes—perhaps because of a change in climate—individual organisms with hereditary make-up different from that of most other members of the species may have a slight advantage in competing for survival, a principle that Darwin called *natural selection*. The descendants of the favored members may undergo further natural selection, and over a sufficient period of time the accumulated changes may result in the development of a new species.

The history of the peppered moth in Britain illustrates a modern example of natural selection. Prior to 1850, before industry was

widespread, most peppered moths were light in color with dark speckles. They blended with the tree lichens on which they rested and became inconspicuous to predator birds. A small proportion of the moths were of a dark variety. After the rise of industrialization, air pollution killed many of the lichens and darkened the trees with soot, so that in industrial areas the dark moths were less conspicuous to birds. By 1900, almost all peppered moths around industrial centers were dark (see Figure 1.11).

Figure 1.11 (left) The photograph shows the light and dark forms of the peppered moth at rest on a lichen-covered tree located in a nonindustrial area in England. The light form, which is far less visible to predator birds than the dark form, is the dominant type in this region.

(right) The photograph shows the light and dark forms at rest on a tree that is free of lichens and darkened by industrial pollution. In this region the dark form is the dominant type because it is less visible and has a greater chance of survival than the light form.

Interaction of Life With the Environment

The appearance of green plants 3×10^9 years ago—at 5:00 P.M. on the 24-hour clock—set new processes in motion that literally changed the face of the earth. The first plants were confined to the shallow layers of the sea, where water shielded them from the

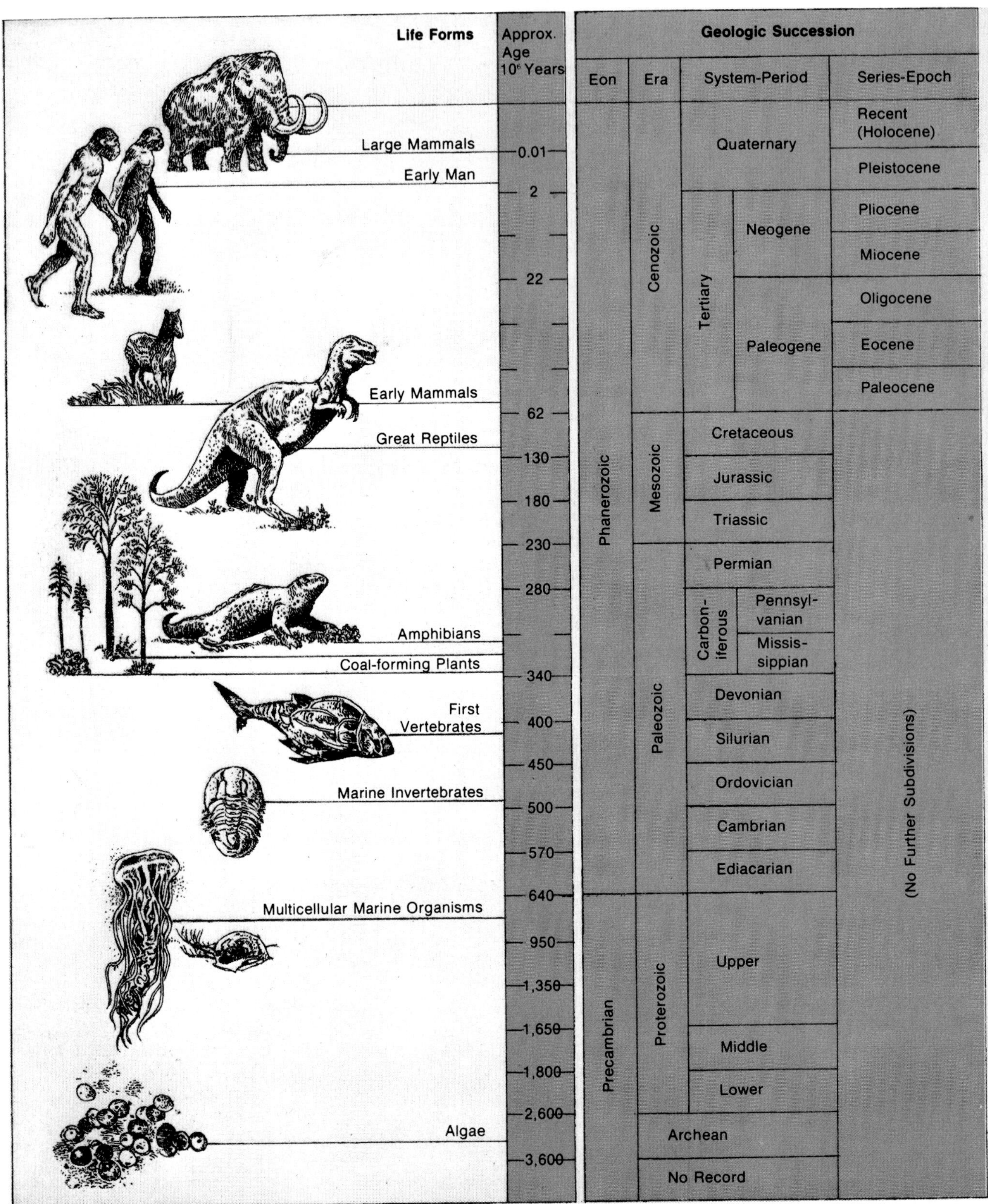

Life Forms
Approx. Age 10⁶ Years
Geologic Succession
Eon
Era
System-Period
Series-Epoch
Large Mammals
Early Man
Early Mammals
Great Reptiles
Amphibians
Coal-forming Plants
First Vertebrates
Marine Invertebrates
Multicellular Marine Organisms
Algae
0.01
2
22
62
130
180
230
280
340
400
450
500
570
640
950
1,350
1,650
1,800
2,600
3,600
Phanerozoic
Precambrian
Cenozoic
Mesozoic
Paleozoic
Proterozoic
Quaternary
Tertiary
Neogene
Paleogene
Cretaceous
Jurassic
Triassic
Permian
Carbon-iferous
Pennsyl-vanian
Missis-sippian
Devonian
Silurian
Ordovician
Cambrian
Ediacarian
Upper
Middle
Lower
Archean
No Record
Recent (Holocene)
Pleistocene
Pliocene
Miocene
Oligocene
Eocene
Paleocene
(No Further Subdivisions)

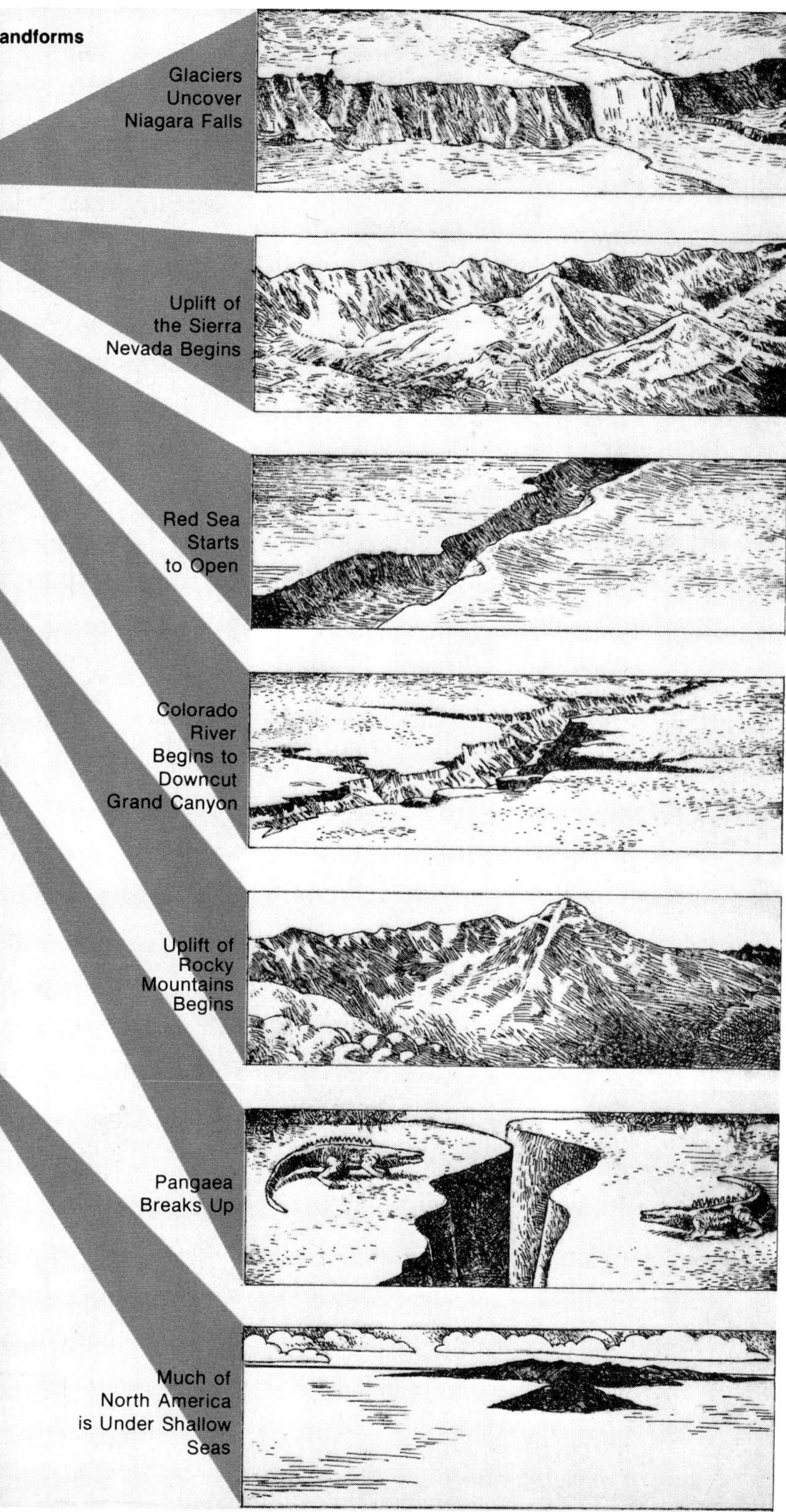

.andforms
Glaciers Uncover Niagara Falls
Uplift of the Sierra Nevada Begins
Red Sea Starts to Open
Colorado River Begins to Downcut Grand Canyon
Uplift of Rocky Mountains Begins
Pangaea Breaks Up
Much of North America is Under Shallow Seas

Figure 1.12 The geologic ages can be distinguished from one another and put into the correct time sequence by examining the fossil record in stratified rocks. The Precambrian, which spans the first 85 percent of the earth's history, corresponds to the period before higher forms of life evolved, when the fossil record is absent or indistinct. Some of the ages are named after the regions where their identifying fossils were first studied; the Jurassic, for example, is named after the Jura Mountains of western Europe. The radioactive dating of rocks had been used to assign an absolute time scale in years to the geologic ages. Note that the time scales on the figure are not evenly spaced but emphasize the most recent 600 million years.

To the left of the age chart are shown selected life forms and the approximate times when they first appeared or were dominant. A general trend from less complex to more complex life forms reflects the processes of organic evolution. Evolution did not proceed along a single course, however; there were many side branches and dead ends such as the great reptiles.

On the right of the age chart, selected events in the earth's geologic history are shown. Some events, such as the rifting of continents and the uplift of mountains, are caused by the motion of the earth's crust. Other events, such as the erosion of rocks by flowing water and the onset or retreat of glaciation, are controlled largely by climate. (After *Adventures in Earth History,* edited by Preston Cloud. W. H. Freeman and Company. Copyright © 1970.)

sun's damaging ultraviolet light. The atmosphere was predominantly carbon dioxide, nitrogen, and water vapor, and the green plants used carbon dioxide dissolved in the water to produce sugar by *photosynthesis*.

In the course of photosynthesis, water molecules are split and oxygen gas is released. Plant cells use some of the oxygen for respiration and give off the remainder. Because oxygen is chemically reactive, most of the oxygen the early plants released probably did not remain in the atmosphere long; it combined chemically with iron minerals in the earth. Eventually, when the chemical reactions of minerals with oxygen were complete, oxygen gas began to accumulate in the atmosphere. The earth's atmosphere today is 20 percent oxygen; green plants produced most of it through geologic time.

Life and the Atmosphere

Part of the oxygen produced by early plants dissolved in the sea water. The availability of oxygen for respiration and of plants for food opened up new possibilities for the development of higher forms of life. The early fossil record contains evidence of plant remains only. But beginning about 600 million years ago—at 10:30 in the evening on the 24-hour clock—the fossil record testifies to a sudden expansion of complex, shelled, multicellular marine animals that could use the oxygen in water for respiration.

The fossil record is silent concerning the transitional forms that led from plants to complex marine animals. Major evolutionary adaptations seem to occur comparatively rapidly—perhaps within a few million years or so, which is too short a time interval to leave a clear record on the fossil calendar.

Plants incorporate carbon from carbon dioxide into the sugar formed during photosynthesis. The early marine animals that fed on plants incorporated carbon and other materials into their hard shells. The amount of carbon dioxide that was originally in the atmosphere gradually decreased as carbon became locked into minerals. Today carbon dioxide constitutes only a small fraction of the earth's atmosphere; there are only about 300 molecules of carbon dioxide for every million molecules of air.

The presence of oxygen in the earth's atmosphere led to the formation of *ozone*. Ozone, which has a molecular structure consisting of three oxygen atoms, forms a layer high in the present atmosphere that absorbs much of the incoming ultraviolet radiation from the sun. The early atmosphere contained little or no oxygen, so no ozone layer was present to prevent ultraviolet light from reaching the earth's surface. Once the lethal radiation was blocked, the land surfaces became accessible to life. Forests of giant seed ferns spread across the land 400 million years ago, only 1 hour before midnight on the 24-hour clock, and amphibious animals appeared.

Life and the Landscape

When plants moved onto the land, new processes of change began to operate, and the landscape took on a new appearance. The chemical attack by plant acids and the mechanical pressure of growing roots accelerated the fragmentation of rocks. The fine rock particles mixed with organic matter to form soils, and a cover of soil and vegetation softened the harsh slopes of barren rock. The plant cover helped protect the land from the erosive action of rain and flowing water, so the rate of sediment production probably decreased in many regions.

Living organisms also helped to form concentrations of minerals. Vigorous plant growth in swampy areas led to the accumulation of thick layers of plant remains, which compacted first to peat and later to coal. Coal formation, which began 350 million years ago and lasted 50 million years, caused a further decrease in the carbon dioxide content of the atmosphere.

Present-day deposits of limestone, such as the White Cliffs of Dover in England, are largely formed from the remains of shelled marine organisms. When the animals died, their shells were deposited on the ocean floor, and after compaction they became limestone. Subsequent uplift of the earth's crust raised many of the deposits above sea level. Most of the limestone beds on the earth today were laid down 70 to 100 million years ago, about 15 minutes before midnight on the 24-hour clock.

The interactions between living organisms and the physical environment altered the environment by introducing new processes of change. Evolutionary adaptation to changing climates and conditions also altered the early plants and animals. For a time, great reptiles dominated the land. But about 50 million years ago—7 minutes before midnight—for an unknown reason, the dinosaurs became extinct, and modern mammals appeared.

Man's Place on Earth

Man is a relative newcomer on the earth. The earliest known manlike remains appeared about 2.6 million years ago—22 seconds before midnight. But man is not an intruder on an alien earth. No less than other organisms, he has been molded by processes as natural as those that built the mountains, and his interaction with the physical environment influenced his development. He is adapted to breathing the gases of the earth's atmosphere, for example, and his eyes respond only to the wavelengths of sunlight that penetrate the earth's atmosphere. His interaction with the environment will continue to affect his development as a species.

Continental Drift and Sea Floor Spreading

The geologic record shows that plants, animals, and landforms have changed with time, and recent evidence indicates that the continents themselves have shifted restlessly over the earth's surface. In the first decades of the twentieth century, the German meteorologist Alfred Wegener and others suggested that portions of the earth's crust have drifted long distances in geologic time. Wegener argued that the presence of extensive Pennsylvanian coal deposits formed from tropical vegetation implies that Pennsylvania was once located near the Equator. As further evidence, Wegener drew attention to the close fit between the east coast of South America and the west coast of Africa. He suggested that these continents were at one time next to one another, perhaps even part of the same landmass. As Figure 1.13 shows, all of the continents fit together fairly well if the "edge" of a continent is taken to be its outline in shallow ocean waters at a depth of 2,000 meters (6,500 feet). Wegener called the original supercontinent *Pangaea*. He speculated that Pangaea broke up into the present continents, which then began drifting slowly across the surface of the earth. Wegener's ideas were not accepted for many years, partly because of conflicting evidence and partly because there seemed to be no mechanism that could account for continental drift.

In the 1960s it was suggested that the sea floor is spreading apart in some places, widening the separation between some of the continents. Somewhat earlier, oceanographers had discovered that the ocean floor is characterized by long, broken oceanic ridges, such as the Mid-Atlantic Ridge, and by a system of deep trenches, such as those that ring the Pacific Ocean basin. The theory of sea floor

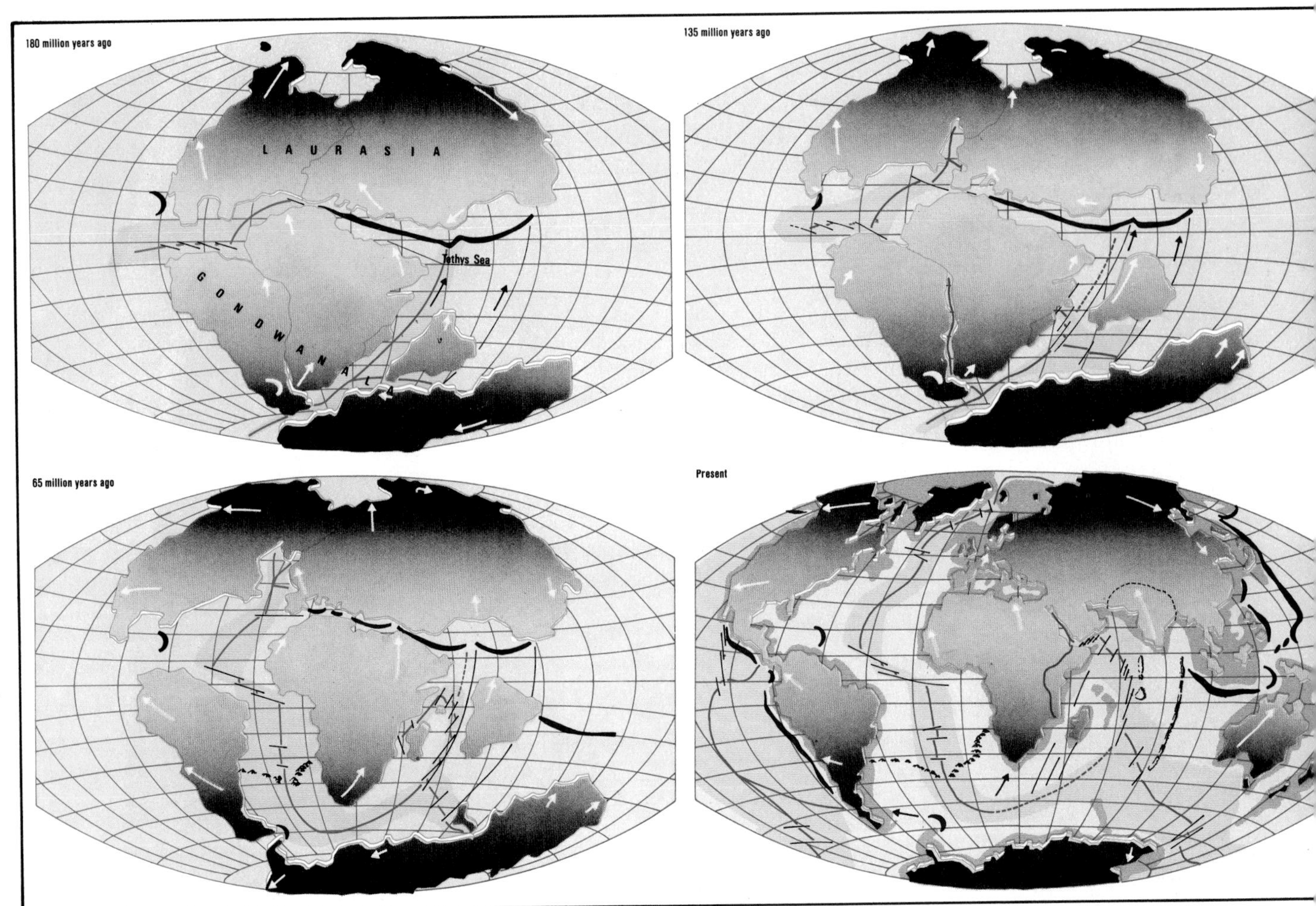

180 million years ago
L A U R A S I A
Tethys Sea
G O N D W A N A L
135 million years ago
65 million years ago
Present

Figure 1.13 Approximately 200 million years ago the landmasses of the earth were grouped together, forming Pangaea, a supercontinent. Pangaea subsequently rifted into two continents, Laurasia in the Northern Hemisphere and Gondwanaland in the Southern Hemisphere. The rifting continued, with the crustal plates moving in the directions indicated by the arrows. The plate that carries India, for example, split from Antarctica and drifted northward until it collided with the Asian plate.

The drifting of the continents is associated with the formation of spreading centers along oceanic ridges such as the Mid-Atlantic Ridge. The crustal plates move away from one another as new crust is formed along the oceanic ridges. The opening of the Atlantic basin, which can be followed in this sequence of diagrams, was accompanied by the drift of the Americas westward away from Europe and Africa. The margins of colliding plates are indicated in the diagrams as heavy lines. At some colliding plate boundaries, such as those along the western edge of the Pacific basin, old crust plunges into the mantle and is remelted. At boundaries such as those along the west coasts of the Americas, the collision of the plates results in uplift of mountain ranges.

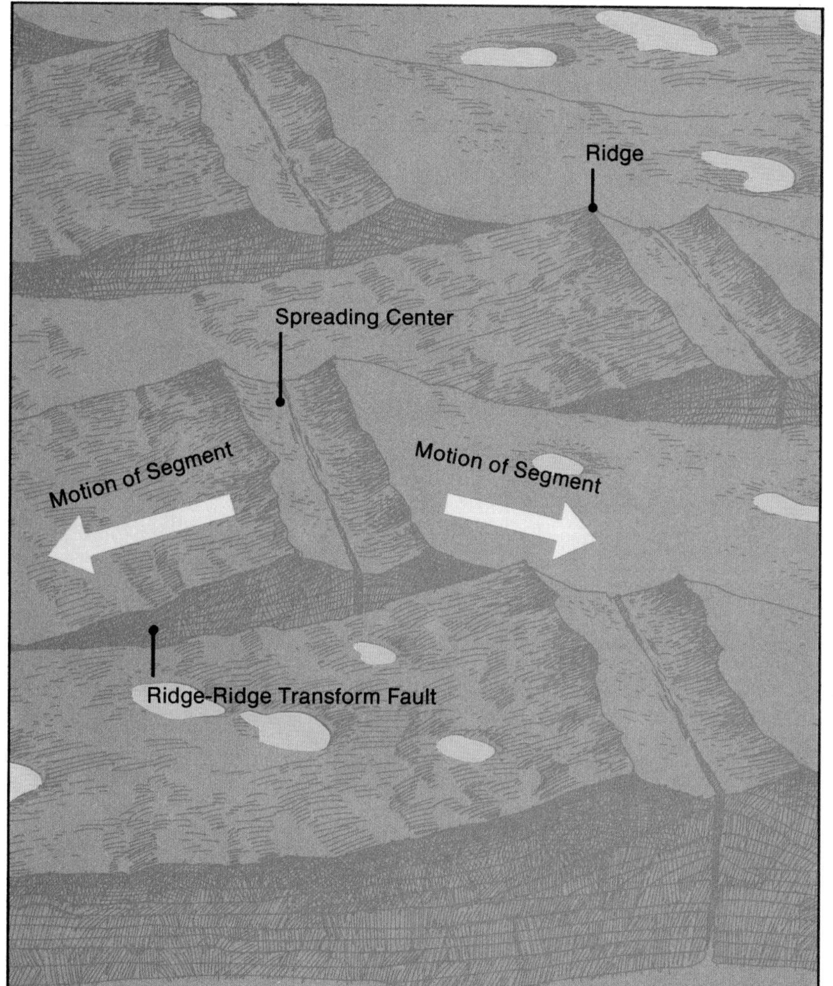
Ridge
Spreading Center
Motion of Segment
Motion of Segment
Ridge-Ridge Transform Fault

Figure 1.14 The structure of a typical oceanic ridge is shown here with exaggerated vertical scale for clarity. As new crust wells up in the spreading center, the segments of the ridge move apart. Ridge-ridge transform faults break the ridge at nearly right angles, which allows the lithospheric plates to fit the contours of the earth as the plates move apart. Earthquakes may occur when segments rub against one another near the offset spreading centers.

spreading suggests that molten rock slowly wells up through deep cracks near the oceanic ridges and creates new ocean floor.

Sea floor spreading has been investigated experimentally. Research ships such as the *Glomar Challenger* have drilled into the floor of the Atlantic basin with hollow drills to bring up samples of sediment. No sediment older than 140 million years has been found, which suggests that the Atlantic Ocean is relatively young. Furthermore, the thickness and age of the sediments increase with distance from the Mid-Atlantic Ridge, because the floor nearest the ridge is the "newest" and has had the least time to collect sediment.

The model of continental drift now generally accepted considers the outer portion of the earth, the *lithosphere*, to be broken into eight or ten major plates and a number of smaller plates. The theory of *plate tectonics* suggests that the plates, perhaps 100 kilometers (60 miles) thick, carry parts of continents and of ocean floor and that they glide on the layer of molten rock constituting the outer part of the mantle. The plates move apart at oceanic ridges where new floor is forming and then slide back into the mantle at the

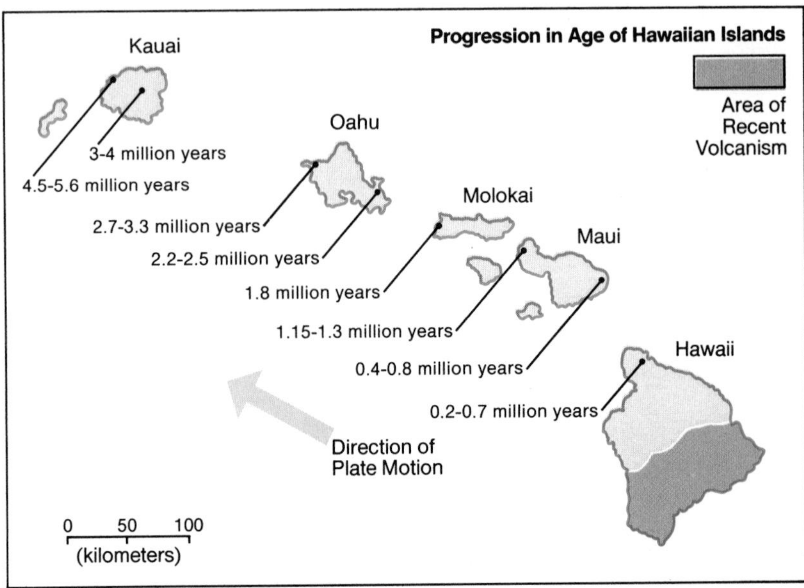
Progression in Age of Hawaiian Islands
Area of Recent Volcanism
Kauai
3-4 million years
4.5-5.6 million years
Oahu
2.7-3.3 million years
2.2-2.5 million years
Molokai
1.8 million years
Maui
1.15-1.3 million years
0.4-0.8 million years
Hawaii
0.2-0.7 million years
Direction of Plate Motion
0 50 100
(kilometers)

Figure 1.15 The ages of the Hawaiian Islands, as determined by the potassium/argon radioactivity dating method, show a general progression from old to young along the chain. These observations were not explained until the development of the theory of plate tectonics, which attributes the formation of the Hawaiian Islands to a plume of volcanic activity fixed in the earth's mantle. As the plate of oceanic crust drifts over the location of the plume, periods of volcanic activity cause the formation of islands of volcanic rock. Can you verify from the map that the measured ages of the Hawaiian Islands are consistent with a drift rate of several centimeters a year? (After McDougall, 1971)

oceanic trenches. Mountain building and island arc formation occur at the colliding boundaries of plates, notably at the edges of the Pacific plates. The formation of the Himalaya Mountains of Asia is attributed to the collision between the plate that carries India and the Asian plate.

Isolated spots of volcanic activity deep in the mantle remain fixed in position, and as the plates glide by, the hot spots erupt, forming volcanic islands on the moving plates. The Hawaiian islands appear to have been formed in this way because they exhibit a definite progression in age from one end of the island chain to the other. Drift rates estimated from these age measurements are a few centimeters per year, so the breakup of Pangaea must have begun about 200 or 250 million years ago to account for the present positions of the continents.

At present, the Atlantic Ocean basin is gradually widening, and North and South America seem to be drifting westward. Southern California, however, is on a northward-moving plate, and earthquakes in that part of the world are associated with the stresses that plate movements set up. The plate that carries Africa and the plate that carries Europe are moving toward each other, and they will eventually squeeze the Mediterranean Sea out of existence.

The Scale of Time

Before 1800, people generally believed that the landscape of the earth had been formed by a series of violent catastrophes that had lasted only a few thousand years. Toward the end of the eighteenth century, James Hutton took a fresh look at the mountains and streams of Scotland and interpreted what he saw as the products of erosion and mountain uplift. He realized that given enough time, these slow processes could have produced the landscape he saw. He concluded that the eras of geologic time must be much longer than had been thought. Charles Darwin realized that evolution also required time spans on the order of hundreds of millions of years.

To check the ideas of Hutton and of Darwin, scientists in the nineteenth century attempted to develop an accurate way to measure the duration of geologic time. The fossil record gave the correct succession of the geologic eras, but it did not indicate their actual ages. So the rate of sediment deposition was measured to calculate how long it would have taken to build up the known thicknesses of sedimentary rock layers. The method of measurement failed because the rate of sedimentation has varied with time and because the record contains too many gaps.

Radioactivity and Time

The accurate natural timekeeper the scientists sought was found early in the twentieth century. They learned that the passage of time could be measured using the decay rates of radioactive elements. A radioactive element decays spontaneously with a certain probability into another element. The rate of decay is characterized by the *half-life*, which is the time required for half of an initial amount of radioactive atoms to decay. The half-life is unaffected by changes in temperature, pressure, or other external factors.

Atoms of the same chemical element may have differently structured nuclei with different degrees of stability and different masses. Such atoms, called the *isotopes* of an element, can be designated by the name of the element and a whole number representing the mass of each of its atoms, such as uranium 238 and uranium 234.

How can radioactivity be used to determine the age of a substance? Suppose a rock contained an initial amount of a radioactive element such as uranium 238 when it solidified from the molten state. Uranium 238 decays with a half-life of 4.5×10^9 years through a chain of steps to the stable end product lead 206. Each uranium atom that decays eventually results in the formation of one atom of lead. Later measurement of the relative numbers of uranium 238 and lead 206 atoms in the rock allows the elapsed time to be computed, because the rate of decay is known. For example, if the rock contains equal amounts of uranium 238 and lead 206, the elapsed time since the rock solidified must have been one half-life, or 4.5×10^9 years.

If the rock initially contained lead 206, the uranium 238/lead 206 method will not be accurate. It is therefore necessary to use more than one radioactive element to establish a date. The most useful radioactive clocks for geologic time are listed in Table 1.1.

Table 1.1. Half-Lives of Radioactive Elements

Radioactive Clock	Half-Life (years)
Rubidium 87/Strontium 87	5.0×10^{10}
Uranium 238/Lead 206	4.5×10^9
Potassium 40/Argon 40	1.3×10^9
Uranium 235/Lead 207	0.7×10^9
Carbon 14	5730

Source: Gray, Dwight E (ed.). 1972. *American Institute of Physics Handbook*. 3rd ed. New York: McGraw-Hill.

Dating the Earth

Among the oldest rocks on earth are some found in Rhodesia in Africa. Their age, determined by radioactivity, is about 3.5×10^9 years. This number represents the youngest the earth might be, because erosion may have worn away older rocks. Radioactive dating of meteorites gives the earth an age of 4.6×10^9 years, assuming that meteorites and the earth were formed at the same time. A direct estimate of the earth's age from measurements of the total uranium and lead content of the crust also leads to an age of 4.6×10^9 years.

Carbon 14 Dating

Radioactive dating by carbon 14 has been especially important for dating events in the past 40,000 years or so. Carbon 14, which is continuously formed in the earth's atmosphere by the action of cosmic rays on nitrogen 14, makes up a definite proportion of all the carbon ingested by plants and animals. The amount of carbon 14 relative to the total amount of carbon in an organism steadily decreases after the organism dies, because it no longer takes in carbon. Measuring the ratio of radioactive carbon to ordinary carbon allows the age of the sample to be inferred. Dating wood from trees that were killed by the advance of glacial ice has helped fix the date of the last ice age, for example.

Other Dating Methods

A number of other dating methods are useful, particularly for measuring relatively short time spans of thousands of years. Most of the methods have been used to determine climatic conditions in the recent geologic past. One method uses the variation in the annual growth rings of trees to estimate average weather conditions. Trees more than 4,000 years old have been found among the bristlecone pines of California.

Another dating method uses *varves*, which are sediment layers deposited in lakes by streams issuing from melting glaciers. Varves have characteristic annual layers because coarse sediment is deposited during heavy summer melting, and fine sediment is deposited later in the year when the waters are still. In Scandinavia, varve analysis has been extended more than 10,000 years into the past, and correlation of varve patterns from different localities has shed light on the course of glacial recession in northern Europe.

Summary

The earth has been a scene of change since its formation about 4 or 5×10^9 years ago. Change continues on the surface of the earth today because of interactions among the atmosphere, rock, water, soil, vegetation, and animals, including man. Physical geography is the study of the changes and interactions on the earth's surface.

Throughout geologic time, dynamic processes have acted on the earth, so the earth today is far different from the earth of billions of years ago. The advent of life, possibly by chance molecular combinations, introduced new processes of change. The action of photosynthesizing plants significantly modified the early atmosphere by giving off oxygen and thereby providing the chemical means for animal respiration.

The landforms on the earth's surface were first shaped by the chemical and physical action of rain as it flowed over the land. Later, when life that could exist on the land had evolved through biological processes of change, vegetation and soil covered the bare rock. Radiant energy from the sun powers most processes on the earth's surface, although the earth's internal energy also produces changes—changes that are evident in volcanic action, mountain building, the drift of continents, and the opening of ocean basins.

Investigations using the fossil record and radioactive dating have traced many of the events of the earth's history. Evidence shows that major changes in life forms and in landforms require long periods of time. Although some changes are too slow to be apparent within a single human lifetime, the interactions that cause these changes occur continuously at the earth's surface. In the remainder of this book, you will read about the interactions and processes at work on the earth today.

2 The Dynamic Planet

Units of Measure: The Metric System

Forms of Energy
Energy and Work
Heat Energy
Units of Energy

Water: An Energy Converter
Changes of Phase and Changes of Energy
Physical and Chemical Properties

The Atmosphere: An Energy Transporter

Systems: Energy Exchange

Budgets of Energy and Matter

Cycles of Materials: Synergy
The Hydrologic Cycle
Man's Input to Cycles

The Classification of Systems

Everything on the earth—its hills and valleys, its trees and rivers, its farms and cities—is unified and driven by energy. Larger swirls push the atmospheric processes; smaller eddies shape the land; the ceaseless flow of energy continually rearranges the tiniest particles of matter. Energy from the sun and from within the earth flows through the earth's systems in dynamic processes and patterns.

The Starry Night by Vincent Van Gogh, 1889. (Collection, the Museum of Modern Art, New York; acquired through the Lillie P. Bliss bequest)

The Dynamic Planet

The sun is always rising somewhere on earth and is always setting somewhere. Man adjusts his life to the cycle of the day—awaking, eating, working, playing, sleeping. As the sun rises, not only man's activities but the tempos of the earth's processes begin to quicken. The sun's radiant energy provides the power to drive most of the geographic processes important on the human time scale of days or years, such as the growth of vegetation and the movement of moisture through the atmosphere by weather systems. With the help of energy from the sun, green plants manufacture sugar from water and carbon dioxide. Heat from the sun's rays lifts water molecules from the ocean's surface into the atmosphere, where they may form clouds and rainstorms. Solar energy warms the air and land, and the heated air circulates in wind and weather patterns. Energy from sunlight also supplies the motive force to transport water from place to place on the earth's surface, an action with significant consequences for vegetation, climate, soil, and erosion.

The sudden shock of an earthquake or the appearance of a new volcanic island emerging from the ocean in an interval of a few months is a spectacular reminder that the earth's internal energy is also at work producing change. Most of the changes induced by internal energy, such as the raising of mountains or the widening of ocean basins, produce important effects only over long spans of geologic time. Erosion processes also tend to be slow, because rock offers great resistance to the action of wind and water. Many of the earth's landforms are shaped both by the earth's internal energy and by external energy from the sun.

Without energy, the earth would be lifeless and unchanging, because energy is the basis for all movement and all change. It is the muscle that makes things go, and no process can operate without the expenditure of energy. But energy alone is insufficient to promote change; there also must be something for the energy to act on. The moon receives radiant energy from the sun, but little change occurs on the moon because it possesses no atmosphere, water, or life to use energy in processes of change.

Physical geography is concerned with the ways energy works on the thin skin of air, water, rock, soil, and plants that covers the earth's surface. How different systems such as vegetation, the atmosphere, and the oceans respond to energy and how they interact by transferring energy to one another are major themes of this book. To set the stage for later chapters, this chapter discusses energy—where it comes from, the forms it takes—and it points out the general ways systems respond to energy. The properties of water are also described, because many systems exchange energy by means of moisture. In later chapters you will read about the earth's principal systems—from green plants to ocean waves—and you will see how energy and moisture interact with particular systems to produce change.

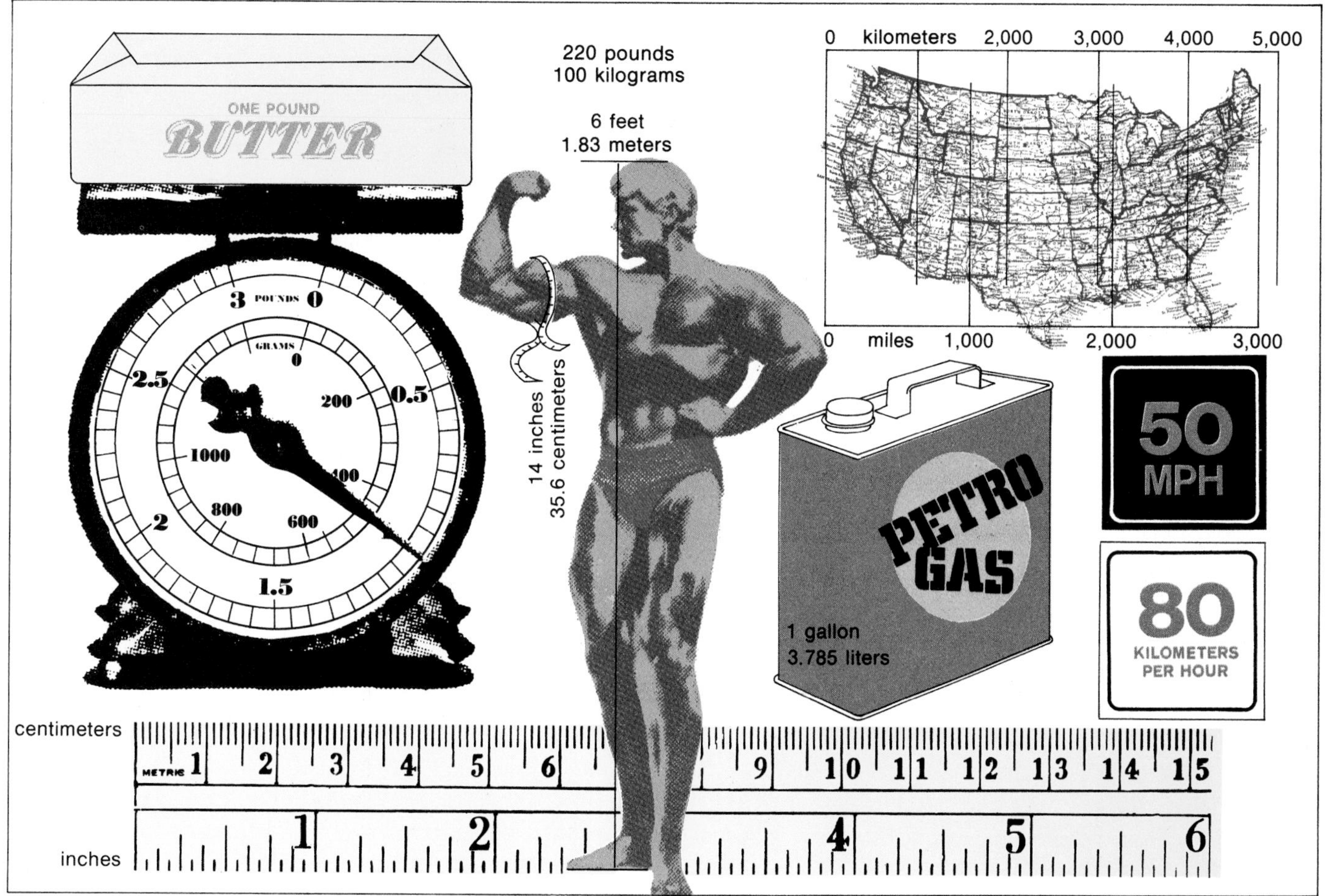
ONE POUND
BUTTER
220 pounds
100 kilograms
6 feet
1.83 meters
14 inches
35.6 centimeters
0 kilometers 2,000 3,000 4,000 5,000
0 miles 1,000 2,000 3,000
PETRO GAS
1 gallon
3.785 liters
50
MPH
80
KILOMETERS PER HOUR
centimeters
inches

Figure 2.2 The metric system of units is used in science, and it is the legal system of measure in most countries of the world. This illustration compares some common quantities expressed in English units with their corresponding metric equivalents.

Units of Measure: The Metric System

Before examining the interactions among systems and the role of energy in these interactions, it is helpful to become familiar with some of the basic language and units of measure employed in science. The length, volume, mass, or temperature of an object must be expressed according to a definite system of units if the measurement is to be meaningful to others. You are probably accustomed to specifying the length of a room in feet or the weight of a bag of sugar in pounds. In the *metric system*, the system of units used in science, length is commonly expressed in meters or centimeters (1 meter = 100 centimeters). One meter is about 3 inches longer than a yard, and a centimeter is approximately 3/8 inch. One kilometer (1,000 meters) is approximately 0.6 mile. Mass in the metric system is expressed in kilograms or grams (1 kilogram = 1,000 grams). One kilogram is equivalent to 2.2 pounds, and 1 gram is equal to about 0.035 ounce. A description of the metric system and a table of conversion factors are given in Appendix I.

Two scales are commonly used in science to measure temperature. On the *Celsius* scale (formerly called the Centigrade scale),

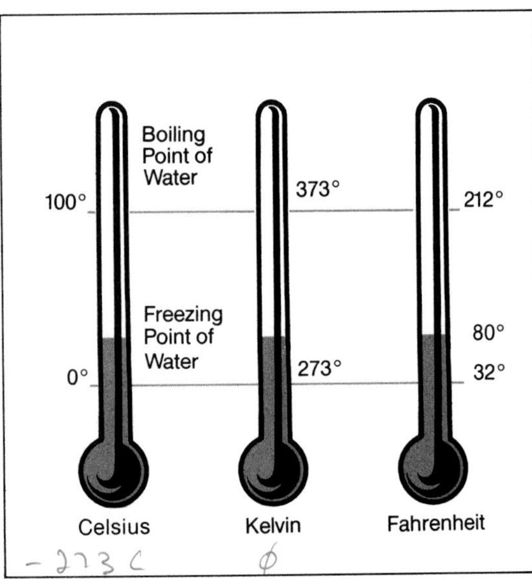
Boiling
Point of
Water
Freezing
Point of
Water
100°
0°
373°
273°
212°
80°
32°
Celsius
Kelvin
Fahrenheit
-273 C
ϕ

Figure 2.3 The Celsius and Kelvin scales are used to report temperatures in scientific work. The freezing point of water corresponds to 0° on the Celsius scale, approximately 273° on the Kelvin scale, and 32° on the Fahrenheit scale. The temperature range between the freezing point of water and its boiling point is divided into 100 equal degrees on the Celsius and Kelvin scales, so that the boiling point of water can be expressed as 100°C (Celsius) or 373°K (Kelvin). The temperature of boiling water on the Fahrenheit scale is 212°F. Can you work out the Celsius and Kelvin temperatures corresponding to 80°F? Conversion tables and formulas are listed in Appendix I.

There are 100 degrees of temperature difference between the freezing and boiling points of water on the Celsius and Kelvin scales and 180 degrees of difference on the Fahrenheit scale. A temperature change of 1°C therefore corresponds to a change of 1.8°F (9/5°F).

the freezing point of pure water under normal conditions is called 0°C, and the boiling point of water, 100°C. The interval from freezing to boiling is divided into 100 equal degrees. Normal room temperature of 72° Fahrenheit is equivalent to approximately 22°C, and normal body temperature is equivalent to 37°C.

The second temperature scale, called the *absolute*, or *Kelvin*, has a different zero point from the Celsius scale. The temperature of a gas is a measure of the kinetic energy of its molecules; as a gas is cooled, the molecules move more slowly and have less energy. Theoretically, there is a temperature at which all motion would cease, and this point is taken as 0 on the Kelvin scale. Zero on the Kelvin scale is equivalent to approximately -273°C (-459°F), so that the freezing point of water on the Kelvin scale is 273°K. Many scientific laws are simpler expressed in absolute temperature because of its fundamental significance on a molecular level.

Forms of Energy

Energy comes in a bewildering variety of forms. The sun transmits radiant energy to the earth by means of sunlight. A teakettle of hot water has heat energy. A stretched rubber band has elastic energy. A tree uses stored chemical energy to grow. A baseball flying through the air has energy of motion. Any moving object possesses energy of motion—the higher the speed, the greater the energy of motion. For a given speed, the energy of motion, or the *kinetic energy*, is proportional to the mass of the object.

The principal reason energy takes part in so many different processes is that one form of energy can be transformed into another. You feel warm when you stand in the sunlight because radiant energy from the sun is being converted to heat energy in your skin. When you run, you use stored chemical energy to produce energy of motion.

Man sometimes makes use of a long chain of energy transformations to accomplish his purposes. The chain begins with the sun's energy falling on green plants. Part of the energy of sunlight is stored as chemical energy by the process of photosynthesis. If a person eats a plant he will obtain chemical energy from it. Or, a plant may decompose and eventually become a fossil fuel. Fossil fuels—coal, oil, and gas—represent radiant energy from the sun stored as chemical energy; they were produced by green plants that grew about 100 million years ago. The heat from burning a fossil fuel or the wood of a plant can be used to drive a steam engine or an electric generator, yielding electric energy. The electricity can then be used to light a light bulb, to turn an electric motor, or to heat a stove. Ultimately the energy came from the sun.

Energy and Work

Energy is the capacity for doing work, and any form of energy can do useful work if it is harnessed properly. When water flows down a hill after a heavy rain, for instance, it does work by carrying pebbles and soil particles. A form of energy associated with altitude, called *gravitational energy*, is often encountered in everyday life, such as when you speed down a hill on a bicycle without pedaling. You gain speed as your altitude decreases because your gravitational energy is being converted to energy of motion. Gravitational energy is sometimes called *potential energy*, because it represents the potential ability to do work. For objects near the surface of the earth, gravitational energy can be considered to be proportional to altitude. Gravitational energy is also proportional to mass. A large rock near the top of a mountain has more gravitational energy and will make a bigger splash in a lake below than a smaller rock located part way down, both because of its greater height and because of its greater mass.

You may view numerous processes in the physical world in terms of work and energy. When the sun's energy causes a molecule of water to leave the surface of the ocean and enter the atmosphere, work is done to break the molecule away from the water and to give it gravitational energy as it rises. Later, when the molecule falls to the earth in a raindrop, its gravitational energy is converted to kinetic energy. If the raindrop happens to fall upon a lump of soil, part of the raindrop's energy of motion may be used to break off bits of the soil.

Although energy can be changed from one form to another, the total *amount* of energy remains constant. One of the fundamental physical principles of the universe is that energy can be neither created nor destroyed. No way has ever been found to do useful work except by expending some form of energy, but the gain of one kind of energy must be accompanied by the loss of another kind. The gain of electric energy generated by a coal-burning power plant, for example, is accompanied by the loss of stored chemical energy in coal. Your body cannot continue to function unless it is supplied with energy from food.

To do work, man relies heavily on energy stored in fossil fuels rather than on the daily input of energy from the sun. The fossil fuels constitute the prime energy source for modern industrial society. It took many tens of millions of years for the earth's coal deposits to accumulate, but in the past 100 years a significant fraction of the known fossil fuel deposits has been consumed to meet the energy needs of industrialization. At the present rate of withdrawal, the known coal reserves will be depleted in less than 800 years. The most promising alternative power source is nuclear energy, which is the excess energy released when one kind of atomic nucleus is changed into another kind that has less internal energy. Nuclear energy represents energy that has been stored in the structure of atoms since the big bang 10×10^9 years ago when the elements were formed. But the supply of nuclear energy is ultimately finite. And because nuclear energy powers the sun, the supply of solar energy is limited. In a few billion years, when the supply of suitable atoms is exhausted, the sun will no longer radiate energy.

Heat Energy

Although energy is never created or destroyed, it is frequently degraded into forms that cannot be used efficiently. Imagine throwing

6
8
9
7
4
10
13
12
11
16
17
15
14

2
1
5
3
19
18
J. DAVIS

Figure 2.4 This painting interprets some of the forms of energy important in physical geography, and it illustrates ways in which energy interacts with systems on the earth.

Processes of change on the earth are powered by two sources of energy: the radiant energy from the sun (1), and the internal heat energy of the earth (14). Radiant energy drives the motion of the atmosphere (2) by causing temperature differences on the earth's surface. The wind in turn transfers energy to the ocean, producing ocean currents and waves (3). Ocean currents and the moving atmosphere help distribute energy from the hot equatorial regions of the earth to the cool polar regions. The energy of waves actively shapes coastlines (4) by eroding rocks and by transporting sand to and from beaches.

On the seas and on the land, energy from the sun frees water molecules to enter the atmosphere (5). The evaporated water is transported by the moving atmosphere and returns to the earth's surface as precipitation (6). Some of the water falling on the land runs into streams (7). As a stream flows toward lower altitudes, its gravitational potential energy is transformed to the ability to do work. Flowing water is a powerful agent of change on the landscape because of its power to erode the land and to transport and deposit sediment. In cold climates or at high altitudes, precipitation may take the form of snow (8). Accumulated snow produces glaciers (9), which are thick sheets of moving ice that scour the landscape.

Vegetation (10) absorbs solar radiant energy and transforms it to stored chemical energy by the process of photosynthesis. Some of the stored energy is passed on to animals (11), which ultimately depend on green plants as their source of food energy. Vegetation, moisture, and rock materials interact to form soils (12). The energy in part of the vegetation becomes stored as fossil fuels, which constitute the main power source for modern industrial society (13).

The energy released in radioactivity (15) contributes significantly to the internal heat energy of the earth. The temperature of the earth increases with increased depth below the surface, and the geothermal energy obtained by tapping this source of heat may become an important source of power. The plastic, partly molten rock underlying the earth's crust allows heat currents to move and deform the crust. On the local scale, deformations of the crust can result in breaks or faults (16). On the global scale, motion of the crust results in the drifting of continents, the opening of ocean basins, and the uplift of mountains (17).

At a boundary where an oceanic crustal plate converges upon a continental plate (18), the oceanic plate may be deflected downward into the deeper layers of the earth. Stresses cause deformation of the plates; deep fissures allow molten rock to come to the surface in active volcanoes (19).

a snowball at a wall. Your muscular energy gives the snowball energy of motion, but when the snowball hits the wall, it comes to a stop and loses its energy of motion. Where does the energy go? Before the snowball strikes the wall, each of its atoms has the same average speed as that of the snowball. After it hits the wall, the atoms are jostled about. Their motion is transformed into a random vibrating motion, with each atom moving in a different direction.

The random vibrating motion of atoms in a solid object or the random motion of atoms in a liquid or a gas represents heat energy. On the atomic scale, heat energy is simply the energy of motion of atoms and molecules. The hotter an object is, the more vigorously its atoms jostle about. The various forms of energy, such as chemical, electric, or radiant energy, are readily converted to heat energy, but because heat energy is associated with random disorderly motion, it is impossible to transform heat energy to other forms with perfect efficiency. The generation of heat energy accompanies all natural processes involving the transfer of energy. Some of the energy being transferred is inevitably transformed to heat, which is usually dissipated and serves only to warm the surrounding area. The conversion of some energy to dissipated heat in an energy transfer means that useful energy can never be transferred with perfect efficiency.

Units of Energy

The measurement system for energy employs a number of different units—ergs, joules, foot-pounds, calories, and British Thermal Units. The unit of energy most frequently encountered in physical geography is the *gram calorie*, often called simply the *calorie*. The calorie is defined as the amount of energy which if converted entirely to heat energy would raise the temperature of 1 gram of water from 14.5°C to 15.5°C. Although the term "calorie" is often used, the energy content of food is actually measured in kilogram calories. One kilogram calorie equals 1000 gram calories. A 100-watt light bulb consumes energy at the rate of approximately 25 calories per second. The sun delivers approximately 2 calories of radiant energy per square centimeter per minute to the top of the earth's atmosphere. (There are approximately 6.5 square centimeters per square inch of area.)

Water: An Energy Converter

Water plays a fundamental role in many physical systems by helping to convert energy from one form to another. Water flowing down a slope, for example, loses gravitational energy but gains the ability to do work in transporting rock debris.

Water is one of the most important substances on the earth. Without water to support the primordial organic soup, life would probably not have arisen. Ancient and medieval philosophers viewed water—along with earth, air, and fire—as one of the four fundamental elements making up the world. According to modern chemistry, water is not an element but consists of the elements hydrogen and oxygen in the ratio of two atoms of hydrogen for every atom of oxygen (H_2O).

Changes of Phase and Changes of Energy

The three physical forms that a substance such as water can take—solid, liquid, and vapor—are called *phases*. Important energy changes occur when a substance changes from one phase to another. Under normal conditions, the melting point of ice is 0°C (32°F), which is the same temperature as the freezing point of liquid water. But a solid at its melting point will not become liquid unless enough energy is added to break the molecules away from the strong mutual attraction they have for one another in the solid. The energy required to melt the ice is called the *latent heat of fusion*; for water, the latent heat of fusion is 80 calories per gram. Conversely, to freeze liquid water at 0°C requires the removal of 80 calories of energy per gram, which is why ice is such an effective cooling agent.

Energy is also required to remove individual molecules, or vapor, from a liquid. This energy, called the *latent heat of vaporization*, has a value of 540 calories per gram for water at the boiling point of 100°C (212°F), and 590 calories per gram for water at outdoor temperatures of 15°C (59°F) or so. The latent heats of fusion and vaporization are much greater for water than for any other common substance.

Water in the atmosphere in the form of vapor represents a large amount of stored energy, and the transport of water vapor through

Figure 2.5 Significant energy changes occur when water changes from one physical state to another. Energy must be added to change water from a physical state in which the water molecules are tightly bound to a state in which they are more loosely bound, even if the temperature of the water remains the same. To change 1 gram of water from solid ice directly into gaseous water vapor requires approximately 680 calories of energy, called the latent heat of sublimation. The change from solid to liquid requires 80 calories per gram, called the latent heat of fusion. The energy required to change 1 gram of liquid water to gaseous vapor, the latent heat of vaporization, varies slightly with the temperature of the water. For water at 15°C the latent heat of vaporization is 590 calories per gram, and for water at 100°C it is 540 calories per gram.

Energy must be removed to change water from a state in which the molecules are loosely bound to a state in which they are more tightly bound. The amount of energy that must be removed is numerically equal to the latent heat; for example, 590 calories must be removed from 1 gram of water vapor at 15°C to form 1 gram of liquid water at the same temperature.

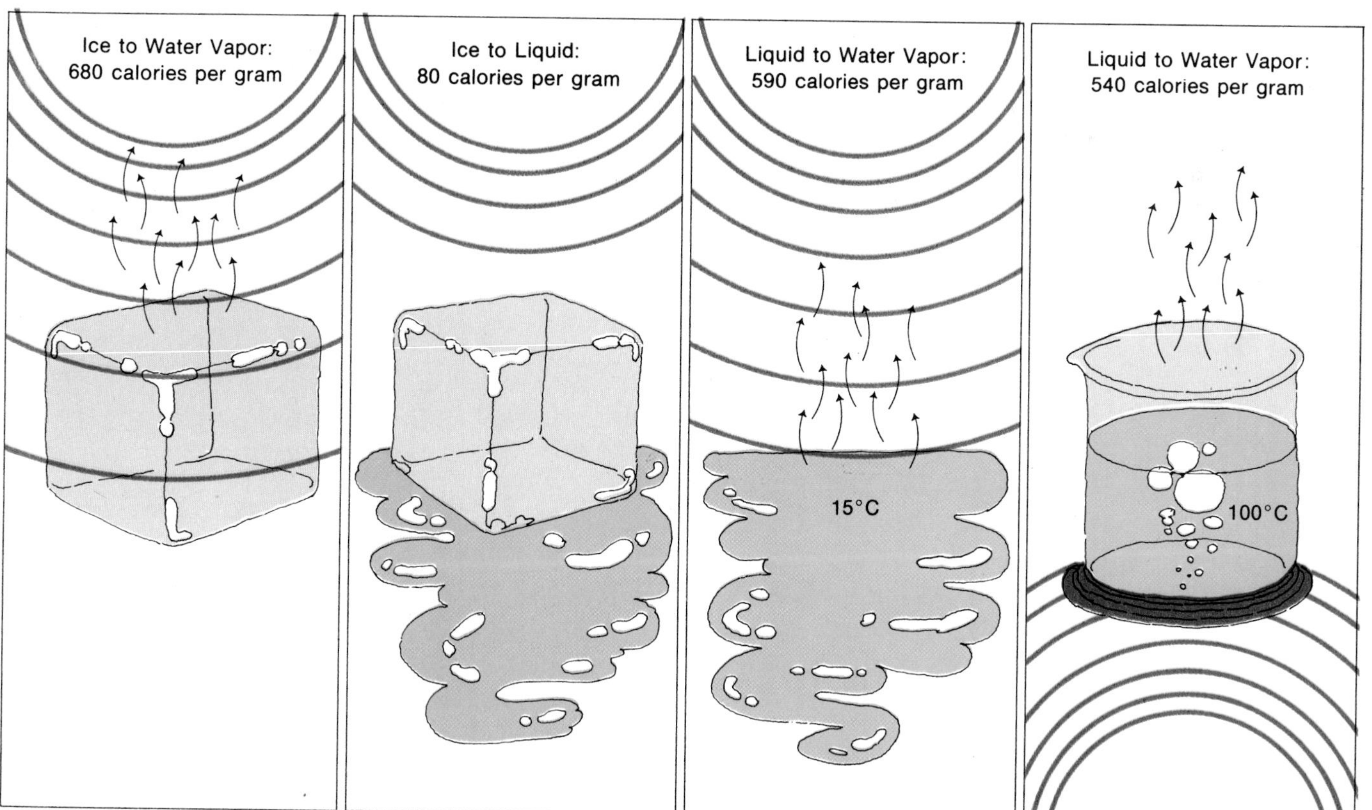

Ice to Water Vapor:
680 calories per gram
Ice to Liquid:
80 calories per gram
Liquid to Water Vapor:
590 calories per gram
15°C
Liquid to Water Vapor:
540 calories per gram
100°C

Figure 2.6(a) The pressure, volume, and temperature of a fixed amount of gas are interdependent; a change in one of the three quantities is always accompanied by a change in one or both of the others. Suppose, for example, that a gas is confined in a sealed box so that its volume is constant. Changing the temperature of the gas changes the pressure the gas exerts on the walls of the box; the gas molecules move more rapidly at high temperatures and therefore collide more forcefully and more frequently with the walls. For gases under normal conditions pressure increases essentially linearly with increased temperature if the volume is held constant.

(b) A parcel of air in the atmosphere is unconfined, and under steady conditions its pressure becomes the same as the external pressure of the surrounding air. If a parcel of air under such conditions is heated, its volume expands until its internal pressure is once more in balance with the external pressure. The volume of the parcel increases linearly with increased temperature for a gas held at constant pressure. The greater the volume occupied by the gas molecules in the parcel, the lower the density of the gas. Density therefore decreases with increased temperature for a gas at constant pressure.

(c) The transfer of energy to and from parcels of air in the atmosphere can produce motion. A parcel of air that is hotter than the surrounding atmosphere is less dense than the atmosphere; in such a situation, the upward buoyant force on the parcel exceeds the downward weight force, and the parcel rises. Conversely, if a parcel of air is cooler than the surrounding atmosphere, it is denser. The downward weight force in this case exceeds the upward buoyant force, and the parcel descends.

the atmosphere can be viewed as the transport of energy. Latent heat is transferred vertically from the surface to the atmosphere, and then it is transported horizontally through the atmosphere until it is released during condensation. When molecules of water vapor form raindrops, 590 calories are released per gram of water. This energy is the principal source of power in thunderstorms and hurricanes. Similarly, the evaporation of water from moist ground or from the ocean by the sun's radiant energy represents an expenditure of energy; 590 calories are required to change each gram of water from liquid to vapor. Water can also change directly from solid ice to vapor, at an energy cost of nearly 700 calories per gram.

Physical and Chemical Properties

In addition to its ability to change phase at the temperatures ordinarily encountered on the earth, water possesses other properties of importance in physical geography. One of water's outstanding qualities is its excellence as a reservoir of heat energy. The amount of heat energy required to raise the temperature of 1 gram of a substance at normal atmospheric pressure by 1°C is known as the *specific heat* of the substance. The specific heat of liquid water is nearly equal to 1 calorie per gram of water per degree Celsius over the temperature range from freezing to boiling. The specific heat of water is about 5 times greater than the specific heat of soil or air; that is, the amount of heat that will raise the temperature of a body of water by 1°C (1.8°F) will raise the temperature of an equal weight of soil by about 5°C (9°F). Similarly, if equal

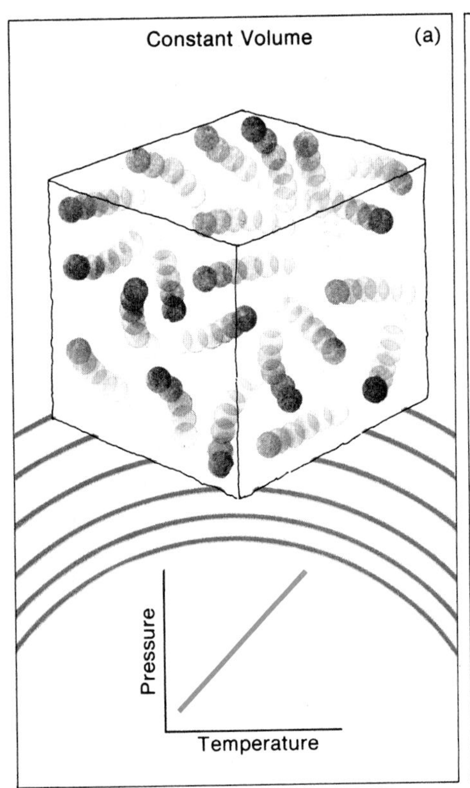
Constant Volume
(a)
Pressure
Temperature

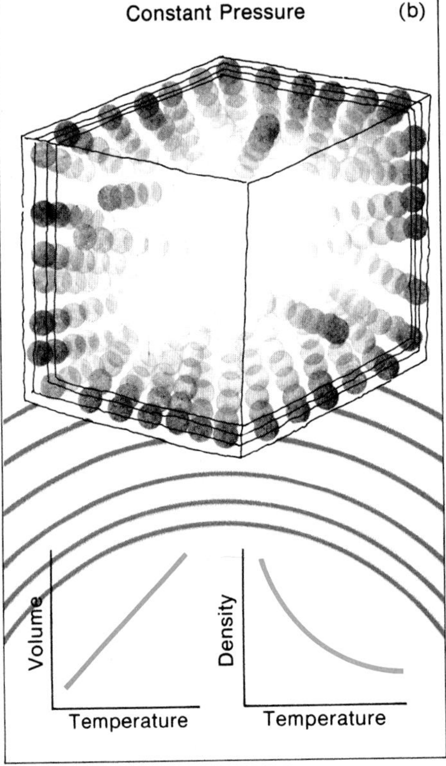
Constant Pressure
(b)
Volume
Temperature
Density
Temperature

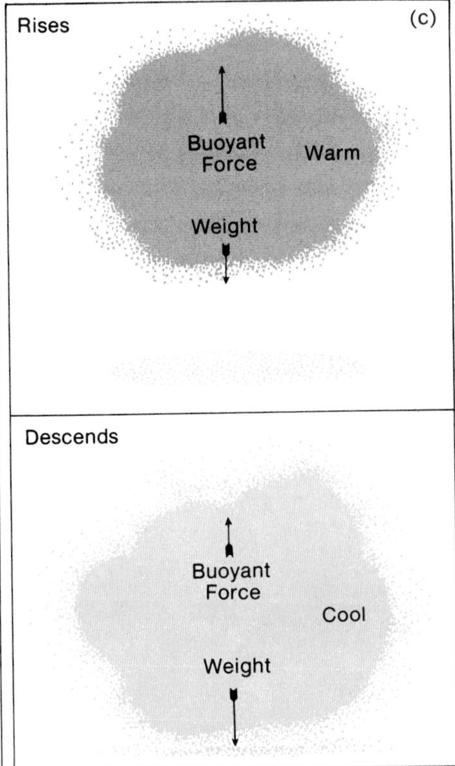
Rises
(c)
Buoyant Force
Warm
Weight
Descends
Buoyant Force
Cool
Weight

weights of soil and water lose the same amount of energy, the soil will cool much more than the water. This is one of the reasons why the temperature fluctuates much less in the oceans than it does on land.

Water also has remarkable *chemical properties*. Nearly a universal solvent, water dissolves almost anything to some extent. Water molecules produce strong electric forces that hold substances dissolved in the water in solution, but the water molecules do not react chemically with the dissolved substance. Because water can carry materials in solution without affecting them chemically, it is an ideal fluid for transporting substances through the circulatory systems of plants and animals.

Water plays an important part in *weathering*, which is the breakdown and alteration of rock materials near the earth's surface. Weathering reduces rocks to particles that can be swept away by wind or water. Weathering proceeds by physical and chemical processes, sometimes with the help of living organisms, and often with the help of water. For example, the pressure water exerts when it freezes in a crack and tries to expand can be great enough to split a rock. Such frost weathering is a major cause of rock disintegration in climates where the temperature range allows repeated thawing and freezing. Water can weather rocks chemically as well as physically. Limestone is particularly susceptible to chemical weathering by water, and areas with extensive limestone beds are noted for the caves and underground channels that water created.

The Atmosphere: An Energy Transporter

A second fundamental substance, the air, is also an essential part of the earth's physical systems. The earth's atmosphere is central in physical geography because it is the medium that transports energy in the form of water vapor over the earth's surface. Because the atmosphere is a mixture of gases, familiarity with the behavior of gases provides a foundation for understanding the detailed discussions of atmospheric phenomena and atmospheric circulation that appear in Chapters 4 and 5.

Unlike the molecules in liquids or in solids, the molecules that constitute a gas are not strongly bound to one another. They are constantly moving, and they frequently collide with one another and with neighboring surfaces. When a gas molecule collides with a surface, it exerts a push on the surface. The total force molecular collisions exert on a unit area of surface at any given time is called the *pressure* of the gas.

The pressure the earth's atmosphere exerts decreases rapidly with height; at an altitude of 5 or 6 kilometers (3 or 3.6 miles), the atmospheric pressure is only half the pressure at sea level. The molecules at sea level must exert enough of an upward push or force to sustain the weight of the entire atmosphere, whereas at an altitude of 5 or 6 kilometers, the molecules must sustain the weight of only half as much air. Standard atmospheric pressure at sea level is specified as 760 millimeters (29.9 inches) of mercury, which is the height of the mercury column that can be supported by standard atmospheric

Figure 2.7 Atmospheric pressure varies with altitude. The pressure exerted by air in the atmosphere decreases rapidly with increased altitude. As the graph indicates, pressure at sea level is approximately 1,000 millibars, but at an altitude of 5 kilometers (3 miles) it has fallen to about 540 millibars, about half the sea level value. Note that the pressure on the graph is not evenly spaced in pressure units but is instead marked off in successive powers of ten. Such a *logarithmic scale* is useful when graphing a quantity that varies over a wide range; for instance, the logarithmic scale on this graph indicates a pressure range from less than 10 millibars to 1,000 millibars. The pressure at sea level can be thought of as the force per unit area exerted by air molecules at sea level to support the weight of a column of the earth's atmosphere. The graph shows the standard value of atmospheric pressure at each altitude; actual atmospheric pressure at a particular location on a particular day can be a few percent higher or lower than the standard value. If the air is denser than usual, the pressure is higher than normal, and if the air is less dense, the pressure is lower than normal.

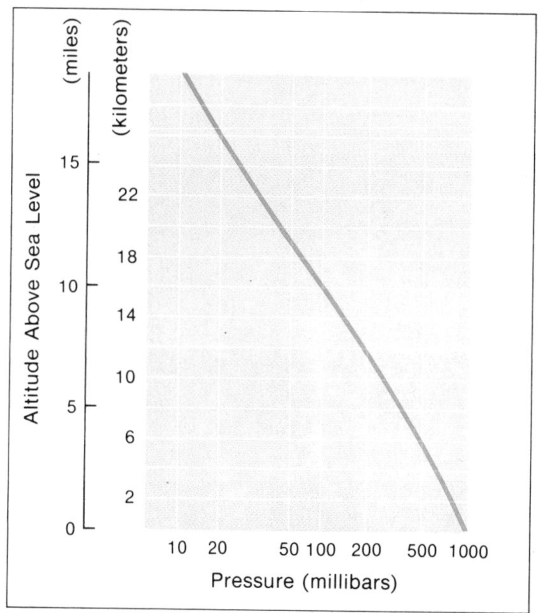

Altitude Above Sea Level
(miles)
(kilometers)
15
10
5
0
22
18
14
10
6
2
10
20
50
100
200
500
1000
Pressure (millibars)

pressure. In meteorology, pressures are usually quoted in *millibars.* Standard atmospheric pressure is equivalent to approximately 1,000 millibars.

When a gas is heated in a closed container, the pressure increases because the molecules move more rapidly at higher temperatures and therefore collide more frequently and more forcefully. If a parcel of unconfined air in the atmosphere is heated, the parcel's volume increases and its pressure decreases until the pressure is equal to the pressure of the surrounding cooler air. Because the air in the heated parcel then has fewer molecules in a given volume than does the cooler air, the parcel begins to rise like a hot air balloon. A parcel of air that is warmer than the surrounding air tends to rise, and a parcel that is cooler than the surrounding air tends to descend. By heating or cooling air, the transfer of energy in the atmosphere can cause movement of the air.

Systems: Energy Exchange

Both the water of the earth and the air are parts of *systems* that interact with one another and with the environment. Energy in its various forms drives the interactions within and among the systems. The earth as a whole can be considered a single system, but it contains so many objects interacting in such a wide variety of ways that the task of understanding them all at one time is extremely complicated. Fortunately, our understanding can be simplified by mentally subdividing the world into smaller systems, such as systems of soil, of vegetation, and of landforms.

Consider several streams flowing down a mountain slope into a basin. Some of the interactions involved in this landform system are the effect of gravity on the motion of the water, the erosive action of water on the underlying rock of the stream bed, the movement of sediment, and the influence of surrounding vegetation and soil on the rate of water runoff from the mountainside into the stream. The system in this example is an *open* system—that is, it receives material or energy, called *inputs,* from outside. In the stream system, the rain that falls on the slope is an input. The end effects a system produces in response to the inputs are called *outputs.* In a stream-slope-basin system, one of the outputs could be the rate at which sediment is deposited in the basin and another could be the peak rate of streamflow under heavy rain conditions.

Ultimately, all systems on the earth are linked by their interactions. The output from one system represents the input to another system. Rainfall is an output of the atmospheric system of water delivery and an input to vegetation and landform systems on the earth's surface.

Although most systems are only partially understood at present, viewing physical geography in terms of systems is a valuable way to organize knowledge, because all systems exhibit general modes of behavior in response to inputs or other changes. Although energy is required to produce change in systems, many systems tend toward an *equilibrium* state in which continued input of energy produces essentially no change in one or more outputs. Consider again a stream flowing toward the sea. A system in equilibrium may show short-term changes, but its average behavior remains effectively constant. A stream in flood may rapidly deepen its bed by erosion, but when the flow returns to normal, the stream channel returns to its original shape and size by depositing sediment to fill the eroded portions of the bed. The average form of the channel remains much the same over long time periods, although there are short-term fluctuations.

The approach toward the equilibrium state takes different amounts of time for different systems. A system's *response time* is a measure of the time required to reach equilibrium. Suppose a volcano erupts and covers the ground with volcanic mineral ash. The chemical composition of the ash gradually changes in response to weather conditions. When plants begin to grow and organic matter becomes mixed with the ash, the composition changes further. Over a sufficient period of time, the ash is transformed to a true soil in equilibrium with its surroundings.

The response time of a system is also a measure of the time required for a system to change from one state of equilibrium to another in response to changed inputs. If a river is dammed, the amount of sand and silt the river transports to the sea is often significantly reduced. A beach on the coast near the river's mouth then receives less sand to replace the sand carried away by ocean waves. The shape and extent of the beach will change for a few years or decades after the dam is built until a new equilibrium is achieved.

Some systems respond to inputs of energy or material by *storing* some of the energy or material. Rocks deep in the earth can steadily store energy from continental drift, much as a compressed spring stores energy. An earthquake occurs when the store of energy is suddenly released.

Many natural systems are vulnerable to inputs that result from the increased tempo of man's industrial and agricultural activities. By understanding the principles and interactions of a system, you can see where a system is most sensitive to change and what the effect of altered inputs will be.

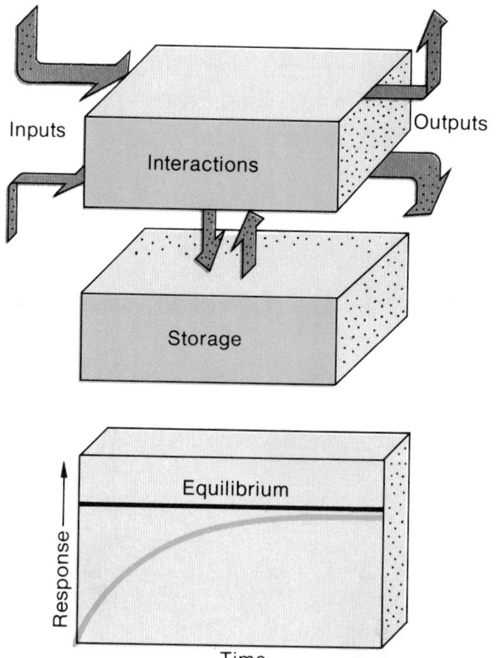
Inputs
Interactions
Outputs
Storage
Equilibrium
Response
Time

Figure 2.8 A system is any collection of interacting objects. Physical geography is concerned with a wide variety of systems, including landforms, soils, vegetation, and the atmosphere. Thinking about the world in terms of systems is a useful concept because systems, whatever their nature, have many features in common, some of which are illustrated in this schematic diagram. The inputs to a system represent material and energy received from outside the system's boundaries. The inputs are transformed by system interactions to new forms that are either placed in storage for a time or transformed to other systems as outputs of material and energy. Consider a field of corn as an example of a system: the inputs include radiant energy from the sun, carbon dioxide from the atmosphere, and moisture and nutrients from the soil. Part of the input is stored as chemical food energy in the process of photosynthesis. Outputs include the water vapor returned to the atmosphere by plant transpiration, the free oxygen released in photosynthesis, and the food energy that eventually is used to feed animals.

The lower diagram illustrates schematically one of the ways that systems respond in time to an input. In this mode of systems behavior, the output does not reach full strength immediately upon application of an input. Time is required for a steady condition to be achieved. Some systems, such as the atmosphere, respond rapidly to new inputs; other systems, such as soils, may require hundreds or thousands of years to come to equilibrium.

Budgets of Energy and Matter

Inputs, outputs, storage, and balance constitute parts of the *budget* of a system, a concept that has proved useful in physical geography for visualizing the distribution of energy or water in complex systems. The word "budget" suggests a procedure of financial accounting that involves balancing income, savings, and expenses.

A simple illustration of the budget concept can be seen in a system that consists of a small pond. A pond without an outlet gains water from rainfall (input) and loses water through evaporation into the atmosphere (output). The principle of the water budget is the recognition that because all the water must be accounted for, the difference between input and output over a given time span represents the net amount of water added to the pond or the net amount of water lost. A detailed knowledge of interactions is necessary to make the budget a quantitative tool, and no interactions should be overlooked. If the bottom of the pond is porous, for instance, water that leaks out of the pond should be included as output in the budget. Similarly, water that drains into the pond from surrounding fields should be included in the input. A detailed discussion of the water budget, which describes the distribution of moisture that falls on the land, is presented in Chapter 6.

Budgets can also be drawn up for energy. All the energy that enters a system either is stored or appears as output. In physical geography, the most important application of energy budgets is to the sun's radiant energy input to the earth.

Cycles of Materials: Synergy

Small changes occur over the world from one year to the next, but they add up to large changes over billions of years if other changes do not occur to restore the balance. Rivers and streams carry water from the land to the oceans, and if water did not *cycle* through the atmosphere from the oceans back to the land, eventually there would be no water on the land. The passage of materials through complete cycles on the earth makes renewal and change possible without a continual supply of new material (see Figure 2.9). In this sense fresh water, for example, is a renewable resource; the water used in agriculture is not lost but eventually returns to the oceans after passing through several systems.

The ability of systems to store materials affects the operation of cycles. On the average, water that evaporates into the atmosphere from the land or from the oceans remains stored in the atmosphere for a week or two, which is enough time for the water to be transported long distances. Water that falls as snow on the Antarctic ice cap, however, may be incorporated into glacial ice for thousands or millions of years before it returns to the oceans.

The Hydrologic Cycle

Where is the water of the earth located? Not all of it is available as free water; an undetermined amount is chemically bound in minerals within the earth. From time to time, volcanic activity releases some of this water to the surface in the form of vapor. The salt water in the ocean basins constitutes 97 percent of the free water on the earth. The next largest store of water is in the glaciers and the polar ice caps. Most of the remaining water is stored in porous rock not far below the surface of the ground, forming the groundwater that is tapped by wells. Lakes, ponds, and rivers contain only a fraction of 1 percent of the total free water on the earth.

The quantity of water contained as vapor in the atmosphere is small, but it plays a crucial role in the *hydrologic cycle*. The fundamental processes of the hydrologic cycle are the *evaporation* of fresh water from the ocean and from the land, the *transport* of water vapor through the atmosphere, and the *return* of water to the earth's surface by *precipitation*.

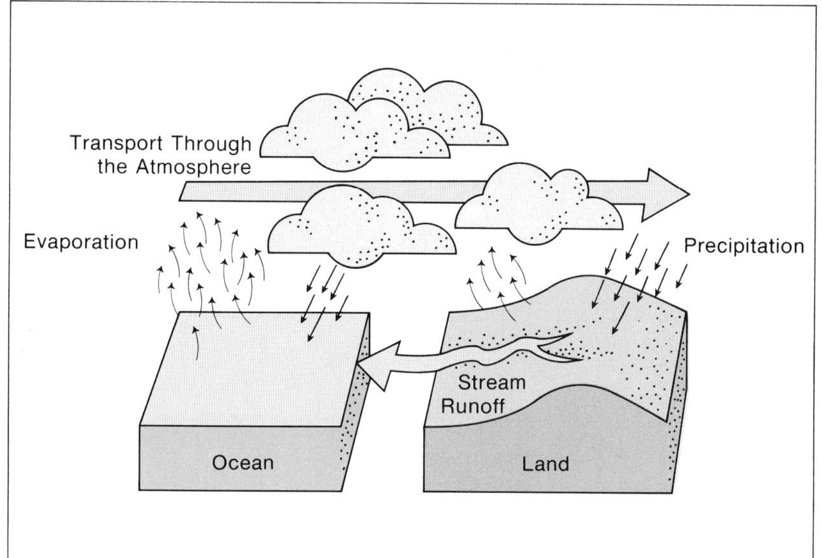
Transport Through the Atmosphere
Evaporation
Precipitation
Stream Runoff
Ocean
Land

Figure 2.9 The hydrologic cycle is a key element in physical geography, because water has important interactions with systems such as vegetation and landforms. The processes involved in the cycling of water include the evaporation of water from the oceans, the evaporation and transpiration from the continents, the transport of water vapor through the atmosphere, and the return of water from the atmosphere to the surface as precipitation. The annual loss of water from the oceans exceeds the amount they gain from precipitation, but the deficit is made up, on the average, by streams flowing from the continents into the oceans.

The average amount of water evaporated from the oceans annually exceeds the amount received from precipitation. Ocean levels are constant, however, because rivers carrying runoff water from the land restore the balance. Averaged over the entire land surface of the earth, precipitation exceeds evaporation. Some of the water falling on the land seeps through the soil and into porous rock to become groundwater. Many cities depend on groundwater for their water supplies even when large rivers are nearby.

Man's Input to Cycles

Man's activities have grown to such a scale that he is now a significant influence in the operation of a number of cycles. In some areas groundwater is being pumped out more rapidly than the hydrologic cycle can replenish it. In Houston, Texas, for example, excessive withdrawal of water from porous clay beds under the city has resulted in compression of the clay and a consequent sinking of the ground level by as much as 1 meter (40 inches).

The carbon cycle provides another example of man's influence in the operation of cycles. The extensive burning of fossil fuels for energy increases the carbon dioxide content of the atmosphere. Study of the carbon cycle shows that part of the carbon dioxide from the atmosphere eventually dissolves in deep ocean water, where it remains in storage for many years. However, the ocean water takes up carbon dioxide slowly, and at present the atmosphere's gain of carbon dioxide from industrial activities exceeds the loss to other systems.

The Classification of Systems

The study of systems simplifies understanding of all the interactions that occur at the earth's surface, and the delineation and classification of regions on the earth facilitates understanding of soils, vegetation, and climates. The earth exhibits such a variety of landscapes

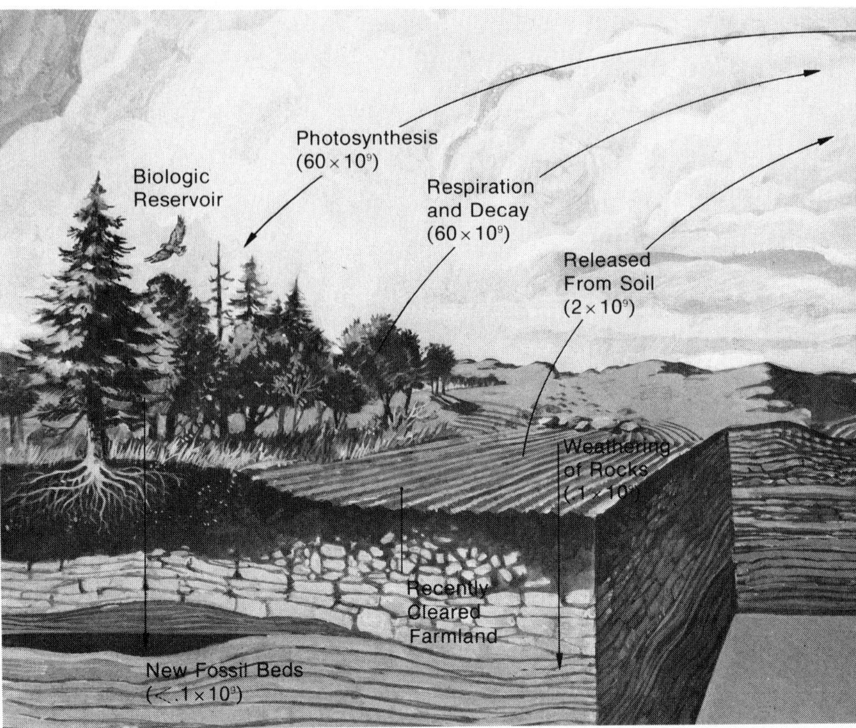

Biologic
Reservoir
Photosynthesis
(60 × 10⁹)
Respiration
and Decay
(60 × 10⁹)
Released
From Soil
(2 × 10⁹)
Recently
Cleared
Farmland
New Fossil Beds

Figure 2.10 This diagram illustrates the large number of systems through which carbon can pass during its cycle. The numbers in parentheses are estimated values of the tons of carbon dioxide released or absorbed annually in each process or the total amount stored in each reservoir; 1 ton of carbon dioxide contains approximately 550 pounds of carbon atoms.

Transfer of carbon dioxide to and from the atmosphere is an essential part of the carbon cycle. Carbon dioxide enters the atmosphere primarily from the respiration and decay of organisms, from the burning of fossil fuels, and from the carbon dioxide dissolved in the oceans. Photosynthesizing green plants and the ocean absorb carbon dioxide from the atmosphere. Some of the carbon absorbed by plants is locked into long-term storage as fossil fuels, and some of the carbon absorbed by the ocean is stored in sediment beds of marine shells.

The transfer of carbon dioxide to and from organisms and to and from the ocean is nearly in balance. Much of the net amount of carbon dioxide added to the atmosphere each year is a result of man's industrial activities, but approximately half of the net increase is taken up by the oceans. The carbon dioxide content of the atmosphere is increasing by several percent each decade. The possible effects of this increase on the earth's utilization of solar radiant energy is discussed in Chapter 3. (After "Carbon Dioxide and Climate" by Gilbert N. Plass, copyright © 1959 by Scientific American, Inc. All rights reserved)

and climates that it would be impossible to study them in detail without some sort of classification system. Consider, for example, the differences between the Appalachian Mountain country in the eastern United States and the desert country of the Southwest. The forested hills and abundant moisture in the Appalachians contrast with the sparse clumps of grass and the barren, rocky landscape of the desert. Differences can be noted in the amount of rainfall, in the temperature, and in the soil, vegetation, and landforms.

When the earth is considered as a whole, similarities as well as differences among regions become apparent. The southwestern deserts in the United States are similar in some ways to the deserts of central Asia. One of the tasks of physical geography is to classify regions and systems according to their similarities. Systems that are subjected to similar inputs make similar responses; the amount and distribution of energy and of moisture in a region profoundly affect the response of systems in the region.

In later chapters you will see examples of several classification schemes. Because classification schemes represent a human attempt to simplify nature, they cannot possibly take into account all the subtle variables in the physical environment. Any attempt to draw a boundary between one region and another neglects the continuous grading of properties that are actually observed in the world. Southern California, for example, receives only a few inches of rain each year and is "dry"; northern California receives many inches of rain each year and is "moist." Where is the dividing line between California's "dry" and "moist" regions? There is no clear boundary between one region and another, and classification schemes can never

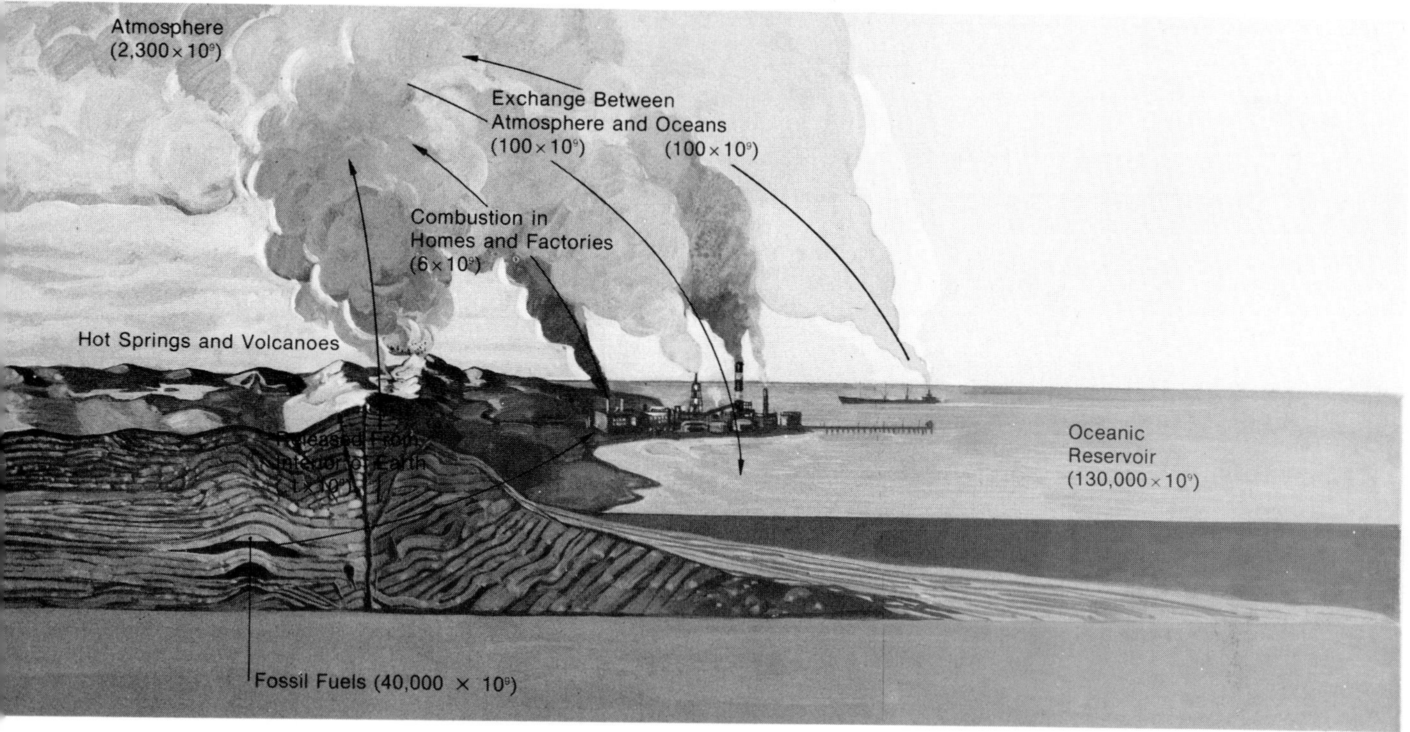
Atmosphere
(2,300 × 10^9)
Exchange Between
Atmosphere and Oceans
(100 × 10^9)
(100 × 10^9)
Combustion in
Homes and Factories
(6 × 10^9)
Hot Springs and Volcanoes
Oceanic
Reservoir
(130,000 × 10^9)
Fossil Fuels (40,000 × 10^9)

embody the infinite degree of variation actually present on the earth's surface. Nevertheless, they form a useful framework for thinking about the world.

Summary

Physical geography is concerned with change, and change can be accomplished only by the expenditure of energy. Energy takes a wide variety of forms; the forms especially important in physical geography include radiant energy, heat energy, gravitational energy, kinetic energy, and the energy associated with water's changes of phase.

Water is important in physical geography as the sustainer of plants and animals and as the sculptor of landforms. Water is able to do work because of energy ultimately derived from the sun. The atmosphere transports energy in the form of moisture from place to place and makes it available as inputs to various systems on the earth's surface. Some of the outputs, or responses, of systems tend toward a state of equilibrium if the inputs remain relatively constant. Time is required for a system to reach a state of equilibrium, and inputs may change before equilibrium is achieved.

Although materials such as water are cycled through systems and can be used over and over again, the atmospheric delivery of water is not uniform over the surface of the earth. Differences in the amounts of energy and moisture delivered to different regions cause systems in each region to follow various lines of development, which leads to a regional differentiation of climate, vegetation, soils, and landforms on the earth's surface.

Graphic Essay

Understanding Nature

The world around us is a wonderfully complex mosaic, the patterns of which have tantalized our curiosity perhaps from the very beginnings of humankind. What is the sky, the earth, the ocean? Why do rivers flow as they do, and why does the land take the forms that it does as it sprawls out in extraordinary configurations to the horizon? And how do the unique bits of the mosaic fit together to create our impressions of an ever-changing natural world?

In our attempts to sort through all the myriad impressions we have of the world, we tend to construct *models* in our mind of what things and events are and how they are interconnected. With models we find a way of perceiving order in nature, of seeking deeper understanding of why and how things happen as they do. Consider the word "tree." For one of us a pine tree may come to mind as the model for "tree"; for another, an elm, or a centuries-old redwood. Even though there are differences among the trees, they nevertheless have enough similarities to be understood by all of us as falling within that category, and we share an understanding of that model.

Now consider the botanist, whose model may be a little more complex: he views a "tree" as a marvelous microenvironment, where different creatures occupy different niches from roots to treetop, where an intricate energy exchange goes on between the inhabitants and the surrounding environment. His model explains much more because

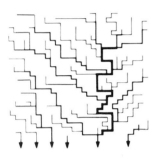

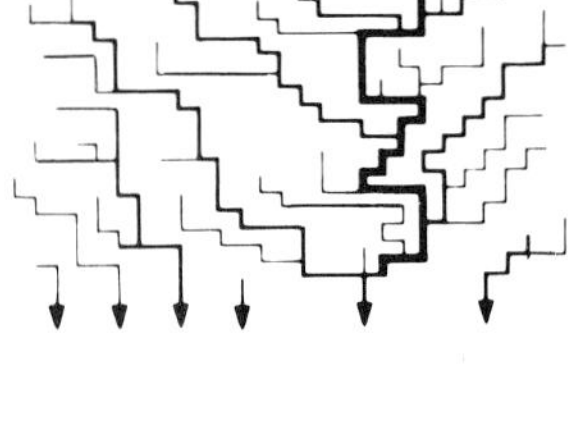

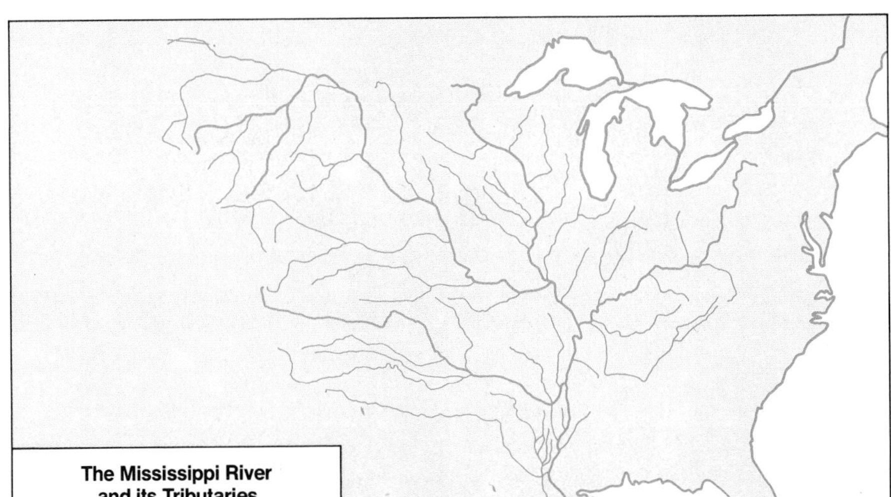

The Mississippi River and its Tributaries

1 | 2 | 3

1. Wherever rainfall runs off the land it dredges its own drainage system. This map, which is an example of a *graphic model,* shows the land area drained by the Mississippi River system (see also Chapter 12). Graphs, photographs, and diagrams are types of graphic models used to aid our conceptualization of the physical world.
2. Drainage patterns may also be simulated by computers, as this example of a *mathematical model* shows. This model is used to study the effects of various types of earth surface on the development of stream patterns and to predict changes that will occur in underground supplies of water as water is added or withdrawn. Computer models can also be used to simulate changes in variables such as precipitation or runoff to see the effect on a total system. In this illustration, the curved lines represent contours of

he has probed a little deeper into a seemingly simple part of the world and has entered still another, miniature world in its own right.

Like all other fields of inquiry, physical geography claims many models that are useful in explaining the natural world on many levels, from many vantage points. The words "cycle" and "system" and "budget" evoke fascinating models in the mind of the physical geographer, models that you will continue to read about in this book. As you come across explanations of them, however detailed they may be, keep in mind that they are the physical geographer's perceptions of the very same world you live in—a world that you can come to know that much better through his eyes.

For example, man has devised innumerable ways to look at and understand the nature of one of the earth's most significant and most common substances, water. Water covers three-fourths of the earth's surface; it is also beneath the surface, permeating the soil and rock, and it is in the atmosphere, circulating in the form of vapor. The water cycle mentioned earlier and described more fully in later chapters is a model to explain how all the earth's original supply of water is continually recycled. The same water is pumped time and time again from the land and oceans into the air; it then falls onto the land and the excess is transported by streams back to the sea. The essence of all life, water supports people, plants, and animals; it shapes the landscape; it influences global climates and local weather; it affects soil formation; and it serves as a source of power.

Different types of models describe different properties of water. A chemist uses a molecular model to explain water's fundamental make-up of two hydrogen atoms and one oxygen atom. A physical scientist is interested in water's unique ability to form three states of matter. A physical geographer is concerned about water as a transporter of heat energy through the atmosphere and as an agent of change on the earth's surface. So he has developed models that describe how water flows across the earth in streams, in rivers, and even in glaciers; he has conceptions of how it circulates in the atmosphere and ocean and how it seeps through pores in the ground. Several models of flowing water are illustrated here as examples of the kinds of models that are useful in physical geography.

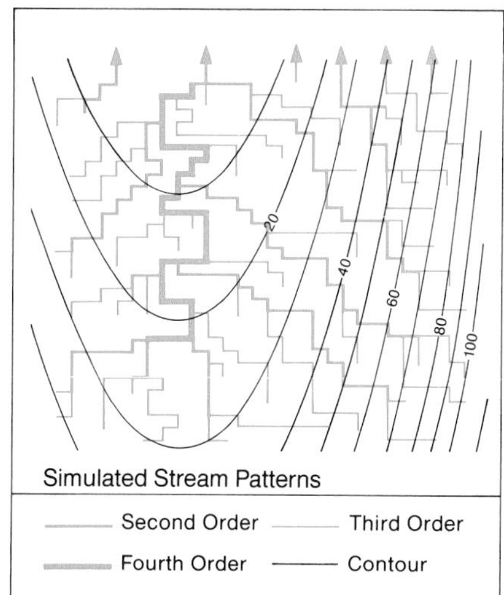
20
40
60
80
100
Simulated Stream Patterns
Second Order
Third Order
Fourth Order
Contour

elevation and the other lines show the development of a drainage pattern on homogeneous material (see Chapter 12). The water flows down the slope in the direction of the arrow. Computer models are used in the oceanic and atmospheric systems as well (see Case Study, Chapter 5) and in other branches of science where complex systems with many variables are involved.

3. An actual replica to scale of the Mississippi River system is an example of a *physical model.* It gives engineers answers to real-life emergencies that arise in flood conditions. During the flood of 1973 (see Case Study, Chapter 12), this model was a valuable addition to computer models in calculating the amount of water-rise that would accompany the opening of a spillway. By using automatic instruments to make graphic recordings of water levels at critical points, inputs and outputs can be tested on the model and the effects of proposed engineering projects can be predicted. Built and operated by the United States Corps of Engineers, the accurate miniature model of the river basin is laid out on 220 acres of land near Clinton, Mississippi. The portion shown in this photograph is the Atchafalaya River basin in southern Louisiana.

3 Energy and Temperature

The Earth-Sun Relationship
The Earth's Tilted Axis
Solar Altitude and Azimuth

Solar Energy Input to the Earth
Total Daily Solar Radiation at the Top of the Atmosphere
The Electromagnetic Spectrum and the Quality of Sunlight
Interaction of Radiation With the Atmosphere

Energy Balance in the Earth-Atmosphere System
Reflection and Albedo
Energy Loss and the Atmosphere
Radiant Energy Balance
Conduction, Turbulent Transfer, and Evapotranspiration
Energy Balance at the Earth's Surface
Heating and Cooling of Land and Water
Temperature Regimes
Modification of the Energy Balance

The sun radiates light and heat, providing the sustenance for all life processes by complex exchanges of visible and invisible forms of radiant energy. The heat of the tropics and the cold of the polar zones are consequences of the earth-sun relation in space and of energy transfers on the earth.

Number 8 by Mark Rothko, 1952. (From the collection of Mr. and Mrs. Burton Tremaine, Meriden, Connecticut)

Energy and Temperature

Basking in the warm sun is a pleasure both people and animals enjoy, and a clear, sunny day seems to refresh everyone's spirits. Radiant energy from the sun is by far the most important source of energy for the processes important on man's time scale of days and years. Radiant energy drives the passing clouds, supplies energy to green plants, lifts water from the oceans to the atmosphere, and keeps the earth at a comfortable temperature.

The amount of energy the sun delivers to a particular location on the earth depends on the position of the sun in the sky. When the sun is low in the sky, as at sunset or during a northern winter, its rays have less heating power. The atmosphere also influences the amount of energy received from the sun. The summer sun at midday brings sunbathers out to the beach, but if a passing cloud blocks the sun's rays, a chill sweeps over the beach.

What happens to the radiant energy that reaches the earth? Energy cannot be destroyed, so all the energy input can be accounted for. If the earth continually received solar energy and did not dissipate any energy, the oceans would boil and the land would be scorched. But the average temperature of the earth remains much the same from year to year, because the earth returns as much energy to space as it receives from the sun, on the average.

This chapter examines the interactions that influence the earth's gain and loss of radiant energy. Because human activities can influence many of these interactions, man may be exerting a significant degree of control over the climate of the earth. But the first step toward understanding the earth's radiant energy balance is to consider the astronomical factors that determine the amount of sunshine that reaches each location on the earth.

The Earth-Sun Relationship

Like every planet, the earth follows a fixed path, or *orbit*, around the sun. The average distance from the earth to the sun is 149.5×10^6 kilometers (92.9×10^6 miles), but the true distance never differs from the average by more than 1.8 percent because the earth's elliptic orbit is nearly circular. The earth is closest to the sun early in January and farthest from the sun early in July. It completes one orbit in a year, so a year is the natural period for discussing the variation of sunlight and energy input at a particular location. Superimposed upon the annual variation is the more rapid daily cycle caused by the rotation of the earth on its axis. The earth completes approximately 365¼ rotations for each revolution about the sun. Because the earth is essentially spherical in shape, the sun illuminates only half of the globe at any time. The boundary line between light and dark halves is the *circle of illumination* (see Figure 3.2).

The Earth's Tilted Axis

As the earth rotates on its axis, each part of its surface becomes light, then dark, and then light again. If the earth's axis of rotation were perpendicular to the plane of its orbit, every place on the earth would have 12 hours of daylight and 12 hours of darkness each day of the year, and the seasonal variation from long summer days to brief winter days would not occur. But the axis of the earth is not perpendicular to the plane of the earth's orbit; it tilts at an angle of 23½° from the perpendicular, so the duration of daylight varies throughout the year at a given location.

Figure 3.2 illustrates that as the earth journeys around the sun, the orientation of the earth's axis remains essentially fixed. For this reason, a location at a given latitude receives different amounts of sunlight through the year. About December 21 or 22, for example, the sun appears to be directly overhead at latitude 23½°S. This parallel is called the *Tropic of Capricorn*, and the particular instant when the sun is directly overhead at this parallel is known as the *winter solstice*. The sun never appears directly overhead at any latitude south of the Tropic of Capricorn. In the northern latitudes at winter solstice, the sun is low in the southern sky, and from any location south of the Tropic of Capricorn, the sun always appears toward the north. The European explorers who navigated south of the Tropic of Capricorn for the first time were surprised to see the sun on the right at midday as they sailed toward the west.

Although half of the earth is illuminated at any given time, the circle of illumination rarely falls along a meridian because of the tilt of the axis. The proportion of each parallel of latitude that lies in the illuminated part of the globe therefore varies from north to south, as Figure 3.2 illustrates. You can estimate the length of daylight from the fraction of the parallel that is illuminated. Consider the situation at winter solstice, when the 0° parallel at the Equator lies half in the illuminated zone, so that day and night are of equal duration. North of the Equator, proportionately less of each parallel lies in sunlight, which means that nighttime is longer than daytime. In northern latitudes, the late rising and early setting of the sun in late autumn is a familiar fact. North of latitude 66½°N, which is the latitude of southern Greenland, the parallels lie completely in darkness and the sun is not visible at all. South of the Equator, proportionately more of each parallel lies in the illuminated zone, which means that daytime is longer than nighttime. For latitudes south of 66½°S, in Antarctica, there is no nighttime at all—the sun is always above the horizon at winter solstice.

As the earth traverses its orbit, the relation of its axis to the sun changes. Figure 3.2 shows that about June 21, 6 months after winter solstice, the sun is overhead at latitude 23½°N, the parallel known as the *Tropic of Cancer*. The instant the sun is overhead is called the *summer solstice*. At that time of year, the extreme north receives constant illumination, while latitudes south of 66½°S are in constant darkness.

At two intermediate times—about March 21 and September 21—the sun is overhead at the Equator. The instant the sun passes over the Equator is called the *vernal equinox* (March) or the *autumnal equinox* (September). Only at these times of the year does the circle of illumination pass through the poles and coincide with meridians

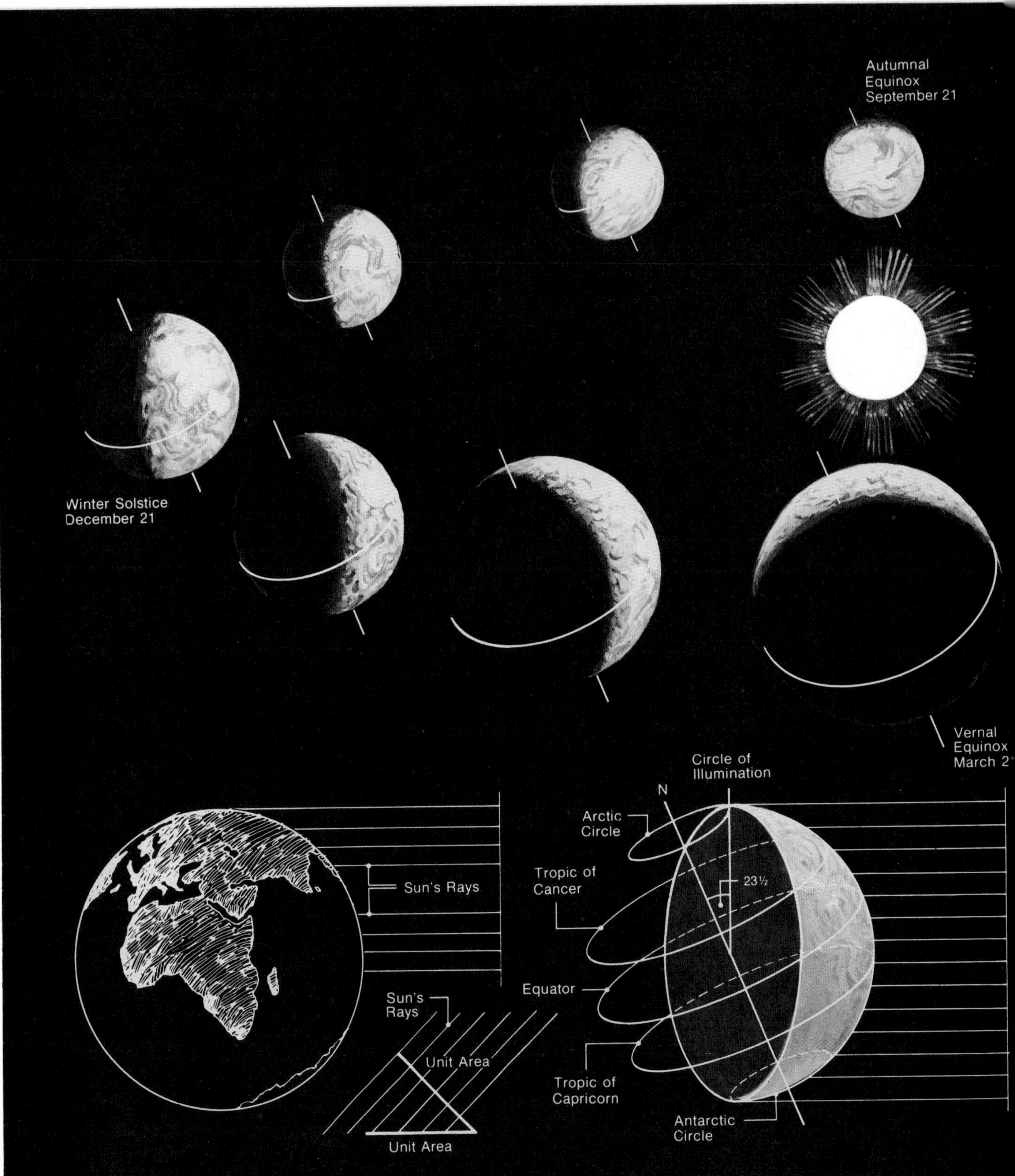

Autumnal
Equinox
September 21
Winter Solstice
December 21
Vernal
Equinox
March 2
Circle of
Illumination
N
Arctic
Circle
Tropic of
Cancer
23½
Sun's Rays
Equator
Sun's
Rays
Unit Area
Unit Area
Tropic of
Capricorn
Antarctic
Circle

Summer Solstice
June 21
N
90
80
70
60
50
40
30
20
10
Latitude
Longitude
W 90
80
70
60
50
40
30
20
10
0 E
Prime
Meridian
Parallels
Meridians

Figure 3.2 (top) The amount of solar energy that reaches a given location at the top of the atmosphere depends on the distance between the earth and the sun and on the orientation of the earth. The top half of the diagram shows the earth at twelve different times during the year. The maximum distance of the earth from the sun is 152×10^6 kilometers (94.4×10^6 miles) and occurs early in July. The minimum distance is 147×10^6 kilometers (91.4×10^6 miles) and occurs early in January. Because solar energy input to the earth varies inversely as the square of the earth-sun distance, the amount of solar radiation the earth receives is 1.07 times greater in early January than in early July. The earth's axis makes an angle of 23½° with a perpendicular to the plane of the earth's orbit. As the earth circles the sun, the orientation of the axis remains the same. At the time of the winter solstice, the sun is overhead at a point on the Tropic of Capricorn in the Southern Hemisphere, and at the time of the summer solstice, it is overhead on the Tropic of Cancer in the Northern Hemisphere. At the time of the vernal and autumnal equinoxes, the sun is overhead at the Equator.

(bottom left) The amount of solar radiant energy that a unit area on the earth's surface intercepts depends on the angle between the sun's rays and the plane of the area. The sun is so far from the earth that the rays near the earth can be considered parallel. As the sketch shows, a unit area intercepts the greatest amount of radiant energy when the area is perpendicular to the rays.

(bottom center) The solar radiant energy input to a location on the earth during a given 24-hour period depends partly on the duration of daylight. At any instant of time, one-half of the earth is illuminated by the sun. The circle of illumination, which is the idealized boundary between the light and dark regions of the earth, is a *great circle*—a circle whose plane passes through the center of the earth. Because of the tilt of the earth's axis, the duration of daylight is in general different at different locations. The diagram illustrates the situation at winter solstice.

(bottom right) The surface of the earth has in principle been divided into two sets of intersecting grid lines to enable locations on the earth to be specified accurately and without ambiguity. The *parallels of latitude* are circles parallel to the plane of the Equator. A particular parallel of latitude is specified by its angular distance north or south of the Equator, with the Equator designated as latitude 0°. The North Pole has latitude 90°N, for example.

The second set of grid lines, the *meridians of longitude,* are great circles drawn with the earth's axis as a diameter. One meridian, chosen to be the *prime meridian,* is designated as 0° longitude. In the United States, and in many other countries, the prime meridian is taken to be the meridian on which the astronomical observatory in Greenwich, England, is situated. Any other meridian is specified by its angular distance east or west of the Prime Meridian. The angle of longitude is therefore limited to the range 0° to 180°.

For precision, a degree can be divided into 60 minutes of angular measure, and a minute can be divided into 60 seconds. The latitude and longitude of Washington, D.C., for example, can be written 38 degrees 54 minutes North, 77 degrees 2 minutes West, or in abbreviated form as 38°54′N, 77°02′W.

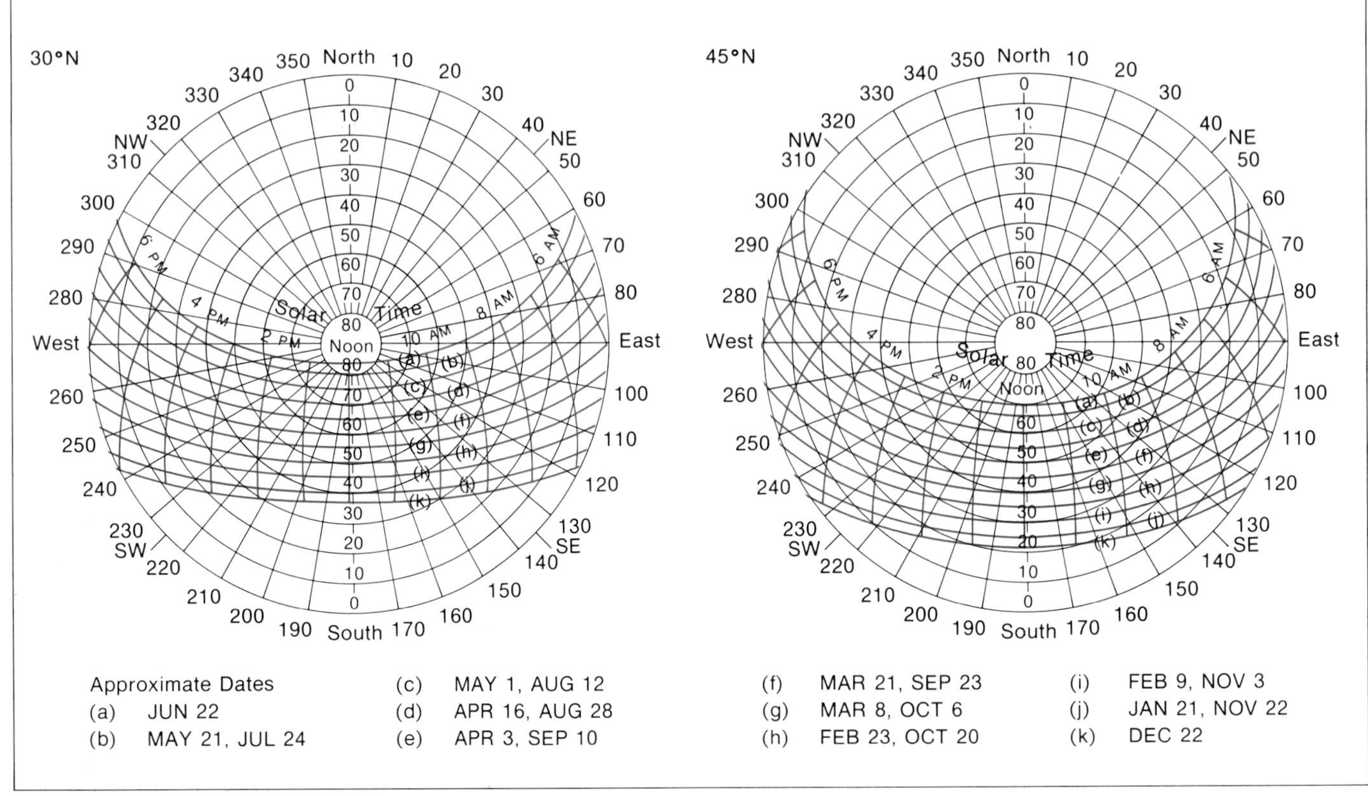
30°N
North
NE
East
SE
South
SW
West
NW
Solar Time
Noon
6 AM
8 AM
10 AM
2 PM
4 PM
6 PM
45°N
Approximate Dates
(a) JUN 22
(b) MAY 21, JUL 24
(c) MAY 1, AUG 12
(d) APR 16, AUG 28
(e) APR 3, SEP 10
(f) MAR 21, SEP 23
(g) MAR 8, OCT 6
(h) FEB 23, OCT 20
(i) FEB 9, NOV 3
(j) JAN 21, NOV 22
(k) DEC 22

Figure 3.3 The solar altitude and azimuth chart on the left is for latitude 30°N, the latitude of New Orleans, Louisiana, and the chart on the right is for latitude 45°N, the latitude of Minneapolis, Minnesota. To find the sun's position in the sky at a given time, first select the grid line corresponding to the date, and then find the point of intersection with the line representing the time of day. Intermediate dates and times can be estimated by interpolation. The sun's altitude can be read from the scale of concentric circles, and the sun's azimuth can be read from the scale of radial lines. On March 21 at latitude 30°N, for example, the sun's altitude at 11:00 A.M. is 57° above the horizon, and its azimuth is approximately 150° from true north.

The duration of daylight can be estimated from solar position charts by finding the times on a given day when the sun's altitude is greater than 0°. On December 22 at latitude 45°N, the sun has an altitude greater than 0° between 7:45 A.M. and 4:15 P.M. Because these charts show the geometric position of the center of the sun, the duration of sunlight is a few minutes longer than these times suggest. Note that the time of day used in these charts is *true solar time* rather than local standard time. The relation between true solar time and standard time can be determined according to the method outlined in Figure 3.4. (After List, 1971)

of longitude (see Figure 3.2). At the time of the equinoxes, daytime and nighttime are of equal duration everywhere on earth.

Solar Altitude and Azimuth

The position of the sun in the sky at any time and at any location can be calculated by taking into account the position of the earth in its orbit, the tilt of its axis, and its spherical shape. The results of such calculations can be charted for each latitude. The two solar position charts shown in Figure 3.3 give the sun's position in terms of its *altitude* and *azimuth*, a system for specifying location in the sky analogous to the system of latitude and longitude on the earth.

Suppose you stand facing north with your arm outstretched, pointing north, and then turn your body and raise your arm until you are pointing at the sun. The angle through which you turn your body is the sun's azimuth, and the angle through which you raise your arm is the sun's altitude. Azimuth is specified as an angle between 0° and 360°, starting from north and going clockwise. Altitude is specified as an angle between 0° and 90°, starting from the level horizon. Thus an azimuth of 90° and an altitude of 45° means that the sun is due east and is halfway up in the sky.

One of the things you can read from a solar altitude and azimuth chart is the maximum height of the sun above the horizon. Using the charts in Figure 3.3, you can verify that at summer solstice the maximum altitude of the sun above the horizon is 83½° in New Orleans, Louisiana, and 68½° in Minneapolis, Minnesota. At winter solstice, the maximum altitude of the sun is 36½° in New Orleans and only 21½° in Minneapolis.

Solar Energy Input to the Earth

The sun emits radiant energy at the same average rate in all directions. Because the earth is relatively small compared to the radius of its orbit, it intercepts only a small fraction of the total energy the sun emits. Dividing the total radiant energy from the sun into a billion equal units, the earth intercepts only 1/2 of 1 unit. But this small fraction is still a huge amount of energy, amounting to 2.6×10^{18} calories per minute. The solar radiant energy intercepted by the earth in 1 minute is equal to the total electric energy generated on the earth in 1 year.

The earth's atmosphere considerably modifies the solar radiation that passes through it to the earth's surface. These modifications are discussed later; for now the discussion applies only to the radiant energy input to the top of the atmosphere.

You know that the duration of daylight is different at different latitudes and varies through the year at the same latitude. This variation has long determined the rhythm of man's agricultural practices. But duration of daylight is not the only factor that influences radiant energy input to a given location. The distance of the earth from the sun varies slightly through the year, causing the radiant energy input to vary as well. But the range of fluctuation is no greater than 7 percent, with the greatest amount of energy received early in January when the earth is closest to the sun.

Another important factor influencing the radiant energy input to a given location is a geometric effect associated with the sun's altitude. You know that the winter sun, low on the horizon, warms the earth far less than the high summer sun. Imagine holding a square of cardboard in a stream of sunlight. If you hold the card perpendicular to the sun's rays, the card intercepts the maximum possible amount of energy. If you turn the card edge on toward the rays, the card receives little or no energy. At intermediate angles, intermediate amounts of energy strike the card. A farmer concerned about his wheat crop wants to know how much energy falls on his particular acres, not how much total energy the earth receives. The farmer's field is analogous to the card. When the sun is low in the sky, less energy is delivered to a given area than when the sun is high. The relative amount of energy falling on a particular area depends on the angle at which the incoming rays strike the area. Sunbathers recognize this principle by propping themselves in lounge chairs to be more perpendicular to the sun's rays. The strawberry fields in Switzerland and the vineyards in wine-producing countries are planted on steep valley slopes for the same reason. When the sun is 60° above the horizon, a patch of ground on a steep 30° slope facing the sun receives nearly 15 percent more energy input than a horizontal patch of the same area.

The appropriate measure for discussing the distribution of radiant energy is the amount of energy that falls on an area of standard size. In physical geography distribution of energy is specified in

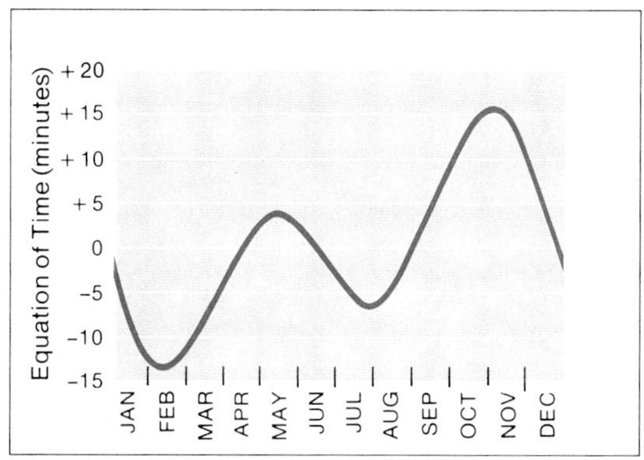

Equation of Time (minutes)
+20
+15
+10
+5
0
-5
-10
-15
JAN
FEB
MAR
APR
MAY
JUN
JUL
AUG
SEP
OCT
NOV
DEC

Figure 3.4 The *equation of time* shown in this graph is required to convert local standard time to true solar time. For convenience, the surface of the earth is divided into 24 time zones, each of which spans approximately 15° of longitude. In each time zone, there is a standard meridian, a multiple of 15° longitude east or west of the prime meridian. To find the *average solar time* at other locations, add 4 minutes of time to the standard time for every degree of longitude the location is east of the standard meridian, and subtract 4 minutes for every degree west of the standard meridian.

True solar time differs slightly from average solar time through the year, because of astronomical factors such as the systematic variation of the speed of the earth in its orbit. To convert average solar time to true solar time, add (+) or subtract (–) the correction shown in the graph to the average solar time. Note that the horizontal scale of the graph begins at January 1 and ends at December 31. If local standard time is 1:00 P.M. on the first of August at a location 2° east of the standard meridian, for example, average solar time is 1:08 P.M. The correction read from the graph is approximately –6 minutes, so that true solar time is 1:02 P.M. (After List, 1971) units of calories per square centimeter. Because of its importance, the unit of energy distribution has been assigned a special name, the *langley*:

1 langley = 1 calorie per square centimeter

The unit *langley per minute* combines the ideas of energy distribution over space and energy flow over time.

Radiant energy from the sun strikes a unit area at the top of the earth's atmosphere at a rate known as the *solar constant*. The definition of the solar constant assumes that the unit area is perpendicular to the sun's rays and that the earth and the sun are separated by their average distance. The accepted value of the solar constant is 1.94 calories per square centimeter per minute, or 1.94 langleys per minute, but values up to 2.04 langleys per minute have been reported. Solar radiation, measured near the earth's surface,

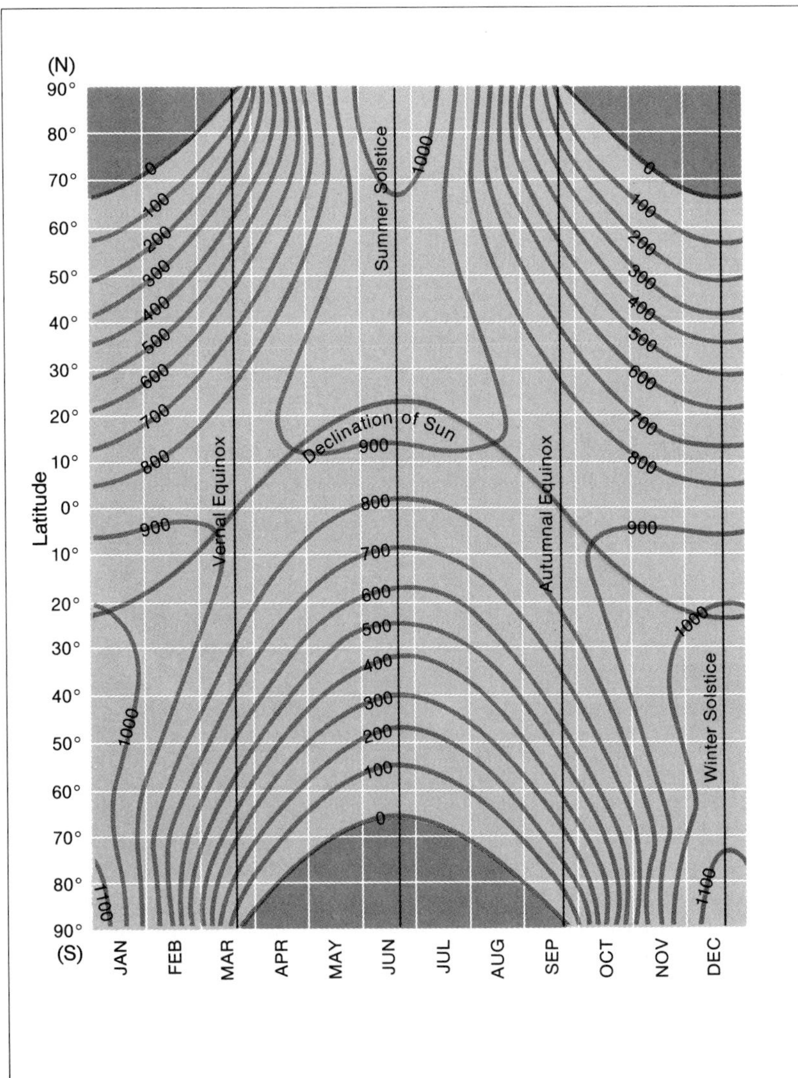

(N)
90°
80°
70°
60°
50°
40°
30°
20°
10°
0°
10°
20°
30°
40°
50°
60°
70°
80°
90°
(S)
Latitude
JAN
FEB
MAR
APR
MAY
JUN
JUL
AUG
SEP
OCT
NOV
DEC
Vernal Equinox
Summer Solstice
Autumnal Equinox
Winter Solstice
Declination of Sun
0
100
200
300
400
500
600
700
800
900
1000
1100

Figure 3.5 This graph shows solar radiant energy input to the top of the atmosphere. The horizontal scale is marked off in days of the year, from January 1 at the left end of the scale past December 31 at the right. The vertical scale lists latitude north and south of the Equator. The curved contour lines give the solar radiant energy input to the top of the atmosphere in units of langleys per 24 hours, based on a value of 1.94 langleys per minute for the solar constant. On the Equator at summer solstice, for example, the radiant energy input is approximately 800 langleys in 24 hours. The shaded areas of the diagram poleward of latitude 66½° represent times of perpetual darkness when there is no solar radiant energy input.

The diagram is not perfectly symmetrical between the Northern and Southern Hemispheres. At summer solstice in the Northern Hemisphere (June 21), locations at latitude 15°N receive 940 langleys per 24 hours, but at summer solstice in the Southern Hemisphere (December 21), locations at latitude 15°S receive 1,000 langleys per 24 hours. The reason is that the earth is nearer the sun in January than in July, and the energy input to the earth is 7 percent higher in January than in July.

is less than the solar constant because of atmospheric particles and dust.

Total Daily Solar Radiation at the Top of the Atmosphere

You have read how the sun's position in the sky can be determined for any time of day at any location on earth. Consider now a unit area of 1 square centimeter at the top of the atmosphere. The amount of radiant energy striking the area depends on the value of the solar constant, the duration of daylight, the sun's altitude, and the true distance between the earth and the sun. The total energy striking the area in 1 day can be found in principle by calculating the position of the sun for every minute of the day and then using the angle appropriate to each time to find the amount of energy delivered to the square centimeter of area in each minute. The total energy delivered per day is the sum of the energies received during each minute.

Figure 3.5 shows the total daily solar radiation input to a horizontal square centimeter at the top of the atmosphere for any latitude and any day of the year. The number on each curve represents energy input in langleys per 24 hours. For example, the energy input at latitude 40°N on November 1 is approximately 450 langleys in 24 hours. The shaded areas on the chart represent times of continuous darkness.

The amount of energy input varies more at the poles throughout the year than at the Equator, where the input is relatively uniform from day to day. Locations at the Equator never receive as much as 900 langleys in 24 hours, but the energy input near the South Pole sometimes exceeds 1,100 langleys in 24 hours. The reason for this is that at winter solstice the South Pole receives continuous sunlight the whole day, while at the Equator daylight never lasts longer than 12 hours. Averaged over a full year, however, locations at the Equator receive nearly 2½ times as much energy as areas near the South Pole.

Figure 3.6 This graph shows the solar radiant energy input per 24 hours to the top of the atmosphere through the year at three selected latitudes. The data needed to construct this graph can be read from Figure 3.5. At the Equator, latitude 0°, the energy input is high and comparatively constant through the year; the slight decline in July occurs because the earth is farthest from the sun at that time. At higher latitudes, the maximum energy input in a 24-hour period exceeds that at the Equator because of the long duration of daylight toward the pole in the summer. During the 4 months from April 15 to August 15, the energy input at latitude 60°N is comparable to the energy input at the Equator. However, at other times of the year the energy input at high latitudes is markedly smaller than the energy input at low latitudes. (After Robinson, 1966)

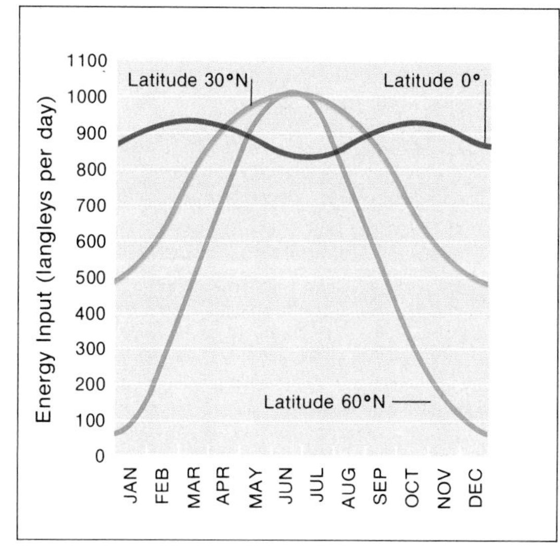
Energy Input (langleys per day)
1100
1000
900
800
700
600
500
400
300
200
100
0
Latitude 30°N
Latitude 0°
Latitude 60°N
JAN
FEB
MAR
APR
MAY
JUN
JUL
AUG
SEP
OCT
NOV
DEC

The Electromagnetic Spectrum and the Quality of Sunlight

During a passing rainstorm, water droplets in the air sometimes break up sunlight into its component colors—the pure colors we see in a rainbow. Light travels in waves, and every pure color can be characterized by its *wavelength*. A typical wavelength for red light is 6.5×10^{-5} centimeter (2.6×10^{-5} inch); purple has a shorter wavelength of about 4×10^{-5} centimeter (1.6×10^{-5} inch). The wavelengths of other colors fall between these limits in a range called the *visible spectrum* of light. Wavelengths are often specified in *microns*, where

$$1 \text{ micron} = 10^{-4} \text{ centimeter} = 3.937 \times 10^{-5} \text{ inch}$$

Thus, the wavelength of red light is 0.65 micron, and the wavelength of purple light is 0.4 micron.

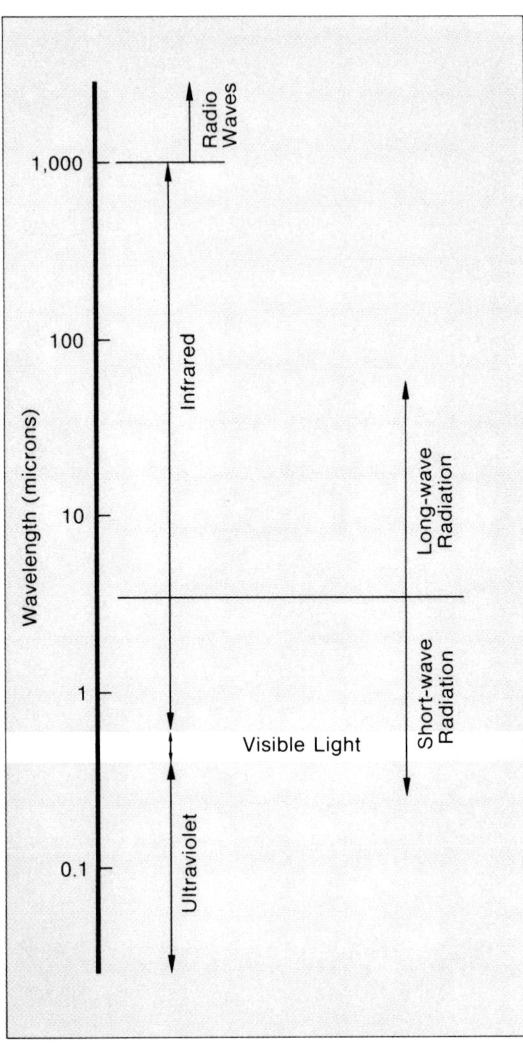
Radio Waves
1,000
100
Infrared
Wavelength (microns)
10
Long-wave Radiation
Short-wave Radiation
1
Visible Light
Ultraviolet
0.1

Figure 3.7 The electromagnetic spectrum is conventionally divided on the basis of wavelength. The divisions most important in physical geography are infrared radiation, visible light, and ultraviolet radiation. Infrared radiation consists of wavelengths intermediate between the wavelengths of radio waves and those of red light. Visible light consists of wavelengths between approximately 0.7 and 0.4 micron, and ultraviolet light corresponds to wavelengths shorter than 0.4 micron. At extremely short wavelengths, the ultraviolet portion of the spectrum merges into the x-ray and gamma ray divisions.

If infrared radiation has a wavelength longer than 4 microns, it is called long-wave infrared radiation; if the wavelength is shorter than 4 microns, it is called short-wave infrared radiation. The distinction is useful in physical geography because solar radiant energy input to the earth is at wavelengths shorter than 4 microns and the radiant energy loss from the earth is at wavelengths longer than 4 microns (see Figure 3.8).

The spectrum of visible light from the sun is only one of many forms of *electromagnetic radiation*. Although you cannot sense them with your eye, there are important forms of electromagnetic radiation with wavelengths longer and shorter than the wavelengths of visible light. Radio waves, for example, consist of electromagnetic radiation in the range of wavelengths from 0.1 centimeter or so to hundreds of meters. Figure 3.7 shows how the electromagnetic spectrum is conventionally divided into a number of sections according to wavelength. Radiation with wavelengths shorter than radio waves but longer than the wavelength of red light is called *infrared radiation*. Radiation with wavelengths shorter than purple light is called *ultraviolet radiation*. The visible spectrum constitutes only a small part of the entire electromagnetic spectrum.

Every form of electromagnetic radiation carries energy, and a portion of the solar energy that reaches the earth lies outside the visible range. The radiant energy emitted by the sun consists of radiation with wavelengths that range from radio waves to x-rays. The spectrum of radiation the sun emits depends primarily on the surface temperature of the sun. Similarly, when you turn on a heating element of an electric stove, the coil remains dull black while it is warming up; it emits mostly invisible infrared radiation, which you can feel as heat. As it becomes hotter, the coil begins to glow dull red and eventually becomes bright red. If the coil could be heated further without melting, it would become yellow-white like the sun, then blue-white like the winter star Sirius.

Every object, whether warm or cold, emits electromagnetic radiation, called *thermal radiation*, because of the jostling of its molecules. Thermal radiation is emitted over a range of wavelengths depending on the temperature of the object. For an object at room temperature (293°K or 68°F), the emitted thermal radiation consists of radio waves and long-wave infrared radiation with wavelengths greater than 4 microns or so. The surface of the sun, at a temperature of 6000°K (10,800°F), emits most of its thermal radiation at short wavelengths of less than 4 microns or so, and a considerable fraction of the emitted radiation falls within the visible spectrum.

The intensity of the thermal radiation an object emits depends on the temperature of the object; the intensity increases rapidly with increased temperature. The nature of the surface also affects the rate of emission. For a given temperature, objects with dull black surfaces, such as soil or blacktop roads, are relatively good emitters of thermal radiation, whereas shiny objects such as clouds are relatively poor emitters. The hypothetical, ideal object that emits the maximum intensity of thermal radiation possible at a given temperature is called a *black body*.

The thermal radiation from the surface of the sun is similar to the radiation that a black body would emit at a temperature of 6000°K. The sun emits strongly in the infrared, visible, and ultraviolet portions of the spectrum, but the radiation is modified as it passes through the outer layers of relatively cooler gases surrounding the sun. These gases selectively absorb certain wavelengths from the radiation. Ultraviolet light is also weakened in intensity as it passes through the sun's outer layers.

Interaction of Radiation With the Atmosphere

The atmosphere plays a central role in determining the earth's response to solar energy input. The atmosphere consists primarily of nitrogen, oxygen, and argon gas, in the proportions listed in Table 3.1. Argon, an inert, chemically unreactive gas, is far more abundant in the atmosphere than other inert gases, such as neon and krypton. Most of the argon seems to have been produced by the decay of radioactive potassium in the earth's crust. The amount of air per unit volume and the pressure of the atmosphere decrease rapidly with increased altitude. One half of all the gas in the atmosphere lies below an altitude of 5 kilometers (3 miles). Above that altitude, the oxygen content of the atmosphere is insufficient for human survival.

Table 3.1. Principal Gases in the Earth's Lower Atmosphere

Gas	Molecular Formula	Number of Molecules of Gas per Million Molecules of Air	Proportion (percent)
Nitrogen	N_2	7.809×10^5	78.09
Oxygen	O_2	2.095×10^5	20.95
Water Vapor	H_2O	variable	variable
Argon	Ar	9.3×10^3	0.93
Carbon Dioxide	CO_2	330.0	0.03
Neon	Ne	18.0	1.8×10^{-3}
Helium	He	5.0	5.0×10^{-4}
Krypton	Kr	1.0	1.0×10^{-4}

Source: adapted from: List, Robert (ed.). 1971. *Smithsonian Meteorological Tables*. 6th rev. ed. Washington, D.C.: Smithsonian Institution Press.

The nitrogen, oxygen, and inert gases in the atmosphere are well mixed and have the same relative proportions at 25 kilometers (15 miles) altitude as they do at sea level. Other constituents of the atmosphere, such as carbon dioxide and water vapor, vary with location, altitude, and time. Water vapor is confined largely to the lowest 10 kilometers (6 miles) of the atmosphere in a layer called the *troposphere*. The next layer, the *stratosphere*, is practically free of water vapor. The *ozone layer* occurs at an altitude of about 25 kilometers (15 miles). The atmosphere also carries a variable load of dust, salt crystals, pollen grains, and debris in the form of industrial smoke and automobile exhaust fumes.

Electromagnetic radiation from the sun interacts with gas molecules in the atmosphere by *absorption* and by *scattering*. In absorption, which is usually a selective process dependent on wavelength, a molecule takes up radiant energy and converts the energy to a different form. Scattering, which can occur for any wavelength, redirects radiant energy from a beam of sunlight into other directions. Because of scattering, the earth receives light from the sky

Figure 3.8 The solar radiant energy received by the earth is distributed over a band of wavelengths. Because of the high surface temperature of the sun, 6000°K (10,800°F), energy from the sun is emitted primarily at wavelengths shorter than 4 microns, with much of the energy concentrated in the visible region of the spectrum. In contrast, the thermal radiation emitted by the earth is confined to wavelengths longer than 4 microns and has a broad peak at about 10 microns, because of the earth's comparatively low average surface temperature of 285°K (54°F).

The lower portion of the figure indicates the degree to which atmospheric gases, primarily carbon dioxide and water vapor, absorb electromagnetic energy near the earth's surface. Wavelength bands of strong absorption are shown in red, and bands of relative transparency are indicated by yellow. The lower atmosphere is relatively opaque to long-wave infrared radiation, so that much of the radiant energy emitted by the earth's surface is absorbed in the troposphere. The atmosphere is relatively transparent to electromagnetic radiation in the band from 8.5 to 11 microns, and radiant energy in this band can escape to space if the sky is clear. (After Dobson, 1963)

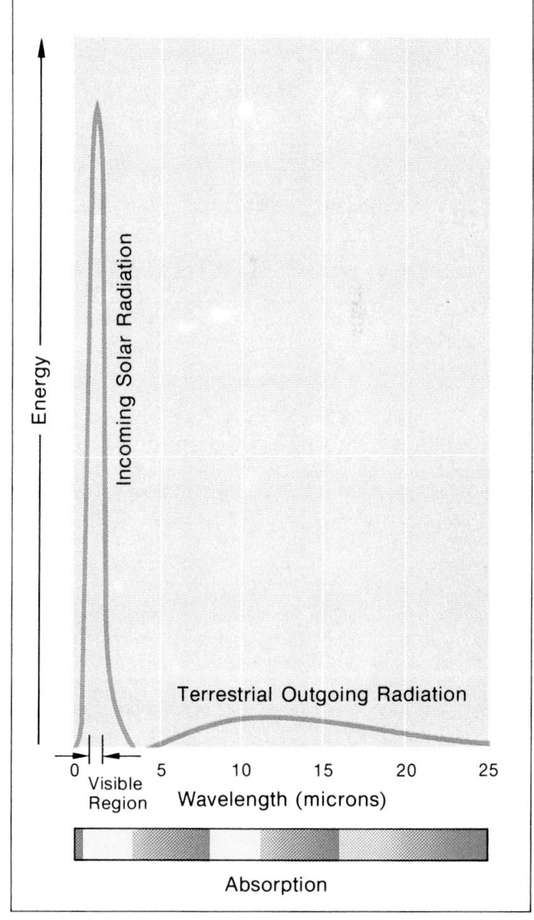

Energy
Incoming Solar Radiation
Terrestrial Outgoing Radiation
0
5
10
15
20
25
Visible Region
Wavelength (microns)
Absorption

Figure 3.9 The atmosphere is conventionally divided into layers, primarily according to the variation of temperature with altitude. The *troposphere,* which is the lowest layer of the atmosphere, decreases in temperature with increased altitude at the average rate of 6°C per kilometer (3.3°F per 1,000 feet). The troposphere is warmest at the earth's surface, on the average, because it is heated primarily from below by the transfer of heat energy from the surface. Air in the troposphere contains water vapor, and most clouds and weather phenomena are confined to this layer. Temperature increases with increased altitude in the *stratosphere,* largely because atmospheric gases such as ozone absorb a portion of the radiant energy incident from the sun. Air in the stratosphere is nearly devoid of water vapor, so clouds seldom form there.

The temperature curve in the diagram is made up of straight line segments and corresponds to the values assumed for the 1962 United States standard atmosphere. A standard atmosphere is meant to represent the average state of the atmosphere, but the actual temperatures and pressures on a given day may differ from the standard values.
(After The United States Air Force, 1965)

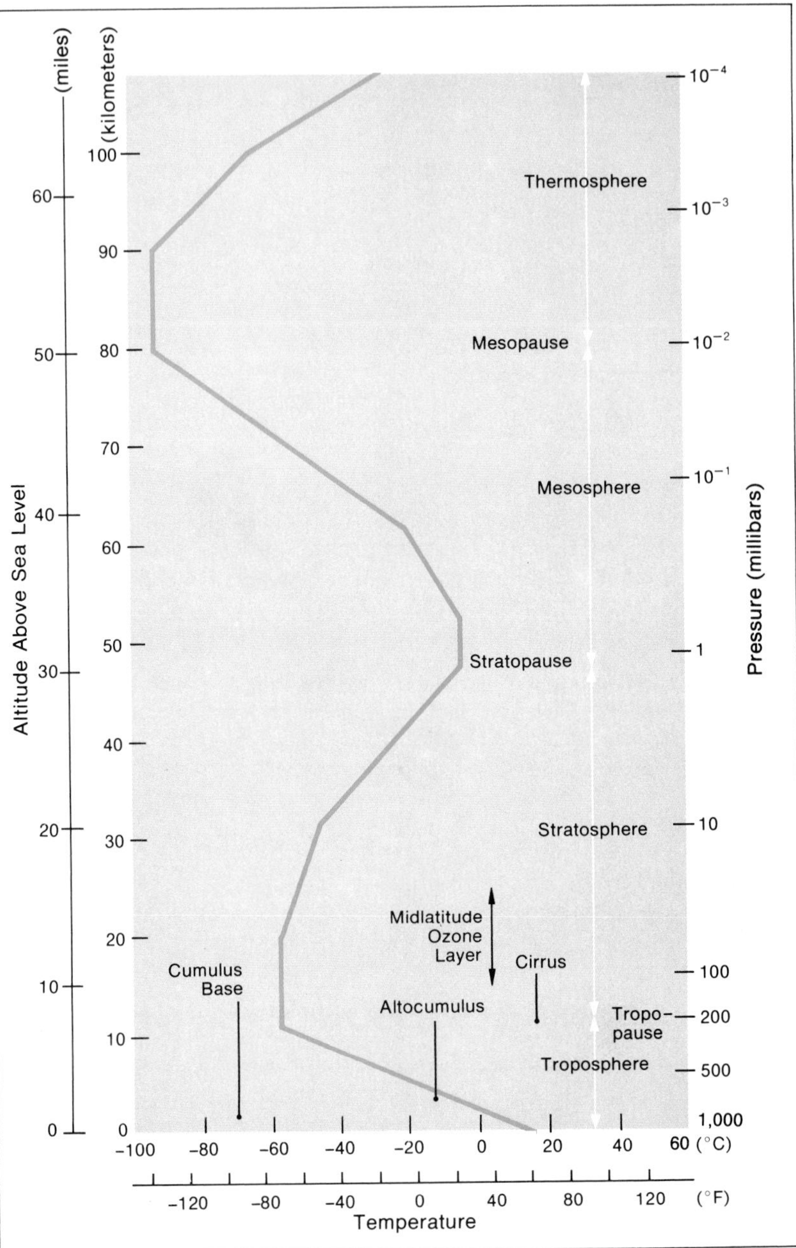

Altitude Above Sea Level
(miles)
(kilometers)
0
10
20
30
40
50
60
0
10
20
30
40
50
60
70
80
90
100
Thermosphere
Mesopause
Mesosphere
Stratopause
Stratosphere
Midlatitude Ozone Layer
Cirrus
Cumulus Base
Altocumulus
Tropo-pause
Troposphere
Pressure (millibars)
10^{-4}
10^{-3}
10^{-2}
10^{-1}
1
10
100
200
500
1,000
-100
-80
-60
-40
-20
0
20
40
60 (°C)
-120
-80
-40
0
40
80
120
(°F)
Temperature

in addition to the light coming directly from the sun; the moon, in contrast, has no atmosphere to scatter sunlight, which is why shadows on the moon are so harsh. Gas molecules in the earth's atmosphere scatter blue light much more effectively than red light, so most of the light scattered from the sky is blue.

Infrared, visible, and ultraviolet radiation differ in their interactions with matter. They are all scattered to some extent by the gases, dust particles, or water drops in the atmosphere, but each kind of radiation behaves differently in absorption. Infrared radiation that passes through a gas is absorbed by causing the atoms in the gas molecules to vibrate back and forth. When the vibrating gas molecules collide, the energy is transformed to heat energy in the form of thermal motion. Energy transfer from infrared radiation to gas molecules is selective. For each kind of gas molecule, the transfer occurs strongly only for certain bands of wavelengths. Because the sun's surface temperature is so high, most of the infrared radiation from the sun is at wavelengths shorter than 4 microns, which is too short for any molecules in the atmosphere to be effective absorbers. Water vapor and carbon dioxide molecules in the atmosphere are the principal absorbers of long-wave infrared radiation with wavelengths greater than 4 microns.

The wavelengths of visible light are even shorter than those of infrared radiation, so no molecule in the earth's atmosphere can absorb visible light effectively. Visible light therefore travels long distances through a clear atmosphere without losing much energy through absorption.

Ultraviolet radiation is energetic and often transfers sufficient energy to a molecule to rearrange its structure or even to break the molecule apart. Ultraviolet light is harmful to living organisms, because it can disrupt the complex molecules required for life processes. Fortunately, little ultraviolet radiation actually reaches the surface of the earth; most of it is filtered out in the upper atmosphere. In the outer layers of the earth's atmosphere, molecular oxygen strongly absorbs ultraviolet light with wavelengths shorter than 0.2 micron. The ozone layer strongly absorbs ultraviolet light with slightly longer wavelengths (between 0.2 and 0.3 micron). The absorption of energy in the upper atmosphere and in the ozone layer heats these layers to temperatures comparable to ground level temperatures. You can see from Figure 3.9 that the temperature of the atmosphere decreases with height in the troposphere but increases with height in the stratosphere, reaching a relative maximum at the altitude of the ozone layer.

Energy Balance in the Earth-Atmosphere System

As solar radiation penetrates the earth's atmosphere, gas molecules may scatter or absorb it, or clouds may reflect it. Part of the scattered solar radiation is directed upward and may be lost to space. Or further scattering may direct some of the energy back toward the earth.

Reflection and Albedo

In the troposphere, where clouds form, part of the incoming solar radiation may be reflected from clouds. In the process of *reflection*, radiant energy is not absorbed; it simply is redirected. With the exception of totally black objects, every object reflects light to some extent. The fraction of incident light an object reflects is called its *albedo*. The albedo of a shiny mirror is nearly 100 percent. The albedo of a thick, puffy summer cloud can be as high as 90 percent—perhaps you have seen the dazzling brightness of a cloud layer from an airplane. Light reflected from a cloud may return to space, or it may be further scattered or reflected. Of the solar radiation that reaches the upper atmosphere, about 33 percent is

Table 3.2. Albedo of Various Surfaces

Surface	Albedo (percent)
Fresh Snow	80-95
Dense Stratus Clouds	55-80
Ocean (Sun Near Horizon)	40
Ocean (Sun Halfway up Sky)	5
Bare Sandy Soil	25-45
Desert	25-30
Dry Steppe	20-30
Meadow	15-25
Tundra	15-20
Green Deciduous Forest	15-20
Green Fields of Crops	10-25
Coniferous Forest	10-15
Bare Dark Soil	5-15

Sources: List, Robert (ed.). 1971. *Smithsonian Meteorological Tables.* 6th rev. ed. Washington, D.C.: Smithsonian Institution Press. Table 15.4, pp. 442-443.

Budyko, M. I. 1958. *The Heat Balance of the Earth's Surface.* Nina A. Stepanova (tr.). Washington, D.C.: United States Department of Commerce.

Figure 3.10 The average solar radiation received at the ground in the United States during January and July is shown in units of langleys per day. (After *The National Atlas of the United States of America,* 1970)

The average solar radiation received globally at the ground-sea surface during December and June is shown in units of langleys per day. (After Löf, Duffie, and Smith, 1966)

During December and January, when the sun is overhead in the Southern Hemisphere, the solar radiation received in the Northern Hemisphere decreases rapidly with increased latitude. Southern Florida, for example, receives more than 300 langleys per day during January on the average, but northern Minnesota receives only 150 langleys per day. Solar radiation input is small at high latitudes during the winter because of the short duration of daylight and because of the low altitude of the sun in the sky. The solar radiation input to a given location is larger when skies are clear than when skies are cloudy; for this reason the southwestern United States receives more solar radiation than cloudy locations receive at the same latitude.

During the summer months (June to September in the Northern Hemisphere, December to March in the Southern Hemisphere), the solar radiation input exhibits little variation with latitude. Locations in northern Canada receive the same input as locations in Texas. The increased duration of daylight with increased latitude compensates for the lower altitude of the sun toward the pole. The variation in solar radiation input between different locations during the summer is caused primarily by differences in the degree of cloudiness of the atmosphere. The dashed isolines indicate areas where data are missing or incomplete. Data for Greenland are unavailable for June.

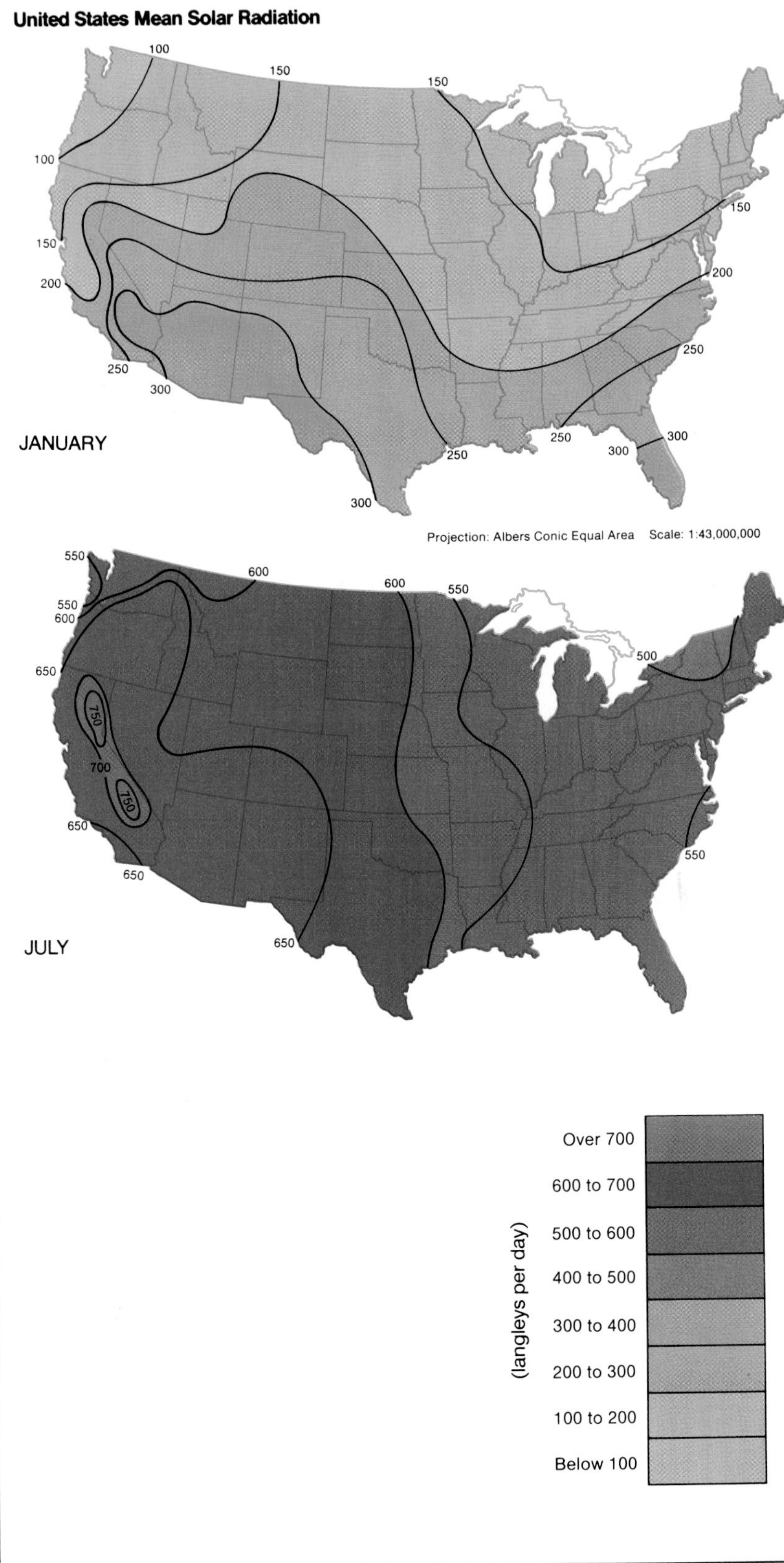

United States Mean Solar Radiation
JANUARY
100
150
200
250
300
Projection: Albers Conic Equal Area Scale: 1:43,000,000
JULY
500
550
600
650
700
750
(langleys per day)
Over 700
600 to 700
500 to 600
400 to 500
300 to 400
200 to 300
100 to 200
Below 100

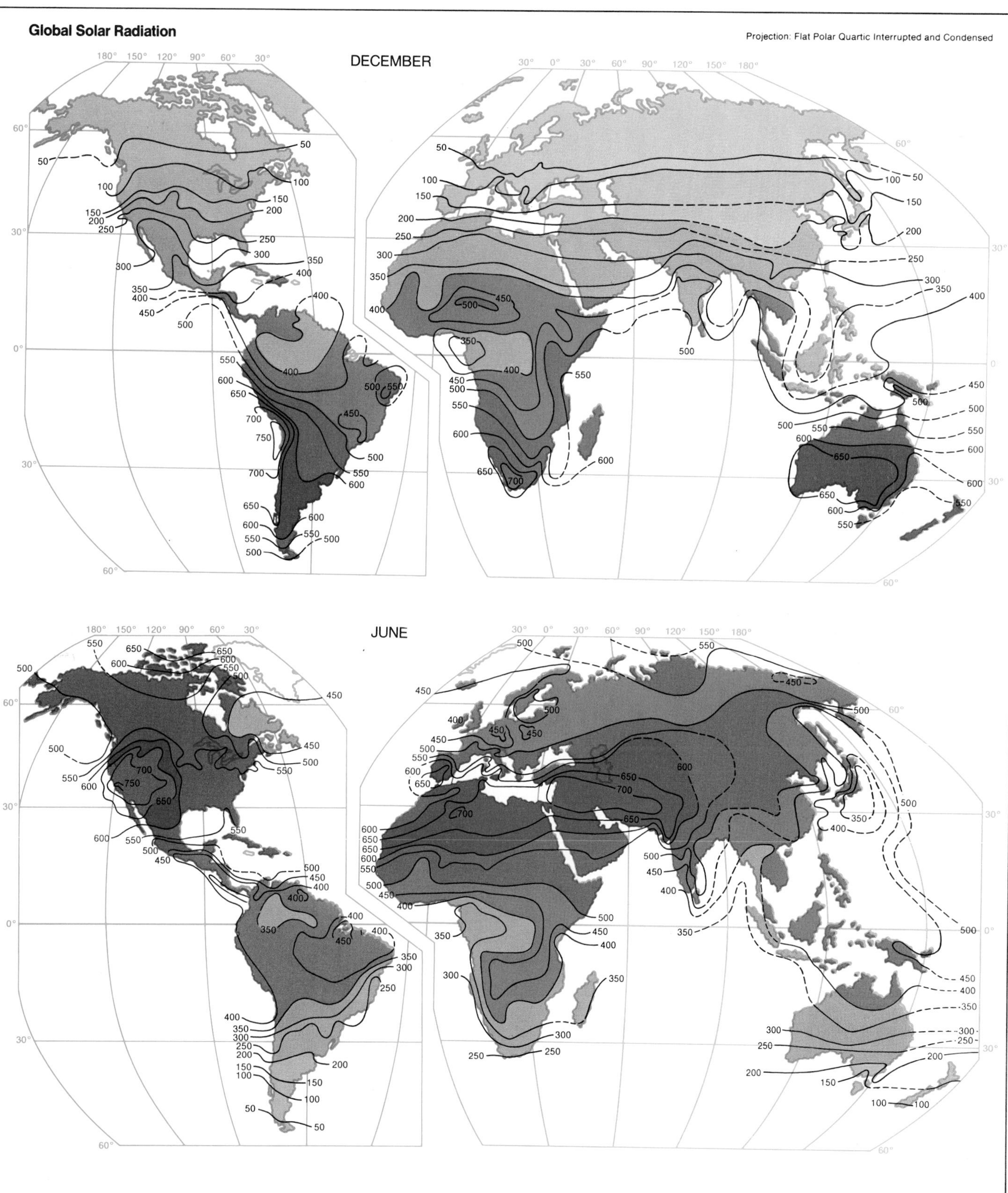
Global Solar Radiation
Projection: Flat Polar Quartic Interrupted and Condensed
DECEMBER
JUNE

scattered or reflected by clouds or by the ground and returned to space. Atmospheric gases and clouds absorb about 22 percent. The remaining 45 percent is absorbed by the ground.

Clouds have a significant effect on solar energy input to the earth's surface. Figure 3.10 shows the average solar radiation input to a horizontal surface on the ground for the United States and for the globe in winter and in summer. The amount of annual energy input to the top of the atmosphere depends only on latitude. The irregularities in the pattern of energy reaching the ground are caused primarily by differences in cloudiness. The southwestern deserts in the United States are farther north than Florida but receive more energy because of the lack of clouds. China and the deserts of North Africa are both at latitude 25°N, but the African deserts receive more than 1½ times as much solar energy input over the course of a year because of extensive cloudiness over China.

A portion of the radiant energy reaching the ground is immediately reflected without transfer of energy. The albedo of the ground can vary from 90 percent for a fresh snow cover to a few percent for a blacktop road. Furthermore, the albedo at a given location varies from season to season as snow comes and goes and as bare fields become covered with crops (see Table 3.2).

Energy Loss and the Atmosphere

The earth's surface itself is warm—285°K (54°F) on the average—and it emits thermal radiation continuously. Most of the radiant energy the earth emits is long-wave infrared radiation with wavelengths greater than 4 microns. If the earth had no atmosphere, all of the emitted radiant energy would escape into space, and the unlit half of the earth's surface would rapidly become cold after sunset.

The atmosphere helps the earth's surface retain much of its heat energy. Water vapor and carbon dioxide molecules in the lower atmosphere are efficient absorbers of long-wave infrared radiation, so that a large part of the thermal radiation from the earth's surface is absorbed in the lower atmosphere. The absorbed radiant energy heats the lower atmosphere to a temperature comparable to the temperature of the earth's surface. According to the laws of thermal radiation, the atmosphere also radiates long-wave infrared radiation; some of this radiation is scattered upward and is lost to space, but most of it returns to the surface of the earth. Although the earth's surface loses energy by thermal radiation, it regains an average of nearly 80 percent of the lost energy by reradiation from the lower atmosphere. Because water vapor and carbon dioxide molecules are the only molecules in the lower atmosphere capable of absorbing long-wave radiation, these minor constituents of the atmosphere play a major part in the earth's utilization of radiant energy.

Clouds contribute to the heating of the earth's atmosphere by absorbing long-wave radiation emitted from the earth's surface. They contribute to the heating of the earth by returning some of the thermal radiation to the surface. Like the atmospheric gases, clouds emit long-wave radiation, part of which is intercepted by the ground. The earth's surface cools more rapidly on a clear night

Figure 3.11 (opposite) This diagram traces 100 units of solar radiant energy incident on the top of the earth's atmosphere as they interact with the atmosphere and with the ground. The numbers represent average global values. The interactions of ultraviolet, visible, and short-wave infrared radiation are shown on the left, and the interactions of long-wave infrared radiation are shown on the right. All of the indicated interactions occur at a given location during daylight hours, but at night when there is no solar radiant energy input, only the long-wave interactions occur.

The earth as a whole is neither gaining nor losing radiant energy; for every 100 units of solar radiant energy received, an average of 33 + 67 = 100 units are returned to space. Similarly, energy is in balance at the earth's surface, on the average. The surface receives 45 + 98 = 143 units of energy and loses 113 + 30 = 143 units. Can you use the diagram to show that the average gain and loss of energy by the atmosphere are also in balance?

Note that the 98 units of long-wave radiant energy returned to the ground by the atmosphere are important components in the energy balance at the ground. The ground absorbs 45 units of short-wave solar radiant energy and emits 113 units of long-wave thermal radiation. If it were not for the energy contributed by the atmosphere, the ground would cool to the temperature where the radiated energy was in balance with the absorbed energy. In the absence of an atmosphere, the average temperature of the earth would be approximately −20°C (−4°F). (After London, 1957) than it does when there is a thick cloud cover.

None of the gas molecules in the atmosphere is an efficient absorber of radiant energy in the wavelength range from 8.5 to 11 microns, so the atmosphere is transparent to this band of wavelengths. Radiation emitted in this band escapes to space through a sort of *atmospheric window* when the sky is free of clouds.

Radiant Energy Balance

Figure 3.11 traces the course of solar radiant energy as it passes from the top of the atmosphere toward the earth's surface. Except for the absorption of some ultraviolet radiation by the ozone layer,

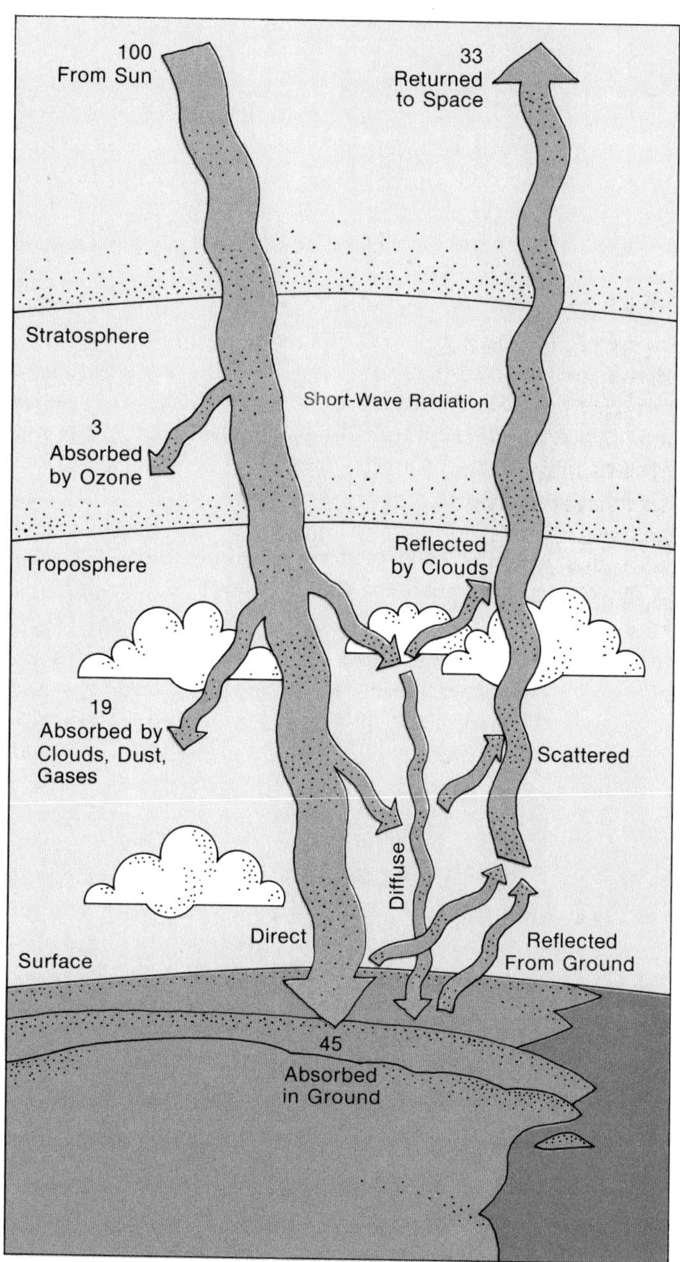
100
From Sun
33
Returned
to Space
Stratosphere
Short-Wave Radiation
3
Absorbed
by Ozone
Troposphere
Reflected
by Clouds
19
Absorbed by
Clouds, Dust,
Gases
Scattered
Diffuse
Direct
Reflected
From Ground
Surface
45
Absorbed
in Ground

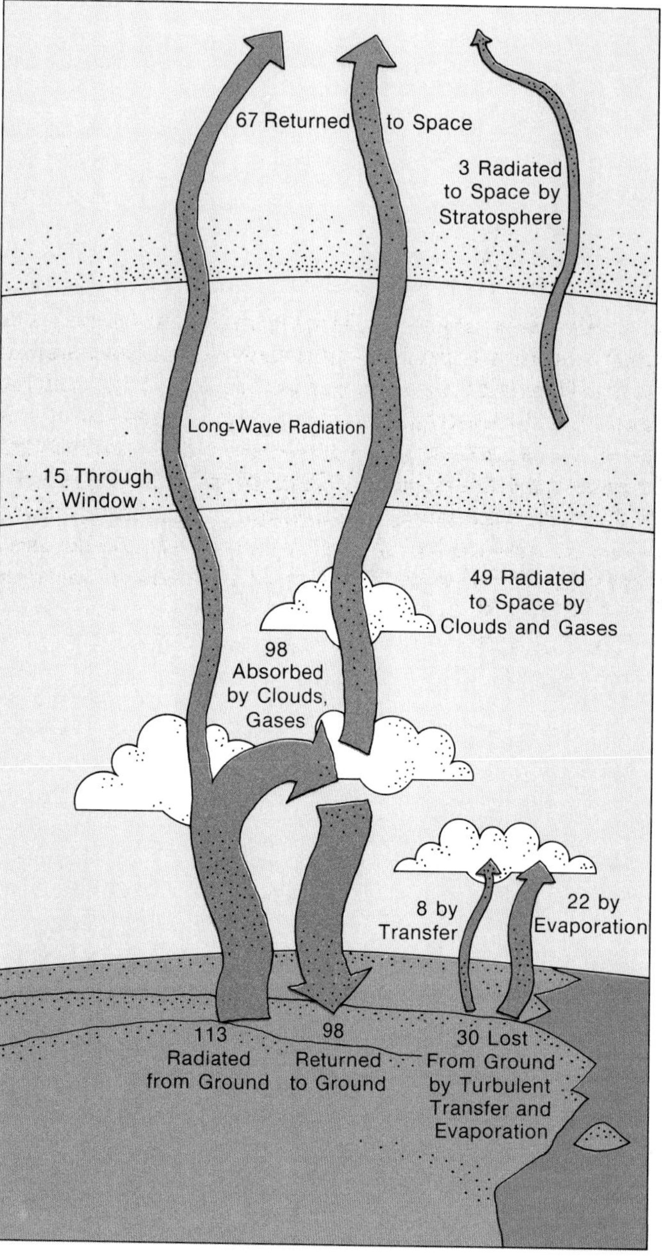
67 Returned to Space
3 Radiated
to Space by
Stratosphere
Long-Wave Radiation
15 Through
Window
49 Radiated
to Space by
Clouds and Gases
98
Absorbed
by Clouds,
Gases
8 by
Transfer
22 by
Evaporation
113
Radiated
from Ground
98
Returned
to Ground
30 Lost
From Ground
by Turbulent
Transfer and
Evaporation

Figure 3.12 (opposite top) The radiation balance and its principal components during a 24-hour period are shown for Hancock, Wisconsin, a moist midlatitude location, and for Tempe, Arizona, an irrigated desert location. The radiation balance *R* is the difference between the rate at which solar radiant energy is absorbed at the ground and the net loss of long-wave thermal radiation in the exchange between the surface and the atmosphere. During most of the day, the radiation balance is positive, which indicates a net flow of radiant energy toward the surface of the ground. During the night *R* is negative because the net flow of radiant energy is away from the surface.

Energy cannot accumulate at the surface, and the excess or deficiency of radiant energy flow to the surface can be accounted for as the sum of three components: *LE*, the energy removed by the evaporation and transpiration of water; *H*, the energy flowing from the ground to the air by turbulent transfer; and *G*, the energy that flows into the soil from the surface of the ground. At Hancock and Tempe at midday, *LE* is positive because of evaporation and plant transpiration during photosynthesis. The small negative value of *LE* before sunrise at Hancock indicates the condensation of water vapor on the ground as dew. At Hancock, *H* is positive at midday, which indicates that heat from the ground is warming the air. At Tempe, rapid evaporation from the irrigated plot keeps the ground cooler than the air so that heat flows from the air over the dry landscape to the irrigated ground, making *H* negative. (After Sellers, 1965)

Figure 3.13 (opposite bottom) The diagram on the left shows the average rates at which the earth and the atmosphere absorb solar radiant energy and emit long-wave thermal radiation to space. During the winter, absorption exceeds emission only at latitudes near the Equator, but during the summer the rate at which energy is absorbed exceeds the rate of emission at every latitude. Much of the excess absorbed energy warms the oceans and is stored there.

The diagram on the right shows the average surface temperature of the atmosphere at a height of approximately 2 meters above the surface. The temperature difference between low and high latitudes is greater during the winter than during the summer. The temperature differential helps to power atmospheric motion and a poleward flow of energy from the tropics. (After Riehl, 1972)

the atmospheric gases do not strongly absorb sunlight, because the wavelengths represented in sunlight are shorter than 4 microns or so. The atmosphere and the earth's surface absorb radiant energy from the sun during daylight hours. But the atmosphere and the ground emit thermal radiation at all times, both day and night. Part of the thermal radiation is exchanged as long-wave infrared radiation between the earth's surface and the atmosphere, and part is lost to space. The surface emits an average of 113 units of radiant energy for every 100 units of solar energy incident at the top of the atmosphere. The atmosphere strongly absorbs long-wave radiation from the surface; some of the absorbed energy is lost to space and some is radiated back to the surface.

The atmosphere's blanketing effect over the earth's surface has been compared to the functioning of a greenhouse. Short-wave sunlight passes as easily through the glass of the greenhouse as through the atmosphere. Because glass is opaque to the long-wave radiation from the warm interior of the greenhouse, it hinders the escape of energy.

As a planet, the earth is not warming or cooling appreciably on the average, because it loses as much radiant energy as it gains. Figure 3.11 shows that for every 100 units of solar energy incident on the atmosphere, 100 units of radiant energy are lost, primarily as long-wave thermal radiation. Similarly, the surface of the earth neither gains nor loses energy on the average. Of the 143 units of radiant energy it gains, the surface loses 113 units by long-wave thermal radiation; the excess 30 units of energy are lost from the surface by the mechanisms discussed in the following section.

Conduction, Turbulent Transfer, and Evapotranspiration

The ground absorbs radiant energy at the surface of the earth and converts it to heat energy. When two objects at different temperatures are in contact, heat flows by *conduction* from the warmer to the colder object. When the top layer of the ground is warmer than the deeper layers, as at midday, heat energy flows into the ground. Late in the day or at night, when radiation losses have cooled the top layer, heat energy flows from the warm, deeper layers toward the surface.

Part of the excess energy at the ground is transferred directly to the atmosphere by conduction. The air's turbulent motion helps to increase the efficiency of this process. When the air is warmer than the ground, heat is conducted from the air to the ground. But on the average, heat flows from the ground to the atmosphere to dissipate the excess energy at the ground.

Another important process in the energy balance involves the transformation of liquid water to vapor. Approximately 590 calories are required to vaporize 1 gram of water at a temperature of 10°C, which is the average temperature of the earth's surface. On land, plants play a key part in converting liquid water to vapor. They absorb water from the soil through their roots and pass water vapor out through openings in their leaves in a process called *transpira-*

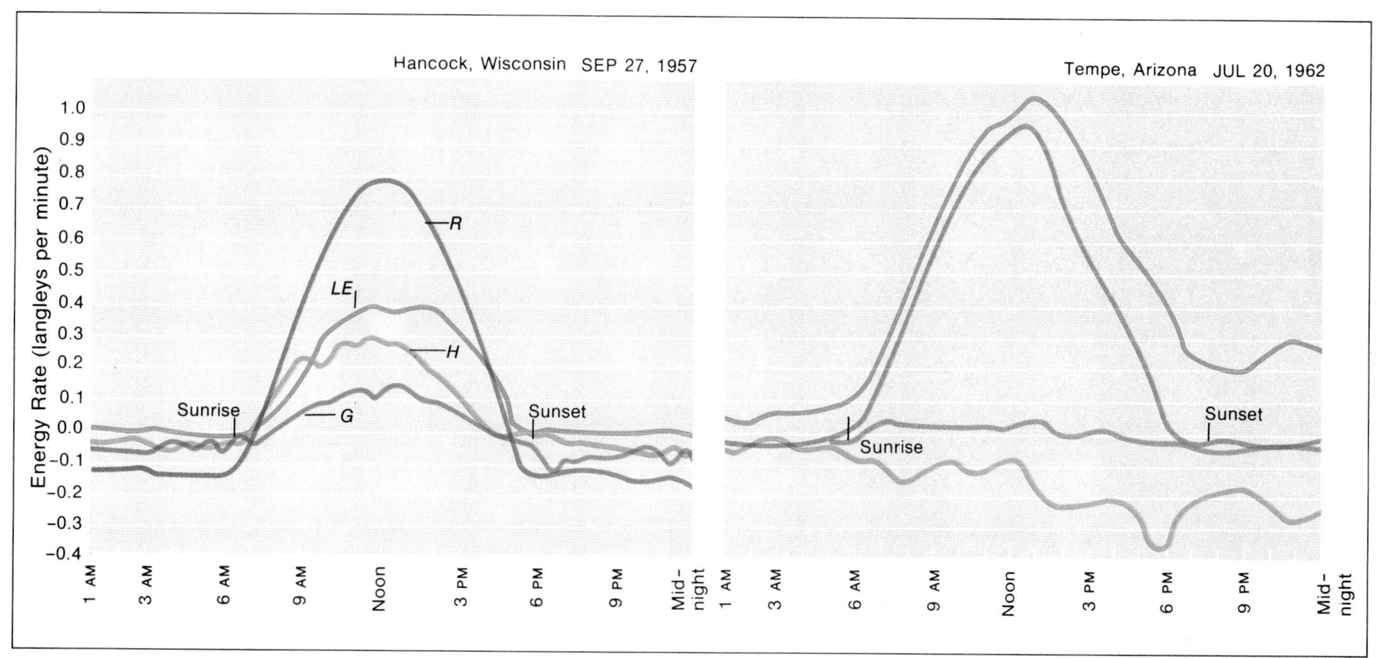
Hancock, Wisconsin SEP 27, 1957
Tempe, Arizona JUL 20, 1962
Energy Rate (langleys per minute)
1.0
0.9
0.8
0.7
0.6
0.5
0.4
0.3
0.2
0.1
0.0
-0.1
-0.2
-0.3
-0.4
R
LE
H
G
Sunrise
Sunset
Sunrise
Sunset
1 AM
3 AM
6 AM
9 AM
Noon
3 PM
6 PM
9 PM
Mid-night
1 AM
3 AM
6 AM
9 AM
Noon
3 PM
6 PM
9 PM
Mid-night

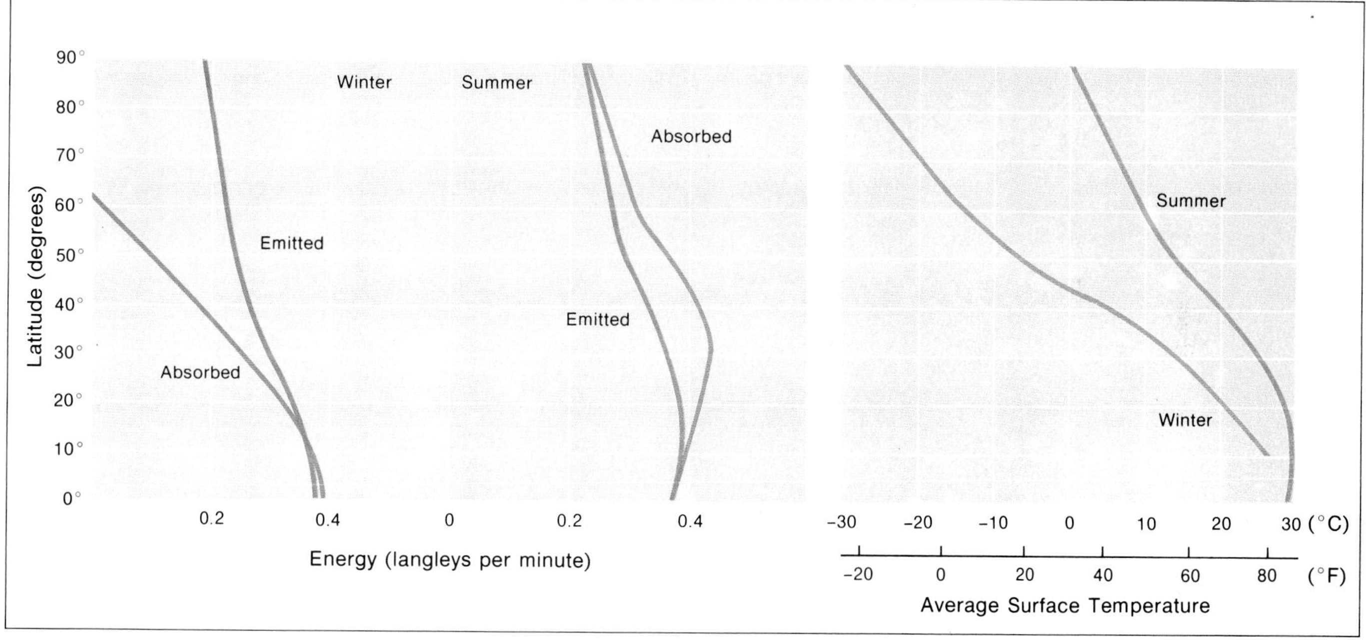
Latitude (degrees)
90°
80°
70°
60°
50°
40°
30°
20°
10°
0°
Winter
Summer
Emitted
Absorbed
Absorbed
Emitted
0.2
0.4
0
0.2
0.4
Energy (langleys per minute)
Summer
Winter
-30
-20
-10
0
10
20
30 (°C)
-20
0
20
40
60
80
(°F)
Average Surface Temperature

tion. Evaporation of water from the soil and transpiration of water from plants are difficult to measure independently, so the two processes are considered jointly as *evapotranspiration.* The energy for evapotranspiration ultimately comes from part of the excess radiant energy at the ground.

Energy Balance at the Earth's Surface

On an annual world-wide average, energy input to the earth's surface is equal to the energy lost from the surface. At any particular

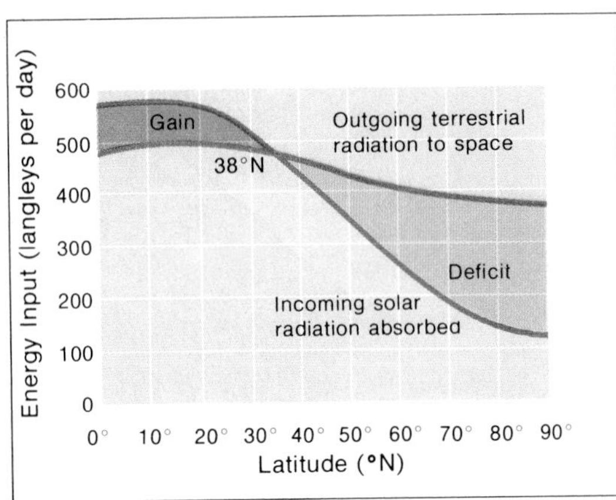
Energy Input (langleys per day)
600
500
400
300
200
100
0
Gain
Outgoing terrestrial radiation to space
38°N
Deficit
Incoming solar radiation absorbed
0° 10° 20° 30° 40° 50° 60° 70° 80° 90°
Latitude (°N)

Figure 3.14 The graph shows average annual global values of absorbed solar radiant energy and emitted long-wave thermal radiation at each latitude in the Northern Hemisphere for the earth's surface and the atmosphere together. Between the Equator and latitude 38°N, absorption exceeds emission, and there is a net input of energy to the earth. Poleward of 38°N, emission exceeds absorption, and there is a net loss of energy. The temperature in the lower latitudes therefore rises until the rate of heat flow toward the poles is sufficient to carry away the excess energy. (After Hare, 1966)

instant or location during a sunny day, more radiative energy is received than is radiated back to the atmosphere and space, resulting in a net radiative gain. The radiant energy input is allocated in varying proportions to evapotranspiration and to heating the air and soil; a very small proportion is utilized directly for photosynthesis. The energy balance equation of the surface shows that the net radiative gain minus evapotranspiration, heating of air and soil, and photosynthesis equals zero. At night, more radiative energy is radiated away than is received from the atmosphere. Thus, the radiation budget is negative. Energy flows from the warmer atmosphere to the cooler ground and from warmer layers in the ground toward the surface. On the average, heat flows from the surface into the atmosphere, heating it from below. This vertical transfer of energy provides one of the driving forces for the weather.

In addition to vertical energy transfer, there is horizontal energy transfer over the earth. Figure 3.14 shows that the earth-atmosphere system in regions near the poles loses more energy in a year than it gains, whereas the tropics have an annual energy excess. To maintain the temperature equilibrium of the earth, energy must on the average flow from the equatorial regions toward the poles. The difference in energy input between the tropics and the poles sets up temperature and pressure differences that drive the circulation of the atmosphere and of the ocean. Warm air and water move poleward from the tropics, and cooler air and water move toward the tropics.

Heating and Cooling of Land and Water

The top fraction of a centimeter of soil almost completely absorbs radiant energy falling on it. Soil has a small specific heat, so a moderate amount of absorbed energy can raise the soil temperature

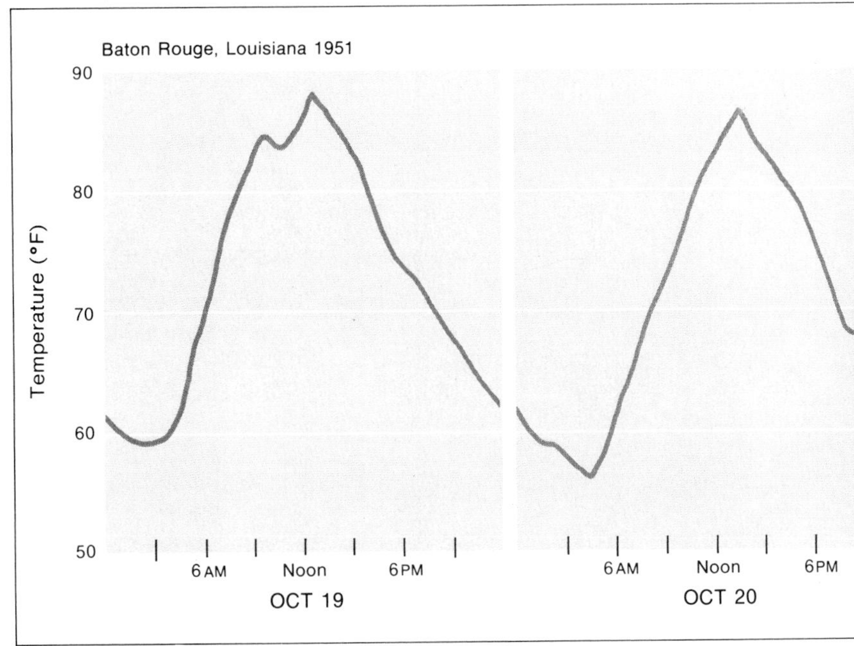
Baton Rouge, Louisiana 1951
Temperature (°F)
90
80
70
60
50
6 AM
Noon
6 PM
OCT 19
6 AM
Noon
6 PM
OCT 20

markedly. Conduction carries some heat energy from the hot surface to the cool deeper layers, but the process is relatively slow because the air space in the soil acts as an insulator.

On a clear summer day at noon, the surface of soil exposed to the sun can easily reach temperatures of 40°C (104°F) or more. Dry beach sand is often too hot for bare feet in the summer. Normally, heat energy in the hot soil flows into the atmosphere, into the deeper layers of the ground, and into evapotranspiration.

At night, the soil cools rapidly because of radiation. If the sky is not cloudy, the amount of radiation returned to the ground is relatively small, and the soil surface temperature can drop to low values. Cold soil cools the atmosphere at immediate ground level. If there is no wind to mix the air, the air temperature near the ground may be several degrees cooler than the air a few meters higher. Even during the day, the soil cools rapidly if shielded from the sun. The daytime temperature of the soil under a cover of tall plants may be as much as 10°C (18°F) cooler than the temperature at the top of the plants.

The surface temperature of the oceans remains much more constant than the temperature of the land surface. Radiant energy in the ocean can penetrate to a depth of tens of meters, distributing energy through a large volume of water. Ocean currents mix the water and help to distribute the energy. In addition, the large heat capacity of water tends to prevent the absorbed energy from raising the water temperature significantly.

Temperature Regimes

The physical principles underlying the radiant energy balance may be applied on a local scale to help explain temperature variations at a given location. Figure 3.15 shows temperature records for a

Figure 3.15 The temperature records at Baton Rouge, Louisiana, for October 19 to 23, 1951, illustrate the daily variation in air temperature as the ground heats and cools. The temperature rises in the morning, when the radiant energy balance becomes positive, and reaches a peak in the early afternoon. The radiant energy balance declines in the afternoon and becomes negative at night, causing the temperature to drop.

On October 19 to 21, skies were clear, but on October 22 and 23, skies were cloudy. On a cloudy day the daily variation in temperature is less extreme than on a clear day, because clouds reduce solar energy input during the day. Clouds also reduce heat loss caused by long-wave thermal radiation, so that temperatures do not fall as low on a cloudy night as on a clear one.

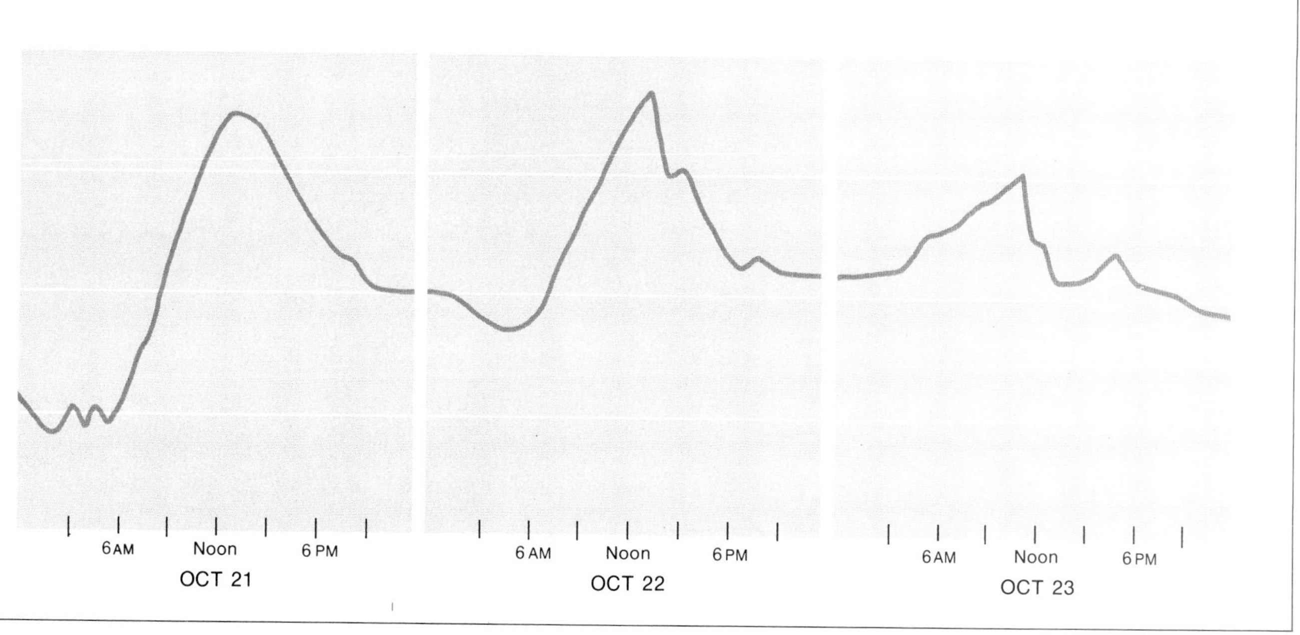
6AM
Noon
6 PM
OCT 21
6 AM
Noon
6 PM
OCT 22
6AM
Noon
6 PM
OCT 23

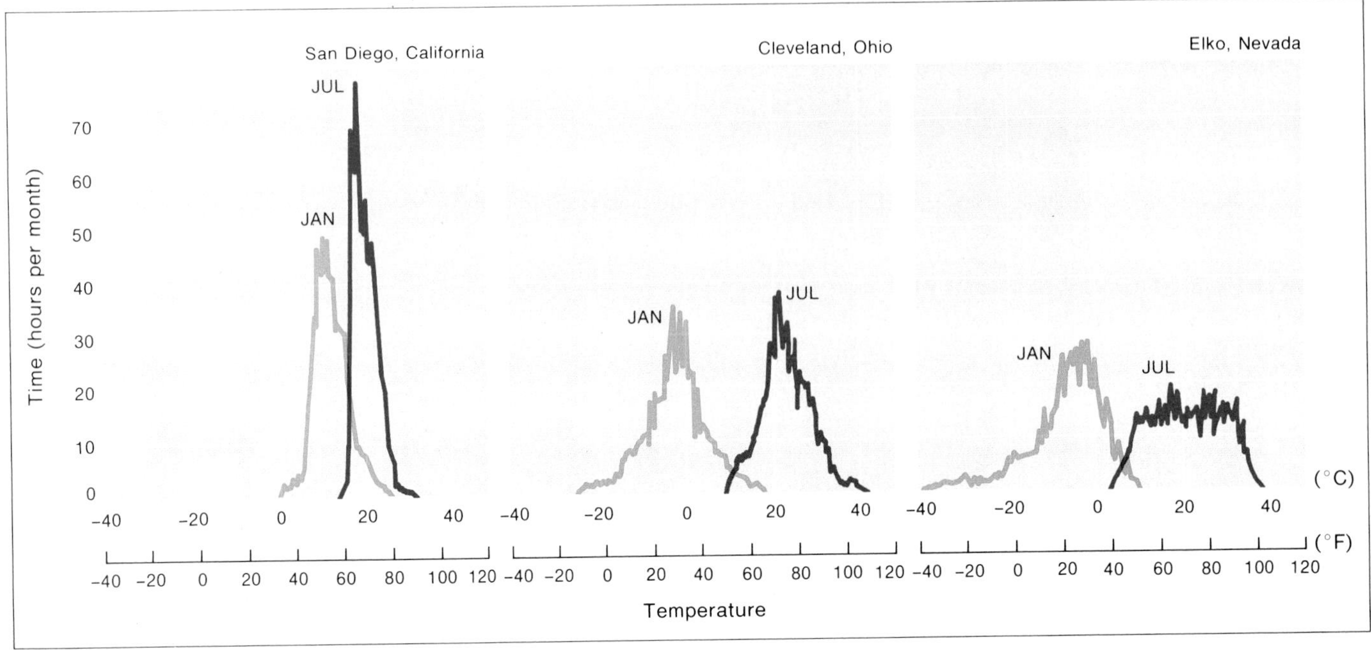
San Diego, California
Cleveland, Ohio
Elko, Nevada
JUL
JAN
Time (hours per month)
Temperature
(°C)
(°F)

Figure 3.16 The average temperature distributions during January and July are shown for three cities in different climatic regions over the 4-year period 1935 through 1939. Each point on a temperature distribution tells the number of hours during the given month that the specified temperature was recorded; in San Diego in July, for example, a temperature of 20°C (68°F) was recorded in 57 separate hours.

A narrow temperature distribution, which San Diego, a coastal city, shows, means that the temperature remains comparatively constant. At Elko in July, in contrast, the temperature distribution is broad, reflecting the hot days and cool nights of the desert. The distributions are moderately broad for Cleveland, a a midcontinent city with a moist climate (After Court, 1951)

Figure 3.17 (opposite) Global temperature distributions are shown for January and July. Temperatures have been reduced to approximate sea level values by applying a correction for the decrease in temperature with increased altitude. In the winter hemisphere (January in the Northern Hemisphere, July in the Southern Hemisphere), temperatures generally depend markedly on latitude and decrease regularly from the Equator toward the pole. In the summer hemisphere, temperatures depend less strongly on latitude. These trends reflect the behavior of solar radiant energy input to the earth (see Figure 3.10).

Note that blue tones on this map indicate below-freezing temperatures and "warmer" colors indicate above-freezing temperatures. The extreme poles are not shown on the flat polar quartic map projections used in this book. (After *Goode's World Atlas*, 1970, and Trewartha, 1968)

period of several days at Baton Rouge, Louisiana. The air temperature near the ground tends to follow the temperature of the ground. In clear weather, the air and ground temperatures rise markedly during the day and fall at night. The daily rise in temperature begins an hour or so after sunrise, when the radiant energy input finally surpasses the energy losses. The air and ground continue to warm up while the energy balance is positive. Cooling begins in the late afternoon when energy losses exceed energy input.

During cloudy weather, the temperature changes very little. Clouds reduce the amount of incoming radiation during the day and increase the amount of return radiation at night, which smooths out extreme variations in temperature. Proximity to a large body of water also tends to smooth out variations in temperature, because the temperature of a large body of water changes little from day to day. Energy transfer between the water and the atmosphere helps to maintain a uniform temperature at coastal cities compared to cities in the heart of the continent.

Monthly temperature records also may be interpreted according to the physical principles that govern the radiant energy balance. Figure 3.16 shows temperature records for January and July at San Diego, California, a coastal city; Cleveland, Ohio, a midcontinent city with frequent cloud cover; and Elko, Nevada, a midcontinent city in the desert. The figure shows the *distribution* of temperature through each month, which is a more meaningful way to describe temperature variation than to quote simple average or extreme values. The graphs are made by measuring the temperature every hour for a month and then plotting the number of times a certain temperature occurs versus the temperature.

Because the energy input is reduced in winter, the distributions in all three cities for January fall at generally lower temperatures

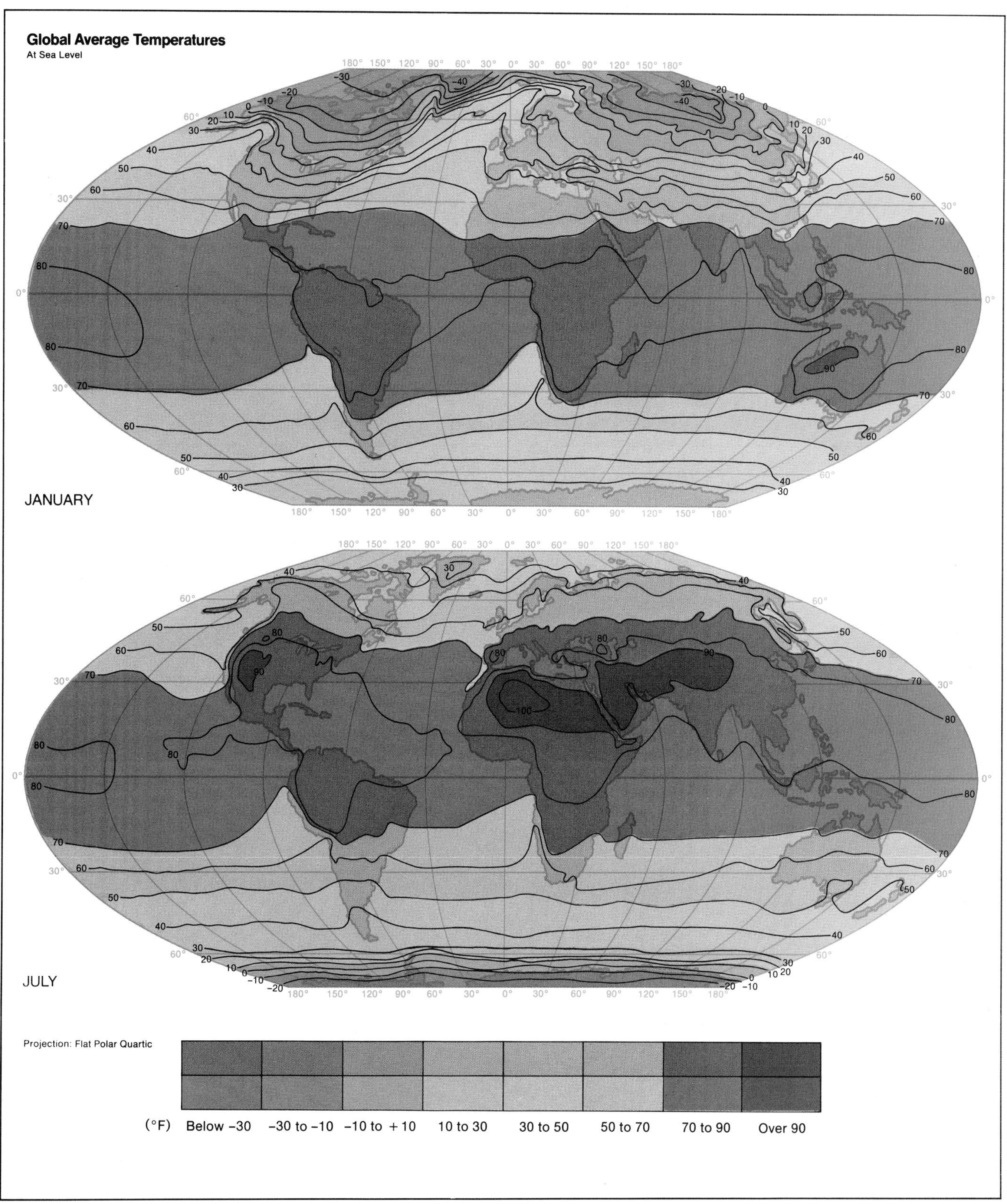
Global Average Temperatures
At Sea Level
JANUARY
JULY
Projection: Flat Polar Quartic
(°F) Below -30 -30 to -10 -10 to +10 10 to 30 30 to 50 50 to 70 70 to 90 Over 90

than the distributions for July. The January and July distributions for San Diego show the most overlap, which reflects the uniformity of coastal temperatures. The monthly distributions for San Diego are the narrowest for the same reason. The January–July differences for Cleveland and for Elko are typical of midcontinent cities. The monthly distributions are relatively wide because of the daily rise and fall of temperature. The extreme width of the July distribution for Elko is attributable to the lack of cloud cover during summer in the desert. The January distribution at Elko and the Cleveland distributions are narrowed by the presence of cloud cover.

Modification of the Energy Balance

The complex interaction of the earth-atmosphere system with radi-

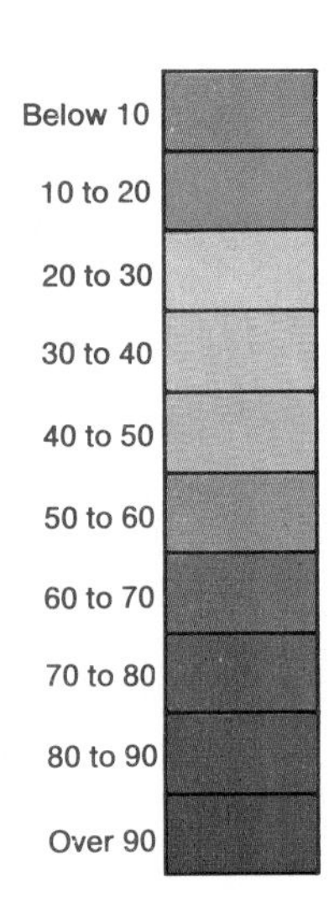

Below 10
10 to 20
20 to 30
30 to 40
40 to 50
50 to 60
60 to 70
70 to 80
80 to 90
Over 90

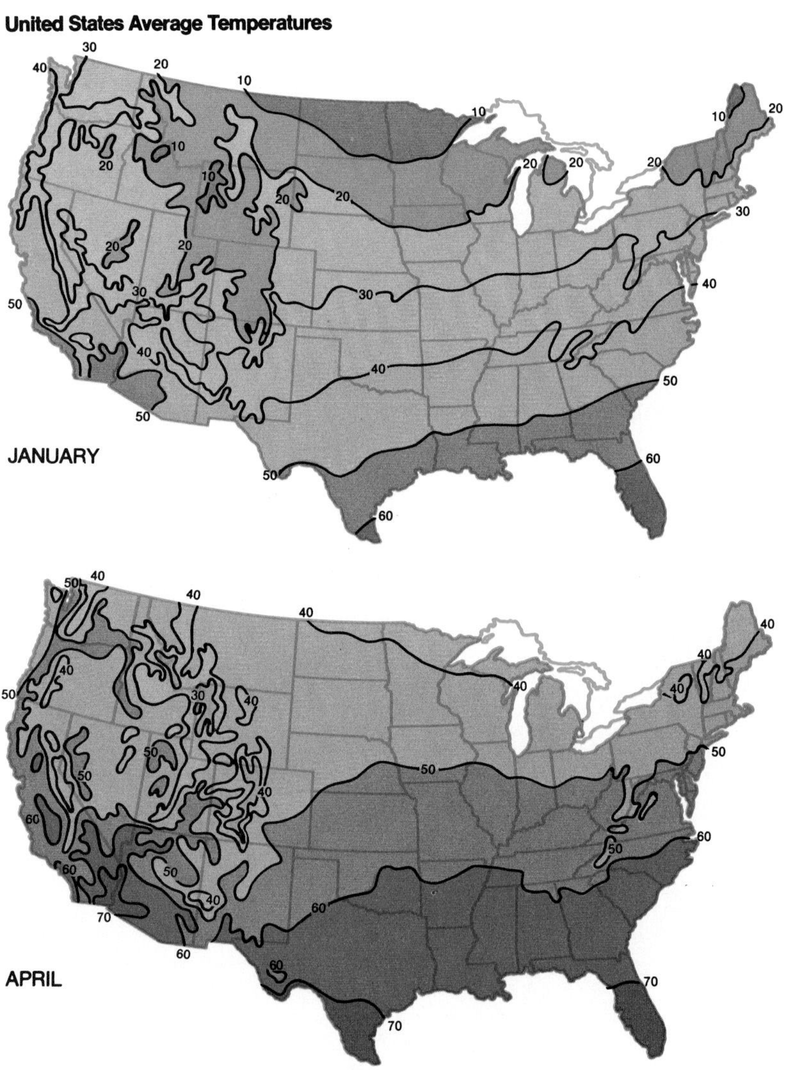

United States Average Temperatures
JANUARY
APRIL

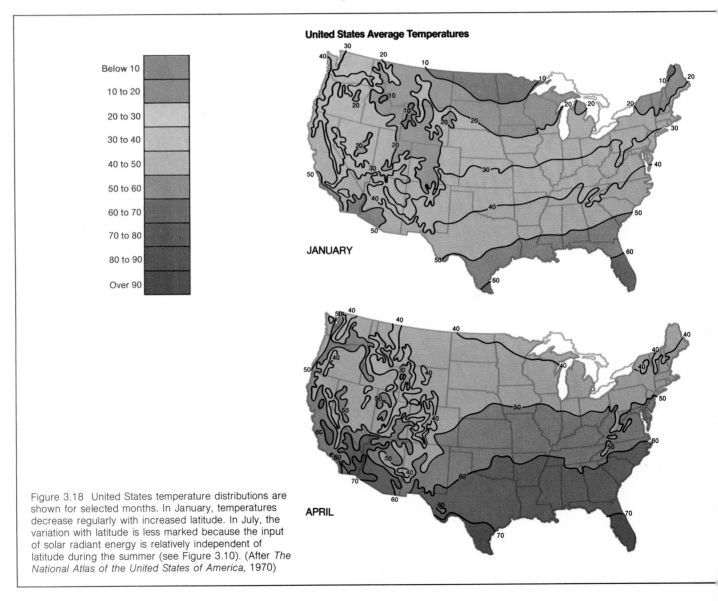

Figure 3.18 United States temperature distributions are shown for selected months. In January, temperatures decrease regularly with increased latitude. In July, the variation with latitude is less marked because the input of solar radiant energy is relatively independent of latitude during the summer (see Figure 3.10). (After *The National Atlas of the United States of America,* 1970)

ant energy depends on a wide variety of factors. The composition of the atmosphere, the cloudiness of the sky, the albedo of the ground—all influence the utilization of radiant energy. A change in any of these factors can alter the energy balance. Man's activities, both purposeful and accidental, have contributed to such alterations on a world-wide scale as well as on a local one.

On the local scale, gardeners and orchard growers are concerned with ways to prevent frost damage to plants. On clear nights, there are no clouds to return radiant energy to the earth's surface, so the return radiation from the sky is minimal. The ground rapidly loses heat by radiation, and temperatures near the ground can drop to below the freezing point. Garden plants and flowers can be protected by covering them with insulating material to reduce radiative

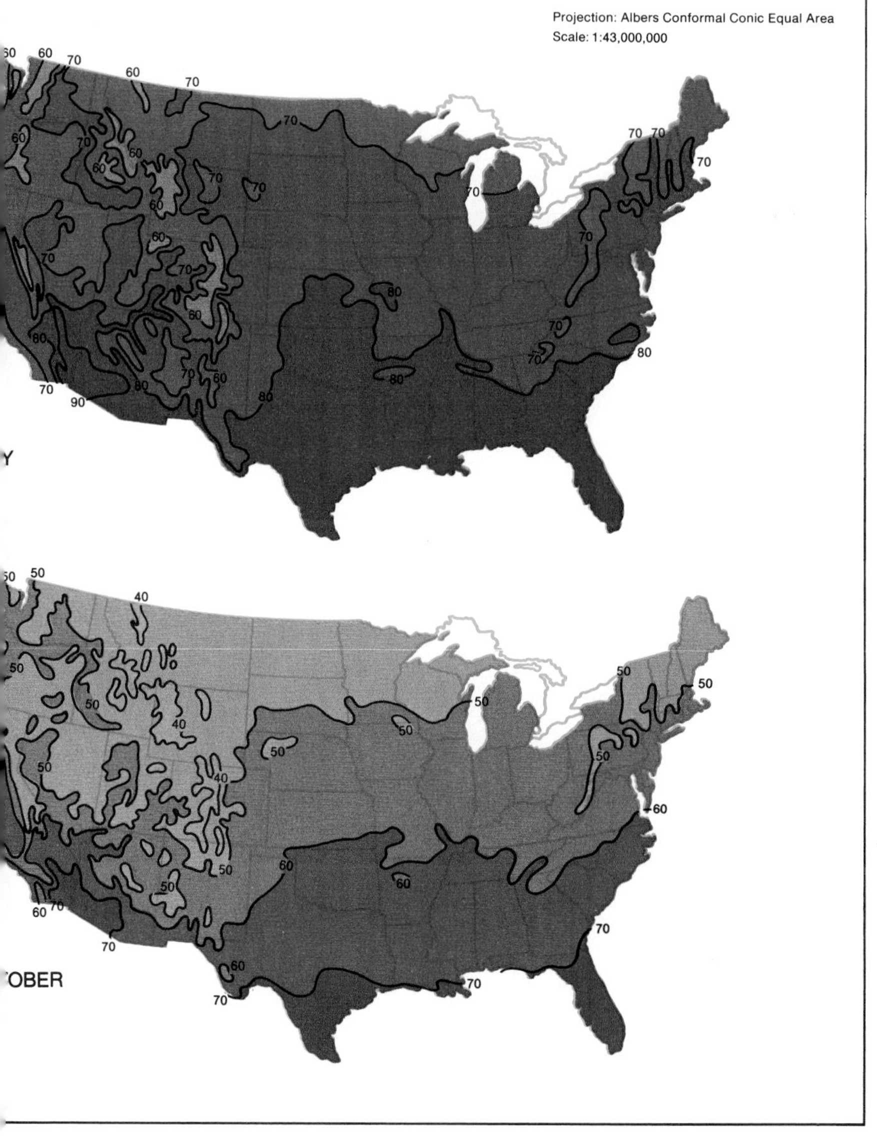
Projection: Albers Conformal Conic Equal Area
Scale: 1:43,000,000

Figure 3.19 The graphs show *R*, the radiant energy balance at the surface of the ground, through the year at four stations. Also shown are the three components of *R*: *LE*, the rate of heat flow from the ground because of evaporation and transpiration; *H*, the rate of heat flow from the ground to the air by turbulent transfer; and *G*, the rate of heat flow from the surface downward into the soil.

West Palm Beach and Yuma are south of latitude 35°N, and Astoria and Madison are north of latitude 40°N. The values of *R* during the summer are similar at all four locations. During the winter, however, *R* is much smaller at Astoria and Madison than at West Palm Beach and Yuma, because of the lower altitude of the sun and the shorter duration of daylight at the more northern locations. At Madison, *R* is negative on the average during the early winter, but cloudy skies at Astoria reduce heat loss at night and *R* is small but positive there during the winter.

LE is usually the largest component of *R* during the growing season. At West Palm Beach, *LE* is especially large because of soil moisture availability to the vegetation. *LE* is zero at Madison during the winter because the low temperatures inhibit evaporation and plant transpiration. (After Sellers, 1965)

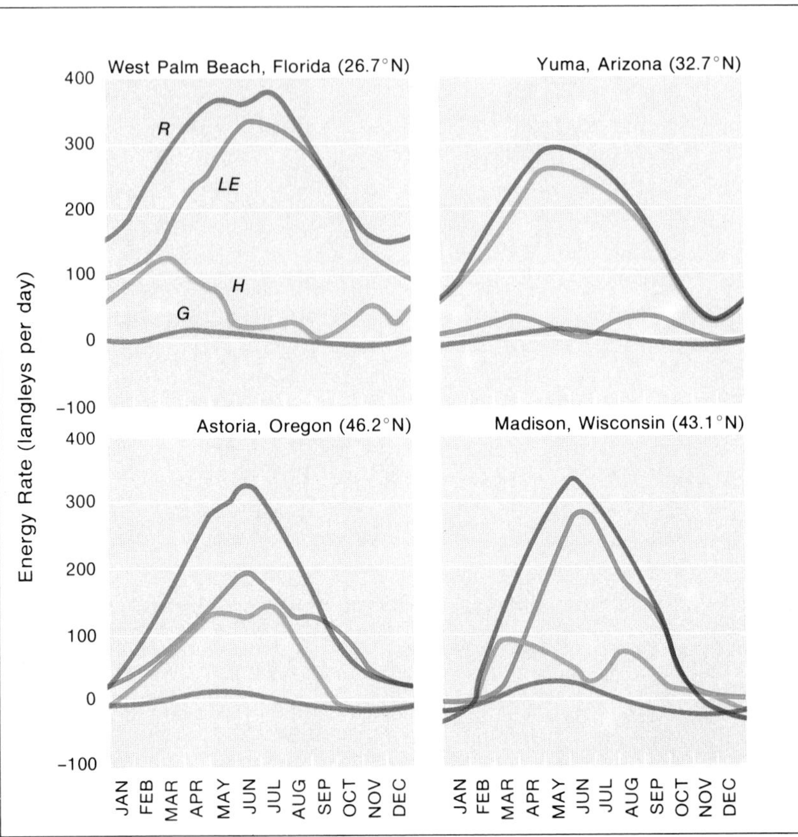

West Palm Beach, Florida (26.7°N)
Yuma, Arizona (32.7°N)
Astoria, Oregon (46.2°N)
Madison, Wisconsin (43.1°N)
Energy Rate (langleys per day)
400
300
200
100
0
-100
R
LE
H
G
JAN FEB MAR APR MAY JUN JUL AUG SEP OCT NOV DEC

energy loss from the surrounding ground. Orchard growers have employed large fans or helicopters to displace the layer of cold air next to the soil by warmer air from above. Smudge pots, which are small heaters that produce a dense blanket of smoke to reflect infrared radiation back to the ground, are generally less effective protection. Spraying vineyards with water throughout the night can protect them from frost. As the water freezes, 80 calories of heat energy are released per gram of water. Even though ice forms on the plants, the temperature is held to 0°C (32°F), which most plants can endure.

Large cities can produce their own distinctive environments, partly by modifying the radiation balance. Cities can be as much as 7 or 8°C (12 or 14°F) warmer than the surrounding countryside, because the towering city buildings receive considerable radiant energy even when the sun is low in the sky, and the canyonlike streets tend to catch reflected radiation. The dense concrete and bricks of cities have a higher heat capacity than soil, and they also cool less rapidly. Energy released from fuel burning in homes, factories, and automobiles augments the radiant heat input in cities. If the air over a city is dusty or smoky, the radiant energy balance is affected. The opacity of the air, called the *turbidity*, affects the energy balance in two opposing ways: by blocking incident radiant energy, turbidity decreases the energy input to the ground, but by increasing the return infrared radiation to the ground, turbidity also decreases the heat loss.

Man's modification of the radiant energy balance did not begin with the industrial revolution. Ancient agricultural practices, such as overgrazing and excessive forest cutting, have altered the albedo and evapotranspiration rates over perhaps 20 percent of the land surface. Removal of plant cover has allowed the wind to blow soil particles into the air, which has increased the turbidity of the atmosphere.

The industrial age has intensified man's impact. Jet aircraft deposit exhaust products and make condensation trails high in the atmosphere, which results in greater cloudiness and greater reflection of short-wave and long-wave radiation. Industrial activities have changed the scattering and absorption of radiation in the earth's atmosphere because they have increased the turbidity and the carbon dioxide content of the atmosphere. The construction of artificial lakes and of roads (roads cover nearly 1 percent of the surface of the United States) has altered albedos and water evaporation patterns.

Data from satellite surveys of key parameters such as albedo, land use and plant cover, and degree of cloudiness over the entire surface of the earth are required to assess the effects of these changes. The launching of the Earth Resources Technology Satellite, ERTS-1, in the summer of 1972 was an important first step in this program. From its orbit between the poles, ERTS-1 surveys bit by bit nearly the entire earth rotating beneath. The significance of such surveys

Figure 3.20 The different colors in this *thermogram* of downtown New York City correspond to different temperatures, with red indicating the warmest temperature. The generation of heat within cities and the comparatively high heat capacities of building materials turn cities into *heat islands* that are several degrees warmer than the surrounding countryside. Washington, D.C., for example, is frost-free for 1 month longer than neighboring rural areas because of man's modifications of the local climate.

Figure 3.21 Man's activities sometimes induce undesirable interactions with environmental systems on the earth's surface. When Los Angeles, California, is free from smog, Mount Baldy 40 miles away can be seen clearly, as shown in the photograph on the left. During smog conditions, as shown in the photograph on the right, the mountain is invisible and the buildings downtown are obscured. Much smog is produced by the action of sunlight on automobile exhaust emissions.

is that every kind of weather together with every kind of land or water surface makes unique energy balance combinations. The energy balance in turn affects the formation of weather systems.

Summary

The sun provides the energy for all life processes on the earth. Solar energy warms the earth's surface and drives the circulation of the air and of the oceans, it powers the hydrologic cycle and the movement of weather systems, and it provides the energy for photosynthesis. The processes described in this chapter underlie the weather and climate systems discussed in later chapters.

The sun's location in the sky can be described by altitude and azimuth angles in a similar way that a pair of angles can specify locations on the earth according to the grid system of parallels of latitude and meridians of longitude. The amount of daylight possible at a particular location depends on the latitude and on the time of year, because of the tilt of the earth's axis of rotation. The solar constant expresses the intensity of the sun's radiant energy output at the top of the atmosphere. The average solar energy delivered to a particular area on the earth depends on latitude because of the tilt of the earth's axis and because of the angle between the sun's rays and the area.

The sun transmits energy in the form of electromagnetic radiation, which has a spectrum of wavelengths. Radiation from the sun consists of infrared, visible, and ultraviolet radiation. Sunlight passing through the earth's atmosphere is modified by absorption, scattering, and reflection.

The earth receives solar energy and it radiates energy back to the atmosphere in the form of long-wave infrared thermal radiation. Water vapor and carbon dioxide molecules in the lower atmosphere absorb much of the thermal radiation the earth emits, thus heating the atmosphere from below. The long-wave radiation reemitted by the atmosphere is a major energy input to the earth's surface, and it significantly reduces energy loss from the earth's surface.

Therefore, the earth's surface gains energy from sunlight and from long-wave radiation. It loses energy by thermal radiation, conduction in the soil, turbulent transfer of heat to the atmosphere, and evapotranspiration. The daily variation of air temperature near the ground can be interpreted in terms of the energy balance between gain and loss of energy. The energy balance can be modified by cloudiness, nearness to bodies of water, and by man's activities. The next chapter explains how the sun's energy drives precipitation processes in the atmosphere.

4 Moisture and Precipitation

Transfer of Water to the Atmosphere
Saturation Vapor Pressure and Evaporation Rate
Evapotranspiration

Moisture in the Atmosphere
Absolute, Relative, and Specific Humidity
Humidity and Temperature
Formation of Dew
Vertical Distribution of Water Vapor

Condensation Processes
The Coalescence Model of Rainfall
The Bergeron Ice Crystal Model of Rainfall
Formation of Hail
Formation of Fog
Formation of Clouds

Adiabatic Processes

Rainmaking: Artificial Condensation Processes

Modification of Harmful Weather Conditions

Water enters the atmosphere through evaporation, is transported in clouds, and falls again to the earth as rain, snow, sleet, or hail. Condensation and precipitation are the results of a number of complex physical interactions that involve energy and moisture.

Seascape Study With Rain Clouds by John Constable, *c.* 1824–1828. (Royal Academy of Arts, London)

Moisture and Precipitation

In the spring of 1972, rain clouds gathered over the Black Hills of South Dakota; rain poured steadily and dumped more than a foot of water on several areas in a 6-hour period. The mountain streams became swollen as they coursed down the hillsides to the land below. In Rapid City, normally placid streams overflowed their banks in a rampaging torrent that swept houses off their foundations. Buildings and automobiles blocked a dam upstream from Rapid City, causing the dam to overflow. The additional water released contributed to the flood wave of death and destruction that passed through the city.

What happens in the atmosphere to cause a deluge that results in a disastrous flood on the earth's surface? The water that fell on the Black Hills gathered as molecules of water vapor over the Gulf of Mexico, thousands of miles south of South Dakota. Moving masses of air carried the water vapor northward across the farmland of Kansas and Nebraska. Along the way, some of the molecules of water vapor coalesced to form droplets of water a few microns in diameter, which were too small to fall to the ground as rain. Over the Black Hills, the droplets coalesced further to form raindrops. The return of water to the earth's surface in the form of rain completed the chain in which a phenomenal number of molecules—about 10^{22} separate molecules of water vapor—joined to form each raindrop.

Rainfall is the result of several complex physical interactions. The conditions that lead to rainfall are so narrowly determined that it should be considered an unusual phenomenon rather than a commonplace event. Although half the earth is covered by clouds at any time, only a small fraction of the clouds possess the necessary qualifications for generating rain.

The farmer who watches rain clouds pass over his parched fields and deposit their rain on the next county probably wishes that rain were under man's control. In recent years, scientists have directed considerable effort toward understanding precipitation mechanisms, with the hope that greater knowledge will lead to ways of controlling rainfall. Although scientists have learned a great deal about precipitation processes, they have met with only limited success in the practical control of rainfall. Adequate rainfall or irrigation at the required times is one of the most important conditions for successful agriculture; prolonged lack of moisture can have far-reaching social and economic effects on a country. Unusually dry weather in the Soviet Union in 1972 had political implications when extensive crop failures led the Russians to buy large quantities of grain from the United States.

The *hydrologic cycle* involves the transfer of water from the earth's surface to the atmosphere and the eventual return of the water to the surface (see Figure 4.3). This chapter describes the processes involved in the atmospheric part of the hydrologic cycle:

Figure 4.2 Man depends on the moisture delivered by the atmosphere to sustain his food crops. Because precipitation occurs only when certain conditions are present, the amount of moisture received by a given region may vary markedly from year to year. The coming of rain, or the lack of it, is a central concern of farmers each year.

(top) People in India rejoice in the coming of the monsoon season rains, which supply the country with much of its annual precipitation.

(bottom) This farm in Oklahoma was abandoned in the 1930s when a succession of drought years made farming impossible. The dry soil was transported by the wind, forming drifts of soil and turning the region into a dust bowl for a period of several years.

Energy
Condensation
Evaporation
Evaporation and Transpiration
Precipitation

Figure 4.3 The entry of water into the atmosphere, its transport through the atmosphere, and its delivery to the surface of the earth are important steps in the hydrologic cycle. Energy from the sun frees water molecules from the land and the ocean, and lifts them into the atmosphere in the form of gaseous vapor. The vapor is transported by the atmosphere, and if certain conditions are met, the water vapor condenses to droplets of liquid water that form clouds. Under special circumstances the droplets coalesce into raindrops, which fall from the atmosphere as precipitation.

evaporation, the conversion of liquid water to vapor; condensation, the conversion of water vapor to water droplets; and precipitation, the coalescence of water droplets to raindrops, snowflakes, or hailstones that are large enough to fall to the ground.

Transfer of Water to the Atmosphere

Water is transferred from the earth's surface to the atmosphere by the process of evaporation. When water evaporates, individual water molecules leave the liquid and enter the atmosphere as water vapor. In a liquid, molecules are bound together by attractive forces, but the molecules have thermal energy of motion and continually collide with one another. The higher the temperature of the liquid, the more energy the molecules have and the more forcefully they collide. Especially forceful collisions near the surface of the liquid may give some of the molecules enough energy to break away from the liquid and enter the air. Because the escaping molecules carry energy with them, evaporation tends to be a cooling process. For this reason, you sometimes feel a chill when you step out of a swimming pool even though the air may be warmer than the water. Considered another way, energy must be supplied to a liquid to keep its temperature constant as it evaporates; this energy is the latent heat of vaporization discussed in Chapter 2. On the earth's surface, the radiant energy from the sun ultimately supplies the energy required to evaporate water. To evaporate a layer of water

1 centimeter thick from a pan or a puddle requires the expenditure of 590 calories for every square centimeter of surface area.

Saturation Vapor Pressure and Evaporation Rate

If a jar partly filled with water is left open in a dry room, the water in the jar eventually will evaporate completely. The water molecules that leave the liquid are moving rapidly, and they disperse to all parts of the room; only a few molecules return to the jar. The rate at which water molecules leave the liquid greatly exceeds the rate at which they return, so the water gradually evaporates. Heat from the room supplies the energy for the evaporation.

If the jar is tightly sealed, water molecules continually evaporate from the liquid under the driving force of heat. The gas above the liquid gradually becomes a mixture of air and water vapor. Because the jar is sealed, the molecules in the vapor cannot disperse widely, and so they strike the surface of the water from time to time. In some of the collisions, the water molecules lose so much energy that they become bound to the liquid again. Eventually a condition is reached in which the number of water molecules leaving the liquid is balanced by the number of water molecules entering. In this state of equilibrium, the air is said to be *saturated*. It contains as much water vapor as it can hold at its particular temperature. The pressure exerted by the water molecules in saturated air is called the *saturation vapor pressure* of the air.

When the temperature is increased, the water molecules move more energetically and a larger proportion of the molecules gain enough energy to leave the liquid. The rate of evaporation increases rapidly with increasing temperature, so the saturation vapor pressure of warm air is greater than that of cold air. Figure 4.4 shows the saturation vapor pressure of air over liquid water for the temperature range from -30°C to 40°C (-22°F to 104°F). The steep increase in the vapor pressure with higher temperature reflects the rapid increase in the evaporation rate with temperature. Except in fog or in clouds, the air in the atmosphere is not saturated; it contains less than the maximum amount of moisture allowed at its particular temperature.

The rate of evaporation of water from an exposed wet surface is sometimes less than the maximum rate set by the prevailing temperature. In still air, for instance, water molecules that leave the surface of liquid water travel only a short distance before they collide with air molecules. Because repeated collisions prevent most of the evaporating water molecules from leaving the vicinity of the liquid water, a layer of nearly saturated air forms directly above the surface of the liquid. Many water molecules from this layer of moist air return to the liquid, so that the net rate of evaporation from the water is reduced to the value required to replace the comparatively few molecules that wander away into the air.

When the wind blows across a moist land or water surface, however, water molecules are carried off as soon as they leave the liquid,

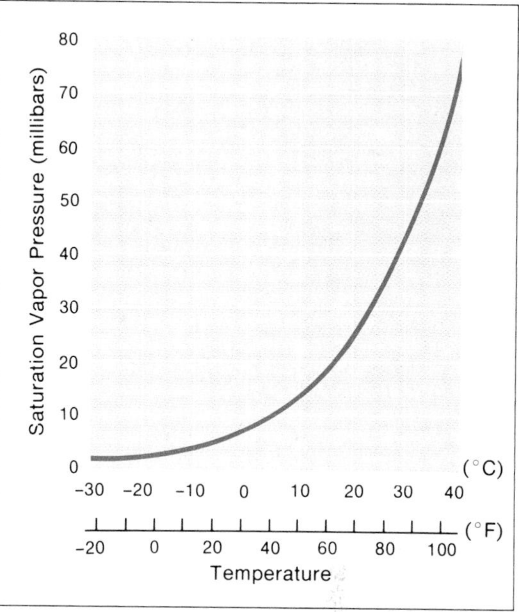
Saturation Vapor Pressure (millibars)
80
70
60
50
40
30
20
10
0
(°C)
-30 -20 -10 0 10 20 30 40
(°F)
-20 0 20 40 60 80 100
Temperature

Figure 4.4 The curve shows that the saturation vapor pressure of air increases rapidly with increased temperature. The data extend to temperatures lower than 0°C (32°F), the normal freezing point of water, because small droplets of water can remain liquid at temperatures as low as -40°C (-40°F). (After List, 1971) so that a layer of nearly saturated air cannot form. That is why wet clothes dry more rapidly on a windy day than on a still day. Under windy conditions, the evaporation rate of water reaches the maximum value permitted by the water's temperature.

The *temperature*, the *wind*, and the *amount of moisture already contained in the surrounding air* affect the actual rate of evaporation from an exposed wet surface. If the air is nearly saturated, the net rate of evaporation is slow because water molecules from the air enter the liquid water nearly as often as molecules leave the liquid. Air moving over the earth's surface sometimes covers large regions with moist air. On a muggy day the air is nearly saturated, and evaporation from skin perspiration occurs too slowly to give a comfortable cooling effect. Electric fans do not help much because the moving air itself is close to saturation. Conversely, on a hot, dry, windy day in the desert, the rate of evaporation is maximized, and in a few hours the body can lose enough moisture to endanger health.

Evapotranspiration

Over lakes and oceans, evaporation occurs directly from the surface of the water. On land where plants are growing, a small part of the water vapor entering the atmosphere comes from direct evaporation of soil moisture, but most comes from water vapor released through openings, called *stomata*, in plant leaves. Plants absorb soil water with their roots. The water and the dissolved nutrients from the soil are then transported up the stem of the plant, and most

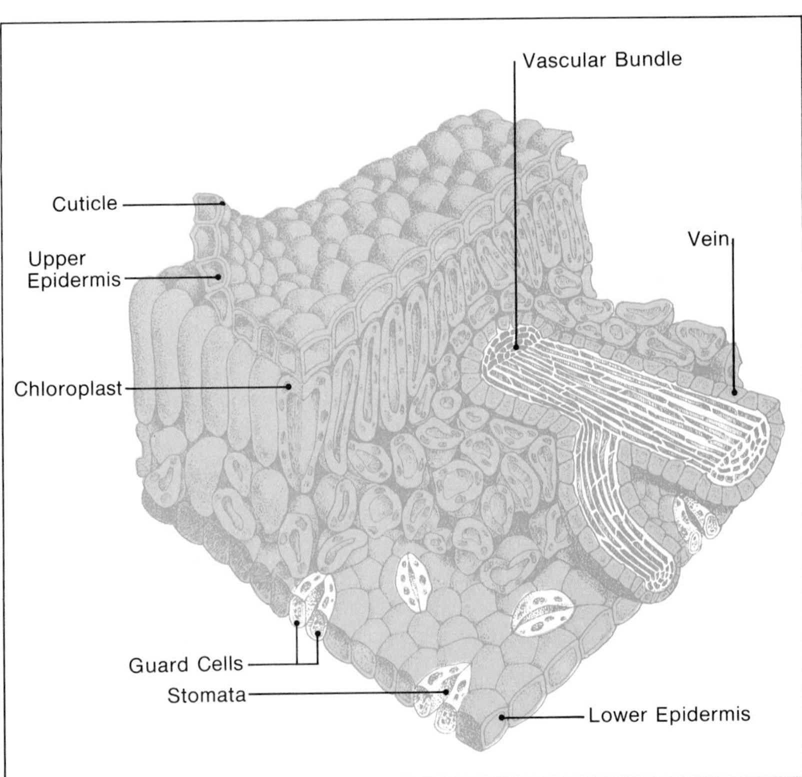
Vascular Bundle
Cuticle
Vein
Upper Epidermis
Chloroplast
Guard Cells
Stomata
Lower Epidermis

Figure 4.5 The leaves of green plants possess openings known as *stomata* in the bottom layer of protective epidermis. When the leaf is exposed to light, photosynthesis occurs in the chloroplasts and the stomata are open to allow the entry of carbon dioxide and the exit of water vapor. Soil moisture absorbed by the roots is transported to the leaves through the vascular bundles in the veins. The transpiration of water vapor is an essential process in green plants, and most of the water vapor entering the atmosphere in heavily vegetated regions is from plant transpiration rather than from evaporation.

The guard cells around the stomata close the openings if the plant lacks moisture and begins to wilt. In plants adapted to hot, dry climates, the guard cells keep the stomata closed during the hottest part of the day when water loss by transpiration would be greatest. In some plant species adapted to dry climates, the covering, or cuticle, is particularly thick and waxy to prevent loss of moisture through cell walls.

of the water is eventually transpired through the stomata as vapor. The flow of water through a plant is an essential part of its life processes. During the day, when a plant is photosynthesizing, the stomata are open to allow the entry of carbon dioxide. Simultaneously, water vapor escapes by transpiration. In the dark, the stomata are closed and there is little transpiration. The energy requirement for transpiration is the same as that for evaporation: 590 calories are required to transpire 1 gram of water at ordinary temperatures.

Evapotranspiration, which is the combination of evaporation from the soil and transpiration from plants, increases rapidly with temperature; it is greatest in hot, dry, windy weather and negligible on freezing winter days. Plants transpire almost all of the water that rainfall and irrigation supply to them. Evaporation from the soil is small once the surface of the soil becomes dry, so most of the water that enters the atmosphere from forests and fields is water that plants transpire. In an area with dense plant cover, water loss by transpiration is at least 2 or 3 times greater than water loss by evaporation.

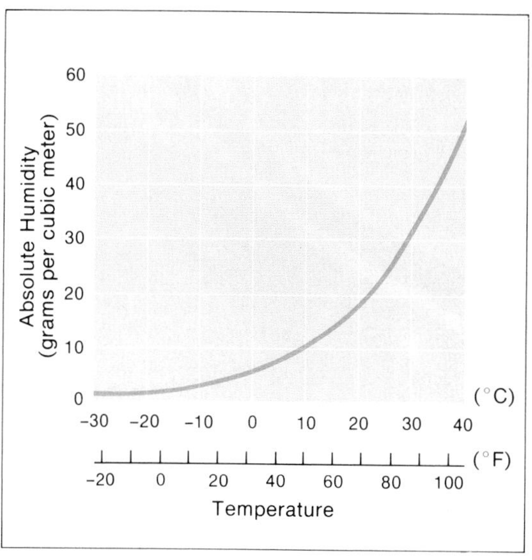

Absolute Humidity (grams per cubic meter)
60
50
40
30
20
10
0
(°C)
-30 -20 -10 0 10 20 30 40
(°F)
-20 0 20 40 60 80 100
Temperature

Figure 4.6 The curve shows the maximum amount of water vapor that can be contained in a cubic meter of air at a given temperature. If the air is not saturated, the absolute humidity lies below the curve. The absolute humidity of saturated air rises rapidly with increased temperature, so that warm air is able to contain more water vapor per unit volume than cool air can contain. The shape of the absolute humidity curve is similar to the saturation vapor pressure curve shown in Figure 4.4. Over a limited range of temperature, the pressure of water vapor in the air depends primarily on the amount of water vapor per unit volume of air and only secondarily on the temperature. (After List, 1971)

Moisture in the Atmosphere

The air above the surface of the earth always contains moisture and it is always moving. The entry of water vapor into the atmosphere by evaporation presents an opportunity for the atmosphere to transport water and deposit it elsewhere.

Absolute, Relative, and Specific Humidity

There are several ways to specify the amount of water present as vapor in a quantity of air. The most direct measure of the air's moisture content is the *absolute humidity*, which is the weight of water present in a given volume of air. Absolute humidity is usually expressed in units of grams of water per cubic meter of air. (A cubic meter is approximately equal to 35 cubic feet.) Saturated air contains the maximum amount of water possible at a given temperature and under normal conditions. As Figure 4.6 shows, the absolute humidity of saturated air is strongly dependent on temperature. At a given temperature, air may contain any amount of water from none up to the maximum value set by saturation. As you can see from the graph, air at 12°C (53.6°F), for example, can contain up to 10 grams of water vapor per cubic meter. Ten grams, which is the weight of two American 5-cent pieces, may not seem like much water, but rain clouds that contain 10 grams of water per cubic meter have the potential to produce rainfall several centimeters deep on the ground below.

Relative humidity is the ratio of the amount of water actually present in a quantity of air to the amount that could be held by the same air if it were saturated. To say it another way, relative humidity is the ratio of the *actual vapor pressure* of water vapor in the air to the *saturation vapor pressure*. Usually expressed as a percentage, the relative humidity ratio does not by itself indicate the actual moisture content of the air. As Figure 4.6 implies, hot

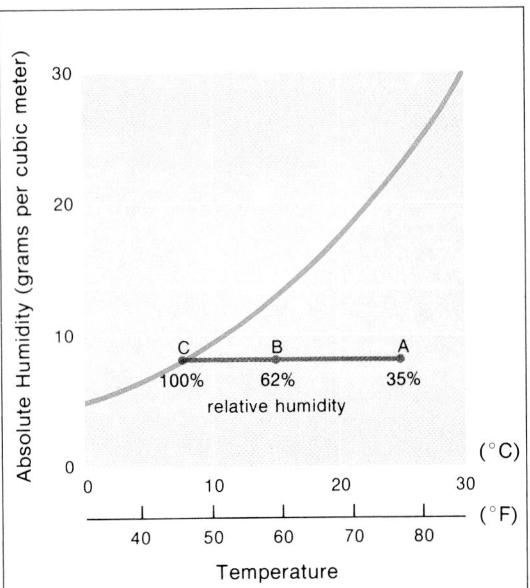

Absolute Humidity (grams per cubic meter)
30
20
10
0
C
B
A
100%
62%
35%
relative humidity
(°C)
0
10
20
30
(°F)
40
50
60
70
80
Temperature

Figure 4.7 The rapid decrease in the absolute humidity of saturated air with decreased temperature has important implications for condensation processes. The curve in this graph is a portion of the absolute humidity curve plotted in Figure 4.6.

Consider a given sample of air that has a relative humidity of 35 percent at a temperature of 25°C. The maximum absolute humidity of air at 25°C is 23 grams per cubic meter according to the curve, so that air with a relative humidity of 35 percent has a moisture content of 23 × 0.35 = 8 grams per cubic meter. The sample is represented by point *A* in the figure.

Suppose that the sample of air is cooled to 15°C, point *B*. The actual moisture content of the air remains at 8 grams per cubic meter. Because the maximum absolute humidity at 15°C is 13 grams per cubic meter according to the figure, the relative humidity is now 8/13 = 0.62, or 62 percent. If the air is cooled further to point *C*, where the maximum absolute humidity is 8 grams per cubic meter, the relative humidity becomes 100 percent and the water vapor is able to condense if conditions are favorable. (After List, 1971)

air at 40°C (104°F) and 50 percent relative humidity contains 25 grams of moisture per cubic meter, but cool air at 20°C (68°F) and 80 percent relative humidity contains only 14 grams of moisture per cubic meter. The cool air in this example has a lower absolute humidity but a higher relative humidity than the hot air. If both relative humidity and temperature are known, the absolute humidity can be found with the aid of Figure 4.6.

Two other units that are sometimes used to characterize the moisture content of air are the *specific humidity* and the *mixing ratio*. In a kilogram of moist air, a small portion of the mass is water vapor and the remainder is air. The specific humidity is the number of grams of water vapor per kilogram of moist air. The mixing ratio is the ratio of the weight of the water vapor to the weight of the dry air. Because a kilogram of air seldom contains more than a few grams of water vapor, the numerical values of the specific humidity and the mixing ratio are nearly the same.

Humidity and Temperature

Temperature determines the maximum amount of moisture the air can hold. When the temperature of moist air changes, the relative humidity of the air changes also, because air at different temperatures has different capacities for holding moisture. Relative humidity is therefore less useful than absolute humidity as a measure of the water content of air. If a parcel of air containing a fixed amount of water vapor is heated, the relative humidity decreases because the saturation vapor pressure increases much more rapidly with temperature than does the actual vapor pressure of the water vapor.

As shown in Figure 4.7, air that is not fully saturated can be represented on a graph of absolute humidity versus temperature by a point corresponding to the actual moisture content and temperature of the air. Point *A* on the graph represents air with 35 percent relative humidity at 25°C (77°F). If the temperature of the air is lowered to 15°C (59°F), no water is lost and the moisture content of the air remains the same. But the cooler air represented by point *B* is closer to saturation than the warmer air, so the relative humidity

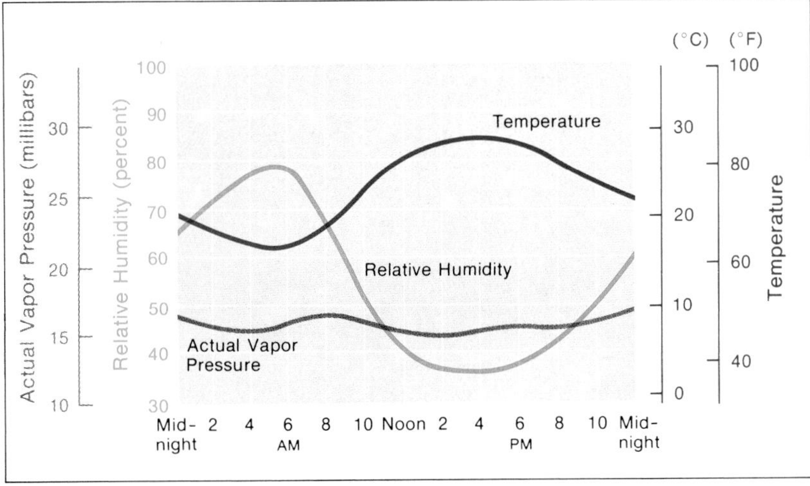

Actual Vapor Pressure (millibars)
30
25
20
15
10
Relative Humidity (percent)
100
90
80
70
60
50
40
30
Temperature
Relative Humidity
Actual Vapor Pressure
Mid-night
2
4
6 AM
8
10
Noon
2
4
6 PM
8
10
Mid-night
(°C)
30
20
10
0
(°F)
100
80
60
40
Temperature

Figure 4.8 During stable weather conditions, the pressure of water vapor in the air near the ground remains comparatively constant through the day, as the graph indicates. However, the relative humidity changes markedly through the day as the air temperature changes, because the saturation vapor pressure of air is strongly dependent upon temperature.

Figure 4.9 Dew forms on plants that cool to the dew point of the surrounding air.

is higher at the lower temperature. If the temperature of the air drops further, to point C, the relative humidity becomes 100 percent.

Formation of Dew

The temperature at which the relative humidity becomes 100 percent is called the *dew point*. If the air cools to the dew point and then continues to cool, the air cannot remain in stable equilibrium with the amount of moisture it contains; some of the water vapor must condense into liquid. A glass of ice water rapidly becomes covered with condensed moisture in humid summer weather because the air near the glass is chilled to the dew point, and it deposits its moisture on the glass. The same process accounts for the formation of *dew*. Soil and plants cool rapidly during the night by radiation, and if they cool to the dew point, moisture from the air condenses on them. The lowest soil temperatures are reached an hour or so after sunrise, before the energy balance becomes positive, and dew is therefore most likely to form early in the morning. The sidewalk on the shady side of a street is protected from solar heating and sometimes becomes cool enough in the early morning to collect a layer of dew while the sidewalk on the sunny side remains dry.

Condensation of water vapor to liquid makes available approximately 590 calories of heat for every gram of water vapor that condenses. The heat energy made available when dew forms tends to keep the soil from cooling as much as it otherwise might. The latent heat of condensation sometimes helps melt snow cover. Ordinarily, snow melts slowly because it has a high albedo and reflects much of the incident radiant energy from sunlight. In addition, the cool air of late winter usually cannot transfer much heat energy to the snow. The surface of slowly melting snow stays near the melting point, 0°C. If the air is so humid that the dew point is 0°C or higher, condensation takes place on the snow surface. Every gram of water vapor that condenses makes enough latent heat available to melt more than 7 grams of snow, so rapid melting of snow can occur even on a cool, overcast day by this process. Condensation melt can supplement the melting that occurs as cool rain falls on snow, and it can produce widespread winter snowmelt floods in areas where the air is often humid in the winter, such as near the Great Lakes and in New England.

Air can contain water vapor at temperatures below 0°C, as Figure 4.6 shows. If condensation occurs when the dew point is lower than the freezing point of water, the water vapor begins to condense as ice crystals. Further deposition of water vapor takes place on the already existing ice crystals, forming long, feathery crystals of *hoar frost*.

Vertical Distribution of Water Vapor

Water vapor normally moves upward into the atmosphere from the earth's surface. During a summer day, when the ground is warm, evapotranspiration is rapid and the absolute humidity near the ground is higher than it is a few tens of meters above the ground. At a height of 100 meters (330 feet) on a clear summer day, the absolute humidity is typically 10 or 15 percent lower than it is at ground level, and water vapor moves from the ground to the atmosphere. Sometimes in the early morning, however, the ground is cool enough to promote the formation of dew, and under these conditions, water vapor moves from the atmosphere to the ground. The water returned to the ground in the form of water vapor is only a small fraction of the water that leaves the ground as vapor in evapotranspiration. Water returns to the ground primarily as rain or snow; fog is not a significant precipitation event, except in coastal

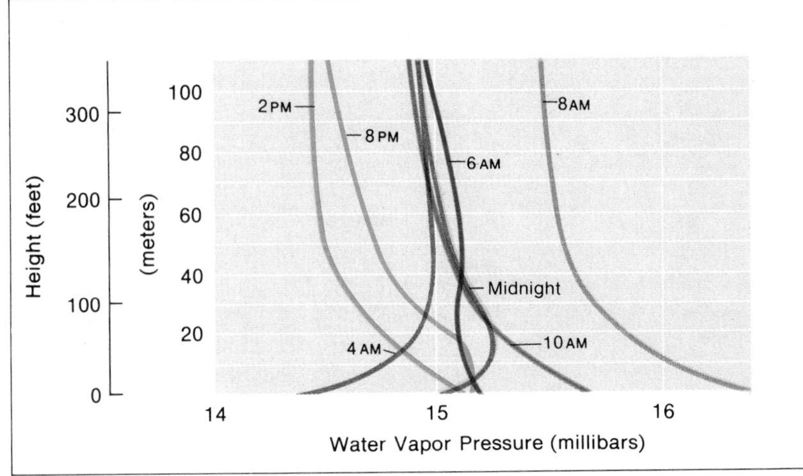

Height (feet)
300
200
100
0
(meters)
100
80
60
40
20
2 PM
8 PM
6 AM
8 AM
Midnight
4 AM
10 AM
14
15
16
Water Vapor Pressure (millibars)

Figure 4.10 Each curve on this graph shows how the water vapor pressure in the atmosphere near the ground varies with height at a particular time on a clear summer day. At 8:00 A.M. the water vapor pressure decreases rapidly with increased height, indicating a flow of water vapor from the surface into the atmosphere. The water vapor pressure decreases from 8:00 A.M. to 2:00 P.M. because the air becomes heated during the morning and early afternoon, causing the onset of turbulent mixing and a consequently more efficient removal of moist air from near the surface. Much of the water vapor remains in the lowest 50 meters of the atmosphere, forming a humid blanket. Water vapor continues to flow upward from the surface during the day while plants are transpiring. By 8:00 P.M. the water vapor pressure is nearly constant through the lowest 15 meters of height, indicating that little water vapor is leaving the surface. Later at night, the vapor pressure immediately above the surface increases with increased height, indicating a flow of water vapor from the atmosphere toward the ground. By 6:00 A.M., after sunrise, the flow is once again from the surface to the atmosphere. (After Geiger, 1965) forests where the trees collect water droplets from the fog as it is blown by. The sequoia forests along the northern coast of California obtain moisture primarily from fog during the summer months when rainfall is light.

Condensation Processes

On the average, the amount of water that leaves the atmosphere is equal to the amount that enters it. Some of the water vapor in the air condenses directly onto cool surfaces as dew or frost. But the condensation of water vapor into masses of microscopically small water droplets, or clouds, is of much greater importance. A small proportion of clouds somehow generates water drops large enough to fall to the earth as rain.

When moist air is cooled to its dew point, the water vapor in the air no longer remains in stable equilibrium; instead, it tends to condense. The condensation process does not occur easily. If moist air free from dust is cooled, condensation does not occur until the vapor pressure of the water is 3 or 4 times the saturation value. Air containing more than its normal saturation amount of water vapor is *supersaturated.*

In clean air that is moderately supersaturated, tiny droplets of liquid water form momentarily, only to be torn apart again by molecular collisions. If the air contains dust particles, however, condensation of vapor to water droplets occurs as soon as the air becomes even slightly supersaturated. The particles, called *condensation nuclei*, act as collection centers for water molecules, and they promote the growth of water droplets to a size large enough to be stable.

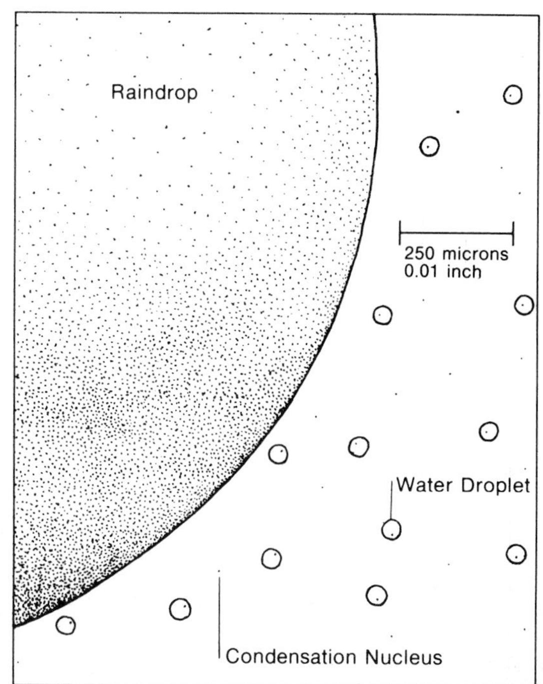

Raindrop
250 microns
0.01 inch
Water Droplet
Condensation Nucleus

Figure 4.11 This drawing shows to scale the relative sizes of condensation nuclei, water droplets, and a raindrop. Most condensation nuclei are less than 1 micron in radius; water droplets are usually less than 20 microns in radius, and raindrops have radii of 1,000 microns or greater. Each water droplet forms in saturated air by condensation of water vapor on a condensation nucleus, and other processes cause millions of water droplets to coalesce and form a raindrop.

Condensation nuclei are microscopically small. Very few exceed a radius of 1 micron (1 micron = 10^{-4} centimeter) and most have radii smaller than 0.1 micron. Polluted air over land contains many thousand such particles per cubic centimeter of air. Giant condensation nuclei—those with radii greater than 1 micron—number only about 1 per cubic centimeter. Principal sources of condensation nuclei in the atmosphere are vapors and smokes from industry and forest fires, salt crystals from sea spray, clay particles from blown soil, and chemicals (particularly sulfates) formed by reactions of trace materials in the air. In regions such as the northwestern United States, where discharges from pulp mills release particles into the air, rainfall appears to be increasing. Perhaps the increased rainfall is a result of the increased number of cloud condensation nuclei from industrial sources.

Fine particles can remain in the atmosphere for days; when the volcano Krakatoa near Java exploded in 1883, the dust content of the global upper atmosphere was greater than normal for more than a year afterward. The force of gravity attracts small particles in the air toward the earth, but the drag force of the air retards their movements. The drag force increases with the speed of the particle. A falling particle soon reaches a steady maximum speed, called the *settling velocity.* The smaller the particle, the slower the settling velocity. The settling velocity of a 1-micron particle is only a few

Figure 4.12 (opposite) The mechanisms that cause water droplets to coalesce to drops large enough to fall from clouds have been studied intensively since the 1940s. The principal models that successfully explain precipitation from clouds are the coalescence model and the Bergeron ice crystal model; the main steps of each process are outlined in this figure.

The coalescence model of precipitation is dominant in the tropics, where many clouds are too warm to contain supercooled water droplets and ice crystals. The Bergeron process dominates at higher latitudes, where the upper portions of many clouds are considerably colder than the normal freezing point of water, even in summer.

The type of precipitation that falls from a cloud depends on the mechanism involved and on the variations of temperature in the atmosphere. Precipitation is formed as snow in the Bergeron process, and if the atmosphere beneath the cloud is below freezing, the precipitation reaches the ground as snow. If the snow crystals fall into warm air, however, the precipitation arrives at the ground in the form of rain. The coalescence process cannot produce snow, but either process can produce sleet or ice pellets if raindrops fall into a layer of cold air near the ground and freeze.

microns per second. Small particles, such as those in tobacco smoke, remain effectively "suspended" in air because random air currents carry such small particles upward faster than the particles can fall. Rain tends to wash particles out of the air, but natural processes and industrial activities renew the supply.

Air in the atmosphere is normally dusty enough for water vapor to condense readily to droplets, forming a cloud, when a parcel of air is cooled to the dew point. The droplets grow to a radius of 10 or 20 microns before condensation of additional water molecules on the droplets ceases and the droplets stop growing. A typical cloud contains several hundred such droplets per cubic centimeter, and the average water content of a cloud is about 1 gram of water per cubic meter of air.

A water droplet with a radius of 10 microns has a settling velocity of a fraction of a centimeter per second. Such small droplets can never fall to the ground as rain; they would evaporate in the drier air below the cloud after falling only a few centimeters. A water drop must have a radius of at least 100 microns in order to fall from a low cloud to the ground. Because volume increases with the cube of the radius, thousands of 10-micron droplets must combine to produce one 100-micron drop, and millions of droplets are required to form an average raindrop, which has a radius of 0.1 centimeter (1,000 microns).

The process by which millions of tiny droplets can coalesce to form a raindrop is the key to understanding, and perhaps controlling, precipitation. A closely related problem is why most clouds remain as tiny droplets and never form rain. Condensation from water vapor in a cloud does not significantly increase the size of a droplet after it has grown to a radius of 10 or 20 microns. Furthermore, theoretical and experimental studies have shown that droplets smaller than about 20 microns in radius cannot collide and coalesce. As such droplets approach one another, they are forced apart by the air flow between them and they cannot merge.

The Coalescence Model of Rainfall

Raindrops seem to form in clouds either by the coalescence of droplets or by a mechanism involving the formation of ice crystals. Droplets grow by coalescence in clouds that are too warm to allow the formation of ice crystals; such clouds are most frequently found in the tropics. Although average droplets cannot grow by merging with other droplets, oversized droplets larger than 20 or 30 microns can grow by coalescing with smaller drops. The larger the embryo raindrop grows, the faster it falls through the air, and as it falls, it sweeps up the droplets with which it collides. Rising currents of air carry the drops upward faster than they can fall out of the cloud. When the growing drop becomes so large that air currents can no longer support it, it falls from the cloud as rain. A drop requires about 1/2 hour to grow to raindrop size by coalescence, and rain clouds must be about 1 kilometer (0.6 mile) thick for the growing drops to remain in the cloud long enough to become raindrops. Thinner clouds limit the growth of drops but may produce

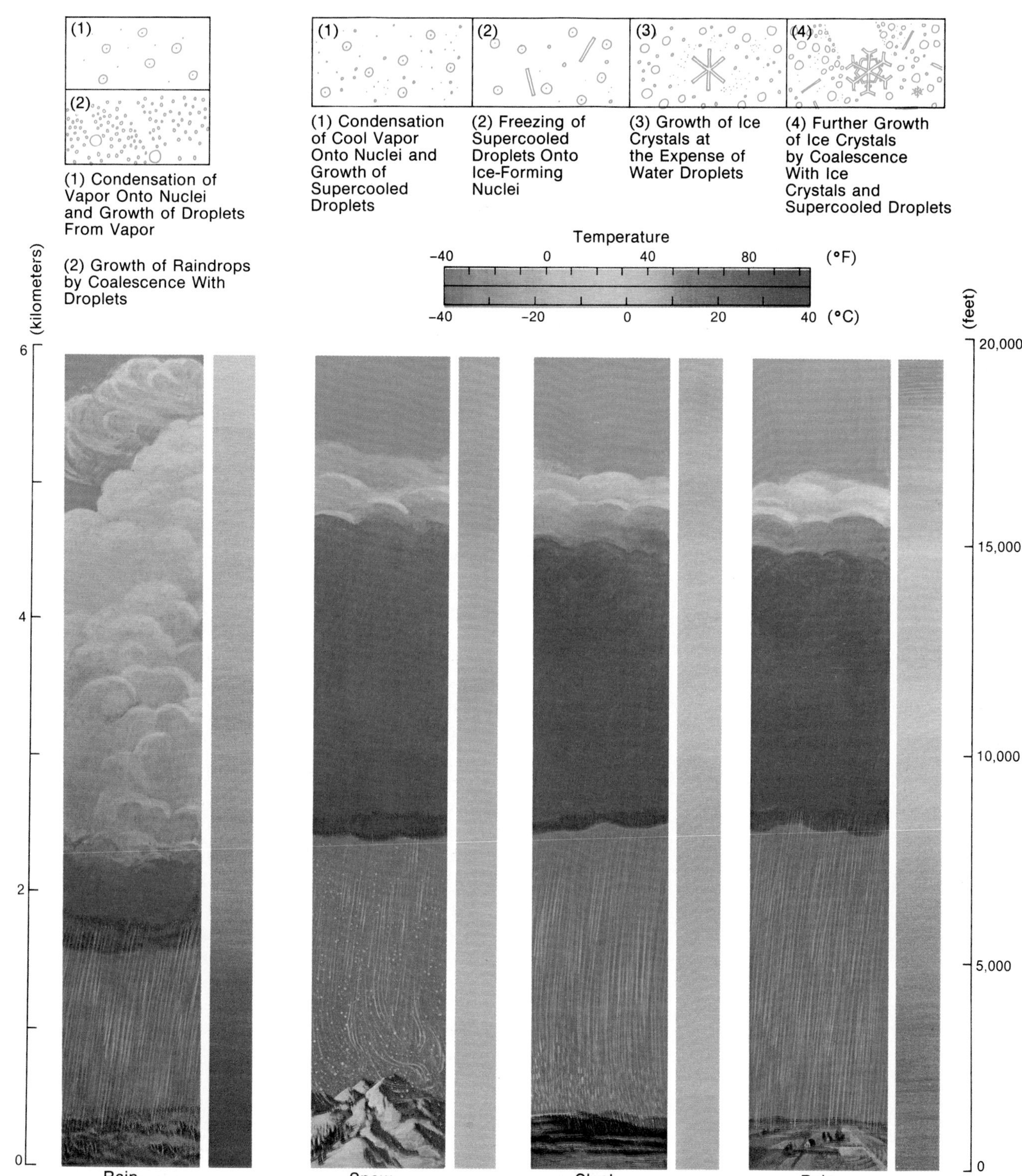
(1)
(2)
(1) Condensation of Vapor Onto Nuclei and Growth of Droplets From Vapor
(2) Growth of Raindrops by Coalescence With Droplets
(1)
(2)
(3)
(4)
(1) Condensation of Cool Vapor Onto Nuclei and Growth of Supercooled Droplets
(2) Freezing of Supercooled Droplets Onto Ice-Forming Nuclei
(3) Growth of Ice Crystals at the Expense of Water Droplets
(4) Further Growth of Ice Crystals by Coalescence With Ice Crystals and Supercooled Droplets
Temperature
-40
0
40
80
(°F)
-40
-20
0
20
40
(°C)
Altitude
(kilometers)
6
4
2
0
(feet)
20,000
15,000
10,000
5,000
0
Rain
Snow
Sleet
Rain

drizzle, a form of precipitation that consists of drops smaller than average raindrops.

Each cubic meter of a typical cloud contains about a million water droplets. Because about a million droplets are required to form an average raindrop, the generation of rain requires the presence of only one initially oversized droplet for every million small droplets. The production of rain by the mechanism of coalescence is therefore surprisingly dependent on the rather rare occurrence of oversized droplets, which form only on giant condensation nuclei. Salt particles 1 micron or more in size make particularly effective condensation nuclei for oversized droplets. One reason that salt particles work so well as condensation nuclei is the *hygroscopic* nature of salt—its natural tendency to take up water vapor. Even far inland, salt particles blown in from the ocean can play an important role in precipitation.

The Bergeron Ice Crystal Model of Rainfall

Coalescence of oversized water droplets accounts for precipitation from clouds that are everywhere warmer than the freezing point of water. But many rain clouds, such as those in the middle latitudes, extend upward to altitudes where the temperature is well below freezing. In 1933 Tor Bergeron, a Norwegian meteorologist, proposed a mechanism of droplet growth that explains rainfall from cool clouds.

Water in a relatively large quantity, such as the water in an ice cube tray, freezes at 0°C. But water dispersed as fine droplets can remain liquid at temperatures as low as -40°C (-40°F). Water that remains liquid at a temperature below the normal freezing point is called *supercooled*. Because of supercooling, water droplets in a cloud can remain liquid even when the upper part of the cloud is in air colder than 0°C. Certain particles in the air, known as *ice-forming nuclei*, promote the freezing of water droplets. Once ice crystals have formed on nuclei, they can continue to grow by adding water molecules from the vapor. Water molecules are more strongly bound to ice than to supercooled water droplets, so ice crystals can grow at the same time that droplets are losing molecules by evaporation. Ice particles can also grow by colliding with supercooled droplets, which freeze onto the ice. As the large ice crystals fall through the lower, warmer parts of the cloud, they may melt and grow larger by coalescing with water droplets. Finally, they fall from the cloud as rain. Ordinary summer raindrops in the middle latitudes often begin life as ice and snow in the upper parts of towering clouds and then melt as they fall. In the winter, they fall as snow if the air is too cold to melt the ice crystals.

Although effective ice-forming nuclei are relatively rare in nature, only a hundred or so per cubic meter are needed to trigger rainfall in a thick cloud. The formation of ice crystals usually occurs only in parts of the cloud supercooled to -15°C (5°F) or lower. In addition, the size of the cloud and the speed of the air currents must be favorable to the growth of large drops if rain is to reach the

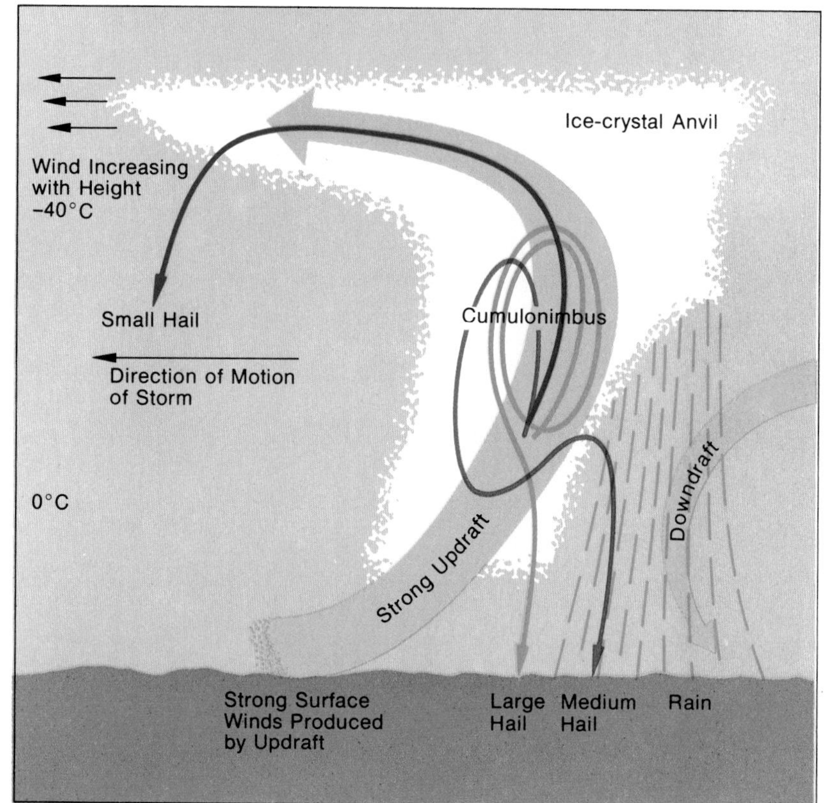

Ice-crystal Anvil
Wind Increasing
with Height
-40°C
Small Hail
Cumulonimbus
Direction of Motion
of Storm
0°C
Strong Updraft
Downdraft
Strong Surface
Winds Produced
by Updraft
Large
Hail
Medium
Hail
Rain

Figure 4.13 This cross section of an intense thundercloud illustrates the cyclic process involved in the formation of large hailstones. Powerful currents of air in the central column of the cloud carry small ice particles upward. Particles that are blown far from the central column fall from the cloud and usually reach the ground in the form of rain. Large hailstones are formed when ice particles cycle through the cloud many times. A falling ice particle may be swept up in the central updraft because of the forward motion of the storm cloud. On each passage through the cloud, the particle gains a new layer of ice. Clouds with strong updrafts are able to support hailstones through many cycles until they fall from the cloud or become too heavy for the air currents to lift. (After Flohn, 1969)

ground. The numerous factors involved and their interplay are not yet fully understood.

Formation of Hail

Hailstones are spherical balls of ice that fall from tall, vigorous clouds such as summer thunderheads. The size of hailstones ranges from miniscule grains to balls the size of a grapefruit. When cut open, hailstones show an onionlike structure of clear and opaque shells of ice. The layering implies that a hailstone is cycled through a cloud many times during its growth (see Figure 4.13).

According to the accepted model of hailstone growth, a hailstone is carried upward by powerful air currents in the central column of a tall cloud. As it rises, the hailstone can grow by colliding with a few small supercooled droplets, which freeze on impact and form a layer of cloudy ice around the hailstone. If the hailstone rises through a region that contains a large number of supercooled droplets, a wet layer may form. The subsequent freezing of this layer produces a layer of clear ice.

Near the top of the cloud, the hailstone emerges from the central column of uprushing air and falls downward in slower air currents in the outer parts of the cloud. As it nears the base of the cloud, however, the advancing cloud can sweep the hailstone back into the central column to repeat the cycle. After a number of cycles, the hailstone grows too large to be supported by the air currents,

Figure 4.14 (right) This photograph shows San Francisco Bay and Oakland, California, shrouded in fog, with Mount Diablo in the distance to the northeast. Low-lying advection fog frequently occurs in the coastal region around San Francisco as moist air moving eastward over the Pacific Ocean cools to its dew point.

Figure 4.15 (below) This map shows the average annual number of days of fog in the conterminous United States. Fog occurs most frequently along the Pacific coast and along the coast of New England, where moist air is cooled by cold ocean waters. Fog also occurs frequently in the Appalachian Mountains of the east central United States. It seldom occurs in the warm, dry air of the western deserts. (After Court and Gerston, 1966)

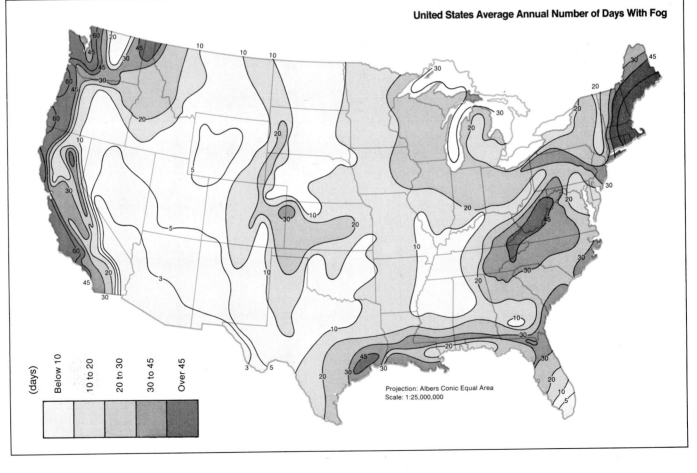

United States Average Annual Number of Days With Fog

(days)

Below 10

10 to 20

20 to 30

30 to 45

Over 45

Projection: Albers Conic Equal Area
Scale: 1:25,000,000

and it falls out of the cloud to the ground. Large hail is associated with violent, turbulent thunderstorm clouds, because vertical air currents with speeds of 20 to 30 meters per second (45 to 65 miles per hour) are required to carry a large hailstone upward.

Formation of Fog

Fog occurs when moist air near ground level is cooled to its dew point. There are several ways such cooling can occur: depending on the mechanism involved, fogs can be classified as *radiation fogs, advection fogs,* or *orographic fogs.*

Radiation fog is usually produced at night, when the ground loses heat by radiation. The air near the ground also becomes cooled, by contact with the cool ground. If it is windy, the cool air near the ground mixes with warmer upper layers of air, and the dew point may not be reached. So radiation fog is most likely to occur on comparatively still nights when the air is clear and when the air near the ground is moist. Clear, still nights facilitate radiational cooling, and if the air near the ground happens to be moist, the dew point is high, and fog is likely to form.

Advection refers to the horizontal motion of air across the earth's surface. When warm, moist air passes over a cool surface such as snow or cool ocean water, the air may be cooled to its dew point and form advection fog. Fogs in the North Atlantic Ocean are advection fogs that form as warm, moist tropical air passes over the cold ocean waters.

The temperature of the surface of the land generally decreases with altitude, because long-wave radiation heats the atmosphere from below. The upper slopes of hills are therefore usually cooler than the bases. When the wind pushes a layer of moist air from lower altitudes up a slope, the air may be cooled to its dew point and form upslope, or orographic, fog.

Formation of Clouds

Clouds are a visible sign that a mass of moist air above ground level is cooled to its dew point. If a large mass of moist air is uplifted

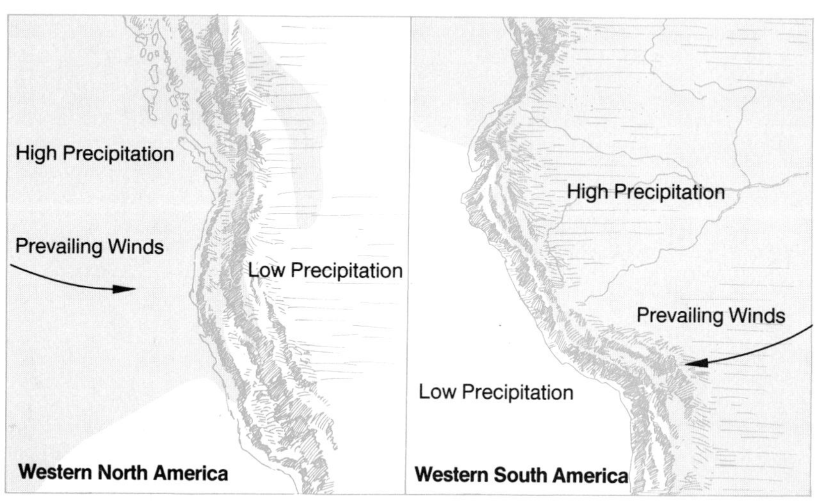
High Precipitation
Prevailing Winds
Low Precipitation
Western North America
High Precipitation
Prevailing Winds
Low Precipitation
Western South America

Figure 4.16 Mountains can exert a strong local influence on precipitation because they can force moisture-laden air to rise to cooler high altitudes where condensation can occur. (left) The mountain ranges along the Pacific coast of the United States intercept moisture transported from the Pacific Ocean by prevailing westerly winds, so that coastal ranges are moist and inland regions are dry. (right) In the South American tropics, prevailing trade winds carry moisture westward from the Atlantic Ocean over the land. The Andes Mountains along the west coast block the moisture, so that the coastal region on the west slope of the Andes is arid, whereas the inland regions to the east are moist.

as it passes over a mountain range, for example, orographic cooling can cause clouds to form above the upslope. In the western United States, air masses tend to advance from west to east, so clouds and precipitation are more prevalent on the western slopes of mountain ranges than on the eastern slopes. Along the Atlantic seaboard, however, moist air masses advance at various times from both east and west, and mountain ranges along the eastern coast receive precipitation on all sides.

A spectacular example of how a mountain can cause rainfall differentiation is found on the island of Kauai in the Hawaiian chain. The average rainfall on one side of Mount Waialeale is 1,170 centimeters (460 inches) a year, and it is only 81 centimeters (32 inches) a year on the other side. The exceptionally heavy rainfall is produced by an orographic cloud that is maintained by steady moisture-laden winds. In addition, the many salt nuclei blown in

Figure 4.17 (top) *Cumulonimbus* clouds, or "thunderheads," rise to great heights and are associated with violent weather such as thunder showers, hailstorms, and tornadoes. They can form near warm or cold fronts when conditions in the atmosphere allow warm, moist air to ascend into colder air of the upper troposphere.

(center) *Fair weather cumulus* clouds, shown here in Wyoming, form on vertical columns of rising warm, moist air. Such clouds tend to be short-lived; they usually evaporate into the surrounding drier air within an hour or so.

(bottom) An *altocumulus,* or "mackerel sky," cloud takes the form of a layer of patchy cloud puffs. It forms at moderate altitudes and often heralds the approach of a warm front and accompanying precipitation.

(opposite top left) *Cirrus* clouds are feathery wisps of ice crystals that form at high altitudes. They frequently are the first indication in the sky that a weather front is moving into a region.

(opposite bottom left) *Cumulonimbus mammatus* is a comparatively rare type of cloud that forms a layer marked by pendulous bulges on its lower surface. It is often associated with severe storms.

(opposite top right) As seen from above, dense stratus clouds present a highly reflective surface.

(opposite bottom right) This photograph of altocumulus clouds over Georgia and Florida was taken on the Apollo 6 mission in 1968. The lines of *cloud streets,* which are a hundred miles or so in extent, demonstrate a remarkable regularity of atmospheric processes.

from the nearby ocean provide condensation nuclei that promote precipitation.

Uplift of a large air mass can occur even in the absence of mountains, because warm air is less dense than cold air and tends to rise over cold air, like a block of wood floating on water. Weather systems move masses of air hundreds of kilometers in length and 5 to 10 kilometers (3 to 6 miles) thick across the surface of the earth. A mass of warm, moist air that is uplifted over a mass of cold air produces widespread cloudiness and gentle precipitation in the lower atmosphere.

The white, puffy *cumulus clouds* of summer, which are only a few kilometers in diameter, are formed by condensation of the water vapor in rising updrafts of moist air. Cumulus clouds show a definite pattern of development during a summer day. Early in the day, the sky is usually clear, but later, scattered wisps of cloud begin

to appear at an altitude of a few kilometers. The wisps increase in size and eventually become large cumulus clouds. What happens is this: the positive radiant energy balance at the earth's surface during the day gradually heats the air near the surface and the heated air rises. The base of the cumulus clouds marks the altitude at which the rising moist air has cooled to its dew point.

Cumulus clouds mark the upper section of a column of rising air. A downflow of air between the clouds replaces the air that rises from the earth's surface. For this reason, cumulus clouds never cover the sky completely. Soaring birds and glider pilots use the rising air under cumulus clouds to gain altitude effortlessly. In fact, observations of the flying habits of birds provided some of the first detailed knowledge of vertical air motion in the lower atmosphere.

Adiabatic Processes

A rising parcel of warm air cools as it rises through the atmosphere. Usually the parcel of air mixes very little with the surrounding cooler air as it rises, and the drop in temperature of rising air is primarily caused by a process known as *adiabatic cooling*.

The pressure of the atmosphere decreases with altitude. As a parcel of air rises from the ground, the higher-pressure air in the parcel pushes outward against the surrounding air until the parcel has expanded enough to make its internal pressure equal to the pressure of the atmosphere. The parcel expands as it rises, and as it expands, it does work pushing outward against the surrounding air. The work the expanding gas does is accomplished at the expense of its internal energy, because no other sources of energy are available. So as the parcel expands, its temperature falls. The decrease in temperature that an expanding gas experiences in the absence of external energy sources is called adiabatic cooling. Similarly, when a parcel of air is compressed as it descends, the surrounding air does work pushing inward on the parcel and the temperature of the parcel increases because of *adiabatic heating*. The rate of adiabatic cooling of air in the absence of condensation is called the *dry adiabatic lapse rate*. Its value is 10°C for every kilometer of increasing altitude (5.6°F per 1,000 feet).

Figure 4.18 When a parcel of dry air rises, it expands and does work against the surrounding atmosphere. The expenditure of energy causes the temperature of the parcel to fall. The figure compares the temperature of a typical parcel of rising dry air with the temperature of the air in the surrounding static atmosphere. The temperature of the parcel and the temperature of the static air both decrease with increased altitude, but in general the temperature of the parcel falls more rapidly, as indicated in the figure. The parcel experiences a net upward force, causing it to continue rising, as long as the temperature of the parcel exceeds the temperature of the static air. The equilibrium height of the parcel is therefore the height where the temperatures are equal.

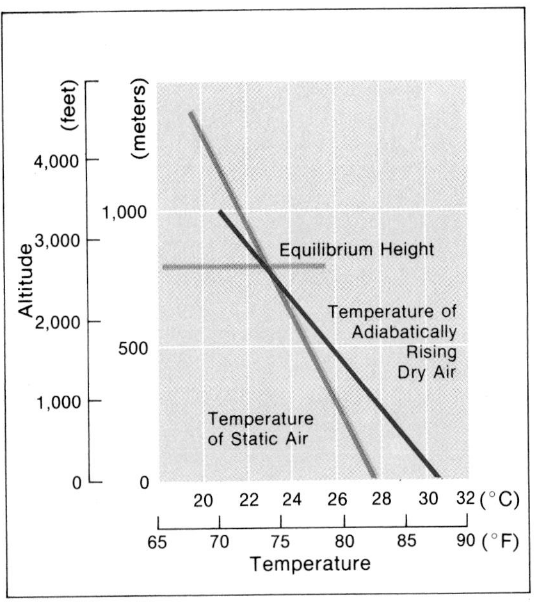
Altitude
(feet)
(meters)
4,000
3,000
2,000
1,000
0
1,000
500
0
Equilibrium Height
Temperature of Adiabatically Rising Dry Air
Temperature of Static Air
20 22 24 26 28 30 32 (°C)
65 70 75 80 85 90 (°F)
Temperature

The temperature of the lower atmosphere varies with altitude according to weather conditions. Normally the temperature of the atmosphere decreases with increased altitude; a typical rate of decrease is 6°C per kilometer (3.3°F per 1,000 feet). A rising parcel of dry heated air cools at the dry adiabatic lapse rate of 10°C per kilometer (5.6°F per 1,000 feet) and eventually reaches an equilibrium altitude where its temperature is equal to the temperature of the surrounding air, as illustrated in Figure 4.18.

When the temperature of part of the atmosphere increases with increased altitude, a *temperature inversion* is said to occur. A warm parcel of air at the surface cannot rise into the warmer, less dense air above it, so inversions trap air at the surface. In large cities, heavy smog frequently accompanies inversions, because pollutants accumulate in the air near the ground when the vertical motion of the air is stopped.

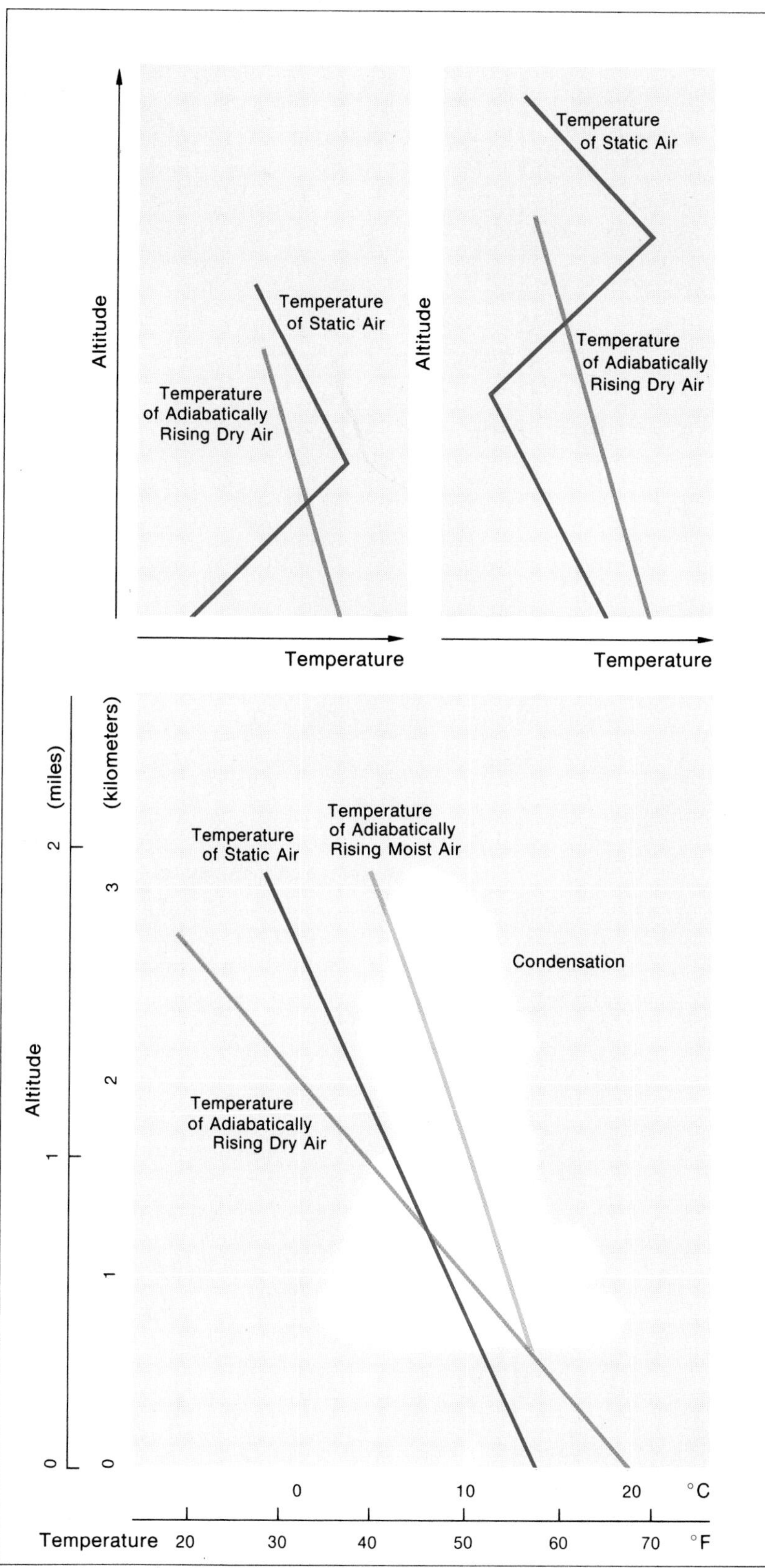

Temperature
of Static Air
Temperature
of Adiabatically
Rising Dry Air
Altitude
Temperature
Temperature
of Static Air
Temperature
of Adiabatically
Rising Dry Air
Altitude
Temperature
Altitude
(miles)
(kilometers)
0
1
2
0
1
2
3
Temperature
of Static Air
Temperature
of Adiabatically
Rising Moist Air
Condensation
Temperature
of Adiabatically
Rising Dry Air
0
10
20
°C
Temperature
20
30
40
50
60
70
°F

Figure 4.19 During a condition of inversion, the temperature of a portion of the atmosphere increases with increased altitude. (top left) The inversion is near the ground. (top right) The inversion occurs at a high altitude. Because the equilibrium height of a rising parcel of air occurs where the temperature of the parcel becomes equal to the temperature of the surrounding atmosphere, the effect of an inversion is to decrease the equilibrium height considerably. Pollutants accumulate when an inversion near the ground traps rising air containing waste products. (bottom) When a rising parcel of air contains water vapor, the release of latent heat of condensation can reduce the rate of cooling of the parcel and enable it to rise to high altitudes. In the example shown in the figure, the adiabatically rising dry air has an equilibrium height of 1.2 kilometers, but a moist parcel with the same initial temperature has an equilibrium height well above 3 kilometers.

No condensation occurs in a rising parcel of moist air until its temperature becomes low enough to saturate air in the parcel. The altitude where condensation first occurs marks the base of clouds, as the figure indicates. Not all of the vapor in the parcel condenses at once. Condensation continues as the parcel rises, and at any given altitude the parcel contains just enough water vapor to keep the air in the parcel saturated. This mechanism can produce clouds that tower to great heights.

(inches)

Below 16	16 to 24	24 to 32	32 to 48	48 to 64	64 to 80	80 to 96	Over 112

Projection: Albers Conic Equal Area
Scale: 1:25,000,000

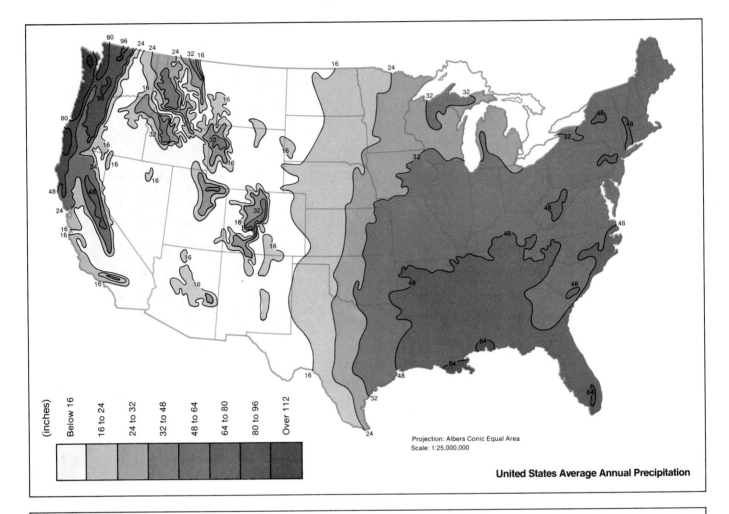

United States Average Annual Precipitation

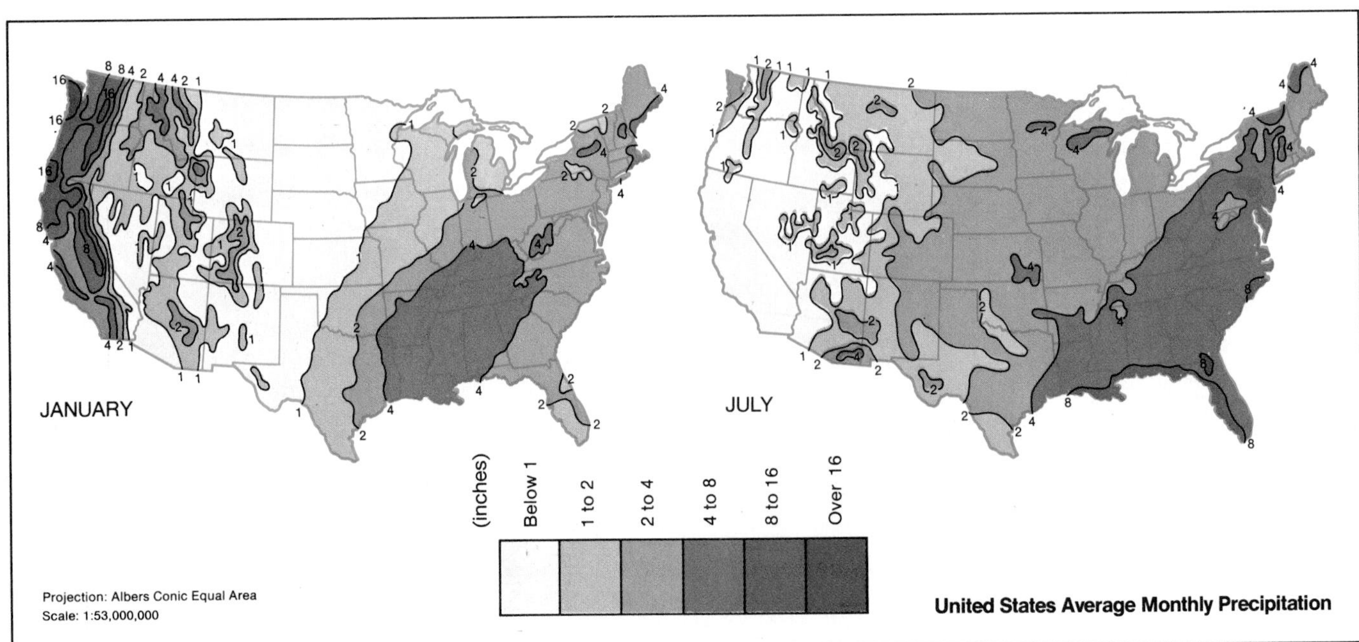
JANUARY
JULY
(inches)
Below 1
1 to 2
2 to 4
4 to 8
8 to 16
Over 16
Projection: Albers Conic Equal Area
Scale: 1:53,000,000

United States Average Monthly Precipitation

A rising parcel of air containing water vapor may cool to its dew point as it rises. As some of the water vapor in the parcel condenses to droplets, latent heat of condensation is made available to warm the air in the parcel. The release of heat promotes further ascent of the parcel and further condensation, which makes the formation of towering clouds possible. The *moist adiabatic lapse rate* varies with the moisture content of the air, and it can be even less than 5°C per kilometer (2.8°F per 1,000 feet).

Rainmaking: Artificial Condensation Processes

The adage has it that although people talk about the weather, no one does anything about it. Some people complain they have too much rain; others do not have enough. People who wanted more rain once argued that planting extensive stands of trees in the Midwest or flooding the deserts of southern California would promote precipitation by increasing the transfer of water vapor to the atmosphere. However, water added to the air in one location may travel as vapor thousands of kilometers with the atmosphere until conditions favorable to precipitation are achieved.

The comparative rarity of precipitation as an atmospheric event means that the amount of precipitation in a particular location is highly variable from year to year. A succession of "wet" years with

Figure 4.20 (opposite top) The pattern of average annual precipitation in the conterminous United States displays some broad regularities. The eastern half of the United States is moist, on the average, because of the moisture transported by winds from the Atlantic Ocean and the Gulf of Mexico. In the central United States, annual precipitation decreases toward the west. Moist air from the Gulf of Mexico seldom flows westward, and moist air from the Pacific Ocean is blocked by mountain ranges. The inland West therefore tends to be dry. The Pacific coast, and particularly the Pacific Northwest, receives moisture from Pacific Ocean air.

(opposite bottom) The eastern United States receives precipitation from moist ocean air during both winter and summer. The central United States is dry in winter because of the influx of cold, dry air from the north. The Pacific coast receives much of its annual precipitation during the winter, and the summers there are dry, particularly in California.

(below) This global map of average annual precipitation shows that precipitation is greatest in equatorial regions and where mountain ranges intercept moist air. The driest regions include the deserts near latitudes 30°N and 30°S, certain inland areas such as central Asia, and regions where mountains block moist winds, such as the inland western United States. (After *Goode's World Atlas,* 1970)

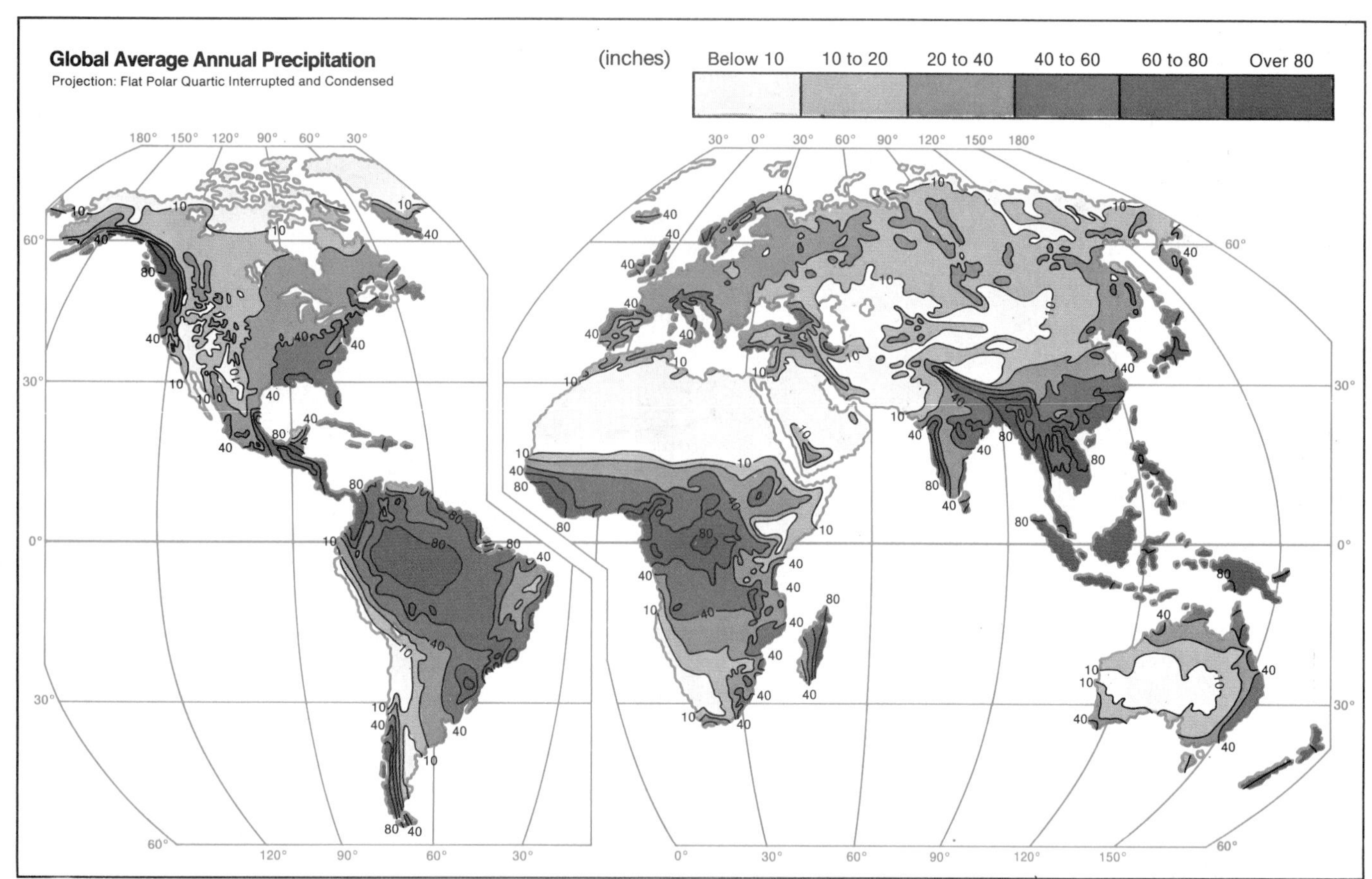
Global Average Annual Precipitation
Projection: Flat Polar Quartic Interrupted and Condensed
(inches)
Below 10
10 to 20
20 to 40
40 to 60
60 to 80
Over 80

above-average precipitation may well be followed by "dry" years with below-average precipitation. In general, precipitation exhibits the highest degree of variability in normally dry regions where the conditions for precipitation are seldom achieved.

The ability to control precipitation would be of considerable economic benefit to farmers, ski resort managers, and city governments faced with the annual expense of snow removal. The key to modern rainmaking is the realization that many clouds possess all the conditions necessary to produce precipitation except a means to trigger the growth of water droplets into raindrops.

The strong dependence of precipitation processes on comparatively rare events suggests that a cloud could be induced to give rain with a small expenditure of effort and energy. In warm clouds, only a few oversized droplets per cubic meter are required to initiate coalescence into raindrops. In supercooled clouds, a few ice crystals or ice-forming nuclei per cubic meter are enough to trigger the growth of ice crystals.

In 1946, Vincent Schaefer of the General Electric Research Laboratories discovered that he could stimulate the Bergeron process and change droplets to ice crystals if he dropped a piece of Dry Ice (solid carbon dioxide at a temperature below -40°C) into a chamber containing supercooled water droplets. Field tests by Schaefer and Irving Langmuir showed that a few pounds of crushed Dry Ice dropped from an airplane onto the top of a supercooled

Figure 4.21 (left) This photograph shows the first conclusive field test of cloud seeding, carried out in 1946 by Irving Langmuir and Vincent Schaefer. The supercooled stratus clouds were seeded by Dry Ice pellets dropped from an airplane flying around an oval track. In less than 1 hour after seeding, the seeded area cleared because of the induced precipitation.

(right) In this early attempt at rainmaking, electricity from a hand-cranked generator was pumped into clouds in the belief that the electricity of a lightning flash was somehow responsible for rain.

layer cloud could cause light precipitation from the base of the cloud. In numerous experiments since then, the summits of cumulus clouds have been seeded with Dry Ice. Precipitation is usually produced if the cloud is thick enough, if its top is cold, and if the cloud is long-lived—in other words, if the cloud already has the conditions necessary for producing rain.

The tiny crystals in the smoke from burning silver iodide also make excellent ice-forming nuclei. Supercooled droplets will form ice crystals on silver iodide particles at temperatures as high as -4°C (24.8°F); no other known material is as effective. One feature of silver iodide seeding is that the required number of nuclei can be obtained by burning only a few ounces of silver iodide in burners located on the ground. Commercial rainmaking operators use chains of burners that are lit whenever promising clouds appear. Because silver iodide particles rapidly lose their effectiveness when exposed to sunlight, the effects of seeding remain localized.

The effectiveness of rainmaking operations is difficult to assess. Experiments on the seeding of selected clouds have definitely produced precipitation as well as dramatic changes in cloud structure. For the practical requirements of agriculture, however, the most suitable clouds are seldom available, and cloud seeding is not reliable enough for day-to-day operations. Experience in the mountainous areas of the western United States suggests that cloud seeding has increased rainfall by 10 or 15 percent. Such increases are economically significant, but it is difficult to assess whether a particular rainfall was initiated by seeding or whether it would have occurred naturally. Naturally occurring rainfall is itself highly variable; variations of 40 percent from one year to the next are common in dry climates where rainmaking is most practiced, and a change of a few percent is difficult to attribute exclusively to seeding. Furthermore, the overall effects of weather modification are unpredictable. Increased rainfall on a valuable crop could be beneficial, but increased rainfall on a river drainage system might result in high waters and flooding.

Modification of Harmful Weather Conditions

Rainmaking is not the only aspect of weather modification. Some forms of weather—severe rainstorms, fog, hailstorms, and hurricanes, for example—can be costly to human life and property. Understanding of precipitation processes has been applied in an effort to disrupt conditions that lead to unwanted weather.

Because airports and highways become dangerous or even unusable in heavy fog, economic methods of fog dispersal are being sought. If the droplets in fog are supercooled, seeding with dry ice can dissipate the fog as snow crystals. Most fogs are warm, however, and no effective and inexpensive way of clearing a warm fog has been found. Heating the air to a temperature above the dew point will clear up a warm fog, but the energy input required makes the method too costly for general application.

Hailstorms cause so much damage to crops that efforts to disrupt

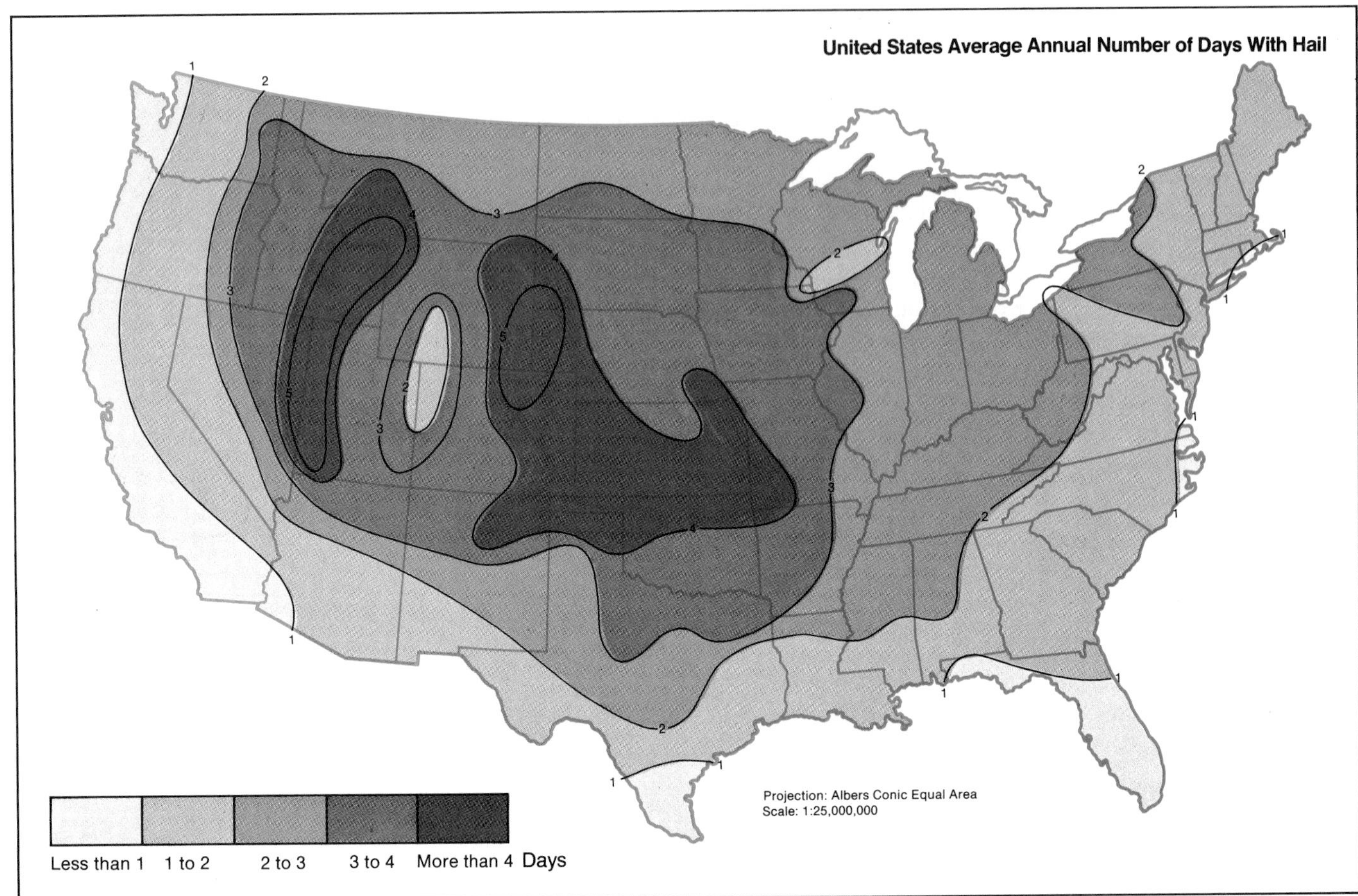
United States Average Annual Number of Days With Hail
Projection: Albers Conic Equal Area
Scale: 1:25,000,000
Less than 1
1 to 2
2 to 3
3 to 4
More than 4 Days

Figure 4.22 The map shows the average number of days per year with hail at locations in the conterminous United States, based on weather records from 1899 to 1938. Because hail is damaging to crops and poultry, hailstorms do hundreds of millions of dollars of damage to United States agriculture each year. The frequency of hailstorms is greatest in southeastern Wyoming and northeastern Colorado. Known as Hail Alley, the region is the meeting place where moisture from the Gulf of Mexico, rising air from the hot prairies, and strong winds from the Rocky Mountains spawn intense storms. Some farmers in Hail Alley average only one hail-free crop every 5 years; experiments are in progress to seed thunderclouds with silver iodide to reduce the chance of hail formation. (After *Yearbook of Agriculture,* 1941)

the formation of hail in turbulent clouds are usually worthwhile. Seeding clouds with silver iodide seems either to reduce the size of hailstones or to cause the cloud to produce only rain. Possibly introducing silver iodide into the cloud converts water droplets to ice crystals, which interrupts the growth cycle of the hailstones. Because hailstorms tend to be concentrated in certain parts of the country, it is feasible to set up a chain of silver iodide smoke generators in particularly susceptible areas. The Soviet Union, where artillery shells and rockets are used to shoot silver iodide directly into clouds, reports a considerable reduction in crop damage caused by hail.

Some efforts have been directed toward modifying the behavior of hurricanes, but the naturally erratic movement and growth of hurricanes makes it difficult to know whether a seeding experiment has reduced or intensified the force of the storm. Experiments on Hurricane Debbie in 1969 seemed to indicate that seeding had caused a reduction in the speed of the hurricane's winds, but seeding of Hurricane Ginger in 1971 had little effect.

Summary

The moisture that the atmosphere transports carries energy and has great potential for doing work on the surface of the earth. Moisture enters the atmosphere by evaporation from the ground or from the oceans and by transpiration from vegetation. Moisture condenses in the atmosphere if the temperature of the air is lowered to the dew point, because there is a maximum quantity of water that air at a given temperature can hold as vapor. Phenomena such as dew, fog, and clouds are formed when moist air is cooled to its dew point and when water vapor condenses on condensation nuclei.

The condensed water in clouds usually remains in the form of tiny droplets. Occasionally, however, special processes cause the coalescence of millions of droplets to raindrops large enough to fall to the ground. In warm clouds, a comparatively small number of giant droplets grow to raindrop size by sweeping up smaller droplets in collisions. In cold clouds, ice crystals form on ice-forming nuclei and grow by competing with the water droplets for water molecules.

The movement of the atmosphere carries moisture, and the energy it represents, from place to place on the earth's surface. The general circulation of the atmosphere, which is discussed in Chapter 5, is important in physical geography as a moisture delivery system that brings different amounts of moisture to different regions of the earth. Chapter 6 discusses how systems on the land share the water from precipitation.

5 Atmospheric and Oceanic Circulation

Forces and Atmospheric Motion
The Force of Friction
The Pressure Gradient Force
The Coriolis Force
The Geostrophic Wind

General Circulation of the Atmosphere
Winds and Pressure on a Uniform Earth
Observed Pressure and Wind Patterns
Circulation in the Upper Atmosphere

Midlatitude Secondary Circulations
Forces in Cyclones and Anticyclones
Air Masses
Weather Fronts
Jet Streams—The Atmospheric Vacuum Cleaner
Tornadoes

Tropical Secondary Circulations
Monsoons
Easterly Waves
Tropical Cyclones

General Circulation of the Ocean
The Gulf Stream
Upwelling
The Deep Circulation

A furious storm at sea emphasizes the complex interactions of the atmospheric and oceanic systems. Winds drive the surface ocean currents, and the ocean basins influence the circulation patterns of the atmosphere.

Snow Storm—Steam Boat off a Harbour's Mouth Making Signals in Shallow Water and Going by the Land by William Turner. (The Tate Gallery, London)

Atmospheric and Oceanic Circulation

Almost everyone living in a midlatitude country—the United States is a familiar example—takes the weather into account in his or her daily activities. Picnics are called off because of rain; Christmas trips home are delayed by blizzards; golf tournaments and football games are hindered by soggy fields. A farmer's work schedule is strongly influenced by the trend of the weather, and even a city dweller glances out of the window or checks a weather forecast before deciding what to wear.

The outstanding feature of weather is that it always changes. A sunny morning in Boston, Massachusetts, may give way to a gray and rainy afternoon. And a sultry summer day in Minnesota may see the build-up of the ominous clouds that bring severe thunderstorms and hail.

Despite the notorious changeability of weather in the United States, many regularities are apparent in weather patterns. The transition from clear, sunny weather to overcast, rainy skies often occurs slowly over a period of several days. In the midwestern United States, weather changes usually come from the west; today's weather in South Dakota is an indicator of tomorrow's weather in Minnesota. Regularity is also apparent in weather conditions in which a number of weather factors, such as wind and temperature, appear together. In Boston, a summer wind blowing from the southwest normally is accompanied by warm temperatures and high humidity, whereas winds from the northwest usually bring cooler, drier air to Boston.

Weather in North America is among the most variable in the world, so the changing weather patterns characteristic of the middle latitudes do not always represent those of other regions of the world. Near the Equator, for instance, the day-to-day weather is comparatively monotonous in its regularity. Weather there usually does not exhibit the successive fair and stormy periods, each lasting a few days, that sweep across the United States.

A passing summer shower that covers an area of a few square kilometers and lasts only a few minutes is a local event on the scale of the earth's atmosphere. On a larger scale are the weather systems, or *secondary circulations*, that move through the middle latitudes and control the patterns of local weather. Such features range up to a few thousand kilometers in size and endure for days before they die. On even larger scales of space and time are the air patterns of the *general circulation*, which are comparable in size to the earth and which are persistent features of the atmosphere, changing only slightly from season to season. This chapter focuses on the systematic behavior of the general and secondary circulations and examines the influence of atmospheric and oceanic circulation on local weather conditions.

Atmospheric circulation plays a central role in physical geography because the atmosphere is the vehicle that transports moisture and heat from place to place over the earth's surface. The focus in upcoming chapters is on the response of vegetation, soil, and landform systems to the inputs of energy and moisture. The circulation of the atmosphere delivers water in fairly reliable patterns, but the distribution of water is far from uniform across the earth's surface. Different regions receive different amounts of energy and moisture, and the systems and organisms in different regions respond in characteristic ways. The landforms, vegetation, and soils of the Arizona desert, for example, present a striking contrast to the woody slopes of New England. Because of such differences, the earth can be divided into climatic regions that reflect the patterns of energy and moisture delivery established by the general circulation. Climatic regions are discussed in Chapter 7.

Forces and Atmospheric Motion

The atmosphere is in ceaseless motion, whether churning with a furious storm or puffing an occasional summer breeze. The motion of any parcel of air is determined by the forces acting on it. Although the details of atmospheric motion are complex, some of the general features of the motion can be understood by looking at the forces involved.

A force may be thought of as a push or a pull, analogous to the pushes and pulls you exert with your muscles. According to the fundamental laws of motion, which the English physicist Isaac Newton developed in the seventeenth century, an object's speed or direction of motion cannot change unless a force is made to act on the object. Furthermore, horizontal motion is influenced only by forces pushing or pulling in the horizontal direction, and the vertical motion of an object is influenced only by forces acting in the vertical direction. The force of gravity pulls every parcel of air downward toward the earth, for example, but gravity has no direct effect on the horizontal motion of air across the earth's surface. The principal horizontal forces that act on a parcel of air in the atmosphere are *friction,* the *pressure gradient force,* and the *Coriolis force.*

The Force of Friction

The force of friction is inevitably present for any kind of motion on the earth. Friction is exerted whenever one surface rubs on another, as when the wind drags across the surface of the ocean or the land. Because friction opposes the motion of the air, additional forces are required to keep the atmosphere moving when friction is present. If the forces that drive the atmosphere ceased to act, friction would cause the atmosphere to come to a halt in a week or two. The effect of friction on atmospheric motion is greatest near the ground: friction slows winds near the surface of the earth more than it slows winds at higher altitudes.

The Pressure Gradient Force

Pressure is force per unit area; if the pressure on one side of a parcel of air is greater than the pressure on the other side, the parcel will be pushed from the higher pressure toward the lower pressure. The greater the pressure difference between the two sides of the

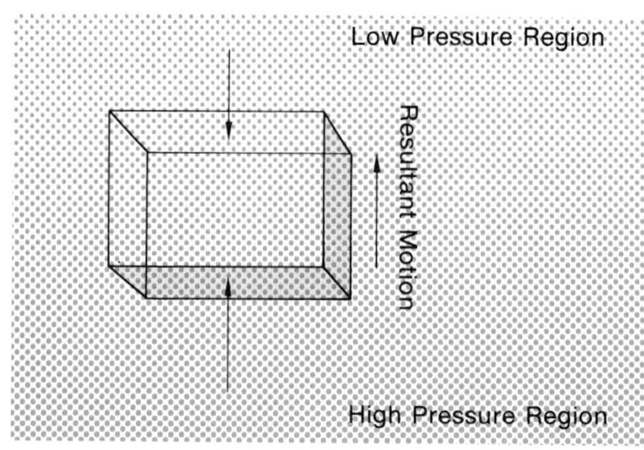
Low Pressure Region
Resultant Motion
High Pressure Region

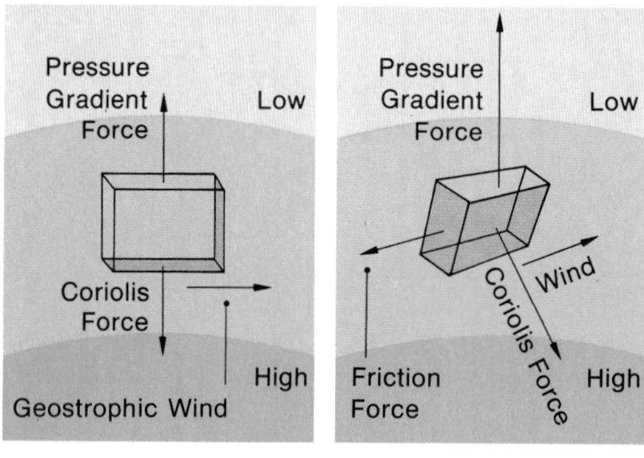
Pressure Gradient Force
Low
Coriolis Force
High
Geostrophic Wind
Pressure Gradient Force
Low
Wind
Coriolis Force
Friction Force
High

Figure 5.2 (top) The pressure gradient force acts in a direction perpendicular to isobars and pushes a parcel of air away from high pressure toward low pressure. If no other forces were present, the parcel would move toward the low-pressure region.

(above left) A parcel of air moving horizontally on the rotating earth experiences a Coriolis force and a pressure gradient force. The Coriolis force acts at right angles to the direction of motion of the parcel. This diagram shows how the Coriolis force can balance the pressure gradient force so that the parcel can move parallel to the isobars with constant speed. Air motion essentially parallel to isobars is known as a geostrophic wind; the speed of a geostrophic wind is highest where the pressure decreases most rapidly with distance. This diagram is drawn for the Northern Hemisphere. How would it appear for the Southern Hemisphere?

(above right) Friction acts on air parcels moving near the ground. The diagram shows that when the retarding effect of friction is taken into account, the forces on a parcel of air can be in balance only if the parcel has a component of motion away from high pressure and toward low pressure. In the upper atmosphere, where friction forces are weak, the winds blow nearly parallel to isobars.

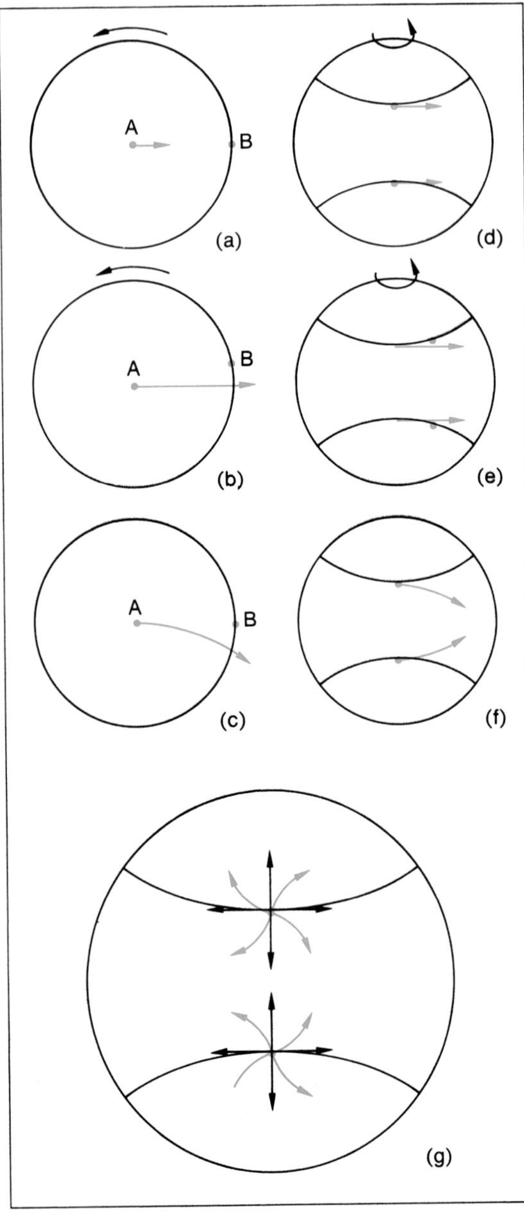
A
B
(a)
(d)
A
B
(b)
(e)
A
B
(c)
(f)
(g)

Figure 5.3(a) Imagine that a child, *A*, on a rotating merry-go-round throws a ball toward another child, *B*. (b) As seen from the ground, the ball flies straight. (c) As seen by *A* and *B*, the ball appears to curve. The motion is attributed to the Coriolis force, a sideways deflecting force.

(d) (e) The diagrams show two objects on the earth traveling in straight lines as seen from outer space. (f) As seen by observers on the rotating earth, the objects follow curved paths; the deflection is attributed to the Coriolis force.

(g) In general, objects moving along the surface of the earth are deflected to the right in the Northern Hemisphere and to the left in the Southern Hemisphere.

parcel, the greater the net push, which is called the pressure gradient force. The pressure gradient force is strongest where the pressure is changing most rapidly with distance.

The measured values of atmospheric pressure from place to place across a region are usually displayed on a map by connecting locations having the same pressure with a curve called an *isobar.* As indicated in Figure 5.2, the direction of the pressure gradient force is at right angles to the isobars, from higher toward lower pressure.

The Coriolis Force

The rotation of the earth complicates the motion of the atmosphere. To understand the principle involved, consider two children throwing a baseball back and forth on a rotating merry-go-round. One child is near the center and the other is near the rim. If the child near the center aims a ball straight toward the child at the rim, the ball will pass to one side of the catcher because the merry-go-round continues to rotate during the ball's flight. From the point of view of the children on the merry-go-round, the ball moves as though a sidewise force were acting to deflect it. The force is called the Coriolis force, after Gaspard Coriolis, the nineteenth-century French engineer who introduced the concept.

The earth is a rotating system analogous to the merry-go-round, and a parcel of air moving horizontally through the atmosphere of air veers to one side as observed by someone standing on the earth, as though a force were pushing the parcel to the side. The Coriolis force always acts sidewise on objects moving horizontally on the earth. As Figure 5.3 indicates, the deflection is toward the right of the parcel's direction of motion in the Northern Hemisphere and toward the left of its direction in the Southern Hemisphere. The strength of the Coriolis force on an object is proportional to the speed of the object and depends on latitude (see Figure 5.4). The Coriolis force has no effect on horizontal motion at the Equator and has maximum effect at the poles.

The Geostrophic Wind

In the middle latitudes, horizontal wind speed and direction are controlled primarily by the pressure gradient force and by the Coriolis force. If the earth were not rotating, air would flow directly from higher to lower pressure because of the pressure gradient force. But the earth's rotation introduces a Coriolis force, which modifies the action of the pressure gradient force.

A parcel of air moving under the action of both a pressure gradient force and a Coriolis force eventually moves parallel to isobars, because this is the only way the two forces can come into balance (see Figure 5.2). Winds that move along isobars are called *geostrophic winds.* The direction of a geostrophic wind is expressed by *Buys Ballot's law,* which states that an observer in the Northern Hemisphere standing with a geostrophic wind at his back will have high pressure to his right and low pressure to his left.

The geostrophic wind is an idealized situation in which the friction force on the moving air is neglected. When friction is taken into account, the direction of the wind is no longer exactly parallel to the isobars. As Figure 5.2 illustrates, the wind tends to move toward the region of lower pressure and away from the region of higher pressure.

General Circulation of the Atmosphere

The earth's low-latitude regions receive a net influx of radiant energy during the year. Part of the available energy is used to drive the *general circulation* of the atmosphere, which carries heat energy from low-latitude regions toward the poles.

According to the model of atmospheric circulation that is generally accepted by meteorologists today, air that is warmed near the Equator rises and begins to move through the upper atmosphere toward the poles. Radiational cooling, however, prevents the upper air streams from remaining warm enough to stay aloft; as they cool, the streams sink to the ground, returning to the surface at about latitudes 30°N and 30°S. The return of the air to the Equator as a cool surface flow closes the loop of circulation. A vertical loop of circulation is called a *convective cell,* or a *Hadley cell,* after John Hadley, who proposed a model of atmospheric circulation in the year 1735.

If the earth were not rotating, the general circulation in low-latitude regions would consist of flow along the surface toward the Equator and flow through the upper air toward the poles. Surface winds in the low-latitude regions of the Northern Hemisphere would blow from the north, and winds would tend to follow meridians of longitude.

The earth's rotation, however, causes winds to be deflected sidewise and to gain a significant component of motion along the parallels of latitude. In the Northern Hemisphere, air that is flowing toward the Equator is deflected toward the west, because the Coriolis force acts to the right of the direction of motion. The surface winds in the low-latitude regions of the Northern Hemisphere therefore tend to blow from the northeast. In the Southern Hemisphere, deflection is to the left, and surface winds in the low-latitude regions of the Southern Hemisphere tend to blow from the southeast.

Winds and Pressure on a Uniform Earth

Continents, oceans, and mountain ranges significantly influence the motion of the earth's atmosphere, so it is easier to understand the general features of the circulation if you imagine an idealized, featureless earth entirely covered by water. On a uniform earth, the surface winds would form belts stretching around the world along parallels of latitude. The rotation of the earth induces patterns of air circulation that cause climate to vary with latitude from the tropics to the polar regions. The belt of winds in the low latitudes north of the Equator on the uniform earth would consist of winds from the northeast, and the low-latitude belt south of the Equator would consist of winds from the southeast. Winds in these belts are often called *trade winds.* The northeast trades and the southeast

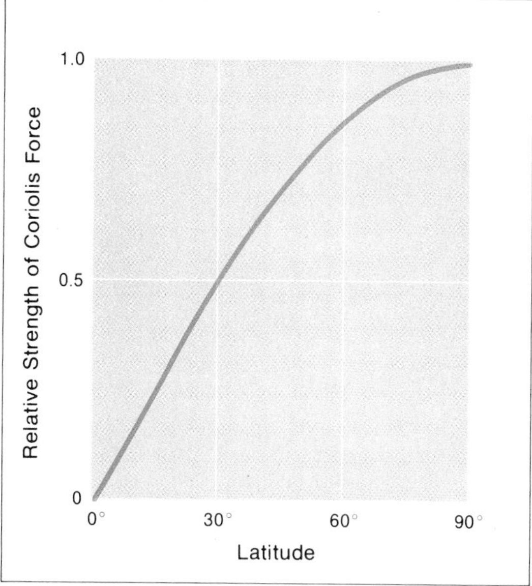

1.0
0.5
0
Relative Strength of Coriolis Force
0°
30°
60°
90°
Latitude

Figure 5.4 For an object such as an air mass moving horizontally on the earth's surface, the strength of the Coriolis deflecting force varies with latitude, as the graph shows, and is proportional to the speed of the object.

Figure 5.5 The general circulation of the atmosphere in the low latitudes is largely governed by convective, or Hadley, cells. In each Hadley cell warm air from the equatorial zone rises and moves poleward. The air cools by emitting thermal radiation and sinks to the surface near latitude 30°. The descending air becomes warmer by adiabatic heating, and tends to be warm and dry at the surface. It then flows along the surface toward the equatorial zone, completing the convective loop. For clarity, the vertical height of the Hadley cells is greatly exaggerated in this diagram; the height of the cells is no greater than 10 or 15 kilometers (6 or 9 miles).

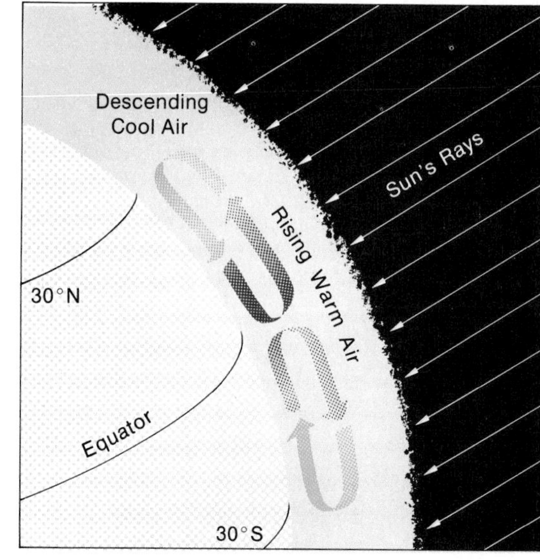

Descending
Cool Air
Sun's Rays
Rising Warm Air
30°N
Equator
30°S

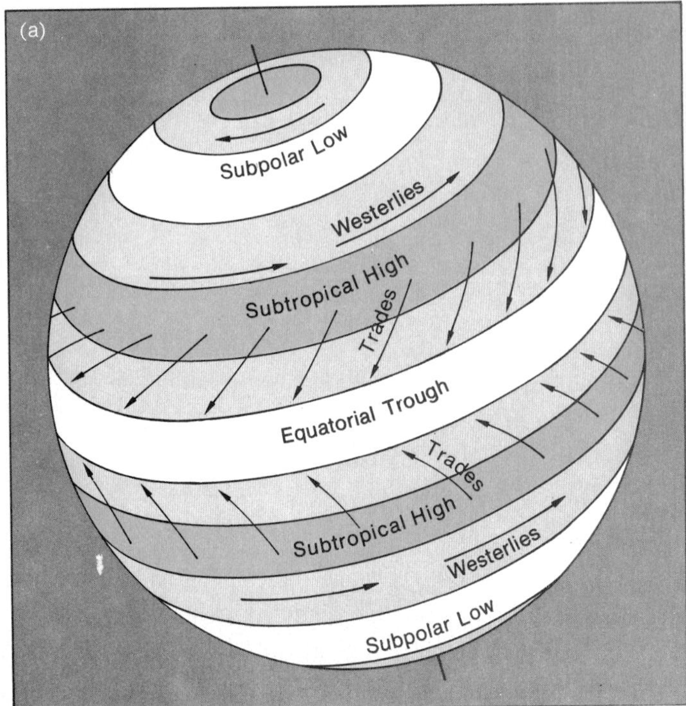
(a)
Subpolar Low
Westerlies
Subtropical High
Trades
Equatorial Trough
Trades
Subtropical High
Westerlies
Subpolar Low

(b)

Figure 5.6(a) If the earth were completely covered with water, orographic effects and effects due to differences in the unequal heating of sea and land would be absent. On a uniform rotating earth the winds and atmospheric pressure regions would form regular belts around the earth, as shown in this diagram. The winds on a uniform earth tend to flow from regions of high pressure to regions of low pressure; winds in the tropics are not fully geostrophic because of the weakness of the Coriolis force at low latitudes. In the midlatitudes, where the Coriolis force is strong, the prevailing westerly winds are geostrophic between the subtropical high and the subpolar low in each hemisphere. The winds on the earth are always a mixture of easterly and westerly flow because the motion of the atmosphere must average to zero relative to the earth.

(b) The swirling cloud streamers in this photograph of the earth from space illustrate how complex the atmospheric circulation on the earth actually is.

trades meet in a zone called the *intertropical convergence zone.* On a featureless earth, the intertropical convergence zone would be continuous around the earth.

The pattern of surface winds on a uniform earth is closely associated with the vertical circulation of the atmosphere and with global patterns of air pressure. In the low latitudes, vertical motion of the air occurs in the tropical Hadley cells north and south of the Equator. As air rises from the surface in the vicinity of the Equator, the removal of air produces a trough of low air pressure at the surface. The pressure gradient force pushes air along the surface toward the low-pressure trough; this movement of air constitutes the trade winds. The intertropical convergence zone lies within the equatorial low-pressure trough.

Air moving through the upper atmosphere in the circulation of the tropical Hadley cells subsides to the earth's surface near latitudes 30°N and 30°S. These regions are therefore zones of high air pressure at the surface, because dense, heavy, cool air is added from above. The pattern of surface air pressure in the low-latitude regions of a uniform earth consists of a low-pressure belt near the Equator and high-pressure belts in the subtropics near latitudes 30°N and 30°S.

The air moving out of the subtropical highs toward the Equator forms the trade winds. Air from the subtropical highs also tends to move poleward, and the sidewise deflection of these air streams produces belts of surface westerly winds in both the Northern and the Southern Hemispheres. Each belt of westerlies extends from about latitude 30° to about latitude 60°. On a featureless earth, there would probably be troughs of low pressure at latitudes 60°N and 60°S and caps of high pressure at the poles.

Observed Pressure and Wind Patterns

The actual average patterns of pressure and wind at the earth's surface are more complicated than the simple belts associated with a uniform earth model. The presence of continents considerably influences the circulation of the earth's atmosphere. Rock and soil have a smaller specific heat than water has, so landmasses become much warmer than the ocean in the summer and much cooler in the winter. The extreme heating or cooling of air over the continents affects the global circulation patterns. In addition, mountain ranges on the continents disturb the flow of air.

Figure 5.8 shows the average global distribution of pressure corrected to sea level for January and for July. Between latitudes 45°S and 65°S, the earth's surface is practically devoid of landmasses, as assumed in the water-covered model discussed previously. The pattern of isobars over the ocean surrounding Antarctica exhibits a regularity similar to the uniform belt pattern on a uniform earth model.

Outside of the southern subpolar region, however, the pattern of pressure distribution consists of a number of separate cells of high and low pressure. The pressure cells shown in Figure 5.8 are average, long-lasting features of the dynamic general circulation and are distinct from the fast-moving, short-lived pressure cells of the secondary circulation that are shown in daily weather maps.

Figure 5.8 for January shows an intense high-pressure cell over Siberia. The Siberian high is a characteristic feature of the general circulation during winter in the northern latitudes; it is associated with cold, dense, heavy air over the Asian continent. Low-pressure cells form in winter over the relatively warm waters of the North Pacific Ocean and the North Atlantic Ocean. During summer in the northern latitudes, the continents become warmer than the oceans, and a low-pressure cell over southern Asia replaces the Siberian high. High-pressure cells form over the relatively cooler ocean waters.

The observed pattern of pressure cells on the earth exhibits traces of the pressure distribution pattern associated with the uniform earth model. Study of the global map shows that the equatorial region is generally a low-pressure region, reminiscent of the equatorial trough belt on the uniform earth. The observed intertropical convergence zone is not continuous around the actual earth, however, and at some places or in some seasons, it exists only weakly, if at all. The equatorial trough as well as the intertropical convergence zone tend to be situated north of the Equator when it is summer in the Northern Hemisphere and near the Equator or slightly south of it when it is summer in the Southern Hemisphere. The north-south shift of the general circulation's pressure patterns with the seasons may result in a seasonal shift in the type of weather that a particular location experiences.

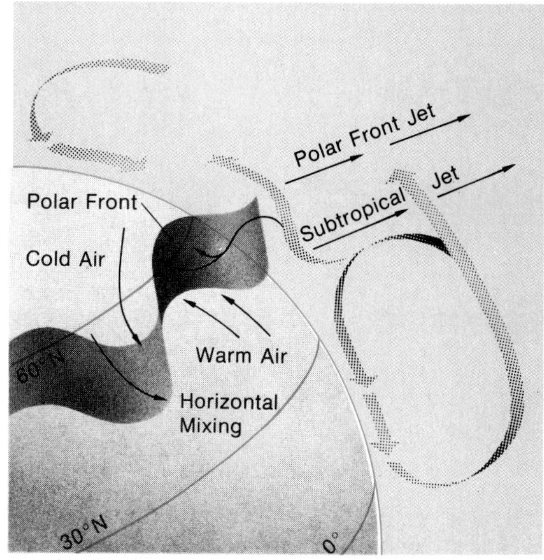
Polar Front Jet
Subtropical Jet
Polar Front
Cold Air
Warm Air
Horizontal Mixing
60°N
30°N
0°

Figure 5.7 This diagram shows a conception of the general atmospheric circulation for the Northern Hemisphere in winter. The circulation in the low latitudes is dominated by the tropical Hadley cell, but the circulation at higher latitudes is not fully understood. The poleward movement of warm air from the tropics is an important mechanism for transferring surplus heat energy to higher latitudes. Much of the transfer appears to occur in the horizontal eddies where warm and cool air mix. The polar front, which is the boundary between warm tropical air and cool polar air, is therefore not uniform around the earth.

This diagram also shows the general location of the westerly jet streams, which are ribbons of fast-moving air in the upper atmosphere. Jet streams, which influence the development of weather systems, are discussed in greater detail later in this chapter. (After Chorley and Haggett)

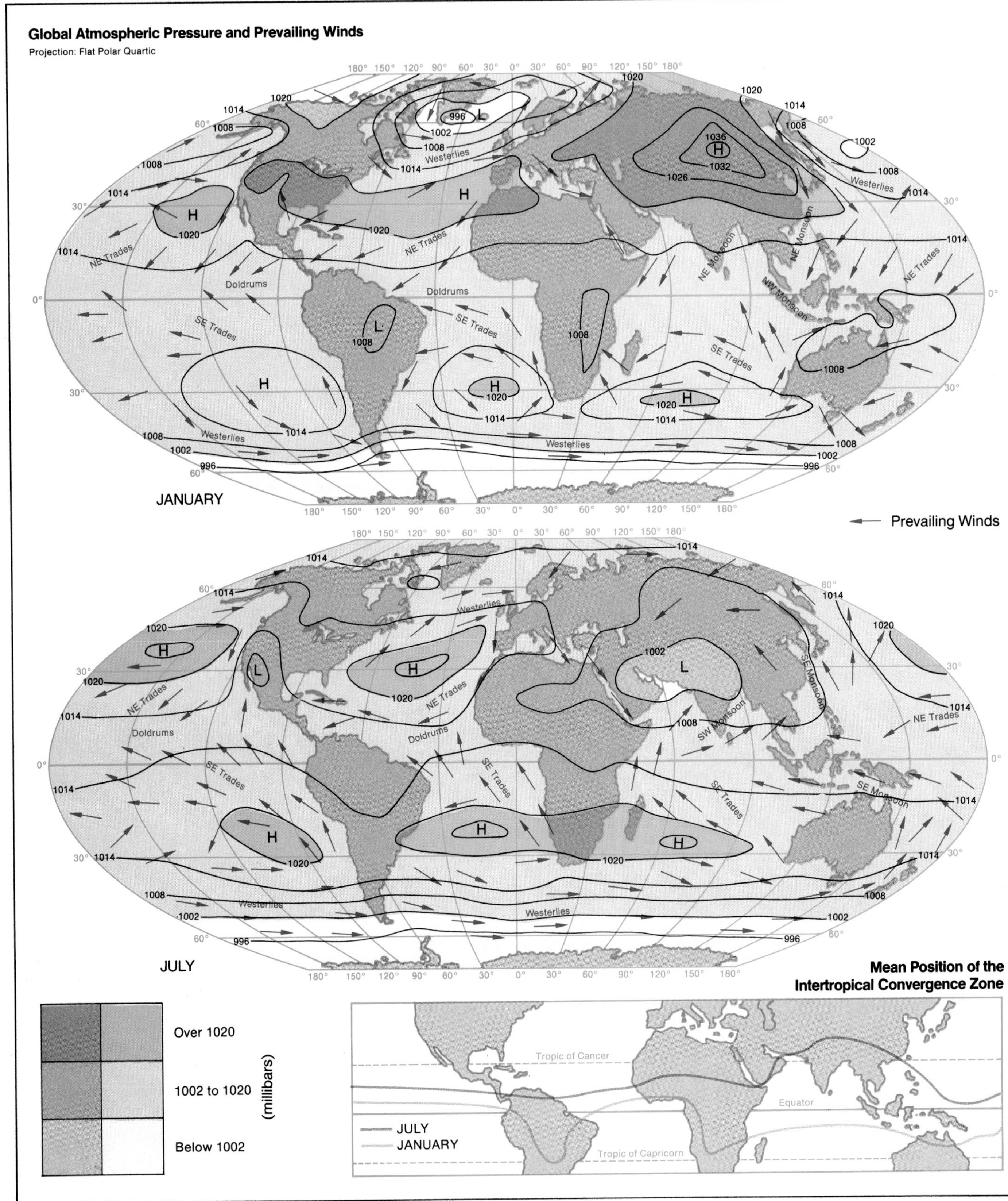
Global Atmospheric Pressure and Prevailing Winds
Projection: Flat Polar Quartic
JANUARY
JULY
Prevailing Winds
Westerlies
NE Trades
SE Trades
Doldrums
NE Monsoon
NW Monsoon
SW Monsoon
SE Monsoon
Over 1020
1002 to 1020
Below 1002
(millibars)
Mean Position of the Intertropical Convergence Zone
Tropic of Cancer
Equator
Tropic of Capricorn
JULY
JANUARY

Atmospheric Pressure in Polar Regions

Projection: Stereographic

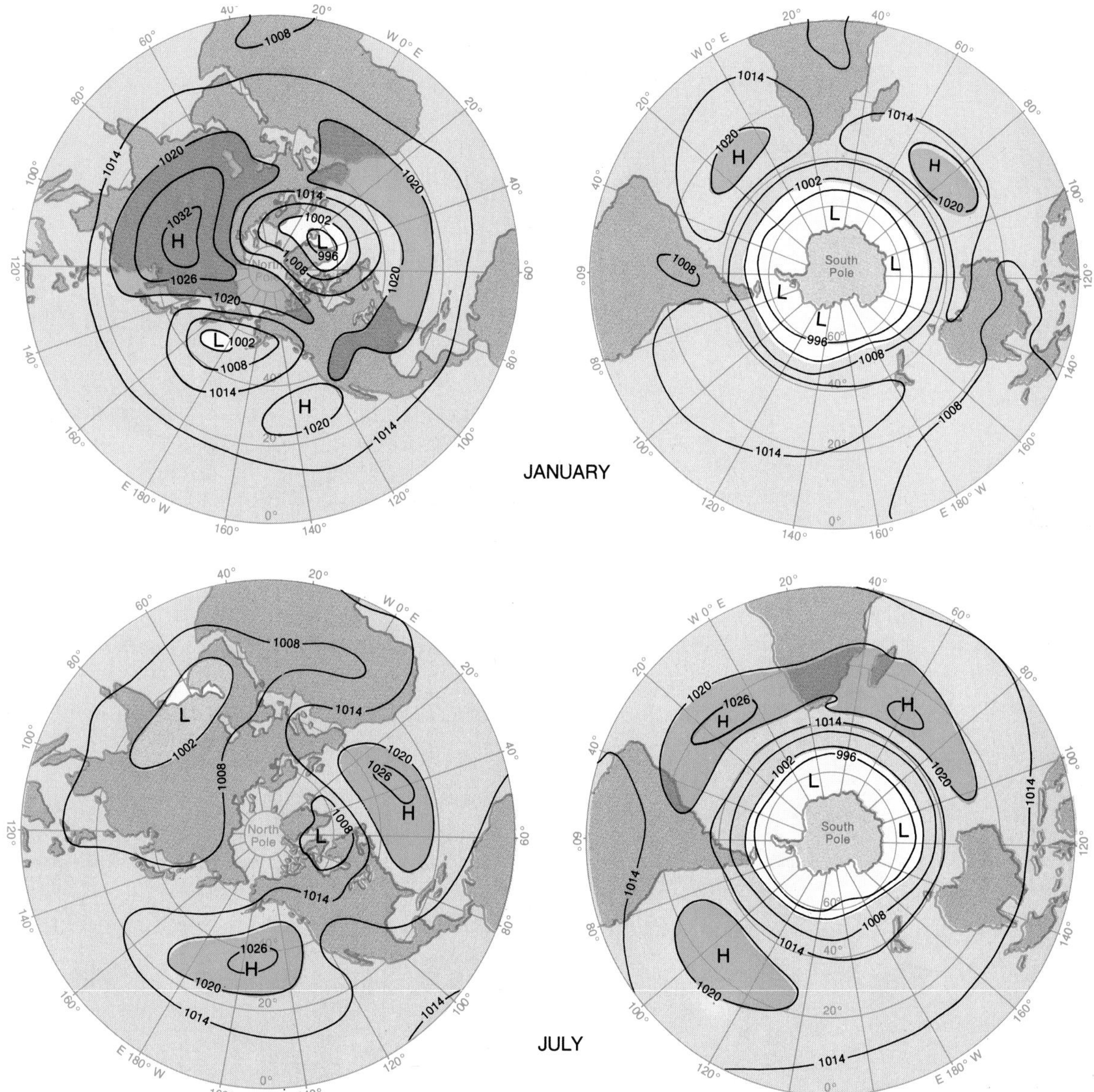

JANUARY
JULY
North Pole
South Pole
H
L
1008
1014
1020
1026
1032
1002
996
W 0° E
E 180° W

Figure 5.8 Global maps show the average direction of surface winds and the average atmospheric pressure at the surface for January and July. Pressure readings are in millibars. Near the Antarctic Circle, where there is open ocean around the entire globe, the isobars form a continuous belt. Elsewhere, the presence of landmasses causes the regions of high and low pressure to be broken into individual pressure cells. The cells tend to be distributed along bands of latitude that are analogous to the pressure belts that would form on a uniform earth. The formation of pressure cells is affected by temperature differences between the oceans and the continents. (After the United States Weather Bureau)

The average distribution of surface winds is closely related to the pressure distribution. Winds tend to flow outward from high-pressure regions and inward toward low-pressure regions. In many areas, the winds follow a pattern of geostrophic flow modified by friction with the earth's surface.

The distribution of pressure cells generally shifts to the north in July and to the south in January. Over the poles, the high-pressure cells are more diffuse and less numerous in the summer than in the winter. The resulting seasonal shift in wind patterns generates a seasonal variation in the weather of many regions. The winds on the east coast of central Africa, for example, change from the northeast trades in January to the southeast trades in July. The inset map shows how the average position of the intertropical convergence zone, which is the meeting place of the trade winds, shifts between January and July. (After Schwerdtfeger, 1970, and Flohn, 1969)

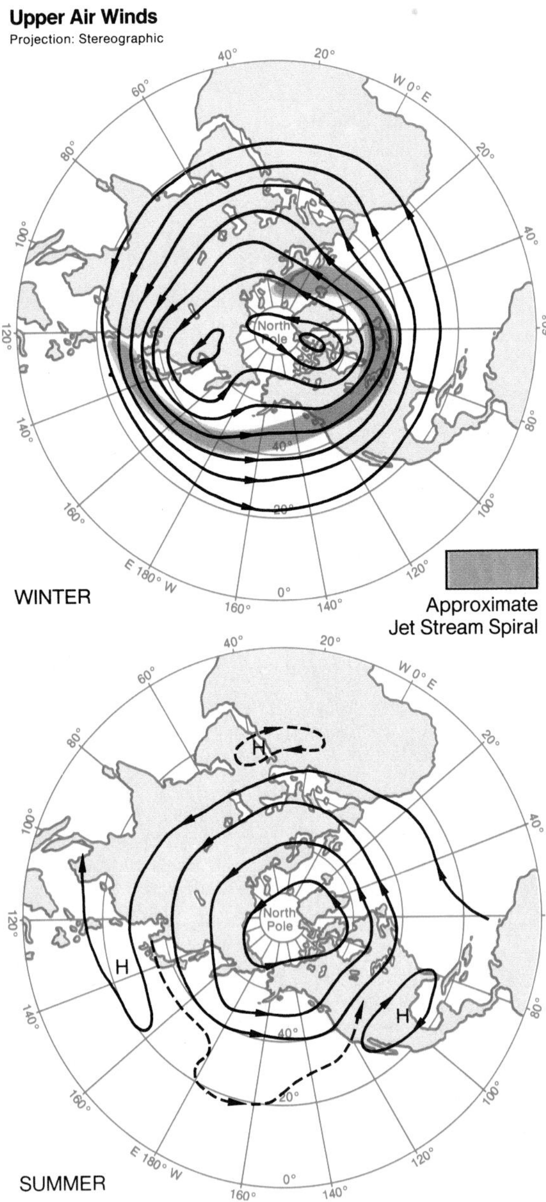
Upper Air Winds
Projection: Stereographic
North Pole
WINTER
Approximate
Jet Stream Spiral
H
SUMMER

Figure 5.9 The maps show winter and summer wind patterns for the Northern Hemisphere in the upper atmosphere near an altitude of 5.5 kilometers (18,000 feet), where the atmospheric pressure is approximately one-half the pressure at sea level. Wind directions and speeds in the upper atmosphere are usually determined by tracking freely drifting radio-equipped balloons that transmit meteorological data to the ground. In winter the upper air winds are strong and form a well-defined pattern of circulation with several undulations. The shaded region indicates the average location of fast-moving jet streams. Many winter storms in the northern latitudes appear to be generated with the help of jet streams. The flow of upper air winds tends to be weaker during summer than during winter. Dashed lines indicate particularly weak flows. (After Riehl, 1972)

The zonal string of high-pressure cells near latitudes 30°N and 30°S are considered to be associated with the subsidence of air from the tropical Hadley cells. The high-pressure cells form an almost beltlike arrangement in the Southern Hemisphere, which has only half the land mass of the Northern Hemisphere. The map also indicates traces of subpolar lows and polar highs. As the Northern Hemisphere becomes warmer in summer, the patterns move northward; six months later, they move southward.

The complex pattern of pressure distribution on the earth's surface makes the surface wind patterns complex as well, because the pressure gradient forces and the effects of the earth's rotation largely determine the motion of air. Figure 5.8 shows the observed global distribution of average surface wind directions for January and for July. Only a trace of the ideal wind belts of the uniform earth model can be seen on the real earth. Winds over the Atlantic and Pacific Oceans in the low latitudes generally exhibit the easterly flows of the trade winds, and poleward of about latitude 40°, the winds are generally westerly.

The wind patterns in many regions do not conform to the simple wind belts of the uniform earth model. The high-pressure cell that cold, dense air forms over Asia in the winter causes an outward flow of air in a variety of directions; the westerlies do not prevail in this region. As another example, the winds over India and over the Indian Ocean blow from the north in winter and from the south in summer. A seasonal reversal of wind direction is known as a *monsoon*. The northerly flow of air from the Gulf of Mexico over the central United States in summer and the southerly flow from Canada in winter give a hint of monsoon flow.

Circulation in the Upper Atmosphere

Prior to the 1930s and 1940s, most weather observations were made near ground level, and meteorologists focused their attention on pressure and wind systems at the earth's surface. The availability of balloons carrying weather instruments and the use of high-flying aircraft opened the upper atmosphere to study. Meteorologists found that phenomena in the upper atmosphere, at altitudes up to 10 or 20 kilometers (6 or 12 miles), exert significant control over weather at the earth's surface. The details of how such control is exerted remain a topic of active investigation by satellites today.

Observations show that above an altitude of 10 kilometers or so, upper air winds nearly all over the earth have a strong component of flow from the west. As Figure 5.9 indicates, the circumpolar westerlies in the upper atmosphere tend to exhibit undulations that are global in scale. Embedded within the circumpolar westerlies at altitudes of 10 or 15 kilometers (6 or 9 miles) are narrow ribbons of fast-moving air known as *jet streams*, which are discussed later in this chapter.

Midlatitude Secondary Circulations

Almost all daily weather maps for the United States show cells of high or low pressure. The isobars around a pressure cell usually form asymmetric oval shapes. The patterns that form around a central region of low pressure are called *cyclones*, or *lows*. Patterns that form around high-pressure regions are called *anticyclones*, or *highs*. The horizontal differences in pressure associated with large highs or with large lows usually amount to a pressure difference of less than 10 millibars over a distance of 100 kilometers (60 miles), only a small fraction of the normal sea level atmospheric pressure of 1,013 millibars.

Like geostrophic flow, the winds that accompany cyclones and anticyclones tend to follow the isobars. In the Northern Hemisphere, the upper winds around a cyclone circulate counterclockwise as viewed from above, and the winds around an anticyclone circulate clockwise. In the Southern Hemisphere, the directions of circulation are reversed.

Forces in Cyclones and Anticyclones

The arrangement of forces differs in cyclones and anticyclones. In cyclones, the pressure gradient force pushes inward, toward the central low pressure, and the Coriolis force pushes outward. In anticyclones, the pressure gradient force pushes outward, away from the central high pressure, and the Coriolis force pushes inward.

As discussed earlier in this chapter, friction with the surface of the earth turns the direction of geostrophic winds away from the isobars; it causes the surface winds to blow more toward the region of low pressure, so that air blows in toward the center of a low and out from the center of a high. Near the ground, air *converges* into a low and *diverges* out of a high.

Air Masses

An *air mass* is a large, quasi-uniform body of air moving together as a unit, usually in association with high-pressure or low-pressure systems. The circulation of the atmosphere moves surface pressure cells from place to place on the earth. Rotating cyclones and an-

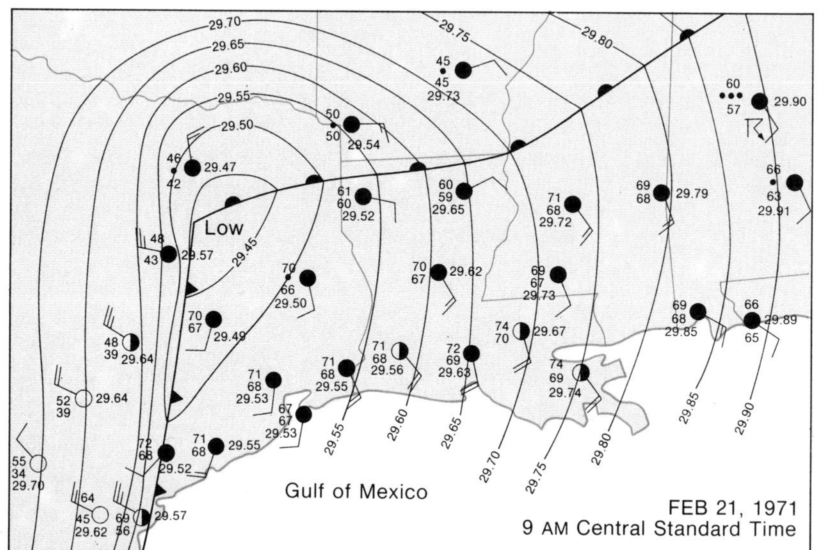

29.70
29.65
29.60
29.55
29.50
29.45
29.75
29.80
29.85
29.90
Low
Gulf of Mexico
FEB 21, 1971
9 AM Central Standard Time

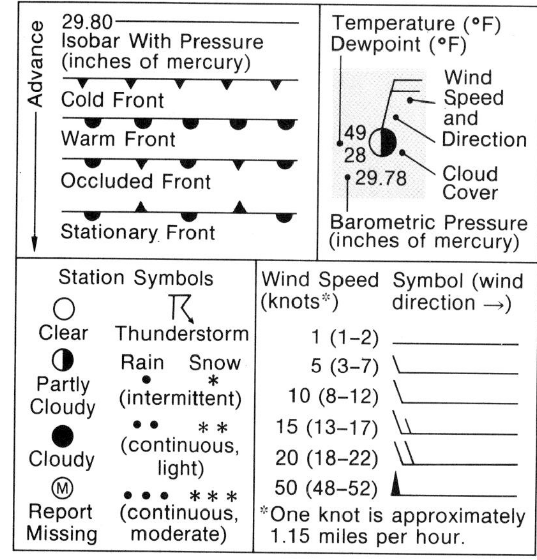

Advance
29.80
Isobar With Pressure (inches of mercury)
Cold Front
Warm Front
Occluded Front
Stationary Front
Temperature (°F)
Dewpoint (°F)
Wind Speed and Direction
49
28
29.78
Cloud Cover
Barometric Pressure (inches of mercury)
Station Symbols
Clear
Thunderstorm
Partly Cloudy
Rain
Snow
(intermittent)
Cloudy
(continuous, light)
Report Missing
(continuous, moderate)
Wind Speed (knots*)
Symbol (wind direction →)
1 (1–2)
5 (3–7)
10 (8–12)
15 (13–17)
20 (18–22)
50 (48–52)
*One knot is approximately 1.15 miles per hour.

Figure 5.10 The weather map portrays weather conditions in the south central United States on February 21, 1971, based on ground station reports at 9:00 A.M. Central Standard Time. The information on the map can be read with the help of the accompanying symbol table. The map shows, for example, isobars labeled with barometric pressure readings in inches of mercury, a cold front advancing eastward through Texas, a warm front advancing northward, and numerous station reports. The station symbols indicate temperature and dew point in °F, barometric pressure in inches of mercury, wind direction and speed in knots, cloud coverage, and precipitation events. Note that a station in Alabama, near the eastern edge of the map, reports a thunderstorm and heavy rain.

The station symbols in the weather maps prepared by the United States Weather Service carry more detailed information than do the station reports in the simplified weather map shown here. Among other additional station data, the Weather Service maps present cloud type, height of the cloud base, change of barometric pressure in the previous 3 hours, and weather conditions in the previous 6 hours.

This weather map shows an intense low-pressure system over eastern Texas. Note that most of the reported winds around the low tend to be parallel to the isobars and that to the west of the cold front winds are strong, with a speed of 25 knots. During the afternoon of February 21, the low-pressure area generated thunderstorms and tornadoes in Mississippi. Damage from the tornadoes was severe, and many people were injured or killed. (After Muller, 1973)

Figure 5.11 In a region of high pressure (left), air descends from the upper atmosphere and diverges outward along the surface of the ground. The air streaming outward from highs therefore tends to be dry. In a region of low pressure (right), air converges inward along the surface of the ground and ascends to the upper atmosphere. If the air streaming into a low is warm and moist, condensation and cloud formation occur as the air rises and cools. The directions of circulation are shown for the Northern Hemisphere.

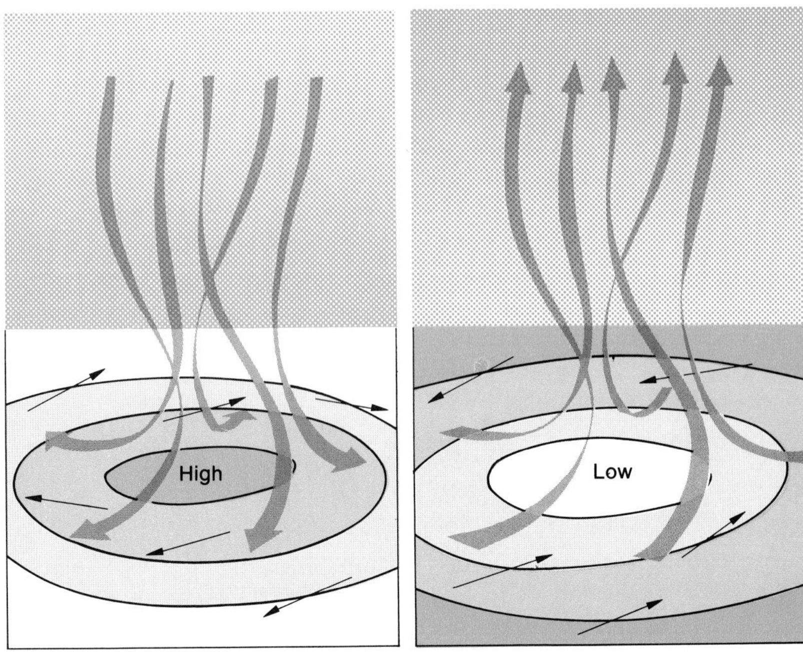
High
Low

ticyclones are remarkably stable and often travel across an entire continent at a speed of 800 kilometers a day (500 miles a day) or so without substantial changes in their circulation patterns. A sequence of daily weather maps for the United States frequently shows pressure cells drifting eastward across the United States under the influence of westerly winds aloft.

Air masses are significant to weather forecasting because they retain the characteristic temperature and humidity of the air where they are first formed. Air masses with polar or tropical characteristics may move long distances, bringing cool, dry polar air or warm, humid tropical air into midlatitude regions.

Air masses are classified into any of four general categories, depending on whether they originate in polar (cold) or in tropical (warm) regions and whether they originate over land (continental, dry) or over oceans (maritime, moist). Extreme cold winter weather in the United States is associated with the intrusion of continental polar air from northern Canada. The cold, dry air mass, usually in the form of an anticyclone, moves easily across the central plains of North America and often penetrates deep into the South.

Although they retain many of their original characteristics, air masses are modified somewhat as they move across the country. The weather that accompanies a polar continental air mass, for example, depends on the path the air mass follows. As polar continental air passes into North Dakota, it brings clear, cold, dry weather accompanied by a temperature inversion in which air near the surface is colder than air higher in the atmosphere. The inversion is at low altitudes, and air near the ground is unable to rise buoyantly through the atmosphere; hence no thick clouds are formed. If a polar continental air mass travels toward the east over the relative-

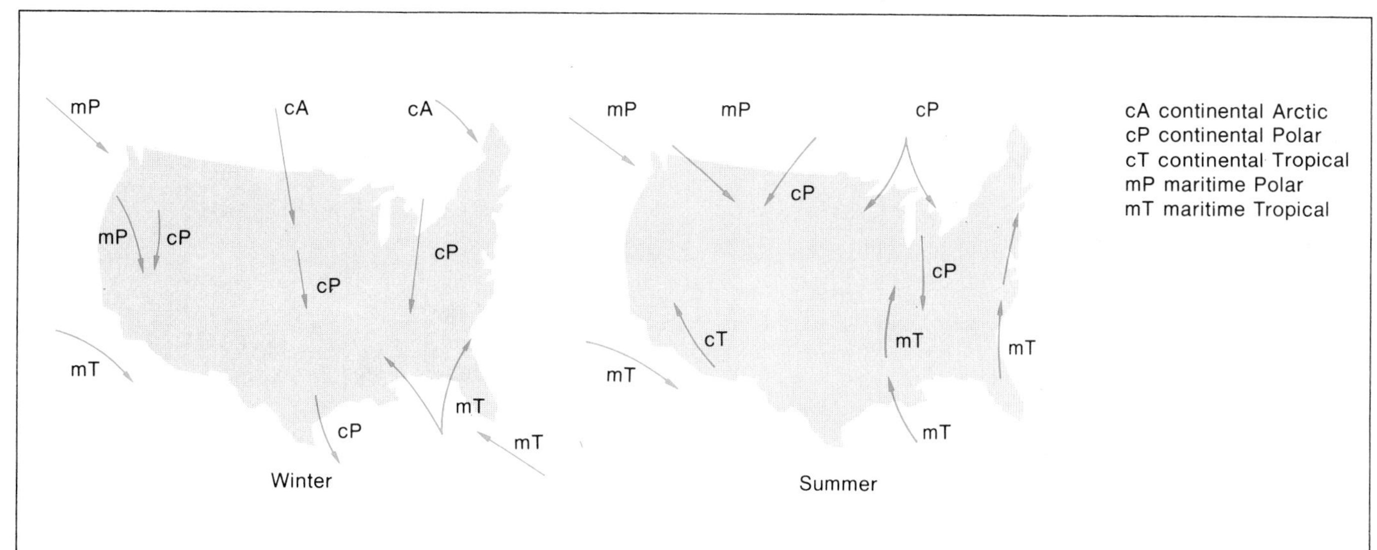
mP
cA
cA
mP
cP
cP
cP
mT
cP
mT
mT
Winter
mP
mP
cP
cP
cP
cT
mT
mT
mT
mT
Summer
cA continental Arctic
cP continental Polar
cT continental Tropical
mP maritime Polar
mT maritime Tropical

Figure 5.12 The air masses that enter a region can strongly influence weather conditions there by replacing the existing air with air of different temperature and humidity. Maritime air tends to be humid, and continental air tends to be dry. The maps in this figure depict the average movements of air masses over the conterminous United States during the winter and the summer. During the winter much of the northern half of the United States is invaded by cold polar air, whereas during the summer the northward flow of tropical air dominates the weather in most regions.
(After Brunnschweiler, 1957)

ly warm waters of the Great Lakes, the air becomes warmer and more humid near the earth's surface. The rising of the warm, moist air and the consequent condensation can produce heavy snow at the eastern ends of the Great Lakes. An anticyclone, or high, can bring either fair or cloudy weather, depending on how it is altered in its travels.

Weather Fronts

In the nineteenth century, Heinrich Wilhelm Dove and Robert Fitzroy first studied the boundaries, or *fronts*, between different bodies of air. Fronts are most distinct between a cold polar air mass and a warm tropical air mass because of their contrasting qualities. Temperature and winds can differ markedly from one side of a front to the other, perhaps a distance of only a few kilometers.

A *warm front* occurs when a warm air mass moves into a region occupied by cooler air. As Figure 5.13 indicates, the boundary between the two air masses at a warm front slopes gradually and gently as the warmer, less dense air overrides the cool air. The slope is usually gentle, and the front at an altitude of 1 kilometer may be several hundred kilometers ahead of the front at ground level. Daily weather maps show the ground-level locations of fronts.

The particular weather conditions at a given location depend on many factors, such as the detailed history of an intruding air mass and the presence or absence of mountains. Nevertheless, a characteristic sequence of weather conditions associated with an incoming warm front can often be recognized, despite the modification of weather by local conditions. The lifting of warm air over cool air at a warm front is somewhat similar to the lifting of air at a mountain slope. At altitudes where the air is cooled to the dew point, condensation and cloud formation occur. Because the front slopes gently upward, the coming of a warm front is heralded by the appearance of high cirrus clouds a day or more before the ground-level

Figure 5.13 A warm front is formed when a mass of warm air overrides a denser mass of cool air. The diagram shows a vertical cross section of a warm front, together with the ground level symbol as it would appear on a weather map. As the uplifted warm air becomes cooler, condensation and cloud formation may occur. The vertical section shows that the resulting precipitation arrives at a location ahead of the ground level warm front itself. (After Flohn, 1969)

Figure 5.14 A cold front occurs where a mass of cool air pushes into a region of warmer air. The vertical cross section of the front shows the warm air ahead of the front rising upward over the incoming cool air. Cloud formation and precipitation may occur as the uplifted warm air cools. The vertical section shows that precipitation normally arrives at a location after the ground level front has passed. (After Flohn, 1969)

Figure 5.15 An occluded front involves three air masses of different temperatures, and it combines some of the features of both warm and cold fronts. The vertical cross sections show that the mass of incoming cold air forces warm air to rise above both regions of cool air. In the first vertical section, the occlusion is not complete and the warm air is in contact with the ground in close proximity to a cold front and a warm front. In the vertical section at the rear, the occlusion has formed and the warm air is lifted completely above the ground. Precipitation may occur on both sides of an occluded front, as the diagrams indicate. (After Flohn, 1969)

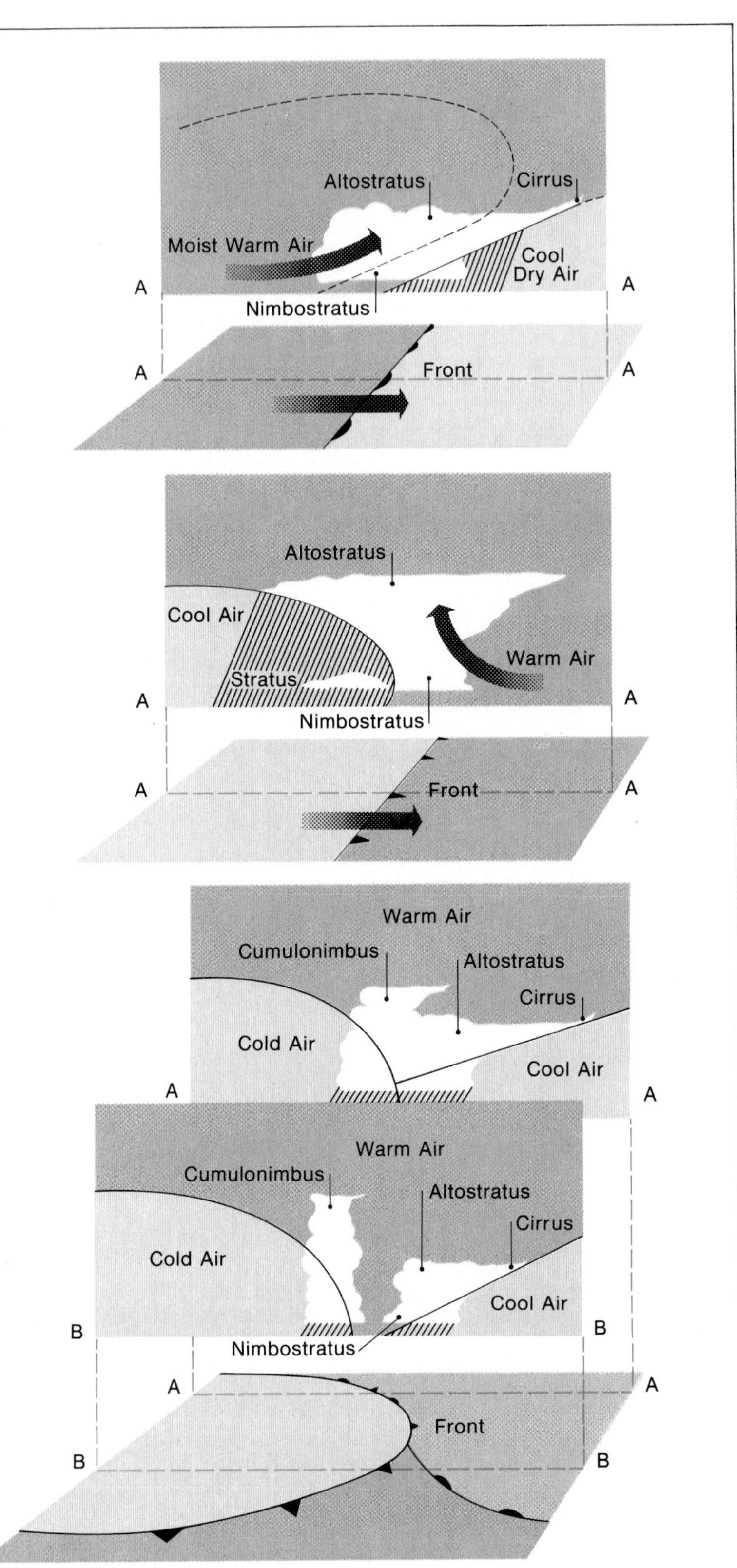

Altostratus
Cirrus
Moist Warm Air
Cool
Dry Air
A
A
Nimbostratus
A
Front
A
Altostratus
Cool Air
Warm Air
Stratus
A
A
Nimbostratus
A
Front
A
Warm Air
Cumulonimbus
Altostratus
Cirrus
Cold Air
Cool Air
A
A
Warm Air
Cumulonimbus
Altostratus
Cirrus
Cold Air
Cool Air
B
B
Nimbostratus
A
A
Front
B
B

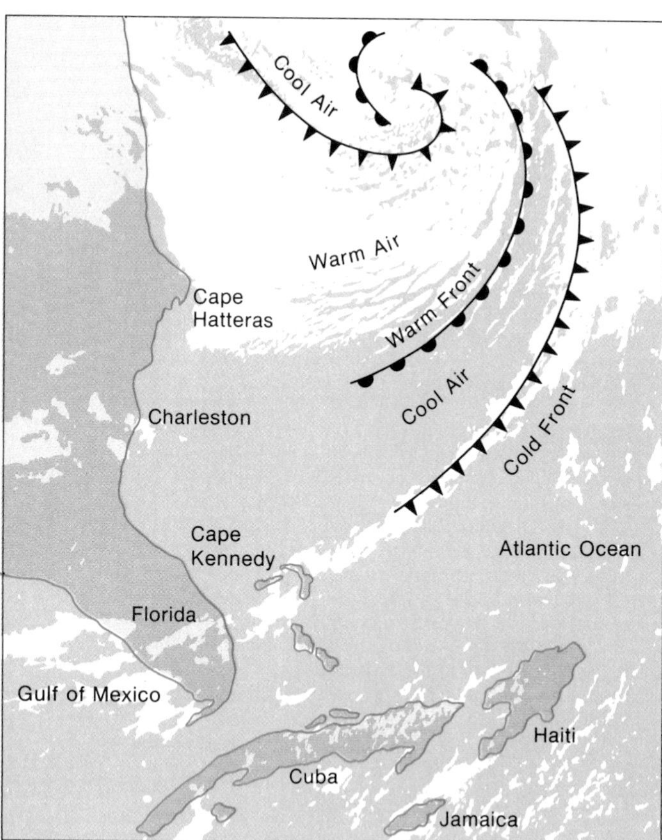
Cool Air
Warm Air
Warm Front
Cool Air
Cold Front
Cape
Hatteras
Charleston
Cape
Kennedy
Florida
Gulf of Mexico
Atlantic Ocean
Cuba
Haiti
Jamaica

Figure 5.16 The photograph shows the Atlantic Ocean off the east central coast of the United States as seen from the Apollo 8 spacecraft shortly after it left Earth orbit. The principal feature exhibited by the atmospheric circulation is a large cyclonic region of low pressure having a diameter of 1,500 kilometers (900 miles). The boundaries, or fronts, between regions of warm and cold air shown on the accompanying map are suggested by the cloud patterns; the warm air has a denser pattern of clouds than the cold air because it is moister. Note also the cloud cover that forms over land areas such as Cuba because of the ascending moist air from the heated land.

front itself arrives. As the ground-level front approaches, the clouds become thicker and lower, forming sheetlike stratus clouds accompanied by a broad band of precipitation.

A *cold front* occurs when a mass of cold air moves into a region occupied by warmer air. The structure of a cold front differs from that of a warm front; cold fronts are steeper than warm fronts, as indicated in Figure 5.14. Contact with the advancing cold air forces the warm air to rise steeply, which induces the rapid formation of vertical clouds. Perhaps you have experienced the break in hot summer weather when a cold front passes by; the sudden appearance of a line of thunderstorms and a rapid drop in temperature sometimes accompany the passage of a cold front.

More complicated fronts form when three distinct air masses are involved. An *occluded front* has some of the characteristics of both a warm and a cold front. A cold occlusion, for example, forms when a mass of colder air overtakes a warm front separating warm and cool air. As the colder air moves in, all of the warm air is forced to rise over the two cooler air masses, and thick clouds are generated. A new front, similar to a cold front, forms between the incoming cold air and the cool air. If rain forms in the warm air in a cold occlusion, it must fall through the cooler air masses to reach the ground. When the temperature of the cooler air is below freezing, raindrops freeze on cold ground, trees, and power lines.

Fronts are instrumental in the initiation and development of cy-

Figure 5.17 (opposite) Cyclones in the midlatitudes often follow a characteristic pattern of development involving the interaction of warm and cold fronts to form an occluded front. (a) The process begins at a *stationary front,* where a stream of cool air and a stream of warm air are moving past each other. (b) A stationary front is comparatively unstable, and a bulge, or *wave,* usually develops within 1 or 2 days. If the wave is 500 kilometers or more in size, there is a good chance that the disturbance will continue to grow. (c) The continued warping of the stationary front leads to the formation of a well-developed warm front on one edge of the wave and a cold front on the other. Associated with this structure is a developing counterclockwise circular flow characteristic of a cyclone. The cyclone reaches a mature state within 4 to 5 days after the wave first appears on the stationary front. (d) In maturity, the cold front overtakes the warm front, lifting the warm air off the ground and forming an occlusion. The thorough mixing of the air in the mature stage eliminates the temperature differences that drive the circulation, and the cyclone soon dies away. (e)The sequence of development can be traced on the three daily weather maps for March 5, 6, and 7, 1973. As a cyclone matures, a new wave disturbance may be initiated farther back on the stationary front. One front may support a cyclone family of three to five members in various stages of development. Later cyclones are weak, because the mixing of the warm and cool air by already mature cyclones reduces the energy available for the growth of younger cyclones.

clones. The idealized sequence of events in the formation of many midlatitude cyclones is described in its main outlines in Figure 5.17.

Jet Streams—The Atmospheric Vacuum Cleaner

The model of cyclone development shown in Figure 5.17 was originated in the 1920s and does not incorporate recent discoveries concerning the role of the upper atmosphere in cyclone formation. Surface air flowing into the low-pressure center of a young cyclone must be constantly removed to maintain the low pressure. It appears that *jet streams* may be the atmospheric vacuum cleaner that extracts air from the center of a cyclone.

Jet streams were first observed in the upper atmosphere with the development in the 1940s of aircraft that could fly at altitudes of higher than 10 kilometers (6 miles). Jet streams occur at altitudes of between 10 and 15 kilometers and are thousands of kilometers long, hundreds of kilometers wide, and several kilometers thick. The wind speed along the central core of a jet stream may occasionally exceed 300 kilometers per hour (200 miles per hour), fading out to much slower speeds near the edges.

Jet flow exhibits a certain degree of instability, and a jet stream may give rise to the growing wave disturbance described in Figure 5.18. Cyclones at the earth's surface seem to form by interaction with jet streams aloft. A jet stream extracts air from the earth's surface, which produces a region of low pressure favorable to the formation of a cyclonic weather system.

Tornadoes

A *tornado* is a highly localized circulation of the atmosphere consisting of a column, or funnel, of rapidly whirling air. The column is seldom more than a few hundred meters in diameter, and the speed of air in the column is estimated to be well in excess of 300 kilometers per hour (200 miles per hour). Tornadoes move across the land with an average speed of about 30 kilometers per hour.

The interior of a tornado column is a region of low pressure and rapid updraft. The pressure in a tornado is sometimes 50 to 80 millibars less than the pressure of the surrounding atmosphere, and the pressure differential can do great damage to the area the tornado passes through. Houses literally explode from the sudden drop in outside pressure, and tornadoes have sucked ponds dry and lifted and carried automobiles several blocks.

Several hundred tornadoes are reported in the United States each year, mainly in the central plains states. Tornadoes are known to occur in association with violent thunderstorms, so that tornado warnings are issued when strong cold fronts generate the atmospheric instabilities that lead to thunderstorm activity.

Tropical Secondary Circulations

Much less is known about tropical weather patterns than about midlatitude weather patterns. The region between the Tropic of Cancer and the Tropic of Capricorn has few observing stations, and weather data from the extensive tropical ocean surface is espe-

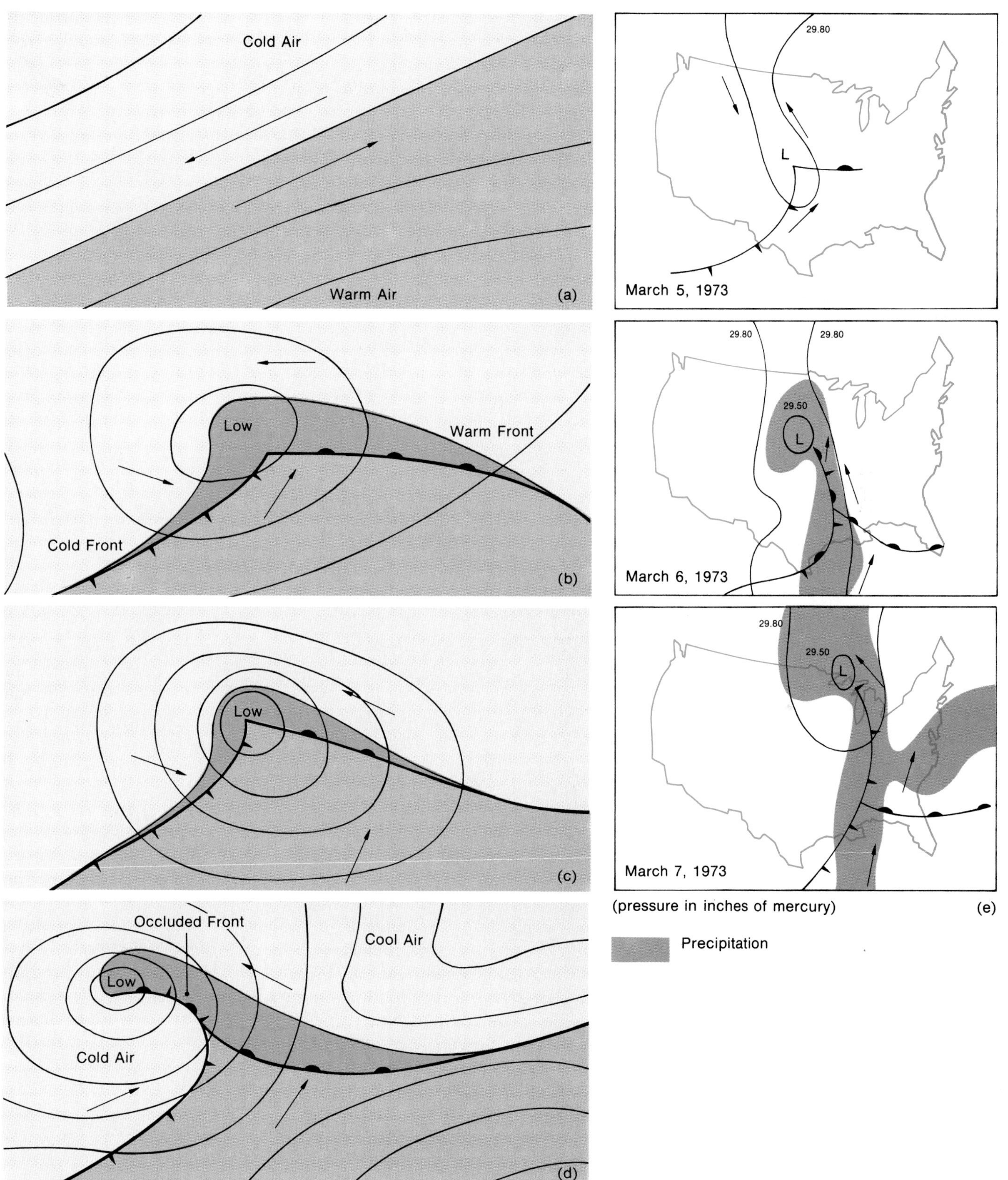

Cold Air
Warm Air
(a)
Low
Warm Front
Cold Front
(b)
Low
(c)
Occluded Front
Cool Air
Low
Cold Air
(d)
29.80
L
March 5, 1973
29.80
29.80
29.50
L
March 6, 1973
29.80
29.50
L
March 7, 1973
(pressure in inches of mercury)
(e)
Precipitation

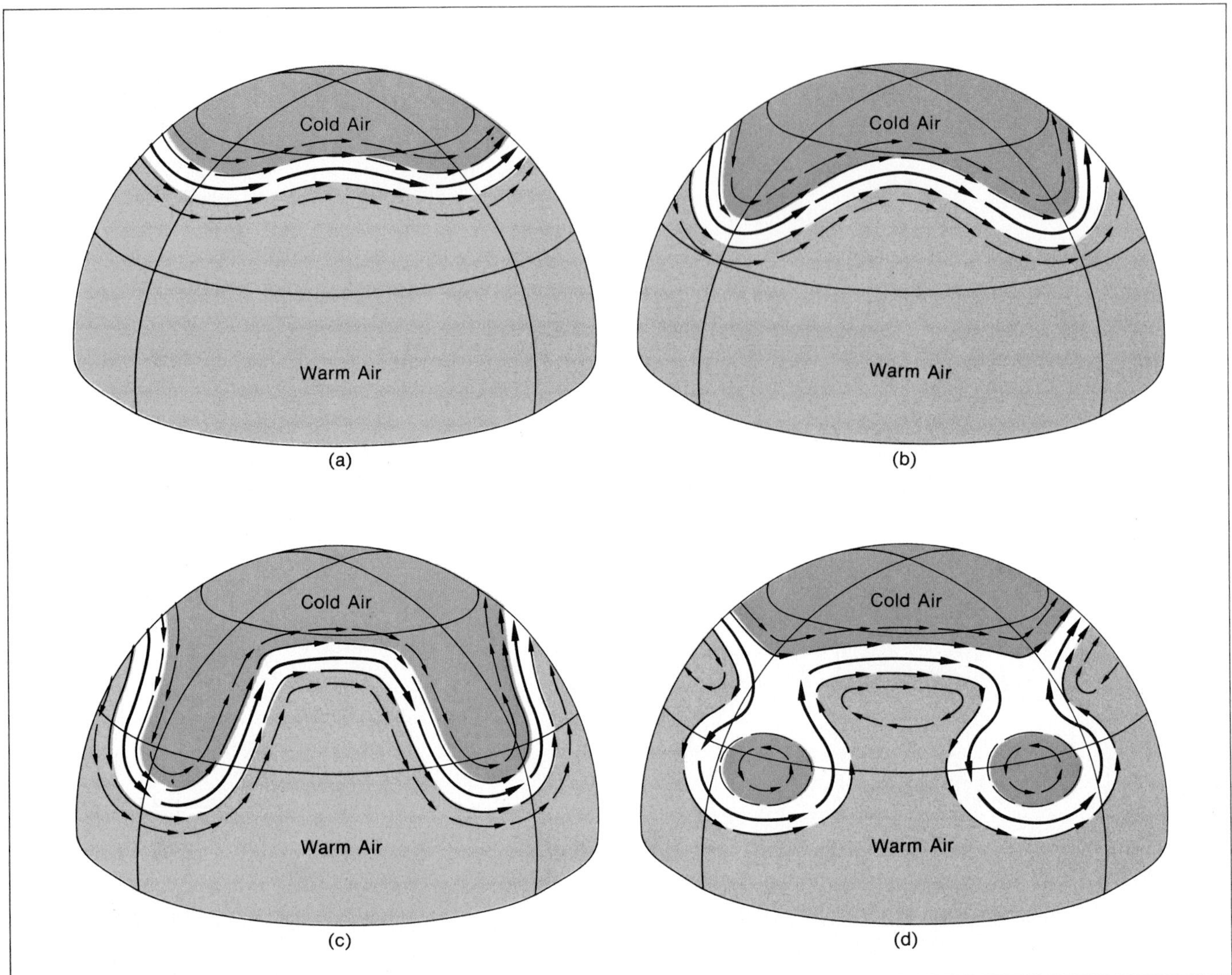
Cold Air
Warm Air
(a)
Cold Air
Warm Air
(b)
Cold Air
Warm Air
(c)
Cold Air
Warm Air
(d)

Figure 5.18 Circumpolar jet streams circle the earth at an altitude of 10 to 15 kilometers (30,000 to 45,000 feet) and gain much of their energy from the temperature difference between warm tropical and cold polar air. (a) The jet streams appear in conjunction with large, slow-moving undulations, or waves, in the atmosphere. There are typically three to six waves in a complete pattern around the earth. The waves may undergo increasingly severe oscillations (b) (c), until cells of rotating warm and cold air are formed (d). The undulations then die away until the pattern once more resembles the pattern shown in (a); the entire cycle is completed in 4 to 6 weeks. This mechanism results in the movement of warm air toward the poles and cold air toward the tropics and helps to maintain the poleward flow of energy from equatorial regions. Jet streams and upper air waves also appear to generate weather systems, but the details of the process are not fully understood. (After "The Jet Stream" by Jerome Namias. Copyright © 1952 by Scientific American, Inc. All rights reserved)

cially sparse. The Coriolis force is small near the Equator, so winds are not geostrophic there. Rotating cyclones and anticyclones do not form near the Equator, and frontal activity is absent. In the middle latitudes, the interaction of different air masses dominates the weather. In the continental tropics, the daily cycle of heating and cooling of a comparatively homogeneous mass of warm air is the principal determinant of the weather. The temperature variation during a single day in the tropics often exceeds the variation in the average temperature through the year. In spite of a lack of variety in weather patterns, the tropics do exhibit characteristic surface circulations not found elsewhere in the world.

Monsoons

The seasonal shift in the direction of surface winds over India and over the Indian Ocean strongly affects weather there. The southwest summer wind picks up moisture as it crosses the sea toward India,

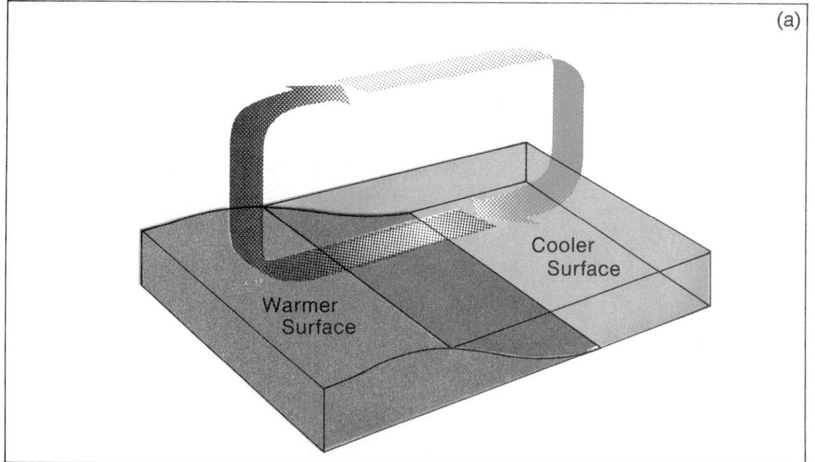
(a)
Cooler
Surface
Warmer
Surface

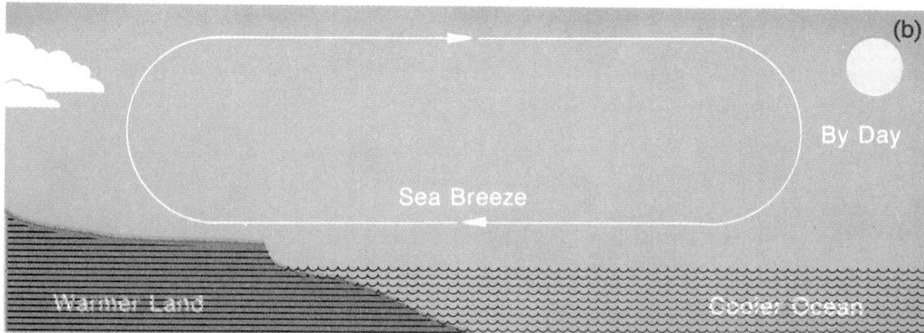
(b)
By Day
Sea Breeze
Warmer Land
Cooler Ocean

By Night
Land Breeze
Cooler Land
Warmer Ocean

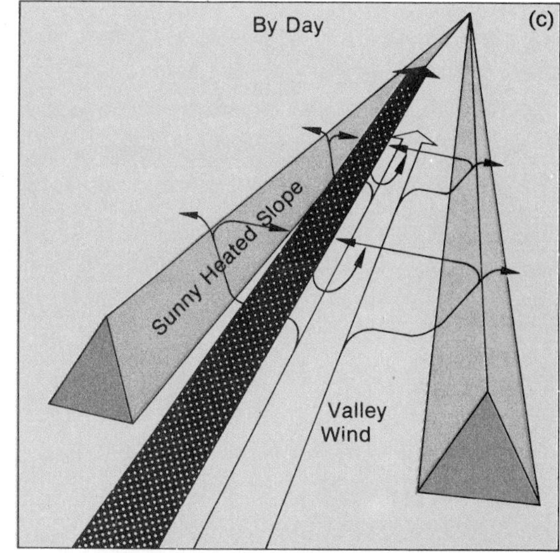
By Day
(c)
Sunny Heated Slope
Valley
Wind

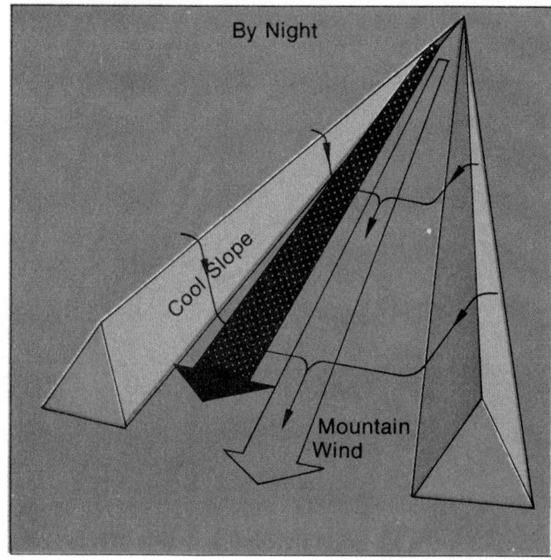
By Night
Cool Slope
Mountain
Wind

Figure 5.19 Small-scale circulations of air can significantly modify local weather conditions. (a) This diagram shows a small convective cell generated near the boundary between a warm and cool region.

(b) Land and sea breezes develop because of the difference in temperature between the ocean and the land. During the day, the land heats more rapidly than the ocean, and a surface sea breeze develops from the ocean toward the land. At night, the land cools rapidly and the surface flow of air is a land breeze from the land to the ocean.

(c) Mountain and valley winds develop because of the temperature difference between valleys and mountain slopes. During the day, the slopes are warm, and as air rises up the slopes, a surface valley wind flows up the valley to replace the ascending air. (d) At night, cool air descends the slopes and flows down the valley out of the mountains.

and precipitation from this air accounts for 70 percent of India's annual rainfall. The mountains of India help induce orographic precipitation from the moist southwest winds; some mountain locations in India receive more than 1,000 centimeters (400 inches) of rain in a year from this source.

The change between the northeast winter winds and the southwest summer winds occurs gradually across India. The summer monsoon begins to migrate into southern India late in May but is not felt in northwest India until July. In northern India, the northeast winter winds begin in September but do not arrive in southern India until January.

Upper air phenomena in the tropics also seem to be connected with the monsoon. A jet stream blowing from the west lies at an altitude of 16 kilometers (10 miles) above the south slope of the Himalaya Mountains in northern India. It shifts north of Tibet at the start of the summer monsoon, while a jet stream blowing from

Figure 5.20 The kink in the isobars denotes the presence of an easterly wave. Easterly waves move toward the west with the trade winds. Ahead of an easterly wave, to the west of the trough, cool, dry air descends from the upper atmosphere, and the weather is fair. Behind an easterly wave, to the east of the trough, moist air from the surface ascends to great heights and severe thunderstorms may be generated. Each year a few easterly waves increase in intensity and develop into severe tropical cyclonic storms such as hurricanes. (After Flohn, 1969)

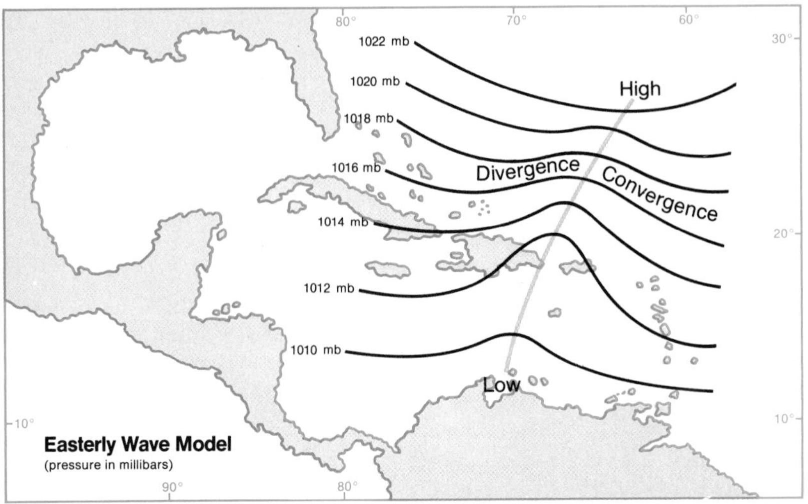

80°
70°
60°
30°
1022 mb
1020 mb
High
1018 mb
1016 mb
Divergence
Convergence
1014 mb
20°
1012 mb
1010 mb
Low
10°
10°
Easterly Wave Model
(pressure in millibars)
90°
80°

the east appears over northern India. The dynamics of the monsoon, however, are not fully understood.

Easterly Waves

The meeting of the trade winds at the intertropical convergence zone sometimes produces a weak front and a line of clouds when the air from the north and the air from the south have slightly different temperatures and moisture properties aloft. Away from the intertropical convergence zone, the weather in the tropics varies little and is usually characterized by long stretches of hot, sunny days. Nevertheless, certain dynamic atmospheric instabilities occur in the tropics.

Sometimes a weak trough of low pressure several thousand kilometers long and running roughly north and south forms in the path of the trade winds and drifts slowly westward. The disturbance, called an *easterly wave*, forms most often in the summer hemisphere. In the tropics, a warm layer of air usually overlies the surface layer, which forms a weak temperature inversion and prevents the surface air from rising to higher altitudes. The inversion is often temporarily destroyed behind an easterly wave, and the rising moisture can form thunderclouds of great height and intensity.

Tropical Cyclones

A surprising feature of the usually monotonous tropical weather is the development of intense cyclonic disturbances. If its wind speed exceeds 120 kilometers per hour (75 miles per hour), a cyclone is classified as a *tropical cyclone*. Tropical cyclones that form near the east or west coasts of North America are called *hurricanes*, and those forming in the western Pacific are called *typhoons*. Both words mean "big wind"—in Arawak and Chinese, respectively.

The "extratropical" cyclones that regularly sweep across the middle latitudes are usually more than a thousand kilometers in diameter, have winds of moderate intensity, and are driven by the horizontal temperature differences between different air masses. In contrast, a tropical cyclone is compact and intense, often with a diameter of only a few hundred kilometers and with pressure changes that can exceed 30 millibars per 100 kilometers. The winds in a tropical cyclone may have velocities greater than 150 kilometers per hour (90 miles per hour). The energy for tropical cyclones comes from condensing water vapor.

Tropical cyclones rarely form within 5° latitude of the Equator, probably because of the weakness of the Coriolis force at low latitudes. And because of the relative lack of water vapor over cold water, tropical cyclones always form over warm oceans. Hurricanes do not often form at latitudes greater than 30°N off the east coast of North America.

Many hurricanes begin as a rotating tropical storm associated with an intense easterly wave. Why some storms die out and others continue to build to hurricane strength is not known. The characteristic structure of a fully formed tropical cyclone consists of rapidly rotating walls of cloudy, moist air surrounding a nearly calm, relatively cloudless central "eye." The height of a tropical cyclone is typically 10 or 15 kilometers (6 or 9 miles), and the eye may be several tens of kilometers in diameter. Dry air from above descends into the eye and becomes warmer than the surrounding air because of adiabatic compression. The moist air in the cloud walls is warm and follows an ascending pattern.

Hurricanes usually follow unpredictable paths. Some sweep into the Gulf of Mexico, cross the Gulf coast, and travel inland. Others curve more sharply and move along the east coast of the United

Figure 5.21 This cross section of a hurricane shows the central column, or eye, and the swirling cloud bands that give a hurricane its characteristic appearance as seen from above. A tropical cyclone in the Northern Hemisphere has an intense counterclockwise rotation. Wind speeds near the center may exceed 200 kilometers per hour (120 miles per hour). The central eye is a region of low atmospheric pressure, and the dry air descending into the eye from above makes weather in the eye clear and calm.

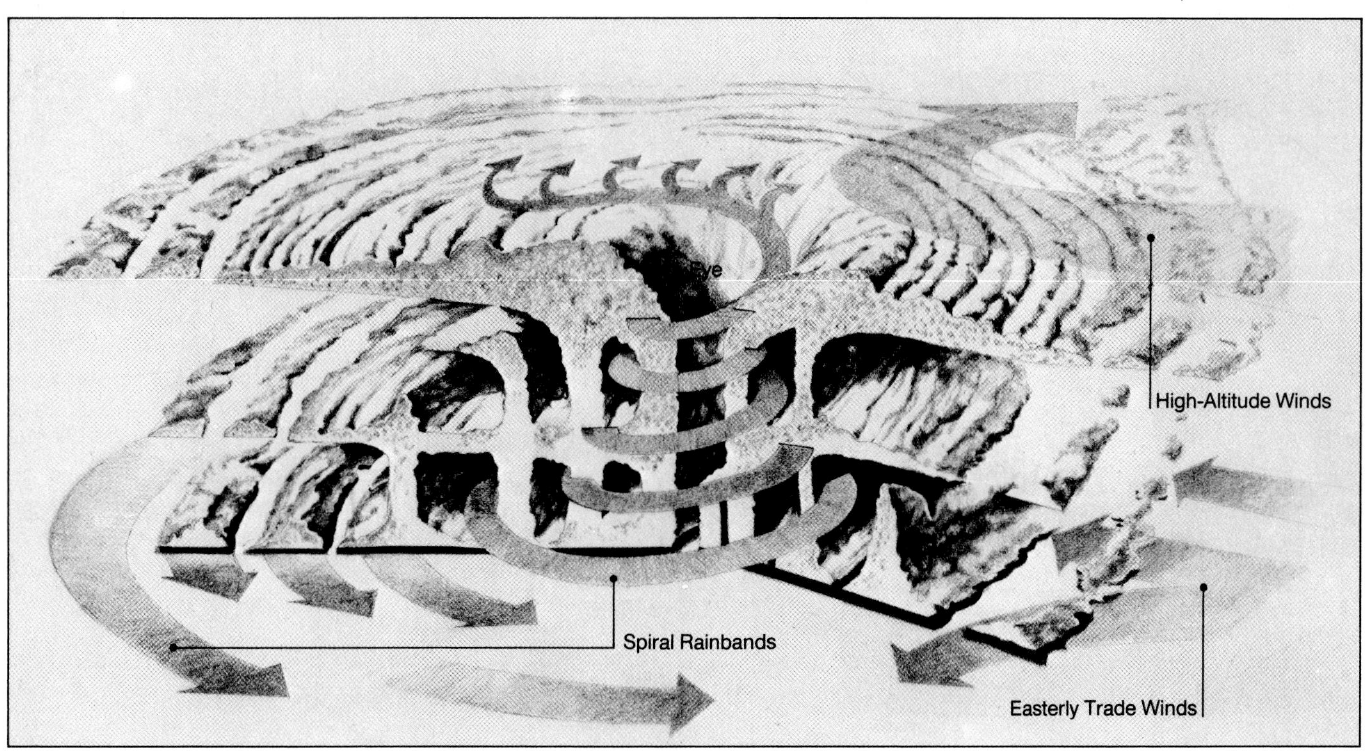
Eye
High-Altitude Winds
Spiral Rainbands
Easterly Trade Winds

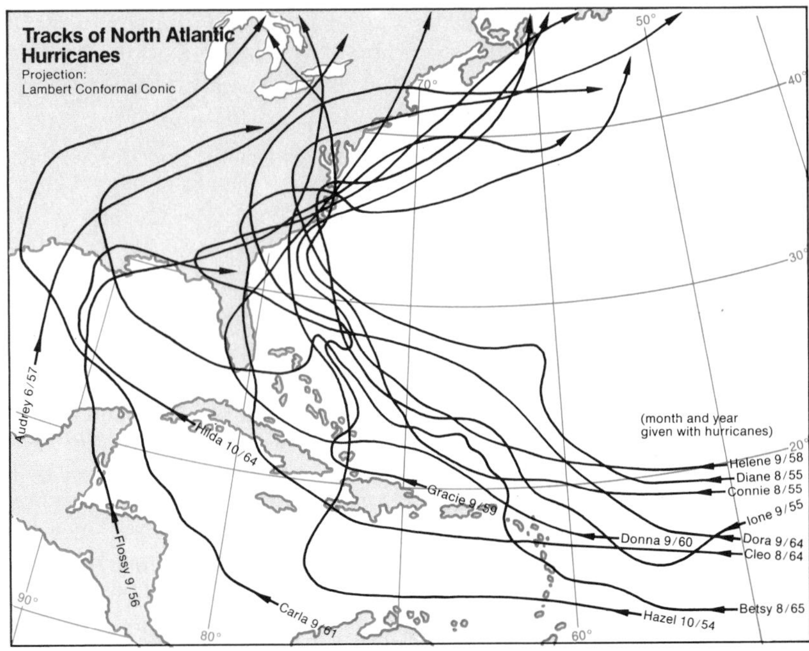
Tracks of North Atlantic Hurricanes
Projection:
Lambert Conformal Conic
50°
40°
70°
30°
20°
90°
80°
60°
(month and year given with hurricanes)
Audrey 6/57
Hilda 10/64
Flossy 9/56
Carla 9/61
Gracie 9/59
Helene 9/58
Diane 8/55
Connie 8/55
Ione 9/55
Donna 9/60
Dora 9/64
Cleo 8/64
Betsy 8/65
Hazel 10/54

Figure 5.22 This map shows the tracks of some devastating North Atlantic hurricanes for the years 1954 through 1965. The path of an individual hurricane tends to be erratic, but as the map indicates, many of the hurricanes generated in the Caribbean Sea follow the same general course westward and northward. Some hurricanes turn northward early and skirt the east coast of the United States; others enter the Gulf of Mexico before heading north. Hurricanes tend to lose strength over land because of friction and lack of water vapor, but some hurricanes, such as Hilda, 1964, have traveled long distances across the United States. A hurricane traveling over land often brings heavy rains that cause rivers to overflow. (After *The National Atlas of the United States of America,* 1970)

Figure 5.23 This photograph of hurricane Gladys, 1968, was taken from the Apollo 7 spacecraft. The open central eye and the counterclockwise circulation of the hurricane are clearly visible.

States. Once a tropical cyclone begins to travel over land or over cold water, it is cut off from a supply of energy in the form of water vapor, and its strength diminishes. Tropical cyclones in the open ocean travel only a few tens of kilometers per hour and increase in speed as they invade middle latitudes. Tropical cyclones are dangerous storms; when forceful hurricane winds cause the ocean to surge up onto low coasts, the water may rise several meters above the normal high-water mark and cause extensive flooding. A rainfall of more than 25 centimeters (10 inches) is not unusual as a hurricane passes by, and if the hurricane moves over land, heavy runoff from the rains soon brings rivers to the flood stage. Fortunately, tropical cyclones are relatively rare. The number of hurricanes generated in the Atlantic and the Caribbean varies from one to two per year to ten or eleven. Only three hurricanes formed off the east coast of North America in 1972, a year in which ocean water temperatures were much below normal.

Because the distinctive cloud formations of a tropical cyclone are easy to see on weather satellite photographs, a moving storm can be accurately tracked in time to give advance warning to threatened areas. Improved warning and communication systems have steadily reduced the loss of life resulting from hurricanes in the United States, even though those who are in danger do not always take the warnings seriously. When a typhoon in the Bay of Bengal struck East Pakistan in 1970, the almost total lack of warning, together with ineffective communication and transportation, resulted in the loss of an estimated 250,000 lives.

General Circulation of the Ocean

The atmosphere and ocean systems interact with one another. The motion of the air drives the surface ocean currents, and the ocean basins affect the circulation patterns of the atmosphere. Wind blowing along the surface of the ocean exerts a push on the water, but because friction forces in water are much stronger than they are in air, the speed of the water is only a small fraction of the wind speed. Typical ocean currents have speeds ranging upward from a few kilometers per day to a few kilometers per hour. Friction causes the speed of fast ocean currents to decrease rapidly with depth below the water's surface. Most fast currents are confined to the upper hundred meters or so of the ocean.

The response time of large ocean currents to the atmospheric circulation is many months. The motion of a large ocean current cannot respond rapidly to changes in the wind, and even the waves that strong winds produce on the surface of the ocean require many hours to reach their full heights. Because ocean currents reflect average wind conditions over a period of a year or more, the general circulation of the ocean is closely related to the general circulation of the atmosphere.

If water entirely covered the earth, the winds would form well-defined belts and the ocean currents would move around the earth in belts under the influence of the prevailing winds. However, only the ocean around Antarctica has relatively unhindered passage around the globe. The largest oceans, the Atlantic and the Pacific, are mostly confined to basins bounded by continents on the east and on the west, so free flow of their currents around the earth is impossible.

The presence of continents strongly influences the motion of the wind-driven currents. Because a current cannot easily leave its basin, the idealized general circulation in the ocean basins consists of closed loops of circulation, or gyres. These gyres correspond to the major belts of wind; both the Atlantic and the Pacific basins have three gyres north of the Equator and three to the south. To understand the ideal gyre pattern shown in Figure 5.24, think of the wind directions. The trade winds, for example, drive the low-latitude currents of the subtropical gyres, and the prevailing westerlies drive the high-latitude return current from the west. The actual circulation of the oceans is shown in Figure 5.25. The main gyre patterns can be seen in the Atlantic and Pacific Oceans, except perhaps near Antarctica, where the westerly winds establish a westerly global ocean current.

Figure 5.24 The idealized oceanic circulation in an ocean basin consists of loops, or gyres, which are driven by prevailing winds. (After Weyl, 1970)

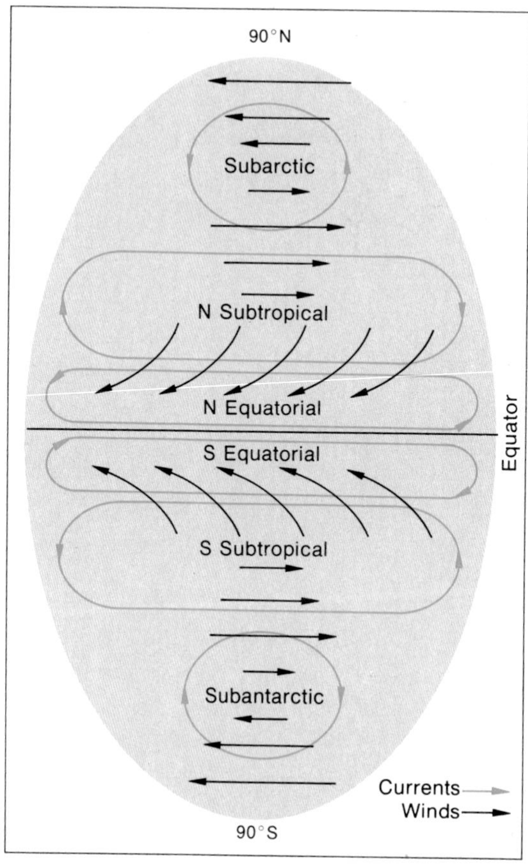

90°N
Subarctic
N Subtropical
N Equatorial
S Equatorial
S Subtropical
Subantarctic
Equator
Currents
Winds
90°S

The Gulf Stream

A general northward current of warm equatorial water flows at the west edge of each ocean basin in the Northern Hemisphere, and cooling southward currents flow at the east edges. The Gulf Stream is the warm northward current in the Atlantic basin; its counterpart in the Pacific basin is the Kurushio off Japan. Both are fast, narrow currents moving several kilometers per hour. By contrast, the southward currents on the east edges of the basins are sluggish and poorly defined.

The amount of water the Gulf Stream transports is more than 30 times as great as the amount of water that flows in rivers and

Case Study: Expanding Horizons in Atmospheric and Oceanic Research

As we learn more about the interactions of global systems we see that an event in one part of the world can have far-reaching effects in another part of the world. The development of sophisticated technology for observing and analyzing phenomena in the air and water has contributed to our knowledge of the interactions of systems such as the atmosphere and ocean.

For example, computers at NASA (National Aeronautics and Space Administration) "painted" the portrait shown here of 1969 Hurricane Camille. A special computerized technique has been devised to study hurricanes by processing data obtained from weather satellite observations. The colors signify the temperatures in different parts of the hurricane. This computerized color image has seventeen shades of color, each one indicating a different cloud temperature.

The photograph of Camille was taken by a satellite from an altitude of about 1,435 kilometers (900 miles) just after the hurricane had passed Bay St. Louis in Mississippi. The hurricane was traveling in a northwesterly track and had surface winds up to 296 kilometers per hour (185 miles per hour). The color used for this print covers a temperature range of 100°C with intervals of 5° in brightness. Purple represents the ground or sea surface temperature of 27°C (81°F), or the temperature of low clouds at an altitude of 0.5 kilometer (0.3 mile). Yellow signifies cooler clouds at about 6 kilometers (3.7 miles) in altitude, and blue indicates an even cooler temperature and higher altitude. The highest clouds, found at approximately 15 kilometers (10 miles), are shown by gray colors and are located over the eye of the hurricane near New Orleans, Louisiana.

The implication of this computer technique is that weather data can be obtained and analyzed rapidly, thus making weather predictions and storm warnings more rapid and more accurate. Studies of the temperature distribution in hurricanes may lead to ways of abating the fury of such storms.

Information about the atmosphere and the surface of the earth is also being obtained by another remote sensing device called ERTS-1, Earth Resources Technology Satellite (see also Chapter 3 and Appendix II). The mission of the satellite, which passes over a given point on the earth every 18 days, is to survey the growth and decay of natural vegetation and crops; to locate mineral deposits, geothermal power sources, and underground reserves of fresh water; and to observe the effects of human activities such as industry and fires.

Television cameras in the satellite scan strips of the earth's surface, thereby providing systematic, repetitive global coverage. The cameras detect both visible and infrared radiation reflected from the surface, and the transmitted imagery is processed into false-color photographs at NASA facilities on the ground.

One of the implications of ERTS for atmospheric and oceanic systems is that it can survey and monitor air and water pollution. Algae-choked lakes, for example, show up

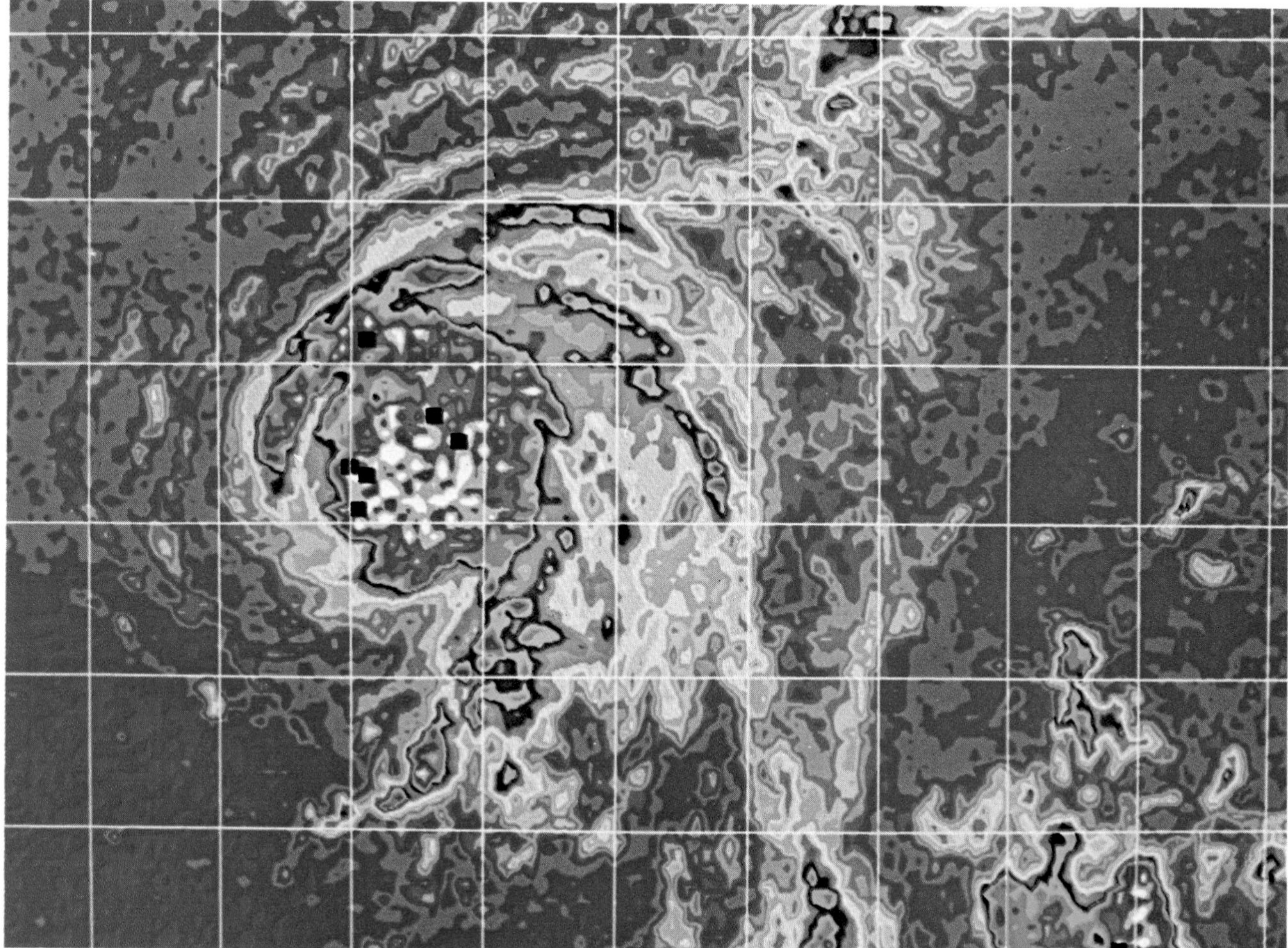

clearly in false-color photographs. Suspended sediment and turbulence in rivers and lakes are also readily observed, and in many cases the wastes discharged from large plants are visible.

Such earth watches can have social and legal as well as geographic implications. The availability of detailed data makes possible informed policies for the management of limited resources of energy and materials. Governmental agencies are considering such topics as the governing of ocean resources, the boundaries of fisheries, the management of coastal zones, the impacts of weather modification, and the interactions of the atmosphere and ocean.

The interdependent relation between the ocean and the atmosphere has profound ramifications over the entire earth, because the same physical processes determine the operation of both systems. Compared to the response time of the atmosphere, however, the ocean responds slowly to outside perturbations. The ocean's importance as a source of stored heat energy, and its ability to interact with the atmosphere, lead to the hypothesis that changes which occur in the ocean may be the precursors of long-term shifts in the general circulation of the atmosphere. Because observing certain key features and processes in the oceans may lead to prediction of atmospheric phenomena, scientists are engaged in a number of study projects to increase our knowledge of the ocean and its interaction with the atmosphere.

Some of the significant results from research on air-sea interactions are the discovery of feedback mechanisms between the two systems that enhance abnormal weather regimes and that steer major atmospheric systems; the suggestion that little-understood regional climatic shifts are influenced by air-sea interactions; and the observation that abnormalities in one part of the North Pacific basin extend their influence over large parts of the Northern and (perhaps) Southern Hemispheres. These earlier research findings have led scientists to ask the question of how the sea may affect climatic fluctuations over time.

Scientists continue to ask questions and suggest hypotheses about the mechanisms involved in physical systems. As technology advances, some of these questions may be answered, but the new technology may also uncover problems for further exploration.

from glaciers. The Gulf Stream proper runs from Cape Hatteras, North Carolina, to near the Grand Banks of Newfoundland, where it merges into the broader, slower North Atlantic Drift. The warm water that the North Atlantic Drift carries to western Europe significantly raises the average winter temperature there.

Upwelling

Winds can induce vertical flow, or upwelling, in the ocean. The wind force and the Coriolis force can combine at a seacoast to cause surface water to move away from the coast out to sea. A flow of cool water from a depth of a few hundred meters then moves up to replace the water that has blown seaward.

Coastal upwelling of cool water occurs on the west coasts of all continents, but it is particularly pronounced off California, Peru, Senegal, and southwest Africa. Upwelling brings nutrient-rich cool water to the surface, which fosters the development of sea life. The upwelling waters off the coast of Peru fail to appear about once every 7 years. When this unexplained event occurs, many forms of marine life are deprived of the nutrients they depend on for life. The guano birds that feed on marine life must migrate or starve.

The Deep Circulation

In the deep ocean, below the surface currents, the motion of the water is slow. The driving force for the motion comes primarily from density differences. The density of sea water varies with both temperature and salt content. The densest water is cool and highly saline; such water is produced off Antarctica when surface water freezes, leaving its salt content in the remaining unfrozen water. The ocean water from the region around Antarctica slowly slides down toward the ocean floor. Because the deep circulation is slow, water masses tend to move with little mixing of different waters. According to carbon 14 dating, water in the deep sea may remain there for hundreds of years before returning to the surface.

Summary

Solar radiant energy input to the earth provides the energy to drive the motion of the earth's atmosphere. The rise of hot air and the descent of cold air set up a general circulation of the atmosphere that carries energy from the tropics toward the poles. The earth's rotation causes the circulation to be deflected so that the movement of air tends to follow parallels of latitude.

On an ideal, smooth earth, the global distributions of surface air pressure and of surface winds would follow well-defined belts extending around the earth. But the differences in temperature and topography over the earth's surface from continents, oceans, and mountain ranges cause the simple belt pattern to be modified into cells. Traces of the ideal circulation pattern exist, however, particularly in the Southern Hemisphere, where landmasses are small.

The general circulation drives air masses associated with weather systems from place to place on the earth, transporting energy and

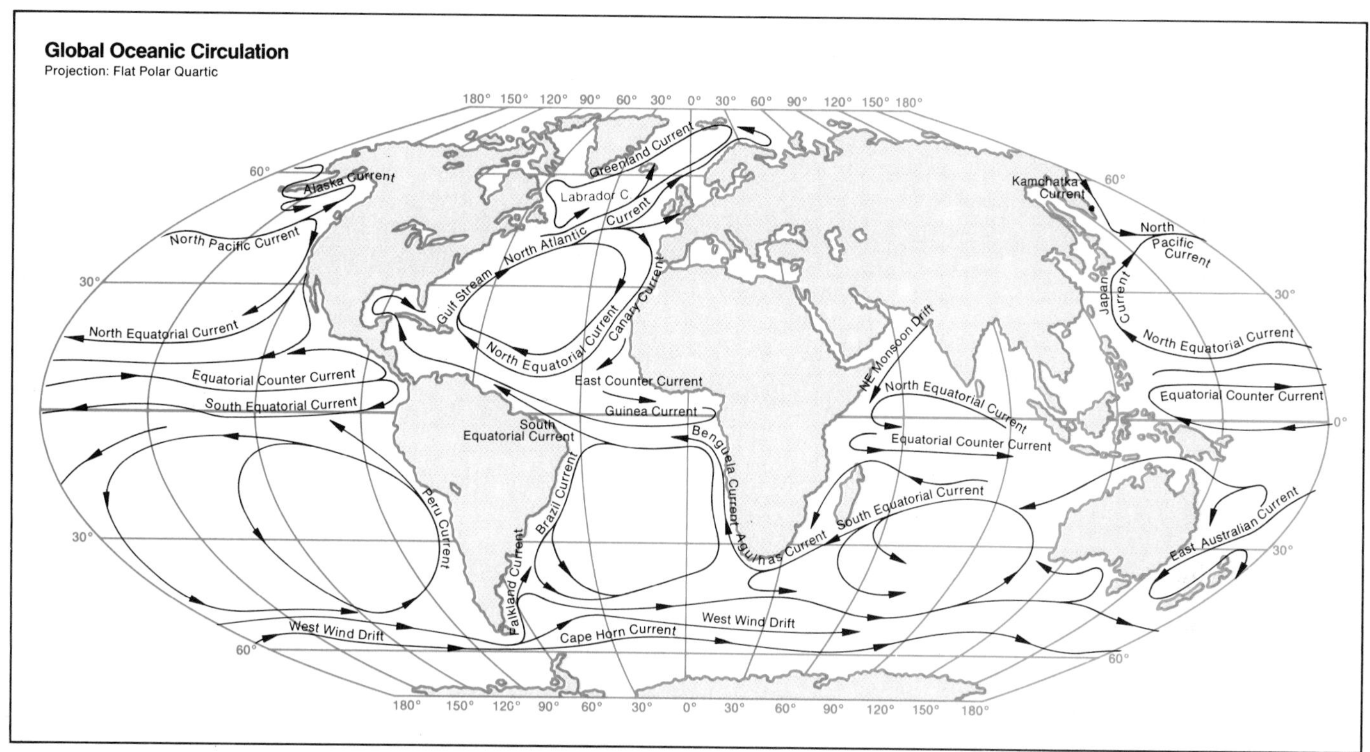

Global Oceanic Circulation
Projection: Flat Polar Quartic
Alaska Current
North Pacific Current
North Equatorial Current
Equatorial Counter Current
South Equatorial Current
Peru Current
West Wind Drift
Greenland Current
Labrador C.
Labrador Current
North Atlantic
Gulf Stream
Canary Current
North Equatorial Current
East Counter Current
Guinea Current
South Equatorial Current
Benguela Current
Brazil Current
Falkland Current
Cape Horn Current
West Wind Drift
Agulhas Current
South Equatorial Current
NE Monsoon Drift
North Equatorial Current
Equatorial Counter Current
Kamchatka Current
North Pacific Current
Japan Current
North Equatorial Current
Equatorial Counter Current
East Australian Current

moisture. The secondary circulation of weather systems at the surface of the earth in the middle latitudes consists of large cells of high or low pressure. The air masses often retain the characteristic temperatures and humidities of their places of origin, and by moving across the earth, they bring changes of weather. When air masses of different temperatures meet, warm, moist air rises over cool air, and a frontal system, often marked by the growth of clouds, is formed.

Weather systems and fronts of the type found in the middle latitudes seldom form at low latitudes, although intense cyclonic storms sometimes develop in the tropics; these storms can cause great flood damage if they travel to a populated coast.

Each region of the earth has a characteristic climate determined by radiant energy input, by the overall controls of the general circulation, and by the influences topography exerts. The connection between climate and the general circulation is discussed in detail in Chapter 7. The transport of moisture by the atmosphere is a key element in the working of the hydrologic cycle. The return of moisture to the earth's surface from the atmosphere exhibits irregularity and variation, and Chapter 6 discusses how systems at the local scale share available moisture.

Figure 5.25 This map shows the principal oceanic currents in the surface layer of the oceans. The currents have a tendency to form loops of circulation that are marked by strong currents on the perimeters and relatively little movement internally (compare with Figure 5.24). Ocean currents move warm water poleward and cold water toward the tropics, which helps to equalize the distribution of incident radiant energy over the earth.

The principal ocean currents along the east coast of the United States include the Florida Current and Gulf Stream, which carry warm waters northward, and the Labrador Current, which carries cool waters southward past the coast of New England. (The Florida Current is the name given to the portion of the Gulf Stream from Florida to Cape Hatteras, North Carolina.) The principal ocean current along the west coast of the United States is the southward-flowing California Current. (After Leet and Judson, 1965)

6 The Hydrologic Cycle and Local Water Budget

Water Use and Water Management

Water on the Earth

Principles of Hydrology
Interception, Throughfall, and Stemflow
Infiltration and Soil Moisture
Groundwater and Aquifers
Runoff and Streamflow

The Local Water Budget: An Accounting Scheme for Water
Potential and Actual Evapotranspiration
Measuring Evapotranspiration and Potential Evapotranspiration
The Water Balance Equation

A Local Water Budget: Baton Rouge

Improving the Water Budget Model

The sun's radiant energy provides the power for the atmospheric circulation to deliver energy and moisture to the earth's surface. On the land the water interacts with the rocks, soils, and vegetation. In time, most of it returns to the atmosphere in a never-ending cycle of renewal.

Rainy Season in the Tropics by Frederic E. Church, 1866. (Fine Arts Museum of San Francisco)

The Hydrologic Cycle and Local Water Budget

Inland from the gently rolling lowlands surrounding Puget Sound in Washington, the landscape changes abruptly into a spine of high mountains that bisects the state. This is the Cascade Range, cloaked in thick stands of dark green fir and spruce and laced with foaming waterfalls that tumble downward in their search for the sea. It is a region of contrasts. To the west of the mountains rain is present much of the year, falling from storm clouds that sweep in from the North Pacific only to meet the mountain barrier. Here on the western slopes of the Cascades the waters accumulate in streams and rivers that drain toward the coast. To the east lies a vast region of rolling, dry grasslands, protected by the mountains from much of the heavy rain characteristic of the western part of the state. Here are the sun-drenched orchards of Yakima and the Wenatchee Valley, the wheat fields of Spokane. And here man must turn to irrigation practices in order to sustain his crops.

These contrasting regions illustrate only a few of the variables at work in determining what happens to water when it returns to the land, and how man has learned to adjust his activities to accommodate the variation. So far you have read about the precipitation processes that convert water vapor in the atmosphere into clouds of droplets and then into large drops that fall as rain. And you have studied the system of circulation that drives moist air masses thousands of kilometers through the atmosphere to deposit their moisture on the earth's surface. This chapter discusses what happens to water that is cycled from the earth's surface to the atmosphere and back to the surface.

Precipitation processes and the atmospheric delivery system condition man's use of water. The *hydrologic cycle*, introduced in Chapter 2, traces water's evaporation from the earth's surface, its storage and transport by the atmosphere, and its subsequent return to the surface by precipitation (see Figure 6.2). The global processes of evaporation and precipitation occur on a local scale as well, and they are traced by the *local water budget*, which is discussed later in this chapter.

Water Use and Water Management

A gallon of water is an ample daily supply of drinking water for one person, but water use in modern society extends far beyond this essential minimum. Washing a car or taking a bath requires 30 or 40 gallons of water, and watering a lawn may use 250 gallons of water or more. Indirect use of water is even greater. Large quantities of water are used in manufacturing and in agriculture. To manufacture 1 pound of paper, man uses 30 gallons of water, and to grow 1 pound of cotton, cotton plants transpire 1,000 gallons of water!

Average daily use of water in the United States amounts to 1,500 gallons of water per person, not including water that is momentarily diverted for hydroelectric use. The public directly utilizes less than 10 percent, and agriculture uses about 40 percent. Fossil fuel electric power plants also use about 40 percent, primarily for cooling, and the remaining 10 percent is used in industry.

Most of the water man uses at home, on farms, and in industry is not consumed but is returned to the hydrologic cycle. Often, however, water is taken from a convenient supply and discharged to a less convenient one. The water used in irrigation may come from a river or from groundwater, but evapotranspiration returns most of it to the atmosphere, which transports the moisture long distances. The water used in homes may come from a convenient supply of groundwater, but the water may be returned to streams by the sewage system and may ultimately flow to the sea. The total amount of fresh water received by the earth as precipitation each year is more than sufficient for man's needs, but the water is not always conveniently available.

Water is an essential resource, and in many parts of the world, its proper utilization is the key to expanding agricultural production. Like any commodity, fresh water has a price. The cost of transporting water where it is wanted in the quantity wanted is a factor often overlooked in normally moist regions where water is abundant. The country of Kuwait, near the head of the Persian Gulf, possesses one-fourth of the world's oil reserves, but its natural fresh water supply is limited to a few inches of rain a year. Income from oil has made it feasible to pay the high price of 1 dollar per 1,000 gallons for distilling fresh water from sea water in quantities large enough to support agriculture by irrigation. In Israel, irrigation is economically practical if water can be supplied at 30 cents or less per 1,000 gallons, but in the southwestern United States, the price must be no more than a few cents per 1,000 gallons for irrigation to be profitable.

Throughout history men have engaged in extensive construction projects to ensure a supply of water that is adequate to meet continuing demands. The simple clay dams built 5,000 years ago to retain water from the Nile River have their modern counterpart in the immense Aswan High Dam on the Nile. The amount of water stored in the lake formed behind the Aswan Dam exceeds the annual flow of the Nile. The dam was built to generate power, to control floods, and to provide water for irrigation throughout the year. The aqueducts in California that carry water hundreds of miles to Los Angeles have historical parallels in the aqueducts that brought ancient Rome a water supply that was ample even by modern standards. More ambitious water control projects are being contemplated; there are plans to dam the Congo River in Africa and to dam the Amazon River in South America, which would create huge fresh-water lakes.

Water on the Earth

The oceans cover about 70 percent of the earth's surface and contain 97 percent of the earth's total water supply. Ocean water contains dissolved minerals—primarily salt—and is unfit for human consumption or for agriculture. When water evaporates from the ocean,

Net Horizontal Transport of Water Vapor From the Oceans to the Land (7)
Condensation
Precipitation Over the Land (23)
Evaporation and Transpiration From the Land (16)
Storage as Ice and Snow
Interception by Plants
Temporary Surface Storage
Surface Runoff
Infiltration
Soil Moisture Storage
Percolation
Storage in Rivers and Lakes
Groundwater Storage
Groundwater Runoff to Streams

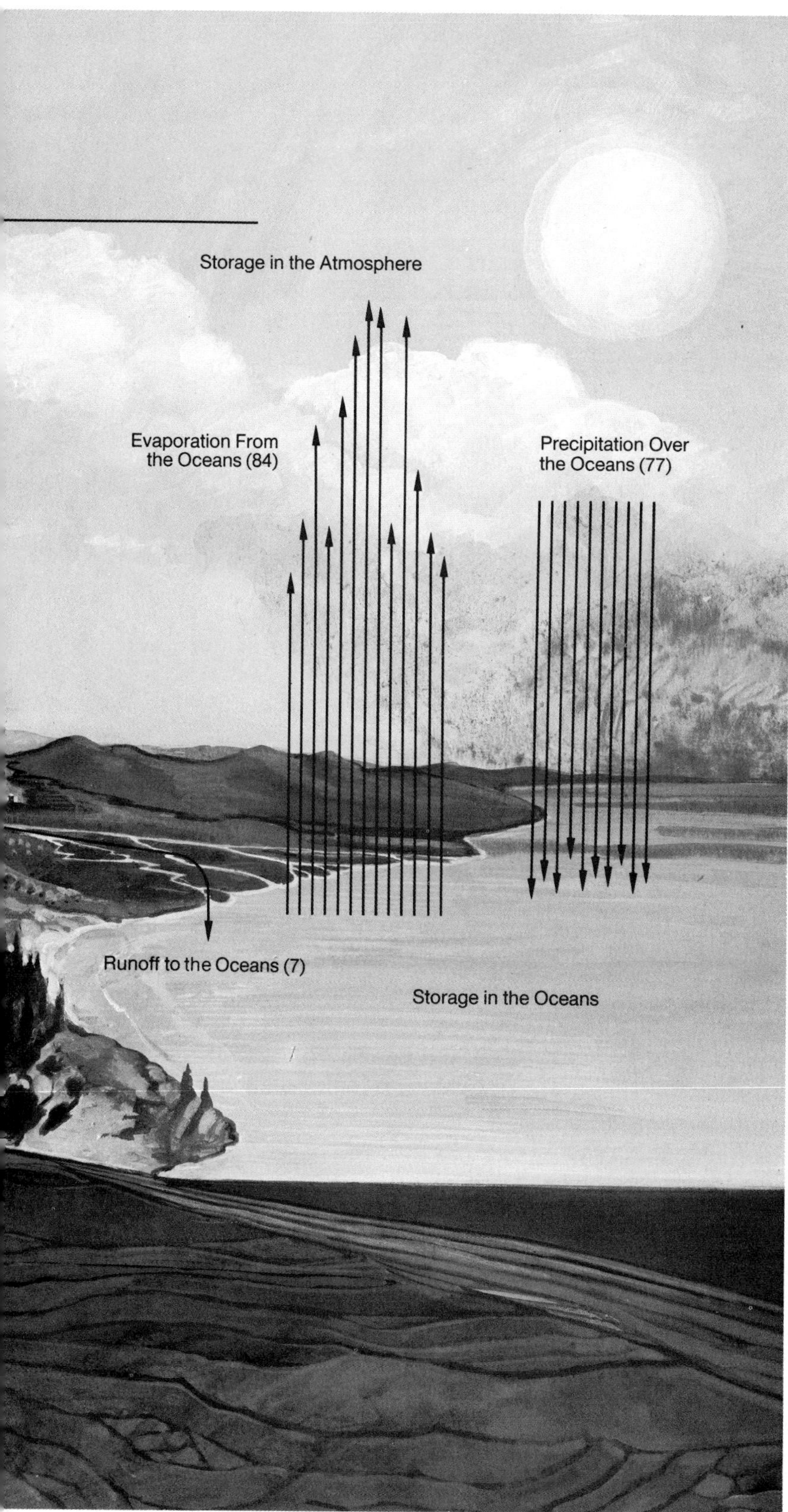
Storage in the Atmosphere
Evaporation From the Oceans (84)
Precipitation Over the Oceans (77)
Runoff to the Oceans (7)
Storage in the Oceans

Figure 6.2 The movement of water through the hydrologic cycle involves numerous interactions and storage processes, as shown in this illustration. Water enters the atmosphere by evaporation from the oceans and by evaporation and transpiration from the land. The numbers in parentheses give the relative number of units of water involved in each process annually, with 100 units equal to the average annual global precipitation of 86 centimeters (34 inches). As the figure shows, 84 units of water enter the atmosphere from the oceans each year compared to 16 units from the land.

Water vapor in the atmosphere condenses into clouds and eventually returns to the surface as precipitation. On the average, 100 units of water enter the atmosphere each year and 100 units leave it as precipitation. Over the oceans, however, more water leaves by evaporation than returns as precipitation; each year 7 units of water are transported by the atmosphere from the ocean to the land. Surplus water from the land returns to the ocean by stream runoff to balance the cycle.

When precipitation falls on the land, a portion of the moisture is intercepted by vegetation and evaporates from temporary storage on leaves. The moisture that reaches the ground either infiltrates into the soil, runs off across the surface, or evaporates from temporary storage in pockets and depressions. Some of the water that infiltrates the soil is stored as soil moisture, and a portion percolates deeper into the ground and enters groundwater storage. The flow of streams is maintained both by direct surface runoff and by underground runoff from groundwater.

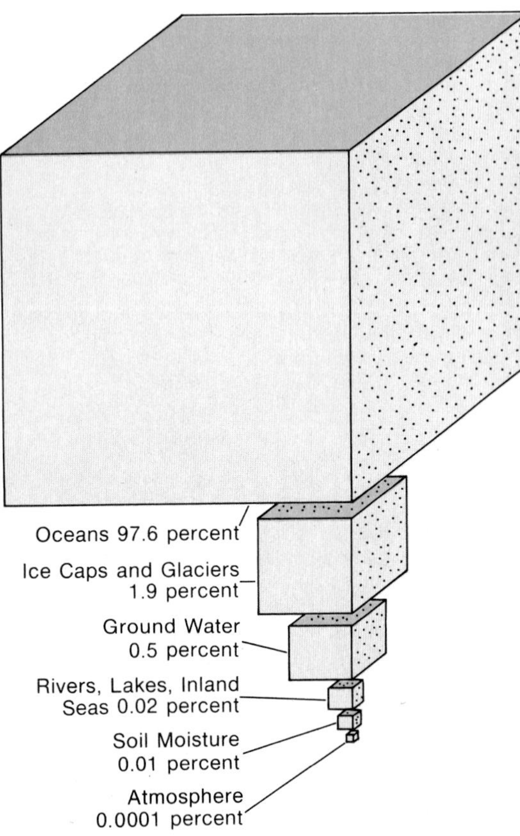
Oceans 97.6 percent
Ice Caps and Glaciers 1.9 percent
Ground Water 0.5 percent
Rivers, Lakes, Inland Seas 0.02 percent
Soil Moisture 0.01 percent
Atmosphere 0.0001 percent

Figure 6.3 The volumes of the cubes show the relative amounts of free water in storage on the earth. Nearly 98 percent of the water is stored in the oceans, which contain an estimated volume of 1.3×10^{18} cubic meters (46×10^{18} cubic feet) of water. Glaciers and ice caps contain the largest store of fresh water, but the turnover is too slow to be important except on the geologic time scale of ice ages. Most of the readily available fresh water is stored in porous rock beds as groundwater. The amount of water stored in the atmosphere is relatively small, but because it is actively transported it plays a key role in the hydrologic cycle. (After Nace, 1969)

the dissolved salts are left behind, and the water that condenses from the atmosphere is fresh. Man relies almost entirely on the fresh water produced by the natural desalination process, because the operation of desalination plants requires costly amounts of fuel.

About three-fourths of all the world's fresh water is stored as ice and snow, primarily in the ice caps covering Antarctica and Greenland. The ice caps receive an annual snowfall equivalent to about 10 centimeters (4 inches) of liquid water, but some of the water returns to the ocean when coastal ice melts or when icebergs break off. For the most part, ice caps represent water in long-term storage and are not important in the year-to-year utilization of water.

Evaporation of water into the atmosphere is an essential part of the hydrologic cycle because the atmosphere transports water from one part of the earth to another. The amount of water present in the atmosphere at any given time is not great: if all the water in the atmosphere were to fall uniformly over the earth as rain, it would form a layer only a few centimeters deep. In the course of a year, however, the atmosphere delivers and recycles enough precipitation to cover the earth with a layer of water 86 centimeters (34 inches) deep.

The evaporation and precipitation balance over the ocean differs from the balance over the continents. Over the ocean, the average quantity of water evaporated exceeds the quantity precipitated. Over the continents, precipitation exceeds evaporation. There is a net transport of water through the atmosphere from the oceans to the land by means of air masses hundreds or thousands of kilometers in diameter, as discussed in Chapter 5. The net transport of water from the ocean to the land is balanced on the average by the flow of water from the land through streams back to the ocean. Before its return to the ocean or to the atmosphere, water can be involved in a number of transport and storage processes on the land. The study of water over, on, and under the land surface is called *hydrology*.

Principles of Hydrology

What happens to water that falls on the land? You are probably aware of some of the principles of hydrology from your own experience and observation. Consider, for example, the various routes rain can take as it falls to the earth.

Interception, Throughfall, and Stemflow

Not all of the rain that falls reaches the soil. In cities, some rain falls on buildings and on paved areas. If the rain is heavy, sidewalks and parking lots become flooded and part of the rain eventually runs off to the surrounding earth. The water that remains on pavements and buildings evaporates and returns to the atmosphere.

Vegetation cover also prevents some rain from reaching the soil. Except in arid regions, plant growth covers most of the ground during the growing season. Leaves are likely to *intercept* raindrops falling on dense plant cover. At the beginning of a rainstorm, leaves

Figure 6.4 Some of the precipitation that is intercepted by vegetation and held in temporary storage by leaves returns to the atmosphere by evaporation and does not reach the soil.

catch and hold a considerable amount of water, and not much rain reaches the soil. For the first few minutes of a rain, the ground under a tree remains fairly dry, but if the rainfall is prolonged, the capacity of the leaves to retain water will be exceeded and water will start to drip from the leaves to the soil as *throughfall.* As the rain becomes heavier and more prolonged, the tree becomes less effective as a shelter. A small amount of water also reaches the soil as *stemflow* by trickling along branches and down the trunk or stem to the ground.

After rain stops falling, water may continue to drip from leaves onto the ground for a time. Nevertheless, some water always remains on the leaves and evaporates back into the atmosphere. The amount of water that actually reaches the ground is less than the amount that falls as precipitation. Because of interception, the amount of water reaching the soil depends not only on the total amount of water that falls but also on its pattern and timing. Compare the effects of interception on a continual, steady rainfall totaling 1 centimeter with that on a rainfall of ten brief showers falling at the same intensity and totaling 0.1 centimeter each. In steady rain, the leaves' capacity to retain water will soon be exceeded and throughfall will carry water to the soil. But between brief showers, the leaves will have a chance to dry, and a significant portion of the rain from each shower will be intercepted and will subsequently return to the atmosphere.

Infiltration and Soil Moisture

When you water your garden after a dry spell, the water readily soaks into the soil. Puddles begin to collect on the surface only after prolonged heavy watering. A few days after watering, the surface of the soil may have dried out, but you will find moist soil if you dig down a few centimeters, which demonstrates that soil can store water.

The *soil moisture* stored in the upper few tens of centimeters of soil, in the zone penetrated by plant roots, is essential to plant life because it is the only water that plants can utilize easily. Water enters a plant through its root system, and it is therefore important in agriculture to maintain sufficient soil moisture. When soil moisture becomes too depleted, the cells of the plants lose fluid and the plants wilt, which decreases crop yields significantly. Too much moisture is also harmful to plants. Because roots require a source of oxygen and space to eliminate carbon dioxide waste from root metabolism, the soil must contain air as well as water.

Soil consists of fine grains of mineral and organic matter, which vary in size from one type of soil to another. For instance, clay soils have much smaller grains than sandy soils. Soil moisture is stored in the spaces, or *pores*, between the grains. If the pores are interconnected, the soil is *permeable*. Water can move downward through permeable soil because of gravity or upward by capillary action.

Water is held in the pores near the surfaces of soil particles by forces of molecular attraction. A thin film of water only a few

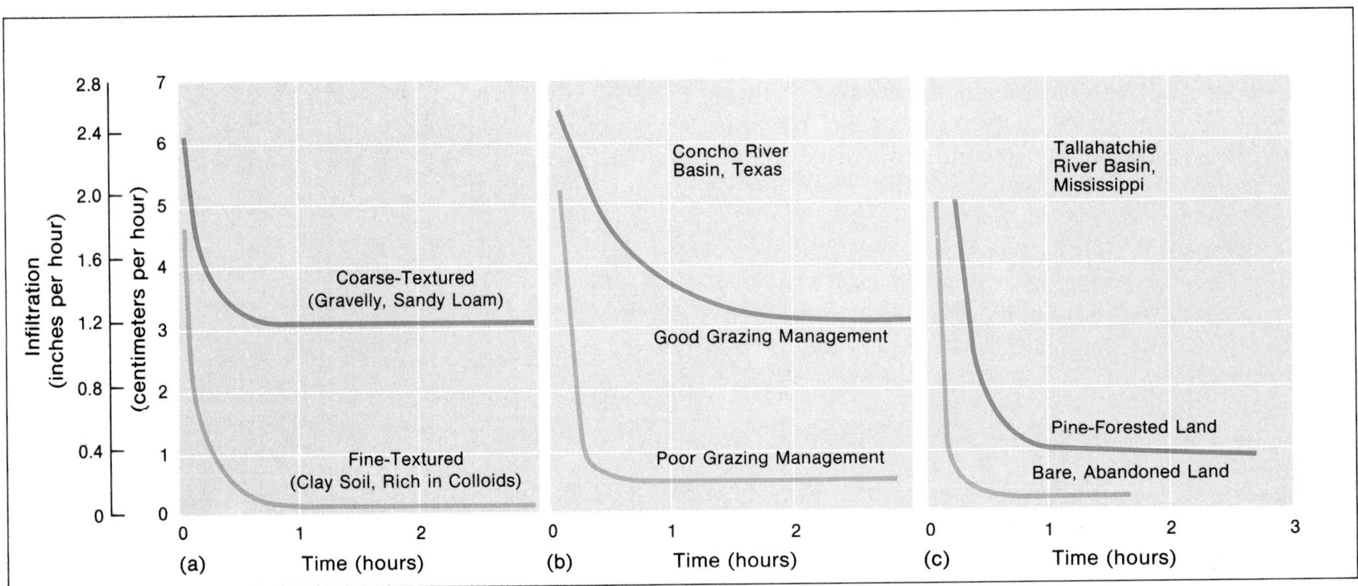
Infiltration (inches per hour)
(centimeters per hour)
2.8
2.4
2.0
1.6
1.2
0.8
0.4
0
7
6
5
4
3
2
1
0
Coarse-Textured (Gravelly, Sandy Loam)
Fine-Textured (Clay Soil, Rich in Colloids)
Concho River Basin, Texas
Good Grazing Management
Poor Grazing Management
Tallahatchie River Basin, Mississippi
Pine-Forested Land
Bare, Abandoned Land
0
1
2
3
(a)
(b)
(c)
Time (hours)

Figure 6.5 The infiltration rate of water into a soil depends on such factors as the texture, porosity, permeability, and moisture content of the soil and on the condition of the surface layer. The graphs show the infiltration rates of various soils measured from the time at which water is added to their surfaces. The infiltration rate falls sharply at first in all cases as the top layer of soil becomes well moistened; then a constant rate of infiltration is attained.

(a) Water infiltrates a coarse, permeable soil more easily than it does a dense clay soil, in which the water passages are small.

(b) The surface of well-managed grazing land retains an open texture and has a high infiltration rate. The infiltration rate is lower on poorly managed land because overgrazing exposes bare soil, which allows the surface to become compacted by raindrops and by animal hooves.

(c) Soil has a low infiltration rate when there is no vegetation cover to protect the surface of the soil from compaction by the impact of raindrops. (After Foster, 1949) molecules thick is held by molecular attraction directly to the surfaces of the soil grains. This water is too tightly bound to the grains to be absorbed by plants. Somewhat thicker layers of water are held by *surface tension*, a molecular attraction that water molecules in a volume of liquid exert on water molecules at the surface of the volume. Surface tension forces are too weak to support large, heavy volumes of liquid, so that soil moisture can be held in soil only if the pores are so small that the water films are thin. A bed of coarse gravel has such large pore spaces that water passes through almost immediately and little water is stored.

Water enters, or *infiltrates*, soil from the surface of the ground. The maximum rate at which a given soil can absorb water is called its *infiltration rate*. If water arrives at the surface at a rate less than the infiltration rate, all of the water will soak into the soil. Once the upper layers of soil are saturated with moisture, additional water can enter only as the water already in the soil slowly seeps down to lower levels. A dry, porous soil can absorb water at an initial rate of about 10 centimeters per hour (4 inches per hour), but the infiltration rate rapidly decreases as the soil becomes wetter. Soil is said to be at *field capacity* when it contains the maximum possible amount of moisture down to the depth of the roots of the vegetation growing in it.

Infiltration rate also depends on the state of the surface of the soil. If the pores are open, infiltration proceeds easily. But if the soil grains have been compacted by vehicles or by animal hooves, the soil may not be able to absorb moisture at all. Rain falls with enough force on bare soil to decrease the infiltration rate noticeably. A layer of vegetation absorbs the impact of the drops and helps the soil maintain a high infiltration rate. If animals such as sheep have grazed an area too heavily, the land will have a sparse vegetation cover. Restoration of overgrazed land to agricultural productivity is difficult because the compacted soil does not absorb moisture readily.

Soil moisture returns to the atmosphere by evaporating directly from the soil or by transpiring through plant leaves, as you read in Chapter 4. Several factors work to reduce evaporation from the soil. Once the top few millimeters of soil have dried out, evaporation from the soil decreases because water vapor diffuses upward through soil very slowly. Furthermore, plant cover reduces both the temperature of the soil and the wind speed at the surface of the soil, reducing evaporation from the soil. Transpiration from plants produces a blanket of humid air near the ground, which further reduces the rate of evaporation from the soil.

Groundwater and Aquifers

The top kilometer of the earth's land crust contains a vast and accessible store of fresh water equal to 15 percent of the amount of fresh water stored in the ice caps. The water is stored in underground deposits of gravel or of porous rock, such as sandstone. These deposits, called *aquifers*, supply the water for wells and springs (see Figure 6.6). The water in the fully saturated portion of an aquifer—*groundwater*—constitutes some 30 times more fresh water than the amount of water found in all the rivers, lakes, and streams of the world. Groundwater is found within 100 meters (330 feet) of ground level in many locations—even in arid regions such as Israel, where there is little surface water. The thin layer of soil moisture is not considered to be groundwater.

The quality of groundwater is often excellent. The porous rock in the aquifer filters the water and removes suspended particles and bacteria, although contaminants seep into aquifers occasionally. Natural springs, frequently noted for the purity of their water, are fed by groundwater. Many aquifers are extensive enough to serve as permanent water supplies for large cities, so some cities rely on comparatively pure groundwater rather than use artificially purified water from a badly polluted river. The temperature of groundwater taken from depths of 10 to 20 meters (30 to 60 feet) in midlatitude locations is usually 1 or 2°C higher than the average annual temperature at the pumping station, so groundwater can supply cool water in the summer. The relatively constant temperature of groundwater makes it suitable for manufacturing processes and for industrial cooling.

Several different kinds of subsurface materials are sufficiently porous and permeable to make suitable aquifers. The majority of aquifers now being drawn from in the United States are underground beds of sand and gravel. Some of these beds were deposited long ago by receding glaciers; others are the products of ancient erosion. Sand and gravel aquifers typically are 50 meters (160 feet) thick and cover an area as large as several thousand square kilometers. The cross section of the aquifer in Figure 6.6 shows sand and gravel beds. Clay deposits, although porous, have such fine pores that water can be extracted from them only with difficulty. Also, removing water from clay may cause the clay to compress and become impermeable. Sandstone and limestone form deep, extensive aquifers. A limestone aquifer 300 meters (1,000 feet) thick underlies a large area of the southeastern United States, and a lime-

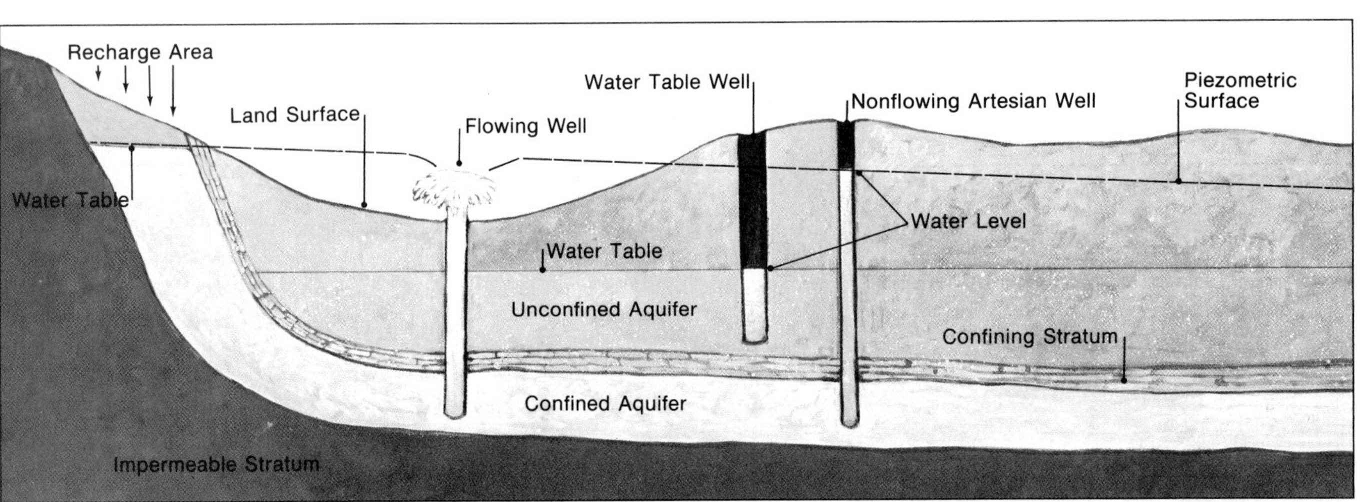
Recharge Area
Land Surface
Flowing Well
Water Table Well
Nonflowing Artesian Well
Piezometric Surface
Water Table
Water Level
Water Table
Unconfined Aquifer
Confining Stratum
Confined Aquifer
Impermeable Stratum

Figure 6.6 This diagram shows the principal features of aquifers in schematic form. If the rock above an aquifer is permeable enough to allow the vertical movement of water, the aquifer is said to be unconfined. The water table, or the water level in an unconfined aquifer, is the level to which water will rise in a well sunk into the aquifer. If the rock above an aquifer is impermeable, the aquifer is confined and must be replenished from a recharge area, which is permeable to water from above. When a well is sunk into a confined aquifer, the level to which the water rises is called the *piezometric surface*. The piezometric surface can be a considerable height above a confined aquifer, particularly in the lower portion of a sloping aquifer, where the water pressure is high. (After Kazmann, 1972) stone aquifer in England is being tapped to supply water for London. Limestone is fissured with long channels and cracks that allow water to flow through it easily.

If an aquifer rests on a layer of relatively impermeable rock or compacted clay, water is prevented from draining out of the aquifer. If an aquifer is covered with rock permeable to water, upward movement of water in the aquifer is not confined, and the water in the saturated portion of the aquifer reaches the *water table*. Because the water table represents the level to which water naturally rises, the depth of the water table below the surface of the ground is the minimum depth a well must be sunk in order to draw water from the aquifer.

The water table is not necessarily horizontal but tends to follow the slope of the land. Water in an aquifer need not reach a horizontal level, as free water in a pond does, because the water in a sloping aquifer is not in stationary equilibrium but slowly flows through the porous rock from higher to lower elevations. Unlike the swift flow of a mountain stream down a steep slope, the flow in a sloping aquifer is gradual, because the fine pores of the aquifer bed offer resistance to the water.

Almost all of the water stored in aquifers comes originally from precipitation. Rain water that infiltrates the soil seeps, or *percolates*, slowly downward through the ground to the underlying groundwater. Similarly, water may seep through lake bottoms or stream beds into aquifers. The surface region over which water for an aquifer collects is called its *recharge area*. Because water travels horizontally through aquifers, the principal recharge area for an aquifer may be hundreds of kilometers away from the point at which water is being withdrawn. Local precipitation at the point of withdrawal does not necessarily govern the recharge rate of an aquifer.

Sometimes part of the upper surface of an aquifer is bounded, or *confined*, for some distance by a layer of impermeable rock. The concept of the water table does not apply to a confined aquifer, because the impermeable upper layers may slope downward through the ground. The difference in height between the higher and lower parts of a confined aquifer may result in a considerable pressure differential. If a well is sunk into a low section of a confined aquifer, the water in the well rises some distance above the local height of the aquifer bed.

The speed at which water flows through aquifers varies from a few meters a year to several kilometers a day. Water is stored for a long time—many hundreds of years—in large aquifers that have a low rate of flow. Careful management of water entails ensuring that the amount of water withdrawn from the aquifer not exceed the net flow into it. Many aquifers are pumped at rates greater than the rate of recharge, which depletes the amount of stored water and lowers the water table.

Approximately 50 percent of all the groundwater extracted in the United States is used for irrigation in Texas, Arizona, and California. Excessive withdrawal lowered the water table in northwest

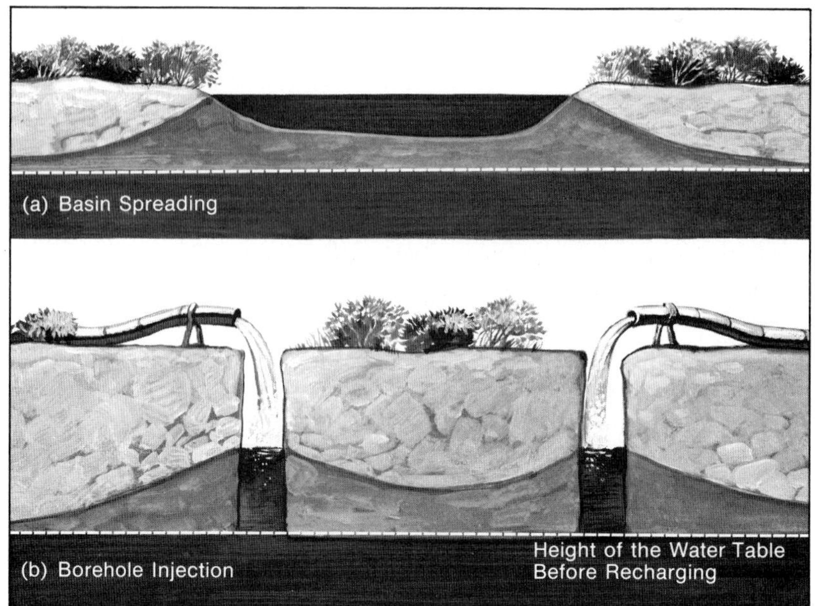
(a) Basin Spreading
(b) Borehole Injection
Height of the Water Table
Before Recharging

Figure 6.7 This diagram illustrates two methods that have been used to recharge aquifers and raise the level of the water table in regions where supplies of groundwater have been depleted by excessive withdrawal.

(a) Water pumped into shallow surface depressions in the natural recharge area of an aquifer seeps through the permeable rock into the aquifer.

(b) Water pumped into boreholes sunk into an aquifer seeps into the permeable rock to recharge the aquifer.

Texas so seriously that artificial recharge was tried as a remedial measure to restore some water to the aquifer. During the rainy season in Texas, water collects in shallow depressions. Some of this water, which would ordinarily evaporate during the dry season, was pumped to the recharge area of the aquifer to help replenish it. Artificial recharge has also been used in coastal regions when the extensive mining of water from an aquifer has allowed sea water to seep in and contaminate wells.

Runoff and Streamflow

Rainfall that reaches the surface of the ground initially infiltrates the soil, but as soon as the infiltration rate of the soil drops below the rate at which water reaches the surface, some of the excess water collects on the surface. Surface irregularities and vegetation, such as the dense underbrush of a forest, store some of the water for later infiltration. Once the capacity of the surface to hold water is exceeded, water begins to flow across the ground in trickles or sheets under the influence of gravity. Surface runoff increases from small rivulets at the beginning of a rainstorm to a steady torrent if the rain lasts long enough to saturate the storage mechanisms in and on the soil. Toward the end of a prolonged rain, nearly all of the water that reaches the ground joins the surface runoff. A rain of long duration produces more surface runoff than several brief showers of the same intensity separated by dry periods.

Moist regions are laced with a network of small streams fed by surface runoff. The small streams carry water into a few large streams, which in turn supply major rivers. In this way, the water carried by a river is collected over a large drainage basin. The water that rivers carry from the land to the sea balances the net flow of water through the atmosphere from the oceans to the land.

The time lag between a rainstorm and the arrival of water from surface runoff at a major river varies with the size of the drainage area. In general, the time lag is only a few days. In the middle latitudes, a considerable amount of precipitation is stored on the ground as ice and snow, and during spring thaw, large amounts of liquid water are released in a short time. Most of the melted snow contributes to surface runoff because the frozen soil has little capacity for absorbing moisture. For this reason, flood danger is at a maximum during rapid spring thaws and when heavy rains fall on a river's drainage basin. The amount of water a river must carry increases rapidly soon after the heavy surface runoff drains into the streams.

Rivers do not depend only on surface runoff for their water. Major rivers continue to carry large amounts of water even after a long, dry spell has completely eliminated surface runoff for weeks, because runoff from groundwater supplies the *base flow* that maintains rivers through periods of drought. In contrast to the rapid variation of surface runoff in response to precipitation, groundwater runoff provides a much steadier flow of water to streams. And groundwater flow changes only slowly through the year, because seepage through permeable rock is a slow process. The water a river carries may be visualized in two parts: one part, from groundwater runoff, is comparatively steady from day to day; the other part, from surface runoff, can show marked changes from day to day because of rainstorms and snow melting.

The Local Water Budget: An Accounting Scheme for Water

Water is not always available in the desired amounts when and where we need it. The principles of hydrology discussed earlier in this chapter show how soil moisture and runoff depend on the amount and timing of precipitation. But precipitation is a specialized and localized event, subject to marked variations from month to month and from place to place. The variability of precipitation shows up in the variability of water available for soil moisture and for runoff to streams.

The water budget is the local version of the hydrologic cycle; the hydrologic cycle represents the sum of the local water budgets for all regions of the earth. The local water budget has practical applications for evaluating water needs. Using it makes it possible, for example, to estimate the amount of soil moisture and the amount of runoff throughout any year for which temperature and precipitation records are available. Such studies are valuable guides for forecasting the probable irrigation requirements in a given location and for determining streamflow where no direct measurements on the stream exist. The local water budget has also been used to gauge the suitability of unpaved terrain for vehicle passage.

The local water budget takes into account four principal components of water distribution: precipitation, soil moisture, evapotranspiration, and runoff. When appropriate, snow accumulation and melting may also be included. Of all these components, evapotranspiration is the most difficult to treat properly, because of its com-

Figure 6.8 (a) This schematic diagram shows the effect of an upstream rainstorm on the amount of water carried by a stream, measured from the time of the storm. From A to B, water from the rain has not reached the downstream region, and the discharge, or rate of flow, is due to the *base flow* supplied by groundwater. From B to C, the discharge rises rapidly as direct surface runoff from the upstream drainage basin reaches the downstream region. From C to D, the discharge falls slowly as the last of the surface runoff, including the runoff retarded by the vegetation cover, makes its contribution. From D to E, the discharge is due primarily to groundwater supplies, which have been newly recharged by the rain. The discharge of the stream gradually decreases as the aquifers become depleted. (After Ward, 1967)

(b) This hydrograph shows the discharge of the Mississippi River during the summer of 1935. The peaks are due to rainstorms over the river's drainage basin; note that the discharge peaks tend to exhibit the rapid rise and slow decrease shown schematically in (a). The base flow steadily decreases through the comparatively dry summer months, but the groundwater supplies become replenished by precipitation during the winter and spring. (After Barnes, 1939)

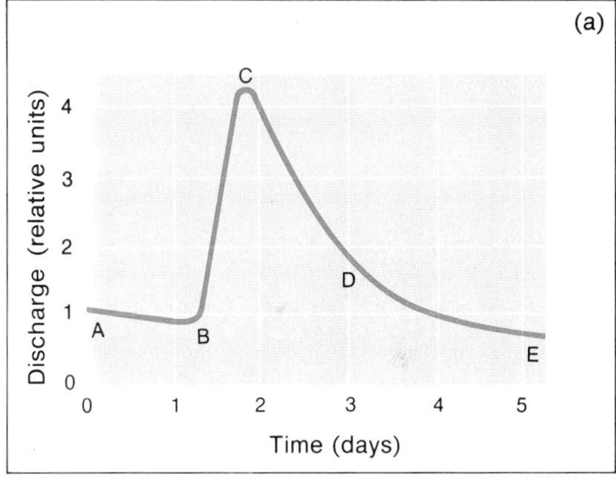
(a)
Discharge (relative units)
0
1
2
3
4
A
B
C
D
E
0
1
2
3
4
5
Time (days)

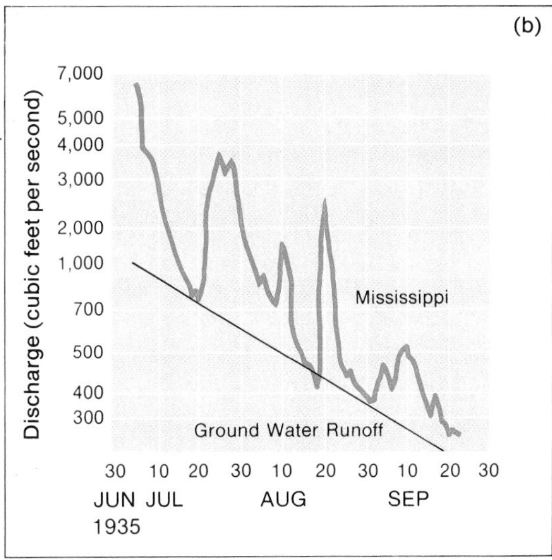
(b)
Discharge (cubic feet per second)
7,000
5,000
4,000
3,000
2,000
1,000
700
500
400
300
Mississippi
Ground Water Runoff
30 10 20 30 10 20 30 10 20 30
JUN JUL AUG SEP
1935

Case Study: The California Water Plan

(Below left) This simplified map shows the existing and planned channels of the California State Water Project. Now under construction, the project will extend nearly the length of California, bringing water from the north to the more densely populated southern portion.

(Below right) In this photograph, a portion of the California Aqueduct is shown winding through the San Joaquin Valley.

Shasta Lake
Lake Oroville
Sacramento River
Feather River
Lake Tahoe
Sacramento
North Bay Aqueduct
South Bay Aqueduct
San Francisco
California Aqueduct
San Luis Reservoir
Los Angeles Aqueduct
Coastal Branch
Castaic Lake
Silverwood Lake
Lake Perris
Los Angeles
Salton Sea
Colorado River
San Diego
Colorado River Aqueduct

The management of fresh water is often a controversial subject. According to one view, there is plenty of water on the earth; it simply needs to be redistributed. Others argue that man should not plan his activities where there is a shortage of water; he should live where water is naturally available. However, man has been moving water about for thousands of years to turn barren land into fertile farmland.

A case to consider is golden Southern California, which has become greener and more productive by the importation of water from the north. In the early 1900s, it became clear that the growth of Los Angeles—then a small coastal town—would be stimulated by irrigating the surrounding land. For a source of water, Los Angeles turned to the Owens River Valley, 400 kilometers (250 miles) to the north at the eastern foot of Mount Whitney. The farmers of the Owens Valley had hoped for the construction of an irrigation system along their own valley, but the rights to the water were given to Los Angeles. One reason why Los Angeles received the water was that the climate in the Owens Valley could not support the highly profitable citrus groves that grow well around Los Angeles. Southern California has the stable climate that is conducive to productive agriculture if adequate water is available.

The Los Angeles aqueduct from the Owens Valley to Los Angeles was completed in 1913. Soon more water was needed, however, and aqueducts from the Colorado River were built to supply Los Angeles and San Diego. Now more than 100 cities rely on water from the Colorado—a supply

that has recently been supplemented by water provided by a state project. Rather than continuing to rely on local and regional plans, the state of California has prepared a water resources development plan for the entire state.

The California Water Plan is founded on the recognition that most of the people and most of the irrigated land are found in the southern part of the state, whereas most of the water is in the north. The plan details the construction of the aqueducts, dams, reservoirs, and power stations needed to transport water from the north to the south. The principal source of water for the present stage of the plan is the Sacramento delta 100 kilometers (60 miles) east of San Francisco. The delta is formed by a confluence of streams from the west slope of the Sierra Nevada as they drain toward the San Francisco Bay through the only opening in the coastal range.

Recently a major project in the plan, Perris Dam, was completed. The dam is located southeast of Los Angeles and is one part of the effort to move an immense amount of water from one area of the state to another. When full, Perris Lake traps water that flowed from more than 1,000 kilometers (600 miles) away, where the project begins at massive Oroville Dam on the Feather River north of Sacramento. Never before has one of man's projects sent so much water flowing. Some sections of the concrete-lined aqueduct are as wide as a major river. At one point in its journey southward, the water is lifted nearly 600 meters (2,000 feet) through the Tehachapi Mountains—a record lift for so much water. Five huge dam-fed power plants provide the energy supply to lift the water in stages from sea level at the delta to its highest point, the Tehachapi crossing.

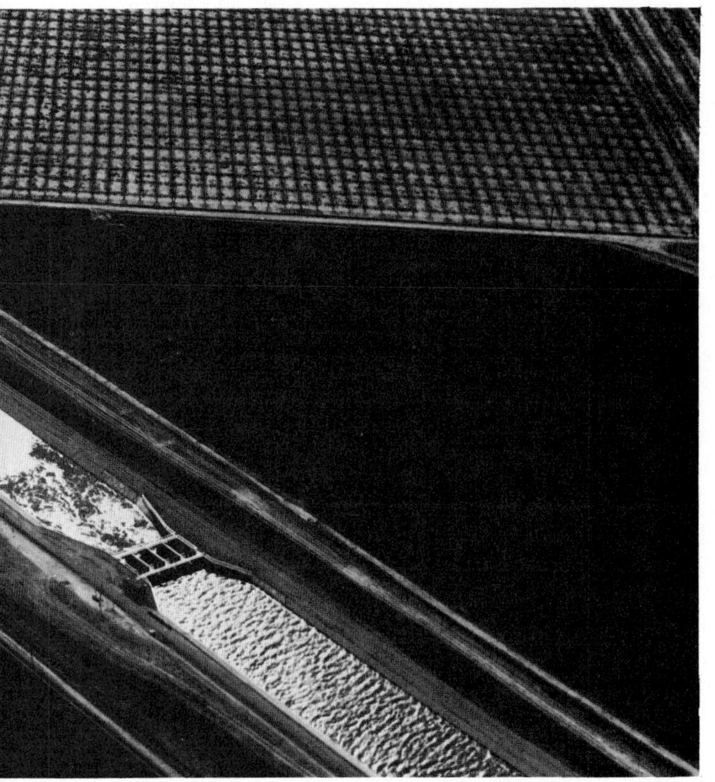

In addition to supplying water for southern regions, the aqueducts provide irrigation for the dusty, windblown San Joaquin Valley. Eventually new lands made available for agriculture will amount to 1 million acres.

Conceived as the Feather River Project, the water development program has spanned about 30 years. It has been one of the most controversial programs ever undertaken in California. Adoption of the plan instigated an angry sectional feud between the moist "north" and the dry "south." Southern California was accused of trying to steal northern water, but the population concentration in the southern part of the state carried the vote and the issue was approved. Environmentalists still claim that the project has scarred the countryside irreparably, upsetting natural balances of streams, estuaries, vegetation, and wildlife. It is also argued that providing water will promote population growth, which will lead to further urbanization and land development. For instance, perhaps the availability of water in Los Angeles will lead to a larger population, more congestion, and more smog.

In such a controversy, sound arguments come from both sides. Proponents of the water plan say that a rapid population increase is *not* expected. They also contend that a canal from the Sacramento delta will provide water to flush out the estuarine area near the San Francisco Bay.

Many of the questions and controversies over the water plan center on whether the water is really needed. Ninety percent of the water used in California is for irrigation; there is evidently abundant water for domestic and industrial use. If the fraction of water used for irrigation were cut to 80 percent, the amount available for other uses would double from 10 to 20 percent. But if the irrigation supply were reduced, what effect would that have on agriculture and food supplies for the entire country? California is now one of the most productive agricultural areas of the country.

The allocation of water resources is a matter of economics as well as a matter of technology. Water demand is, in reality, often a demand for water at a sufficiently low cost to make irrigation profitable. The emphasis on specialty crops such as fruit and nuts in California agriculture reflects the need for profitable crops to offset the cost of irrigation water. The California water project has been criticized for using public funds to increase the value of privately held farmland. Furthermore, technological advances in desalination plants may give a new dimension—unforeseen when the water plan was devised—to the problem of water resources.

Although more aqueducts, canals, and pumping plants are planned for the 1970s and 1980s, it is uncertain when and if they will be completed. The pace of the water plan is influenced by the projected water needs, the development of new technology, and the ecologic impacts.

plicated dependence on biologic and meteorologic factors. In the 1940s, the American climatologist C. Warren Thornthwaite developed a way of treating evapotranspiration. To overcome the difficulties in estimating the actual rate of evapotranspiration, Thornthwaite introduced the concept of *potential evapotranspiration* and developed a formula for calculating it (see Appendix III). Potential evapotranspiration, a key element in the water budget, is worth examining in detail.

Potential and Actual Evapotranspiration

Potential evapotranspiration is the maximum rate at which water is lost to the atmosphere from a dense vegetation cover that has been supplied with all the moisture it can use. If the soil moisture is partly exhausted, the actual rate of evapotranspiration from a field may fall below the potential rate because plants may be unable to absorb and transpire the full amount of water they could use.

Evapotranspiration is expressed as the equivalent depth of liquid water converted to vapor in a given time. You probably think of precipitation in terms of inches of water; the amount of water involved in evapotranspiration is expressed in the same units. A typi-

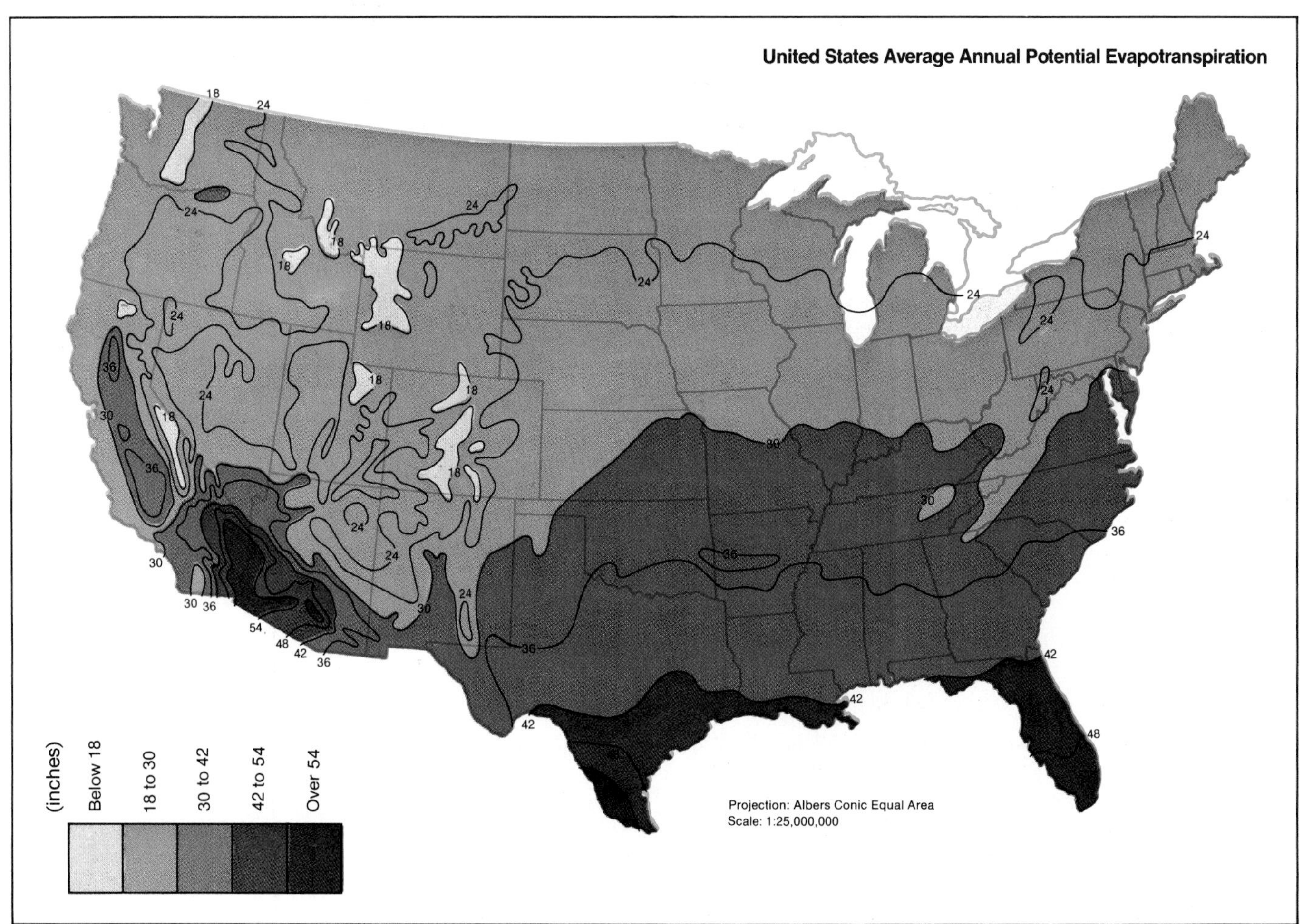
United States Average Annual Potential Evapotranspiration
(inches)
Below 18
18 to 30
30 to 42
42 to 54
Over 54
Projection: Albers Conic Equal Area
Scale: 1:25,000,000

Figure 6.9 This map shows the average annual potential evapotranspiration for the conterminous United States, calculated from Thornthwaite's formula. The Southwest, Texas, and Florida have high values of potential evapotranspiration because of the hot, sunny weather there. These regions are potentially highly productive for agriculture if proper soils and sufficient moisture are available. In northern Maine, however, the potential evapotranspiration is low because of the short growing season and the generally cool and cloudy weather. Most of the important agricultural section through the Midwest has moderate values of potential evapotranspiration that correspond well with available supplies of moisture.

cal value for potential evapotranspiration over a 30-day period in a hot, sunny region is about 15 centimeters (6 inches).

Of all the factors that determine evapotranspiration, the most important is the amount of solar energy the plants absorb. Both evaporation and transpiration of water require the expenditure of 590 calories per gram of water. Evapotranspiration therefore depends on the duration of daylight, the position of the sun relative to the land surface, the degree of cloud cover, and the albedo of the plant cover, because these are the principal factors that determine the amount of energy the plants absorb.

Potential evapotranspiration from a field is essentially independent of the type of plant cover. Imagine looking down on a forest or a field during the middle of the growing season. The plants present a uniform, dense cover of overlapping green leaves. The albedo of almost all green plants is about 15 percent. Therefore, an acre of forest and an acre of cotton field, both densely covered, absorb about the same amount of energy under the same external conditions.

It is significant that potential evapotranspiration depends only on external conditions, such as solar energy input and weather, and not on the detailed characteristics of the plants. The *potential* evapotranspiration for a given region at a given time can in principle be calculated from astronomical and meteorologic data alone. By contrast, *actual* evapotranspiration, or the amount of evapotranspiration that actually occurs, varies with the amount of soil moisture and with the condition of the vegetation cover as well as with external conditions. The average annual potential evapotranspiration for the United States, calculated from Thornthwaite's formula, is shown in Figure 6.9.

Measuring Evapotranspiration and Potential Evapotranspiration

Measurement is fundamental to science. It provides the quantitative checks required to test the validity of a theory and also provides the data required to apply theory to specific situations. In evaluating data, it is important to know how the data were obtained.

Potential evapotranspiration at a particular location can be measured in a device called an *evapotranspirometer*, which consists essentially of several open tanks sunk flush with the ground. The tanks, perhaps 2 feet in diameter and 3 feet deep, are filled with soil and planted with a dense cover of grass. If the grass is well watered, the soil moisture remains sufficient and constant so that the difference between the amount of water added to the tank and the amount of water draining from the bottom of the tank represents the amount of water used by the plants, or the potential evapotranspiration in this case. Actual evapotranspiration is measured in soil tanks called *lysimeters*. In an attempt to simulate the natural environment, some lysimeters are built large enough to carry a stand of trees. Setting up and operating an evapotranspirometer is a major undertaking that requires the time and attention of skilled personnel. Measurements of potential evapotranspiration have therefore been restricted to a few locations over limited periods of time, and for most practical applications, a calculated estimate of potential evapotranspiration, such as Thornthwaite's, must suffice.

Thornthwaite's calculation of the monthly potential evapotranspiration at a particular location depends on three factors: (1) the average air temperature of the given month, (2) the length of day for the given month and latitude, and (3) the average air temperature of each of the 12 months of the year. The average monthly air temperature and the day length represent solar energy input; the same conditions that cause temperature to be high also increase potential evapotranspiration. The yearly factor represents the effects of general weather conditions.

Although other formulas have been developed to estimate potential evapotranspiration, Thornthwaite's formula is the most widely used. It has the advantage of being based on standard meteorologic data and it requires no additional measurements. More important, however, many studies have shown that Thornthwaite's formula gives satisfactory values for potential evapotranspiration in almost all regions of the world. Knowing the potential evapotranspiration of a region is essential to calculating its water budget.

Figure 6.10 This schematic diagram illustrates the principal components of the local water budget. The input to the system of soil and vegetation is the amount of moisture supplied by precipitation. A large portion of the input returns to the atmosphere by evaporation and plant transpiration. Some of the input moisture is stored in the soil. However, the storage capacity, or the field capacity, of the soil is limited, and under some conditions a moisture surplus becomes available to supply surface runoff and groundwater recharge.

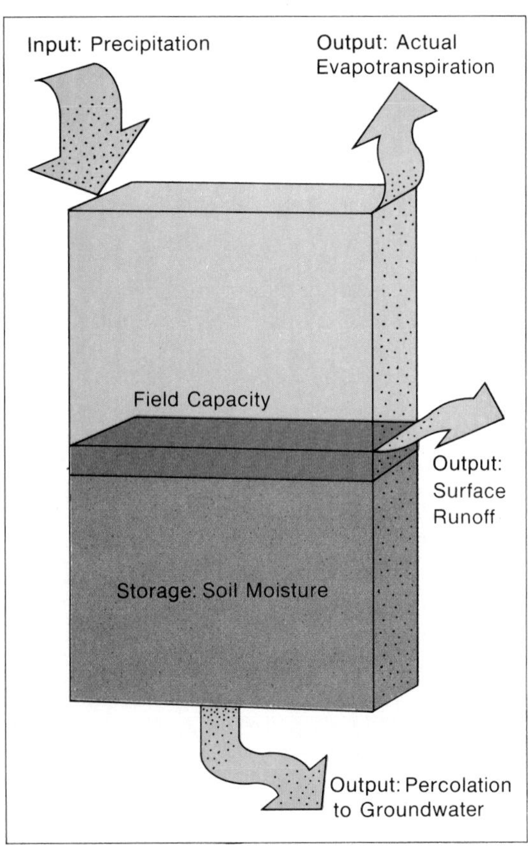
Input: Precipitation
Output: Actual Evapotranspiration
Field Capacity
Output: Surface Runoff
Storage: Soil Moisture
Output: Percolation to Groundwater

The Water Balance Equation

The local water budget is concerned with the various components of water use. The *water balance equation* makes it possible to set up simple relationships between the amounts of water in such components as precipitation and soil moisture. The equation is based on the idea that all the water involved in a process at a particular place can be accounted for. In a typical hydrologic process, a certain initial amount of precipitation, or *income*, is available. Part of the water income may enter storage. The remainder, or *outgo*, represents the amount of water used by plants or the amount that becomes surplus and can be used for other purposes. The water balance equation may be written:

Amount of water income = amount of water outgo + amount of water entering storage

If water leaves storage instead of entering it, the equation becomes:

Amount of water income = amount of water outgo - amount of water leaving storage

A Local Water Budget: Baton Rouge

To illustrate the calculation of a local water budget, consider the example of Baton Rouge, Louisiana, in 1962. In an average year, Baton Rouge receives more than 50 inches of rainfall, a comparatively large amount. Nevertheless, the water budget analysis shows that for several months in 1962 Baton Rouge was "dry" and crops needed watering.

In this example, income, outgo, and storage are considered month by month. Weekly or even daily periods could also be used. The first step in constructing a monthly local water budget is to use meteorologic records to find the monthly precipitation at Baton

Rouge, which is listed in the first row of the chart in Figure 6.11. The values are given in inches, because weather data in the United States are recorded in English units. The total precipitation for the year was 52.9 inches, all of which fell as rain because of the warm winter temperatures.

Involved computations are required to calculate monthly potential evapotranspiration using weather records and Thornthwaite's formula. The results have been worked out and are listed in the second data row of the table. As you can see, potential evapotranspiration is much smaller in the winter months than it is in the summer.

Precipitation is the source of water income. In months when rainfall exceeds potential evapotranspiration, the excess of moisture allows plants to transpire at the maximum rate. The monthly potential evapotranspiration is subtracted from monthly precipitation and is entered in the third data row. The six negative values show that for 6 months of 1962, precipitation was less than potential evapotranspiration.

Aside from snowfall, soil moisture is the principal storage mechanism considered in the local water budget. Because 1961 was a wet year, 1962 began with the soil moisture in Baton Rouge at its full capacity of 6.0 inches. The simple water budget considered

Figure 6.11 This is the local monthly water budget for Baton Rouge, Louisiana, in 1962. See the text for a discussion of how the budget is constructed. (After Muller, 1970)

	JAN	FEB	MAR	APR	MAY	JUN	JUL	AUG	SEP	OCT	NOV	DEC	Total
1. Precipitation (P)	6.4	0.7	3.3	9.7	1.6	11.4	2.0	4.5	4.3	5.2	0.9	2.9	52.9
2. Potential Evapotranspiration (PE)	0.4	1.7	1.3	2.6	5.3	6.1	7.4	6.8	5.2	3.4	1.1	0.6	41.9
3. Precipitation-Potential Evapotranspiration	6.0	−1.0	2.0	7.1	−3.7	5.3	−5.4	−2.3	−0.9	1.8	−0.2	2.3	
4. Change in Stored Soil Moisture, (ΔST)	0.0	−1.0	1.0	0.0	−3.7	3.7	−5.4	−0.6	0.0	1.8	−0.2	2.3	
5. Total Stored Soil Moisture (ST)	6.0	5.0	6.0	6.0	2.3	6.0	0.6	0.0	0.0	1.8	1.6	3.9	
6. Actual Evapotranspiration (AE)	0.4	1.7	1.3	2.6	5.3	6.1	7.4	5.1	4.3	3.4	1.1	0.6	39.3
7. Deficit (D)								1.7	0.9				2.6
8. Surplus (S)	6.0		1.0	7.1		1.6							15.7

Figure 6.12 The graphs show the local monthly water budgets for Baton Rouge, Louisiana, in 1961, 1962, and 1963. Information on each water budget graph is conveyed by three solid lines that denote the measured precipitation, the assumed potential evapotranspiration, and the calculated actual evapotranspiration. Solid colors denote times when precipitation and soil moisture withdrawals do not meet potential evapotranspiration, when soil moisture is being utilized or recharged, and when surplus precipitation is available for runoff or groundwater recharge. The calculation of soil moisture depletion in this model assumes that all stored soil moisture is equally available to plants. During times of moisture surplus, such as in most of 1961, the moisture available from precipitation exceeds the moisture that can be utilized by plants and that can be stored by the soil. Actual evapotranspiration is equal to potential evapotranspiration when a surplus of moisture is available. In times of deficit, such as in August and September, 1962, all of the available precipitation and soil moisture is used for evapotranspiration, and the actual evapotranspiration is less than the potential evapotranspiration.

The average annual precipitation at Baton Rouge is 56 inches. In 1961, a wet year, 74 inches of precipitation were received and surpluses were generated during 8 months. In 1962, an average year, the precipitation totaled 53 inches; 2 months suffered deficits during a summer dry spell. In 1963, a dry year, 42 inches of precipitation fell. Four months suffered deficits, but 3 months had surpluses even during the dry year because of the variability of precipitation from month to month. (After Muller, 1970)

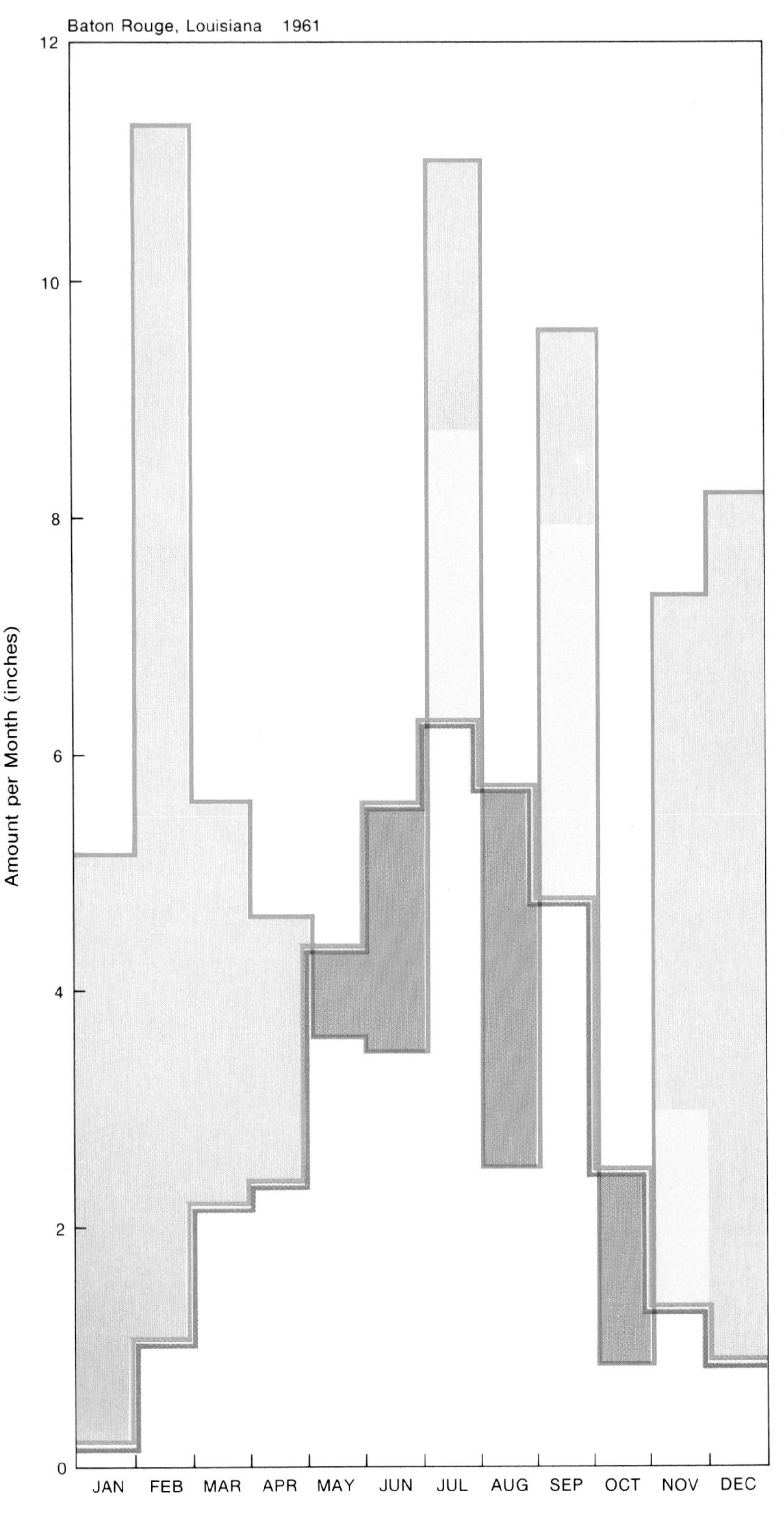

Baton Rouge, Louisiana 1961
Amount per Month (inches)
0
2
4
6
8
10
12
JAN
FEB
MAR
APR
MAY
JUN
JUL
AUG
SEP
OCT
NOV
DEC

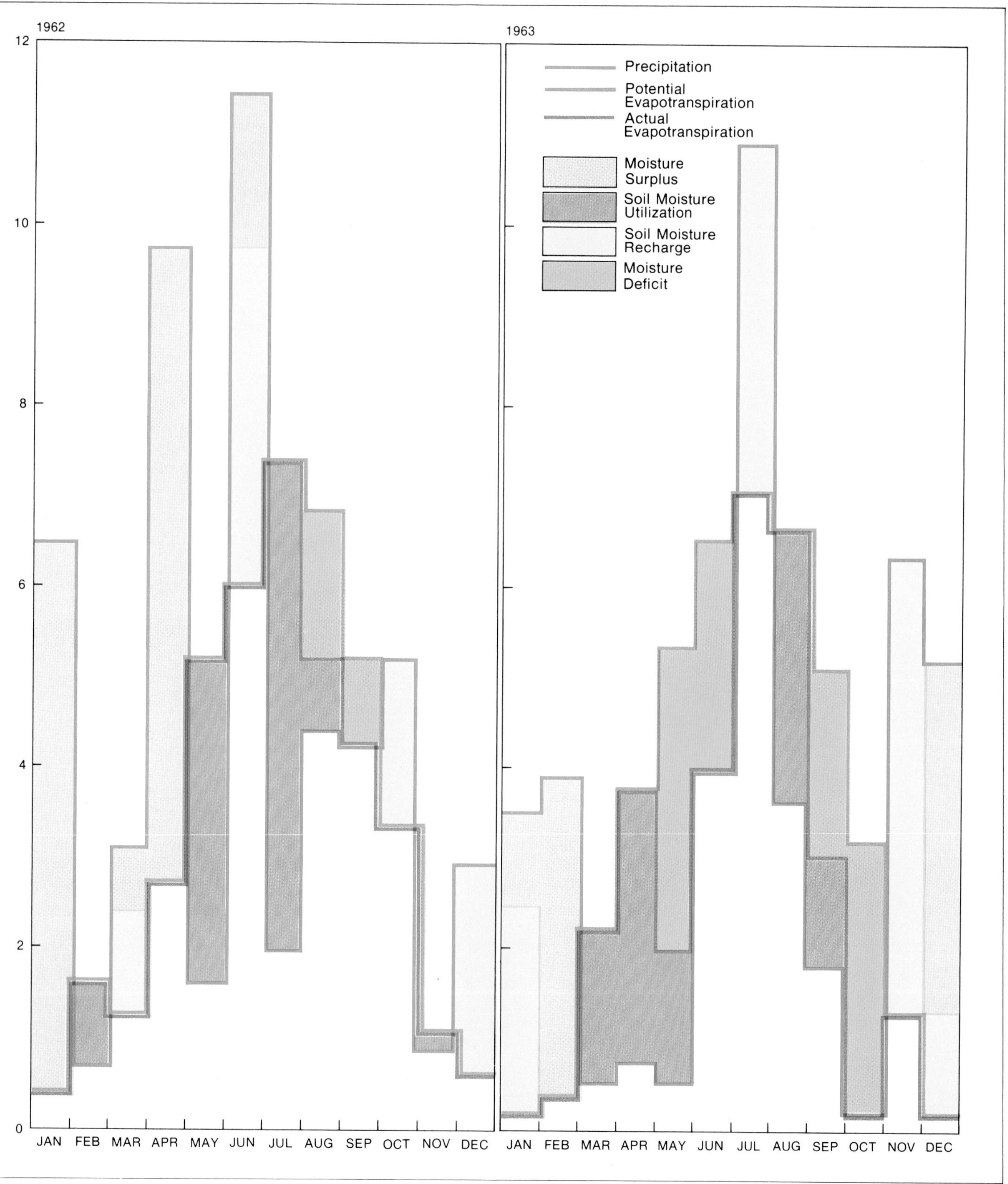
1962
1963
12
10
8
6
4
2
0
Precipitation
Potential Evapotranspiration
Actual Evapotranspiration
Moisture Surplus
Soil Moisture Utilization
Soil Moisture Recharge
Moisture Deficit
JAN FEB MAR APR MAY JUN JUL AUG SEP OCT NOV DEC
JAN FEB MAR APR MAY JUN JUL AUG SEP OCT NOV DEC

here assumes that all of the soil moisture is equally available to plants, even if the soil moisture is below field capacity. The model also assumes that if potential evapotranspiration exceeds rainfall, plants will use soil moisture as needed until the supply is exhausted. If rainfall exceeds potential evapotranspiration, the model assumes that the soil will absorb as much water as it can up to its full capacity of 6.0 inches.

Data row 4 shows the change in soil moisture and row 5 shows the total soil moisture for each month. These quantities must be calculated together, as a few examples will indicate. In January, rainfall exceeded potential evapotranspiration, but the excess was not absorbed because soil moisture was already at full capacity. In February, a rainfall deficiency of 1.0 inch caused 1.0 inch of water to leave soil moisture, which reduced the stored total to 5.0 inches. In March, an excess precipitation of 2.0 inches made water available for soil moisture, but only 1.0 inch could be absorbed because this amount brought the soil moisture back to its full capacity of 6.0 inches. In July, the soil moisture was 0.6 inch; a deficiency of rainfall in August caused the remaining 0.6 inch of soil moisture to be used, which left the soil moisture at 0.0.

Potential evapotranspiration is equal to actual evapotranspiration only when the soil contains some moisture. Row 6 shows actual evapotranspiration equal to potential evapotranspiration for all months in which total soil moisture is greater than 0.0. For September, when soil moisture was exhausted, actual evapotranspiration was equal to the amount of rainfall.

The difference between potential and actual evapotranspiration, called the *deficit*, is listed in row 7. A deficit represents the minimum amount of irrigation water required to restore the crops to full transpiration; it does not represent an actual loss of water. The deficit represents the amount of unavailable water that the plants could have used. Baton Rouge suffered small deficits for August and September of 1962.

Whenever precipitation exceeds actual evapotranspiration, there is an excess of water, part of which recharges the depleted soil moisture supplies. Any excess water that remains after soil moisture is brought to full capacity is called the *surplus*. Row 8 shows the calculated monthly surpluses for Baton Rouge in 1962. Surpluses were present only in January, March, April, and June. In March, rainfall exceeded actual evapotranspiration by 2.0 inches, but of this excess, the soil absorbed only 1.0 inch, which left a surplus of 1.0 inch. Surplus water is important, because it is available for surface runoff and for groundwater recharging. It is the water that enters streams and rivers and that eventually shapes the landscape by erosion and by sediment deposition.

A graph of the tabulated water budget data is shown in Figure 6.12, together with graphs for 1961 and 1963 at Baton Rouge. Each graph shows that surpluses tend to accumulate in the cooler months when transpiration is at a minimum and that deficits occur most often in the summer months. No month in 1961 had a deficit.

The total rainfall for 1961 was 73.8 inches, an above-average value for Baton Rouge. By contrast, a total deficit of 10.4 inches was accumulated in 1963, a "dry" year with only 41.9 inches of rain. Even an area such as Baton Rouge, which normally is considered "wet," can experience "dry" periods. Whether a surplus or a deficit is generated in a particular month depends on the complicated interplay of the timing and amount of rainfall, the transpiration requirements, and the soil moisture conditions. Potential evapotranspiration varies smoothly with solar energy input through the year, but the highly variable nature of precipitation from day to day and from year to year causes marked changes in water use. The monthly water budget illustrates these effects clearly; it is used again in Chapters 7 and 10 to compare conditions in different regions.

Improving the Water Budget Model

The simple water budget model presented here is only an estimate and it neglects a number of details. It does not take into account the change in the infiltration capacity of a soil under prolonged rainy conditions, for example. It also assumes for simplicity that soil moisture is readily available to plants, regardless of the moisture

Figure 6.13 This graph compares the amount of stored soil moisture estimated by the water balance method to the measured amount. The excellent agreement indicates the suitability of the local water balance approach to agricultural problems.

The course of the soil moisture through the year reflects seasonal conditions. The amount of stored soil moisture is maximum in late winter and spring, when precipitation is heavy and transpiration is small due to lack of plant cover. Soil moisture decreases in April and May and reaches a minimum during the summer, when precipitation is low and high temperatures and dense plant cover cause evapotranspiration to be high. The local peaks in the soil moisture are caused by rapid recharge during rainstorms. Can you explain why the average level of soil moisture increases in the autumn? (After Thornthwaite and Mather, 1955)

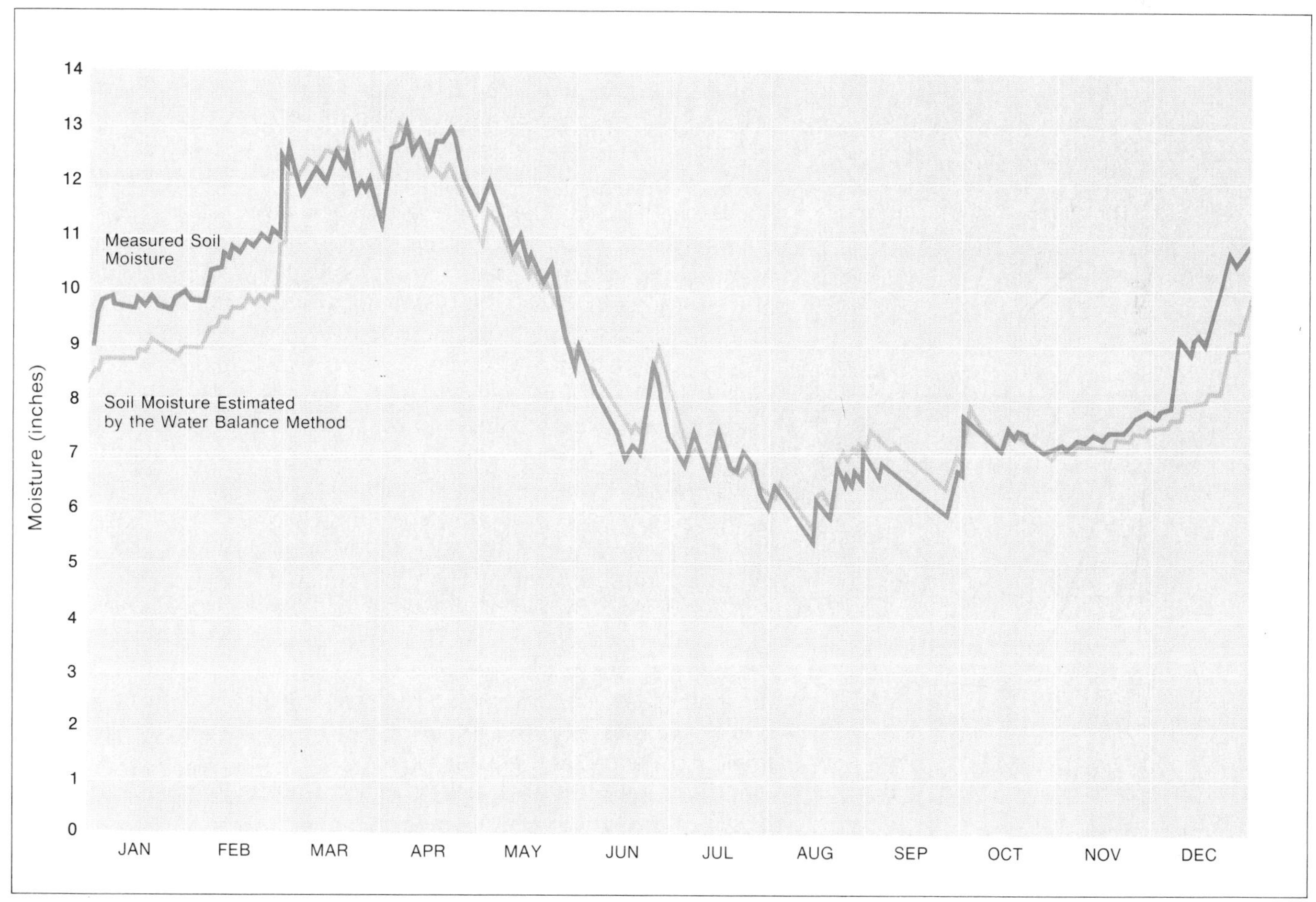
Moisture (inches)
14
13
12
11
10
9
8
7
6
5
4
3
2
1
0
Measured Soil Moisture
Soil Moisture Estimated by the Water Balance Method
JAN
FEB
MAR
APR
MAY
JUN
JUL
AUG
SEP
OCT
NOV
DEC

content of the soil. In reality, the smaller the amount of moisture stored in the soil, the more difficult it becomes for plants to utilize it. More complex water budget models take such decreasing availability into account.

The monthly water budget can be criticized on the more fundamental grounds that potential evapotranspiration, a key component of the budget, is usually estimated from an approximate formula and may be in error. Furthermore, a month is too long a period to represent fluctuations in precipitation adequately. In an average year in Baton Rouge, at least one 24-hour period has a rainfall of more than 3 inches. The water from a heavy rainfall is not distributed in the same way as a gentle steady rain over several days, so a daily water budget would be more representative of actual conditions than a monthly budget. Modern computing facilities have made the preparation of daily water budgets an easy task.

The precipitation data required to construct a water budget are not always sufficiently accurate. Precipitation is conventionally measured by noting the accumulation of water in a standard bucket, or rain gauge, exposed under standard conditions. In the United States, however, there is an average of only one rain gauge for every 500 square kilometers (200 square miles). Because rainfall is highly localized, this rain gauge network cannot give fully representative precipitation data. New instrumentation techniques may eventually improve the quality of the data, however. Urban runoff and storm sewer problems have prompted the city of San Francisco to install thirty rain sensors in the city and on its hills. Every sensor sends a signal to a central computer for each 0.01 inch of rainfall. The computer indicates the progress of a rainstorm across the city and the possibility of local flooding. And if other modern instrumentation techniques made rainfall intensity data available, the effect of the varying infiltration rate during a rainstorm could be incorporated into water budget calculations.

Despite some deficiencies, the local water budget is a valuable tool. With it, important water use components can be estimated using standard meteorologic data. Knowledge of water surplus is an important guide in assessing a region's ability to supply runoff water, and knowledge of water deficit is required to evaluate irrigation needs. If you calculated a water budget for your own locale, you could see at what times it would be wisest to wash your car or water your garden.

Summary

Water is a renewable resource that continually shuttles between the atmosphere and the earth's surface in the hydrologic cycle. Modern civilization requires large quantities of water for agriculture and for industry, and efficient management and major construction works are required to transfer water from regions with a surplus of water to regions where water is wanted.

Radiant energy from the sun and the circulation of the atmosphere deliver energy and moisture to the earth's surface. Part of the water that falls on the land immediately evaporates back to the atmosphere, part enters the ground to be stored as soil moisture or to percolate into groundwater storage, and part becomes surface runoff to streams. Most of the water that falls on the land returns to the atmosphere by evaporation or by transpiration without flowing to the ocean.

The availability of water for storage as soil moisture or for runoff can be estimated using the local water budget, which is the regional version of the hydrologic cycle. Because of insufficient data on evapotranspiration, estimated values of potential evapotranspiration must be calculated from meteorologic data. The water budget reveals how variations in rainfall and how the use of water by vegetation influence the amount of water available for runoff and for groundwater storage. Chapters 7 and 10 use the water budget to interpret the climate of a region in terms of the availability of water for different uses.

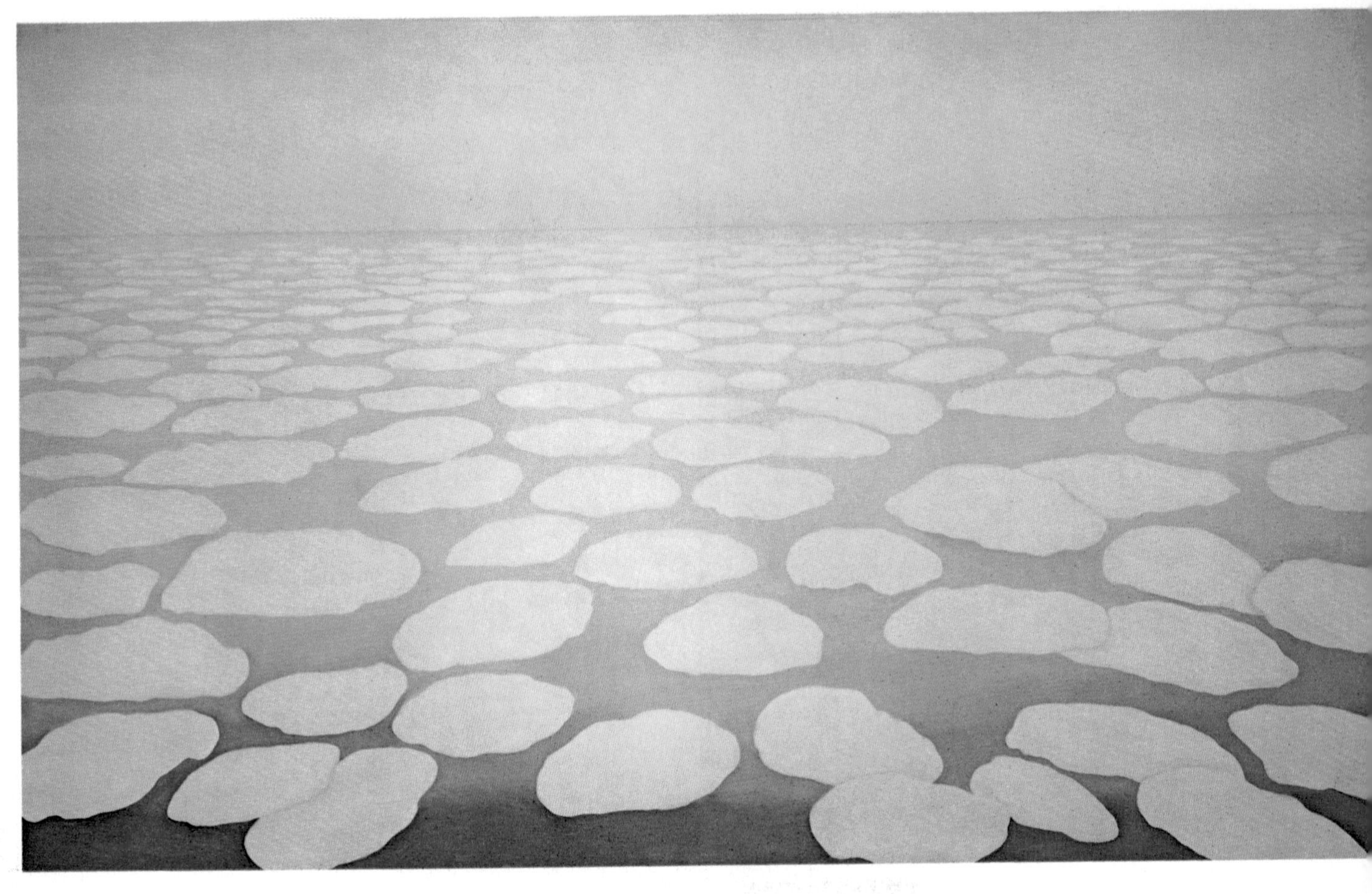

7 Global Systems of Climate

Classification of Climate
Measures of Climate
Time and Space Scales of Climate

One View of Climate: Dynamic Global Delivery System
Distribution of Climatic Regions on a Hypothetical Continent
Distribution of Climatic Regions on the Earth

Another View of Climate: System Interactions at the Surface
Climate and the Water Budget
Distribution of Energy and Moisture

Three Functional Realms of Climate

The Köppen System of Climate Classification

The Thornthwaite System of Climate Classification

Sky Above Clouds II by Georgia O'Keeffe. (Collection of Mrs. Potter Palmer, Lake Forest, Illinois; with permission of Georgia O'Keeffe)

Dynamic processes in the atmosphere cause the weather patterns that give each region of the earth a distinctive climate. Climate can be viewed in two ways: as the delivery system of energy and moisture through the atmosphere, and as the interactions of energy and moisture at the earth's surface with systems of soils, vegetation, and landforms.

Global Systems of Climate

The southwestern part of the United States is a region where potential evapotranspiration far exceeds precipitation. The annual *deficit* of moisture at Phoenix, Arizona, for example, is 40 inches in an average year, and extensive irrigation is needed to support the growth of crops. A plan has been advanced recently that has as its object the radical increase of precipitation in the Southwest, so that moist woodlands would grow where only cactus can exist today. The idea is to build a stupendous dam across the Bering Strait between Siberia and Alaska. The dam would have to be 100 kilometers (60 miles) long, far more massive than any dam on the earth today.

How could a dam at the Arctic Circle affect weather in the Southwest 5,000 kilometers (3,000 miles) away? If such a dam could be built, it would stop the circulation of cold Arctic waters into the Pacific Ocean. The planners anticipate that warming the Pacific Ocean would increase rainfall in the Southwest, but meteorologists opposed to the scheme argue that it is impossible to predict the full consequences of such a drastic global scheme to modify climate. Changing the temperature of the Pacific Ocean would affect the pattern of the general circulation of the atmosphere and would change the temperature of the air above the Pacific basin. And because the general circulation of the atmosphere exerts controls on the climates of the earth, the effects of the dam could be far-reaching indeed.

Climate is the general pattern of weather in a region over a period of years. Although weather at a given location varies from day to day, it exhibits a predictable sameness over time. Understanding the global distribution of the earth's climatic regions is the key to understanding many of the characteristic features of regions, because the climate of a region has a profound influence on its soils, vegetation, and landforms: it determines the suitability of a region for agriculture, and it influences the processes that shape the landscape. This chapter discusses how solar energy input and the circulation of the atmosphere determine the climate of a region. The local water budget is used to show why climate is the active factor in controlling other systems, and two widely used methods of climate classification are described.

Classification of Climate

The general circulation of the atmosphere is an important factor in determining the variety and timing of weather events that can occur at each global location. The comparative stability of the general circulation produces characteristic regional weather patterns that give each region a distinctive flavor. The winter vacationer in Florida, for example, has every reason to expect a succession of warm, sunny days during his stay.

The atmosphere is the vehicle that delivers energy and moisture around the world. It is useful, therefore, to attempt to classify different regions of the earth according to the type of climate they experience, because the distribution of energy and moisture carries implications for other systems in the region. The classification of climates helps reveal general patterns in other systems and serves to organize a wealth of information about the earth's surface.

To be most useful, a classification scheme for a system should have enough categories and distinctions to account for the system's principal features, but not so many that a general understanding is lost in a welter of detail. The classification schemes for climate discussed later in this chapter classify the earth's climates according to about five main categories and one or two dozen subcategories. No two places on earth have exactly the same climate, however, so any classification scheme inevitably submerges individual details. Because climatic regions usually blend into one another gradually, the boundaries of a classification system should usually be interpreted as broad zones of transition rather than as sharp distinctions. On a global scale, however, the transition zones are frequently narrow compared to the size of the climatic regions, so it is meaningful to speak of boundaries if the concept is not pushed too far. Despite the artificial nature of classification systems, the degree of organization they impose can provide a useful framework for perceiving the world.

Measures of Climate

Classification systems for climates generally employ one or more measures of climate, or *climatic indexes*, to distinguish one climatic type from another according to rules devised by the inventor of the system. What constitutes an appropriate index of climate for a climate classification system depends on the intended use of the classification. Someone interested in constructing buildings would want to emphasize wind speed and direction to assess the possibility of wind damage. Someone concerned with transportation would be especially interested in the frequency of fog and icing conditions in various regions. The timing and duration of temperature inversions must be known to assess a region's potential for air pollution.

Climate classification systems in physical geography generally employ temperature and precipitation as indexes of climate because physical geographers are interested in agriculture—how vegetation is distributed and what potential there is for agriculture in a particular region. They do not ordinarily use indexes of wind, inversions, or other atmospheric phenomena, because too many indexes can encumber a system and because for many regions the only available data are temperature and precipitation records. Nevertheless, temperature and precipitation are only indirect measures of energy and moisture inputs to systems. Other measures, such as potential evapotranspiration, may be more suitable indexes of climate, as discussed later in this chapter.

Time and Space Scales of Climate

Weather conditions change constantly. The movement of air masses into a region can change the local air temperature significantly within a few hours. Weather conditions also vary markedly from place to place in the same local region. Passing storm clouds can bring heavy rain to isolated sections while other locations remain dry. Given that the indexes of climate are to be temperature and precipitation, geographers must still decide when, where, and how often to measure these factors. The following example shows how important the time scale—the "when"—of measurement is.

The average annual temperature at Aberdeen, Scotland, is 8.3°C (46.9°F), and at Chicago, Illinois, it is 9.7°C (49.5°F), less than 2°C (3°F) higher. The average annual precipitation at both locations is the same, 84 centimeters (33 inches) per year. Despite these similarities in annual averages, the climates at the two locations differ significantly. The monthly averages give a much more accurate indication of the climate in each city. During January, the coldest month, Aberdeen averages 3.9°C (39°F) while Chicago averages -3.9°C (25°F), a difference of nearly 8°C (14°F). During July, the warmest month, Aberdeen averages 14°C (57°F) and Chicago averages 23°C (73°F), which is 9°C (16°F) higher. The variation of temperature at the two places is markedly different through the year; the climate at Chicago is more extreme than the climate at Aberdeen. The example shows that, in general, the timing of temper-

Figure 7.2 The nighttime temperatures in this Louisiana backyard show a large degree of variation from place to place because of such factors as local movements of air, differences in the amount of heat stored in the ground and buildings, and differences in cooling rates. Climate measured on such a small scale is called the *microclimate* of the area. Such details are lost when the climates of large regions are classified into broad categories. (After Muller, 1973)

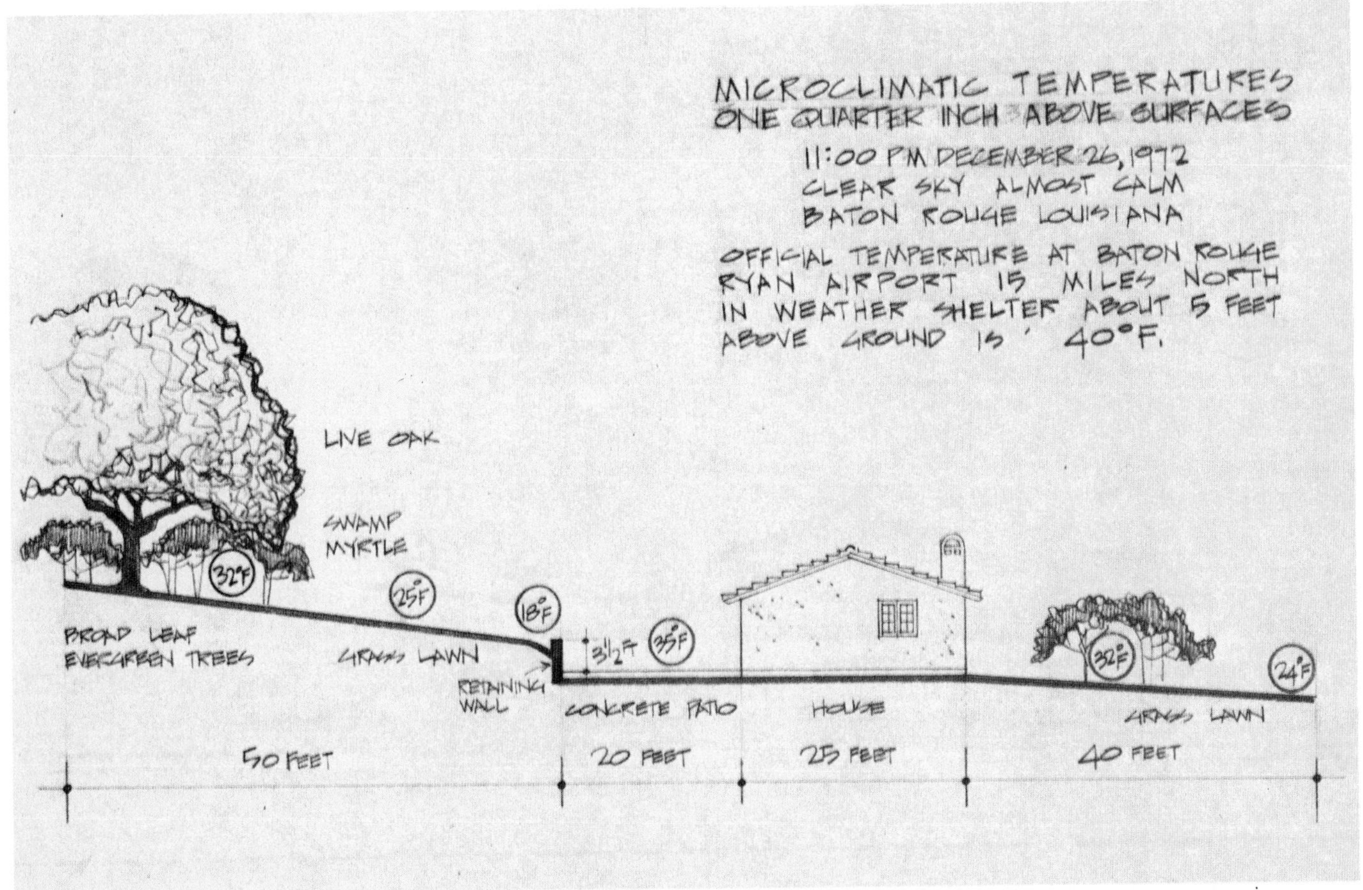
MICROCLIMATIC TEMPERATURES
ONE QUARTER INCH ABOVE SURFACES
11:00 PM DECEMBER 26, 1972
CLEAR SKY ALMOST CALM
BATON ROUGE LOUISIANA
OFFICIAL TEMPERATURE AT BATON ROUGE RYAN AIRPORT 15 MILES NORTH IN WEATHER SHELTER ABOUT 5 FEET ABOVE GROUND IS 40°F.
LIVE OAK
SWAMP MYRTLE
32°F
25°F
18°F
3½ FT
35°F
32°F
24°F
BROAD LEAF EVERGREEN TREES
GRASS LAWN
RETAINING WALL
CONCRETE PATIO
HOUSE
GRASS LAWN
50 FEET
20 FEET
25 FEET
40 FEET

ature and precipitation, as well as the average annual values, is important in characterizing the climate of a region.

The size scale of climatic regions is also important in classification systems. Figure 7.2 shows temperature readings made at 11 P.M. at various points in a backyard located in Baton Rouge, Louisiana. The yard has its own microclimate, with a wide range of temperatures over a small area. The microclimate of the yard is too small to appear on even a countywide climate survey, but climatic conditions in different parts of the yard are important to someone trying to protect subtropical vegetation from a hard freeze.

The best size scale for a climate classification system depends on the intended use of the system. On a global scale, most of the West Coast of the United States is considered to be in a single climatic region. But someone concerned with water resources on the West Coast would need to see a much more detailed picture that showed the gradation from the abundant rainfall in the Pacific Northwest to the dryness of southern California.

This chapter focuses on climate classifications on continental and global scales to give an orientation to the general distribution of climate on the earth. Excessive detail is not desirable on such a large scale. The Köppen system of global climate classification discussed later in this chapter assigns New York City and Nashville, Tennessee, to the same climatic region, for example. On a nationwide scale, the climates of the two cities are obviously different, but on a global scale the climates are more like each other than they are like climates in a tropical rainforest or an Asian steppe.

One View of Climate: Dynamic Global Delivery System

The delivery system of energy and moisture to different regions of the earth follows global patterns imposed by the distribution of solar radiant energy and the movements of the general atmospheric circulation. The positions of the earth's landmasses, oceans, and mountain ranges also influence the global distribution of climatic regions. The gross features of the earth's climates can be understood by looking at how solar energy input and the general circulation control the climate regions of an idealized hypothetical continent.

Distribution of Climatic Regions on a Hypothetical Continent

The hypothetical continent shown in Figure 7.3 embodies many of the features of the actual continents. Like most continents, it is surrounded by oceans. It is broad at the north and extends to high latitudes, like North America and Asia, and it is narrow and ends near latitude 60°S, like South America and Africa.

The division of the hypothetical continent into climatic regions can be accounted for by considering solar energy input, global distributions of winds and pressure, and the seasonal meridional migration of the general circulation. The climatic regions on the hypo-

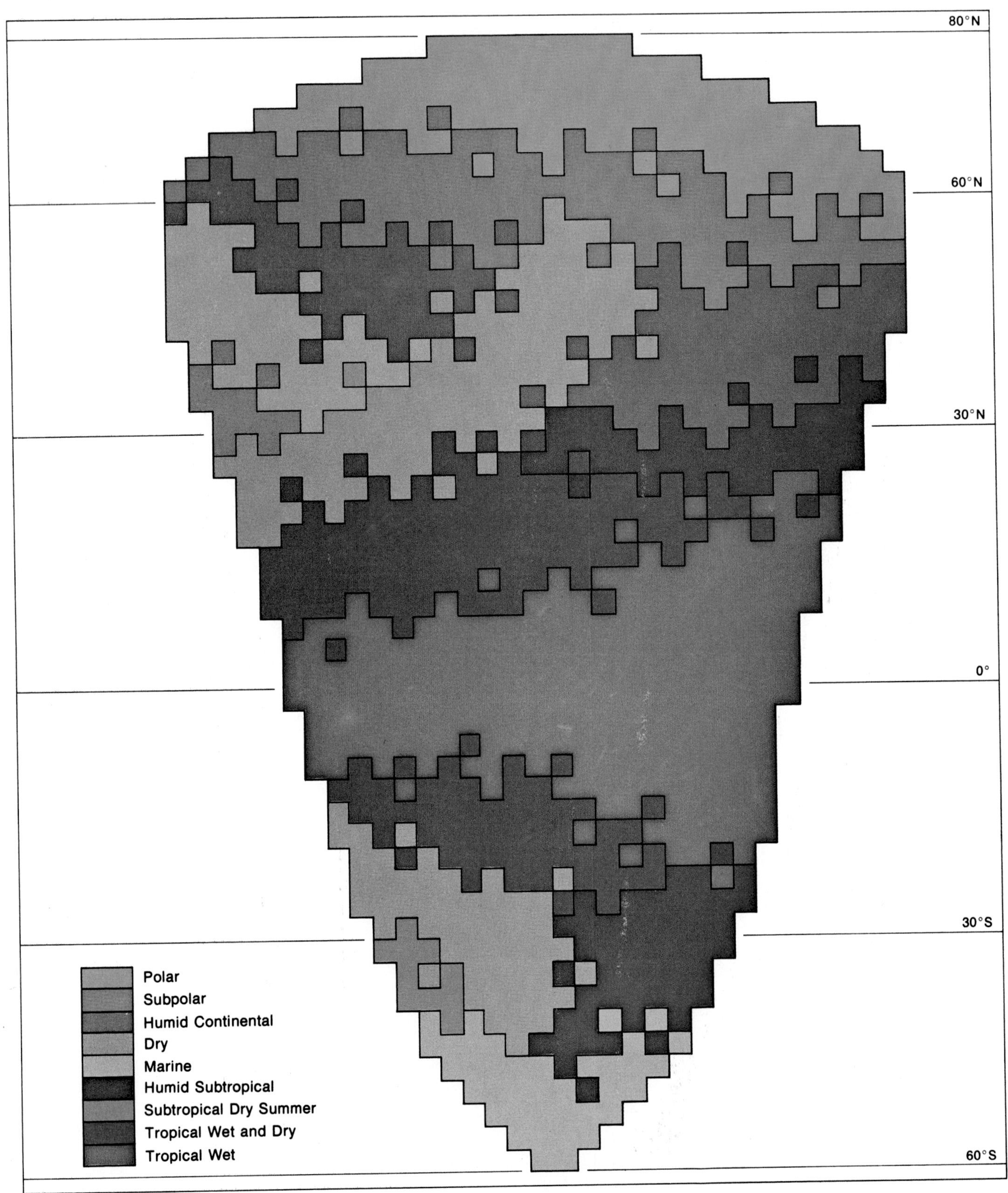
80°N
60°N
30°N
0°
30°S
60°S
Polar
Subpolar
Humid Continental
Dry
Marine
Humid Subtropical
Subtropical Dry Summer
Tropical Wet and Dry
Tropical Wet

thetical continent tend to have a lower average temperature the farther they are from the Equator. The difference between average summer and average winter temperatures tends to be greater farther from the Equator as well. The tropical and subtropical regions near the Equator receive the greatest total amount of solar energy and tend to be the warmest; latitudes between 38°N and 38°S receive an annual surplus of energy from the sun, whereas regions at higher latitudes receive less solar energy and have lower average temperatures. The climatic regions in the extreme north include polar ice and tundra regions. Because of the tilt of the earth's axis, the amount of solar energy the higher latitudes receive is much greater in summer than in winter, and the seasonal variation of temperature is correspondingly large. The nearly uniform input of solar energy to equatorial regions makes temperatures there comparatively constant through the year.

Several mechanisms control the distribution of moisture to the various climatic regions on the hypothetical continent. Winds reaching the continent after crossing the ocean are moisture-laden. The continent lies in the path of the easterly trades, which are between latitudes 30°N and 30°S, so that at low latitudes the eastern section of the continent tends to receive more moisture than the western section. The low-latitude eastern section contains a wide region with a tropical wet climate and regions with humid subtropical climates.

The equatorial trough of low pressure occupies the region near the Equator. Air that converges into this trough is frequently moist from the oceans, and as the converging air rises up into the circulation of the subtropical Hadley cells, condensation occurs. Precipitation tends to be frequent near the Equator, so the climatic region there is classified as tropical and wet.

The climatic regions immediately to the north and south of the tropical wet region are moist during the summer but receive little precipitation during the winter. The intertropical convergence zone and the zone of uprising moist air shift toward the summer hemisphere, bringing moisture to the summer tropical wet and dry region and leaving the winter tropical wet and dry region with much less moisture than it receives in summer. The tropical wet region near the Equator receives moisture almost uniformly through the year.

The dry climatic regions near 30°N and 30°S are located in the zone of subtropical highs, where dry air from the upper atmosphere subsides to the surface. The dry air brings little moisture, and its outflow along the surface prevents moist air from reaching the dry climatic regions. Between latitudes 30° and 65°, the continent lies in the path of the westerlies, so that western sections tend to be wetter than eastern sections. The western sections in the middle latitudes include subtropical and marine climates.

The delivery of moisture to the subtropical and marine climatic regions in the western sections is governed by a high-pressure cell in the ocean to the west of the continent. In winter, the cell is at lower latitudes, and midlatitude cyclones bring moisture to both the subtropical and marine climatic regions. In the summer, the

Figure 7.3 (opposite) The climatic regions represented on this hypothetical continent are determined by the input of solar radiant energy and by the delivery of air masses and moisture by the atmospheric systems. The regions are depicted with overlapping boundaries to suggest that climate changes gradually from one location to another across the surface of the earth. (The term "subarctic" is also used to indicate subpolar regions.)

subsidence associated with the high-pressure cell moves poleward and blocks moist air from reaching the subtropical climatic region. The subtropical climatic region has a wet winter and a dry summer, while the marine climatic region farther poleward receives moisture throughout the year.

In the north middle latitudes, where the continent is broad, there is a humid continental region. The rapid heating of the land in summer and the rapid cooling in winter lead to marked differences between summer and winter temperatures. Air from the ocean, which remains comparatively uniform in temperature through the year, moderates the temperatures of the subtropical and marine climates in the western section. There is no humid continental climatic region in the Southern Hemisphere because the continent is narrow at high southern latitudes, which allows the moderating effect of marine winds from the west to reach all sections.

Distribution of Climatic Regions on the Earth

A global map of the dynamic climates of the earth indicates that many of the climatic regions described for the hypothetical continent are apparent on the actual continents (see Figure 7.5). Near the Equator are tropical wet regions, with tropical wet and dry regions immediately to the north and south. Africa and Australia show the expected tropical dry regions near latitude 30°. The mid-latitude west coasts of North and South America and of Europe show the expected marine climates, and the interiors of the United States and Europe are humid continental climatic regions. Subpolar and polar regions occupy northern Asia and North America.

The distribution of climates on the earth also exhibits the effects of the distribution of continents, oceans, and mountain ranges. The mountain ranges in western North and South America, for example, prevent the west coast marine climates from extending as far inland as they do in Europe. As a further consequence, the regions to the east of the west coast mountain ranges are dry because moist air is blocked from reaching them. Similarly, the west coast ranges in the low latitudes of South America block the easterly trades, which prevents moisture from reaching the west coast of South America between the Equator and latitude 30°S so that the coastal regions there are dry. The monsoon circulation greatly influences the climatic regions of India and Southeast Asia. The exact cause of the monsoons is unknown, but the high-pressure cell over Siberia in winter, low pressure over the Indus Valley in summer, the presence of the Himalaya Mountains, and a jet stream over Asia all seem to play a role.

Another View of Climate: System Interactions at the Surface

Climate as discussed above is sometimes defined as the composite average of such weather conditions as temperature, precipitation, winds, degree of cloudiness, and so forth. This view of climate emphasizes the atmosphere as a delivery system and neglects the interaction of energy and moisture with systems on the surface of the earth. However, systems on the surface of the earth share the energy and moisture that climate delivers; plant growth, for example, depends on the amount of soil moisture available to vegetation and not directly on total precipitation. So to bring out the relationships between climate and other systems, the view of climate can be extended to include the interactions of delivered energy and moisture with the systems on the earth's surface.

Climate and the Water Budget

The local water budget discussed in Chapter 6 is a useful tool for thinking about climate in terms of interactions. The water budget deals with energy by using the concepts of potential and actual evapotranspiration, and it takes into account the availability of soil moisture for plant growth and the amount of moisture available as runoff for streamflow and erosion.

Consider the local water budgets at two locations, Cloverdale, California, and Manhattan, Kansas. Cloverdale receives an average of 100 centimeters (39.4 inches) of precipitation annually, and Manhattan receives 80 centimeters (31.5 inches). Both locations have a total annual potential evapotranspiration of nearly 80 centimeters (31.5 inches), so their energy endowments are essentially equal.

On the basis of annual precipitation, Cloverdale appears to be

Figure 7.4 Cloverdale, California, and Manhattan, Kansas, receive similar amounts of precipitation annually and have nearly the same total annual potential evapotranspiration. The climates at these locations appear to be similar on the basis of the annual endowments of moisture and energy. But their local water budgets indicate that the timing of precipitation compared to the demand for moisture at each location makes the climate different for systems on the earth's surface. At Cloverdale, heavy precipitation and a low demand by vegetation during the winter generate large surpluses of moisture in winter. Deficits occur during the dry summer, so that Cloverdale is moist for runoff and dry for vegetation. At Manhattan, the timing of precipitation through the year is in close accord with the demand by vegetation, so that surpluses and deficits are small. Manhattan is therefore dry for runoff and moist for vegetation. (After Carter, Schmudde, and Sharpe, 1972)

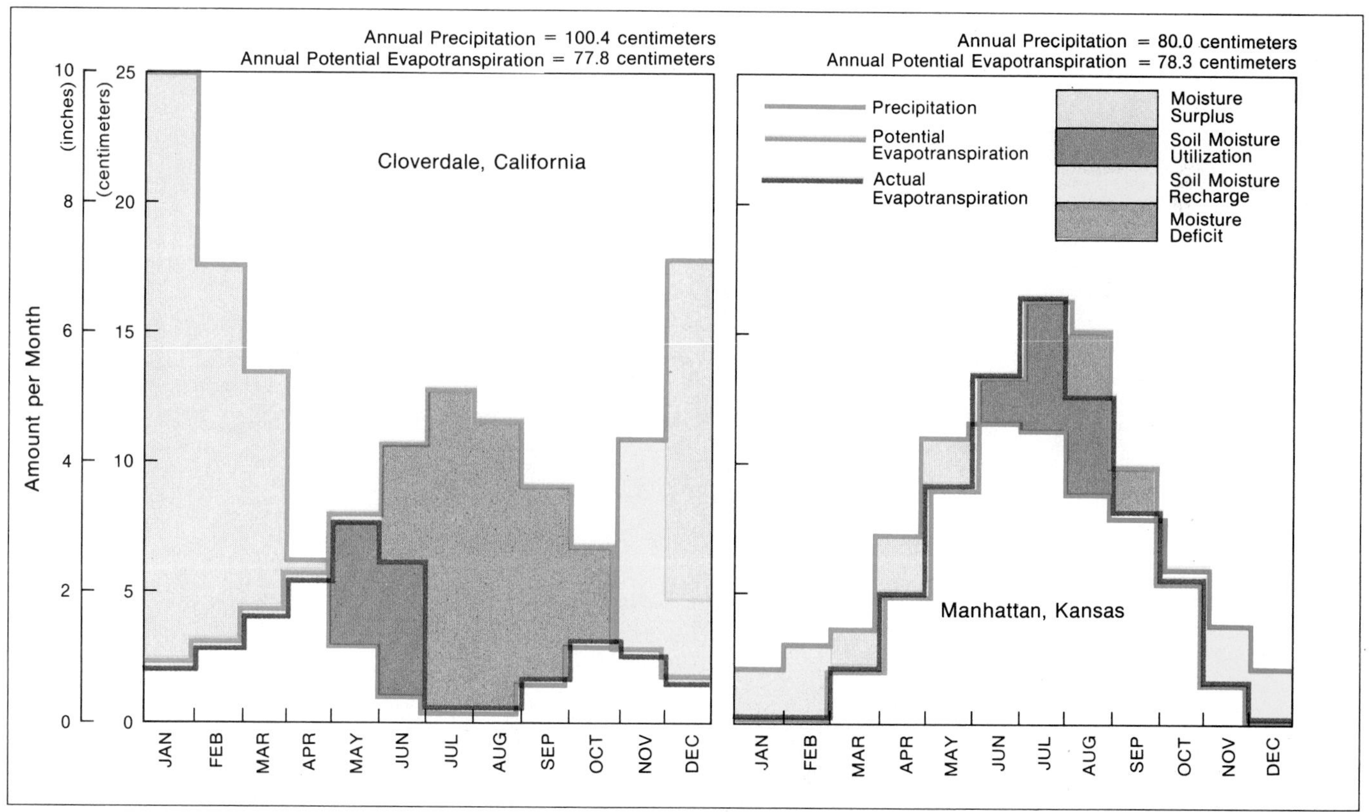

Annual Precipitation = 100.4 centimeters
Annual Potential Evapotranspiration = 77.8 centimeters
Cloverdale, California
Amount per Month
(inches)
(centimeters)
0
2
4
6
8
10
0
5
10
15
20
25
JAN
FEB
MAR
APR
MAY
JUN
JUL
AUG
SEP
OCT
NOV
DEC
Annual Precipitation = 80.0 centimeters
Annual Potential Evapotranspiration = 78.3 centimeters
Precipitation
Potential Evapotranspiration
Actual Evapotranspiration
Moisture Surplus
Soil Moisture Utilization
Soil Moisture Recharge
Moisture Deficit
Manhattan, Kansas
JAN
FEB
MAR
APR
MAY
JUN
JUL
AUG
SEP
OCT
NOV
DEC

Figure 7.5 This global map of climate emphasizes the classification of climates according to the delivery of energy and moisture. Compare it with the distribution of climates on the hypothetical continent shown in Figure 7.3. The landmasses of the high northern latitudes possess polar or subpolar (subarctic) climates like the hypothetical continent, because of the low input of solar radiant energy. Like the hypothetical continent, Africa and South America possess equatorial tropical wet climate regions flanked by tropical wet and dry climate regions. Africa and Australia exhibit dry climate regions near latitude 30°S. The effect of mountains on climate distributions is apparent in the way the tropical wet region of South America is blocked from the west coast by the Andes.

The distribution of continents and oceans on the earth affects the movements of air masses. The southeastern United States, for example, possesses a humid subtropical climate instead of a humid continental climate because of the northward movement of moist tropical air from the Gulf of Mexico and the Caribbean Ocean. The United States and much of Europe lie in the belt of westerly winds. Parts of western Europe and the west coast of the United States possess marine climates. In Europe the marine climate gradually merges into a humid continental climate region, but in the United States the west winds from the Pacific Ocean are blocked by mountains. The region immediately inland from the Pacific coast is dry desert. The north central and northeastern United States possess a humid continental climate, however, partly because of cold winds from the north in winter and warm winds from the south in summer. (After Finch and Trewartha, 1949)

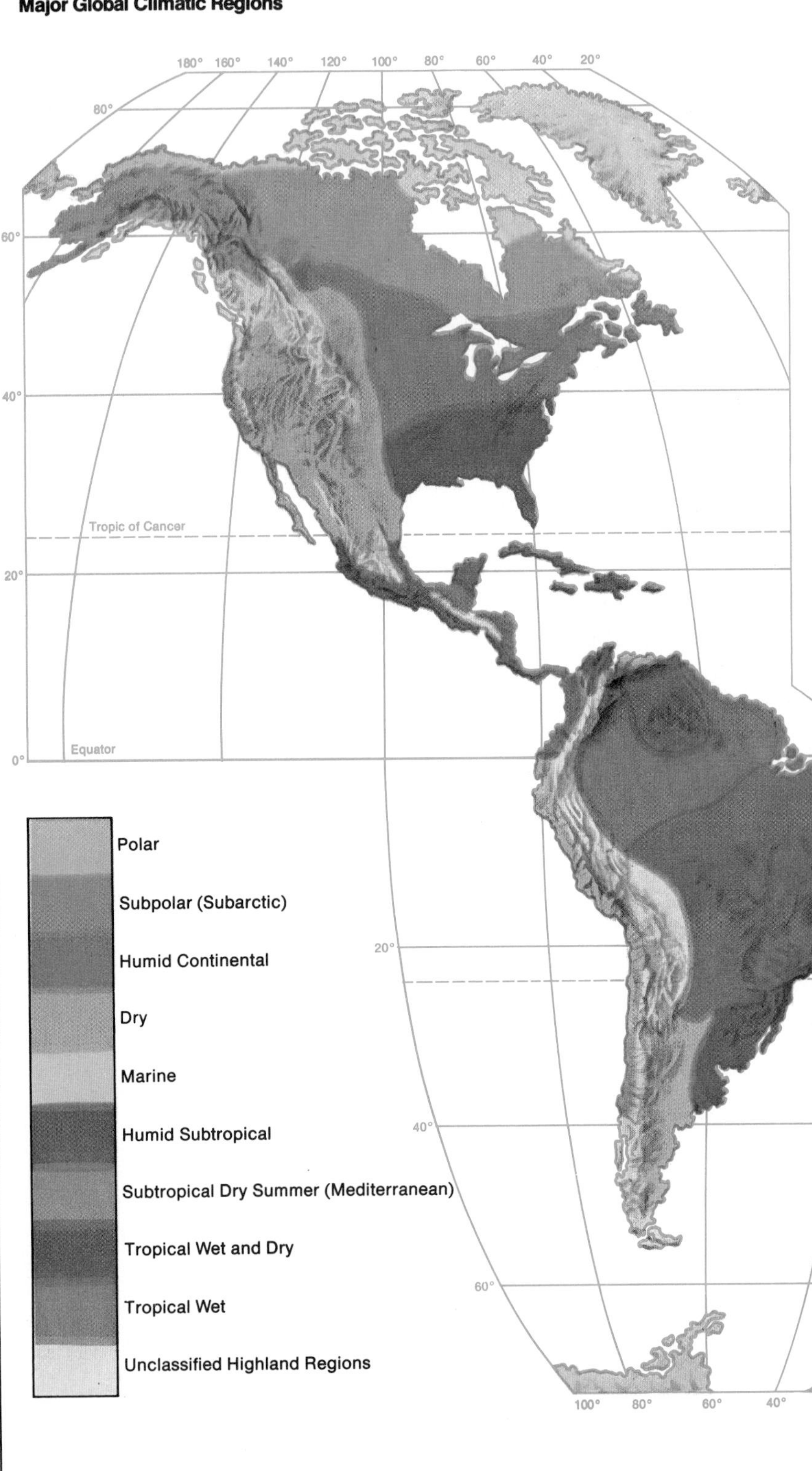

Major Global Climatic Regions
180°
160°
140°
120°
100°
80°
60°
40°
20°
80°
60°
40°
20°
0°
Tropic of Cancer
Equator
20°
40°
60°
100°
80°
60°
40°
Polar
Subpolar (Subarctic)
Humid Continental
Dry
Marine
Humid Subtropical
Subtropical Dry Summer (Mediterranean)
Tropical Wet and Dry
Tropical Wet
Unclassified Highland Regions

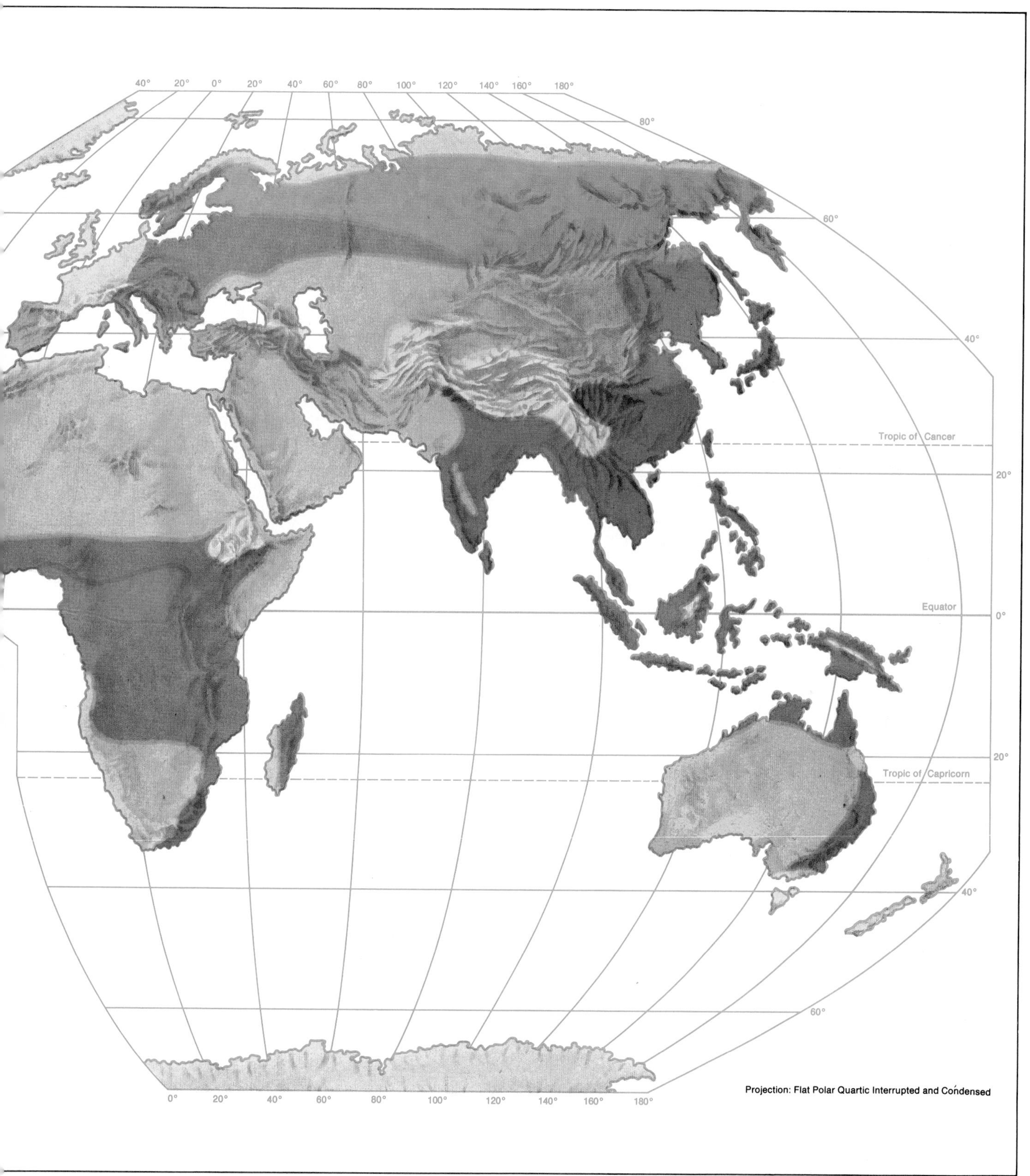

40°
20°
0°
20°
40°
60°
80°
100°
120°
140°
160°
180°
80°
60°
40°
Tropic of Cancer
20°
Equator
0°
20°
Tropic of Capricorn
40°
60°
0°
20°
40°
60°
80°
100°
120°
140°
160°
180°
Projection: Flat Polar Quartic Interrupted and Condensed

more humid than Manhattan. But as you read in Chapter 6, the utilization of moisture at a place depends on the timing of precipitation. Figure 7.4 shows the water budgets for the two cities. At Cloverdale, large deficits of soil moisture are generated in the summer months, when potential evapotranspiration is high and precipitation is low. In terms of interactions, Cloverdale has a dry climate for vegetation during the summer. In the winter and spring, however, heavy precipitation results in a large surplus of water for runoff, making Cloverdale moist for purposes of streamflow.

At Manhattan, the water budget shows that the timing of precipitation tends to follow the variation of potential evapotranspiration through the year. Therefore, only small deficits and surpluses are generated at Manhattan; for vegetation purposes Manhattan is moist, but for streamflow it is dry.

Distribution of Energy and Moisture

The examples of Cloverdale and Manhattan show that when a geographer considers the relationship of other systems to climate, it is not sufficient to consider only the amounts of energy and moisture delivered. The distribution of energy and moisture to systems must also be taken into account. A given location may exhibit different types of climate in terms of different interactions—Cloverdale, for example, is moist for streamflow but dry for vegetation. Components of the water budget can be used as indexes of climate to characterize the interactions between climate and the environment.

Various components of the water budget represent different ways for systems on the earth's surface to share moisture. Surplus in the water budget is, for instance, the amount of water available for surface runoff and groundwater recharge. Surplus is therefore a measure of the availability of moisture for erosion and for the support of streamflow.

Figure 7.6 shows the surplus generated at seven different locations in the United States. The data are presented in terms of frequency distributions over a 30-year period to indicate the degree of variability that can occur. At Blue Canyon, California, for example, a moisture surplus greater than 50 inches was generated half of the time. But in individual years the surplus ranged from over 90 inches to under 25 inches. In 70 percent of the years a surplus greater than 45 inches occurred at Blue Canyon.

Infrequent events such as an exceptionally great surplus can be important for systems like streams that have a short response time. High floods in streams are a consequence of unusually high runoff, and flood conditions can alter the channel of a stream by rapidly scouring the banks and bed. On the other hand, the average value of surplus is sufficient to give a good measure of the erosion rate by flowing water on a rocky outcrop, because the response time of rock to erosion is long.

From the standpoint of moisture surplus, Blue Canyon should be classified as exceptionally humid, Morgan City, Louisiana, as humid, and Phoenix, Arizona, as arid. However, the classification

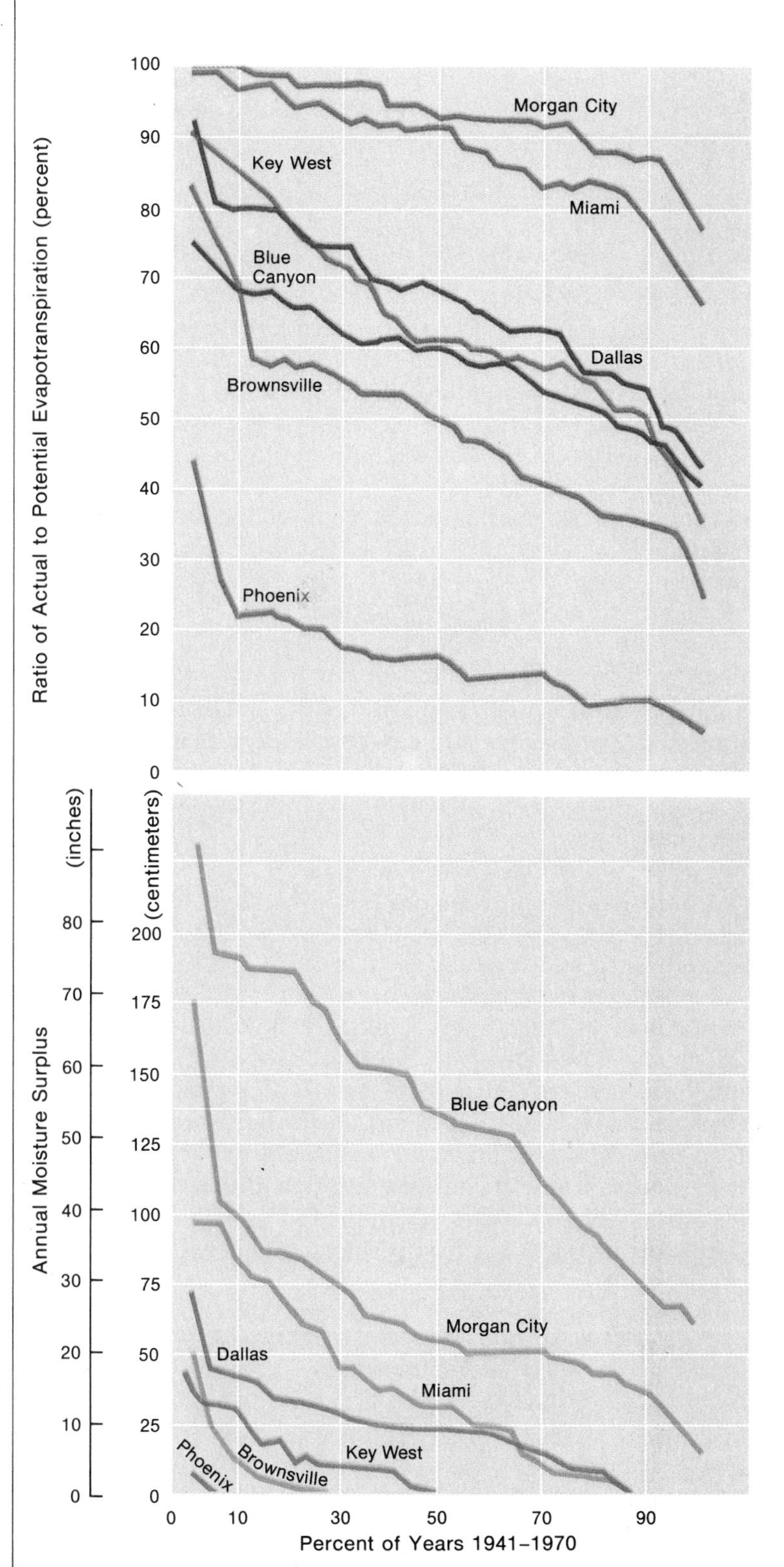

Ratio of Actual to Potential Evapotranspiration (percent)
100
90
80
70
60
50
40
30
20
10
0
Morgan City
Key West
Miami
Blue
Canyon
Dallas
Brownsville
Phoenix
Annual Moisture Surplus
(inches)
(centimeters)
80
70
60
50
40
30
20
10
0
200
175
150
125
100
75
50
25
0
Blue Canyon
Morgan City
Dallas
Miami
Key West
Phoenix
Brownsville
0
10
30
50
70
90
Percent of Years 1941–1970

Figure 7.6 Components of the local water budget can be used as climatic indexes. The graphs show frequency distributions of the ratio of actual to potential evapotranspiration (AE/PE) and the total annual moisture surplus at seven locations for the 30-year period from 1941 through 1970. Locations where the moisture surplus is large for most years are moist for stream runoff, and locations where the ratio of actual to potential evapotranspiration is close to 100 percent in most years are moist for vegetation. Note the degree of variation in the indexes over the 30-year period, which shows that simple averages are not fully representative of the climate at a given place. In Miami, for example, the ratio of actual to potential evapotranspiration was 93 percent on the average, but for 5 years of the 30, the ratio fell below 80 percent. Climate is determined by variations as well as by average values. Perennial plants that require abundant moisture every year might not be able to maintain themselves during drier than average years in a normally moist location such as Miami. (After Muller, 1972) may change when other interactions are used as climatic indexes. Figure 7.6 shows the frequency distribution of the ratio of actual evapotranspiration to potential evapotranspiration at the seven locations. Plants that require abundant moisture can grow where the ratio is close to 100 percent. On the basis of this climatic index, Morgan City and Miami, Florida, are more humid than Blue Canyon.

Three Functional Realms of Climate

The example of the local water budget clearly shows that the total amount of moisture available in a location is not the index that measures the effectiveness of climate for the functioning of other systems. An annual precipitation of 80 centimeters (30 inches) in one location may represent an abundance, and the same amount elsewhere may represent a serious deficiency.

The nineteenth-century German geographer Albrecht Penck recognized that in terms of the interaction of climate with other systems, three distinct *realms of climate* can be designated. The criteria that distinguish Penck's three natural realms of climate can be stated in terms of potential evapotranspiration. The *frozen* realm embraces regions where potential evapotranspiration equals 0. Plants cannot transpire when the temperature is too low, so the realm in which potential evapotranspiration equals 0 is characterized by low temperatures. The *dry* realm includes all regions where precipitation is less than potential evapotranspiration. Typical examples of the dry realm are dry grasslands and deserts, where deficiencies of soil moisture are common. The *moist* realm, in which forests are typically located, includes all regions where precipitation exceeds potential evapotranspiration.

The earth's regions can be classified as frozen, dry, or moist realms according to definite criteria that involve precipitation and potential evapotranspiration. The line where precipitation and potential evapotranspiration are equal marks the boundary between the dry and moist realms. Each realm can be subdivided by imposing criteria connected with energy, most often using some measure involving temperature. However, the resulting climatic types, such as hot-moist, cool-moist, and so forth, prove in practice to be inadequate to characterize the world's climates in sufficient detail, even on a global scale. The dry and moist realms require further division.

Different systems of climate classification divide the dry and moist realms in different ways. The remainder of the chapter discusses how this division is accomplished in two widely accepted systems of climate classification.

The Köppen System of Climate Classification

Wladimir Köppen, a German botanist and climatologist, developed and refined a system of climate classification over a period spanning the first half of the twentieth century. Köppen's classification of climates was strongly influenced by the work of the nineteenth-century plant geographers who mapped the vegetation of the world on the basis of extensive field studies.

Köppen's system employs five principal climatic types, which he labeled *A, B, C, D,* and *E*. The types *A, C,* and *D* are divisions of the moist realm, *B* is the dry realm, and *E* is a polar climatic type that includes the frozen realm. The *B* type is subdivided into the *BW*, or arid desert climatic type, and the *BS*, or semiarid steppe climatic type. Köppen subdivided the *E* climatic type into the *ET*, or moist tundra type, and the *EF*, or perpetual ice type. His definition of the *ET* climatic type illustrates his use of vegetation as an indicator of climate. The *ET* climatic type is represented by climates in which the average temperature of the warmest month falls between 0°C (32°F) and 10°C (50°F). Such a climate can support limited growth of vegetation, and the 10°C limit for the warmest month corresponds approximately to the poleward boundary of forests. By choosing this limit, Köppen ensured that regions with *ET* climates would be essentially treeless.

Köppen devised definite criteria based on temperature and precipitation to distinguish between principal climatic types. A schematic diagram of his method of division is given in Figure 7.9. The *A* climatic types are hot and moist, the *C* types are warm and moist, and the *D* types are cool and moist. The *B* climatic types, however, span a wide range of temperatures and a range of moisture. The boundary line of the *B* climatic type moves to greater precipitation as the temperature increases. The reason for this shift is that the boundary between the dry realm and the moist realm is intended to fall where precipitation and potential evapotranspiration are equal, so the boundary shifts because potential evapotranspiration increases with temperature.

The Köppen system was an outstanding achievement for its time, and it still provides a good orientation to the distribution of the world's climatic regions. One of the strong points of the Köppen system is its flexible terminology, which is outlined in greater detail in Appendix III. The flexibility of the system comes from the use of additional symbols to describe special features of climatic regions. The symbol *f*, for example, means that adequate precipitation falls in every month of the year. Therefore, the symbol *Af* specifies a tropical wet climate, as in the equatorial rainforests of South America. The symbol *Cf* describes the warm, moist climate of the eastern United States. The symbol *w* means essentially that little precipitation occurs in the winter, and *s*, that little precipitation occurs in the summer. Additional symbols are added to characterize special conditions such as frequent fog.

However, Köppen did not possess an accurate means of estimating potential evapotranspiration. He assumed that the transition between the dry *B* climatic type and the moist *A, C,* and *D* climatic types occurred at forest boundaries. By studying temperature and precipitation data for typical forest boundaries, Köppen established empirical relations involving temperature and precipitation that could be used to define the boundary of the *B* climatic types. His assumption that the forest boundary is where precipitation equals potential evapotranspiration is incorrect; precipitation exceeds potential evapotranspiration at most forest boundaries. Because his

$$\text{Moisture Index} = 100\left(\frac{P - PE}{PE}\right)$$

P = Precipitation

PE = Potential Evapotranspiration

Figure 7.7 The value of Thornthwaite's moisture index is positive for locations where precipitation exceeds potential evapotranspiration and negative where precipitation is less than potential evapotranspiration.

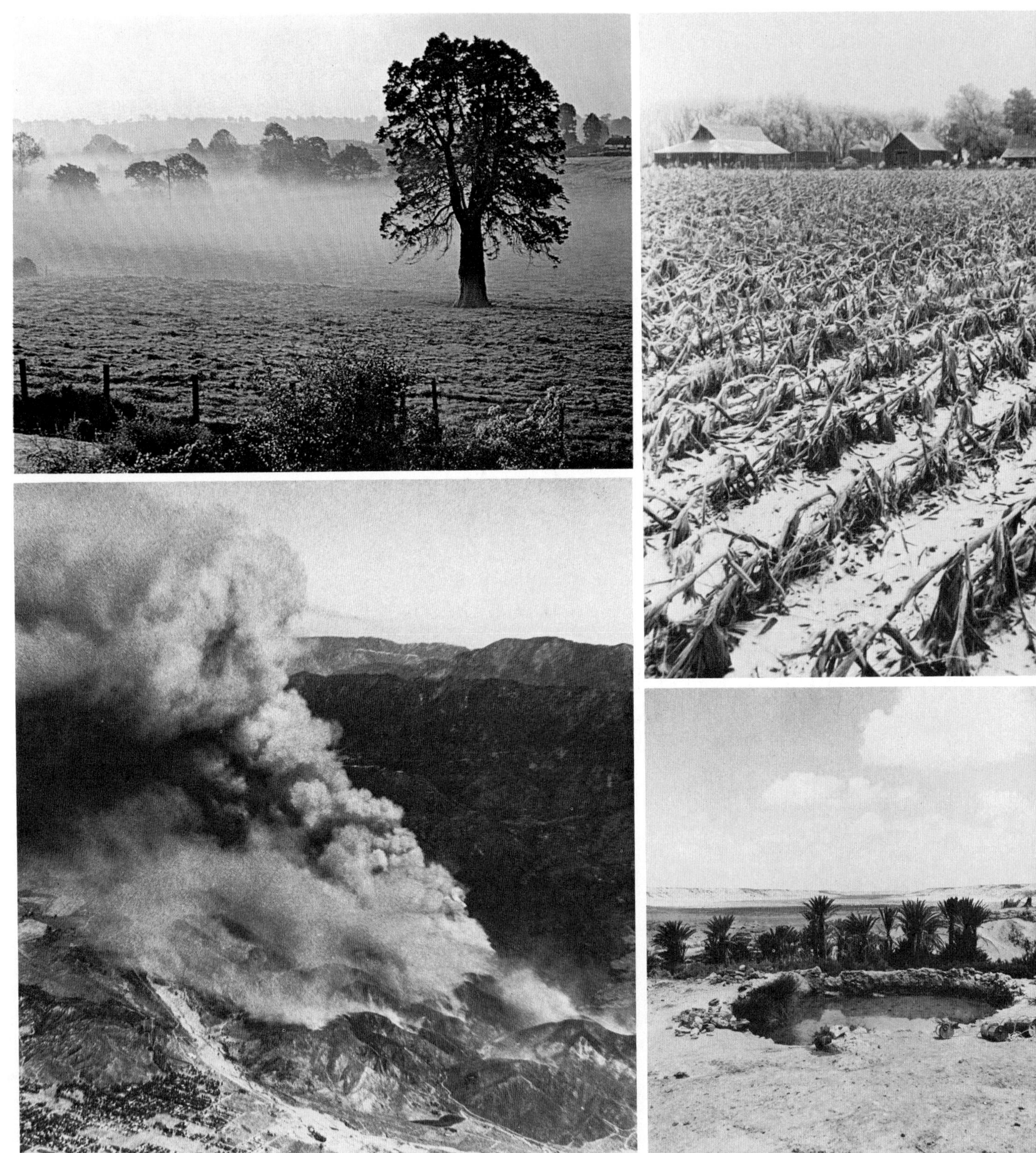

Figure 7.8 (opposite top left) England has a marine climate because of moist westerly winds from the Atlantic Ocean. The climate is moist throughout the year, with moderate temperatures. This early morning mist in Devon, England, formed in a layer of cool air near the wet ground.

(opposite bottom left) The Mediterranean climate of the coastal regions of southern California is moist during the winter and fosters the growth of vegetation. During the dry summer season, brush fires are common among the dry vegetation. The fire shown in the photograph occurred near the mountains east of Los Angeles, California, in early July after a moist winter.

(opposite top right) This snowy cornfield in Iowa exemplifies the seasonal temperature variations of a humid continental climate region. Snow is frequent in Iowa during the cold winter months, but sunshine and high temperatures make the summer an excellent growing season. Moisture is available throughout the year.

(opposite bottom right) Precipitation in arid climates seldom exceeds a few centimeters per year. This oasis in the Sahara depends on groundwater to supply the needs of humans and animals.

(top left) Convection currents rising from the warm tropical island of Fiji in the southwestern Pacific Ocean carry moist air upward. The water vapor condenses at a low altitude to form the fair weather cumulus clouds characteristic of moist tropical regions.

(center left) The climate near San Francisco, California, is moderate and supports the growth of subtropical vegetation. A rare occurrence of freezing temperatures during the winter of 1972 killed these eucalyptus trees on the hills of Berkeley, to the east of San Francisco.

(bottom left) In subarctic climate regions, snow and ice may persist through the year. The photograph shows icebergs breaking off, or *calving,* from a glacier in northern Iceland.

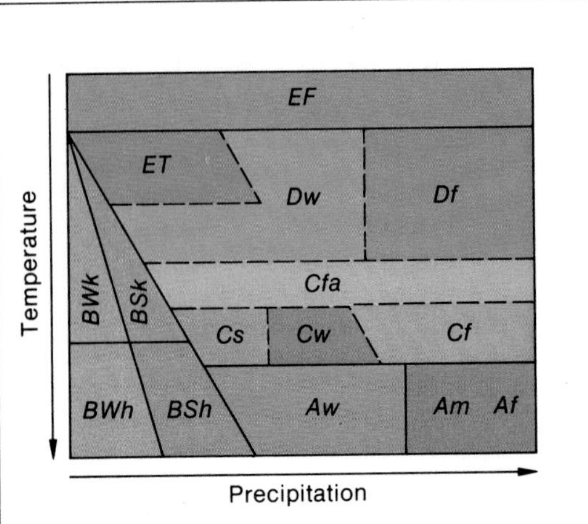
EF
ET
Dw
Df
Cfa
Cs
Cw
Cf
BWk
BSk
BWh
BSh
Aw
Am
Af
Temperature
Precipitation

Figure 7.9 This schematic diagram represents the principal climatic regions of the Köppen classification system in terms of average annual temperature and precipitation. The *E* climates, near the top of the figure, are the coldest, and the *A* climates, near the bottom, are the warmest. Reading from left to right in the direction of increasing precipitation, the *BW* climates are the driest and the *Af* climates are the wettest. The precipitation scale in particular is highly schematic; it shows general trends in each temperature zone, but it does not indicate, for example, that the annual precipitation in *Df* regions is much less than in *Af* regions. The Köppen climatic regions do not easily fit the scheme illustrated, because they are defined in terms of such factors as the seasonality of temperature and precipitation as well as on the basis of annual values.

Köppen-Geiger System of Climate Classification
Tropic of Cancer
Equator
Tropic of Capricorn
EF
ET
Dfc
Dfb
Dfa
Cfa
Cfb
Cfc
Csa
Csb
BSk
BWk
BSh
BWh
H
Aw
Am
Af
Cw

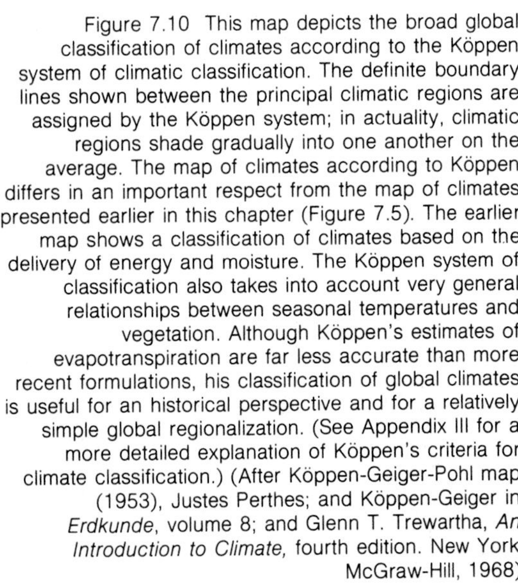

Figure 7.10 This map depicts the broad global classification of climates according to the Köppen system of climatic classification. The definite boundary lines shown between the principal climatic regions are assigned by the Köppen system; in actuality, climatic regions shade gradually into one another on the average. The map of climates according to Köppen differs in an important respect from the map of climates presented earlier in this chapter (Figure 7.5). The earlier map shows a classification of climates based on the delivery of energy and moisture. The Köppen system of classification also takes into account very general relationships between seasonal temperatures and vegetation. Although Köppen's estimates of evapotranspiration are far less accurate than more recent formulations, his classification of global climates is useful for an historical perspective and for a relatively simple global regionalization. (See Appendix III for a more detailed explanation of Köppen's criteria for climate classification.) (After Köppen-Geiger-Pohl map (1953), Justes Perthes; and Köppen-Geiger in *Erdkunde*, volume 8; and Glenn T. Trewartha, *An Introduction to Climate,* fourth edition. New York: McGraw-Hill, 1968)

empirical relations for the *B* climatic type boundary are inaccurate, he sometimes places the boundary hundreds of kilometers from the true boundary between the dry and moist realms.

The Köppen system has been criticized because many of its boundaries were developed to correspond to the boundaries of certain vegetation regions. The climatic regions defined by the Köppen system therefore automatically show a good correlation with regions of vegetation, and an evaluation of the relationship between climate and vegetation cannot be made objectively and independently with Köppen's climatic regions. Another criticism of the system is that the boundary between *Cs* and *Cf* climatic regions, for example, is based solely on precipitation, whereas the boundary of *B* regions depends on both precipitation and temperature in a crude measure of potential evapotranspiration.

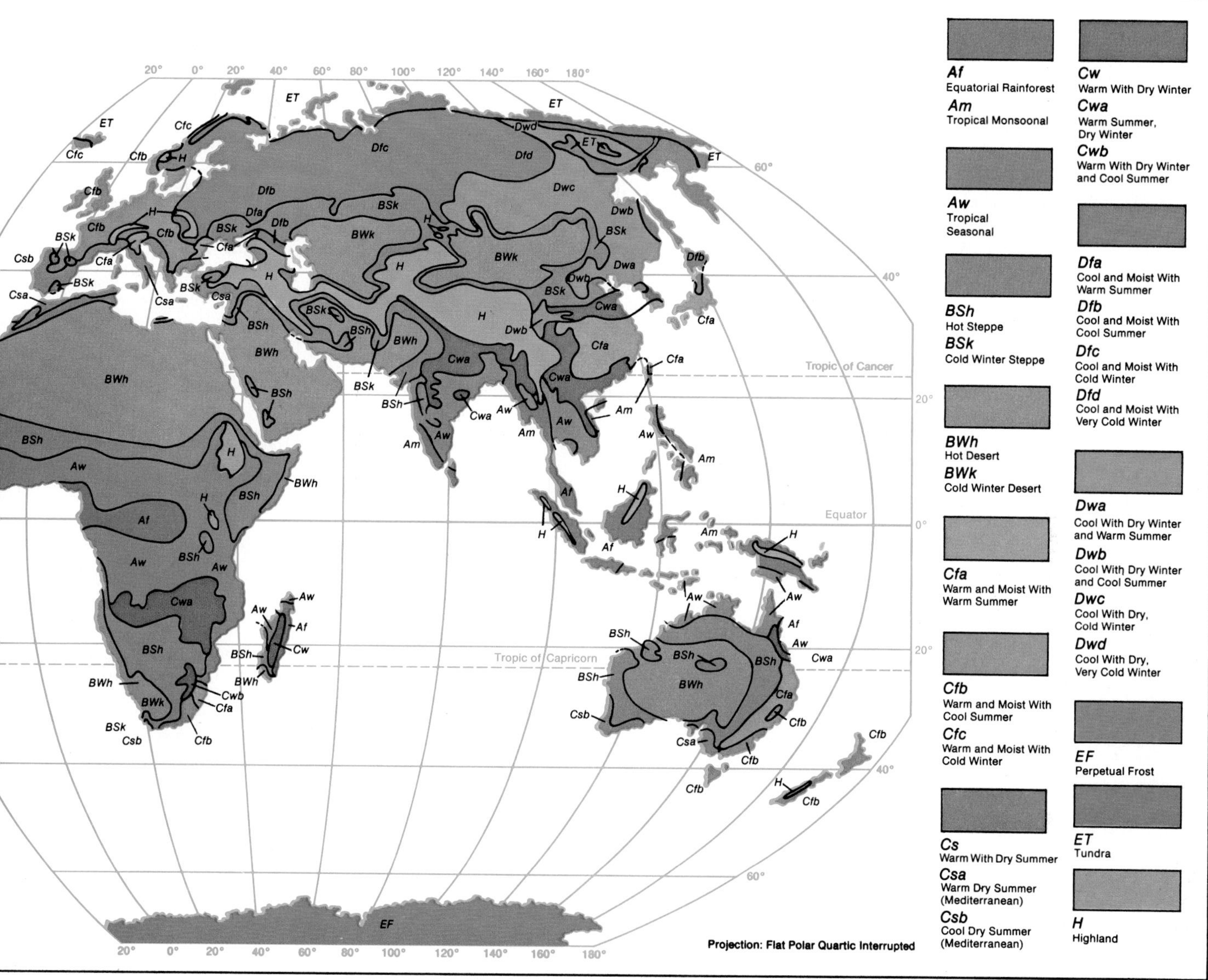
Af
Equatorial Rainforest
Am
Tropical Monsoonal
Aw
Tropical Seasonal
BSh
Hot Steppe
BSk
Cold Winter Steppe
BWh
Hot Desert
BWk
Cold Winter Desert
Cfa
Warm and Moist With Warm Summer
Cfb
Warm and Moist With Cool Summer
Cfc
Warm and Moist With Cold Winter
Cs
Warm With Dry Summer
Csa
Warm Dry Summer (Mediterranean)
Csb
Cool Dry Summer (Mediterranean)
Cw
Warm With Dry Winter
Cwa
Warm Summer, Dry Winter
Cwb
Warm With Dry Winter and Cool Summer
Dfa
Cool and Moist With Warm Summer
Dfb
Cool and Moist With Cool Summer
Dfc
Cool and Moist With Cold Winter
Dfd
Cool and Moist With Very Cold Winter
Dwa
Cool With Dry Winter and Warm Summer
Dwb
Cool With Dry Winter and Cool Summer
Dwc
Cool With Dry, Cold Winter
Dwd
Cool With Dry, Very Cold Winter
EF
Perpetual Frost
ET
Tundra
H
Highland
Tropic of Cancer
Equator
Tropic of Capricorn
Projection: Flat Polar Quartic Interrupted

However, the Köppen system is comprehensive and flexible, and its terminology is recognized by all geographers. Köppen's criteria for establishing the location of boundaries between different climatic regions are objective and easy to calculate from meteorologic data. Figure 7.10 shows the climates of the world according to the Köppen system.

The Thornthwaite System of Climate Classification

In an attempt to overcome the deficiencies of the Köppen system, C. Warren Thornthwaite devised in 1948 a method for classifying climates in which climatic types are defined in an orderly way according to water balance evaluations of energy and moisture. In Thornthwaite's classification system, energy is specified by poten-

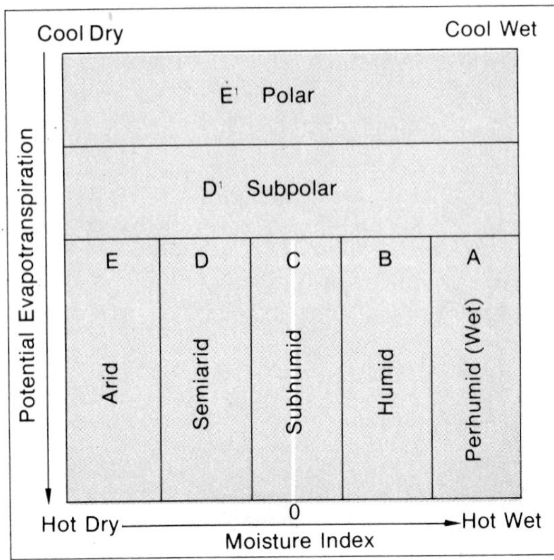

Cool Dry
Cool Wet
E' Polar
D' Subpolar
E
D
C
B
A
Arid
Semiarid
Subhumid
Humid
Perhumid (Wet)
Potential Evapotranspiration
Hot Dry
0
Hot Wet
Moisture Index

Figure 7.11 This schematic diagram represents the principal climatic regions in the Thornthwaite classification system. Potential evapotranspiration is used as a measure of moisture utilization by environmental systems on the earth's surface. The climates run from cool and dry at the upper left corner of the diagram to hot and wet at the lower right. The Thornthwaite system is systematic because the division of the diagram into blocks is done on the basis of orderly ranges of potential evapotranspiration values and ranges of the moisture index. Five principal moisture regions, *A* through *E*, are recognized in this system. Note that the subhumid climate region *C* is centered on the moisture index of 0, where precipitation and evapotranspiration are equal. (After Carter and Mather, 1966)

tial evapotranspiration, and moisture is specified by a *moisture index*.

Thornthwaite's moisture index depends essentially on the difference between precipitation and potential evapotranspiration. The moisture index has the value −100 when precipitation is 0, and may exceed +100 where rainfall far exceeds potential evapotranspiration. At the boundary between the dry and moist realms, the moisture index is 0.

Figure 7.11 shows schematically how climatic types are defined in Thornthwaite's classification system. There are no special criteria such as in the Köppen system, and the divisions are regular and orderly. Each climatic type corresponds to a well-defined range of energy and moisture values, which change proportionately from one climatic type to the next. The Thornthwaite system embodies a more quantitative measure of climate than does the Köppen system.

One of the principal differences between the Köppen and the Thornthwaite classification systems is in the way the dry and moist realms are divided. Köppen divided the moist realm into three principal climatic types, *A, C,* and *D,* and the dry realm into two types, *BW* and *BS*. Thornthwaite also divided the dry and moist realm into five principal climatic types: two in the moist realm, two in the dry realm, and one, the subhumid, that spans the dry and the moist realms. The subhumid climatic type contains the boundary where precipitation and potential evapotranspiration are equal. The soil and vegetation in the dry grasslands of the central United States are so distinctive that they appear to occupy a separate climatic region between the moister regions to the east and the drier regions to the west. When Köppen's criteria are applied to find the boundaries of climatic regions for the United States, the dry grasslands do not emerge as a distinct climatic region. However, the dry grasslands correspond closely to the subhumid climatic type of the Thornthwaite system. Figure 7.12 shows the regions of the United States divided according to Thornthwaite's moisture index alone. The subhumid region occupies the central United States.

Although the Thornthwaite system was not developed by using boundaries between vegetation regions, the system is oriented toward agriculture and forest management by its emphasis on energy and moisture, the factors most important for plant growth. Thornthwaite's method for estimating potential evapotranspiration is more accurate than the crude estimates built into the Köppen system.

Summary

The combined actions of solar radiant energy, precipitation processes, and the circulation of the atmosphere deliver energy and moisture to the earth's surface. The comparative stability of these factors gives a definite climate to each region of the earth. Many of the features of the earth's climatic regions can be understood by looking at the effects of global solar radiation, pressure, and wind distributions on a hypothetical continent. The continents, oceans, and mountain ranges further influence actual climates.

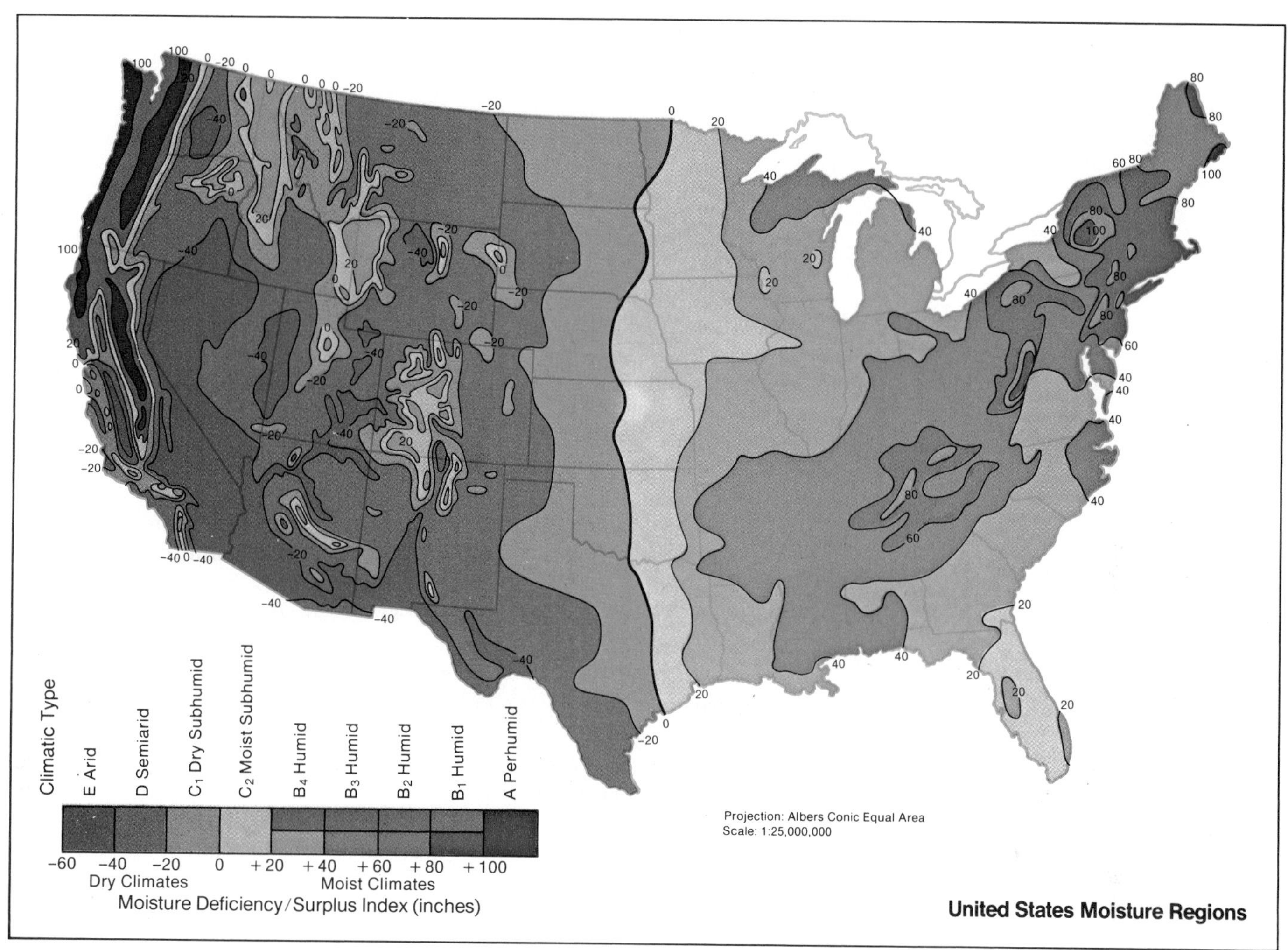
United States Moisture Regions
Climatic Type
E Arid
D Semiarid
C_1 Dry Subhumid
C_2 Moist Subhumid
B_4 Humid
B_3 Humid
B_2 Humid
B_1 Humid
A Perhumid
-60 -40 -20 0 +20 +40 +60 +80 +100
Dry Climates
Moist Climates
Moisture Deficiency/Surplus Index (inches)
Projection: Albers Conic Equal Area
Scale: 1:25,000,000

Because physical geographers are concerned with the response of such systems as soils, vegetation, and landforms to inputs of energy and moisture, they classify climate in terms of the interactions of energy and moisture with other systems. Components of the local water budget can be used as climatic indexes to classify the climate of a region according to a particular interaction. On the basis of these interactions, climate can be divided into three realms—frozen, moist, and dry—depending on potential evapotranspiration and its relation to precipitation. The three realms are subdivided according to moisture and energy in different ways in different systems of climate classification. The Köppen and Thornthwaite systems of classification are the most widely recognized. The following chapters on soil, vegetation, and landforms illustrate how systems on the earth's surface utilize the energy and moisture provided by climate.

Figure 7.12 This map of the conterminous United States shows the principal moisture realms as classified according to the Thornthwaite moisture index. Precipitation exceeds potential evapotranspiration where the index is positive, and it is less than potential evapotranspiration where the index is negative. The line where the moisture index is 0 extends from western Minnesota south to the Gulf of Mexico. Immediately to the east of this line there is a surplus of moisture, on the average, and immediately to the west there is a deficiency, on the average. The moisture index of a region bears a close relation to the form of its vegetation. Forests, for example, can maintain themselves only where the moisture index is positive and has a value of at least 20 or so. Only grasslands and desert vegetation can survive where the moisture index is negative. (After Thornthwaite and Mather, 1955)

8 Global Systems of Soils

Properties of a Soil
Texture and Structure
Chemistry of Solutions
Chemistry of Soils
Horizons and Profiles

Weathering: The Disintegration of Rocks
Physical Weathering
Chemical Weathering

Factors Affecting Soil Development
Inorganic Material
Topography
Climate
Organisms
Rate of Formation

Formation of Soil
The Leaching Process
Climate and Soils

Classification of Soil

Examples of Principal Soil Types
Tundra Soils
Podzols
Gray-Brown Podzolic Soils
Red-Yellow Podzolic Soils
Latosols
Chernozems
Prairie Soils
Chestnut-Brown Soils
Desert Soils
Bog and Meadow Soils

Soil Management

Lime Banks by Andrew Wyeth, 1962. (Collection, Mr. Smith W. Bagley; copyright © by Andrew Wyeth)

A thin layer of soil cloaks the earth's surface, softening its contours and supporting its vegetation. The soils we see today are products of the interactions of living and nonliving materials with energy and moisture over hundreds or thousands of years. The development of characteristic soil types is associated particularly with climate and with plant cover.

Global Systems of Soils

Sometime before the beginning of the ninth century, the desire for religious monasticism led Irish monks to found a monastery on Skellig Michael, a pyramidal bare rock 10 kilometers (6 miles) off the west coast of Ireland. The site offered the solitude and isolation they sought because the stormy Atlantic often prevented passage between the island and the mainland for weeks at a time. Birds and fish were available for food, but no herbs or flowers grew on the bare rock. So the monks laboriously built retaining walls in every niche of the steep crags to hold tiny pockets of soil they brought from the mainland.

Skellig Michael is unusual in having little or no natural soil of its own. The application of energy and moisture and the action of vegetation create soil nearly everywhere, even on the newly formed volcanic island of Surtsey near Iceland. Soil is a complex physical, chemical, and biological system on the earth's surface that consists of fine rock particles and organic material. This chapter discusses the processes that form soil and the properties of some common soil types. The soils you see today are products of the interaction of climate with organic and inorganic materials over hundreds or thousands of years.

A home gardener wants dirt that will grow flowers and vegetables. A farmer is interested in the complex factors that contribute to a soil's productivity. And a construction engineer needs to know if a soil is stable enough to support a highway or a bridge abutment. Knowing the capabilities and limitations of different soils is important to the progress of agriculture—in backyard gardens as well as in huge farming complexes. In agriculture, the soil's primary functions are to support plants mechanically, to store water for plant use, and to supply mineral nutrients to the plants. The soil's ability to perform these tasks varies with its chemical and physical makeup. Without productive soil, a plant cannot make full use of the available radiant energy and water. The Puritans who settled on the east coast of Massachusetts found the soil so poor that they learned to follow the Indian practice of burying fish with the seeds for fertilizer. A fertile soil is so essential to agricultural productivity that some people have chosen to farm productive soils on the slopes of active volcanoes despite the imminent danger of an eruption.

Properties of a Soil

The inorganic matter in soil consists of rock particles from the earth's crust. The chemical elements most abundant in crustal rock—oxygen, silicon, aluminum, and iron—are frequently found in soils. Because oxygen reacts easily with other elements, it occurs only in chemical combination with other elements. Silicon dioxide, a combination of silicon and oxygen, is a major constituent of many soils; both beach sand and quartz crystals are forms of silicon dioxide. Many other elements are represented in the inorganic portion of soil, in quantities that depend on the history of the soil and on the nature of the parent rock materials that formed the soil.

Chemical elements in a soil can combine to form a wide variety of compounds. Some compounds of iron and oxygen give a distinctive reddish color to the soil, as in the soils of the Amazon rainforest. Other iron compounds often give the soil a yellowish color. Sometimes soluble mineral salts of calcium and magnesium accumulate in the pores of a soil, giving it a whitish appearance. However, various compounds of silicon and oxygen are the building blocks of most soils.

Rock particles alone do not make a soil, because soils contain organic matter in addition to minerals. No soils existed until life appeared on the land. When plants and animals die, bacteria of decay reduce them to semisoluble organic matter, which gradually washes down into the soil. A product of decomposition is flaky black organic matter called *humus*. Humus mixes with the mineral matter of the soil and becomes largely inseparable from it. Humus formation is inhibited in waterlogged soils, because most bacteria of decay require oxygen to live.

The nitrogen essential for plant growth comes primarily from the humus in the soil. Among the products of humus decomposition are soluble compounds of nitrogen that the plant roots can readily absorb. In addition, certain bacteria, which live freely in the soil or in the root nodules of certain plants, can convert nitrogen from the atmosphere into nitrogen compounds that the plants can utilize.

Texture and Structure

The *texture* of a soil refers to the distribution of the size of its particles. Particle sizes are conventionally classified into three main groups: sand, silt, and clay. In the United States, particles with diameters between 50 and 2,000 microns are called *sand*, particles from 2 to 50 microns are called *silt*, and those smaller than 2 microns are called *clay*, as shown in Table 8.1. The designations sand, silt, and clay refer only to the sizes of the particles and not to their chemical compositions.

The United States Soil Conservation Service has set up a broad classification of soils according to texture. The texture class to which any soil belongs can be determined by measuring the proportions of sand, silt, and clay in the soil. Figure 8.3 gives the class names for soils of different textures. Texture influences a soil's ability to accept and to store moisture. Coarse-grained sandy soils absorb water easily but cannot store water through long periods of drought. The pores in fine-grained silt and clay soils are efficient in storing moisture, but the small surface pores make infiltration difficult.

Grains of soil join together in larger units called *peds*. The shape and size of the peds characterize the *structure* of a soil. The peds can be platelike, granular, or prismlike. Plants generally grow best in a soil with a surface structure consisting of peds 1 to 5 millimeters (0.04 to 0.2 inches) in size. Smaller peds reduce the infiltration of

Table 8.1. Grades of Soil Texture
United States Department of Agriculture Classification

Grade	Size (millimeters)
Very Coarse Sand	1-2 (0.04-0.08)*
Coarse Sand	0.50-1 (0.02-0.04)
Medium Sand	0.25-0.5 (0.01-0.02)
Fine Sand	0.10-0.25 (0.004-0.01)
Very Fine Sand	0.05-0.1 (0.002-0.004)
Silt	0.002-0.05 (0.00008-0.004)
Clay	less than 0.002 (less than .00008)

*Numbers in parentheses indicate inches.

Source: United States Department of Agriculture. 1951. *Soil Survey Manual.* United States Department of Agriculture Handbook No. 18. Washington, D. C.: United States Department of Agriculture.

Figure 8.2 The particles that constitute the inorganic component of soil are classified according to size. Particles smaller than 2 microns in diameter are called clay, particles from 2 to 50 microns in diameter are called silt, and larger particles are considered sand or gravel.

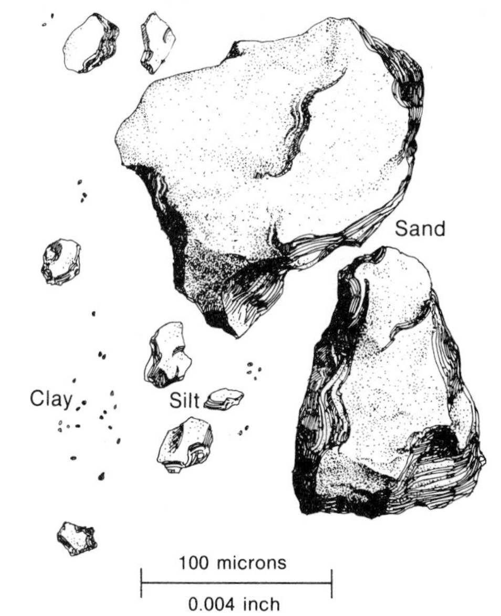
Sand
Clay
Silt
100 microns
0.004 inch

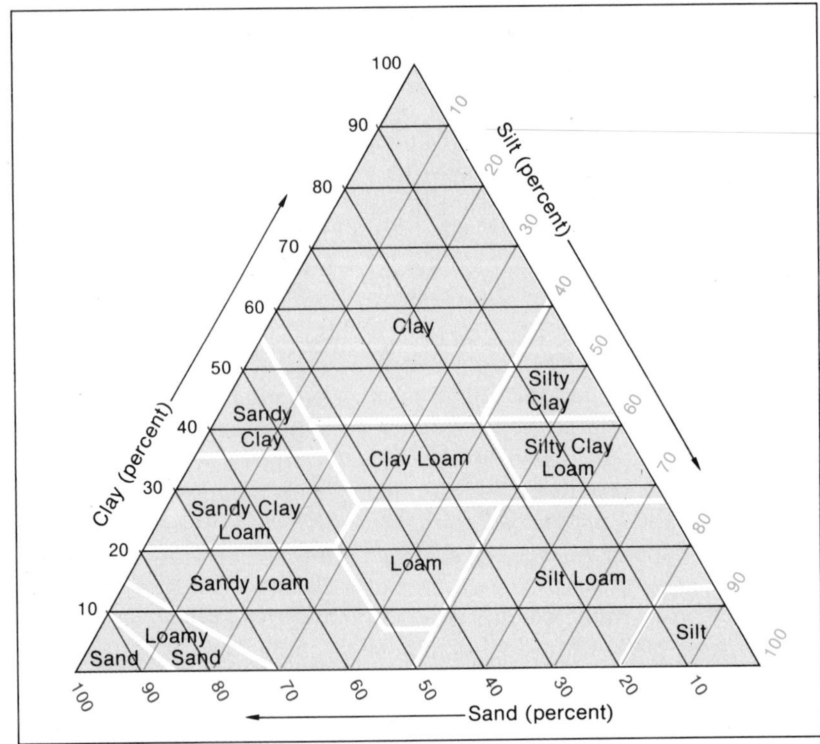
100
90
80
70
60
50
40
30
20
10
Clay (percent)
Silt (percent)
Sand (percent)
Clay
Silty Clay
Sandy Clay
Clay Loam
Silty Clay Loam
Sandy Clay Loam
Loam
Sandy Loam
Silt Loam
Loamy Sand
Sand
Silt

Figure 8.3 The texture of a soil is determined by measuring the proportions of clay, silt, and sand in the inorganic part of the soil. Texture is measured by sifting the soil sample through a series of screens graded from coarse to fine. The soil texture triangle shown in the figure can be used to classify the texture of a soil sample once the percentages of the components are known. If a soil sample contains 30 percent clay and 40 percent sand, for example, it would be classified as a clay loam. (After Bridges, 1970)

air and water into the soil, and larger peds leave too much space around small plant roots. One of the reasons for plowing soil before planting is to make the structure of the soil as suitable for plant growth as possible. A stable ped structure helps reduce the erosive effect of wind on soil.

Chemistry of Solutions

Electric forces bind atoms together in molecules. Many of the simple soluble chemical salts of interest in soil chemistry are held together *ionically*. An ion is an electrically charged atom that has gained or lost one or more electrons. In an ionically bound molecule, one or more of the negatively charged electrons leaves one atom for another. Atoms of metallic elements usually lose electrons when they form ionic molecules. In a sodium chloride molecule, for example, an electron leaves the sodium atom and goes to the chlorine atom. A positively charged ion, which has lost one or more electrons, is called a *cation*. A negatively charged ion, one which has gained electrons, is called an *anion*.

When an ionic molecule is dissolved in water, the electric forces exerted by the water molecules split the ions apart. Because they are electrically charged, ions exert much stronger forces than electrically neutral atoms and are therefore much more reactive chemically. The positively charged hydrogen ion plays a particularly important role in chemistry, because it is always formed when acids are dissolved in water.

The *acidity* of a soil is one of the measures used to characterize a soil. Acidity is measured in terms of the concentration of hydrogen

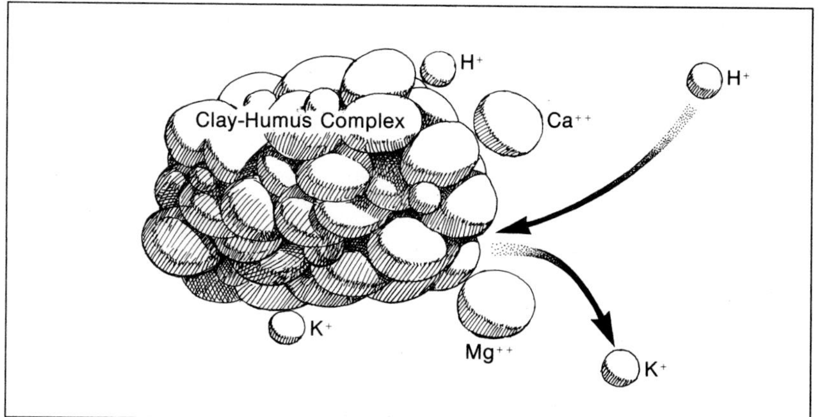
H+
H+
Clay-Humus Complex
Ca++
K+
Mg++
K+

Figure 8.4 The fine inorganic particles and humus in a soil bind together to form the clay-humus complex. On a submicroscopic scale, the particles of the clay-humus complex act like giant molecules with the power to attract ions electrically. The complex performs an important function in a soil by preventing nutrient materials needed by plants from washing out of the soil. An acid soil contains numerous hydrogen ions that can replace other ions on the surface of the complex, as the figure shows. Hence an acid soil soon loses much of its content of inorganic nutrients.

ions according to the pH scale. Pure water, which has been assigned a pH of 7, contains 10^{-7} gram of hydrogen ions per 1,000 cubic centimeters. A weak acid with 10^{-6} gram of hydrogen ions per 1,000 cubic centimeters has pH 6, and so on. The pH scale extends in the opposite direction as well. Values greater than pH 7 indicate basic, or alkaline, soils. Natural soils have pH values between 3 and 10; values between 5 and 7 seem best for growing most crops (see Figure 8.5).

Chemistry of Soils

The chemical behavior of soil is closely connected with its physical characteristics. The fine clay particles in soil become attached to small particles of humus, forming a clay-humus complex that has such small particles that chemically they behave like large molecules. The clay-humus complex plays an essential part in maintaining soil fertility. The salts necessary for plant nutrition are dissolved in soil moisture, where they dissociate into ions. The clay-humus complex attracts ions, particularly cations, and holds them by electric attraction. In this way the complex prevents soil moisture from washing salts out of the soil.

In a productive soil, the clay-humus complex maintains a delicate balance. Its hold on the essential cations must be strong enough to prevent the cations from being washed away, but weak enough to allow the plant roots to absorb cations as required. The clay-humus complex holds different kinds of cations with different degrees of strength. A cation weakly bound to the complex can be replaced by a new cation that will be more strongly bound. This process of *cation exchange* can cause the soil properties to vary with time as one type of bound cation gradually replaces another.

Figure 8.5 The pH value of a soil, or its hydrogen ion concentration, is one of the measures that can be used to estimate a soil's suitability for agriculture. A low pH indicates that a soil is acid and may have lost many of its nutrient salts by exchange with hydrogen ions. A high pH indicates that a soil contains strong alkalis, which may be damaging to plant root tissues. (After Lyon and Buckman, 1943)

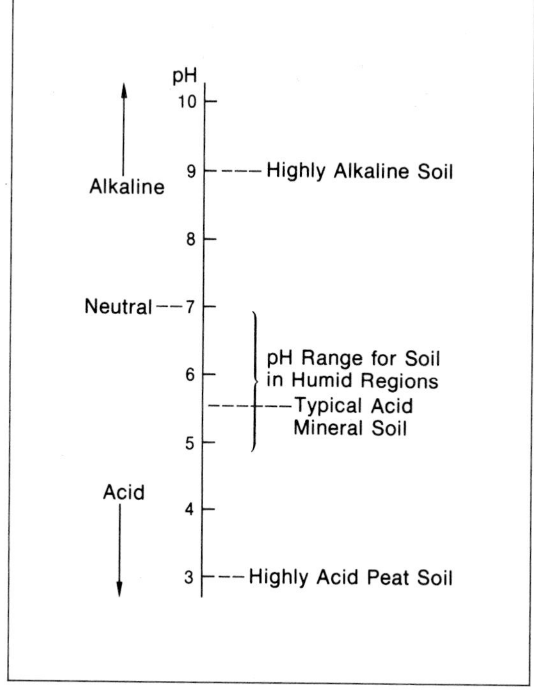
pH
10
9 Highly Alkaline Soil
Alkaline
8
Neutral 7
6
pH Range for Soil in Humid Regions
Typical Acid Mineral Soil
5
Acid
4
3 Highly Acid Peat Soil

Horizons and Profiles

When a digging exposes a vertical section of a soil, several distinct horizontal layers, or *horizons*, are apparent (see Figure 8.6). The downward flow of water through a soil tends to remove substances from the upper parts of the soil and to redeposit them in lower parts, forming distinct layers. Loss of materials from a soil layer by downward movement is known as *eluviation*, and accumulation

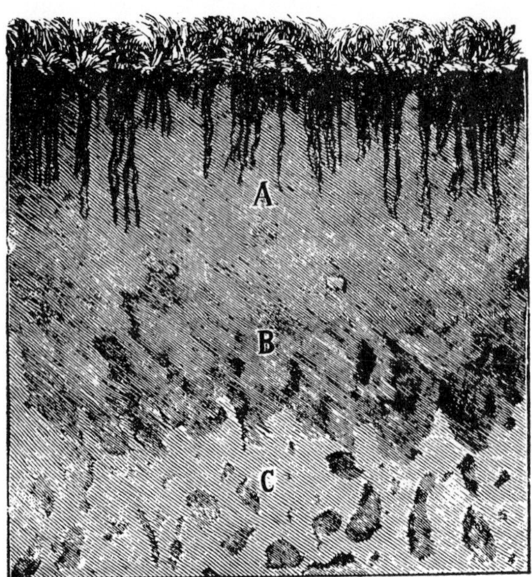
A
B
C

Figure 8.6 V. V. Dokuchaiev, the father of soil science, used this drawing to illustrate the soil profile of chernozem soil, a productive agricultural soil found in Russia and in the United States. The *A* horizon at the top is rich in organic material, which soil animals and percolation of water rapidly carry downward. The *C* horizon in a chernozem soil often contains the burrows of soil animals, lined with lime from the calcium salt concentrations in the soil.

of materials in lower layers is known as *illuviation*. The horizons are labeled *A, B, C,* and so on from the top of the soil down. The uppermost, or *A*, horizon is a zone of eluviation, and the next layer, the *B* horizon, receives material from above. The *B* horizon tends to have less organic matter than the *A* horizon, and it may contain fragments of the parent inorganic material. The *C* horizon largely consists of weathered parent material. In certain types of soils, the horizons are not fully formed.

How can you distinguish one soil from another? A number of factors characterize a soil: chemical content, color, texture, structure, and acidity. Usually the most distinguishing characteristic of a soil is the degree of development of the horizons in the vertical section. This *profile* development is a record of the past and present influences that have acted on the soil.

Weathering: The Disintegration of Rocks

The inorganic component of soil comes from the *weathering* of rocks. Soil may form at the site of weathering if the organic components are available, or wind and water may transport the rock particles to a new location. Weathering of rock can occur in the absence of vegetation, but plants exert physical and chemical effects on rocks that speed the weathering process. Physical processes merely break apart rocks, whereas chemical processes change the nature of the rock substance. Both processes are important in the preparation of rock particles for soil formation.

Physical Weathering

Igneous rocks formed deep underground are subject to great pressure. When geologic processes raise rocks to the surface, the rocks may split into sheets, sometimes with explosive force. Cracks and pores help to induce weathering, because most weathering processes occur wherever rock is exposed to air and to moisture. The

Figure 8.7 Granite rock usually breaks into slablike sheets when it is exposed at the earth's surface. The joints in this granite at Pike's Peak, Colorado, have been enlarged by physical and chemical weathering active at the exposed surfaces.

weathering of igneous rock during geologic time produced sediment that eventually hardened into consolidated sedimentary rock.

Water expands in volume by nearly 10 percent when it freezes. Expansion in a confined space can generate high disruptive pressures, which is why exposed water pipes may split during a hard freeze. Frost, one of the most effective physical agents of weathering, is most active in the moist middle latitudes, especially where extreme temperature changes cause frequent cycles of freezing and thawing. In hot, arid regions, the force exerted by the growth of salt crystals in small cracks and pores of rocks can dislodge rock particles.

The surface of a rock exposed to the sun becomes warm and expands more than the cooler inner portions. Except in isolated cases, however, nonuniform heating of a rock does not appear to produce strains great enough to cause significant cracking. Other minor processes of physical weathering include the abrasive effect of wind-borne sand on rock and the mechanical abrasion of rocks tumbling together in streams.

Chemical Weathering

Chemical weathering processes occur only in the presence of water. In moist climates chemical weathering is more important, on the whole, than physical weathering, and even in arid regions, chemical weathering plays a significant role in breaking down rocks.

Rain water and surface water, which are often slightly acid, react chemically with many of the minerals in rocks and convert them to soluble or weakened forms. The acids plant roots produce tend to promote chemical weathering of rocks. Because chemical weathering occurs at surfaces, it proceeds most rapidly when the weathered material is steadily removed so that fresh rock surfaces are exposed to attack.

Factors Affecting Soil Development

The science of soils, called *pedology*, has been actively pursued in Russia for more than 100 years. Vasilii Vasielevich Dokuchaiev, a Russian soil scientist and the leading pedologist of the nineteenth century, enumerated several factors that influence the development of a soil: (1) inorganic parent material, (2) aspects of topography, such as slope and drainage, (3) climate, and (4) plants and animals. In addition, many soils require long time periods to come to equilibrium under given conditions.

Inorganic Material

As you have read, the inorganic parent rock material determines which elements are present in soil. In addition to the common elements oxygen, silicon, aluminum, and iron, soils frequently contain the alkaline metals sodium and potassium and the alkaline earth metals magnesium and calcium. From an agricultural standpoint, calcium is a desirable component of soils. Acid water moving through the soil tends to remove alkaline materials from the clay-humus complex by exchanging the alkaline cations with hydrogen

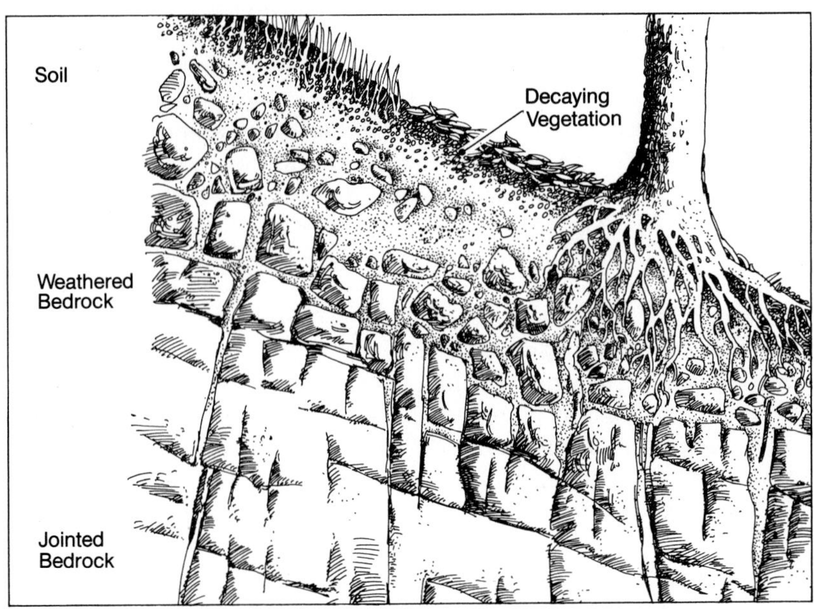
Soil
Decaying
Vegetation
Weathered
Bedrock
Jointed
Bedrock

Figure 8.8 Chemical and physical weathering processes break down massive rock into the small particles that form the inorganic component of soil. In warm, moist climates, weathering proceeds actively along the exposed surfaces of jointed rock. Some of the mineral nutrients supplied to the soil by the parent rock are incorporated into growing vegetation. The nutrients are returned to the surface of the soil when plants decay, and rain washes them downward into the soil.

ions. The soil therefore becomes more acid and less suitable for agriculture. If the mineral portion of the soil is rich in calcium, however, calcium compounds can steadily enter the soil moisture to replace the calcium that has been carried away.

Many other elements, such as phosphorus and titanium, are found in soils; some of them are present only in small *trace* amounts. Traces of elements such as boron, sulfur, and copper are necessary for plant nutrition because they allow a plant to produce the chemicals required for photosynthesis and growth. Plants require so little copper that the application of only a few pounds of copper compounds per acre may be sufficient to restore copper deficiencies. The trace elements that humans require for proper nutrition come indirectly from the soil through plants in the diet.

Although the parent material of a soil determines which elements are in a soil, the original rock does not necessarily exert a dominant effect on soil formation. Many other soil properties, such as texture and structure, depend primarily on particle size and humus content. A single type of parent rock can therefore give rise to soils that have dissimilar properties, and soils from two different kinds of parent rock may have similar properties.

Topography

The soil at the base of a steep hill often differs from the soil on the slopes because flowing water and gravity transport material down the slope toward the base. Fine weathered particles tend to accumulate at the base, leaving coarser material on the slope. Flowing water may also carry away upper layers of soil, so that the soil layer on the slope is thinner than at the base. The plant communities on a slope are usually not the same as the vegetation on level ground nearby because of different exposures to the sun, different temperature conditions, and different rates of water drainage. The organic content of soils on a slope therefore differs from the organic content of soils on a lowland. If drainage water collects at the base of a hill, an acid, peaty soil, rich in organic material, may be formed because plants decay slowly in watery soils.

Climate

Climate strongly influences the formation of soil. On a global scale, soils show a close relationship to climatic zones and to vegetation zones. The energy and moisture the atmosphere delivers affect the rate of vertical movement of soil moisture and dissolved salts in soils, the rate of erosion and weathering, and the rate of chemical reactions in soils. In addition, soils often show a close association with vegetation types, which are in turn influenced by climate. Russian pedologists showed that over time, soils that start from reasonably similar parent materials eventually reach forms that are primarily determined by climate. The soil-forming processes and soil types discussed later in this chapter are closely related to climate (see Figure 8.9).

Organisms

Life is essential to the formation of soil. Plants and animals supply organic matter, which bacteria and fungi break down to forms plants can use. Bacteria also make nitrogen available to plants by converting atmospheric nitrogen to soluble salts. The lichens clinging to a rock in the Antarctic produce a tiny amount of "soil" by pulling flakes of rock away and dissolving them with plant acids.

Soils contain numerous organisms in addition to bacteria. The largest soil animal is the earthworm—an acre of typical midlatitude soil contains earthworms that equal the weight of a cow. Earthworms improve soil by ingesting and breaking down plant debris, but more important, they improve soil texture by dispersing soil grains and by opening up channels for air and water in the soil. In soils where improper use of pesticides has destroyed the earthworm population, or in acid soils where earthworms cannot live, the mixing of new humus into the soil takes longer than it would if earthworms were present.

Rate of Formation

The rate of soil formation has been estimated in a few special cases from knowledge of geologic events. After the explosion of the volcano Krakatoa, for example, soil became established in the volcanic ash at the rate of about 1 centimeter per year (0.4 inch per year), an exceptionally rapid rate of formation. Volcanic ash in Central America developed soil 30 centimeters (1 foot) thick in 1,000 years.

Soils tend toward a state of equilibrium with climate and with the controls of the environment. However, significant climatic and environmental changes may occur more rapidly than soil can respond. So soils seldom reach a completely stable form; they continually undergo slow change toward new conditions. In parts of

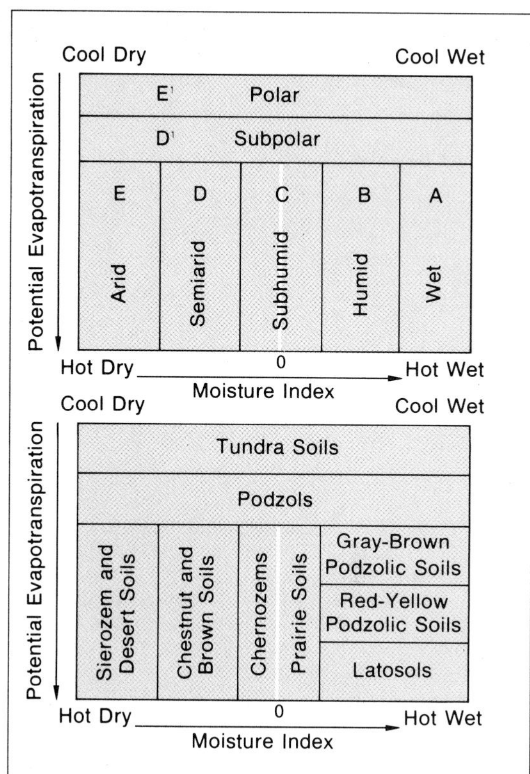

Cool Dry
Cool Wet
Potential Evapotranspiration
E'
Polar
D'
Subpolar
E
D
C
B
A
Arid
Semiarid
Subhumid
Humid
Wet
Hot Dry
0
Hot Wet
Moisture Index
Cool Dry
Cool Wet
Potential Evapotranspiration
Tundra Soils
Podzols
Sierozem and Desert Soils
Chestnut and Brown Soils
Chernozems
Prairie Soils
Gray-Brown Podzolic Soils
Red-Yellow Podzolic Soils
Latosols
Hot Dry
0
Hot Wet
Moisture Index

Figure 8.9 Climate is one of the important factors in soil formation. The top diagram shows the principal climate regions according to the Thornthwaite system. Potential evapotranspiration, a measure of energy, increases downward from cool to hot on the vertical scale. The moisture index, a measure of the moisture available for systems, increases from dry to wet along the horizontal scale. The bottom diagram classifies the principal soil types according to the same method.

The similarities between the two diagrams show that some soils form primarily in a particular climatic region. Tundra soils, for example, are found in polar climates, whereas chestnut and brown soils are found in semiarid regions. Chernozems and prairie soils occur in the subhumid climate region, where the moisture index is small or 0. In the humid and wet climate regions, the soils arrange themselves primarily according to energy rather than to moisture, partly because of the important effect of vegetation type on soil development.
(After Blumenstock and Thornthwaite, 1941)

Figure 8.10 The first stage in the formation of a soil may be the dissolution of massive parent rock by the attack of plant acids from lichens. When a suitable base has been established, rapidly growing plants can invade the soil and accelerate the process of soil formation.

Figure 8.11 In the leaching process shown here schematically, rain water, which has become acidic from dissolved carbon dioxide or from organic acids on the ground, infiltrates the soil. The hydrogen ions in the acid replace alkali and alkaline earth ions from the clay-humus complex, causing a downward movement of nutrient salts. If the acidity of the soil becomes high, the clay particles disintegrate chemically, usually into compounds of aluminum and iron. (After Bridges, 1970)

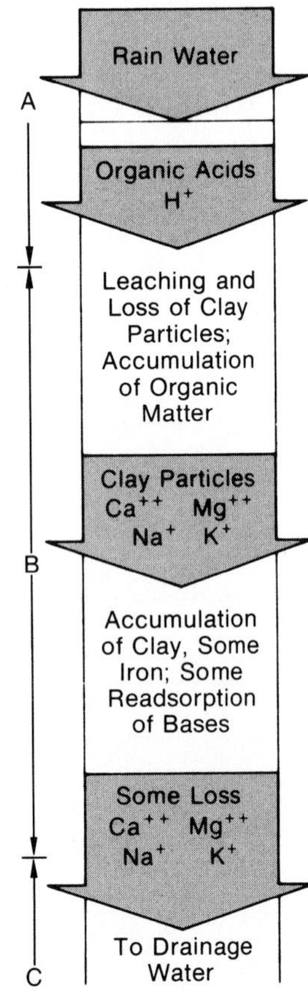
Rain Water
A
Organic Acids
H⁺
Leaching and Loss of Clay Particles; Accumulation of Organic Matter
Clay Particles
Ca⁺⁺ Mg⁺⁺
Na⁺ K⁺
B
Accumulation of Clay, Some Iron; Some Readsorption of Bases
Some Loss
Ca⁺⁺ Mg⁺⁺
Na⁺ K⁺
To Drainage Water
C

Europe, subsoils that were subject to glaciation 10,000 years ago show similarities to arctic soils, but upper soil layers reflect the warmer climate of more recent times.

Formation of Soil

There are a number of chemical and mechanical processes that produce distinctive soils or that alter the nature of a soil. Because these processes depend on environmental controls, soil types are often associated with particular climatic and vegetation zones.

The Leaching Process

Leaching is a process that occurs in many types of soils. When weakly acid solutions from rain water and from plant acids are washed downward through a soil, the hydrogen ions in the acid displace other cations from the clay-humus complex. The displaced cations include sodium, potassium, magnesium, and calcium, all of which form alkaline compounds. Thus in leaching, alkaline materials move downward in the soil, while the upper layers become more acid from the increased hydrogen ion concentration, which Figure 8.11 shows. The white salty material that often appears on the outside of clay flower pots is evidence of leaching in the soil.

Climate and Soils

The chemical processes that operate in soils differ in different climatic realms, so that soils often exhibit distinct characteristics associated with climate. How differences in temperature and moisture affect the mobility and reactions of ions in the soil is not fully understood. Until fundamental processes of soil chemistry are better known, it is preferable simply to describe the observed properties of soils in different climates.

Different ions move and accumulate in different realms of climate. In dry realms, soluble calcium salts often become concentrated in the soil. In the calcified soils found in semiarid plains and grasslands, a white layer of calcium salts frequently accumulates

Figure 8.12 These ferns have established themselves in a lava bed in Kilauea crater, Hawaii, only 4 years after the lava flow. In time, weathering of the lava and addition of organic matter from the plants will cause a well-developed soil to form.

below the *A* horizon. In extremely dry climates, calcium and magnesium salts concentrate near the surface of the soil, sometimes causing white patches of salts to form on the surface of the ground. The salts may fill the pores of the upper layers of the soil, giving the *A* horizon a characteristic white appearance.

Iron and silica move through soils at different rates in different climatic realms. In moist, *cool* regions, iron is mobile and silica remains fixed. The profiles of such *podzolic* soils have an ash-gray *A* horizon where the soil has become richer in silica. The *B* horizon may be reddish or yellowish because of the presence of iron compounds illuviated from above. In moist, *warm* regions, iron is fixed and silica is mobile. The *lateritic* soils, or *latosols*, formed in such regions are usually red or yellowish red from iron compounds. The abundant energy and moisture in moist, warm regions produce deep weathering, and latosols can be many meters thick.

Where rainfall is abundant, water normally passes through a soil and percolates into the parent rock and groundwater. If the underlying rock is impermeable to water, however, the soil pores fill with water and the soil becomes waterlogged. The bacteria of decay cannot exist in a waterlogged soil because of the lack of oxygen. The top portion of a waterlogged, or *gley*, soil consists of a thick layer of partly decayed, peaty plants. Below this black horizon, there is usually a bluish gray horizon colored by iron compounds.

Classification of Soil

The most useful methods of soil classification attempt to group together soils with similar characteristics. In this way, classification systems are useful for organizing a large amount of information into a compact form. A workable system must possess general categories that emphasize the similarities among soils in the same category and that bring out the differences among soils in different categories. It must also possess highly specific subcategories so that a particular soil can be described precisely and unambiguously.

Classifying soils is far more difficult than classifying climates.

Figure 8.13 Vegetation is an important factor in the formation of soils, and some major types of soil are characteristically associated with certain kinds of vegetation, as this figure illustrates. The separate effects of climate and of vegetation on soil development are sometimes difficult to differentiate, because climate influences the growth of vegetation. Nevertheless, vegetation should be considered an independent factor in soil formation; in the same climate region, for instance, different types of soil develop under different types of vegetation. The influence of the parent rock on soil is usually minor compared to the effects of climate and vegetation. The figure shows several different kinds of soil developing from the same type of parent rock.

Tundra soils. The vegetation above tundra soils consists of mosses, lichens, low shrubs, grasses, and flowering plants. The character of tundra soils is influenced by the presence of a permanent frost layer several feet below the surface, which prevents water from draining freely through the soil. In some places wedges of solid ice remain in the soil throughout the year. (See also Figure 14.16, Cold Climate Landscape)

Podzols develop beneath coniferous forests. The soils tend to be moist because of the shade and protection offered by vegetation cover. The moisture and the surface layer of acid humus promote rapid leaching, which leaves a gray layer of silica after the clay is dissolved.

Tundra Soils

Podzols

Chernozems are formed under dry, short-grass prairie and are classified as pedocals. Grassland soils are well mixed with humus because of the extensive root network. The amount of leaching is small, so that clay minerals remain intact and the soil retains alkaline nutrient compounds.

Prairie soils are the soils of the moist, tall-grass prairie and are classified as pedalfers. The grasses absorb and return large amounts of mineral nutrients, so that despite the abundant moisture the soils retain mineral nutrients and are only slightly acid. The development of some prairie soils may have been influenced by the presence of forest at one time.

Chernozems

Prairie Soils

Gray-Brown Podzolic Soils

Latosols

Gray-brown podzolic soils develop beneath the broadleaf deciduous forests of the moist midlatitudes. Such soils tend to be rich in mineral nutrients, partly because deep roots penetrate the parent rock. The nutrients stored in the vegetation seasonally return to the surface when leaves fall.

Latosols are the typical soils of the rainforest. The heat and abundant moisture cause deep weathering and vigorous leaching, but the steady decay of plant litter on the forest floor returns mineral nutrients to the soil for uptake by vegetation.

Desert Soils

Bog and Meadow Soils

Desert soils are low in humus because of the sparse vegetation, and they tend to be high in alkaline nutrients because of the lack of moisture.

Bog and meadow soils develop in waterlogged land. The vegetation consists of low shrubs, herbs, mosses, and grasses; trees cannot subsist in perpetually wet land. The soil is highly acid from decaying plant material, and the upper horizon may be dark peat. If a changing water level allows air to reach lower horizons, grayish oxides of iron may form.

Climates are described by comparatively few variables—primarily temperature and rainfall distributions. Soils, on the other hand, possess a large number of distinctive characteristics of relatively equal importance, such as texture, structure, character of the parent material, humus content, chemical composition, nature of the horizons, and so forth. Furthermore, climatic zones cover broad areas, whereas soil types may differ within small areas. Soil classification systems should also allow for man's modification of the soil. Two farms in Ohio a mile apart have the same temperature and rainfall, but their soils may be noticeably different, possibly as a result of different agricultural practices.

To be of value to agriculture, a system of soil classification must be comprehensive enough to describe thousands of soil types. However, the generalized global distribution of a dozen or so major soil types shows the general features of soils and their relationships to climate and vegetation. The major soil types are closely associated with specific climatic and vegetation zones because temperature, moisture, and plant cover exert controls on soil-forming processes.

The pedologist Dokuchaiev recognized that many soils have achieved an equilibrium with their local climate and plant cover. Climatic and vegetation zones tend to foster the development of soils of definite characteristics, called *zonal* soils. Zonal soils usually exhibit distinct horizons, and their profiles reflect their climatic and vegetation zones. Soils that have not had time to come to equilibrium with the climate, such as newly formed volcanic ash, are called *azonal* soils. They do not have developed profiles. Soils whose main characteristics are determined primarily by local conditions—a waterlogged bog soil on poorly drained land, for example—are called *intrazonal* soils.

In 1938, the United States Department of Agriculture published a soil classification system that divided soils into three main *orders*: zonal, intrazonal, and azonal. Each order is subdivided into *suborders* based on criteria such as the climatic and vegetation zone and the principal soil-forming process. Each suborder is further divided into forty *great soil groups*, such as tundra soils and chestnut soils. A listing of *the 1938 classification* is in Appendix III.

The 1938 classification does not possess the detail required for classifying all known soils. So in 1960 the Soil Survey staff of the United States Department of Agriculture published a more comprehensive system, known as *the 7th Approximation*. The 7th Approximation, which was developed from earlier stages over a period of years, recognizes ten orders of soils, differentiated according to the processes of formation. Further division into suborders, great groups, subgroups, families, and series makes approximately 7,000 categories available for the description of soils in the United States. An outline of the 7th Approximation is also in Appendix III.

Examples of Principal Soil Types

The 1938 classification system is adequate for describing the distribution of soils on the global scale, and it has the advantage of using common descriptive names for soils. This section describes ten of the most important great soil groups to give a graphic description of major world soil types. Except for the intrazonal bog and meadow soils, all of the soils described here are zonal soils.

Note in each description the close association of zonal soils with particular regimes of climate and vegetation. In the moist higher latitudes, for example, podzolic soils are most widespread. These soils have an accumulation of aluminum and iron in their *B* horizons and are sometimes called *pedalfers*. In drier climates, calcified soils, called *pedocals*, dominate. As the map in Figure 8.14 shows, the approximate boundary between the pedalfers of the eastern United States and the pedocals of western regions runs north and south nearly along the dividing line between the moist and dry regions defined by Thornthwaite's moisture index.

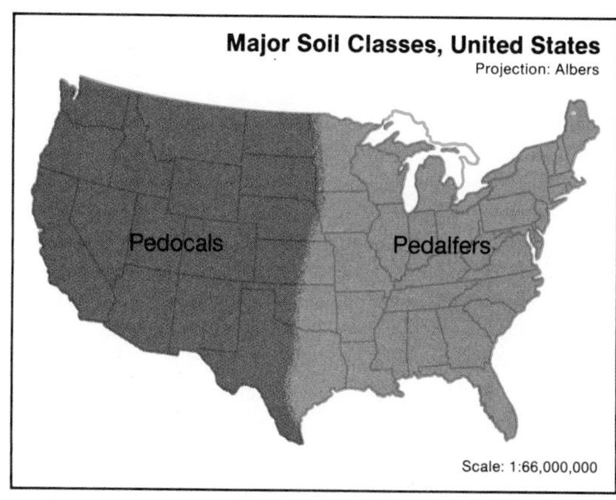
Major Soil Classes, United States
Projection: Albers
Pedocals
Pedalfers
Scale: 1:66,000,000

Figure 8.14 Soils in the conterminous United States can be divided into two major classes, the pedalfers of the East and the pedocals of the West. Pedalfer soils are leached and contain aluminum and iron compounds from the breakdown of clay by acids. Pedocal soils are alkaline and contain calcium compounds. The dividing line between pedalfers and pedocals nearly coincides with the line where Thornthwaite's moisture index is 0. Pedalfers are formed in regions where moisture for leaching is abundant, and pedocals form in dry regions. (After Marbut)

Tundra Soils

Tundra soils are found in the extreme northern portions of Europe, Asia, and North America, north of the forest boundaries. The tundra vegetation in the cold, moist Arctic climate consists of grasses, low shrubs, and lichens. The growing season is short, but the long days in summer help increase plant productivity.

An important feature of the tundra is the subsurface of permanently frozen ground, called *permafrost*. Because the permafrost blocks water drainage, tundra soils tend to become waterlogged gley soils. When winter approaches, the upper layer of soil freezes first, which traps liquid water in the soil between the surface and the permafrost. When water in the next layer finally freezes, the expansion disrupts the soil and prevents the formation of clearly defined horizons. The profile of tundra soil usually shows a black humus layer at the surface and the characteristic blue-gray lower portion of a gley soil. Often large rock fragments are present, which ice moves upward into the soil.

During summer thaw, when the soil above the permafrost is wet and lacks cohesion, the soil tends to creep down slopes. Construc-

Figure 8.15 This *tundra soil* occurs along the coast of northern Alaska and has developed from a parent material of coastal plain sediments. The soil is permanently frozen below a depth of 40 centimeters (16 inches), but the upper portion thaws seasonally. (7th Approximation: pergellic cryaquept)

tion of buildings on tundra soil is difficult because in the summer the soil slips and slides from wetness, and in the winter it shifts and heaves from frost. The agricultural potential of tundra soils is negligible because the severe climate limits crop growth.

Podzols

Podzols, soils of the northern coniferous forests, are found in the upper Great Lakes and in parts of northern New England in the United States. The belt of podzols stretches across Canada, northern Europe, and northern Russia and is bounded on the north by the polar limits of the forest.

The limited rainfall and the cold winter temperatures of the northern forest allow organic debris from needles to accumulate into an acid peaty mat on the surface. The profile of a true podzolic soil shows an ash-gray layer in the lower *A* horizon from the accumulation of silica. The thickness of the layer may range from a centimeter to several meters. Illuviation of the *B* horizon with clay, a cementing agent, sometimes forms a layer of *hardpan*. A continuous hardpan inhibits water drainage and, in some cases, leads to formation of gley soil in the layers above the hardpan.

Podzols are difficult to bring to productive agricultural use, because they are highly acid in the *A* horizon and lack essential mineral nutrients. To be productive at all, they require considerable treatment with fertilizer and with alkali, such as lime. And even with such additives, regions with podzols are not major producers of food crops and are probably best left as forest land.

Figure 8.16 This *podzol* has developed on sandy glacial outwash from granitic rocks of the Adirondack Mountains of northern New York. (7th Approximation: typic haplorthod)

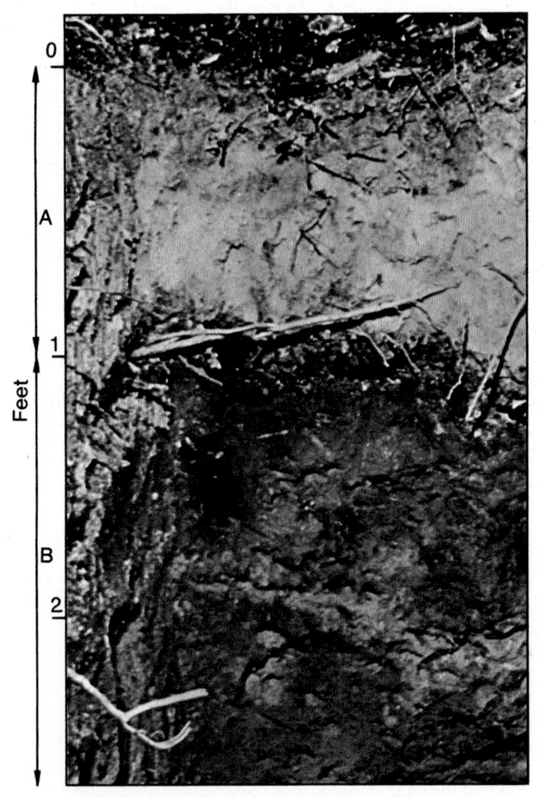
0
A
1
Feet
B
2

Gray-Brown Podzolic Soils

Gray-brown podzolic soils are the most common soils in the productive agricultural land of the Northeast and Midwest. They form in humid continental and in marine West Coast climates, and they are also the common soils of western Europe and of northeastern China. Gray-brown podzolic soils are formed under broadleaf hardwood forests in regions where the annual rainfall is about 100 centimeters (40 inches).

Organic litter from deciduous trees forms humus that is richer and less acid than the humus from coniferous tree debris. The *A* horizon of gray-brown podzolic soil is brownish and lightens in color at lower depths. The *B* horizon is darker than the lower part of the *A* horizon and ranges in color from brownish to yellowish or reddish brown. The combined thickness of the *A* and *B* horizons seldom exceeds 100 centimeters (40 inches).

Gray-brown podzolic soil has an acid *A* horizon, but regular additions of alkali, such as lime, make the soil suitable for a variety of crops. It is good agricultural practice to rotate crops in this type of soil. A planting of alfalfa or clover once every few years helps maintain the texture and organic content of the soil.

Red-Yellow Podzolic Soils

The most common soils in the southern United States are the *red-yellow podzolic soils*. They occur in humid subtropical climates, in southern China, along the coast of eastern Australia, and in southeastern Brazil. Podzolic soils are forest soils. True podzols are formed under coniferous forests, and the gray-brown podzolic soils are formed under temperate broadleaf hardwood forests. The red-yellow podzolic soils are often formed under moist subtropical coniferous woodlands.

The *A* horizon of red-yellow podzolic soils tends to be strongly acid and is usually brownish in its upper portion and yellowish toward the *B* horizon. The *B* horizon, which is colored by iron compounds, is yellowish where soil moisture is abundant and is reddish where conditions are somewhat drier. The composition of the *B* horizon is similar to the clay used to make china.

Red-yellow podzolic soils lack nutrients. They also have a low humus content, because the warm, moist conditions encourage the rapid decay of organic materials on the surface. However, fertilizer applications can make these soils agriculturally productive.

Latosols

Latosols are found in the Amazon Basin, in central Africa, and in southeast Asia, where high temperatures and heavy rainfall make rainforest soils quite different from those of midlatitude climates. The humus content of tropical soils is very low because organic material in the surface of the soil decomposes rapidly in the hot, moist climate, leaving little time for partly decomposed humus material to enter the soil. So most of the nutrients are contained in the vegetation, and mineral nutrients are stored but not bound in

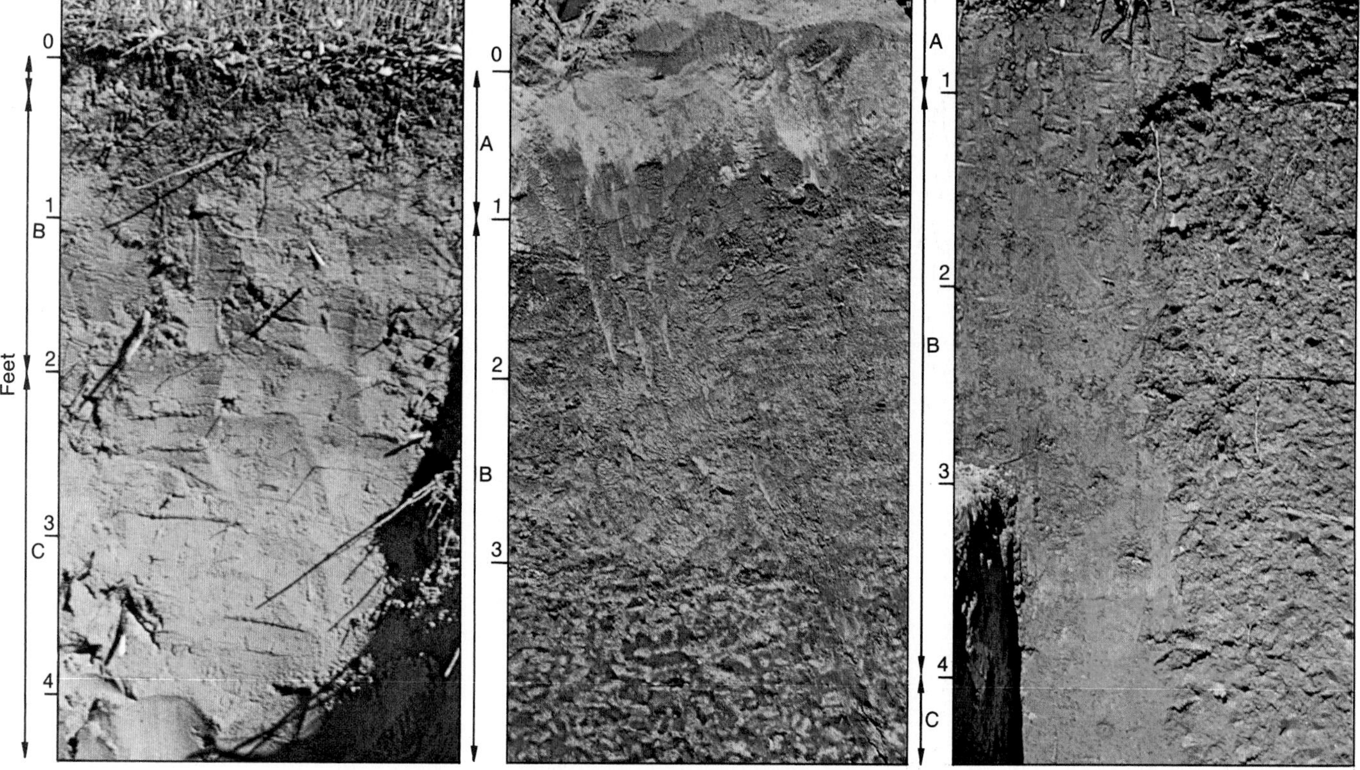

Feet
0
1
B
2
3
C
4
0
A
1
2
B
3
A
1
2
B
3
4
C

Figure 8.17 (left) This *gray-brown podzolic soil* is located in northern Michigan and has developed on sandy glacial outwash. The soil is relatively infertile because of the excessive drainage of water through the sandy parent material.
(7th Approximation: typic udipsamment)

Figure 8.18 (center) This *red-yellow podzolic soil* has developed on coastal plain sediments in central North Carolina. The soil is strongly acid and over 200 centimeters (80 inches) thick.
(7th Approximation: plinthic paleudult)

Figure 8.19 (right) This *latosol* has developed on deeply weathered rock in central Puerto Rico. The soil is low on nutrients but is permeable to water and easily worked in agriculture. (7th Approximation: tropeptic haplorthox)

the soil. Lateritic soils consist of various proportions of iron and aluminum compounds, and the soil is usually red from the iron content. When exposed to air, the heavy iron-aluminum clay in some tropical latosols hardens irreversibly. The temples of Angkor Wat in Cambodia are constructed of laterite.

Chernozems

Chernozems, soils of the midlatitude steppe grasslands and prairies, are found primarily in the Ukraine, Russia's important wheat-growing area. A region of chernozem soil also stretches from Nebraska through eastern North and South Dakota in the United States and extends north into Canada.

The parent material of many chernozems is the fine dust that glaciers deposited and the wind carried. Chernozems are often as much as 2 meters (6 feet) deep. In the comparatively dry climate of the prairie, grasses have deep root systems that give the soil an open texture that is desirable for agriculture. Decayed roots and grasses contribute to the humus content, so the upper horizon is black with humus. Rainfall in these regions is insufficient to produce extensive leaching, and the soils are not acid. On the contrary, the upward movement of calcium salts characteristic of soils in dry climates brings needed alkali material from the parent material into the upper horizons. Chernozems are excellent for growing cereal grains, and they require little treatment with fertilizer unless irrigation leaches nutrients from the upper layers.

Figure 8.20 This *chernozem* has developed in glacial till in southeastern South Dakota. The *A* horizon, which extends to a depth of 25 centimeters (10 inches), is dark from its rich humus content. (7th Approximation: typic argiustoll)

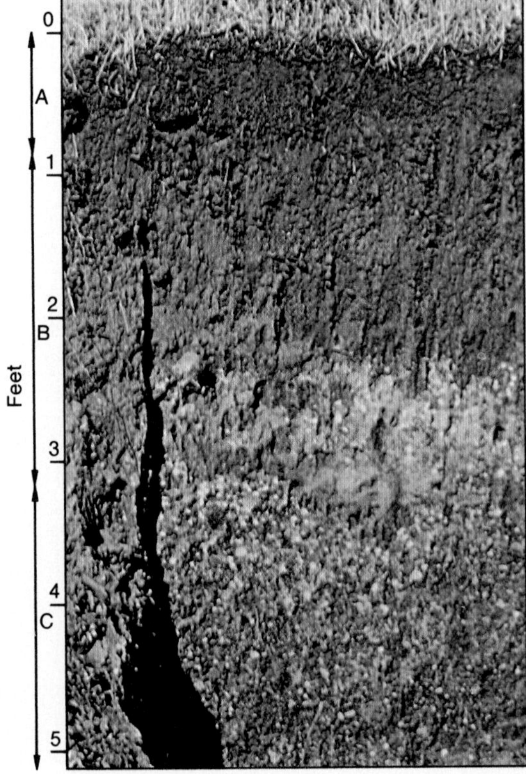

0
A
1
2
B
Feet
3
4
C
5

Prairie Soils

Prairie soils are transitional between the gray-brown podzolic soils of the moist broadleaf forests and the black chernozems of the drier steppe. Prairie soils are also called *brunizems* and, occasionally, gray forest soils. In the United States, prairie soils are found primarily in Iowa and southern Minnesota, between the podzolic soils of the East and the chernozems of the West. The belt of prairie soils in the Soviet Union runs east and west, north of the chernozem region. Except in Uruguay and in southern Brazil, prairie soils do not occur commonly in the Southern Hemisphere. Pedologists do not agree about the type of vegetation responsible for the development of prairie soils. Possibly both tall grass prairie and forest contributed to their formation at different times under different climatic conditions.

Prairie soils have abundant humus. They contain calcium bound to the clay-humus complex, but they do not contain enough calcium compounds to form distinct layers. The *A* horizon is dark with humus and gradually shades into a lighter color, with no sharp demarcation appearing between the *A* and *B* horizons. Prairie soils are fertile and productive, as the cornfields of Iowa testify. Unlike the podzolic soils, they do not require treatment for excess acidity.

Chestnut-Brown Soils

Chestnut-brown soils in the United States are found west of the chernozem soils and east of the Rocky Mountains, in the grassland regions characterized by clumps of short grasses. The chestnut-brown soils occur in numerous dry areas throughout the world, including southern Africa, central Spain, and the dry grasslands stretching across central Asia.

Chestnut-brown soils contain less humus than the chernozems, because the dryness of the grasslands limits the rate at which organic matter is added to the soil. The *A* and *B* horizons are not as dark as the black *A* horizons of the chernozems. The C horizon is often white from calcium salts.

The temptation is great to plant wheat in the chestnut-brown soils that border the highly productive wheat-growing regions of chernozem soils in the United States and in the Soviet Union. But in the dry climate of the short grasslands, crops such as barley and sorghum are usually more productive than wheat. Wheat farming in the zone of chestnut-brown soils is successful only when the soil contains an adequate store of soil moisture at planting time.

Desert Soils

Red and gray soils of the desert cover nearly one-fifth of the earth's landmass. Large belts stretch across northern Africa and Arabia, across central Asia, and across Australia. In the United States, desert soils occupy a large portion of the dry lands between the West Coast mountain ranges and the Rocky Mountains. The gray desert soils, or *sierozem*, are found in the cooler regions, whereas red desert soils are common in the hot deserts of the tropics and of the southwestern United States.

Figure 8.21 (left) This *prairie soil* is located in central Iowa and is used for growing crops such as corn and soybeans. The parent material of this soil is loess, which is fine dust produced by glacier action and transported long distances by wind. (7th Approximation: typic argiudoll)

Figure 8.22 (center) This soil is a brown soil from the category of *chestnut and brown soils*. It is located in eastern Colorado and has developed on loess; it is used primarily to grow small grains. (7th Approximation: abruptic paleustoll)

Figure 8.23 (right) This *red desert soil* is located in southern New Mexico and has developed on sediments washed down into a basin floor from neighboring mountains. (7th Approximation: petrocalcic paleargid)

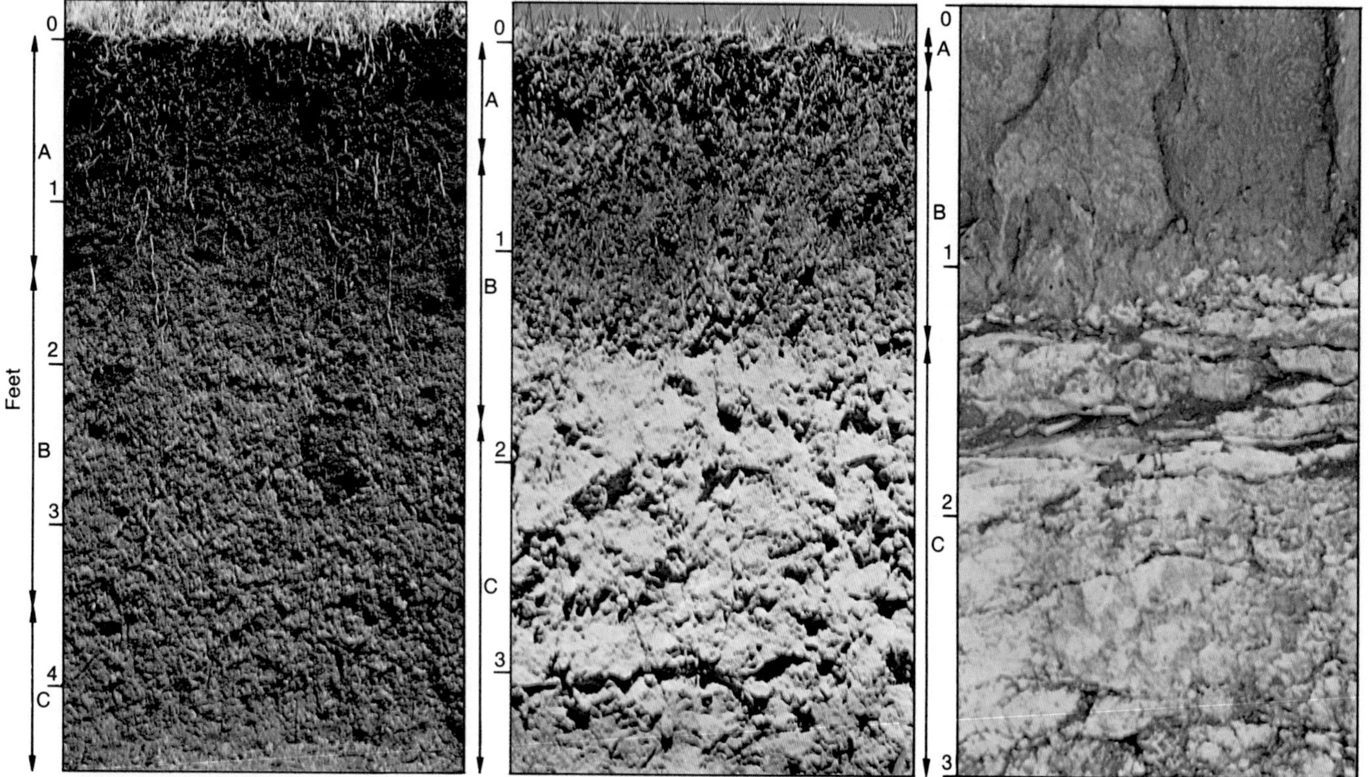
Feet
0
A
1
2
B
3
4
C
0
A
1
B
2
C
3
0
A
B
1
2
C
3

Figure 8.24 The generalized global distribution of major zonal soil types are shown on this map. The map is simplified and should be interpreted as a broad generalization because of the difficulty of classifying a given region in terms of a single soil type. Boundaries between soil regions are purposely left indistinct to suggest that the distribution of soils is not definitely known. Different compilations of world soil distributions differ in detail. The soil categories depicted in this map correspond to the zonal soil types discussed in the text. The symbol colors are not meant to represent the colors of the soils but are only suggestive. The map also shows the distribution of grumusols (7th Approximation: vertisols) and alluvial soils. Grumusols are comparatively dry, deeply cracked dark gray and black soils of the subtropics and tropics. Alluvial soils are those that have developed in sediment; they are found in such locations as the valleys of the Mississippi, Amazon, and Nile Rivers, and in the central valley of California, which receives sediment from neighboring mountains.

The distribution of zonal soils in the United States corresponds closely to regions of climate and vegetation. In the Northeast and Midwest, the soils are largely the gray-brown podzolics of the deciduous forest, merging to true podzols in the coniferous forests of northern New York and northern New England. Soils in the Southeast are the red-yellow podzolic soils of subtropical woodlands. In the central United States, soil regions tend to correspond to moisture regions. A band of chernozems is found in the region of short-grass prairie. To the east of the chernozems, where more moisture is available, lie the prairie soils of the tall-grass prairie. To the west are the chestnut and brown soils of the dry grasslands. The arid western part of the United States is occupied largely by desert soils or by unclassified highland soils. (After Trewartha, Robinson, and Hammond, 1967)

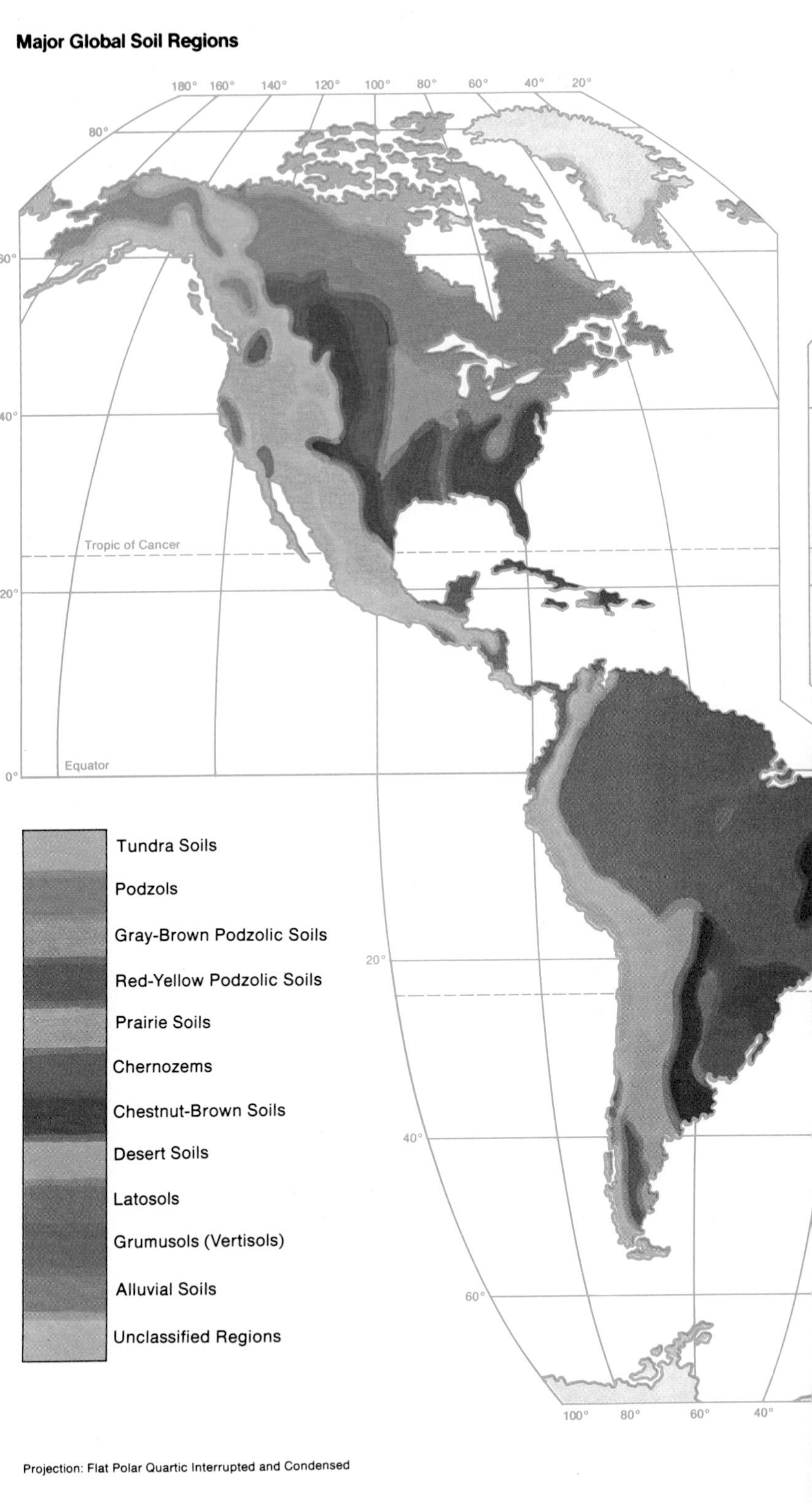

Major Global Soil Regions
180° 160° 140° 120° 100° 80° 60° 40° 20°
80°
60°
40°
20°
0°
Tropic of Cancer
Equator
20°
40°
60°
100° 80° 60° 40°
Tundra Soils
Podzols
Gray-Brown Podzolic Soils
Red-Yellow Podzolic Soils
Prairie Soils
Chernozems
Chestnut-Brown Soils
Desert Soils
Latosols
Grumusols (Vertisols)
Alluvial Soils
Unclassified Regions
Projection: Flat Polar Quartic Interrupted and Condensed

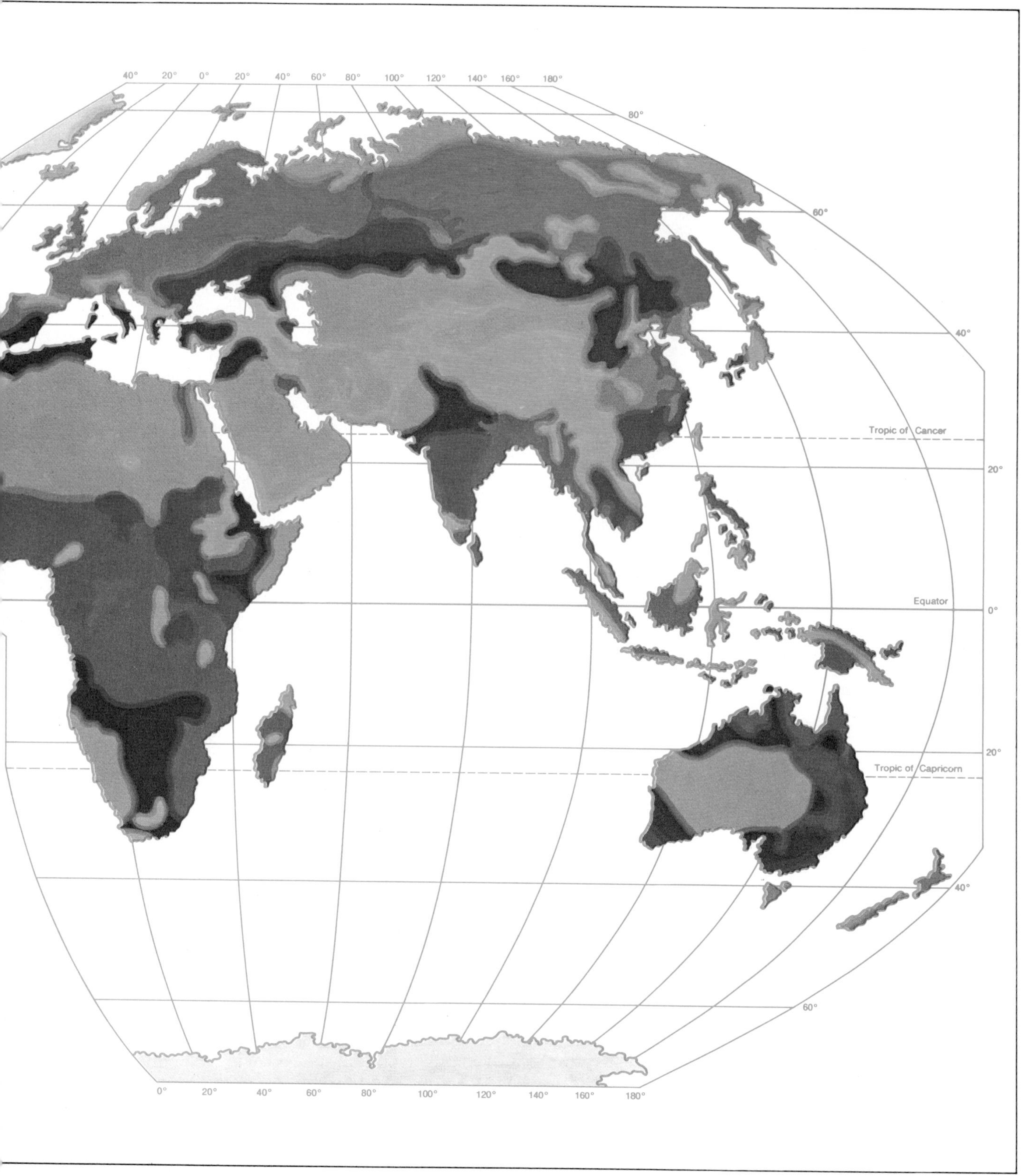

40°
20°
0°
20°
40°
60°
80°
100°
120°
140°
160°
180°
80°
60°
40°
Tropic of Cancer
20°
Equator
0°
20°
Tropic of Capricorn
40°
60°
0°
20°
40°
60°
80°
100°
120°
140°
160°
180°

0
1
2
Feet

Figure 8.25 This waterlogged *bog soil* in southern Michigan consists of a layer of muck over undecomposed peat. (7th Approximation: limnic medisaprist)

Desert soils contain little humus, because vegetation is sparse and grows slowly. Like most soils in dry climates, desert soils have accumulations of calcium salts within the soil; in some regions, salts come to the surface. But the lack of water makes leaching insignificant, so the soils are rich in mineral nutrients. Desert soils have received considerable attention from soil scientists because of the global need for increased agricultural production. To make the soils productive, irrigation and additional humus are required, but satisfactory drainage is an important factor as well. If the soil cannot easily drain the irrigation water, the rapid evaporation of water in the soil will intensify the accumulation of salts. Large areas of irrigated land in West Pakistan, for example, have been removed from agriculture because of salt build-up in the soil.

Bog and Meadow Soils

Bog and meadow soils are waterlogged, intrazonal soils whose properties are determined by local conditions such as a lack of drainage or a high water table. Bog soils are waterlogged because of heavy rains and poor drainage, whereas meadow soils are kept wet by streams. Although bog soils are not common in the United States, some occur on the coast of Florida and along the Gulf of Mexico. England and Ireland have extensive areas of bog.

Both bog and meadow soil profiles are characterized by a thick, black surface layer of partly decomposed plants, an *A* horizon rich in humus, and a bluish gray layer above the parent rock. The layer of organic material, called peat, represents the first stage in the conversion of vegetation to coal. The Irish burn peat to heat their homes and to fuel electric power plants.

Many peat bogs are highly acid from the slow decay of the vegetation, and they have little agricultural value. Peat bogs in low-lying areas are sometimes infiltrated by alkaline water, which neutralizes the acid. When such regions are drained and cultivated, the organic matter becomes exposed to air and decays rapidly to humus, producing a rich fertile soil, as in the fen country of England.

Soil Management

The preceding descriptions of soil types have discussed some of the measures necessary to make soil productive. Because agriculture exposes soil to hazards that are not encountered naturally, proper soil management must consider the future as well as the present. Practices that maintain the fertility of the soil and protect it from erosion should be followed.

Although undisturbed soil tends toward a state of equilibrium, it is a dynamic system. In the equilibrium state, the humus produced annually by the natural vegetation cover of forest or grassland balances the loss of humus from the soil. Clearing the land for agriculture interrupts the process of humus production. Food crops are not allowed to accumulate organic litter on the soil, and in time, the humus content of the soil decreases. In one test, the soil in a plot used solely to grow corn for 30 years experienced a loss of two-thirds of its humus content and a corresponding loss in its nitrogen. Similar plots that used a 3-year crop rotation of corn, wheat, and clover showed little loss of humus and nitrogen. The year devoted to clover pasture allowed organic matter to enter the soil and restore the humus. Adding manure to the soil has a similar effect.

The physical state of a soil is also important to successful soil management. In conventional agriculture, fields are often left bare after a crop has been harvested. Without a plant cover, soil is susceptible to erosion by wind, rain, and running water, so it is wise to leave stubble from the food crop in the soil to help stabilize it. If improper cultivation compacts the surface of the soil, infiltration of water into the soil will be decreased and runoff and erosion will be increased.

In modern agriculture, fertilizer is used to maintain the level of nutrients in the soil. Ancient cultures, without artificial fertilizers, resorted to other methods to keep the soil productive. Heating the soil by fire has been practiced for thousands of years in India to improve the next crop. The heat kills pests and also tends to decompose some of the humus and to drive a small amount of soluble nitrogen salts into the soil.

Summary

Soil is a complex system of inorganic rock material and vegetation matter combined chemically and physically. The physical and chemical processes of weathering produce the rock material for soil formation. In some climates, weathering by frost action is an important process; in others, the action of water plays a central role in weathering.

The organic humus in soil is bound to clay particles, forming a complex with the power to hold cations, particularly hydrogen ions and alkaline cations of calcium and magnesium salts. Energy, moisture, and vegetation influence chemical reactions in soils to give the soils of various regions distinctive characteristics. Soils can be identified by the profile developments of their horizons.

Because there are so many varieties of soils, classifying them is cumbersome. However, many soils proceed toward an equilibrium state that is strongly determined by climate, and on a global scale, soils can be classified into a dozen or so major types with distinctive characteristics associated with climate and plant cover.

Man manages the soil for agriculture by irrigating, fertilizing, adding calcium salts, and plowing. The viability of most soils can be maintained in agricultural use by employing techniques suitable to the special character of the soil.

9 Ecologic Energetics

Energy Use by Green Plants: Photosynthesis and Food Production
Plant Respiration and Net Production
Productivity Measures
Climate and Crop Growth
Potential Photosynthesis and Crop Productivity

Energy Flow Through Ecosystems
Food Chains and Food Webs
Laboratory Ecosystems
Energy Transfer Along Food Chains
Studies of Natural Ecosystems

Agricultural Ecosystems

Problems of World Agriculture

Ecosystems and Man

All interactions among the parts of an ecosystem are united by the energy flow through the system. This intimate interdependence of living things implies a certain stability, a certain dynamic reciprocity, and a certain harmony of nature.

The Waterfall by Henri Rousseau. (Helen Birch Bartlett Memorial Collection; courtesy of the Art Institute of Chicago)

Ecologic Energetics

All life requires energy—energy for growth, energy for movement, energy for change. And the ultimate energy source for human beings and all other animals is the solar energy that green plants convert to chemical energy. Even a blind albino fish that lives in underground streams and is never exposed to sunlight depends ultimately on green plants to provide the food that floats along the stream. And the bacteria that survive on crude petroleum depend on the green plants of prehistoric times from which the petroleum was formed.

Man has developed highly organized agricultural practices to make efficient use of the energy stored in plants. Sometimes he obtains the energy of plants directly, by eating vegetables and fruits, and sometimes he obtains the energy indirectly, by feeding plants to animals such as chickens and cattle.

All animals rely either directly or indirectly on photosynthesizing plants for their food. Most plants live in groupings, or communities, which appear to be mutually beneficial to the members of the community. Animals also live in association with plants and must ultimately interact with plant communities for their food. A natural system that consists of groups of plants and animals interacting with one another and with their environment is called an *ecosystem.*

A system is a collection of structures and the set of interactions between the structures. Depending on how the boundary lines of a system are defined, there may be significant interactions between the system and the world outside the system. It is possible to maintain simple ecosystems that involve only two or three different kinds of organisms under highly controlled laboratory conditions, but the ecosystems found in nature are exceedingly complex. Consider a forest as an example of an ecosystem. A typical midlatitude deciduous forest contains perhaps fifty species of plants and hundreds or thousands of species of animals, primarily insects. Within an Oregon Douglas fir forest are about fifty species of lichens and mosses alone. The task of tracing the interactions of all these organisms with one another and with the wind, rain, sun, and soil is unbelievably immense.

There are enormous complexities in any ecosystem, even if attention is focused only on a single component, such as the feeding habits of one species. Consider an ocean herring, one component in an ecosystem in the sea. An adult herring feeds on eight or ten different species of marine animals, and each of these species feeds on others in intricate relationships.

It is not possible with our present knowledge to give a detailed account of all the interactions that occur in an ecosystem. One way to simplify our understanding of ecosystems is to focus on the energy flow through them. The flow begins with the sun, the ultimate energy source for the earth's surface. Energy—the capacity to do work—can be converted from one form to another; the forms most important to the living organisms in an ecosystem are radiant, chem-

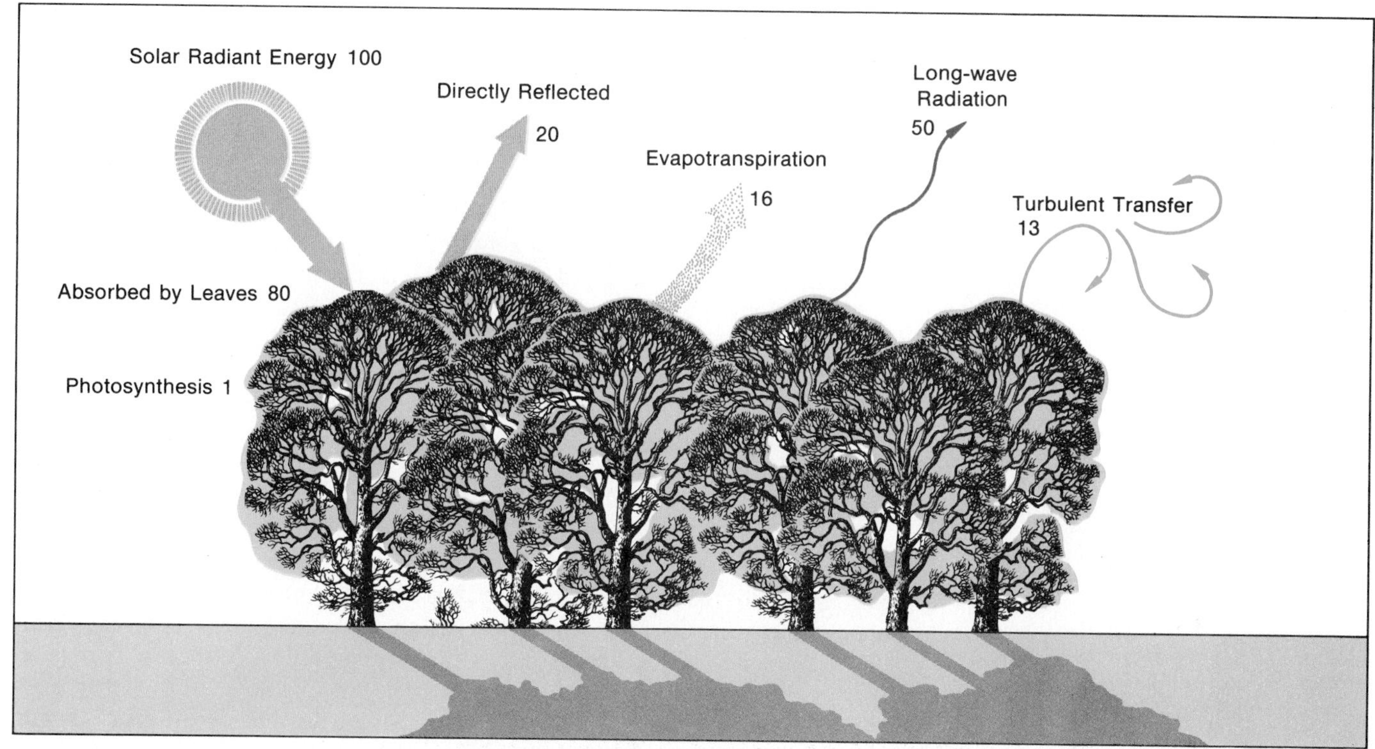
Solar Radiant Energy 100
Directly Reflected
20
Long-wave
Radiation
50
Evapotranspiration
16
Turbulent Transfer
13
Absorbed by Leaves 80
Photosynthesis 1

Figure 9.2 For every 100 units of solar radiant energy incident upon a forest, approximately 1 unit becomes stored energy through photosynthesis. Most of the energy serves to warm the leaves and is returned to the atmosphere by various processes. The diagram shows representative numbers for the heat transfer mechanisms under conditions of moderate temperature, humidity, and wind. The relative importances of the processes differ under different conditions.

ical, mechanical, and heat energy. Radiant energy falls on the leaves of green plants, which convert part of the energy to chemical energy and store it in plant tissue. The plants supply energy to animals that ingest them, and these animals provide energy to other animals in turn.

Looking at an ecosystem only in terms of how energy flows through the system is a simplified but useful approach. Because we know that all energy entering a system can be accounted for, it is possible to set up an energy budget for an ecosystem and trace the energy flow through all the components of the ecosystem. The energy flowing into an ecosystem may be utilized in a variety of ways. For example, not all of the solar radiant energy input to a given plant becomes stored in its tissue; part of the energy is reradiated and part is used to sustain the plant's own life processes. The useful food produced in one ecosystem provides the energy input to other ecosystems. The *efficiency of an ecosystem* is a measure of its ability to pass on some of its input energy to another ecosystem in the form of useful food. Agricultural scientists continually work to increase food production by developing more efficient strains of plants.

Energy Use by Green Plants: Photosynthesis and Food Production

You can begin a study of the energy flow in an ecosystem by looking at the interactions within a green plant. What are the factors that determine how a plant utilizes energy? In the process of photosynthesis, a green plant uses radiant energy to produce stored chemical energy. The detailed chemical processes involved in photosynthesis are quite complex. In terms of energy, however, photosynthesis may be thought of as a process in which simple molecules, carbon dioxide and water, are joined with the aid of solar radiant energy to form a more complex molecule, sugar. The solar radiant energy is stored in chemical form in the sugar. Further chemical reactions use sugar to produce more complex molecules that the plant's cells require for life and growth. In addition to producing sugar, photosynthesis produces free oxygen gas. The production of oxygen is incidental to the synthesis of sugar in photosynthesis, but it is important to the ecosystem as a whole. Without photosynthesis, there would be little oxygen in the atmosphere, and animal life as we know it could not have evolved.

Only a small part of the solar radiant energy falling on a leaf is converted to chemical energy. Some of the incident energy is immediately reflected, and approximately 80 percent is absorbed. Most of the absorbed energy functions to warm the leaf and is then given off as long-wave thermal radiation. And some of the absorbed energy is used in the evaporation of water stored in the plant. Only about 1 percent of the incident energy eventually comes to be stored as chemical energy.

Figure 9.3 Photosynthesizing plants combine carbon dioxide, water, and nutrients with the aid of radiant energy to produce stored food energy. Animals rely on plants for food, and they breathe the oxygen released as a by-product of photosynthesis. Carbon dioxide is returned to the atmosphere by the respiration and decay of plants and animals; nutrients such as organic nitrogen compounds also cycle between organisms and the environment.

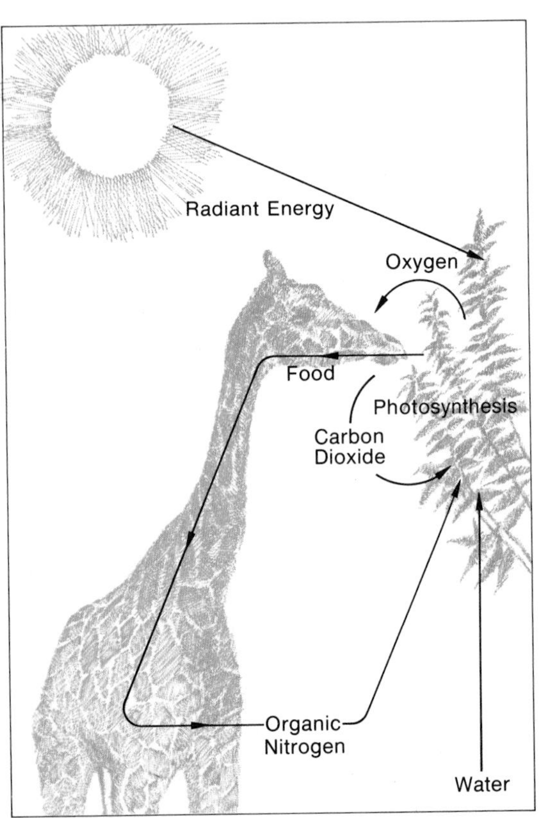
Radiant Energy
Oxygen
Food
Photosynthesis
Carbon
Dioxide
Organic
Nitrogen
Water

A significant question for agriculture is this: How much sugar can a plant produce photosynthetically for a given amount of incident solar energy? In 1926 Edgar Transeau measured the sugar-producing efficiency of a cornfield. He recognized that all the carbon in a plant comes from the sugar produced during photosynthesis, and by measuring the amount of carbon in corn plants, he estimated the amount of sugar produced during the growing season. Because the amount of energy required to produce 1 kilogram of sugar by photosynthesis had already been determined in laboratory studies, Transeau was able to conclude that of the total solar energy incident on a field of corn during the 100-day growing season, only 1.6 percent becomes stored chemical energy through photosynthesis. This percentage is the energy conversion efficiency of corn; all plants have energy conversion efficiencies of a few percent or so.

The *rate* of photosynthesis in a leaf varies with the intensity of the incoming light. If the intensity of light is increased, the rate of sugar production will also increase, up to a maximum value. Further increases of light intensity beyond this point will not result in increased photosynthesis. A leaf in full sunlight receives more than enough light to carry on maximum photosynthesis. Even the leaves that are partially shaded or that receive only indirect light are able to carry on photosynthesis near the maximum rate. Because plants adapted to the tropics receive more solar energy, they generally have higher maximum rates of photosynthesis than plants native to midlatitude regions, where less solar energy is available. Graphs depicting the relationships between photosynthesis and solar radiation are shown in Figure 9.4.

The rate of photosynthesis also depends on the temperature of the leaf, which may differ from the air temperature. For plants in

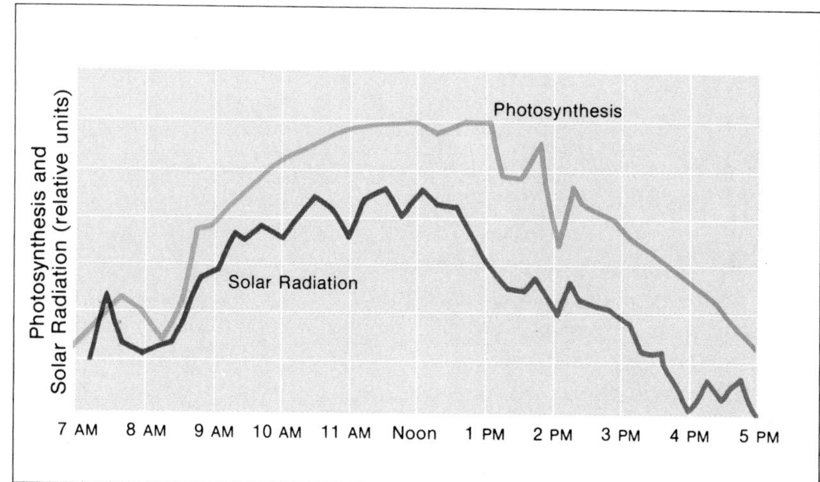
Photosynthesis
Solar Radiation
Photosynthesis and Solar Radiation (relative units)
7 AM
8 AM
9 AM
10 AM
11 AM
Noon
1 PM
2 PM
3 PM
4 PM
5 PM

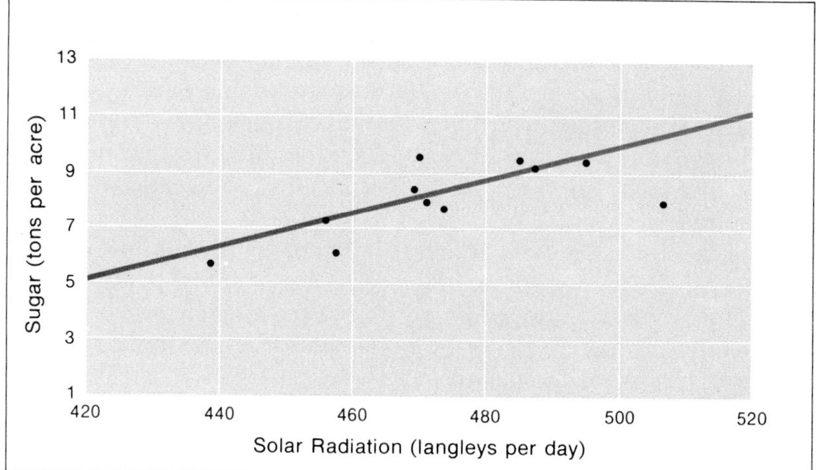
Sugar (tons per acre)
13
11
9
7
5
3
1
420
440
460
480
500
520
Solar Radiation (langleys per day)

Figure 9.4(a) The course of photosynthesis through the day rises and falls with the variations of incident solar radiant energy.

(b) This graph shows the yield of sugar from a Hawaiian sugar-cane plantation during different growing seasons, plotted according to the average daily solar radiation input during the season. The yield is high during seasons when fair weather causes the average radiant energy input to be high and is low during cloudy seasons. (After Jen-Hu Chang, 1968)

midlatitude regions, photosynthesis for a given light intensity reaches a maximum at a leaf temperature of about 25°C (77°F). For arctic plants, maximum photosynthesis occurs at lower leaf temperatures. The rate of sugar production decreases both above and below a plant's optimum leaf temperature, and production stops if leaf temperatures rise above 40°C (104°F) or so. If there is no wind to cool the leaves, leaf temperatures may rise so high that photosynthesis stops during the middle of the day when solar energy input is maximum. Under such conditions, food production is limited to a period in the morning and to a brief period in the late afternoon. Partly for this reason, the warmest regions of the tropics tend to have lower agricultural potential than midlatitude regions.

A number of other factors also influence the rate of photosynthesis. Adequate supplies of water and carbon dioxide are necessary for efficient photosynthesis. The availability of nutrients from the soil, particularly nitrogen and phosphorus, also affects the rate of food production. Phosphorus, a comparatively rare element in the earth's crust, is required for the synthesis of chemical compounds

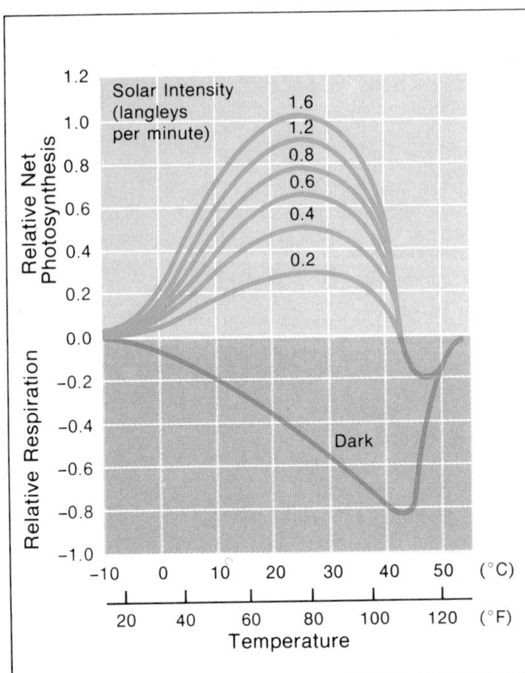
Relative Net Photosynthesis
Relative Respiration
Solar Intensity (langleys per minute)
1.6
1.2
0.8
0.6
0.4
0.2
Dark
1.2
1.0
0.8
0.6
0.4
0.2
0.0
-0.2
-0.4
-0.6
-0.8
-1.0
-10
0
10
20
30
40
50
(°C)
20
40
60
80
100
120
(°F)
Temperature

Figure 9.5 The upper portion of the graph shows the trend of net photosynthesis in a leaf for different leaf temperatures and different rates of radiant energy input. Note that photosynthetic activity is small at both low and high temperatures and that it reaches a maximum at an intermediate temperature. For a leaf at a given temperature, photosynthesis increases with increased radiant energy input, but not proportionately. Doubling the radiant energy from 0.8 to 1.6 langleys per minute increases the photosynthesis by only 30 percent or so. The lower part of the graph shows the energy consumed by respiration at different temperatures for a leaf that is kept in the dark. The respiration increases with increased temperature until the leaf is too warm to function. High temperatures at night therefore promote the use of stored energy for respiration and tend to decrease the net productivity of the plant. (After Gates, 1968)

important in the photosynthetic process. Phosphorus deficiency is often the limiting factor for plant growth in moist climatic regions, in lakes, and in coastal areas. Nitrogen is required for the production of plant proteins.

Plant Respiration and Net Production

Not all of the sugar a plant produces in photosynthesis is available to animals as food. The plant's own cells require a continuous supply of energy, day and night. In the process of *respiration*, sugars combine with oxygen and are reduced to carbon dioxide and water, which releases stored chemical energy for use by the living cells. Photosynthesis stops in the absence of light or when leaf temperatures are too high, but respiration continues both day and night. Cool nights help the plant to conserve the energy produced during the day because a plant uses more energy for respiration when the temperature is higher and when the leaves are exposed to light. The energy released in light-dependent respiration is called *photorespiration*. The most efficient plants, such as sugar cane and corn, tend to have low rates of photorespiration.

One way to find the amount of energy a plant uses in respiration is to seal a plant in a dark box and measure the amount of carbon dioxide it produces by respiratory activity. Studies show that a wide variety of plants, including phytoplankton (floating microscopic green plants), use approximately one-fourth of their stored energy for respiration.

The rate at which a plant converts light energy to chemical energy is called the *gross productivity*. The plant uses some of the chemical energy for respiration, so only a portion of the total chemical energy remains stored in plant tissues. The net rate at which a plant stores energy, exclusive of the energy it uses for respiration, is called the *net productivity*. Net productivity is the quantity important to the consumer of a plant.

Productivity Measures

Productivity is expressed in terms of the amount of energy stored in a particular plant during a given time interval such as a day or a year. The method described for measuring respiration can be extended to the measurement of both gross and net productivity. Suppose a plant is placed in a transparent sealed container. The rate at which the plant gives off carbon dioxide when no light is allowed to enter the container is a measure of the rate at which it is using energy for respiration. When the plant is exposed to light, photosynthesis and respiration both occur. Photosynthesis causes carbon dioxide to be taken up at the same time that respiration causes carbon dioxide to be given off. The net rate at which the plant absorbs carbon dioxide is therefore a measure of net productivity. Gross productivity can be found by adding the rate of energy usage for respiration to the net productivity. For instance, if the net productivity of a carrot plant is 4 grams of sugar in a given time, and it uses the equivalent of 1 gram of sugar for respiration, the gross productivity of the carrot is 5 grams of sugar.

Because it is so difficult to carry out direct measurements of energy flow, productivity is usually measured indirectly. The net productivity of agricultural crops is often estimated by harvesting the complete plants, roots and all, at the end of the growing season. The plants are then dried and weighed, and the productivity is expressed in terms of the dry matter that a given area produces during a specified period of time. Productivity measured by the harvest method can be expressed as grams of dry matter per square meter of field per year.

The energy content of a gram of dry matter differs among plant species depending on the relative amounts of fats, proteins, and carbohydrates in the plant tissues. To correct for this difference, productivities expressed in terms of dry matter can be converted to energy units by burning the dry matter in a device that measures the heat energy released in complete combustion. When the harvest method is applied to natural plant communities, measurements of dry matter must be corrected for losses caused by the shedding of leaves, the depredations of insects, and the death of some plants during the growing season.

Averaged for the world, the net productivities of all important food crops, such as wheat, rice, corn, and potatoes, are of the order of 300 to 500 grams of dry matter per square meter per year. Productivities of 2 or 3 times greater are attained in areas where intensive

Table 9.1. Net Primary Production of Major Ecosystems and of the Earth's Surface

Type of Ecosystem	Area (millions of square kilometers)*	Net Primary Productivity per Unit Area (dry grams per square meter per year) †		World Net Primary Production (billions of dry tons per year)
		Normal Range	Mean	
Lake and Stream	2	100-1,500	500	1.0
Swamp and Marsh	2	800-4,000	2,000	4.0
Tropical Forest	20	1,000-5,000	2,000	40.0
Temperate Forest	18	600-3,000	1,300	23.4
Boreal Forest	12	400-2,000	800	9.6
Woodland and Shrubland	7	200-1,200	600	4.2
Savanna	15	200-2,000	700	10.5
Temperate Grassland	9	150-1,500	500	4.5
Tundra and Alpine	8	10-400	140	1.1
Desert Scrub	18	10-250	70	1.3
Extreme Desert, Rock, and Ice	24	0-10	3	0.07
Agricultural Land	14	100-4,000	650	9.1
Total for Land	149		730	109.0
Open Ocean	332	2-400	125	41.5
Continental Shelf	27	200-600	350	9.5
Attached Algae and Estuaries	2	500-4,000	2,000	4.0
Total for Ocean	361		155	55.0
Total for Earth	510		320	164.0

* One square kilometer is equal to about 0.39 square mile.
† One gram per square meter is equal to about 0.0033 ounce per square foot.
Source: Whittaker, Robert H. 1970. *Communities and Ecosystems.* New York: Macmillan.

mechanized agriculture is practiced. The world average net productivity of sugar cane is high—about 1,500 grams per square meter per year. Productivities of more than 9,000 grams have been measured for sugar cane in some tropical regions, where the growing season lasts the entire year and where the soil is moist and rich in nutrients. As Table 9.1 shows, the average productivity of the land is 730 grams per square meter per year, compared to an average productivity of 155 grams per square meter per year for the oceans. Although the ocean is less productive overall than the land, its most productive parts are the shallow zones near the continents (see Figure 9.6).

The *standing crop*, or *biomass*, is the amount of energy present in plants at any given time, expressed in grams of dry matter per square meter. Biomass should not be confused with productivity, which is the *rate* at which a plant produces food. A redwood forest has a large biomass but a comparatively low productivity. A large amount of energy is stored in the trees, but not much additional energy is added to storage each day. The phytoplankton in the ocean have a high productivity, but their biomass is small because they are continuously consumed by predators. *Yield* is the amount of energy stored during the growing season in the desired portion of a crop, such as the fruit. It is usually determined by weighing the harvested portion of the crop.

Climate and Crop Growth

A plant's basic requirements are water, carbon dioxide, mineral nutrients, and energy. Insufficiency in any of these factors inhibits a plant's growth. Sufficient energy usually is available, and water is often the limiting factor in plant growth. During photosynthesis, a plant absorbs carbon dioxide and releases water vapor. It has been suggested that in a crop ecosystem, actual evapotranspiration is a qualitative measure of plant growth.

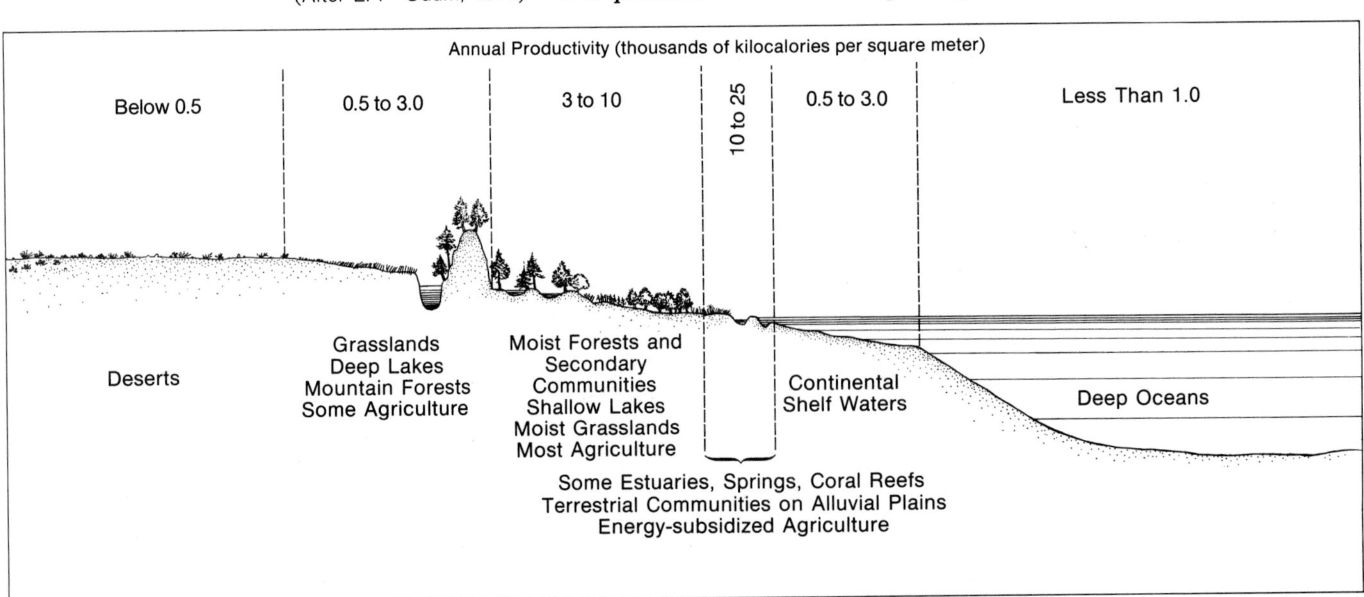

Annual Productivity (thousands of kilocalories per square meter)
Below 0.5
0.5 to 3.0
3 to 10
10 to 25
0.5 to 3.0
Less Than 1.0
Deserts
Grasslands
Deep Lakes
Mountain Forests
Some Agriculture
Moist Forests and
Secondary
Communities
Shallow Lakes
Moist Grasslands
Most Agriculture
Continental
Shelf Waters
Deep Oceans
Some Estuaries, Springs, Coral Reefs
Terrestrial Communities on Alluvial Plains
Energy-subsidized Agriculture

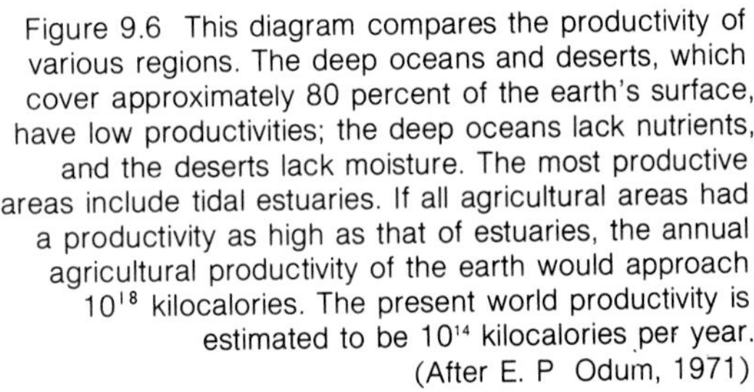
Figure 9.6 This diagram compares the productivity of various regions. The deep oceans and deserts, which cover approximately 80 percent of the earth's surface, have low productivities; the deep oceans lack nutrients, and the deserts lack moisture. The most productive areas include tidal estuaries. If all agricultural areas had a productivity as high as that of estuaries, the annual agricultural productivity of the earth would approach 10^{18} kilocalories. The present world productivity is estimated to be 10^{14} kilocalories per year. (After E. P Odum, 1971)

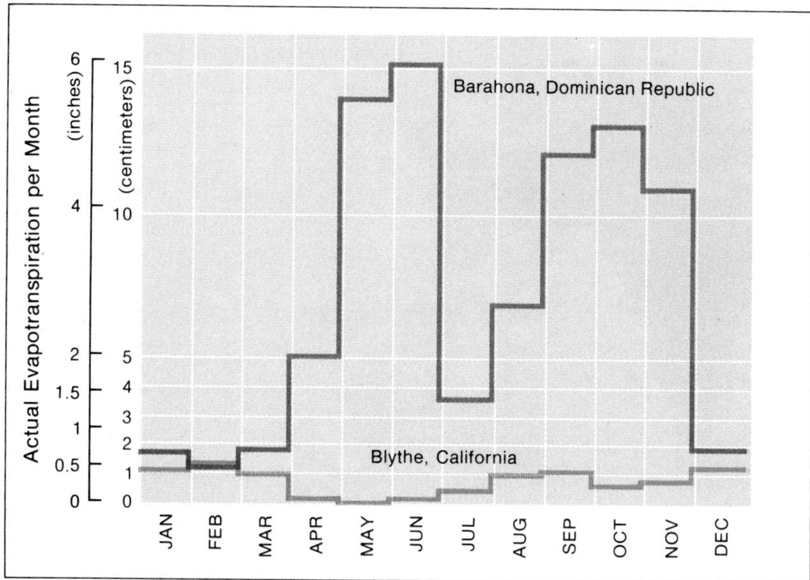

Actual Evapotranspiration per Month
(inches)
(centimeters)
6
4
2
1.5
1
0.5
0
15
10
5
4
3
2
1
0
Barahona, Dominican Republic
Blythe, California
JAN
FEB
MAR
APR
MAY
JUN
JUL
AUG
SEP
OCT
NOV
DEC

Figure 9.7 Actual evapotranspiration, which can be calculated from the local water budget, has been taken to be a qualitative measure of plant activity. The graph shows that in Barahona, where the climate is warm and moist, actual evapotranspiration is high during two periods of the year, with dry periods occurring during the summer and the winter. Plant activity in Barahona is therefore seasonal. In Blythe, a desert location, actual evapotranspiration is small but comparatively constant, so that desert plants maintain steady growth through the year. (After Major, 1963)

Actual evapotranspiration is small when shortage of water inhibits plant growth. The variation of actual evapotranspiration through the year at a particular location indicates when plants are growing actively. Figure 9.7 shows the actual evapotranspiration at a tropical location, Barahona, where there are two dry periods—one in the winter and one in midsummer—and where the growing season is marked by two peaks of activity. But there may be only one long growing season if the amount of stored soil moisture is sufficient or if moisture is added to the soil to help the plants survive the short midsummer dry spell.

In the desert of southeastern California, rainfall occurs primarily in the winter, so the growing season is restricted to the winter months. In the desert of central Arizona, however, some rainfall occurs throughout the year, and the Arizona desert supports plants that cannot survive in the southeastern California desert.

Potential Photosynthesis and Crop Productivity

An important problem for world agriculture is the evaluation of the potential productivity of a given region, assuming that enough water is supplied and that good agricultural practices are followed. *Potential photosynthesis* is the maximum value of a plant's net productivity, given sufficient water and nutrients. Because photosynthesis depends on light intensity and temperature, these factors set a natural limit to the expected rate of energy storage in a particular plant species. If photosynthesis is occurring at its maximum rate, additional water or fertilizer will not increase productivity.

Potential photosynthesis for a particular crop can be estimated from records of temperature and solar radiation. Figure 9.8 shows estimates of the global distribution of potential photosynthesis for a crop of sugar beets. The figure indicates that potential photosynthesis for sugar beets is lowest in the tropics and in polar regions. Although the tropics receive abundant sunlight, high temperatures

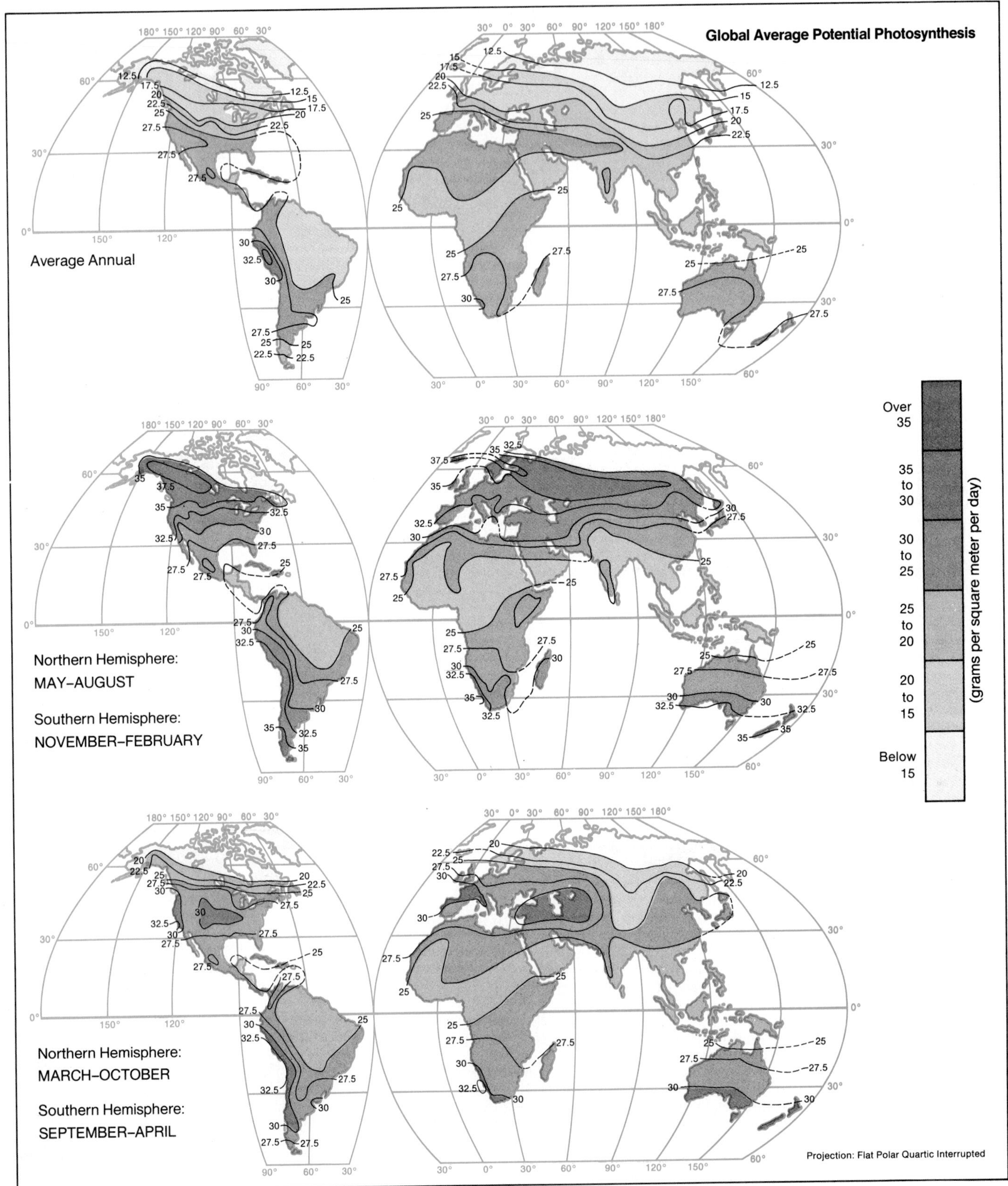
Global Average Potential Photosynthesis
Average Annual
Northern Hemisphere:
MAY–AUGUST
Southern Hemisphere:
NOVEMBER–FEBRUARY
Northern Hemisphere:
MARCH–OCTOBER
Southern Hemisphere:
SEPTEMBER–APRIL
Over 35
35 to 30
30 to 25
25 to 20
20 to 15
Below 15
(grams per square meter per day)
Projection: Flat Polar Quartic Interrupted

limit the rate of photosynthesis. Crops such as sugar cane are better adapted to the tropics and have higher productivity than sugar beets. In the tropics, the lowest values of potential photosynthesis occur near latitude 10°N. Lowland areas at that latitude have the hottest climate on the earth. Productivity at latitude 10°S exceeds productivity at latitude 10°N because temperatures for a given latitude are somewhat lower south of the Equator. Tropical highlands have better agricultural potential than land near sea level because the cooler temperatures at higher altitudes favor a higher rate of photosynthesis.

According to estimates of potential photosynthesis, the world regions with the greatest potential agricultural productivity probably lie in the middle latitudes and have a Mediterranean climate. Regions with the potential for a highly productive 8-month growing season include the central United States, California, southwestern Europe, southern Australia, and New Zealand, all of which are excellent agricultural regions in actual practice. The high estimated productivity near latitude 60°N during the 4-month growing season is of particular interest. The long days of the northern summer and the moderate temperatures favor a high level of plant activity. The Matanuska Valley in southern Alaska, for example, is known for its giant cabbages and other luxuriant crops produced during the short summer growing season.

Figure 9.8 Potential photosynthesis is a measure of a plant's maximum net production if the only limiting factor is solar radiant energy input. The global maps in this figure show calculated potential photosynthesis in units of grams of production per square meter per day for a crop of sugar beets.

(opposite top) Average annual potential photosynthesis is given on this map. The highest values occur in regions that are classified as *Cs* or *Aw* climates according to the Köppen system.

(opposite center) Potential photosynthesis is shown for the 4-month principal growing season (May through August in the Northern Hemisphere, November through February in the Southern Hemisphere). Note the high values of potential photosynthesis that occur in the higher latitudes because of the long duration of daylight during the summer months.

(opposite bottom) Potential photosynthesis is given for an 8-month growing season (March through October in the Northern Hemisphere, September through April in the Southern Hemisphere). Many of the highest values are reached in what are now desert regions, because of the clear skies prevailing in such areas. White areas on the map indicate regions where data are unavailable. (After Jen-Hu Chang, 1970)

Energy Flow Through Ecosystems

Green plants are examples of *autotrophs*, organisms capable of producing food directly from inorganic matter and sunlight. Organisms such as animals that rely on organic matter for food are *heterotrophs*. The autotrophs in an ecosystem are often called *primary producers*, because they can produce food from the inorganic environment, and the heterotrophs are often called *secondary producers*. Food energy flows from autotrophs to heterotrophs in an ecosystem.

Food Chains and Food Webs

The animals that ingest plants are called *herbivores*, and these animals may in turn be ingested by other animals, called *carnivores*. Some animals, such as man, are both herbivorous and carnivorous.

A *food chain* is a sequence of consumption. Some agricultural systems involve a food chain with only a few links. Cattle, for example, eat grass and grain and are in turn eaten by carnivores. A grazing food chain can be symbolized as a linear relationship:

Plant → herbivore → carnivore → second carnivore → top carnivore

However, most feeding relationships in nature do not take the form of simple, isolated chains. In addition to the grazing food chain, there are food chains that involve the feeding of bacteria on dead matter, food chains that involve parasites, and so forth. Many food chains are interconnected, forming complex *food webs*. For that reason, tracing feeding relationships is not a simple task. In a midlatitude forest, for example, there are numerous species

Figure 9.9 This diagram shows the principal members of a food web in a salt marsh near San Francisco, California, during winter. The primary producers are various salt marsh and marine plants, which grasshoppers, snails, fish, and marine invertebrates feed upon. Birds and rodents are intermediate carnivores, and the top carnivores of the web are hawks and owls. (After Smith, 1966, and Johnston, 1956)

of herbivores, each of which may feed on several plant species. Carnivores in the forest may also have a varied diet, feeding on herbivores and other carnivores. One way to analyze a food chain is to take samples of all relevant animal species in the ecosystem and examine the contents of their digestive tracts. An owl's recent diet, for instance, is revealed by the bits of bone and hair in the pellet the bird spits up every few days. The complexity of the food web is apparent when you consider that a few acres of grassland may harbor several hundred different species of insects and numerous other animals.

Laboratory Ecosystems

Because a simple, natural ecosystem contains so many species of plants and animals, quantitative measurements of energy flow are difficult to accomplish. Ecologists have instead turned to laboratory ecosystems made up of a clearly defined plant-herbivore-carnivore chain. Although laboratory ecosystems are artificial, they have the advantage that conditions in them are easily varied. Careful study of the response of laboratory ecosystems to a variety of conditions has led to insight into ecologic principles.

Lawrence B. Slobodkin, for instance, studied a laboratory ecosystem comprising a species of one-celled green algae (Chlamydomonas) as the primary producer, water fleas (Daphnia) as the herbivores, and Slobodkin himself as the carnivore. Slobodkin divided his colonies of water fleas into five groups. He gave each colony in the first group a standard amount of energy each day in the form of algae. Successive groups were given proportionately more food, so that colonies in the fifth group received 5 times the standard portion. In the absence of predation, each colony grew to a size proportional to the rate at which food energy was supplied.

Slobodkin then acted as a predator on each colony. Every fourth day he removed a number of adult water fleas equal to a chosen percentage of the new offspring in the colony. He then compared the energy removed from each colony with the amount of energy supplied as food. As Figure 9.10 shows, the energy output of the colonies drops when predation rates are high. The reason is that if a colony is "overfished," the few remaining animals are unable to make full use of the food supplied.

The most efficient strategy for an intelligent predator to maintain its own food supply is to eat only enough animals so that just the right number remain to use all the food. Slobodkin found that, in using this approach, he could extract by predation no more than 12 percent of the food input energy, irrespective of the feeding level.

Energy Transfer Along Food Chains

Slobodkin's results indicate a significant loss of useful energy at each step of a food chain. Only a fraction of the net production of one stage becomes available as food for the next stage. Food chains in natural ecosystems exhibit approximately the same fractional transfer of useful energy as the laboratory ecosystems show.

Figure 9.10 These graphs summarize the results of L. B. Slobodkin's experiments on a laboratory ecosystem consisting of algae, water fleas, and a predator. Five groups of water flea colonies were fed algae, such that the group with feeding level 5 received five times as much food as the group with feeding level 1. In the absence of predation, each colony reached a population proportionate to its feeding level.

The yield of water fleas that could be extracted from each colony per unit of food energy supplied decreased at high levels of predation because the number of water fleas remaining could not use all of the food.

When colonies are supplied only with the food they will actually consume, the energy yield per unit of food increases with predation level to a maximum energy transfer of 12 percent. (After Slobodkin, 1959)

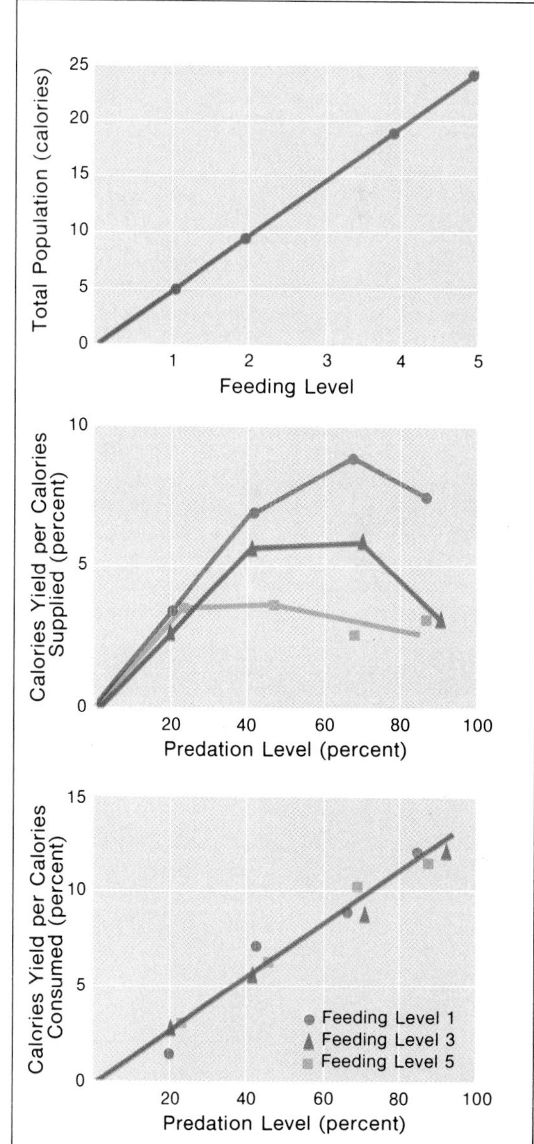

Total Population (calories)
25
20
15
10
5
0
1
2
3
4
5
Feeding Level
Calories Yield per Calories Supplied (percent)
10
5
0
20
40
60
80
100
Predation Level (percent)
Calories Yield per Calories Consumed (percent)
15
10
5
0
20
40
60
80
100
Predation Level (percent)
Feeding Level 1
Feeding Level 3
Feeding Level 5

In a grazing food chain, approximately 10 percent of the energy absorbed during one stage is absorbed during the next stage.

No organism can convert the food it eats into an equal amount of stored energy. An adult man uses most of the energy he obtains from food for body heat, motion, and work. Only a small amount of energy is stored for growth. An infant uses considerable food energy for growth but dissipates energy in heat and motion. One reason for the low efficiency of the energy transfers along food chains is that the chemical reactions required for life are always accompanied by the degradation of energy into forms of heat that cannot be utilized. Grazing food chains also lose energy to decay food chains.

Consider a simple plant-herbivore-carnivore food chain consisting of grass plants, mice, and snakes. The mice obtain approximately 10 percent of the energy absorbed earlier by the grass plants, and the snakes obtain approximately 10 percent of the energy absorbed by the mice. Thus the carnivore receives only 1 percent or so of the energy originally absorbed by the plants. The fraction of the original energy available to a succeeding carnivore stage—a hawk, for instance—is still less. The rapid decrease of available energy along a food chain limits such chains to four or five links. Large carnivores such as lions, which are the last natural link in a food chain, obtain only a small fraction of the energy absorbed by the plants in their habitat. Lions roam over large areas to obtain their food, and one region cannot support many of them.

Studies of Natural Ecosystems

Despite the difficulties involved, the energy flows in several natural ecosystems have been studied in detail. The results of such studies can be conveniently summarized in energy flow diagrams, which show the main energy pathways through an ecosystem.

The energy flow diagram in Figure 9.11(a) represents a typical stage in a food chain. The diagram illustrates how the energy input to the stage is divided among various energy outputs and storage.

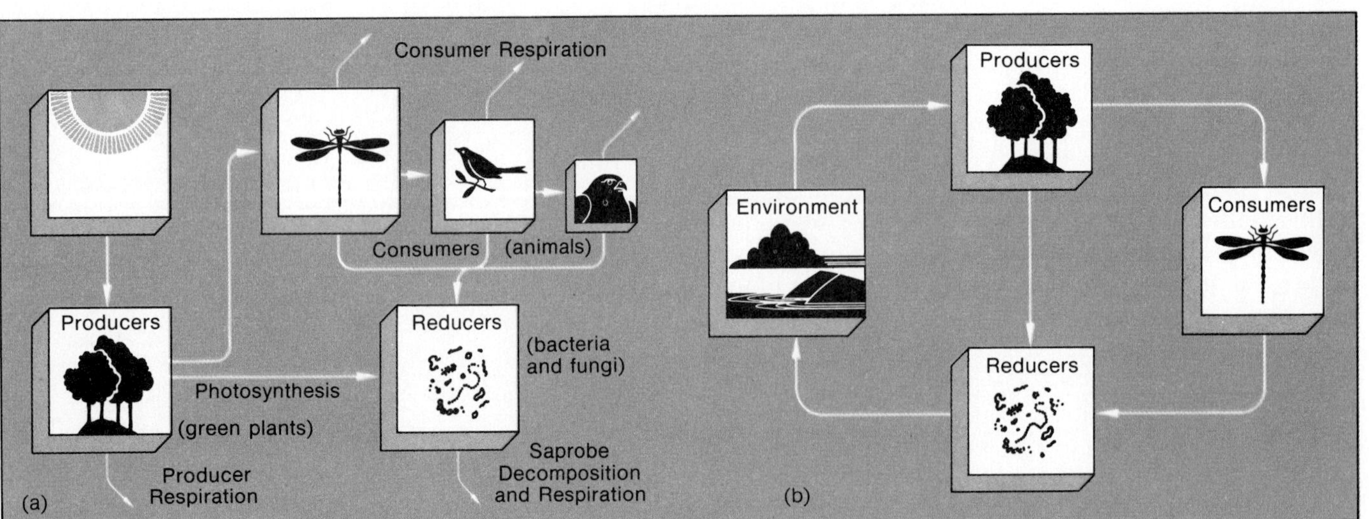

Consumer Respiration
Consumers (animals)
Producers
Photosynthesis
(green plants)
Producer Respiration
Reducers
(bacteria and fungi)
Saprobe Decomposition and Respiration
(a)
Producers
Environment
Consumers
Reducers
(b)

Figure 9.11(a) This schematic diagram illustrates the flow of energy in a natural ecosystem. Following the flow from left to right, solar radiant energy input is converted to food energy by green plants. The plants use some of the energy for respiration, herbivores consume some of it, and organisms of decay convert some of it to products of respiration. Part of the energy in the herbivores is transferred to successive carnivores. However, the amount of transferred energy decreases at each step because energy is used for respiration and other purposes.

(b) This diagram indicates schematically how materials circulate between the principal units of an ecosystem and between the ecosystem and the environment. (After Whittaker, 1970)

The principal outputs include the energy used in respiration, the energy transferred to the next stage of the food chain, the energy transferred to the decay food chain, and the energy utilized for other purposes. In a pond, for example, the energy in plant and animal materials that sink into the bottom sediment is exported from the system. Figure 9.11(b) is a schematic energy flow diagram for an ecosystem consisting of several stages, including a decay food chain.

One of the classic energy flow studies of a natural ecosystem was carried out at Silver Springs, Florida, by the American ecologist Howard Odum and his colleagues. Their results are summarized in the energy flow diagram of Figure 9.12.

In the Silver Springs ecosystem, the groundwater that feeds the spring is essentially pure—it is the radiant energy from the sun that provides almost all the energy input to the autotrophic forms of life in the spring. A small additional contribution of energy is imported through plant and animal debris that enters the system. Only a small fraction of the absorbed radiant energy is employed in photosynthesis.

There are four links in the Silver Springs grazing chain: plant, herbivore, carnivore, and top carnivore. The significant reduction of energy from stage to stage is evident from the flow diagram.

Figure 9.12 As this energy flow diagram for the Silver Springs ecosystem shows, the energy input to the ecosystem is principally solar radiant energy, with a small input of organic matter that has entered the section of stream under study. The energies are expressed in kilocalories per square meter per year. Most of the solar energy incident on the system is dissipated as heat, but a small fraction is converted to food energy by various water plants. Some of the net production is absorbed by herbivores, some is exported in the form of plant matter that is carried downstream out of the system, and some is transferred to the chain of decay. The small energy loop in the decay chain represents the decomposition of decay organisms. Note the rapid decrease in energy transferred along the main food chain. The top carnivores receive only 0.6 percent of the energy absorbed by the herbivores. The *biomass pyramid* for Silver Springs shows that at any given time, most of the biomass of the ecosystem is in the form of primary producers, with only a small fraction in the form of herbivores and carnivores. The *energy flow pyramid* summarizes the energy flow data from the main diagram. (After H. Odum, 1957, and E. P. Odum, 1971)

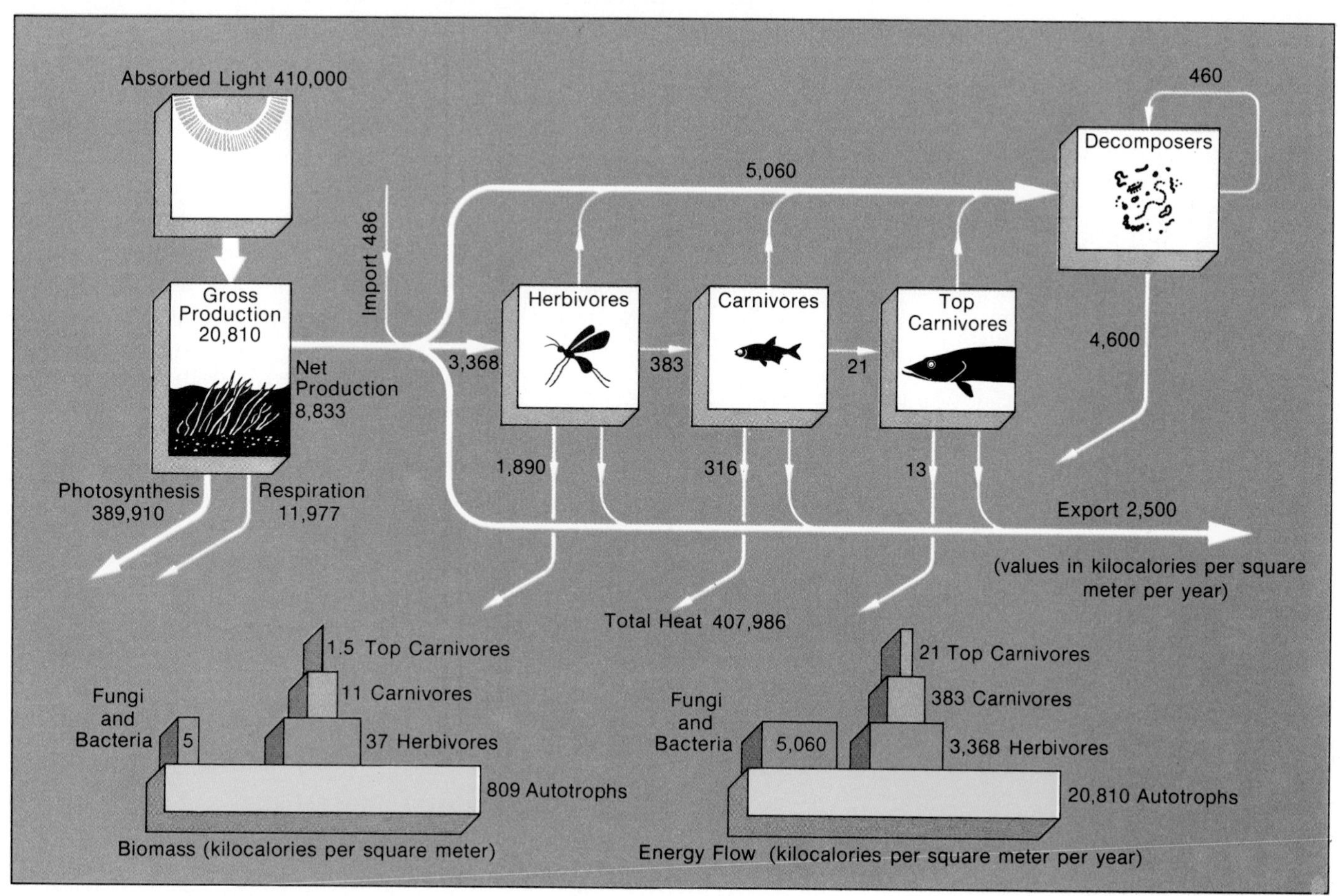

Absorbed Light 410,000
460
Decomposers
5,060
Import 486
Gross Production 20,810
Herbivores
Carnivores
Top Carnivores
4,600
Net Production 8,833
3,368
383
21
1,890
316
13
Photosynthesis 389,910
Respiration 11,977
Export 2,500
(values in kilocalories per square meter per year)
Total Heat 407,986
1.5 Top Carnivores
11 Carnivores
Fungi and Bacteria
5
37 Herbivores
809 Autotrophs
Biomass (kilocalories per square meter)
21 Top Carnivores
383 Carnivores
Fungi and Bacteria
5,060
3,368 Herbivores
20,810 Autotrophs
Energy Flow (kilocalories per square meter per year)

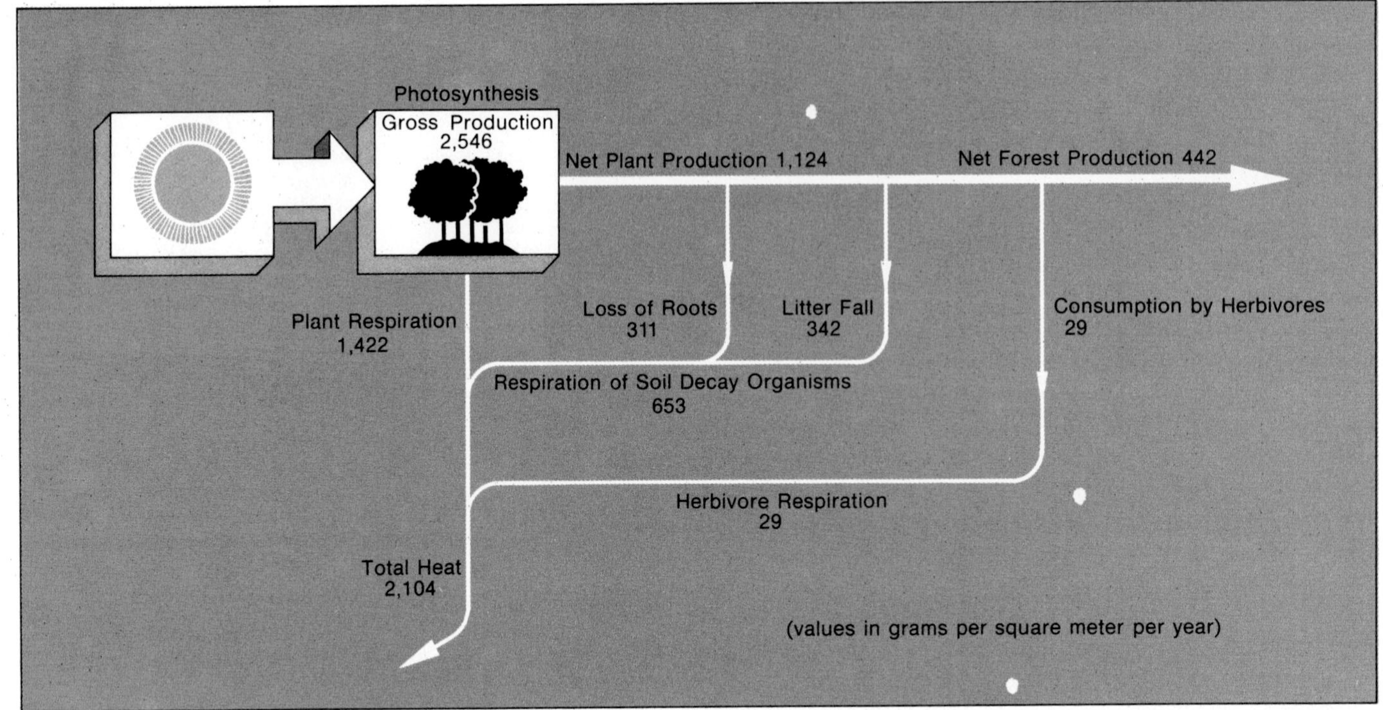

Photosynthesis
Gross Production
2,546
Net Plant Production 1,124
Net Forest Production 442
Plant Respiration
1,422
Loss of Roots
311
Litter Fall
342
Consumption by Herbivores
29
Respiration of Soil Decay Organisms
653
Herbivore Respiration
29
Total Heat
2,104
(values in grams per square meter per year)

Figure 9.13 This diagram traces the energy flow in an oak and pine forest on Long Island, New York, in units of grams per square meter per year. Only 17 percent of the gross productivity is used for net production, indicating that the forest is nearly at the climax stage, the point at which most of the new growth in the forest is balanced by the death and decay of old growth. The gross productivity in this forest is used primarily for the respiration of plants, animals, and decomposers. (After Woodwell and Whittaker, 1968)

Note that the grazing chain is somewhat less important energetically than the decay chain: the energy input to the herbivores is 3,368 kilocalories per square meter per year, but the energy output to the decay chain is 5,060 kilocalories. Plant and animal materials moving downstream out of the region of study account for the exported energy flow.

The energy flow for an oak and pine forest near the Brookhaven National Laboratory on Long Island, New York, has been studied by other researchers. In the course of their work they found that of the annual gross production of dry matter per square meter, only about one-fifth was stored as new growth, litter, and humus. The major part of the gross production was used in respiration.

Plant communities undergo succession toward a stable, comparatively unchanging system that is called *climax* vegetation. From the viewpoint of energy flow, very little energy is used in a climax forest ecosystem to increase the stored energy. The energy released by respiration and by the decay of dead plant and animal tissues balances the energy stored in new growth, so that the total energy stored in a climax forest remains effectively constant. Although an individual tree in a climax forest uses energy for growth, the forest ecosystem as a whole uses essentially all the input energy for respiration. Respiration accounts for 80 percent of the energy used in the oak and pine forest at Brookhaven, which indicates that the forest is in a late stage of succession.

Agricultural Ecosystems

Early human societies represented a hunting and gathering mode of existence; the eventual domestication of animals represented a more effective way of assuring an available food source. Both patterns persist today, and in both it is man who is the top carnivore. However, a hunting or grazing economy cannot support a dense population of top carnivores because so much useful energy is lost at each stage of the long food chains involved. Societies that depend on hunting or grazing are therefore small in size and widely dispersed.

The cultivation of selected plant species in agriculture is more efficient than hunting because the food chain is much shorter. In an agricultural ecosystem, man often plays the part of herbivore, although some agricultural output becomes food for domestic animals. Although people need protein in their diets, only members of affluent societies can afford the cost in energy of eating meat every day—meat is at the end of a long food chain. In the United States, much of the energy available from cereal grains is consumed indirectly as meat because grain is the diet of livestock. The average American consumes directly or indirectly more than 4 times the amount of grain available to the average person in developing countries.

The efficiency of agricultural ecosystems has been remarkably improved in the last hundred years or so. In the Middle Ages in Europe, a return of eight or ten grains from a cereal crop for every grain planted was considered good. Today the average return from a cornfield is 500 grains for each grain planted. There are several reasons for the marked improvement in food production. The principal factors are the development of improved plant varieties; the use of irrigation; the extensive use of fertilizers, weed-killers (herbicides), and insecticides; and the increased mechanization of agriculture.

From 1800 to 1940, corn yields in the United States averaged 25 bushels per acre. Average present-day yields exceed 60 bushels per acre. A primary reason for the improvement was the introduction of high-yield varieties, principally hybrid corn. The recent development of highly productive strains of wheat and rice for use in tropical and semitropical regions has contributed greatly to the food supplies available to countries such as India. The new strains mature early and respond well to fertilizer applications. They also are relatively insensitive to seasonal variation in the duration of daylight, so a new crop can be planted almost any time during the year. This characteristic allows three crops a year to be harvested regularly in tropical countries, which increases the annual productivity of the land.

Figure 9.14 (top) Estimated rice yields in Japan and (center) corn yields in the United States have shown rapid increases in the past few decades because more productive species have been developed and because efficient, energy-intensive agricultural practices have been used. Such a rate of increase cannot be sustained indefinitely for a given crop. (bottom) Wheat yields in the United States increased by 3.5 percent per year from 1950 to 1965, but the projected rate of increase from 1965 to 1980 is 2 percent per year.

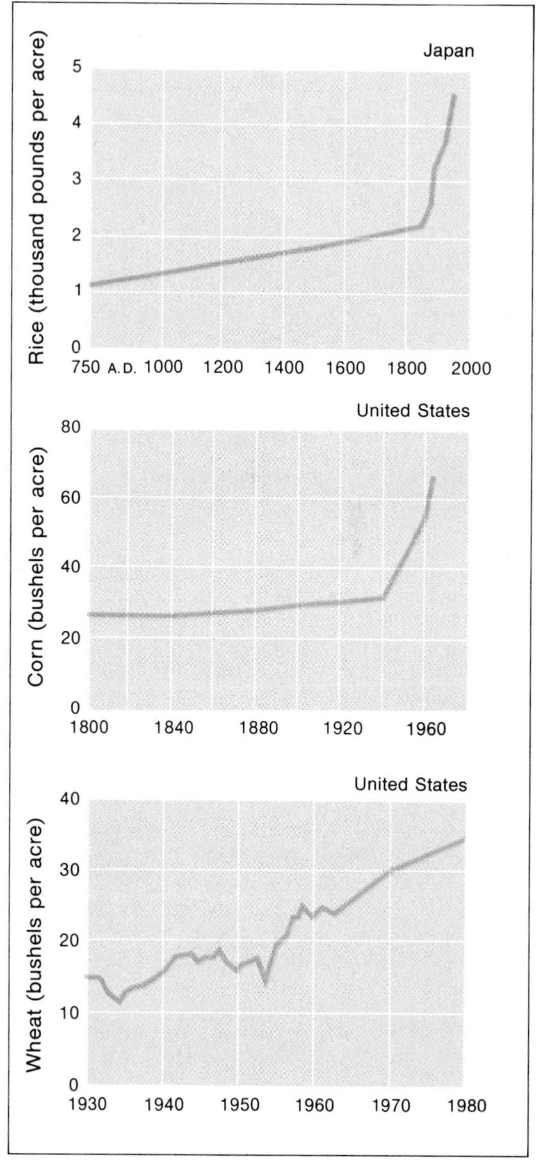

Japan
Rice (thousand pounds per acre)
0
1
2
3
4
5
750 A.D. 1000 1200 1400 1600 1800 2000
United States
Corn (bushels per acre)
0
20
40
60
80
1800 1840 1880 1920 1960
United States
Wheat (bushels per acre)
0
10
20
30
40
1930 1940 1950 1960 1970 1980

Problems of World Agriculture

An agricultural ecosystem is not stable in the way that a climax forest is, a fact well known to anyone who tries to maintain a beautiful green lawn. Man has artificially stabilized agricultural ecosystems by supplying large amounts of external energy to them. In industrialized nations, the stored energy of fossil fuels drives the machinery required to increase agricultural productivity. So the true energy input for food production includes not only radiant energy but also the cost of building tractors, producing fertilizer, training agricultural field agents, and so forth. In highly mechanized agriculture, man may supply more than 1 kilocalorie of energy for each kilocalorie of food energy produced.

The rapid increase in agricultural yields experienced in the last few decades cannot be expected to continue indefinitely, because there are natural limits to agricultural performance. A plant may respond well to moderate applications of fertilizer, but applying 10 times as much fertilizer will not increase the yield tenfold. Although further improvements in productivity will undoubtedly be made, a given plant species cannot be improved indefinitely.

The climatic limits set by potential photosynthesis also affect yields. In recent years, the amount of land devoted to agriculture has declined in the world because of its unsuitability for raising crops. The 100 million acres of virgin land in central Asia put to agricultural use by the Soviet Union is often too dry to produce good harvests, and much of this land is now being released from agricultural activity.

Industrialized nations such as Japan and the United States now grow more food on less agricultural land than they did 20 years ago. It appears that if advanced agricultural methods were introduced into less developed countries, agricultural performance could be significantly improved there. Many social and economic problems are involved, however. There are social and political implications when a society changes from subsistence farming, in which one farm produces barely enough to support its own workers, to modern productive agriculture, in which one farm produces surpluses for other sectors of society. Fewer farm workers are needed, so those forced to leave the land usually congregate in cities to seek new employment. Marketing and transportation systems are needed to distribute farm products. Modern agriculture requires costly equipment and supplies, and the new methods involved require intensive educational programs for farmers.

Figure 9.15 The build-up of grain stocks ceased in 1961, and at present stocks are being depleted to meet world food needs. In developing regions, the food available per person is projected to decrease during the remainder of this century.

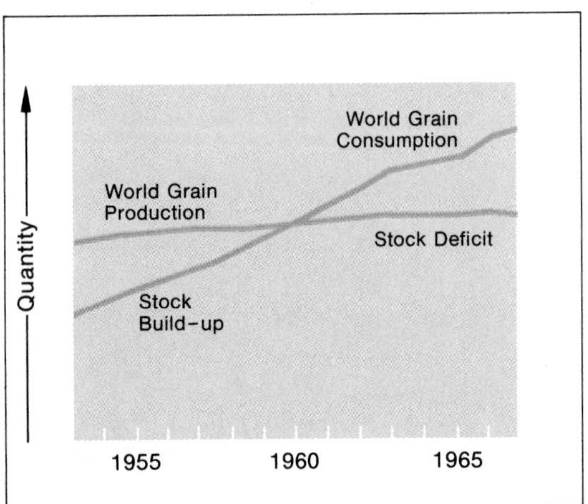
World Grain Consumption
World Grain Production
Stock Deficit
Stock Build-up
Quantity
1955
1960
1965

It is expected that the population of the world will double between 1970 and the end of this century. Agricultural production will not be able to keep pace with such growth indefinitely. The food shortage in the Soviet Union and in a number of other countries after the poor harvests of 1972 shows that food production is not high enough even now. The problem at hand is to feed people adequately until a stable situation is attained. The Food and Agriculture Organization of the United Nations has developed a World Plan for agriculture to meet food needs through 1985. The plan emphasizes cereal production by intensive agriculture and calls for increased production of animals, particularly hogs and poultry, because they are relatively efficient to raise. Humans require protein regularly, and lack of protein is one of the most serious nutritional problems in the world today.

The United Nations plan calls for cereal production to increase by 3.6 percent each year. Reports at the end of 1972, however, indicate that the growth rate has not increased appreciably from

2.6 percent, the value that has been characteristic of the last decade. World food production and consumption are contrasted in Figure 9.15.

Ecosystems and Man

Man has always interacted with natural ecosystems, and he has set up ecosystems of his own to fit his needs. Throughout most of history, and in many countries today, man's thrust in dealing with nature has been to obtain the immediate necessities for survival. Sometimes, however, these activities have made survival more difficult over time.

Man's interaction with nature has been comparatively successful, nevertheless. A large number of people are well-clothed and well-fed today, at least in comparison with earlier ages. It is becoming increasingly clear, however, that human activities in the drive for survival have important repercussions in nature. As we attempt to turn a greater proportion of the earth's productivity to our own uses, we sometimes act in ways that will make that productivity less available to us in the future.

The traditional method of farming in the rainforests of Panama and Brazil has been to clear a section of forest by cutting and burning, to farm the land for a few years, and then to move on to a new section. In a rainforest, nutrients are stored in the natural vegetation rather than in the soil, so after a few years, cleared land becomes less productive. This slash-and-burn, or *milpa*, cultivation

Figure 9.16 The dark-shaded areas on this map sculpted of bread dough indicate counties in the United States where malnutrition is thought to be a serious health problem as of January 1, 1968. The criteria for identifying the critical areas include a high rate of mortality of infants older than 1 month, a high degree of poverty, and a low level of participation in welfare or food stamp programs compared to the number of poor and needy. Lack of proper nutrition during childhood may lead to poor performance in school, stunted growth, and retarded mental development. Diseases associated with nutritional deficiencies debilitate many millions of adults as well.

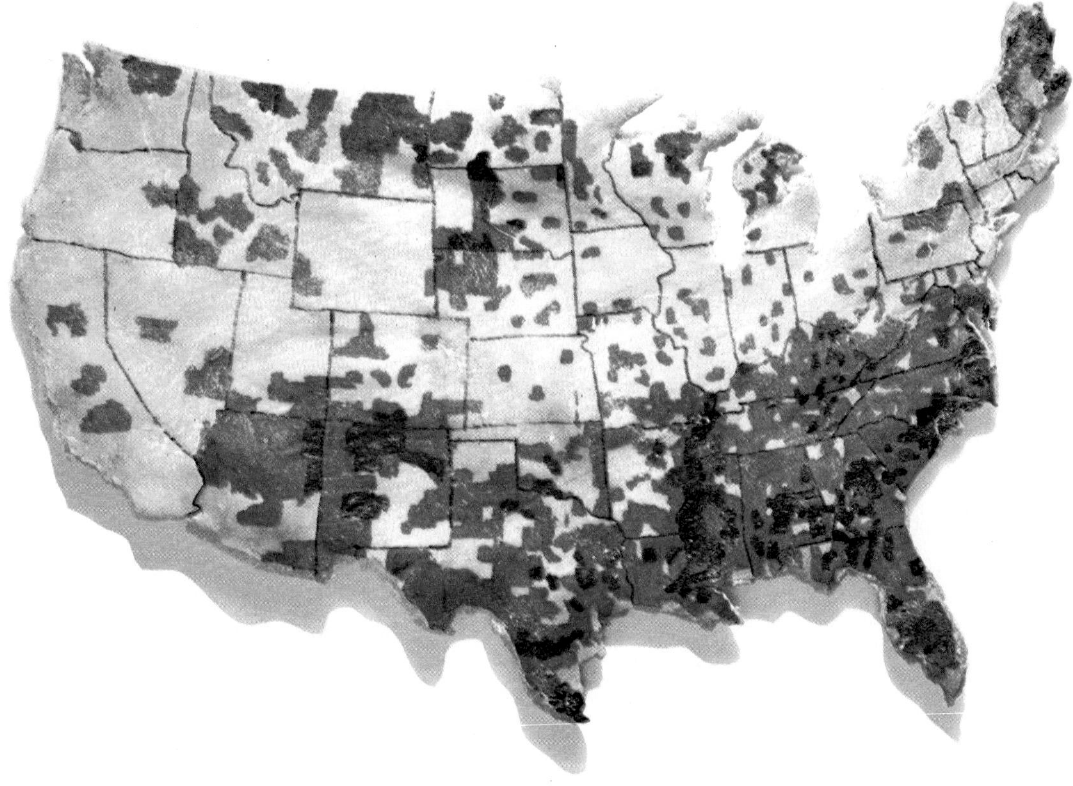

Case Study: The Tragedy of the Whale

Whale species are conveniently placed in two categories, the *toothed* whales and the *baleen* whales. Most toothed whales—porpoises, dolphins, narwhales, and killer whales, for example—are relatively small. The sperm whale, the largest of the toothed whales, is perhaps the best known of all whales; of the five species that sustained the nineteenth century whaling industry, sperm whales were most frequently killed.

The larger baleen whales are so named because of the comblike structure that descends from the roof of their mouths. It is made of closely spaced parallel triangular plates of hornlike substance called baleen. The inner edges of these long, thin slats fray out into a thickly matted net of hair that lines the entire inner surface of the comb so that as the whale swims through fields of plankton or schools of fish it sieves out and swallows whatever food the sea water contains.

The throat and belly of most baleen whales, such as the humpbacks, are deeply furrowed by long pleats from chin tip to navel. These pleats allow the whale's throat to expand so that it can engulf great quantities of water from which the whale strains its food. Baleen whales without these pleats—the right whale, for example—have longer baleen strips and strain their food by swimming, with their mouths part way open, directly into masses of small marine life, thus collecting the tons of krill and animal plankton they feed on.

Most species of whales are social animals, and they move about the oceans in groups that vary in size from small families of a few animals—some of them coming together only during part of the year for pairing and mating—to large groups, such as porpoise schools, that number in the thousands.

Just below the uppermost film of the world's oceans live some of the most gentle, most serene, most immense creatures of the ocean—the whales. But certain species of whales have been severely reduced in number because so many have been killed for commercial use. Whale oil has been used in household paint, in margarine and soap, and in cosmetics. Whale meat, although sometimes eaten by people, is fed primarily to domestic animals—dogs and cats in Europe and the United States and ranch fox and ranch mink in the United States, Canada, and Norway. The remarkable fact about the various products made from the oil and flesh of whales is that all of them could be made from synthetic materials or other, more abundant natural raw materials.

The whale represents the only source of the abundant energy that man has yet harvested significantly in the Antarctic Ocean. Of the utmost importance is the fact that whales are not themselves the basic source of energy from this area but rather they are the "packages" in which the abundant polar energy comes.

It may sound strange to speak of the cooler northern and southern waters as being abundant with energy, but in fact it is the polar seas, not the tropic ones, that are highly productive. Because of the lack of nitrogen cycles in most of the ocean, except in upwelling regions, tropic seas are deserts compared with polar ones. In the polar seas life is so abundant that during the short growing seasons the ocean surface is often discolored with plankton blooms for many square miles. Plant plankton, which carry on the major portion of photosynthesis that occurs in the ocean, use only carbon dioxide, sea water, and certain nutrients to synthesize proteins, fats, and carbohydrates. These plant plankton, which lie at the base of the energy

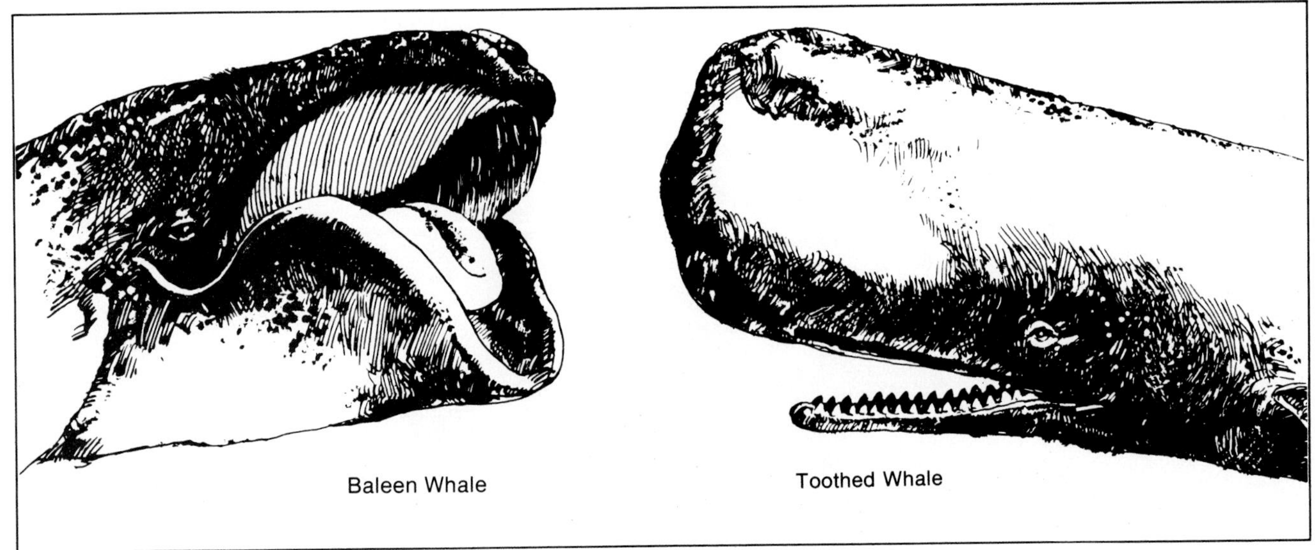
Baleen Whale

Toothed Whale

pyramid, are eaten by a variety of animals ranging in size from single-celled protozoa to crustaceans resembling small shrimp. These are in turn devoured by other, usually larger animals, some of which are eaten by other animals, and so on. In general, 80 to 90 percent of the energy is lost at each step in the food chain.

Baleen whales, which constitute the most abundant group of whale species, feed almost exclusively on the largest plankton animal, krill. Their food chain can be represented as follows:

Plant plankton → animal plankton → whales
(primarily krill)

Because these whales obtain their food from animals that are near the base of the energy pyramid, the amount of energy lost in the food chain is relatively small. Thus, whales can harvest an important part of the ocean's food relatively efficiently, and they store their harvest in neat, easily collected multi-ton packages: themselves.

But the activities of the whaling industry have significantly depleted the whale population in much of the Arctic during the last century and in the North Pacific and the Antarctic during this century. Thus, the whaling industry has destroyed the only access that we have yet used to the abundant plant plankton of the northern and southern seas. The business of whaling, far from being an outdated practice from our history, has until very recently been more extensive than it ever was in previous centuries. The total of whales killed in the 1960s was the largest 10-year kill ever made.

At present, only four species of the "great" whales (fin, sei, Bryde's, and sperm) remain in commercially profitable numbers. Provided present management continues, these should remain at or rebuild to optimum population sizes. (Of the four species, only fin whale stocks are currently below optimum size.)

How many whales can be taken from a population without endangering the species? Scientists are attempting to determine what a viable number of whales might be by studying the rate at which a species can replace itself and thus maintain its population. Under dense population conditions, whales, like all animals, produce more young than are necessary to maintain the whole population, and a fairly high percentage of newborn whales presumably die of natural causes during the first year: predators, disease, inadequate food supply, failure to compete, and so on. When the population of a region is substantially reduced for any reason, a larger proportion of the young survive. Also, the number of offspring produced in a depleted population may increase because of a lowering of the average age of sexual maturity and an increase in the pregnancy rate. Female fin whales, for example, mature sexually at about 6 years in a depleted stock, as opposed to about 10 years in a stable, unexploited stock, and the pregnancy rate increases from about 33 percent to about 50 percent.

From a commercial viewpoint, then, a viable number of whales would be the maximum number of animals that can be taken from a species each year without preventing the species from making up that loss. This number is known as the *sustainable yield*. In order to calculate the sustainable yield of a whale species, one must know, among other things, the size of the overall population, how many young a female bears each year, how long it takes the young to reach reproductive age, how long an individual's reproductive age lasts, and what the life expectancy of each age group is.

Some scientists have claimed that the main schools of whales in the Antarctic have been brought to near extinction. But just what does "near extinction" mean? The whaling industry is concerned only with *commercial extinction*, in which the number of remaining animals is too small to be harvested profitably. Other scientists studying whales are quick to point out that there is little danger that any species will become totally extinct—there will always be at least a handful of surviving members of a species. In terms of ecology, however, neither of these interpretations seems relevant; the few whales that remain at the point of commercial extinction play a very insignificant part in the ecosystems of the oceans. In this sense, the concept of *significant extinction* seems a more relevant one. And it seems apparent that it is the significant extinction of whales that we should strive to prevent.

was carried on for thousands of years with no apparent long-term ill effects, as long as population densities were low and as long as the forest was allowed time to reestablish itself in a region before clearing was repeated.

Today, however, the increased population pressures and the need for food have led the people to shorten the cycle between successive rotations. Furthermore, the cleared land is often used to graze cattle after it has been farmed for a few years. The grazing destroys secondary growth and prevents the reestablishment of the forest. The cattle pack the earth to a nearly impervious layer that seriously diminishes infiltration of rainfall and that increases runoff and erosion. The accelerated pace of milpa cultivation may make large tracts of land in the rainforests economically useless and difficult to restore to productivity (see Case Study, Chapter 10).

The search for food has caused many countries to turn to the sea. The sea as a whole is not a highly productive region. Photosynthesis is limited to the upper 50 to 100 meters (160 to 330 feet) of the ocean, and growth in this layer is usually limited by lack of nutrients in the water. An acre of midlatitude grassland is several times more productive than an acre of ocean: the net production of the ocean is less than half the net production of the land.

The strategy of many nations is to fish as intensively as possible until reduced catches no longer warrant continued investment. For this reason the ocean is being overfished, and many species, including some types of whales, have almost disappeared. In 1935, commercial fishermen in California landed 60,000 tons of Pacific sardines. The catches decreased with time, and by 1961 the sardine industry in California no longer existed. Similar overfishing is evident with Pacific perch, mackerel, and tuna. As Slobodkin showed with his laboratory food chain, the best long-term strategy for an intelligent predator is to leave enough animals to make good use of the available food. However, the desire of many nations for an immediate increase in their living standard becomes more important than the inevitable decline in ocean productivity.

The misuse of rainforest land and the overfishing of the oceans are only two examples that illustrate how immediate benefits can be accompanied by large hidden costs. The farmers and the fishing fleets gain access to an inexpensive source of energy, but restoring the ecosystems that produced the energy will eventually be costly to society as a whole. Because the true costs of his actions are seldom considered, man may eventually suffer the consequences of bankrupt ecosystems.

With his rapid increase in population, man is far from being in a stable long-term relationship to natural and agricultural ecosystems. Andrei Sakharov, a leading Russian physicist, has predicted famine among millions in the next 20 years if production continues to fall further behind population, and he has called upon the affluent nations of the world to use one-fifth of their annual production to help other nations reach higher standards of living. Research can provide the basis for informed actions, but the time available for building a foundation of knowledge is shrinking rapidly.

Summary

A green plant's ability to store solar radiant energy as chemical energy enables it to utilize energy directly from the physical environment. The stored energy that remains after the plant's own energy requirements are met is available as food to other members of natural or agricultural ecosystems.

The climatic factors that condition the productivity of an agricultural crop are solar energy input, temperature, and the amount of moisture available. The distribution of energy and moisture over the surface of the earth again appears as a major factor in the operation of the earth's systems. The transfer of energy from green plants to other organisms takes place through food chains or food webs. Studies of laboratory and natural ecosystems indicate that only about 10 percent of the energy that is absorbed during one stage of a food chain is passed on as food to the succeeding stage. The top carnivore in an ecosystem stands at the apex of an energy pyramid, which the primary producers ultimately support.

Man has developed ecosystems of high agricultural productivity, but the increase in productivity cannot be expected to continue indefinitely. Penalties are incurred when agriculture is expanded into regions that have insufficient energy, inadequate moisture, or unsuitable soils.

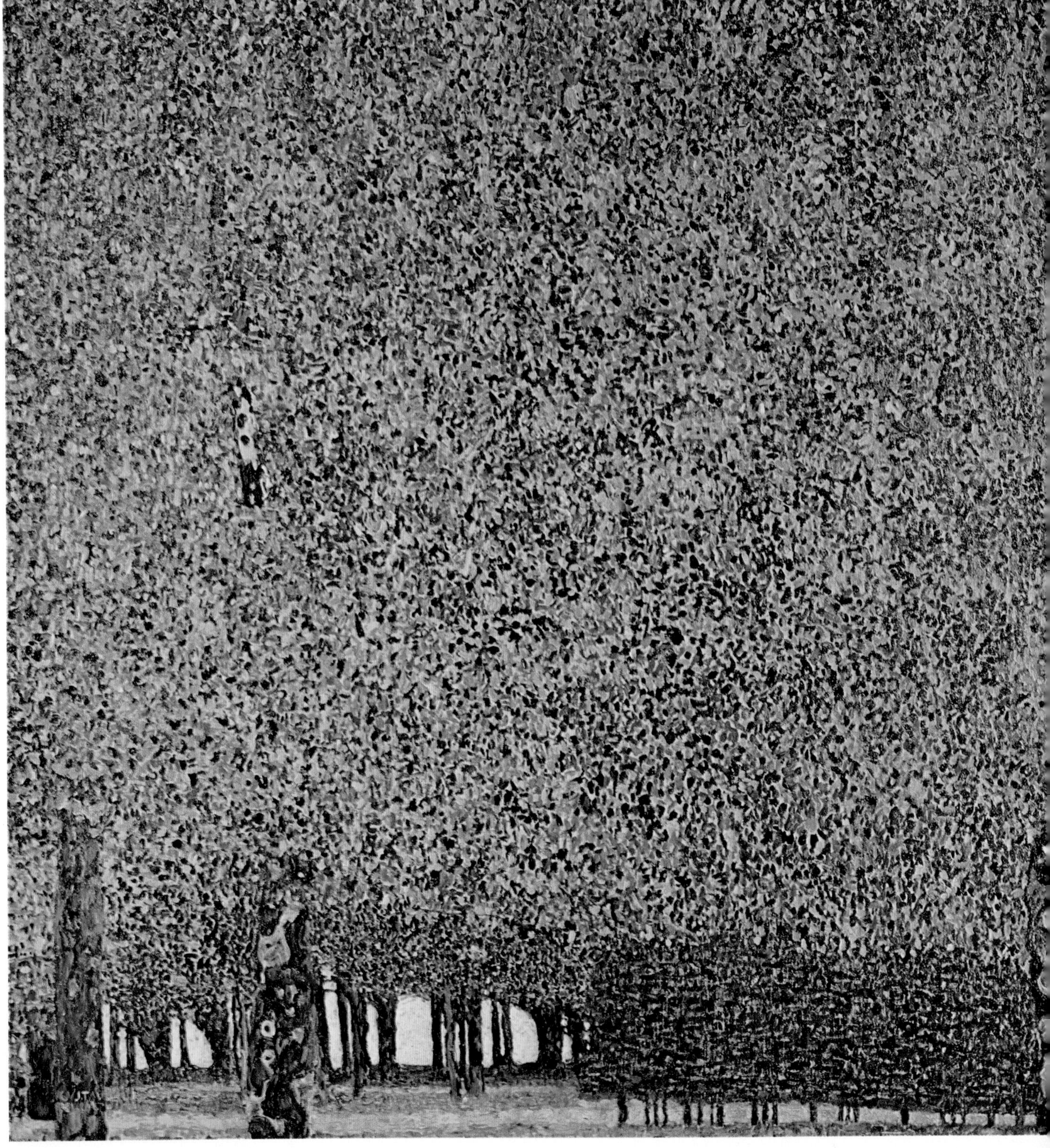

10 Global Systems of Vegetation and Climate

Vegetation and Climate
Plant Communities
Plant Succession and Climax
Natural Vegetation

Principal Vegetation Regions and Their Climates
Equatorial Rainforests
Tropical Seasonal Forests
Tropical Savannas
Tropical Thorn and Scrub Forests
Subtropical Semidesert and Desert Regions
Midlatitude Shrublands
Midlatitude Deciduous Forests
Midlatitude Grasslands
Subarctic Coniferous Forests
Tundra Vegetation

Changes in Climate
Climate of the Early Holocene
Climate in Historic Times
Theories of Climatic Change
Climates of the Future

Sunlight and rain showers give an arboretum its colorful, luxuriant foliage. If the climate in a particular region were to change, the plants in that region would also change because of the close interaction of vegetation with climate.

The Park by Gustav Klimt, 1910. (Collection, The Museum of Modern Art, New York; Gertrud A. Mellon Fund)

Global Systems of Vegetation and Climate

The various regions of the earth have distinctive appearances because climate provides different amounts of energy and moisture to different regions. The sparse, low shrubs of a subpolar tundra, for example, reflect the distribution of energy and moisture they receive in their particular region, as do the luxuriant growth and towering leafy trees of a tropical rainforest.

This chapter describes how vegetation responds to climate in the world's principal climatic regions. It also describes the history of climatic change and it suggests some of the mechanisms that may cause climate to vary.

Vegetation and Climate

Plants go through different phases of development as they grow from seedling to mature plant, and they cannot survive unless temperature and moisture conditions are suitable for each stage. For instance, some plants will not reproduce unless the minimum temperature during the night is sufficiently low compared to the daytime high. Seeds and mature plants are usually more tolerant of variable weather conditions than young shoots and blossoms. In a citrus grove, a late spring frost that does no damage to the leaves or the stem of a plant may kill all of the blossoms and prevent the production of seeds.

Evolutionary adaptation has produced distinct plant forms especially suited to regions with different energy-moisture conditions. Many of the plants adapted to hot, dry climates, such as cactus, tend to have thick, waxy leaves and stems that can store water. To prevent severe moisture loss, special guard cells often close the stomata in such plants during hot weather. Some annual plants are adapted to desert conditions by having seeds that are well protected against moisture loss. The seeds remain dormant in the desert soil for long periods until rain provides enough moisture for the seeds to germinate. Then the desert blooms as the plants quickly mature, flower, and produce seeds to repeat the cycle of growth.

The broadleaf trees, such as maples and birches, that are common to midlatitude forests transpire large amounts of water and require substantial moisture supplies. Forests can withstand comparatively long periods of drought, provided that precipitation is on the average somewhat greater than potential evapotranspiration. Trees have well-developed root systems that can draw on soil moisture during dry periods. The thick ground cover in forests slows runoff from rain and allows time for water to infiltrate the soil and replenish the soil moisture.

For protection against cold weather and freezing temperatures, the broadleaf trees of the middle latitudes are deciduous—they drop their leaves in the fall. Conifers, such as pine or spruce trees, survive more easily than broadleaf trees in high-latitude regions, which have only moderate supplies of moisture and short growing seasons.

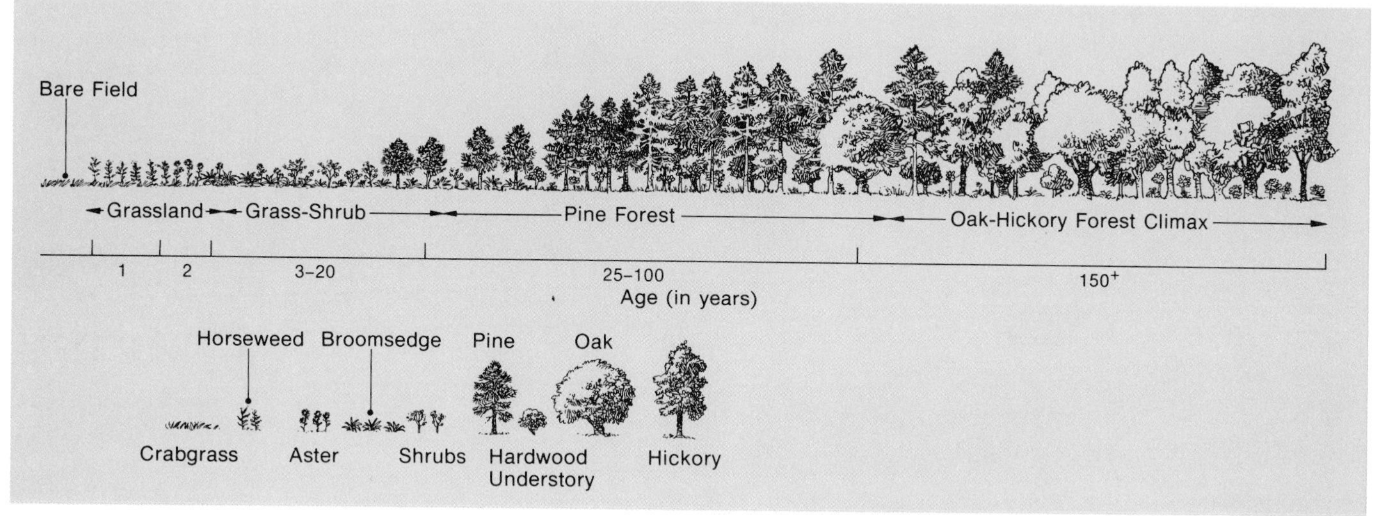
Bare Field
Grassland
Grass-Shrub
Pine Forest
Oak-Hickory Forest Climax
1
2
3–20
25–100
150+
Age (in years)
Horseweed
Broomsedge
Pine
Oak
Crabgrass
Aster
Shrubs
Hardwood Understory
Hickory

Figure 10.2 This diagram illustrates a typical plant succession from open land to a midlatitude deciduous forest. In the initial stages of the succession, low, fast-growing grasses and shrubs dominate. Each stage alters the soil and microclimate of the land, enabling other species to establish themselves and become dominant. The final stage consists of high trees that overshade and force out some of the low shrubs of earlier stages. (After E. P. Odum, 1971)

The thin needles of a conifer are protected from water loss by thick, waxy tissue. Conifers are usually evergreen, and they begin photosynthesis with the first warm spring weather. In contrast to the conifers of the cooler regions, trees in the hot, wet rainforests of the tropics are often broadleaf evergreens. They lose leaves from time to time during the year rather than at one season, so that rainforests are always green.

Plant Communities

In one midlatitude broadleaf forest there are various kinds of trees, shrubs, ferns, grasses, flowers—perhaps fifty or more different species of vegetation—all living together in one *plant community*. A forest, a grassland, or a prairie consists of numerous species of plants living in association with each other and with bacteria, fungi, insects, and other animals. A community affects its local environment by its modification of soils and moisture storage. The individual members of a community interact with one another and respond to changes in a way that gives the community a definite pattern of organization.

The different species in a plant community do not usually compete for energy and moisture because each species has its own special *niche* in the community. In a broadleaf forest low shrubs use the light that filters through the high leafy canopy and do not compete with the towering trees for light. Mosses growing on rocks do not rely on the same moisture supply drawn on by ferns. In central Africa farmers often successfully grow several crops, such as bananas and cassava, in the same field. As long as the crops do not compete for the same moisture and sunlight, the total crop yield per acre can be greater than if the crops were planted separately. If two species in a plant community should happen to compete, eventually one will be driven out by the other, because two competitive species cannot both occupy the same niche indefinitely.

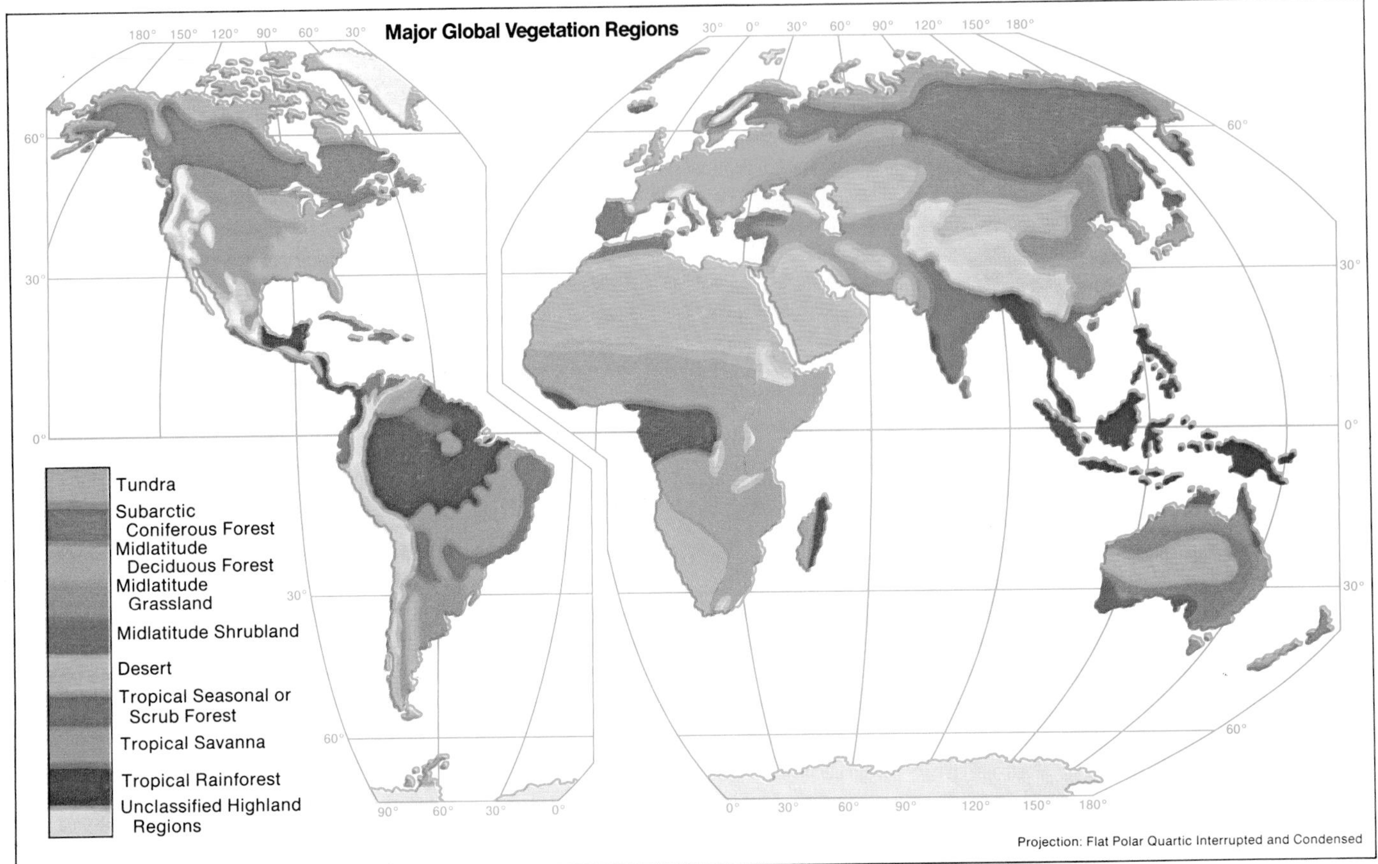

Major Global Vegetation Regions
Tundra
Subarctic Coniferous Forest
Midlatitude Deciduous Forest
Midlatitude Grassland
Midlatitude Shrubland
Desert
Tropical Seasonal or Scrub Forest
Tropical Savanna
Tropical Rainforest
Unclassified Highland Regions
Projection: Flat Polar Quartic Interrupted and Condensed

Figure 10.3 This schematic global map of the distribution of major vegetation types should be compared with the global map of climates (Figure 7.5) and the global map of soils (Figure 8.24). Climate, soils, and vegetation are strongly interacting systems, and their global distributions bear marked similarities. The regions occupied by tropical rainforest, for example, are found in tropical wet climates, and the underlying soils in the rainforest tend to be latosols. Similar broad global generalizations can be made about most of the major vegetation types. (After Finch and Trewartha, 1949)

Plant Succession and Climax

A hiker walking through the woods of New Hampshire sometimes comes upon the foundation stones of farm buildings that were abandoned more than a hundred years ago. Although the land was once cleared for farming, it is now forest again. A forest does not immediately establish itself when cleared land is left to nature. A *succession* of plants occurs, one type of plant giving way to the next, until a stable community is attained. Each community alters the local climate and conditions on the land, and in the new environment, other species become dominant. The particular communities through which a succession advances depend on whether the succession begins in cleared land, filled-in marsh, burned-over forest, or some other condition. They also depend on the kind of stable plant community characteristic of the region. In a succession that changes from cleared land in wooded country to forest, the early communities are dominated by grasses and low shrubs. The grasses give way to communities containing small trees and higher shrubs. The general trend in succession in all plant communities is toward *taller and denser vegetation*, toward *more diverse vegetation*, and toward *greater stability*. As communities become established, the pace of succession tends to slow—large trees, for example, require many years to grow.

The stable community that results at the end of a long, undisturbed succession is called a *climax*. It may take a long time to establish a climax, partly because man's activities, such as farming or logging, frequently interrupt successions. Extensive logging in New York during the nineteenth century selectively removed the softwood coniferous trees useful in industry, which allowed the deciduous hardwoods to expand into areas formerly dominated by softwoods.

Natural Vegetation

In time, a plant community evolves to a climax called the *natural vegetation*. The natural vegetation that grows in an area is a good indicator of a region's potential for agriculture or for tree farming. Figure 10.3 shows a simplified map of the global distribution of natural vegetation. The map shows an idealized system of natural vegetation because once agriculture is introduced, the natural vegetation is not always easy to determine. In Hawaii, a large number of plant species flourish in a small area, but two-thirds of the species were introduced at various times from outside, and a number of native species have been crowded out. Grazing practices also have significantly disturbed the plant communities in Hawaii. By contrast, large tracts of Alaska have lain undisturbed for a long time. The comparatively small number of plant species found in cold regions makes the natural vegetation of Alaska easy to determine; it is primarily spruce and fir forests and tundra plants.

Principal Vegetation Regions and Their Climates

Many of the major vegetation regions described in this section corre-

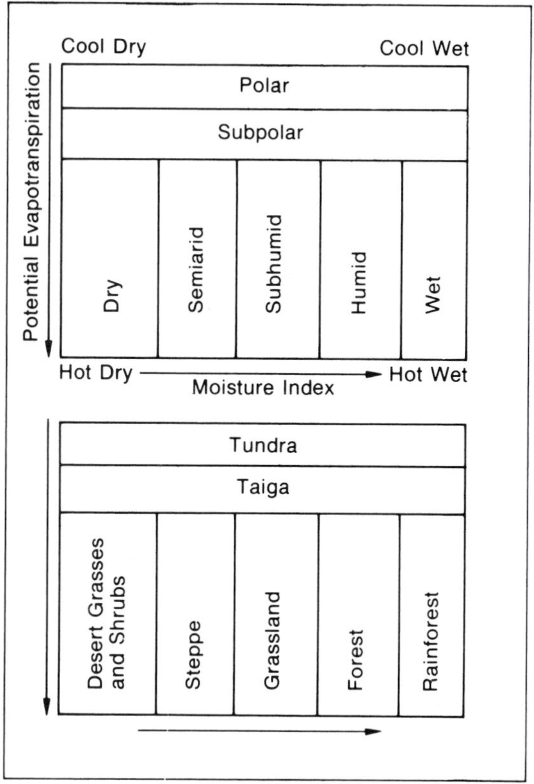

Cool Dry
Cool Wet
Potential Evapotranspiration
Polar
Subpolar
Dry
Semiarid
Subhumid
Humid
Wet
Hot Dry
Moisture Index
Hot Wet
Tundra
Taiga
Desert Grasses and Shrubs
Steppe
Grassland
Forest
Rainforest

Figure 10.4 (above) The top schematic diagram shows the principal climate regions according to the Thornthwaite system of classification. Potential evapotranspiration, a measure of energy, increases downward from cool to hot. The moisture index, a measure of available moisture, increases from dry at the left to wet at the right.

The bottom diagram shows the principal vegetation regions diagrammed according to the same climatic measures. Note that the occurrence of tundra and subpolar (coniferous forest) vegetation is restricted to regions with small energy input. In other regions, the vegetation is determined primarily by the availability of moisture. Forests, for example, can establish themselves only where the moisture index exceeds a certain value. (After Blumenstock and Thornthwaite, 1941)

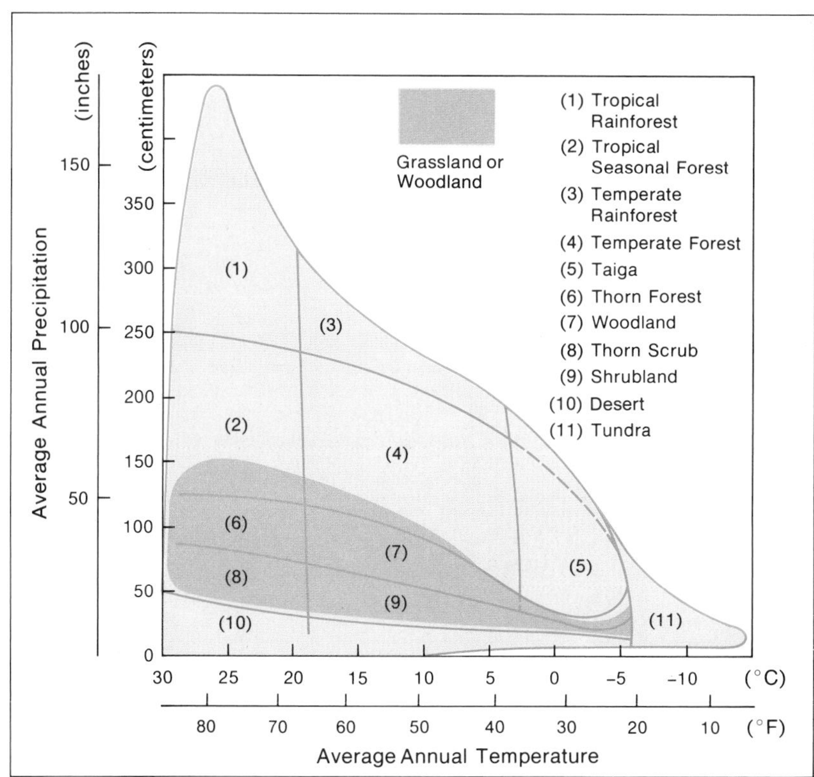

(inches)
(centimeters)
Average Annual Precipitation
150
100
50
350
300
250
200
150
100
50
0
Grassland or Woodland
(1) Tropical Rainforest
(2) Tropical Seasonal Forest
(3) Temperate Rainforest
(4) Temperate Forest
(5) Taiga
(6) Thorn Forest
(7) Woodland
(8) Thorn Scrub
(9) Shrubland
(10) Desert
(11) Tundra
(1)
(2)
(3)
(4)
(5)
(6)
(7)
(8)
(9)
(10)
(11)
30 25 20 15 10 5 0 -5 -10 (°C)
80 70 60 50 40 30 20 10 (°F)
Average Annual Temperature

Figure 10.5 (left) This schematic diagram shows the typical vegetation to be expected in climate regions according to the annual temperature and precipitation. The boundary lines are approximate, because the vegetation in a region also depends on such factors as the seasonality of precipitation. The shaded portion of the diagram indicates the regime of temperature and precipitation that supports either grassland or forest, depending on the history of the region and on man's effects. (After Holdridge, 1947)

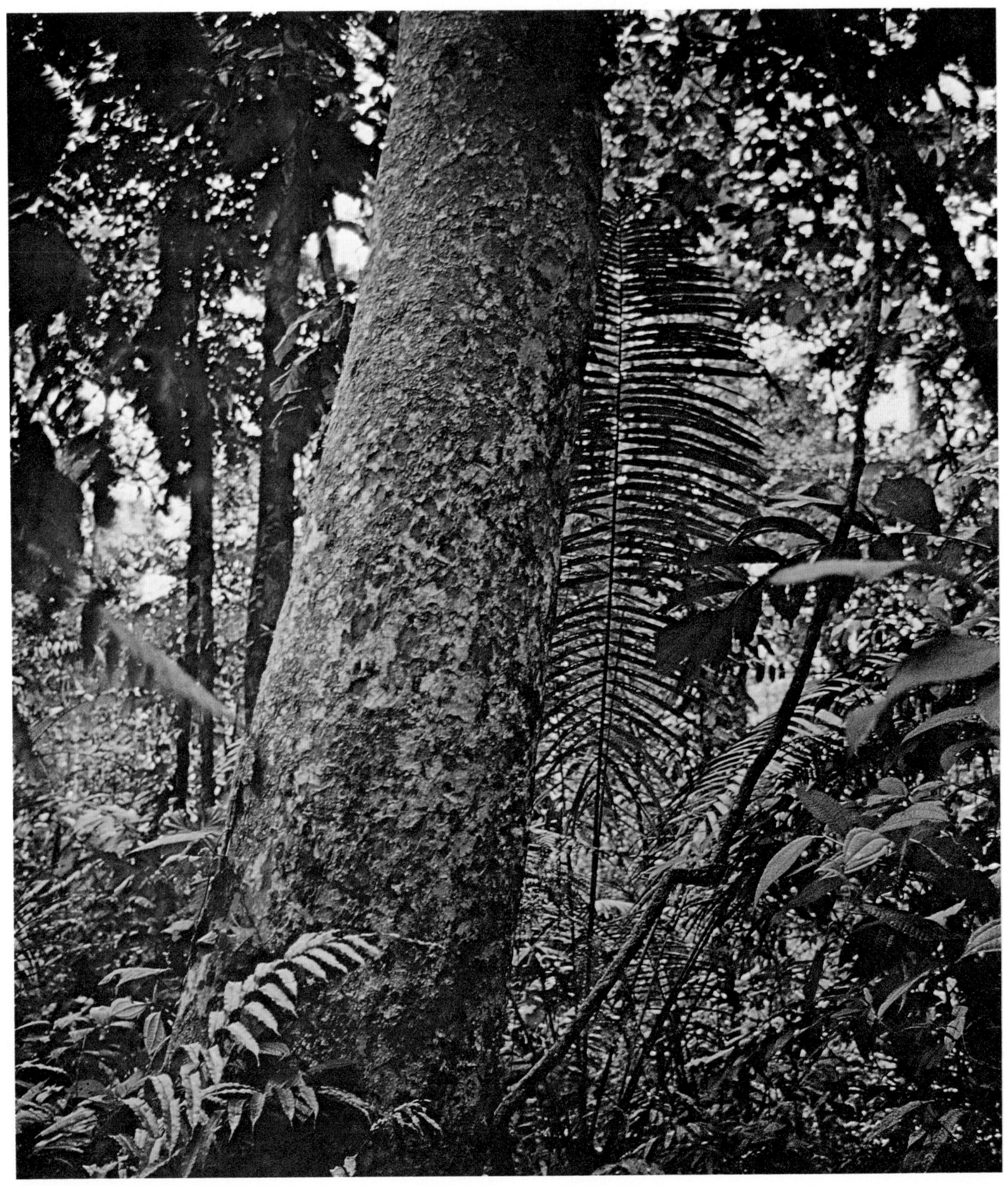

spond closely to climatic regions discussed in Chapter 7. Each region's characteristic plant growth is described and discussed in terms of the amount and distribution of energy and moisture it receives. The correlation between plant communities and climate tends to break down at the artificial boundaries set up between different climatic regions. The boundary lines drawn between climatic regions are based on long-term averages of weather data, but the response of vegetation may depend on short-term fluctuations as well as on averages. If a dry region has a year with above-average moisture from time to time, plants from neighboring moister regions may push their way into the region classified as dry. Some overlap at regional boundaries is therefore to be expected.

Figure 10.6 (opposite) The dense vegetation in this equatorial rainforest in Ecuador often reaches great heights and may have specially adapted forms such as climbing vines.

Equatorial Rainforests

The trees in lowland equatorial rainforests, where energy and moisture are abundant in all months, are largely broadleaf evergreens that shed some leaves throughout the year. Trees that shed all their leaves in one season are seldom found in the rainforest because a tree that remains dormant for part of the year is at a disadvantage competing with evergreens, which maintain photosynthesis and growth in all seasons.

The diverse vegetation of a rainforest presents a layered appearance, because in the competition for light, tree species of different heights fill available niches. The tallest species form a comparatively dense canopy about 30 meters (100 feet) above the ground, and shorter trees grow best where there are breaks in the canopy. Ferns are commonly found in the region of dim light and abundant moisture at ground level. The vegetation of a rainforest exhibits a number of special adaptations to reach the available energy and moisture. The tall trees often have buttresses growing out from their lower trunks to offer additional support. And climbing vines use the trunks of tall trees as supports for growing upward into regions of brighter sunlight.

Singapore is located at latitude 1°N, which is a region classified as a tropical rainforest climate. The temperature through the years shows little variation from season to season, and rainfall exceeds 15 centimeters (6 inches) in every month (see Figure 10.7). There are a variety of ways to characterize such a climate. The *thermoisopleth* diagram in Figure 10.8 shows the variation in air temperature through the day and through the year. Air temperature through the year at a given time of day can be read from a thermoisopleth diagram along a horizontal line. Air temperature through the day can be read along a vertical line. The pattern of the temperature contour lines, or *isotherms*, on a thermoisopleth diagram reflects the climatic features. When the isotherms run primarily along the horizontal direction, temperature variations during a day are greater than variations from season to season. The diagram for Belém, Brazil, located in a tropical rainforest near the mouth of the Amazon River, exhibits such a pattern. The temperature changes by 5°C (9°F) or more during each day but varies by no more than 2°C (3.6°F) through the year at any given hour.

Figure 10.7 Singapore is a station in the tropical rainforest climate (Köppen *Af* region). As the graph shows, the average monthly temperature and precipitation are high and nearly constant through the year. (After Nelson, 1968)

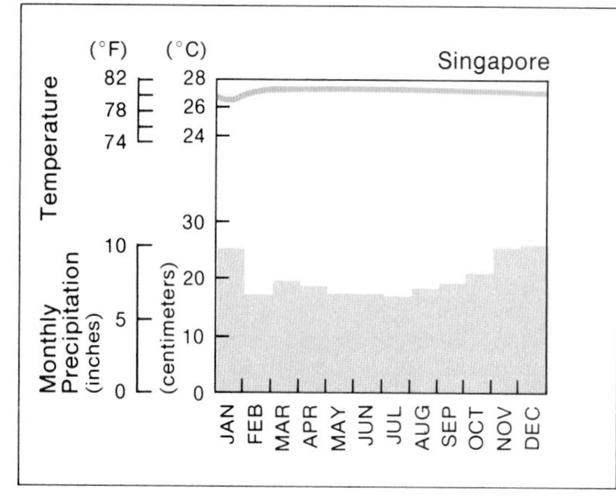
Singapore
Temperature
(°F)
(°C)
82
78
74
28
26
24
Monthly Precipitation (inches)
(centimeters)
10
5
0
30
20
10
0
JAN
FEB
MAR
APR
MAY
JUN
JUL
AUG
SEP
OCT
NOV
DEC

Figure 10.8 Belém, near the mouth of the Amazon River, is located in a tropical rainforest climate (Köppen *Af* region). This thermoisopleth for Belém gives the hourly temperature through an average year. The scale of months begins with July because Belém is in the Southern Hemisphere, so that January and February are summer months. The noon temperature through the year, which can be read by tracing across the diagram from left to right, is between 30 and 31°C (86 and 88°F) until December, when it falls 1 or 2°. The diagram can also be used to trace the temperature through a given day. On January 1, for example, the temperature after midnight remains at about 23°C (73°F) until the early morning, then rises to nearly 30°C (86°F) at midday. The temperature falls again in the evening. In a tropical rainforest, the temperature range through a day is greater than the range of temperature through the year. (After Troll, 1958)

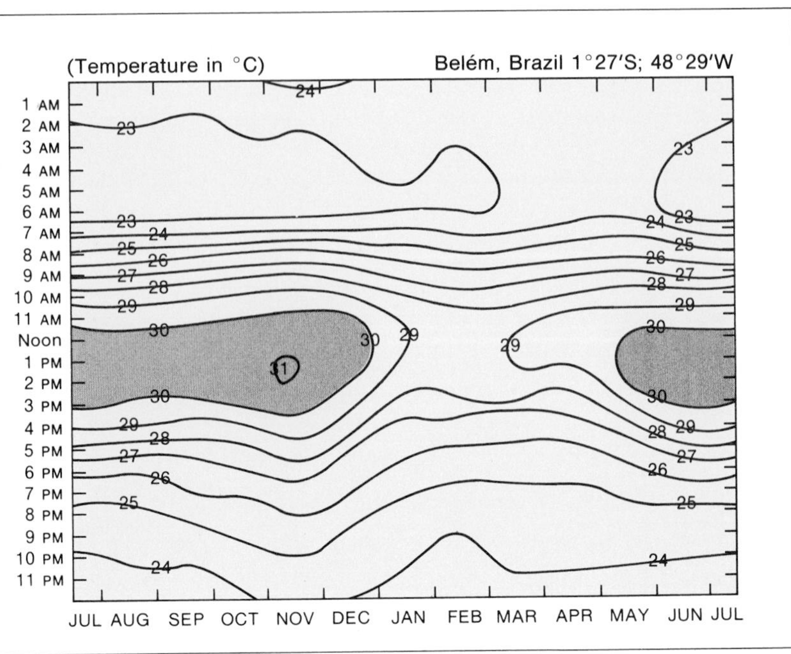

(Temperature in °C)
Belém, Brazil 1°27′S; 48°29′W
1 AM
2 AM
3 AM
4 AM
5 AM
6 AM
7 AM
8 AM
9 AM
10 AM
11 AM
Noon
1 PM
2 PM
3 PM
4 PM
5 PM
6 PM
7 PM
8 PM
9 PM
10 PM
11 PM
JUL AUG SEP OCT NOV DEC JAN FEB MAR APR MAY JUN JUL

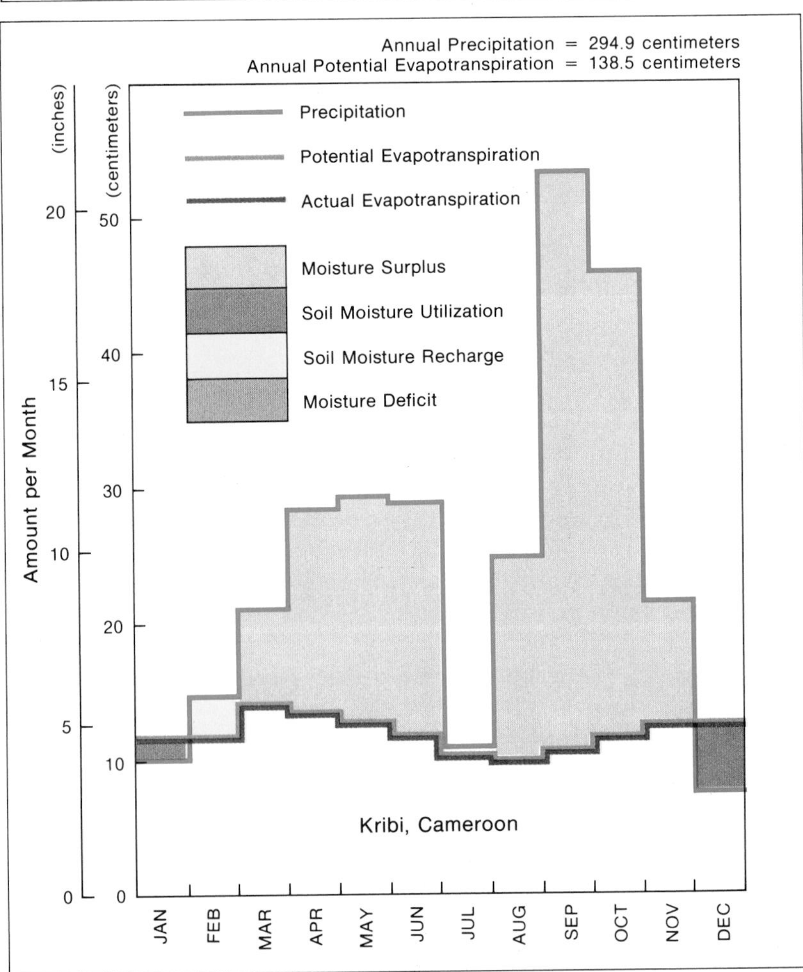

Annual Precipitation = 294.9 centimeters
Annual Potential Evapotranspiration = 138.5 centimeters
Precipitation
Potential Evapotranspiration
Actual Evapotranspiration
Moisture Surplus
Soil Moisture Utilization
Soil Moisture Recharge
Moisture Deficit
Amount per Month
(inches)
(centimeters)
Kribi, Cameroon
JAN
FEB
MAR
APR
MAY
JUN
JUL
AUG
SEP
OCT
NOV
DEC

Figure 10.9 Kribi, Cameroon, is a station in the tropical rainforest of Africa (Köppen *Af* region). In such a climate, the annual precipitation far exceeds the annual potential evapotranspiration, and the water budget for Kribi exhibits large surpluses of moisture in most months. (After Carter and Mather, 1966)

The water budget is another way to characterize climate. Figure 10.9 gives the water budget for Kribi, Cameroon, located in the western section of the tropical rainforest region of Africa. The water budget shows large surpluses of moisture for nearly every month, which reflects the abundant precipitation.

Tropical Seasonal Forests

In tropical regions that have a distinct dry season, moisture is not available to maintain the continuous growth that characterizes the tropical rainforest. Many of the tall trees of the tropical seasonal forest are deciduous and shed their leaves during the dry season, when little soil moisture is available for transpiration. The trees at lower levels are often evergreens. The canopy of a tropical seasonal forest is less dense than the canopy of the rainforest, so more light penetrates to ground level. As a result, the ground layer of vegetation tends to be more varied and luxuriant in tropical seasonal forests than in tropical rainforests.

Figure 10.10 shows monthly temperature and precipitation data at Cuiabá, Brazil, located at latitude 16°S. Like temperatures at locations in tropical rainforests, the temperature at Cuiabá exhibits little seasonal variation. Unlike the rainforest, however, rainfall is seasonally small during the Southern Hemisphere winter from May through September.

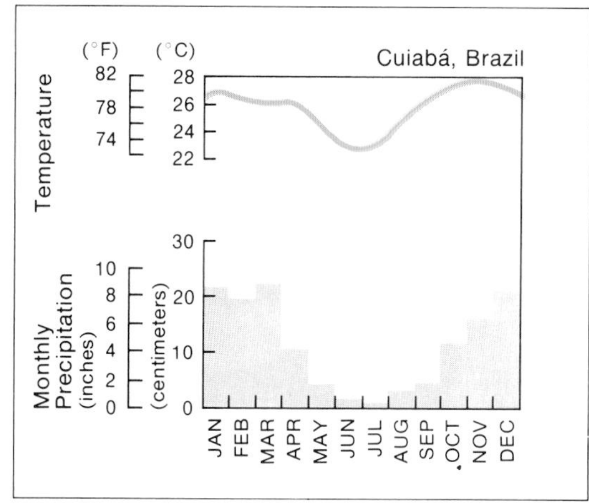
Cuiabá, Brazil
Temperature
(°F)
82
78
74
(°C)
28
26
24
22
Monthly Precipitation (inches)
10
8
6
4
2
0
(centimeters)
30
20
10
0
JAN
FEB
MAR
APR
MAY
JUN
JUL
AUG
SEP
OCT
NOV
DEC

Figure 10.10 Cuiabá, in the Mato Grosso section of western Brazil, has a tropical seasonal climate with a dry winter (Köppen *Aw* region). The graphs of average monthly temperature and precipitation show a dip during the winter months of May through September. Precipitation is nearly 0 during June and July. (After Nelson, 1968)

Tropical Savannas

A savanna is a tropical grassland, often with trees. Savannas lie between tropical wet equatorial climate regions and the dry regions located near latitudes 30°N and 30°S. Although savannas do not

Figure 10.11 These photographs show a tropical seasonal forest in the ancient cities of the Maya, Honduras, during the dry season (left) and the wet season (right).

correspond to a definite climatic region, many of them are found in tropical seasonal climates; extensive regions occur in East Africa.

Savannas vary from open woodland with a ground cover of grass to open grassland with a few isolated trees, and most of the trees shed their leaves during the dry season. The lack of forestation in parts of the savanna is sometimes thought to result from grass fires that destroy young trees rather than from insufficient moisture for tree growth. Man's use of savanna land for grazing also tends to prevent forests from becoming established.

Figure 10.13 shows the water budget for Caracas, Venezuela, located at latitude 10°N in the savanna, or llanos, of Venezuela. The annual precipitation at Caracas is 80 centimeters (32 inches). Savannas are found in regions with annual precipitations as high as 150 centimeters (60 inches), but precipitation as low as 50 centimeters (20 inches) can support a tree savanna if temperatures are sufficiently moderate so that potential evapotranspiration is not too high. The water budget for Caracas shows some deficiency of moisture during the winter months, but soil moisture is never completely exhausted.

The water budgets presented in this chapter are based on a more realistic model of soil-moisture availability than the budget presented in Chapter 6. It is assumed that plants can withdraw only a portion of the total soil moisture in any month, so that both a deficit and a utilization of soil moisture may occur in a given month or in a succession of months.

Tropical Thorn and Scrub Forests

At the dry edge of a savanna where it bounds a desert, precipitation is concentrated in the summer months, and annual totals are considerably less than potential evapotranspiration. The continuous grass

Figure 10.12 The grass savanna in Amboseli Park in Kenya, Africa, is an open grassland dotted with acacia trees.

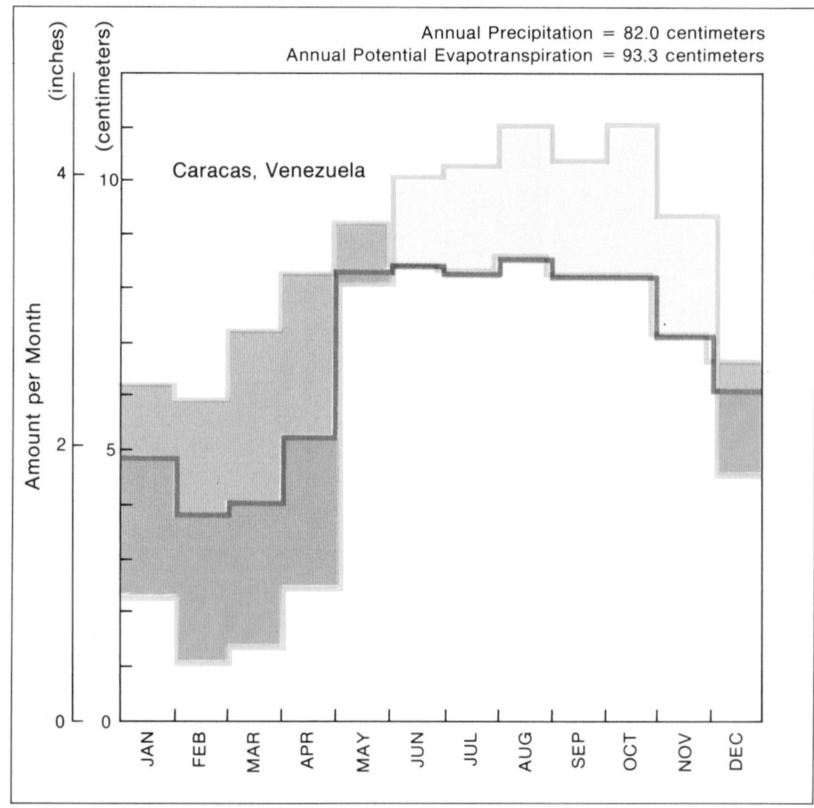

Annual Precipitation = 82.0 centimeters
Annual Potential Evapotranspiration = 93.3 centimeters
Caracas, Venezuela
Amount per Month
(inches)
(centimeters)
4
2
0
10
5
0
JAN
FEB
MAR
APR
MAY
JUN
JUL
AUG
SEP
OCT
NOV
DEC

Figure 10.13 Caracas, Venezuela, has a tropical seasonal climate (Köppen *Aw* region). The average annual precipitation is somewhat less than the annual potential evapotranspiration, but as the water budget shows, the seasonality of precipitation causes moisture deficits to occur during the months of December through May. Soil moisture stores are recharged during the remaining months. This water budget model assumes that only a portion of the stored soil moisture can be drawn on each month by plants during periods of moisture deficiency. (After Carter and Mather, 1966)

cover of the savanna gives way to isolated thorny shrubs or low thorny trees. Thorn scrub and thorn trees have small, hard-surfaced leaves that help the plant resist water loss in the hot climate. Acacia is a typical thorny tree of the tropical desert. Succulent plants with fairly thick, water-filled leaves and hard surfaces are often found growing among thorn scrub.

Subtropical Semidesert and Desert Regions

The warm, dry regions of North Africa, Arabia, and Australia are located under the descending dry air from the subtropical high-pressure cells. Other warm, dry regions include the southwestern United States and coastal regions of South America where moist air is largely blocked by mountain barriers. Potential evapotranspiration is high in warm, dry regions, and the scanty precipitation is insufficient for plants that are not adapted to collect and retain moisture. In some dry regions, such as parts of Saudi Arabia and sections of the Sahara in Africa, precipitation is essentially nonexistent, and large regions support no plant life at all.

Regions with an annual precipitation of 12 to 25 centimeters (5 to 10 inches) are semidesert. Semidesert regions in the United States include the Colorado Plateau of Utah and northern Arizona. Sagebrush, a low, brushy shrub, is characteristically found in Utah. The plants are isolated from one another, so that each plant has a large area of ground from which to draw moisture. Stands of low, well-separated trees, primarily pinyon and juniper, are also found.

Figure 10.14 This scrub region in Australia is analogous to the thorn scrub regions of South America and Africa. The low trees are a species of acacia. The shrubs and trees in thorn scrub forests generally have small leaves.

Figure 10.15 Little or no plant life can survive in this portion of the Sahara in Algeria, because it receives so little moisture. The precipitation must be at least 2 or 3 inches per year for desert-adapted plants such as cactus to subsist.

The vegetation in deserts, where the annual precipitation is less than 12 centimeters (5 inches), is specially adapted to make efficient use of the limited moisture available. Cactus and other succulent plants have thick leaves or stems to store moisture and a thick outer skin, often covered with hairs or spines, to reduce moisture loss. Some plants, such as creosote bush, can withstand the loss of a large proportion of their moisture by lapsing into dormancy during prolonged droughts. Sagebrush and creosote bush have deep root systems that sometimes penetrate to the water table, whereas cactus is usually shallow rooted to take advantage of moisture at the surface from light rain. Ephemeral annual vegetation produces seeds that are well protected against moisture loss and that can survive for several years until sufficient soil moisture becomes available for germination and growth.

Midlatitude Shrublands

Regions with a Mediterranean climate have a hot, dry summer and a somewhat cooler winter during which precipitation is concentrated. The growing season therefore occurs in the winter, when there is abundant moisture and energy and when potential evapotranspiration is not as high as in the hot summer months.

The green of the late-winter growing season gradually becomes brown as grass becomes sere in the dry heat of summer. Trees and shrubs are restricted to species that can withstand a prolonged dry

Figure 10.16 This midlatitude shrubland, or *maquis,* is located in Spain. The plants in such shrublands are adapted to Mediterranean climates and are protected from prolonged drought by thick, leathery leaves. Nevertheless, climate is not the sole factor involved in the establishment of shrublands. Many maquis areas were once forested, and it is believed that land clearing, overgrazing, and fire altered the ecological balance and made it difficult for trees to survive.

season, such as eucalyptus, evergreen oaks, and chaparral. Most of the vegetation is low scrub, and if trees are cleared from the land, scrub may become the climax vegetation.

Figure 10.17 shows the monthly temperature and precipitation at Palermo, Italy, located on the Mediterranean Sea. The temperature is noticeably higher in the summer than in the winter, and most of the annual precipitation falls in the winter.

San Francisco, California, is located in the region of Mediterranean climate on the West Coast of the United States. Its water budget shows a small surplus of moisture in late winter and a marked deficiency of moisture during the summer. The deficiency is not great enough to exhaust the stored soil moisture, however. Figure 10.19 shows the daily temperature and precipitation range at San Francisco for 1954. The average temperature does not change as much as at Palermo because of the moderating effect of the cool Pacific Ocean to the west. The isolated days when temperatures exceed 25°C (77°F) are days when cool air from the ocean does not reach the city, perhaps because of winds from the east. Many days, even days during the moist winter season, show no measurable precipitation.

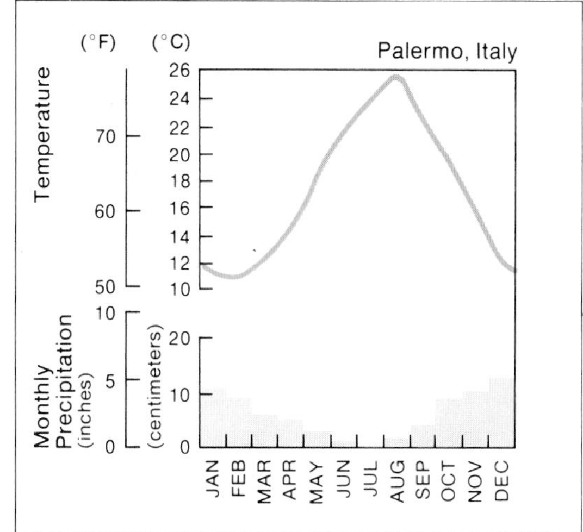
Palermo, Italy
Temperature
(°F)
(°C)
70
60
50
26
24
22
20
18
16
14
12
10
Monthly Precipitation (inches)
(centimeters)
10
5
0
20
10
0
JAN
FEB
MAR
APR
MAY
JUN
JUL
AUG
SEP
OCT
NOV
DEC

Figure 10.17 Palermo, Italy, has a Mediterranean climate characterized by moderate temperatures and a dry summer (Köppen *Cs* region). As the graphs show, the winter in Palermo is cooler and moister than the summer. The average annual precipitation is nearly 80 centimeters (31 inches), but June, July, and August combined receive only 4 centimeters (1.5 inches). (After Nelson, 1968)

Midlatitude Deciduous Forests

Midlatitude deciduous forests constitute one of the dominant plant communities of the eastern United States and of western Europe.

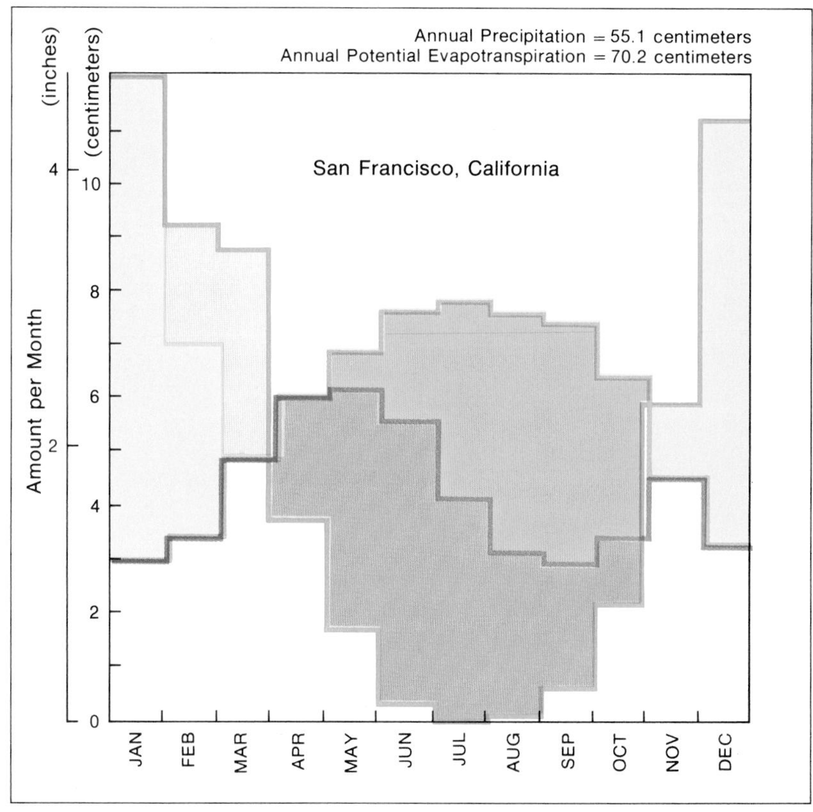
Annual Precipitation = 55.1 centimeters
Annual Potential Evapotranspiration = 70.2 centimeters
San Francisco, California
Amount per Month
(inches)
(centimeters)
4
2
10
8
6
4
2
0
JAN
FEB
MAR
APR
MAY
JUN
JUL
AUG
SEP
OCT
NOV
DEC

Figure 10.18 The local water budget for San Francisco, California, is characteristic of stations with a Mediterranean climate (Köppen *Cs* region). A deficiency of moisture persists through the dry summer months, and plants that live through the summer must draw upon stored soil moisture. From October through March, when most of the precipitation is received, soil moisture stores are replenished, and a surplus is generated in March. (After Carter and Mather, 1966)

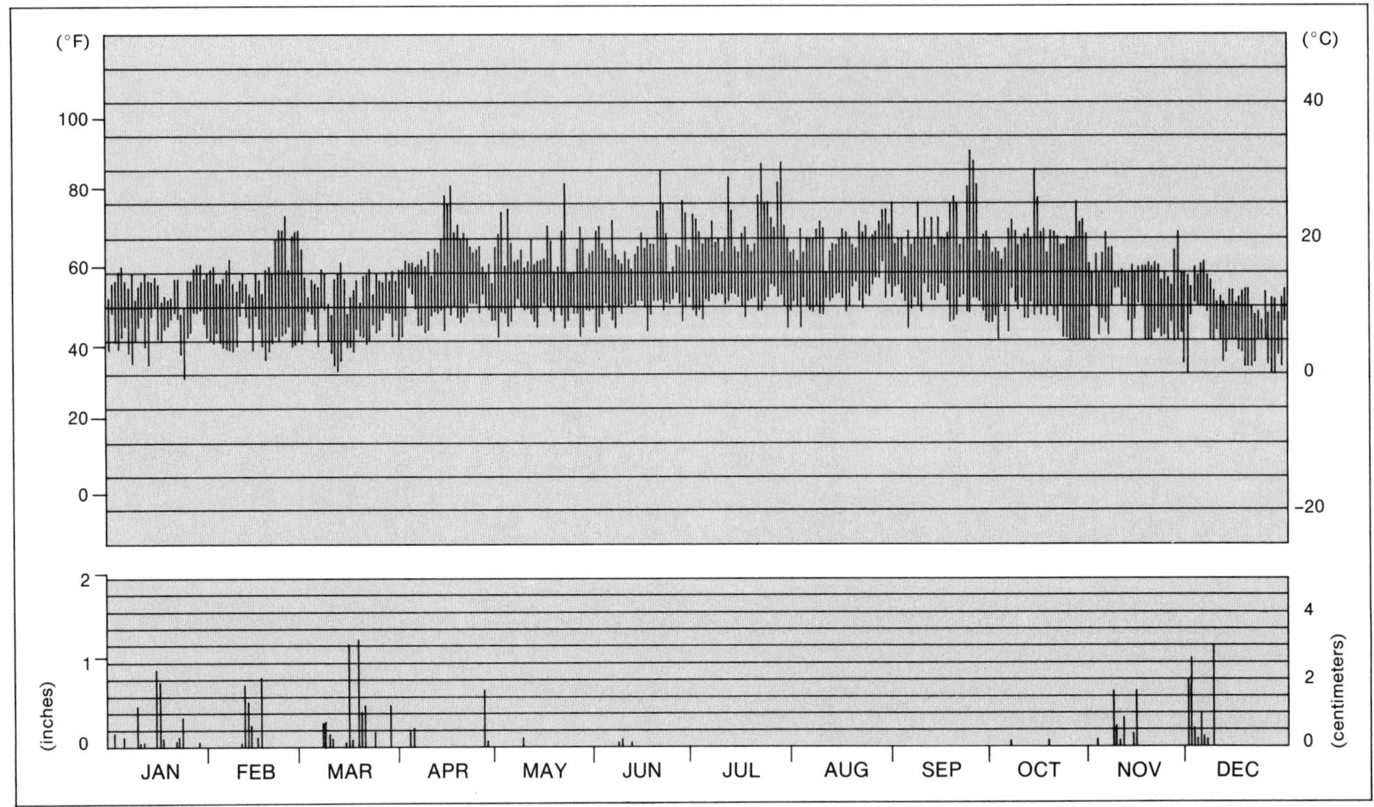

(°F)
100
80
60
40
20
0
(°C)
40
20
0
-20
2
1
0
(inches)
4
2
0
(centimeters)
JAN
FEB
MAR
APR
MAY
JUN
JUL
AUG
SEP
OCT
NOV
DEC

Figure 10.19 The graphs show the daily temperature and precipitation at San Francisco during 1954. Each vertical line on the temperature graph indicates the maximum and minimum temperatures recorded for a given date. Temperatures remain moderate at San Francisco through the year because of the prevailing winds from over the Pacific Ocean. The temperature reached or exceeded 30°C (86°F) on only 5 days in 1954, and freezing temperatures occurred only twice. Note the almost complete lack of precipitation during the summer months. Precipitation during the rainy season tends to occur in episodes of a few rainy days separated by periods of dry weather. (After Hendl, 1963)

Many of these woodlands have been cleared for agriculture because of the desirable properties of forest soils.

The deciduous forests of the eastern United States receive abundant moisture, and many species of trees flourish. Oak, maple, beech, and other large trees of the deciduous forest can grow to heights of 30 meters (100 feet) or more. The trees are usually well separated, and the canopy of leaves is not dense. Low shrubs occupy the niche near ground level, and grasses often cover the ground. In less humid regions, such as the Ozark Mountains near the border of the dry grasslands, well-separated oak and hickory tend to be the dominant trees of the deciduous forests.

Figure 10.21 shows monthly temperature and precipitation data for Pittsburgh, Pennsylvania, located in the humid subtropical climate region. Summer temperatures are much higher than winter temperatures, and rainfall is distributed throughout the year, with a relative maximum in the summer.

Figure 10.22 shows the thermoisopleth for Klagenfurt, Austria, in the humid continental climate region of Europe. The rounded isotherms indicate significant variations in temperature through the day and through the seasons that are characteristic of midlatitude continental climates.

Midlatitude Grasslands

Grasses dominate where trees and shrubs cannot endure. Great areas of grassland are found in the central United States, in eastern

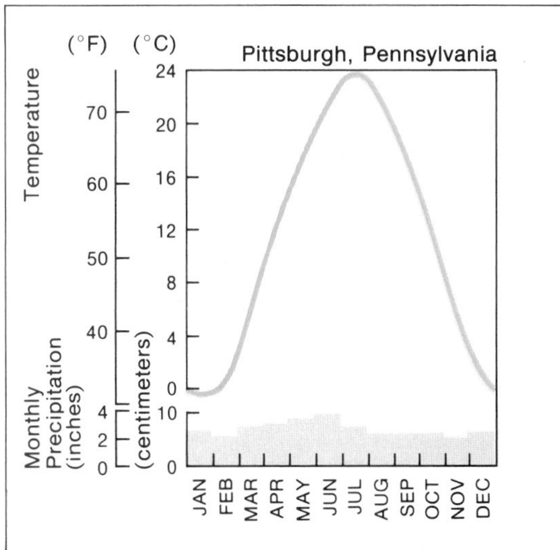

Pittsburgh, Pennsylvania
(°F)
(°C)
Temperature
Monthly Precipitation (inches)
(centimeters)
JAN FEB MAR APR MAY JUN JUL AUG SEP OCT NOV DEC

Figure 10.20 (top left) Birches, as shown in this deciduous forest, are often one of the first tree species to invade cleared land in midlatitude climate regions. The seeds of birches are easily carried long distances by the wind and remain viable under adverse conditions. Birches have successfully displaced coniferous trees in some northern regions.

Figure 10.21 (top right) Pittsburgh, Pennsylvania, has a humid subtropical climate (Köppen *Cfa* region). The graphs of temperature and precipitation show the great range of the average monthly temperature between summer and winter. The monthly precipitation is nearly constant through the year, however, and does not exhibit seasonal behavior. (After Nelson, 1968)

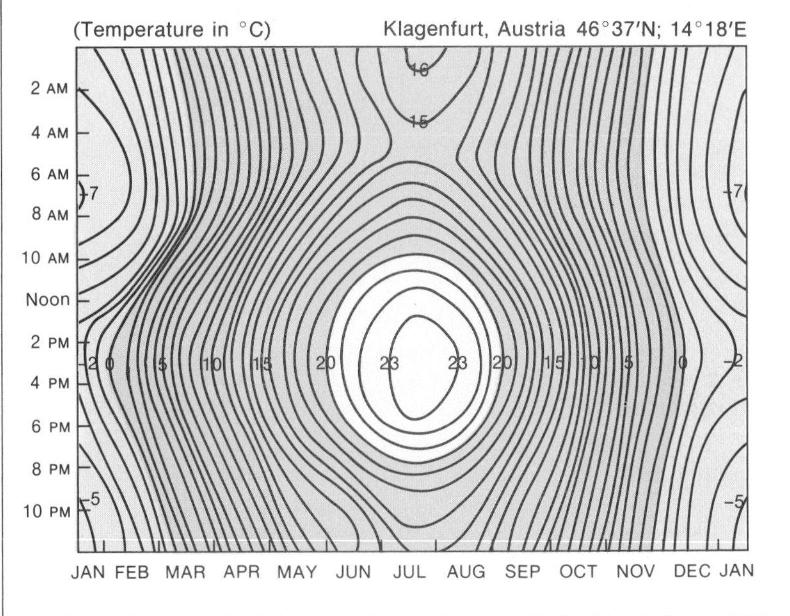

(Temperature in °C)
Klagenfurt, Austria 46°37′N; 14°18′E
2 AM
4 AM
6 AM
8 AM
10 AM
Noon
2 PM
4 PM
6 PM
8 PM
10 PM
JAN FEB MAR APR MAY JUN JUL AUG SEP OCT NOV DEC JAN

Figure 10.22 (left) Klagenfurt, Austria, has a humid continental climate (Köppen *Dfa* region). As the thermoisopleth shows, the range of temperature through the year exceeds the range through the average day. At noon, for example, the temperature in January is approximately −4°C (25°F), whereas in July the noon temperature is 22°C (72°F). Note that in November the temperature through the day varies by only a few degrees, partly because of the moderating effect of cloudy weather. (After Troll, 1965)

Europe, and across central Asia. Although patches of trees may occur in some regions, especially near rivers, grasses are the dominant vegetation. Grasslands contain a large number of plant species. The dominant grasses are usually perennials that lie dormant during the winter and continue their growth in the next growing season. Near the boundary between forest and grassland, where moisture is comparatively abundant, the natural grassland vegetation is usually tall prairie grass. Most such areas in the United States have been converted to agricultural land. Where moisture is less, the dominant vegetation is short prairie grass, and in the driest grasslands the grass becomes clumped and tufted.

Huron, South Dakota, located on the drier western margin of prairie grassland country, receives an annual precipitation of approximately 50 centimeters (20 inches). The water budget for Huron in Figure 10.24 shows that although precipitation is concentrated in the summer months, potential evapotranspiration is especially high then, so there is a deficit of moisture during the summer.

Subarctic Coniferous Forests

Coniferous forests constitute the dominant plant community from approximately latitude 50°N to the northern boundary of the forest. Most conifers are evergreen and do not lose their needlelike leaves

Figure 10.23 This tussocked grassland is in the Canterbury Plains of South Island, New Zealand. Radiocarbon dating of wood fragments indicate that extensive forests occupied some of this region 500 years ago. The conversion to grassland may have occurred as a result of man's use of fire.

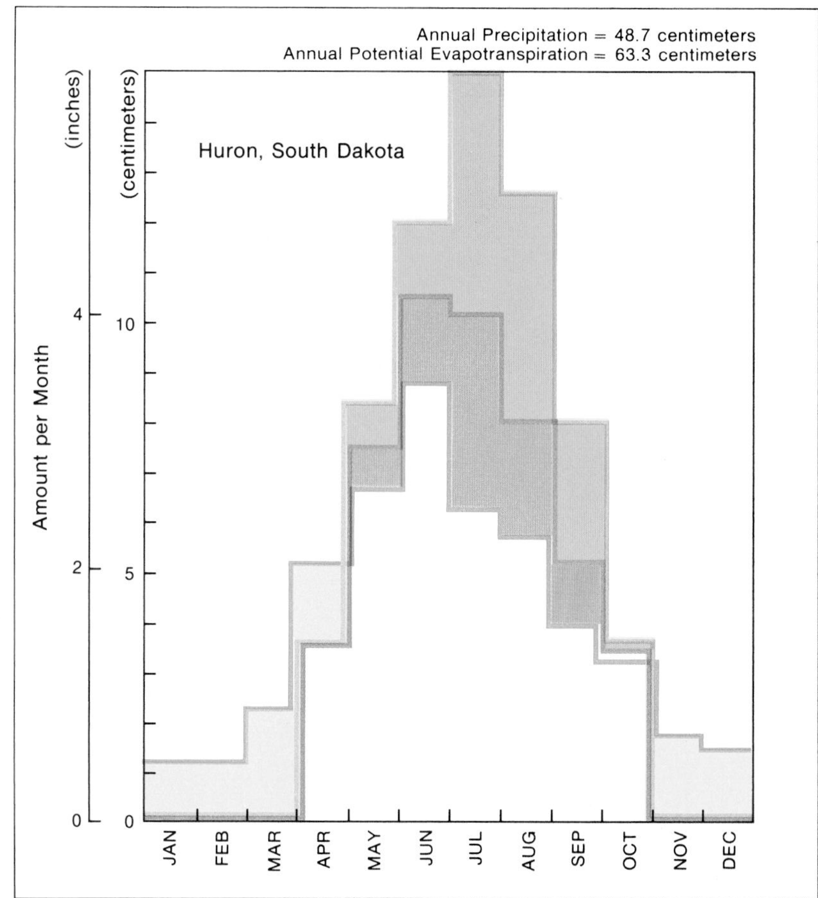

Annual Precipitation = 48.7 centimeters
Annual Potential Evapotranspiration = 63.3 centimeters
Huron, South Dakota
Amount per Month
(inches)
(centimeters)
0
2
4
0
5
10
JAN
FEB
MAR
APR
MAY
JUN
JUL
AUG
SEP
OCT
NOV
DEC

Figure 10.24 Huron, which is located in the prairie grasslands of South Dakota, has a continental climate with moderate moisture and cold winters (Köppen *Df* region). The local water budget for Huron shows that precipitation occurs in all months and is most plentiful during the summer. However, high summer temperatures cause potential evapotranspiration to be high then as well, and moisture deficits normally occur. The soil moisture is recharged during the cooler months, when the vegetation's demand for moisture is small. (After Carter and Mather, 1966)

seasonally. They are therefore able to begin photosynthesis in the spring as soon as the temperature is sufficiently high. Their small, thick-surfaced leaves and thick bark help them withstand the long, cold winters of the north, and their comparatively shallow root systems help them grow in shallow soils and make use of water from snowmelt.

Coniferous forests contain a comparatively small number of plant species. The dense layers of needles allow little light to reach the ground, and the cool temperatures limit plant growth as well. White spruce is common in the coniferous forests of North America. Larch, which has needlelike leaves but is deciduous, is found in Siberia, where conditions may be too severe even for evergreens. In the southern part of the coniferous forests, the trees are tall and densely packed. Farther north, the trees tend to be shorter and well separated from one another.

Tundra Vegetation

Only a few plant species exist in tundra regions. Trees cannot endure unless the temperature of the growing season exceeds 10°C (50°F) or so. In central Alaska and along the northern coast of the Arctic seas, the trees of the coniferous forest give way to grasses, low shrubs, and flowering herbs. Lichens can live on rock surfaces

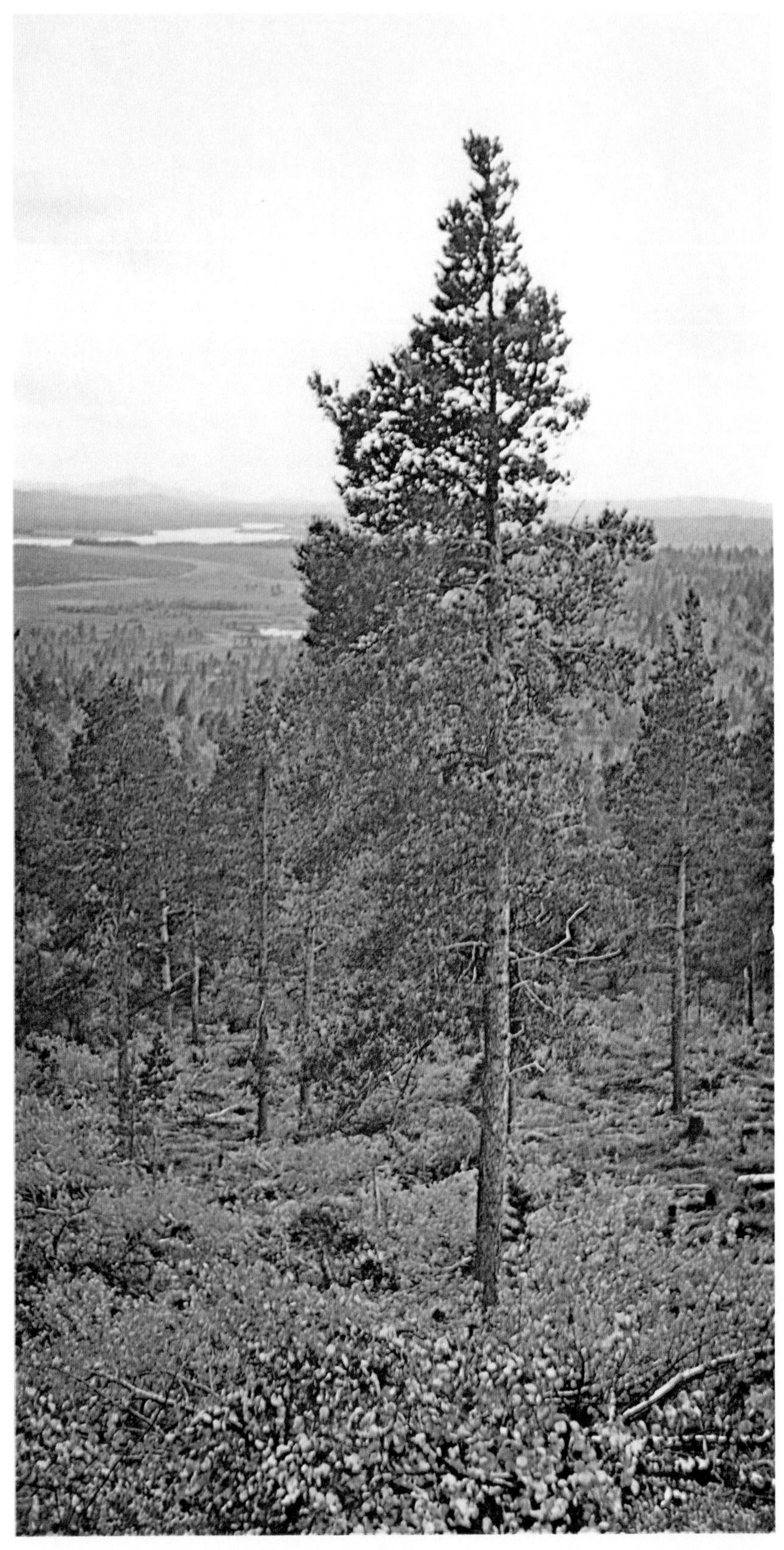

Figure 10.25 Coniferous evergreens, commonly spruce and pine, occupy large regions of the northern middle latitudes and the subarctic. The uniform appearance of this coniferous forest in Norway shows how only a few species of conifers form the dominant vegetation.

without soil, and in sheltered locations they sometimes form layers several centimeters thick. Despite the relatively small precipitation, the tundra is often moist or waterlogged, because moisture cannot penetrate the perpetually frozen layer below the surface of the soil.

Figure 10.27 is the thermoisopleth for Sagastyr, at latitude 73°N, in the tundra region of northern Siberia. The nearly vertical pattern of isotherms shows that the daily variation in temperature is only a few degrees. Seasonally, the midday temperature varies from 6°C (43°F) in July to -40°C (-40°F) in February. The small daily variation in temperature reflects the long duration of daylight in the summer and the long night in winter.

Figure 10.26 Low shrubs, grasses, and flowering herbs are the dominant vegetation types in tundra regions, such as shown here in Norway. Accumulations of water are visible in the background; tundra regions tend to be wet during the period of thaw because water cannot drain through the permafrost layer below the surface.

Changes in Climate

Although there are general similarities in weather within a global climatic region, the weather in a particular location changes from year to year. In your town, one year may be dry with prolonged hot spells, and another year may be wet, with practically no summer vacation weather. An example of annual variations is apparent in the monthly water balance components for the years 1960 through 1967 in Middlesex County, New Jersey, shown in Figure 10.28. The pattern for each year is characteristic of moist midlatitude regions, with surpluses occurring in the cool winter months when transpiration is minimal and deficits occurring in the hot summer months, if at all. The surpluses generated in the period from 1960 to 1967 were approximately equal in amount and timing, but the deficits show a different pattern. For the years 1962 through 1966, the water budget shows a sizeable deficit each summer because rainfall was considerably below average. The years 1962 through 1966 were years of drought in the northeastern United States; in the summer of 1965, particularly, water use had to be curtailed in this normally

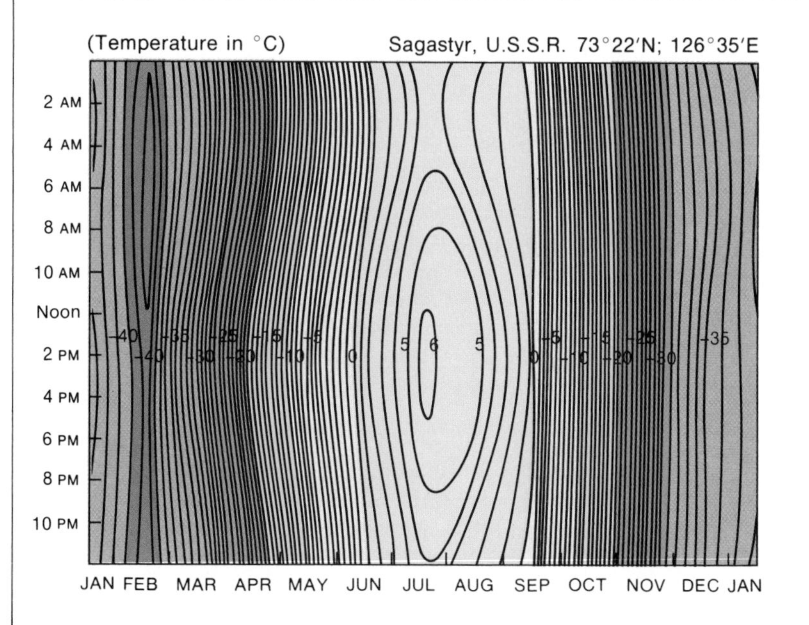
(Temperature in °C)
Sagastyr, U.S.S.R. 73°22′N; 126°35′E
2 AM
4 AM
6 AM
8 AM
10 AM
Noon
2 PM
4 PM
6 PM
8 PM
10 PM
-40
-35
-25
-15
-5
-40
-30
-20
-10
0
5
6
5
0
-5
-10
-15
-20
-25
-30
-35
JAN FEB MAR APR MAY JUN JUL AUG SEP OCT NOV DEC JAN

Figure 10.27 Sagastyr, U.S.S.R., is a station in the tundra region of Siberia north of the Arctic Circle (Köppen *ET* region). The temperature during the year varies through an extreme range because of the lack of sun near the time of winter solstice and the continual daylight at summer solstice. The nearly vertical temperature contour lines imply that the temperature during any given day is essentially constant.
(After Troll, 1965)

Case Study: Slash-and-Burn Agriculture: An Efficient or Destructive Land Use?

In vast areas of the tropics, inhabitants subsist by a method of agriculture largely unknown to midlatitude populations. Slash-and-burn agriculture—also known as shifting cultivation, *milpa* cultivation (Latin America), and *ladang* (southeast Asia)—constitutes the principal method of land cultivation practiced on, for example, approximately one-third of the total land area currently used for agriculture in Southeast Asia. It has been estimated that in some countries, among them the Philippines, up to 10 percent of the country's population depends on shifting cultivation for its food.

In Latin America, typical crops produced by shifting cultivation are maize, manioc, and squash. In Southeast Asia, rice, cucumber, and maize are often planted the first year, followed by second-year crops of cassava, sugar cane, and squash.

The actual technique of slash-and-burn agriculture is essentially just what the name implies. Several months before the region's rainy season begins, the trees and undergrowth on a selected plot of forested land are cut. The vegetation is left on the ground to dry and is later burned. At the beginning of the rainy season, crops are planted, usually by using sticks or hoes to dig holes for the seeds. The crops are later harvested.

The same plot is replanted until the decrease in soil fertility or the growth of weeds and grasses reaches a point at which cultivation is no longer economically feasible. At this point, cultivators abandon the field, leaving it to fallow for several years, and move on to another plot to repeat the process of cutting, burning, planting, and harvesting.

Shifting cultivation is not feasible for a high-density population because of the low crop yields obtained and because of the long fallow period required for the reestablishment of the cleared vegetation. Thus, in a region where slash-and-burn agriculture is the primary method of cultivation practiced, only a fraction of the arable land can be productive at any given time.

The overall effect of slash-and-burn agriculture on the ecology is difficult to evaluate. Many experts argue that shifting cultivation is an efficient and, if properly practiced, an ecologically sound method of utilizing the land of the tropics that would be difficult to cultivate by most other methods. They are also quick to point out that in underdeveloped countries, slash-and-burn agriculture may be necessary to support the people who eke out a subsistence living by practicing it.

Other experts, however, argue that this method of agriculture is a reflection of civilization level rather than of any unavoidable or unalterable physical limitations, and that on the whole the method is wasteful and inefficient. They point out that the clearing of vegetation exposes the already-thin, infertile topsoil of the slopes to the powerful erosive force of

tropical rainfall. The heavy erosion results in an increased silt load in stream channels and often causes flooding of streams. In some regions, slash-and-burn agriculture also is responsible for the depletion of various species of timber.

It seems that most scientists would agree, however, that any damage wrought by slash-and-burn agriculture is minimal—perhaps even negligible—if properly practiced. The amount of available land must be sufficient to allow for adequately long fallow periods, and there must be sufficient seasonal variation in the climate and rainfall of the region so that cut vegetation will dry and newly planted crops will be watered. The method must be used to support a low-density population with a subsistence economy that does not require a large surplus of food for trade, and there must be minimal influence or pressure from external culture groups, such as loggers, that practice economic systems in conflict with the "system" of shifting cultivation.

In addition, certain practices connected with the method itself must be followed: a safety path must be made around the clearing to prevent the spread of fire during the burning process; the same plot must not be used over and over again for the same crop to avoid depletion of soil nutrients; and secondary rather than primary forests must be used as cultivation sites. If these practices are not followed, and if the proper conditions are not present, slash-and-burn agriculture can indeed do great damage to soil, biotic, and water resources, reducing vast areas of land to economic uselessness in a short period of time under the climatic extremes of the tropics.

(top) The natural vegetation on this Honduran slope has been cut, dried, and burned, and replaced by the *milpa* crop shown here. The farmers who cultivated this plot will abandon it for a new area after only a few harvests.

(bottom) Most of this formerly forested area near Oaxaca, Mexico, has been transformed into a wasteland as a result of *milpa* agricultural practices.

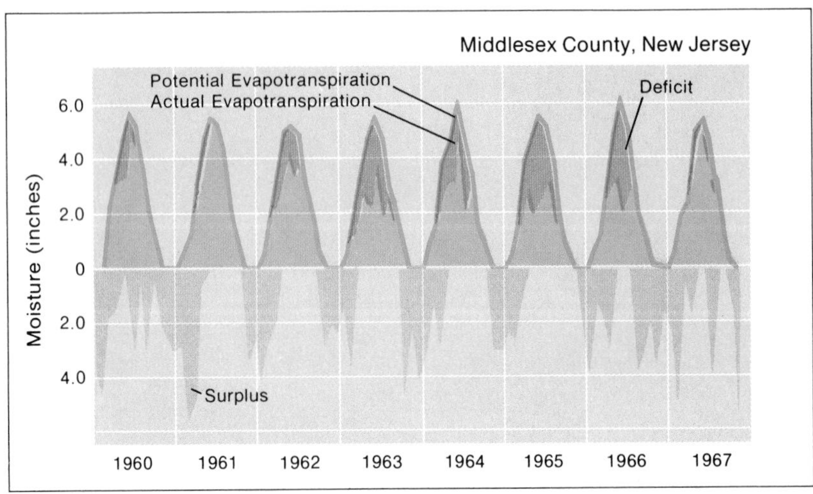
Middlesex County, New Jersey
Potential Evapotranspiration
Actual Evapotranspiration
Deficit
Moisture (inches)
6.0
4.0
2.0
0
2.0
4.0
Surplus
1960
1961
1962
1963
1964
1965
1966
1967

Figure 10.28 These local water budgets for Middlesex County, New Jersey, a normally moist location, show that moisture deficits were generated every summer during the East Coast "drought" years 1962 through 1966. Moisture surpluses were available every winter, but soil and groundwater storage were insufficient to meet water demands during the summer. (After Muller, 1969)

moist region. No one would claim, however, that the climate of the Northeast had changed from moist to dry. The weather trend would have to continue much longer to be considered a climatic shift. The dividing line between a prolonged fluctuation in weather and a shift in climate is difficult to define. Some of the trends observed in historic times have lasted as long as a hundred years or more; these trends certainly could be called changes in climate.

Climate interacts with soils, vegetation, and landforms. If the climate changes, other systems begin to respond to the new conditions and move toward a new state of equilibrium. The response of soils and vegetation systems to climatic change is comparatively rapid. A decade of general warming in subpolar regions would be noticeable in the advance of the forest boundary. Climatic changes are less apparent in landforms. Most landforms have a long response time, and many slopes are still changing from the conditions during the glacial period 10,000 years ago.

Climate of the Early Holocene

In North America and in Northern Europe the last period of major glaciation ended when the climate changed about 11,000 years ago. The first few thousand years of the postglacial, or *Holocene*, period saw the rapid melting back of the glaciers. As thc glacial ice melted, vegetation responded to the improved conditions by growing on the vacated land. The climatic history of this period has been determined primarily by analyzing the fossil pollen grains found in sediment to determine the compositions of early plant communities. The lowest, hence the earliest, layers of sediment found in a given location usually contain pollen from the spruce and fir trees of northland forests. Upper layers include pollen from deciduous trees, which reflects an increasingly hospitable climate. Pollen analysis is providing detailed temperature and moisture information for the Holocene climate of the midwestern United States. In making the analysis, the mix of pollen from a dated sediment layer is compared in composition with pollen from present-day plant communities of different climatic regions until the best match is found. It is assumed that climatic conditions for the ancient plant community, especially the average temperature during the growing season, correspond to conditions of the matched modern community. The July mean temperature found in this way shows a notable rise of 3.3°C (5.9°F) in the 2,000-year period beginning 12,000 years ago. World temperatures continued to rise until about 5,000 years ago, when a *climatic optimum* was reached, with average temperatures several Centigrade degrees higher than at present. The cooler, more moist climates that followed turned many European forests into wet bogs.

Climate in Historic Times

The systematic recording of weather began only 200 years ago, so a wide variety of indirect methods have been employed to determine the climates of earlier times. One method uses the width of tree growth rings as a measure of growing conditions. Extensive studies of tree rings in the southwestern United States show that for the last 2,000 years, the climate there has had long periods of relative stability, interspersed with periods of meteorologic fluctuation lasting several decades. Unusual dryness during a large part of the thirteenth century coincides with the abandonment of large Indian settlements in northeastern Arizona. Water management in early agriculture was apparently unable to cope with prolonged drought.

A secondary climatic optimum occurred in northern Europe in the years 800–1000. Summer temperatures were 1 or 2°C (1.8 or 3.6°F) higher than they are today, and the climate was warm, dry, and free of storms. During this period, the Vikings sailed from Europe to Iceland, Greenland, and North America. The abundance of historical records from the Middle Ages on helps in reconstructing a detailed history of climate in Europe. The major climatic event in northern Europe during historic times was the Little Ice Age, which lasted from about 1550 until 1850. Many travelers' descriptions and engravings from this period show glaciers in the Alps

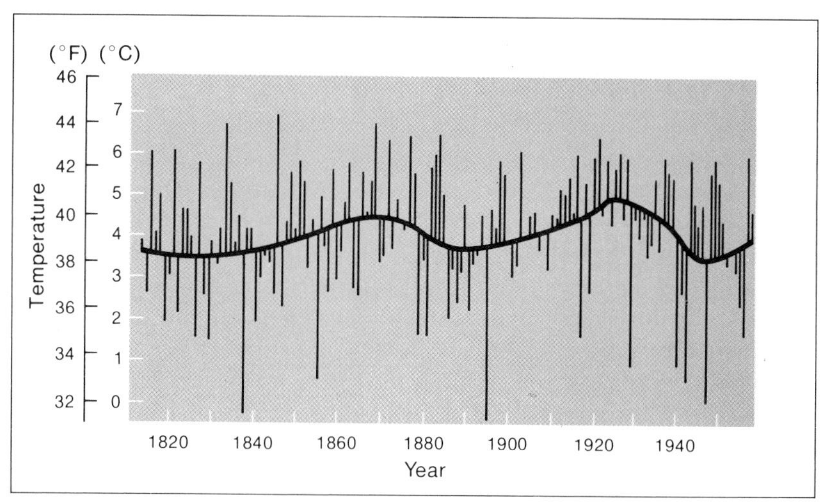

(°F) (°C)
46
44
42
40
38
36
34
32
7
6
5
4
3
2
1
0
Temperature
1820
1840
1860
1880
1900
1920
1940
Year

Figure 10.29 This graph shows the average winter temperature at Oxford, England, for a period of nearly 150 years. The heavy line represents 10-year averages. The tips of the vertical lines indicate the average temperatures for January and February in individual years. The average winter temperature shows a slow variation over periods of decades, reflecting climatic variation. The long-term trend is nearly masked by the fluctuations from year to year. (After Lamb, 1966) far beyond their present-day boundaries. Government records tell of houses being destroyed by the advancing alpine glaciers and of entire villages being abandoned. In Iceland, farmers had to stop growing grain because the climate became too cold and ice covered some of the farmland.

The period from 1850 to 1940 was one of generally increasing temperature, with consequent glacial recession. The average temperature rose 1°C (1.8°F) or so in the Northern Hemisphere. Such a small change is insignificant compared to the normal fluctuation in average temperature from one year to the next, but a shift of a few degrees over a long duration can make the difference between an ice age and a warm period. The average temperature since the 1950s seems to be decreasing, but the long-term trend is difficult to identify.

Theories of Climatic Change

Accounting for climatic change is extremely difficult because many of the interactions of the general circulation are not understood in detail. Furthermore, slight climatic changes can produce disproportionate effects in the environment. If the average temperature were to drop a few degrees, ice and snow might cover a greater proportion of the earth's surface. Because of its high albedo, ice would reflect more radiant energy back to space, which would cause further cooling and further spread of ice and snow.

A number of different mechanisms have been proposed to account for climatic change, although none is entirely satisfactory. Over geologic time, periods of glaciation appear to be associated with periods of mountain uplift, and one theory of climatic change suggests that mountains might provide the conditions required to trigger glacial advance. Because mountains rise above much of the atmosphere, they can easily radiate long-wavelength energy into space, which increases the loss of heat from the earth. Mountains also cause moist air masses to rise, which can result in increased snowfall. Regardless of whether these effects are sufficient to induce glaciation, mountain uplift is too slow to account for the recent ice ages of the last million years.

Another theory of climatic change suggests that the amount of energy radiated by the sun has varied through geologic time, which would cause the climate to vary. Precise measurements of solar radiation over a long period are required to determine whether the sun's energy output varies significantly; the absence of concrete observational evidence makes variable sun theories difficult to evaluate. The earth's motion about the sun exhibits irregularities, which affect the amount of radiant energy that reaches the earth on a time scale measured in tens of thousands of years. There is disagreement, however, as to whether the effects are large enough to cause significant climatic change.

Climatic change has even been attributed to the presence of volcanic dust in the atmosphere, because dust scatters incoming radiation and reduces the amount of energy the earth receives. The effect is significant; for 3 years after the volcano Krakatoa exploded, observatories in France, halfway around the world, measured a 10 to 20 percent decrease in solar energy reaching the earth. A prolonged period of volcanic activity could lower the average global temperature for a significant period of time.

Climates of the Future

Most of the concern for the climate of the near future centers on the effect of man's activities, particularly his use of the atmosphere as a dumping ground. The atmosphere serves as a sensitive controller of climate; cloudiness and dust in the air directly affect the amount of energy reaching the earth, and carbon dioxide in the air affects the amount of energy that is radiated away. Large-scale industry and agriculture add smoke and soil particles to the air, increasing the albedo of the earth and decreasing the amount of energy received from the sun.

The interrelatedness of systems makes it difficult to predict the net effect from a particular change. The amount of carbon dioxide in the atmosphere is expected to increase by at least 20 to 30 percent in the last half of this century, primarily because of industrial activities. The direct effect of the increase in carbon dioxide will be a decrease in radiation losses from the earth's surface, which will raise the average temperature. The higher temperatures will increase evaporation, perhaps leading to increased cloudiness. Because the degree to which each of these changes affects other changes is difficult to determine, the final temperature change caused by the increase of carbon dioxide cannot be predicted.

Small, long-term changes in the atmosphere are important. The present state of the environment is a function of its past history; there is no fresh start each year, because the environment accumulates the effects of minor changes. Orbit wiggles, volcanic dust—these and other small and complicated interactions influence climatic trends. Now man's activities must be added to the list.

Summary

Plants form natural communities, which in a given environment tend to move through a succession to a climax. In the global distribution of vegetation systems, most of the plants of a region are characteristic of the region's climate because of the close interaction between vegetation and the energy and moisture associated with climate. Often plants possess special features adapted to the climatic region in which they flourish.

Climate has changed during the earth's history. Vegetation systems are quick to respond to changes in climate, while soils respond less rapidly, and many landforms respond to changes in climate exceedingly slowly. A change in climate urges systems toward new states of equilibrium. The cause of climatic change is unknown, although several mechanisms have been proposed to account for it.

Graphic Essay

III

Adapting to Climatic Factors

The earth's environmental extremes represent amazingly diverse challenges to its inhabitants. To survive in arctic regions its creatures must resist bone-chilling cold; to survive in the deserts they must find the means to tolerate the blistering sun. Everywhere the earth's creatures must find food and shelter; with varying degrees of urgency they must escape predators; and all must reproduce if they are to assure continuity of their kind. It is through anatomical, physiological, behavioral—and, in the case of man, cultural—adaptation that living creatures have been able to move even into seemingly hostile environments, there to become an integral part of the ecological ebb and flow of a special niche of the world.

Among all the creatures who have adapted to the environment in unique and specialized ways—the polar bear, the camel, the catfish, to name a few—we find one who is all the more remarkable because its success comes not from its ability to specialize but to *generalize* in its adaptations. This creature is man. Through his cultural adaptation he has found a means of condensing the immense spans of time that biological adaptation requires. Why take eons to grow the fur coat needed for warmth in the arctic, when furs already abound? Why accept the waterless deserts as barriers to other lands, when pouches can be made to store water for the journey? Why wander like the wolves with game herds when animals can be bred, conveniently, in captivity? Through his culture, man has moved out of the world's system of environmental checks and balances and he now manipulates not only other

Rotation of crops can prevent depletion of nutrients in the soil. In this field in the flatlands of Texas, the rows of grain ready for harvest alternate with young cotton plants.

creatures but the land as well. It is too soon to say what the consequences of this move toward generalized exploitation will finally be, but even now there are signs that it is having profound impact on the natural environment.

And yet, it is not correct to say that man controls the environment, even as it is not correct to say that the environment still controls man. The influence is and shall continue to be reciprocal. Man has both adapted to climatic factors and he has changed them. Any kind of land use, for instance, involves some kind of environmental control. Contour plowing is a technique that prevents soil erosion. Man moves water around so he can have it where he wants it—the Aswan Dam on the Nile and the California Water Plan are engineering projects that have redistributed tons of water for man's use. In Asia, hill slopes are remodeled to facilitate systematic irrigation and the drainage of rice crops, and in the Netherlands, dike drainage systems operate against the force of gravity. Also in the Netherlands and in other midlatitude areas, man has extended the growing seasons of such crops as tomatoes and cucumbers by controlling heat and light with glass-enclosed greenhouses. He has used similar climate-control measures to maintain his own comfort in homes and office buildings that have their own microclimates regulated by air conditioners or heating devices.

Whether man adapts to his environment or modifies it requires an understanding of nature and an awareness of the implications of using resources of energy, air, water, minerals, and food. In the midst of his success comes foreboding: man's rapid spread and growth in numbers could place excessive demands on the environment that has nurtured him. Is man able to continue controlling his environment in a way compatible with nature?

René Dubos, the microbiologist and scientific humanist, presents a viewpoint that may help answer that question. He writes, "Ecology becomes a more complex but far more interesting science when man is accepted as an integral part of the landscape. . . . From Japan to Italy, from China to Holland, from Java to Sweden, civilizations have been built on a variety of ecosystems, which are almost completely man made. Everywhere man has favored sun-loving plants over shade plants, and has moved them from one part of the world to another. There is hardly one agricultural crop and one decorative plant which has not been adapted from its area of origin to another area with compatible soil and climatic conditions."*

Man and nature are interdependent, not separate entities. In his most successful adaptations man has learned to enhance, not destroy, his natural environment.

*René Dubos, 1972. " 'Replenish the Earth and Subdue It,' Genesis 128: Human Touch Often Improves the Land," *Smithsonian*, 3 (December): 18–28.

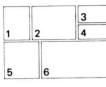

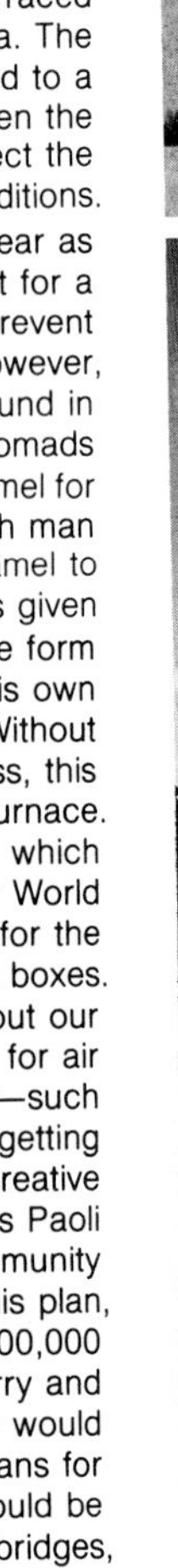

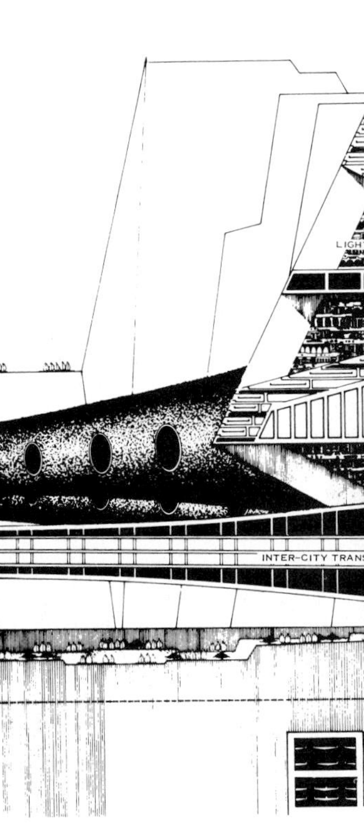
LIGHT
INTER-CITY TRANS

1. Hedgerows and plows tamed the forest that once mantled the English countryside, creating this quilted pattern of mixed agriculture in Taunton, Southwest England. Crops are planted in rotation with pasture so that the humus in the soil can be replenished. The main farming activity in this area is dairying. **2.** Another kind of geometric pattern emerges in the terraced rice paddies of Sumatra, Indonesia. The agricultural practices are adapted to a tropical wet climatic region. Even the people's clothing and housing reflect the warm temperatures and moist conditions. **3.** Buildings in the arctic may appear as bleak as the climate. Except for a predominance of pitched roofs that prevent excessive snow accumulation, however, these houses are similar to those found in midlatitude climatic regions. **4.** Nomads of the Sahara rely on the camel for transport, food, and clothing. Although man is not as naturally equipped as the camel to live in hot, dry climates, culture has given him the means to do so—even in the form of canvas shoes! **5.** Man creates his own microclimates, especially in cities. Without air conditioning and shaded glass, this building would be an intolerable furnace. The photograph shows Lever House, which was built in New York City just after World War II and served as the inspiration for the world-wide generation of black-glass boxes. With questions being raised about our dwindling energy reserves—energy for air conditioning, and light, and power—such specialized architecture is now getting second looks. **6.** Perhaps a creative answer to the glass skyscrapers is Paoli Soleri's imaginative model for a community under a translucent dome. His plan, designed for a community of 100,000 people, would be set in a marble quarry and would overlook a reservoir. The city would have its own power sources and means for climate control. One community would be connected to another by tunnels, bridges, and waterways.

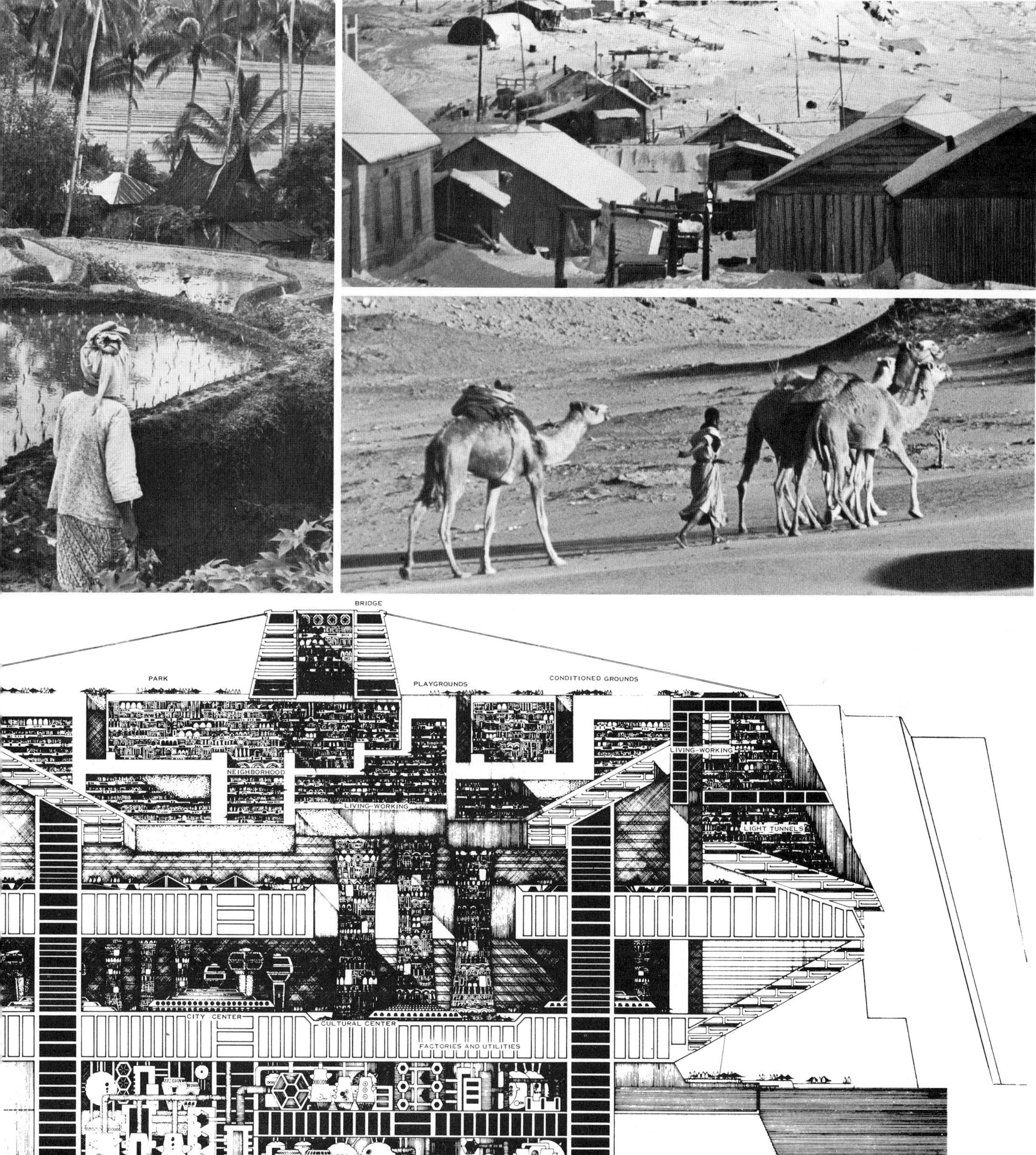
BRIDGE
PARK
PLAYGROUNDS
CONDITIONED GROUNDS
NEIGHBORHOOD
LIVING-WORKING
LIVING-WORKING
LIGHT TUNNELS
CITY CENTER
CULTURAL CENTER
FACTORIES AND UTILITIES
AUTOMATED INDUSTRIES

11 Sculpture of the Landscape

Orders of Relief
First Order of Relief:
Continents and Ocean Basins
Second Order of Relief: Mountains,
Oceanic Ridges, and Plateaus
Third Order of Relief: Hills and Valleys

The Differentiation of Landforms
Geologic Movement
Erosion and Deposition
Diversity of Slopes

Evolution of Landforms
The Cycle of Erosion
Relict Landforms
The Tempo of Geomorphic Change

Like an artist's composition, the natural landscape of a region is made up of many parts. And each of these parts—mountain, boulder, or pebble—was shaped by processes that continue to change the face of the earth. Energy from the sun, acting through climate, soils, and vegetation, combines with energy from within the earth to create unique and inspiring scenery.

Rocky Landscape by Paul Cezanne, 1898. (Reproduced by courtesy of the Trustees, The National Gallery, London)

Sculpture of the Landscape

Some of the earth's landscapes are truly spectacular. A thundering waterfall cascading over a cliff or a massive range of mountains excites awe and wonder. Each year tourists travel long distances to view the fantastic landscape of eroded sandstone pillars and arches in Utah or the impressive skyline of granite slopes in the White Mountains of New Hampshire. But the gentle contours of a pasture are no less interesting than a towering mountain when you pause to wonder how the surface of the earth achieved so many different forms and shapes. Every landform on the earth's surface has a story to tell about the processes that formed it and the processes that work to shape it today.

There is unmistakable evidence that the earth's landforms are in a process of perpetual change. Sometimes the changes are easy to see, such as when a severe flood alters a river channel or a landslide tears away a hillside. Other changes occur gradually over a period of years: the Mississippi River continually alters its course across the land; glaciers slowly advance or recede. Other changes, although significant over the span of geologic time, are too slow to be perceptible in a single human lifetime. Hills and mountains seem rugged and enduring, but after every spring thaw or heavy

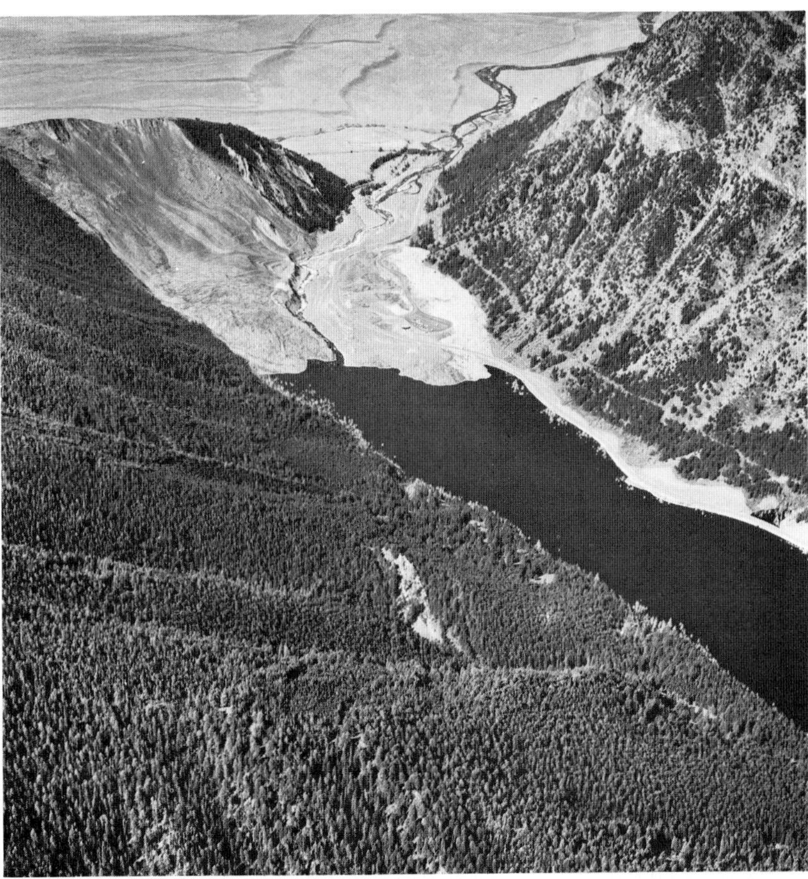

Figure 11.2 Although many processes of geomorphic change are slow, some changes are rapid and dramatic. This photograph shows some of the effects of the Hebgen Lake, Montana, earthquake of August, 1959, which was near Yellowstone National Park, Wyoming. The strong shock caused a portion of the slope shown in the photograph (left background) to slide downhill, which left a broad scar devoid of vegetation. The lake was created when the earth slide dammed the stream in the valley below the slope.

rain, the streams running off a mountain are brown with mud and debris. The mountains are continually being washed to the sea, grain by grain.

Unlike the surface of the moon, the earth's surface is vibrant with activity. A footprint left in the soil of the moon will remain for millions of years, but in a warm, humid climate on earth a footprint disappears in days. The surface of the earth is an interface that receives energy from within the earth and energy from the sun and atmosphere. Wind and rain, aided by the force of gravity, work on the exposed materials of the earth's crust. The internal energy of the earth expresses itself in the work it does lifting, fracturing, and wrinkling the crust. All of these interactions have combined to shape the earth's surface into its multitudinous forms.

Landforms have a past history and a future course of development. The processes that sculpt the landscape take on direct significance as soon as we try to modify the land for our own uses or try to protect ourselves from such hazards as floods and slides. Construction of a dam to regulate a river and to generate power stops the flow of sediment down the river, and beaches miles away may start to erode. If sloping land is cleared for construction, spring rains may cause the slope to slide. In nature's dynamic, complex systems, changing one factor may cause a chain of reactions, many of them unfavorable to man.

The remainder of this book is devoted to *geomorphology*, the study of landforms. This chapter gives an overview of landforms and presents them in a general framework of the geologic structures and climatic processes that shaped them. Many of the ideas discussed earlier concerning energy, hydrology, climate, vegetation, and soils form a background for the study of landforms. Succeeding chapters consider in greater detail the major processes and structures that give the landscape its characteristic appearance.

Orders of Relief

The features of the landscape stand out partly because of differences in vertical elevation, or *relief*. The plains of Iowa, for example, show a relief that differs in altitude by only a few meters over a horizontal distance of several kilometers. In contrast, the land surface in the Rocky Mountains may vary in altitude by 1,000 meters (3,300 feet) over the space of 10 kilometers (6 miles).

First Order of Relief: Continents and Ocean Basins

Relief features occur at all scales on the earth's surface, ranging from a small rain-water channel a fraction of a centimeter deep to the continents and ocean basins themselves. The continents and ocean basins, which constitute the most extensive relief features of the earth's surface, are sometimes referred to as the *first order of relief*.

The continents are great rafts of rock supported from below by a layer of denser, semiplastic rock. According to the concepts of plate tectonics and continental drift discussed in Chapter 1, the

Lena
KOLYMA RANGE
5446
Magadan
-385
2992
KAMCHATKA
PENINSULA
-430
541
-1200
-1100
-90
15584
-13000
6253
5489
5322
8268
2300
SEA OF OKHOTSK
4823
7822
11339
1768
-21500
-2100
7073
4587
SAKHALIN
5958
-28000
Amur
3773
GREATER KHINGAN RANGE
MANCHURIAN PLAIN
SIKHOTE ALIN RANGE
6575
OKHOTSK
ABYSSAL
PLAIN
-14800
-90
(15910)
KURIL TRENCH
-18500
-18500
Vladivostok
7513
HOKKAIDO
CONTINENTAL SLOPE
-34600
9003
-150
BOGOROV
SEAMOUNT
SEA OF JAPAN
-25000
KOREA
Seoul
HONSHU
-21000
YELLOW SEA
-48
Fuji
Tokyo
12388
JAPAN
JAPAN TRENCH
-28000
SHIKOKU
-10000
(6000)
SHATSKY
RISE
-180
KYUSHU
Shanghai
EAST
CHINA
SEA
CONTINENTAL SHELF
-7600
-32800
-19700
-18000
IZU TRENCH
PAREGE VELA RIDGE
-21000
-19800
-18000
1657
-150
Taipei
DAITO
ISLANDS
-18100
RYUKYU ISLANDS
RYUKYU TRENCH
543
-18500
-7300
PAREGE VELA
-220
-22400

Figure 11.3 The ocean basins, which are first-order features of relief, contain second-order features of relief such as deep trenches and extensive undersea mountain ranges. This physiographic diagram of the North Pacific Ocean basin shows the plains, trenches, and sea mounts characteristic of the ocean floor. Heights above and below sea level are given in feet. Note in particular the island arcs along the continental side of the great trenches. Such arcs are located at the boundaries where oceanic crustal plates are plunging downward into the mantle under continental plates, and they are frequently sites of volcanic and earthquake activity. (Courtesy of the National Geographic Society)

earth's crust is divided into a number of thin *lithospheric plates.* The earth's internal energy moves the plates horizontally, perhaps by thermal convection currents set up by radioactive heating in the earth's mantle. The continents and the ocean floor are intimately connected by the process of sea floor spreading, in which molten rock wells up along the great oceanic ridges, building new ocean floor and pushing adjacent plates apart. The Mid-Atlantic Ridge marks the boundary between the plate to the west, which bears North and South America, and the plates to the east, which carry Europe and Africa. As the floor of the Atlantic basin widens, North and South America are drifting westward and Europe and Africa are drifting eastward at a rate of a few centimeters per year.

The Mid-Atlantic Ridge emerges from the ocean's surface in Ice-

ISLAND
-240
CONTINENTAL SHELF
1505
-240
NUNIVAK I.
1675
-90
-1400
BERING SEA
PRIBILOF ISLANDS
-180
ALASKA PENINSULA
Mt. McKinley
20320
ALASKA RANGE
16390
11258
Anchorage
19850
© 1969 BY NATIONAL GEOGRAPHIC SOCIETY
GULF OF ALASKA
15300
Juneau
9500
10002
COAST MOUNTAINS
-2300
PRATT SEAMOUNT
-19700
9387
ALEUTIAN TRENCH
PATTON SEAMOUNT
WELKER GUYOT
-3700
(12300)
PARKER SEAMOUNT
ISLANDS
6920
-23300
5925
4100
BOWIE SEAMOUNT
13104
ROCKY MOUNTAINS
Vancouver
Seattle
Mt. Rainier
14410
COBB SEAMOUNT
-108
15892
CONTINENTAL SHELF
CASCADE RANGE
3506
-18000
-14800
-12100
ALEUTIAN
-18100
ABYSSAL PLAIN
-16400
CASCADIA CHANNEL
12662
13785
7242
9720
14162
3528
-17500
-15500
-17000
FRACTURE ZONE
MENDOCINO FRACTURE ZONE
-16400
-19500
-19200
NORT
Denver
Pikes Peak
14110
San Francisco
Mt. Whitney
14494
Death Valley
-282
-18000
-16400
-18000
-20600
-140
ZONE
MURRAY FRACTURE
FIEBERLING GUYOT
ERBEN GUYOT
-14800
-15000
Los Angeles
6140
San Diego
-1900
JASPER SEAMOUNT
10154
-60
(15940)
18400
-19700
-15800
GULF OF CALIFORNIA
-2100
MIDWAY ISLANDS
-15100
HAWAIIAN ISLANDS
FRACTURE ZONE
-16500
-16100
-16400
HENDERSON SEAMOUNT
-1280
10984
MOLOKAI
-15400
-16400
-18000
-19680
Mauna Kea
13796
NECKER RIDGE
HAWAIIAN RIDGE
HORIZON GUYOT

land, an area characterized by displaced rock masses, open fissures, and volcanic activity. The volcanically active rift valley that cuts across East Africa for thousands of kilometers to the Red Sea is another land feature associated with sea floor spreading. An oceanic ridge adjacent to East Africa runs through the Red Sea and branches into the rift valley. The Red Sea, a comparatively new feature on the surface of the earth, has been widening for only the past 5 or 10 million years. The Gulf of California, which is also widening by the lateral motion of adjacent plates, has opened in the past 15 million years. About 200 or 250 million years ago the earth's landmasses were grouped together and surrounded by sea. The movement of crustal plates slowly separated the continents and opened new shallow seas similar to the present Red Sea.

Figure 11.4 This satellite photograph shows a portion of the east coast of Africa looking toward Saudi Arabia, with the Red Sea at the left and the Gulf of Aden at the right. The opening of the Red Sea is believed to have begun 5 to 10 million years ago when the lithospheric plates carrying Africa and Arabia began to drift apart.

Figure 11.5 (opposite) These maps indicate the probable locations of lithospheric plate boundaries in the earth's crust and the relative motion of the plates. In the upper map, the dark-colored lines represent spreading centers where new crust is being formed, and the light-colored lines are regions where plates are converging. The arrows indicate the general relative directions the major plates are moving. The lower map shows plate boundaries. The seven major plates are identified in bold type and several of the smaller plates in lighter type. Note that the Pacific basin is largely rimmed by subduction zones.

Many features of the earth's crust can be understood in terms of the motion and interaction of plate boundaries. Note, for example, that mountain uplift is active at the plate boundaries along the west coasts of North and South America, where the mountains are known to be young. Along the east coasts, however, there are no plate boundaries, and mountain uplift is inactive there at the present time. As another example, earthquake and volcanic activity is often associated with converging plates such as those around the rim of the Pacific Ocean basin. (top, from John F. Dewey, "Plate Tectonics." Copyright © 1972 by Scientific American, Inc. All rights reserved)

The development of ocean basins has been traced by studying the locations and ages of ancient marine sediments and by using the record of ancient magnetism in rocks. When molten rock extruded at the oceanic ridges solidifies, it becomes magnetized with a strength and orientation that depends on the earth's prevailing magnetic field. Because the earth's magnetic field changes in time, the magnetism in older crust farther from an oceanic ridge differs from the magnetism near a ridge, which is more recently formed crust. The crust in almost all ocean basins exhibits the same pattern of magnetism except for a change of scale that depends on the rate of widening. The data indicate that in the last few million years, the East Pacific basin has been widening at the rate of 10 or 12 centimeters per year (4 or 5 inches per year) compared to 2 centimeters per year (less than 1 inch per year) in the North Atlantic basin.

The Atlantic basin began to form about 200 million years ago, when Africa and North America separated from the original supercontinent Pangaea. Africa remained joined to South America for 80 to 100 million years longer, but by about 100 million years ago, the entire Atlantic basin was open and becoming wider. Australia and Antarctica separated from Africa fairly early but remained joined to each other until 50 or 60 million years ago.

The emergence of new crust in the Atlantic basin has not enlarged the surface area of the earth, because the production of crust in one part of the ocean basins is matched by the destruction of crust elsewhere. The new material that wells up from the oceanic ridges forms denser rock than the material that constitutes the plates carrying the continents. When an oceanic plate and a continental plate drifting toward each other happen to meet, as they do at the western edge of the Pacific basin, the edge of the denser oceanic plate descends into the earth's molten interior. The extensive belt of deep trenches in the floor of the western Pacific basin appears to be a region in which plates descend and crust is destroyed, as shown in Figure 11.6.

Second Order of Relief: Mountains, Oceanic Ridges, and Plateaus

After the continents and ocean basins, the most extensive relief features on the earth are continental mountain ranges, such as the Andes, Alps, and Himalayas; subcontinental plateaus—the Tibetan and African plateaus, for instance; and large depressions, such as the basin of the Caspian Sea. Such features constitute the *second order of relief*. The ocean basins also possess a second order of relief, such as the oceanic ridges and the deep trenches. The oceanic ridges run for tens of thousands of kilometers and rise to heights of 3,000 meters (nearly 10,000 feet) above the ocean floor. They form undersea mountain chains with a continuity unrivaled by any mountains on the land.

Many of the great continental mountain ranges were formed by plate movements. The young, rugged mountain ranges strung along the west coasts of North and South America are being lifted presently, as the Americas drift westward and collide with the oceanic

Global Lithospheric Plates

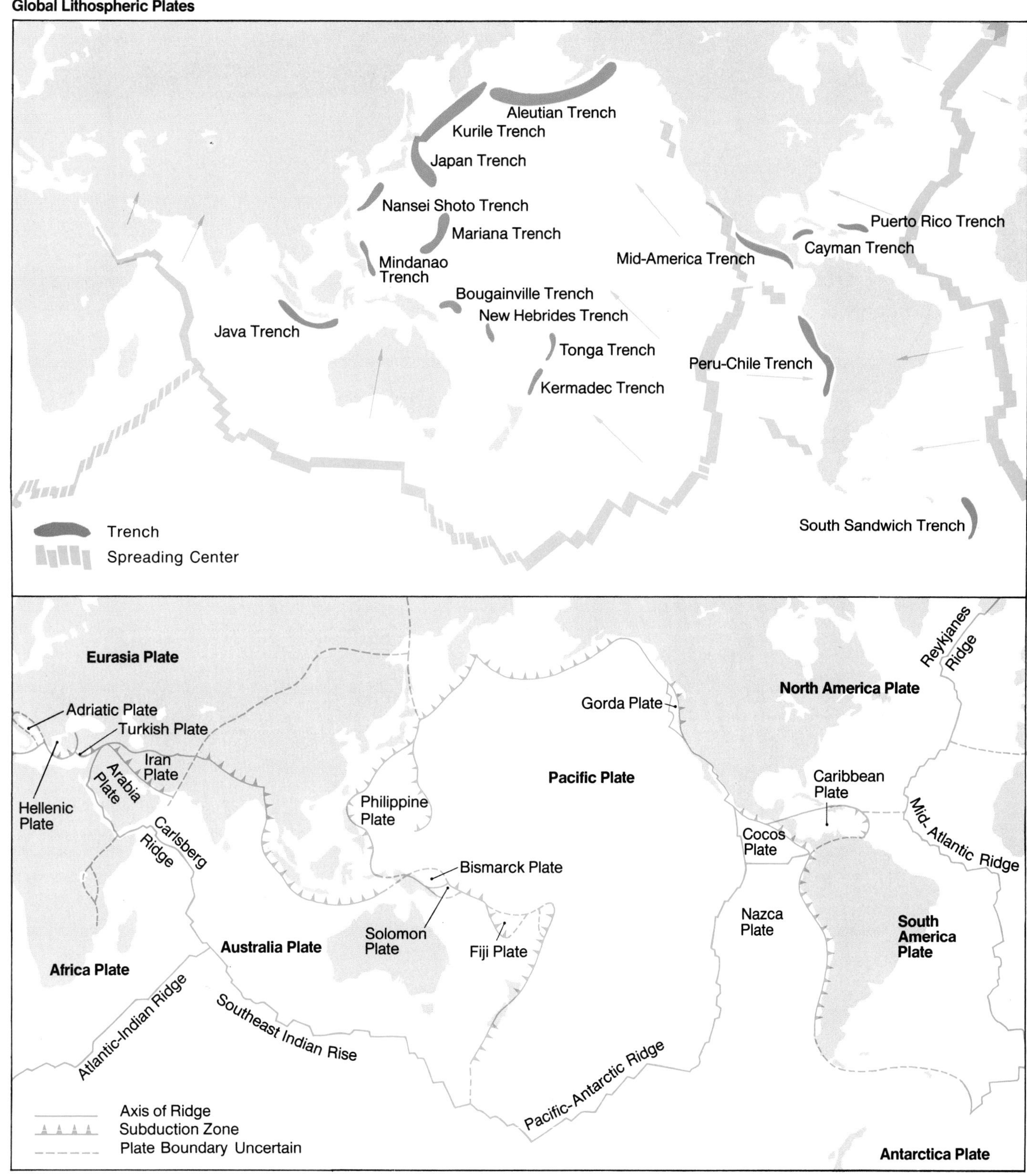

Aleutian Trench
Kurile Trench
Japan Trench
Nansei Shoto Trench
Mariana Trench
Mindanao Trench
Java Trench
Bougainville Trench
New Hebrides Trench
Tonga Trench
Kermadec Trench
Mid-America Trench
Puerto Rico Trench
Cayman Trench
Peru-Chile Trench
South Sandwich Trench
Trench
Spreading Center
Eurasia Plate
Adriatic Plate
Turkish Plate
Iran Plate
Arabia Plate
Hellenic Plate
Carlsberg Ridge
Philippine Plate
Pacific Plate
Gorda Plate
North America Plate
Reykjanes Ridge
Caribbean Plate
Mid-Atlantic Ridge
Cocos Plate
Bismarck Plate
Solomon Plate
Fiji Plate
Nazca Plate
South America Plate
Australia Plate
Africa Plate
Atlantic-Indian Ridge
Southeast Indian Rise
Pacific-Antarctic Ridge
Axis of Ridge
Subduction Zone
Plate Boundary Uncertain
Antarctica Plate

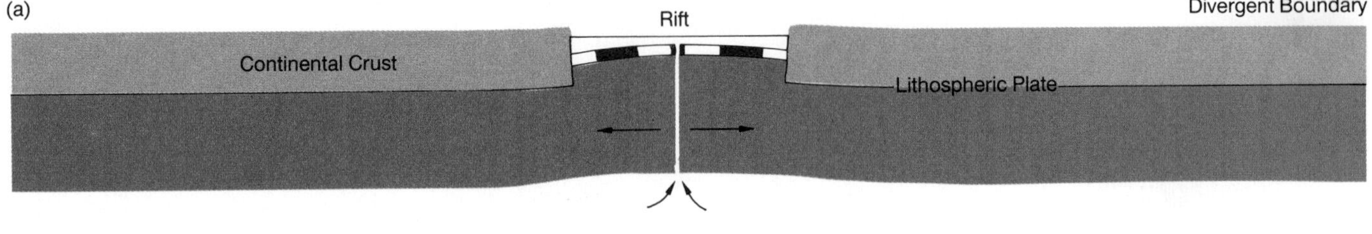
(a)
Rift
Divergent Boundary
Continental Crust
Lithospheric Plate

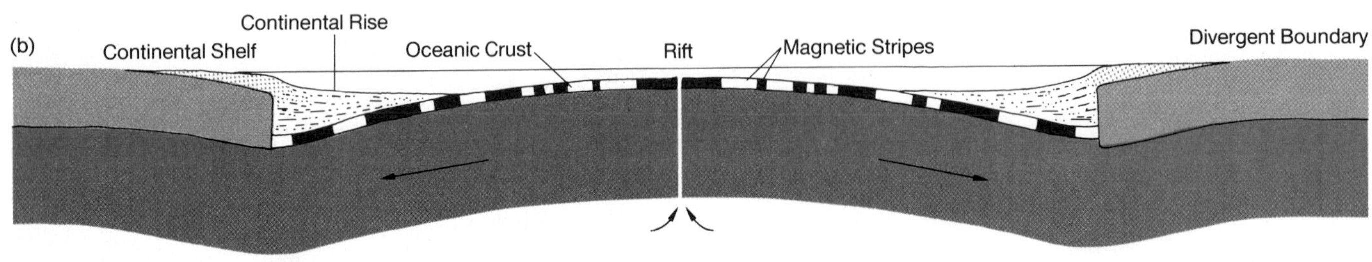
Continental Rise
(b)
Continental Shelf
Oceanic Crust
Rift
Magnetic Stripes
Divergent Boundary

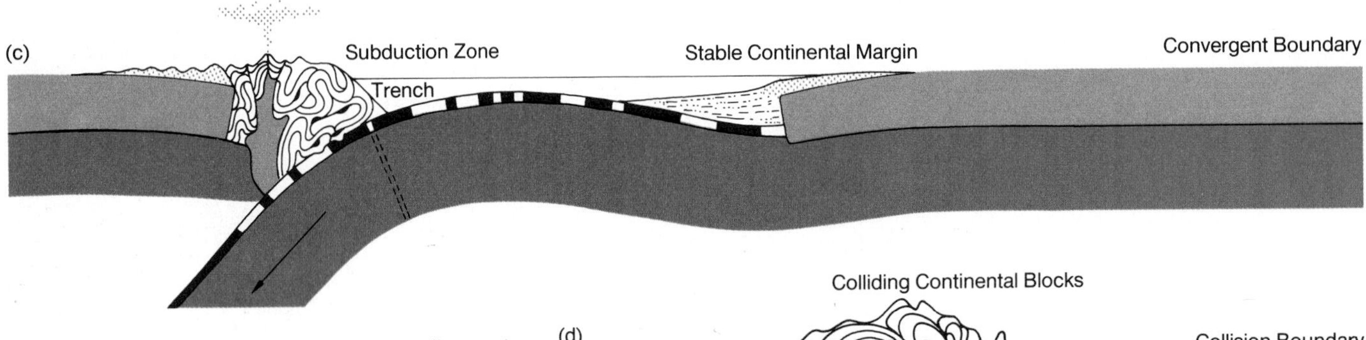
(c)
Subduction Zone
Stable Continental Margin
Convergent Boundary
Trench

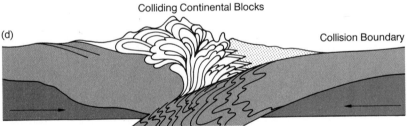

Colliding Continental Blocks
(d)
Collision Boundary

Figure 11.6(a) A newly developed spreading center causes continental plates to rift and move apart (divergent boundary).

(b) The development of new crust at a rifting spreading center causes a sizable ocean basin to form over the course of tens of millions of years. Note that the sediment deposits are thickest near the continental plates, where the oceanic crust is oldest. This schematic diagram represents the Atlantic Ocean basin. The magnetic stripe pattern indicates the record of the earth's magnetism that is frozen into new crustal rock as it cools. The pattern is the same on both sides of the spreading center; such patterns occur in all ocean basins.

(c) Deep ocean trenches, such as those in the western Pacific Ocean, occur where a plate of oceanic crust plunges downward into the mantle beneath either a continental plate or another oceanic plate (convergent boundary). Crustal deformation at a convergent boundary leads to mountain uplift, to the formation of island arcs, and to earthquake and volcanic activity.

(d) A collision between two continents—India and Asia, for example—produces high mountain ranges such as the Himalayas. The continental crust is doubled in thickness and the surficial sediments are contorted and uplifted to form complex overfolds. (From Robert S. Dietz, "Geosynclines, Mountains, and Continent Building." Copyright © 1972 by Scientific American, Inc. All rights reserved)

plates of the Pacific basin. Similarly, the Himalayas were generated when the plate that carries India drifted north into the plate of continental Asia. Along the east coast of the Americas and the west coasts of Europe and Africa, the continents are attached to the oceanic plates, and there are no active plate boundaries at the coasts. Accordingly, these coasts are not centers of mountain building at present.

Mountain ranges significantly affect man's activities, partly through the controls they exert on weather patterns. The grasslands and deserts of the western United States would receive much more rainfall, for example, if the mountain ranges along the western coast did not extract moisture by orographic precipitation from moist air moving eastward from the Pacific Ocean. The Pyrenees and the Alps, which run east and west across Europe, help give the Mediterranean basin its favorable climate because the mountains block the

Figure 11.7 The low hills in the foreground are examples of features having third order of relief. Mount Kilimanjaro, Tanzania, looming in the background, can also be considered a feature of third order; an isolated volcano such as Kilimanjaro covers an area too small to class it with features of second order of relief, such as mountain ranges.

cold polar air associated with cyclonic storms that move with the westerlies across northern Europe.

Third Order of Relief: Hills and Valleys

The individual hills and valleys on the landscape constitute the *third order of relief*. These features are the sculptural details of the mountain ranges and plateaus, and they are the landforms that geomorphology is largely concerned with. Third-order landforms are on man's scale of magnitude—they are features comparable in size to the distance a person can hike in a few hours or a day. Man's more ambitious construction works, such as dams and reservoirs, are also on the scale of third-order relief landforms.

Third-order landforms come in an immense variety of shapes, which geomorphologists have attempted to classify. Sometimes the classification of landforms is based on outward form; for example, a system of streams in a drainage basin may be described as "dendritic," which means "branching like a tree." Sometimes the classification reflects the process assumed to account for a landform, as in the terms "glacially eroded valley" or "wave cut terrace." Or a local name may be used to describe a particular landform regardless of where it occurs—the Gaelic word "drumlin," for example, refers to small, streamlined hills deposited by glaciers.

Many geomorphologists believe that a unified and useful approach to landforms is to view hills, valleys, and other features of third-order relief as a combination of *slopes*. According to this view, the characteristic shape of any landform can be discussed in terms of the processes that determine the development of its slopes. This approach emphasizes the need for quantitative measurement of landforms and the need for knowledge of the processes that form and modify them. However, the study of slopes is a new field, and detailed understanding of slope formation and development is still incomplete.

The Differentiation of Landforms

How is a rock pinnacle in the mountains of northern Italy different from a pinnacle in Monument Valley in the southwestern United States? And why are some landforms in different parts of the world similar to one another? The differences and similarities of landforms around the world are considered in the following chapters. Landforms of first, second, and third order are discussed, for the most part, in the context of the processes that shape them. The work of flowing water, an extremely important process, is covered first, followed by discussions of geologic movements in the crust, climatic influences on landforms, the work of ice on the land, and the action of waves against the coasts.

The landscape is created by two fundamental processes: internal energy and solar energy. The earth's internal energy causes the relative displacement and warping of portions of the crust, which produce local features of relief. In addition, solar energy and moisture redistribute materials across the earth's surface, creating landforms by the erosion and deposition of material. Geologic structures formed by crustal movement are rapidly assaulted by the forces of erosion, and they are rarely seen intact.

Geologic Movement

Crustal movements generate characteristic geologic structures that form the foundation of landscape relief. Sometimes the crust ruptures on the local scale and great blocks of rock slip past one another to produce *faulted* structures. If the displacement is in a horizontal direction, stream beds and chains of hills can be broken and displaced. Vertical displacements can produce severe changes in relief; some cliff faces or steep-sided valleys, such as the valley of the Rhine River in Alsace, are the result of vertical slippage in the earth's crust. Sometimes the materials of the crust react to distorting forces by flexing and folding rather than by rupturing. Many of the characteristic features of the Appalachian Mountains are related to an underlying structure of folded crust. Sometimes the crust is

Figure 11.8 This large, symmetrical ridge of rock in the Kabir Kuh Mountains of Iran is a wrinkle in the earth's crust formed when compressional forces pushed the crust together in this region. The exposed cross section in the foreground shows rock strata that have been folded by intense and prolonged pressure.

elevated in a dome tens or hundreds of kilometers in size. The Black Hills of South Dakota are part of an oval dome 200 kilometers (120 miles) long and 100 kilometers (60 miles) broad.

Erosion and Deposition

The atmosphere interacts strongly with the earth's surface, creating new landforms and modifying the relief forms generated by geologic movement. Weathering processes—frost action, chemical decomposition, the physical and chemical action of plants, among others—break down solid rock into small, movable particles. Flowing water, wind, waves, and ice pick up and transport rock materials in the process called erosion.

Flowing water, usually the most effective agent of erosion, is discussed in detail in Chapter 12. Sheets of water from rainfall and melting snow wash across the ground and carry grains of rock and soil down to stream channels. The rivers of the world ceaselessly carry the highlands of the continents down to the sea. More than 80 percent of the material that the waters of the Amazon River transport comes originally from the Andes Mountains. If it were not for the renewing process of uplift, the surfaces of the continents would eventually become swampy lowlands. Rivers do not necessarily carry all of their debris into the ocean; they deposit some of their load, which builds up characteristic landforms such as floodplains along river channels, deltas at the mouths of rivers, and alluvial fans of rock debris at mountain bases.

Other agents of erosion are ocean waves, winds, and glaciers. Like the flowing water in streams, waves modify the landscape both

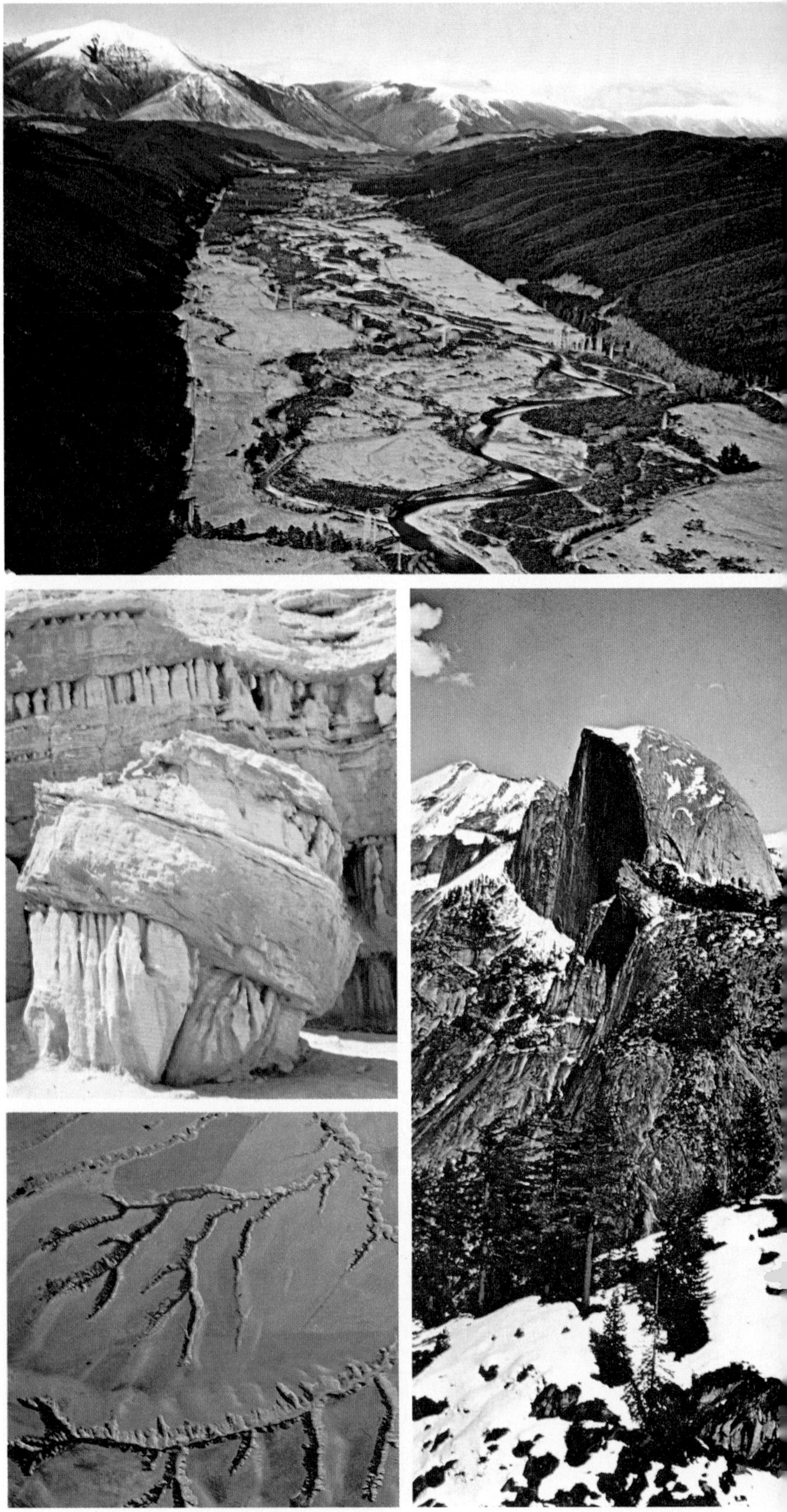

Figure 11.9 Erosion processes acting through time on the varied materials and structures of the earth's crust have generated widely different landforms, as these examples show. (top) Motueka River Valley in New Zealand was carved by a glacier. The sinuous river slowly changes its course by lateral migration, alternately eroding and depositing a veneer of sediment and producing a wide floodplain between the valley walls. (center left) Water acting on these sedimentary rocks in Red Rock Canyon, California, has cut a variety of shapes and patterns in layers of different resistance to erosion. (bottom left) Surface runoff and streamflow are carving a dendritic, or branched, drainage pattern into this landscape north of Christchurch, New Zealand. The relatively uniform resistance of the underlying rock favors the development of a treelike drainage pattern as each of the side tributary branches extends by headward erosion away from the main stream channels. (bottom right) Half Dome, a famous natural feature in Yosemite National Park, California, was originally a complete granite dome, probably shaped by *exfoliation*. When granite is raised to the surface, the mechanical pressure on it is released, and concentric sheets split off, or exfoliate, from the main mass. A glacier later carved away half of the dome.

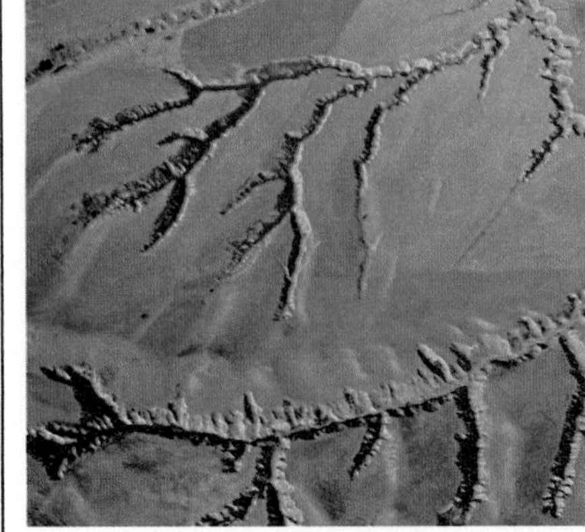

(top left) This natural arch of sandstone is an erosional remnant formed by collapse and by the action of water, even though it is located in the arid southwestern United States where the annual precipitation is only 20 centimeters (8 inches). The less-compacted lower layers of rock weather more rapidly than the upper layers, forming an arch. In time, the entire structure will be weathered away. (center left) The attack of waves at the base of these limestone hills near Dorset, England, has removed the support for the overlying rock. The entire face has become a nearly vertical cliff because of removal of rock from the foot. (top right) The rock spires in Bryce Canyon National Park, Utah, are products of erosion by water of fine-textured sedimentary rocks. They exhibit a steplike structure because the rock layers have different resistances to erosion. (bottom) Pillars have been cut into the soft pumice slopes of an ancient collapsed volcano at Crater Lake, Oregon. Such pillars are formed by rainfall on soft material where boulders or other hard cover prevent portions of the surface from being eroded downward as rapidly as unprotected portions.

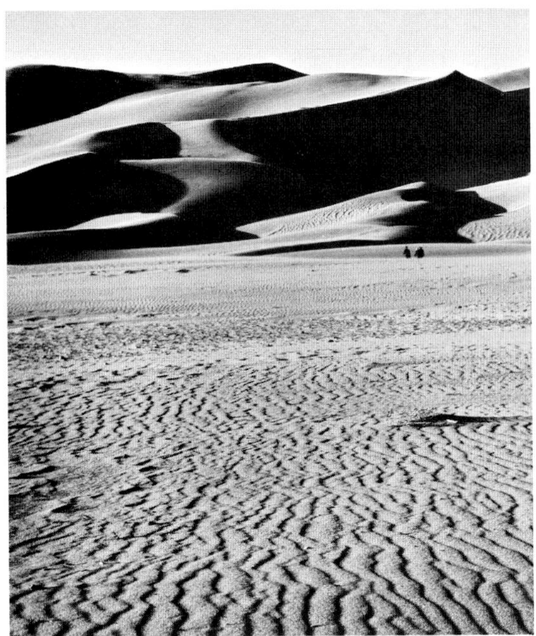

Figure 11.10 In regions where large quantities of sand are available from rock weathering, wind action may cause both erosion and deposition. The wind also shapes the sand into characteristic landforms, as shown here at Great Sand Dunes National Monument in south central Colorado. The surface of the sandy plain in the foreground shows closely spaced ripples transverse to the direction of the prevailing winds. The sand accumulations in the background have been formed into crescent-shaped dunes by the wind.

by erosion and by deposition. The unique landforms of the seacoast are discussed in Chapter 16. Wind also picks up and transports particles and is responsible for depositional landforms such as sand dunes and for accumulations of rock and soil dust such as the steep cliffs of *loess* found along the Mississippi River basin, in China, and in many other locations. Glaciers act by both erosion and deposition. They pluck rocks from the face of a mountain to form deep erosional amphitheaters, and they form hills and extensive plains by depositing the load of debris they carry. The landscapes of the northeastern United States and of the Midwest cannot be understood without reference to the work of the ice sheets that covered these regions in the recent geologic past.

Gravity plays a dominant role in sculpting the landscape by influencing the downward movement of water, ice, and loose materials. Gravity often accelerates the process of erosion. If the forces holding the particles on the slope become too weak—perhaps because of heavy rains—part of the hillside may slide downward. Even on comparatively stable vegetation-covered slopes in midlatitude climates, the soil often slowly creeps down the slopes. Where vegetation is sparse, the rate of soil creep may be as much as several centimeters per year. In cold regions where the summer thaw of frozen soil provides water for lubrication, soil may flow tens of centimeters in a few days until the moisture content diminishes.

Diversity of Slopes

Whenever you look at a ridge of soil, a drumlin, or a mountain range, you can see examples of slopes. Slopes, a basic component

Figure 11.11 Stone Mountain, Georgia, is a massive, barren dome of granite. The domelike form is the result of chemical and mechanical weathering of the granite; large rounded slabs tend to break off as a result of the weathering.

(bottom) This slope consisting of granite boulders overlooks the east face of the Sierra Nevada in California. Weathering along cracks in the granite has produced the rounded forms. The mountains in the background, dominated by Mount Whitney in the center, also consist largely of granite. The steep east face of the Sierra Nevada is the result of faulting and uplift of a massive block of the continental crust.

(opposite left) Vegetation covers steep slopes that are eroded along zones of weak rocks in dead volcanoes on the island of Maui in the Hawaiian Islands.

(opposite right) The rolling slopes of this cow pasture in Devon, England, have developed on hard limestone.

of all landforms, are formed in three main processes: crustal deformation, such as faulting and folding; deposition of new material; and erosion. The Rocky Mountains exhibit the steep, rugged slopes formed by crustal deformation. Depositional slope forms include sand dunes, which are built by the wind, and moraines, which are the piles and ridges of rock debris that retreating glaciers left behind.

Most slopes are formed by erosion, particularly by water flowing down slopes and in channels. The same process that cuts gullies in a farmer's field has also carved out the Grand Canyon of the Colorado River. For many years the role of erosion in valley formation was not understood, and many people thought that a stream simply sought out its course along a valley that already existed. Only for the last century have people recognized that valleys are the result of a stream's work in eroding the land as it flows to lower altitudes.

The processes of deformation and erosion can occur simultaneously, but usually they occur at different rates. Sometimes deformation or uplift is so rapid compared to the rate of erosion that an emerging landform is initially modified only slightly by erosion, although erosion processes continue to shape the landform after its formation. At other times, the crust is deformed or uplifted so slowly that erosion significantly alters the landscape as it is being formed.

Figure 11.11 shows examples of slopes from various parts of the world that illustrate how great a contrast can exist among slopes. The diversity of slopes is due to differences in the processes that

shape them and differences in the response of rock materials. Climate controls many of the processes that shape slopes, such as chemical and mechanical weathering processes. Chemical weathering is most active in climates that have abundant moisture and high temperatures. Mechanical weathering, which takes place primarily through the pressures produced by the freezing of water, is important except where water is lacking or where temperatures remain too high for freezing to occur.

Climate also influences the amount of runoff water available for erosion. As you know from studying the water budget, the amount of water available for runoff in a particular region depends on such factors as the amount and timing of precipitation, evapotranspiration from the vegetation cover, the amount of moisture stored in the soil, and the rate of infiltration into the soil. In arid regions, rare periods of heavy rain can accomplish significant amounts of work on the landscape because water cannot easily infiltrate the compacted soil and a large proportion of the precipitation becomes runoff.

Despite the various climatic influences, there is also a diversity of slopes within a single climatic region. Look again at Figure 11.11 and note the range of materials represented—granite, limestone, lava. A slope's form depends not only on the *processes* that act on it, but also on the *resistance* of its material to these processes. A slope represents the response of particular materials to particular processes. The resistance that each type of rock offers to weathering depends on the rock's chemical properties and physical structure. Sandstone, for example, weathers into individual grains of silica when the cement that holds the grains together is chemically dissolved. A rock with extensive joints and fractures tends to weather initially into large rock fragments; the fragments subsequently suffer chemical and physical attack and are disintegrated to grains. Figure 11.12 shows a *tor* of granite, where erosion along joints has produced a characteristic pile-of-blocks appearance. The debris from weathering is more easily moved by wind or water if it is in the form of small grains. Rock that weathers into large fragments, which are not easily transported, may remain relatively protected from erosion. The climatic influences on slope formation are discussed further in Chapter 14.

Figure 11.12 This granite tor in the Mojave Desert of California developed from a jointed mass of granite under the attack of physical and chemical weathering processes. The closely spaced joints act as channels for water, which attacks the rock. Weathering enlarges the joints and rounds the corners and edges of the rock prisms to produce the characteristic "pile of boulders" appearance of a tor. In some tors, the boulders are not fully separated, although they appear to be so because of the enlarged joints.

Process and rock type help determine the shape of a slope. In a single climatic region, combinations of different processes acting on rocks of various resistances can generate a wide variety of landforms. Furthermore, because of the complex interaction of processes, similar landforms may arise in different climatic regions.

Evolution of Landforms

The interplay of process and resistance generates an immense variety of landforms—mountain ranges, rolling hills, featureless plains. As the earth has evolved through geologic time, its original surface has changed and new features have appeared. A few local changes on the earth's surface—the collapse of a volcanic cone, for instance—have been sudden and catastrophic, but most of the evolution of the earth's landforms proceeds slowly over time by the accumulation of small events. Several theories about the evolution of landforms have been proposed.

The Cycle of Erosion

At the close of the nineteenth century, the American geographer William Morris Davis wrote a series of essays in which he attempted to organize landforms according to a system of evolutionary development. An analogy may clarify Davis' thinking. Imagine a scientist from another solar system visiting the earth for the first time, with only a short time to determine the development of human life from infancy to old age. With no time to follow a person through a full life span, an alternative approach would be to study infants, children, teen-agers, adults, and the aged and then work out a scheme of human development from such observations. Similarly, the pace of geomorphic change is generally much too slow to make the evolution of given landforms apparent over the limited time span of human history. So Davis examined existing landforms and classified them according to stage of development. He then inferred how landforms of one stage might develop into those of another over the course of time.

Davis called his ideal cycle of development the *geographical cycle*, but today it is commonly known as the *Davisian cycle*, or the *cycle of erosion*. To explain the cycle in the simplest way, Davis began with an ideal situation that probably seldom occurs in reality. He visualized a sudden uplift of lowland from near sea level to a higher altitude and assumed that the uplift occurred in a time interval too short for erosion to have a marked effect. Once the uplift was complete, however, erosion was important in shaping the land, according to Davis.

Davis concentrated his attention on the erosive action of flowing water. Rainfall on the uplifted block of land works on the landscape as the water flows down toward the sea and loses gravitational energy. Because the land is initially high, the moving water is an active agent of erosion. Davis pictured the rivers as cutting primarily downward to form deep, steep-sided river valleys with large areas of the original land still remaining between the deep valleys. He called this deeply incised landscape the *youthful* stage of landform evolution.

As the rivers deepen their valleys, they decrease their altitude above the sea, and so the energy available for erosion lessens. Davis called the level toward which rivers descend the *base level of erosion*, because water that has fallen to the base level has no more gravitational energy that can be converted to do the work of erosion. Sea level is normally the ultimate base level, but streams that drain into basins may have a base level higher or lower than sea level, depending on the altitude of the basin. Davis argued that as erosion continued, the descent of rivers would become more gradual and downcutting of the land would give way to horizontal erosion: the widening of the river valleys, and the evolution of floodplains. In the stage Davis called *maturity*, the original uplifted land is reduced

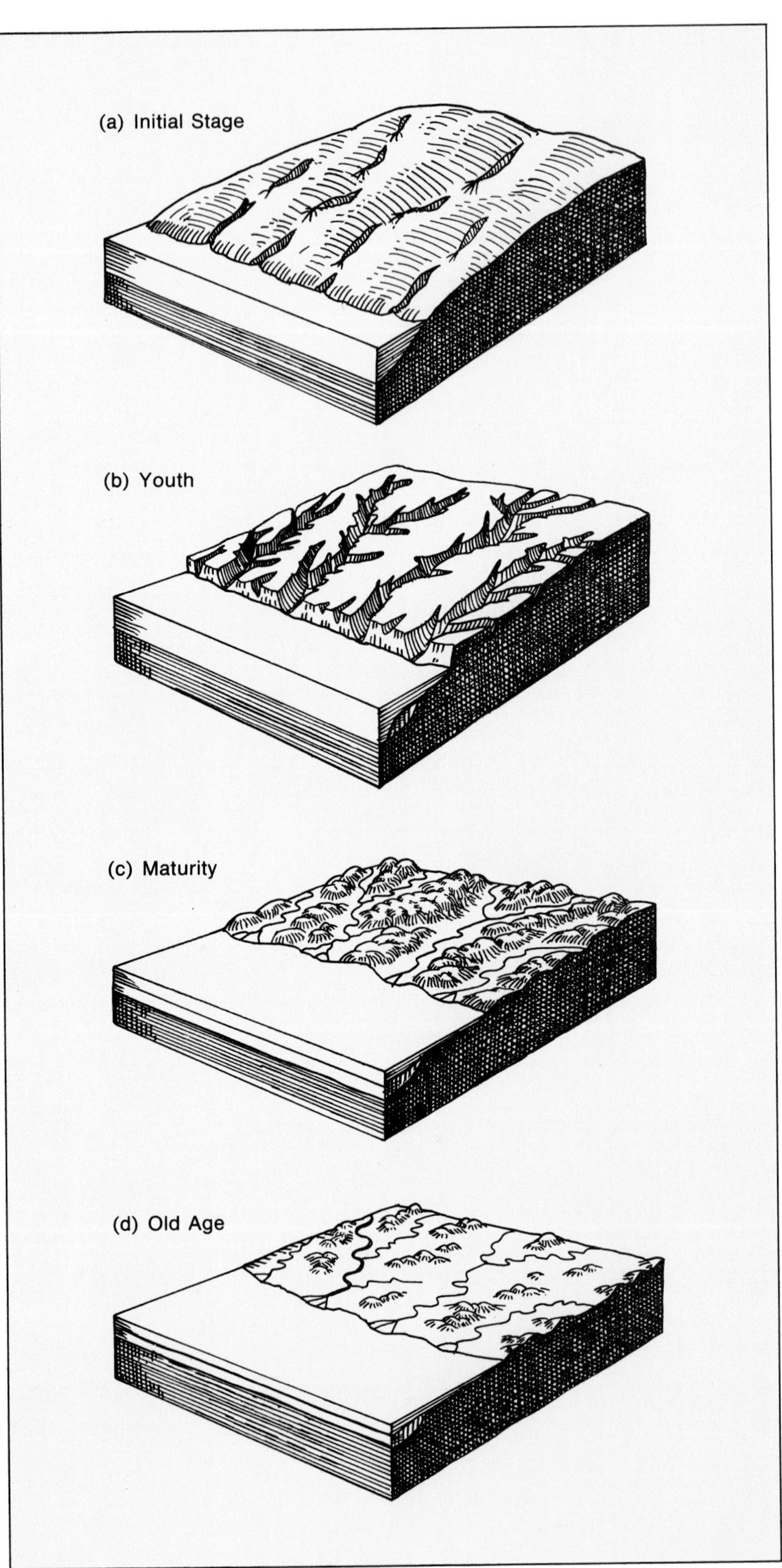
(a) Initial Stage
(b) Youth
(c) Maturity
(d) Old Age

Figure 11.13 This sequence of diagrams illustrates stages of the Davis cycle of landscape evolution. This cycle applies to moist regions where erosion is accomplished primarily by flowing water and where the underlying rock is uniform and exerts no controls over landform development.

(a) The initial stage is a landscape of low relief. After uplift of the region, streams have energy to begin cutting downward. (b) In the stage of youth, the streams have cut comparatively narrow valleys downward, and much of the initial plain remains in the high land between the stream valleys. (c) In late youth and early maturity, runoff water has worn the hill slopes to rounded ridges. The streams have widened their valleys, and little trace of the initial plain remains. (d) In old age, the work of flowing water has worn the region to a plain of low relief (peneplain) dotted with isolated high hills, or *monadnocks.* (After Longwell, Knopf, and Flint, 1941) to rolling hills and wide valleys. Finally, in *old age*, erosion has proceeded so far that the land is once again near sea level. The river valleys in old age are broad plains, and the landscape presents the appearance of a rolling lowland with gentle slopes, called a *peneplain* ("almost a plain"). The land has returned to its original form through a complete cycle of erosion.

Davis realized that his simple model could be modified in reality, perhaps by further uplift during one of the intermediate stages of evolution. He also realized that his model was most appropriate for climates in which flowing water was abundant and that other cycles of erosion would operate in arid or polar regions.

Davis viewed landforms as products of *structure*, *process*, and *stage*. By stage, he meant the stage of evolution. The actual time required for a landscape to reach a certain stage of maturity depends on the nature of the rocks and on the strength of the processes active in erosion. One landscape may be younger than another in terms of the time that has elapsed since uplift, but it can be at a later stage of development because of active erosion or because of a low resistance of the rock structures.

The importance of Davis' work is that it turned geographers' attention from the purely verbal description of landforms toward an effort to understand landform development and the relationships between different landforms. But Davis devised his cycle of erosion at a time when little was known about the processes that form slopes. His ideas were widely adopted as a framework for geomorphology, but as knowledge about slope-forming processes has increased, his ideas have been criticized.

In the 1920s, the German geomorphologist Walther Penck argued that slopes do not inevitably become less steep as the Davisian cycle of erosion describes. Penck suggested the alternate possibility that eroding slopes may not change angle at all but may simply retreat parallel to themselves. Like Davis, Penck did not fully consider all the processes important in controlling slope erosion, but his idea that slopes can erode by parallel retreat as well as by decline has contributed to the present understanding of landforms.

The fundamental concept of an erosion cycle seems less attractive

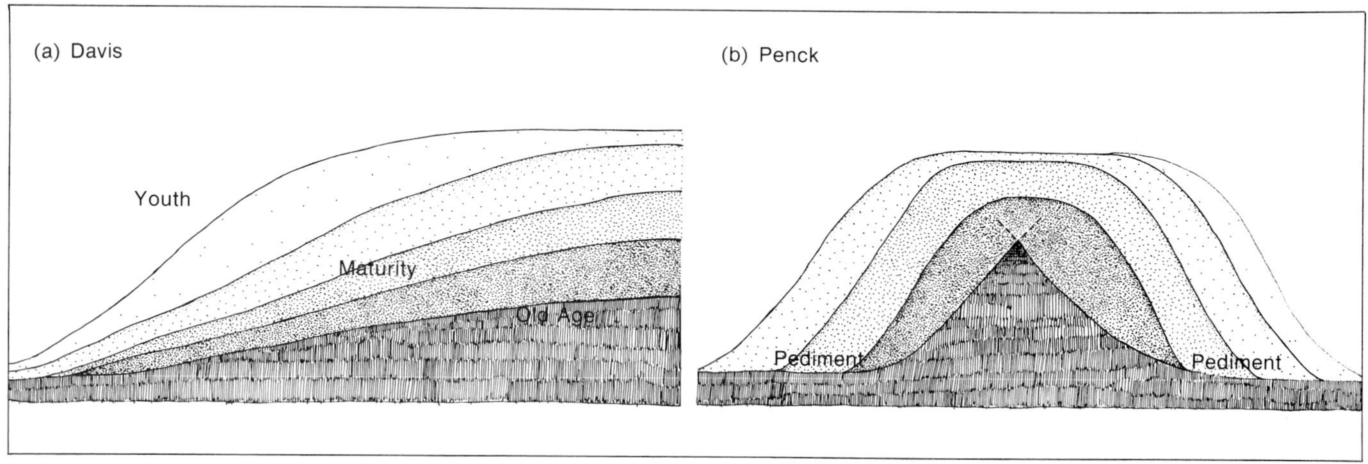

(a) Davis
Youth
Maturity
Old Age
(b) Penck
Pediment
Pediment

Figure 11.14 (a) According to William Morris Davis, the evolution of a slope proceeds through the successive stages shown in this diagram. As the slope retreats, it becomes less steep and more rounded. This type of slope development appears to occur when material is transported down a slope more rapidly than it can be removed from the base of the slope by streams or other agents of erosion.

(b) According to Walther Penck, the evolution of a slope through successive stages occurs by parallel retreat, in which the slope angle remains essentially constant. For this type of slope development to occur, the removal of material from the base of the slope must keep pace with the rate of transport of material down the slope face. (After Holmes, 1955) now than it did earlier in this century. For one reason, it is difficult to find clear examples of peneplains at base levels or near sea level on the earth today. Furthermore, the response time of landform systems to erosion may be long in comparison with intervals between periods of uplift, so that the undisturbed conditions required for the simple Davisian model may not exist. Under conditions of continued change, the concept of a cycle of landform evolution becomes less useful. As a scientific theory, the Davisian cycle of erosion has shortcomings. It is based upon an imperfect knowledge of the processes that actually shape slopes, and it does not lead to an objective check on the conclusions it draws. The Davisian cycle can be regarded as a way to organize a certain amount of descriptive material concerning landforms in moist climates. The terms *youth*, *maturity*, and *old age* have been retained in geomorphology as descriptive terms for certain types of landscapes, but the implication that such terms describe an inevitable succession of evolutionary stages in the development of any landform is no longer accepted. So although the Davisian cycle of erosion provides a framework for thinking about landforms, it cannot tell us at a glance why a given landform in a particular place develops in the way it does.

Relict Landforms

The recent emphasis on process and form in geomorphology has led geomorphologists to seek a deeper understanding of the details of slope formation. One of the problems in studying landforms is

Figure 11.15 Convict Lake, on the east face of the Sierra Nevada, California, lies in a glaciated valley dammed by glacial deposits of rock debris, or by a *moraine.* Although these landforms are about 20,000 years old, they still bear the unmistakable marks of glacier action.

that they respond so slowly to change. Whether a particular hill is changing form, and to what extent, is not a question that can be answered easily with present experimental techniques. Some landforms, such as stream channels, are easier to study because they respond relatively more rapidly to change. Stream channel behavior can also be studied with models in the laboratory.

More resistant slopes, however, still bear the imprint of conditions from the geologic past because they respond slowly to present conditions. In northern Europe and in the northern United States, for example, many landforms show the effects of the ice ages of the past 2 million years, when climate and other conditions were different from those of today. Furthermore, the combinations of climate and material that shaped landforms in the past may no longer exist on the earth. *Relict* landforms are those that have not completely adjusted to the processes active today. Although puzzling and difficult to understand, relict landforms represent an opportunity to learn about the past.

The Tempo of Geomorphic Change

The shape of the landscape is the outcome of the opposition between uplift and erosion. What are the rates of uplift and erosion today? In California, mountain building is active at the present time. Careful and repeated measurements using surveying methods indicate that the rate of uplift in California in active regions of mountain uplift is about 5 to 10 meters per 1,000 years (15 to 30 feet per 1,000 years). The intermountain regions of the western United States are rising a fraction of a meter per 1,000 years.

Most estimates of the rate of erosion are based on measurements of the total amount of sediment carried by all the streams that flow out of a drainage basin. In large drainage basins of moderate relief, such as the basin drained by the Mississippi River, the current rate of erosion is estimated to be about 5 centimeters per 1,000 years (2 inches per 1,000 years). It is known, however, that this is not the rate of erosion that has prevailed over time; rather, it reflects human disturbance of the natural landscape in the past century. In extremely mountainous regions erosion rates may be as high as 1 meter per 1,000 years (3 feet per 1,000 years). In general, the rate of uplift in mountain-building regions is much greater than the rate of erosion. These measurements seem to support Davis' theory of initial rapid uplift followed by slow erosion.

Small increments of change do the major work in shaping the landscape. The uplift of a high mountain range occurs by the accumulation of small upward movements occurring every century or so. The increments occur frequently compared to the millions of years required for the mountains to be built.

On a different time scale, the work done by flowing water is performed mainly during frequent events that recur every few days or weeks. Measurement of the sediment load carried by a stream under different conditions of flow shows that a stream transports most of its annual load of sediment under conditions of moderate flow. Major floods enable a stream to carry a heavy load of sediment

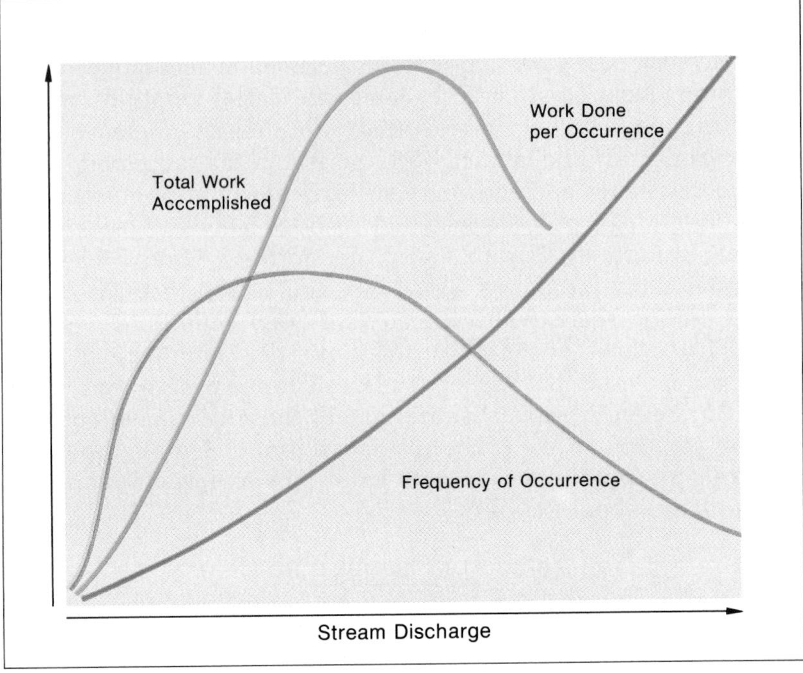
Work Done
per Occurrence
Total Work
Accomplished
Frequency of Occurrence
Stream Discharge

Figure 11.16 This schematic diagram illustrates the principle that most geomorphic change is accomplished by relatively frequent events, each of which performs a moderate amount of work. The work done by a given stream, for example, increases during times of high discharge. However, major floods occur infrequently, so the rising curve showing work done per event is balanced by the falling curve showing frequency of occurrence. Most of the work a stream does in a year is therefore accomplished during the frequent intervals when the discharge is moderately high. (After *Fluvial Processes in Geomorphology* by Luna B. Leopold, M. Gordon Wolman, and John P. Miller. W. H. Freeman and Company. Copyright © 1964)

for a short time, but major floods occur so infrequently that on the average more work is done under conditions of moderately heavy flow that occur comparatively frequently. Figure 11.16 shows this idea schematically.

Some processes of erosion can occur only during events of catastrophic proportions. Rain from an intense storm, for example, may induce a major landslide that significantly modifies the form of a slope. Or a stream in high flood may be able to loosen a large boulder from the stream bed and carry it far downstream. But such events are relatively infrequent. A stream that experiences a flood of certain intensity once every year or two may experience a flood twice as great once every 15 years, and a flood three times as great once in a century. On the whole, however, the modification of landforms is accomplished by events of moderate intensity occurring at moderate frequency.

Summary

The surface of the earth displays features of relief on several scales. The continents and ocean basins are features of first-order relief, and major mountain ranges, plateaus, and depressions are features of second order. The features of third order, such as valleys and hills, are sculptural details on second-order relief forms.

Every feature on the earth undergoes change and development. The drift of crustal plates and the spreading of the sea floor have moved landmasses and opened ocean basins. Uplift of large regions has raised mountain ranges, particularly along the boundaries of colliding plates. Landforms undergo change and development because of internal energy and solar energy. The work of the sun and the atmosphere is carried out primarily by the erosive action of flowing water (Chapter 12), whereas the earth's internal energy expresses itself in displacements and warpings of the earth's crust and in volcanic activity (Chapter 13). Other agents that shape landforms include glaciers (Chapter 15) and ocean waves (Chapter 16).

The *diversity of landforms* arises from the complex combination of rock types, processes, and stages of development (Chapter 14). William Morris Davis, an American geographer, recognized the importance of structure, process, and stage, although his cycle of erosion is inadequate to deal with detailed questions of form and process concerning all landforms. The *response time of landforms* is usually long, so that significant changes in climate sometimes occur before a landform reaches equilibrium. Many landforms in northern latitudes, for example, bear the imprint of glacial action from past ice ages. The *shaping of landforms* usually occurs by the accumulation of small increments of change. The modification of landscapes is largely the work of events of moderate intensity that recur moderately often, although infrequent catastrophic events can produce unique results.

12 Flowing Water and Its Work

Energy of Moving Water

Principles of Water Flow
Turbulence
Transport of Material by Streams
Scour and Fill

The Geometry of Stream Drainage

Anatomy of a River

Stream Channels: Form and Process
Vertical Adjustments
Slopes in Stream Beds
Graded Streams
Knickpoints
Lateral Migration of Stream Channels

Landforms Associated With Rivers
Floodplains
Terraces
Underfit Streams
Deltas
Alluvial Fans

Flowing water is one of the most important agents of change on the landscape, even in arid regions. The energy in moving water empowers it to pick up, carry, and redeposit bits of the land. The landscape is a balance between downwearing due to flowing water and uplift caused by geologic processes.

The Pass Through the Mountains by Fukaye Roshu. (The Cleveland Museum of Art, John L. Severance Fund)

Flowing Water and Its Work

In the 1920s the French government took nomadic desert chieftains on a tour of France. The chiefs were awestruck at their first sight of a mighty river. Thinking of the meager and intermittent waters of their desert home, the chieftains sat on the bank to wait patiently for all the water to flow by. When told that the river had been flowing for untold ages, they declared that the god of the French was good to his people.

For centuries rivers have been important to man for agriculture, for transportation, and for their beauty. Man has relied on the flowing water of rivers to inspire his poetry and songs, to transport his boats, to water his fields, and even to carry away his wastes. From prehistoric times to the present he has built settlements near rivers—Rome on the Tiber, Cairo on the Nile, Paris on the Seine, London on the Thames, New York City on the Hudson; indeed, many cities of significance are situated on rivers.

But streams carry not only water. Before the Egyptian government built the Aswan High Dam, the Nile River overtopped its banks each spring and left a layer of rich sediment on the narrow belts of farmland bordering its channel. The rock and soil particles that the Nile carried into Egypt entered the river thousands of kilometers upstream, where they had been washed down from the mountain slopes of Ethiopia and Uganda by surface runoff and streams. Some of the Nile's load of sand, silt, and clay is deposited in the great fertile delta near the mouth of the Nile. The denudation of the Ethiopian hills supplies the material for building the Nile delta, and flowing water supplies the means of transport. Now that the Aswan Dam is built the delta no longer receives its supply of nutrient-rich sediment. The sediment instead builds up in the region of Lake Nasser behind the dam.

Flowing water has the ability to pick up, transport, and deposit material. From the mightiest river to the tiniest raindrop, water is a powerful sculptor of the earth's surface. Streams are interesting in themselves, particularly because their response times to changing inputs are comparatively short. The study of streams is a field of active investigation for physical geographers, who are attempting to determine the interactions between the speed of streamflow, the shape of the channel, the slope of the bed, and other factors.

Energy of Moving Water

The reason that water can erode rock is that moving water possesses energy. Recall from Chapter 2 that all moving objects have kinetic energy, which can be converted to other forms of energy. When a falling raindrop splatters on the ground, for instance, part of its kinetic energy is converted to heat, but the remainder may be used to loosen rock and push rock particles from one place to another. When a stream descends a slope, its gravitational potential energy is transformed to other forms of energy that enable the water to pick up and carry sand and pebbles. As a stream flows down a

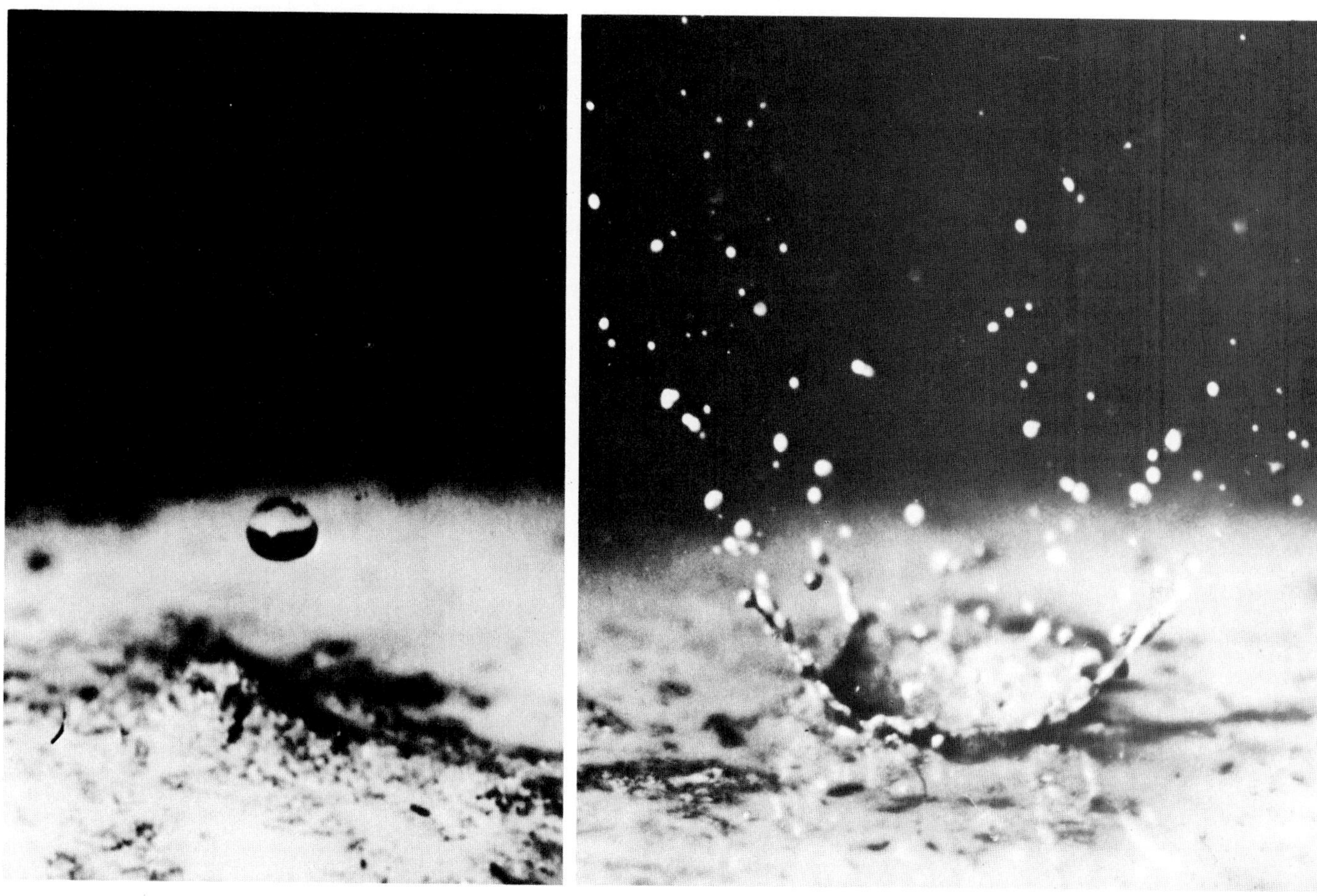

Figure 12.2 These close-up, high-speed photographs of a raindrop splashing into bare soil show graphically the work that can be done by water. In the photograph on the left, the raindrop is 1/8 inch in diameter and is traveling at a speed of 25 miles per hour. (right) When the drop strikes the ground, particles of soil are thrown long distances by the impact. The impact of raindrops on a bare slope tends to move soil particles down the slope, on the average.

slope, the water in a section of the stream experiences a decrease of gravitational potential energy. The amount of decrease depends on the change of altitude and on the amount of water in the section. A stream can do more work when it is in flood than when its flow is normal because under flood conditions there is more water in any given section of the stream.

Principles of Water Flow

To understand some of the principles of water flow, visualize sediment-free water in a long, open cement trough. Water will not flow by itself along the trough unless one end is higher than the other, so that the water can lose gravitational energy and gain energy of motion as it flows along. But the gravitational energy is not converted primarily into energy of motion; it is used largely to overcome friction between the water and the trough walls and friction between water currents moving within the trough. If the trough did not slope, friction would quickly bring the moving water to a halt. For this reason, a stream channel must slope downward, on the average, to maintain flow.

Friction has another important effect on the flow of water in channels. If a drop of water is moving near the bottom or near the bank of a stream, the drop will experience a large retarding effect caused by friction with the bottom or bank. Water near the banks or bottom of a stream channel flows less rapidly than water near the center of the channel, where friction with the channel is less. Figure 12.3 shows measured water velocities at numerous points through the cross section of a stream. The water moves fastest near the center of the stream, away from the channel walls and bottom. Boatmen heading downstream attempt to keep their boats in the center of a river where the water is fastest, but rowing upstream is easiest near the banks, where the water flows more slowly.

The volume of water a stream carries past a given point during a specific time is called the *discharge* of the stream. Discharge,

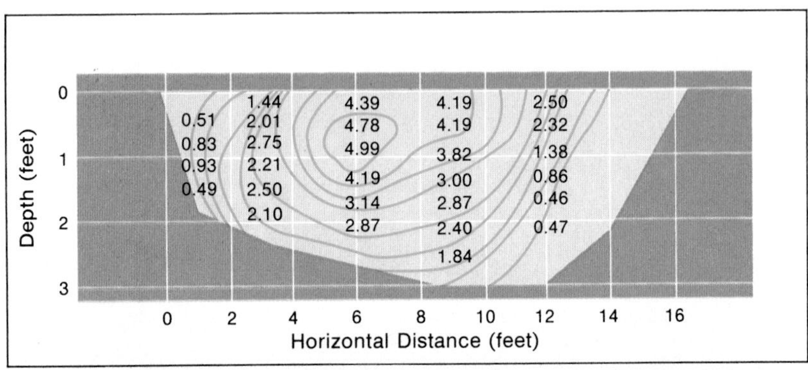

Depth (feet)
0
1
2
3
0.51
0.83
0.93
0.49
1.44
2.01
2.75
2.21
2.50
2.10
4.39
4.78
4.99
4.19
3.14
2.87
4.19
4.19
3.82
3.00
2.87
2.40
1.84
2.50
2.32
1.38
0.86
0.46
0.47
0
2
4
6
8
10
12
14
16
Horizontal Distance (feet)

Figure 12.3 This diagram shows the average measured stream velocities, in feet per second, at various points in a cross section of Baldwin Creek, Wyoming. Note that the velocities tend to be lowest near the sides of the channel; the stream cannot, therefore, actively erode its banks in this section. Velocities are highest near the center of the channel, and they are moderately high near the bed. The stream is therefore able to carry material suspended in its waters and to transport material along its bed. (After *Fluvial Processes in Geomorphology* by Luna B. Leopold, M. Gordon Wolman, and John P. Miller. W. H. Freeman and Company. Copyright © 1964)

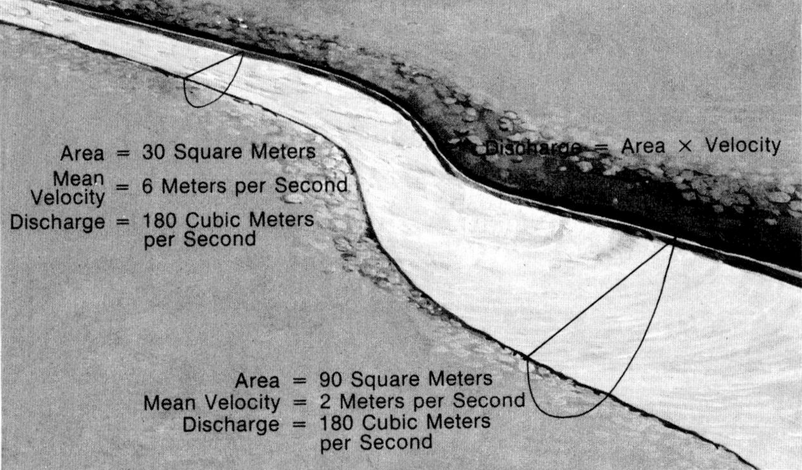

Discharge = Area × Velocity
Area = 30 Square Meters
Mean Velocity = 6 Meters per Second
Discharge = 180 Cubic Meters per Second
Area = 90 Square Meters
Mean Velocity = 2 Meters per Second
Discharge = 180 Cubic Meters per Second

Figure 12.4 As this schematic diagram indicates, the discharge of a stream through any cross section is equal to the area of the cross section multiplied by the average velocity of flow. If the cross-sectional area of a stream increases downstream, as shown in the diagram, the velocity of flow decreases proportionately, assuming that no water is gained or lost by the stream.

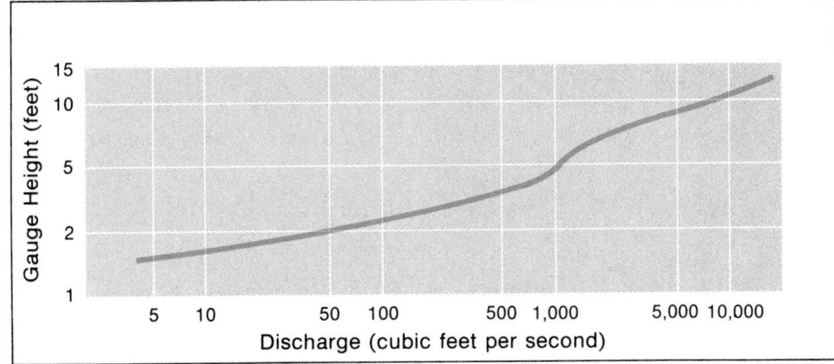

Gauge Height (feet)
15
10
5
2
1
5
10
50
100
500
1,000
5,000
10,000
Discharge (cubic feet per second)

Figure 12.5 This calibration curve for Seneca Creek, Maryland, relates the height of the stream to its measured discharge and allows the discharge to be inferred from a simple measurement of height. The average annual discharge of this stream is 100 cubic feet per second, but discharges of more than 50 times as great have been observed at times. (After *Fluvial Processes in Geomorphology* by Luna B. Leopold, M. Gordon Wolman, and John P. Miller. W. H. Freeman and Company. Copyright © 1964) which can be specified in cubic meters per second, or in cubic feet per second, is directly proportional to the speed of flow and to the cross-sectional area of the stream. The discharge of a stream in steady flow is the same everywhere along the stream if no water is added, lost, or stored. If a stream is forced into a confined channel with a smaller cross section, the average speed of flow must increase to maintain a constant discharge. Figure 12.4 illustrates how the velocity of water in a stream will change if the cross-sectional area is changed and if the discharge remains constant. The section of stream with the larger cross section has smaller flow velocities. The maximum flow speed in any stream seldom exceeds 3 to 5 meters per second (7 to 11 miles per hour).

The United States Geological Survey maintains a network of more than 6,000 gauging stations to measure the discharge of streams in the United States. The data are used to make flood forecasts and to evaluate irrigation systems, dam and bridge designs, water intake systems, and so forth. At most stations the discharge is inferred from the height of the stream's waters; a calibration between height and discharge is previously determined for the section of stream at the location of the gauging station.

Turbulence

The regular and even flow of water running slowly along a smooth channel is called *laminar* flow. In laminar flow, a particle of water moves with constant speed and direction at any given position. If the speed of flow is increased, however, the flow will become irregular, or *turbulent*. The constantly moving *eddies* that make a stream so fascinating to watch are evidence of turbulent flow. The rolling, boiling rapids of a swift stream or the water running along a street after a heavy rainstorm shows irregularities characteristic of turbulent flow, but even placid streams move fast enough to be turbulent. Particles of water moving in an eddy constantly change their speeds and directions of flow. In turbulent flow, particles of water may momentarily flow vertically or across the stream, but the average motion is in the downstream direction.

Friction is present in turbulent flow because particles of water in eddies move with different speeds and directions and rub against each other as they pass. Friction converts some of a stream's energy to heat. Nearly all of the gravitational energy of water in a stream is used to compensate for losses caused by the friction of turbulence and friction with the channel walls; only a few percent of the gravitational energy goes toward picking up and transporting rock debris.

Transport of Material by Streams

Streams are an important link in the hydrologic cycle because they carry excess precipitation from the land back to the sea. As a stream runs over the land to meet successively larger streams, it also carries eroded material from the highlands. Streams running off the Badlands of South Dakota are milky with fine clay. The Mississippi River, which receives sediment from 40 percent of the area of the mainland United States, carries more than 400 million tons of sediment to the sea each year. Table 12.1 lists the discharge, sediment load, and other characteristics of some of the world's rivers.

Streams transport materials by three mechanisms. Part of the material transported is dissolved in the water, forming the *dissolved load*, or *chemical load*, of the stream. The clay and silt particles of the solid load that are carried in suspension are the *suspended load*, and the coarser solid particles that are bumped along the bed of the stream constitute the *bed load*. The bed load cannot always be strictly differentiated from the suspended load because some particles may be lifted completely off the bed to join the suspended load for a time.

The amount of dissolved load a stream carries depends on such factors as the composition of the rocks in the area drained by the stream, the degree of weathering of the rocks, the acidity of the soil moisture, and the effect of temperature on solubility. A major portion of the water flowing in a stream comes from groundwater, which is often high in dissolved minerals.

A stream's ability to carry sediment in suspension depends on the degree of turbulence. Particles in still water eventually settle to the bottom, with the largest particles settling first. In a flowing turbulent stream, upward currents of water keep small, slowly settling particles in suspension, similar to the way that upward currents of air keep water droplets in clouds from settling. Swift

Table 12.1. Characteristics of Selected Rivers

River and Location	Average Discharge at Mouth (thousands of cubic meters per second)	Length, Head to Mouth (kilometers)	Area of Drainage Basin (thousands of square kilometers) ‡	Average Annual Suspended Load (millions of metric tons) §	Average Annual Suspended Load (metric tons per square kilometer of basin)
Amazon (Brazil)	180 (6,400)*	6,300 (3,900)†	5,800	360	63
Congo (Congo)	39 (1,400)	4,700 (2,900)	3,700		
Yangtze (China)	22 (800)	5,800 (3,600)	1,900	500	260
Mississippi (U.S.)	18 (630)	6,200 (3,900)	3,200	310	97
Yenisei (U.S.S.R.)	17 (600)	4,500 (2,800)	2,100		
Irrawaddy (Burma)	14 (500)	2,300 (1,400)	430	300	700
Bramaputra (Bangladesh)	12 (415)	2,900 (1,800)	670	730	1,100
Ganges (India)	12 (415)	2,500 (1,600)	960	1,450	1,520
Mekong (Thailand)	11 (390)	4,200 (2,600)	800	170	210
Nile (Egypt)	2.8 (100)	6,700 (4,200)	3,000	110	37
Missouri (U.S.)	2.0 (70)	4,300 (2,700)	1,370	220	160
Colorado (U.S.)	0.2 (6)	2,300 (1,400)	640	140	210
Ching (China)	0.06 (2)	320 (200)	57	410	7,200

* Numbers in parentheses indicate thousands of cubic feet per second.
† Numbers in parentheses indicate miles.
‡ One square kilometer is equal to about 0.39 square mile.
§ One metric ton is equal to 2,204.6 pounds.

Sources: Holeman, John N. 1968. "The Sediment Yield of Major Rivers of the World," *Water Resources Research*, 4 (August): 737-747.
Fairbridge, Rhodes W. (ed.). 1968. *The Encyclopedia of Geomorphology.* Vol. III, Encyclopedia of Earth Sciences Series. New York: Reinhold.
Espenshade, Edward B. (ed.). 1970. *Goode's World Atlas.* 13th ed. Chicago, Ill.: Rand McNally.

Figure 12.6 The amount and kind of sediment that a stream transports depend on the nature of the material through which the stream flows and on the amount of work the stream is able to do. These merging streams in Central Otago, New Zealand, carry different amounts of suspended material. The Kawera River on the left is heavily loaded with clay and silt; the Clutha River on the right has only a small suspended load and runs clear.

streams can transport larger particles than slow streams, because turbulence increases with increased velocity of flow.

A fast stream has enormous power to transport objects. When a train crossing a trestle in the Tehachapi Mountains of Southern California was caught in a flash flood during a cloudburst, the locomotive and tender were carried a kilometer downstream and buried so completely in gravel that a metal detector had to be used to find them.

Rocks and gravel on the bed of a stream can be pushed along by the water if the flow near the bottom is fast enough. Part of the bed load rolls and slides along the bottom, and part is lifted in and out of the bed flow in a hopping motion called *saltation* (from the Latin word meaning "jump"). As the rocks move along the bed, they collide and grind against one another, abrading and *corrading* each other. Corrasion gradually breaks them into smaller pieces, so that the particles on a stream bed tend to become finer, smoother, and rounder in the downstream direction.

Scour and Fill

Much of the sediment a stream transports comes to it from smaller streams, from unconcentrated runoff, and from stream bank erosion. If the speed of flow near the bottom or banks of the stream channel is great enough, however, the stream can generate sediment by *scouring* particles from its own bed. There is a critical speed below which scouring will not occur because the force of water below that speed is insufficient to detach particles from the bed. The critical speed depends on the size of the particles and on the nature of the bed. Clay particles, for instance, tend to be more strongly bound than sand to a bed.

Figure 12.7 indicates the flow velocities required to move particles of given sizes from a stream bed. The flow velocity near the bottom of the stream must be of the order of 10 to 20 centimeters per second (0.2 to 0.4 miles per hour) to initiate movement. The graph also shows the range of speeds for which a particle of given size will remain in the suspended load. A particle scoured from the bed when a stream is running fast can be carried along even if the speed of the stream subsequently decreases. If the speed becomes too low, however, the particle will fall out of suspension and be deposited on the bed.

When a large dam is built on a major river that carries a suspended load of sediment—the Glen Canyon Dam on the Colorado River, for example—the reservoir that forms behind the dam collects all but a few percent of the sediment because the speed of flow in the reservoir is extremely slow. The rate at which the reservoir fills with silt limits the useful life of the dam. Removal of the sediment also affects the dynamics of the stream below the dam, often causing scouring to increase. Streams tend to carry as much load as their energy permits, and they scour their beds if the load supplied to them is too small.

A stream is a far more complicated system than water flowing in a cement trough, because by scouring and filling, a stream readjusts the elevation of its bed. A stream may scour at one time or place and fill at another time or place, depending on the stream's

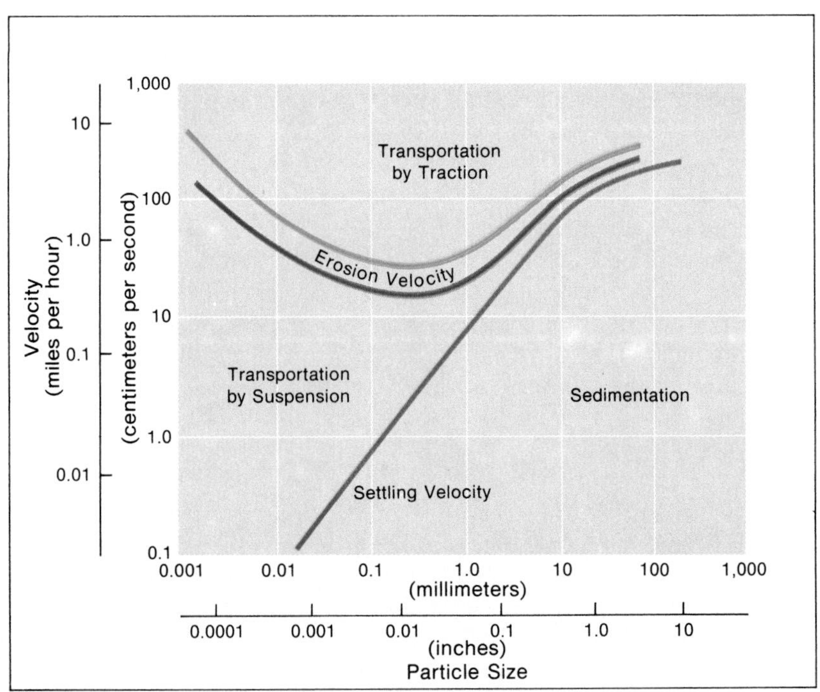
Velocity
(miles per hour)
(centimeters per second)
10
1.0
0.1
0.01
1,000
100
10
1.0
0.1
Transportation
by Traction
Erosion Velocity
Transportation
by Suspension
Sedimentation
Settling Velocity
0.001
0.01
0.1
1.0
10
100
1,000
(millimeters)
0.0001
0.001
0.01
0.1
1.0
10
(inches)
Particle Size

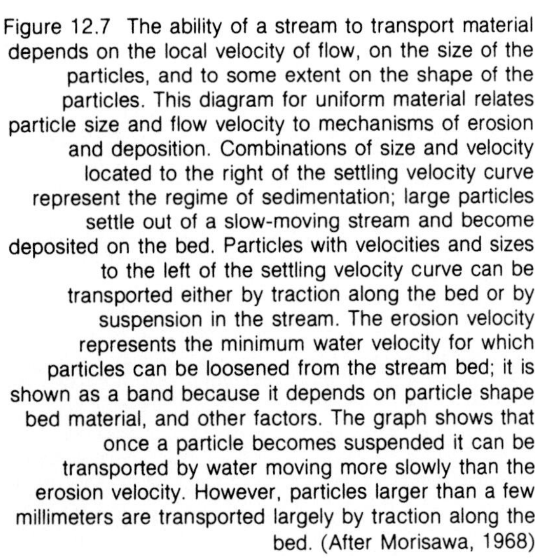
Figure 12.7 The ability of a stream to transport material depends on the local velocity of flow, on the size of the particles, and to some extent on the shape of the particles. This diagram for uniform material relates particle size and flow velocity to mechanisms of erosion and deposition. Combinations of size and velocity located to the right of the settling velocity curve represent the regime of sedimentation; large particles settle out of a slow-moving stream and become deposited on the bed. Particles with velocities and sizes to the left of the settling velocity curve can be transported either by traction along the bed or by suspension in the stream. The erosion velocity represents the minimum water velocity for which particles can be loosened from the stream bed; it is shown as a band because it depends on particle shape, bed material, and other factors. The graph shows that once a particle becomes suspended it can be transported by water moving more slowly than the erosion velocity. However, particles larger than a few millimeters are transported largely by traction along the bed. (After Morisawa, 1968)

speed and on the size of the particles involved. Scouring that continues for a long time is called *degradation*, and steady filling is called *aggradation*.

The Geometry of Stream Drainage

Only a few streams, such as the Nile River, flow for long distances without being joined by other large streams. Most streams are part of a larger geomorphic unit, the *drainage basin*, which is the region drained by a stream and all its tributaries. The scale of a drainage basin may range from a square kilometer or less to a significant fraction of a continent. The pattern of the streams in a drainage basin is called the *drainage net*. Drainage nets differ in appearance to some extent because of differences in texture and pattern. Texture depends on the total number and the total length of streams in a basin compared to the area of the basin. The patterns of drainage nets exhibit a wide variety of forms—barbed, trellislike, dendritic, radial, and so forth.

The feature common to almost all drainage systems is that they consist of numerous little streams feeding into successively fewer, larger streams. The hierarchy of tributaries ranges from streamlets

Figure 12.8 The drainage basin of a large river such as the Mississippi contains a hierarchy of smaller nested drainage basins. The arrows indicate the downhill direction of water flow into the various basins. The Continental Divide along the Rocky Mountains separates basins draining eastward toward the Atlantic Ocean from those draining westward toward the Pacific Ocean.

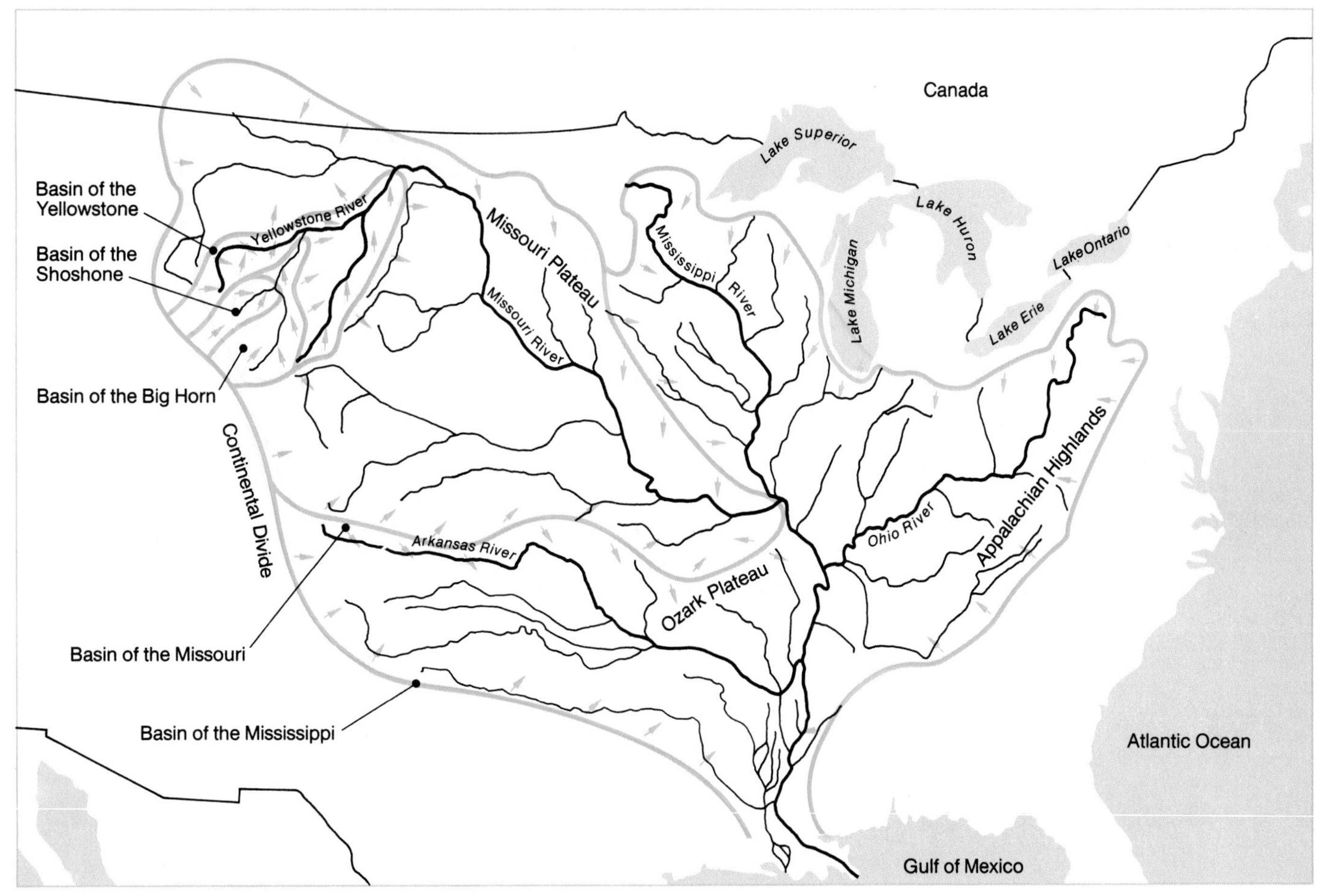
Canada
Lake Superior
Lake Huron
Lake Michigan
Lake Ontario
Lake Erie
Basin of the Yellowstone
Basin of the Shoshone
Basin of the Big Horn
Yellowstone River
Missouri Plateau
Missouri River
Mississippi River
Continental Divide
Appalachian Highlands
Ohio River
Arkansas River
Ozark Plateau
Basin of the Missouri
Basin of the Mississippi
Atlantic Ocean
Gulf of Mexico

too small to be shown on maps up to major rivers. In 1945 Robert Horton, an American hydraulic engineer, devised a quantitative way to study the implications of stream hierarchy. He introduced the idea of *stream order* as a numbering system to indicate the position of a stream in the hierarchy of a drainage net. The smallest streams that have no tributaries feeding them are called first-order streams. The larger streams they join are called second-order streams, and so on.

Horton found simple empirical, quantitative relationships between stream order and factors such as the number of streams and the area of the drainage basin. According to the Horton analysis shown in Figure 12.9, the number of streams of a given order drops off sharply with increased order. The analysis also shows that the total length of a drainage net increases regularly with order, as does the area drained by streams of a given order. The regularities Horton found hold for a large number of drainage basins, from small basins with a few dozen streams to basins with hundreds of thousands of streams draining an area of a thousand square kilometers. Despite the differences in their outward appearances, drainage basins seem to share certain fundamental properties. In most drainage basins, for example, the number of streams in a given order is found to be approximately 3 times greater than the number in the next higher order. A drainage basin with 50 first-order streams should therefore have approximately 15 second-order streams, 5 third-order streams, and 1 or 2 fourth-order streams.

Horton's work had a significant effect on geographers' approach to geomorphology. It turned attention from the purely descriptive aspects of landforms to the possibility of using quantitative measurements to gain deeper insight into the processes that shape the earth's surface.

Anatomy of a River

Streams have characteristic structural features. An overview of these features in the Platte-Missouri-Mississippi river system may help in understanding the processes that shape streams. The drainage basin of the Platte River includes the eastern slope of the Rocky Mountains in Wyoming. During a thunderstorm over the Rockies, the rate of infiltration of rain water into the soil is insufficient to keep pace with the downpour, and a sizable amount of runoff soon develops. At first the water runs as threads over the land; then, where runoff is heavy, the threads combine to form sheets. The *sheetwash* that accompanies runoff, together with the impact of raindrops, is a significant agent of erosion in semiarid and dry climates, particularly over the upper parts of hills. Farther down a slope the water usually cuts a dense, semipermanent drainage pattern consisting of fine parallel channels, or *rills*. Rills, which are a centimeter or so wide and deep and a few centimeters apart, can be seen on most bare slopes, such as those exposed by road construction.

The drainage network of a region progresses from the fine, dense

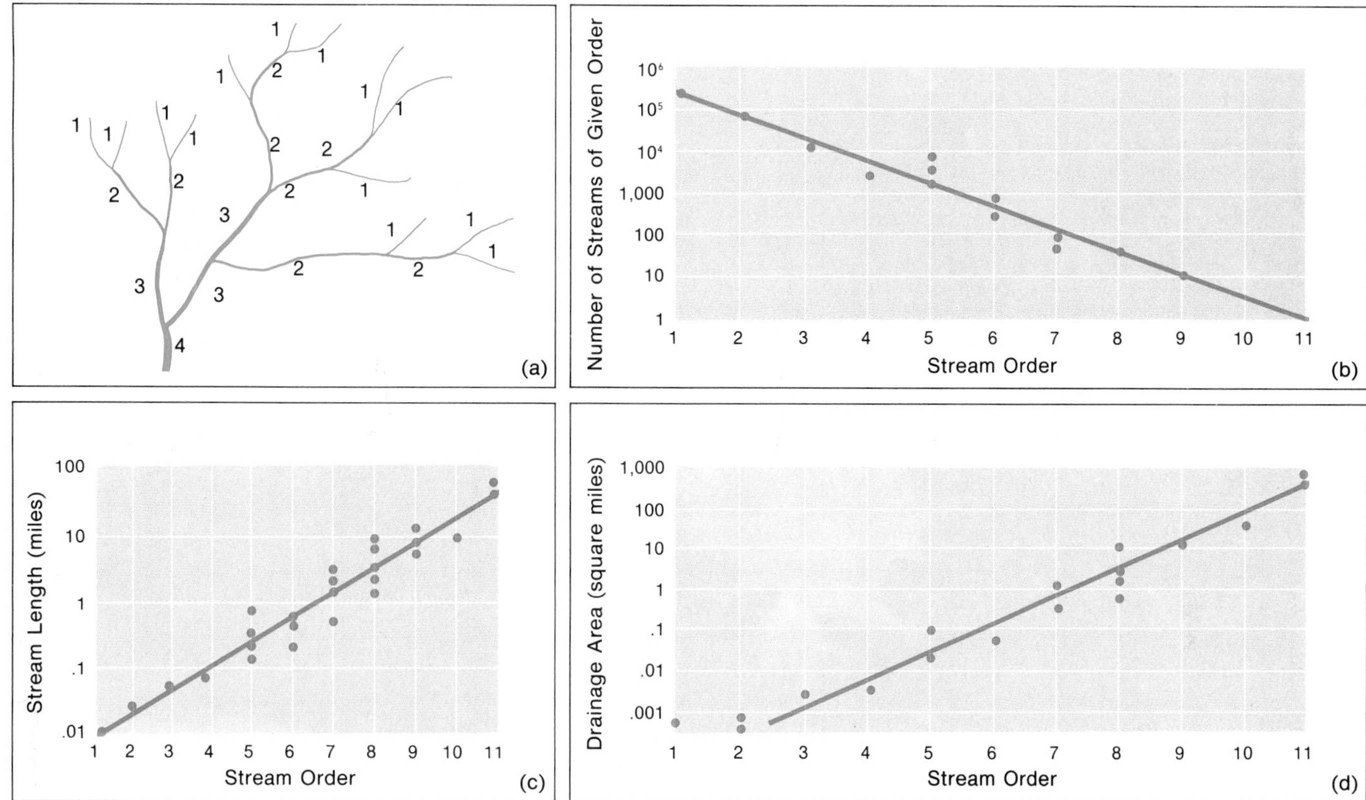
(a)
Number of Streams of Given Order
Stream Order
(b)
Stream Length (miles)
Stream Order
(c)
Drainage Area (square miles)
Stream Order
(d)

pattern of threads and rills to successively larger channels spread out in a coarser pattern. The water from the numerous rills flows into gullies and then is further concentrated into a few streams. In Wyoming water from rain or from the spring thaw may eventually drain into a large stream, such as the Platte River. Eventually the water will reach the mouth of the Mississippi River, thousands of kilometers from Wyoming.

The Platte River arises in the highlands of Colorado and Wyoming and descends from an altitude of more than 2,000 meters (6,500 feet) to an altitude of about 300 meters (1,000 feet) south of Omaha, Nebraska, where it feeds into the Missouri River. The Missouri joins the Mississippi River at St. Louis, where the altitude is only 150 meters (500 feet) above sea level. The slope of the Platte River in Nebraska corresponds to a drop of 1 meter in height for every kilometer in length (5 feet per mile), whereas the slope of the Mississippi at St. Louis is one-tenth as great. On the average, most river beds become less steep downstream over any long stretch of the river. For example, the average slope of the Mississippi River decreases by a factor of 3 over the last 1,500 kilometers (900 miles). Elevation of a river versus distance along the river is called the *longitudinal profile*. The structure of most streams on every scale from rill to major river shows a longitudinal profile that is concave upward, as Figure 12.10 indicates, although the profile steepens locally where streams pass over falls or hard rock.

Figure 12.9 (a) This diagram illustrates the system of stream ordering devised by the geomorphologist Arthur Strahler. A second-order stream arises at the junction of two first-order streams, for example, and remains second order until joined by another second-order stream. (b) (c) (d) These three graphs represent the Robert Horton analysis of a drainage basin near Santa Fe, New Mexico. The graphs show the regularities that are exhibited by the streams in a given basin according to their order. (b) The number of streams of given order decrease regularly with increased stream order; drainage patterns have numerous first-order streams that feed a comparatively few streams of higher order. (c) The length of streams increases regularly with increased order; small first-order streams are much shorter than streams of higher order. (d) The area drained by a stream of a given order is larger with higher-order streams; first-order streams drain only a small local area, whereas a stream of high order may drain an appreciable fraction of a continent. (After *Fluvial Processes in Geomorphology* by Luna B. Leopold, M. Gordon Wolman, and John P. Miller. W. H. Freeman and Company. Copyright © 1964)

Figure 12.10 The examples on these graphs illustrate the longitudinal profiles of eight streams varying in length from a few hundred feet to several thousand miles. The vertical scales have been exaggerated for clarity. In each case the stream tends to develop a longitudinal profile that is concave upward on the average, with the steepest portion upstream. There are local irregularities in the profiles, perhaps because of local differences in rock structure along the bed. Note that the downstream portions of the Rio Grande and the Nile River develop concave longitudinal profiles even though their discharges are effectively constant because of the lack of tributaries. (After *Fluvial Processes in Geomorphology* by Luna B. Leopold, M. Gordon Wolman, and John P. Miller. W. H. Freeman and Company. Copyright © 1964)

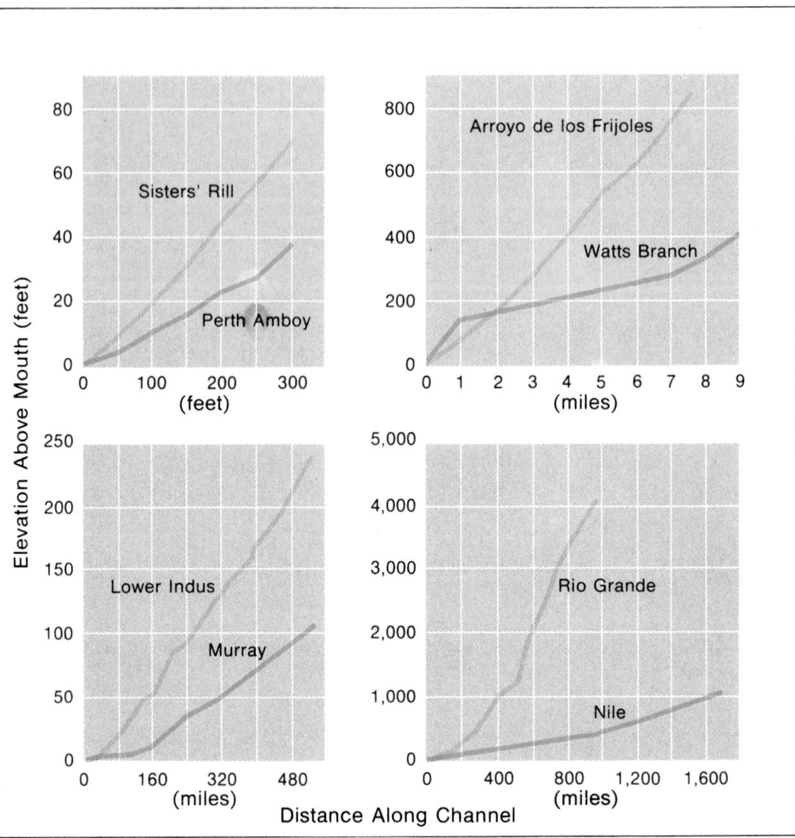
Elevation Above Mouth (feet)
Distance Along Channel
Sisters' Rill
Perth Amboy
(feet)
Arroyo de los Frijoles
Watts Branch
(miles)
Lower Indus
Murray
(miles)
Rio Grande
Nile
(miles)

Seen from above, a stream channel cut into soft material usually displays either a braided or a sinuous pattern. *Braided* sections mark the comparatively shallow channel of the Platte River where the main stream divides into several smaller, interwoven channels separated by bars of sand and gravel. Braided sections form on layers of easily eroded sand and gravel, and the shallow subsidiary channels frequently change course.

It is rare to find a stretch of river that is straight over any appreciable distance unless hard rock walls confine the river. A stream free to cut and fill its banks usually develops a remarkably convoluted and sinuous pattern of partial loops called *meanders*. The structure of both the Platte and the Mississippi, as well as of most other rivers that flow in large open valleys, includes large, well-developed, rapidly curving meanders. The meander pattern itself is not completely stable; new meanders develop and old meanders change their courses. The Mississippi has shortened its course by 800 kilometers (500 miles) in the last 200 years partly by cutting off old meanders. The development of new meanders lengthened the course by approximately the same amount, except in sections between Cairo, Illinois, and the mouth, where the United States Army Corps of Engineers has cut across some of the meanders, shortening and straightening the river considerably. Eliminating meanders makes navigation along the river easier and increases the

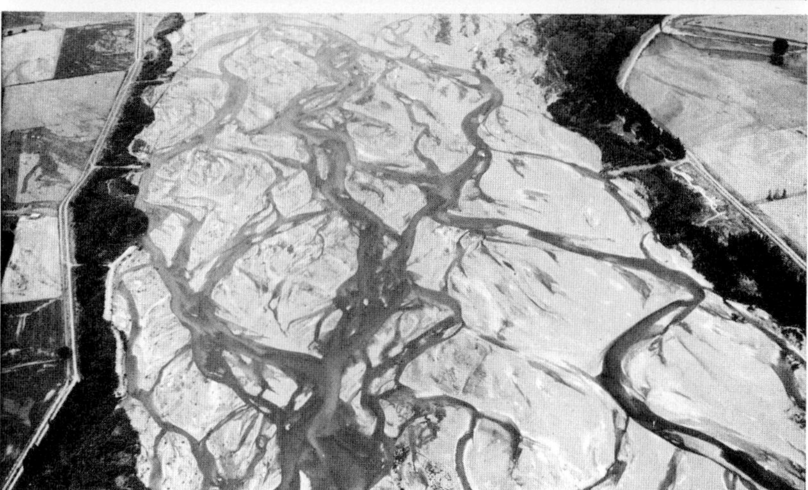

Figure 12.11 (top) The James River near Huron, South Dakota, follows a sinuous, meandering course. Note how the stream's course wanders from side to side, or laterally, across the low-relief floodplain.

(bottom) The Waimakariri River on the Canterbury Plains of New Zealand is a typical braided stream. The pattern changes with time as old channels become filled with sediment and as new channels are scoured out of the loose sand and gravel of the bed.

speed of flow by increasing the average slope of the river per unit section of stream.

Stream Channels: Form and Process

A stream is a complicated, dynamic system because it freely adjusts some of its structural features in response to inputs and to other factors. The input to a given section of a stream consists of the discharge and sediment load from farther upstream. The section responds to its inputs by adjusting its channel shape and size, its slope, the sinuosity of its course, the speed of its flow, and the roughness of its bed.

The discharge and sediment load presented to a section of a stream for transport downstream require the stream to do work. Exactly how a section of a particular stream will adjust to its dis-

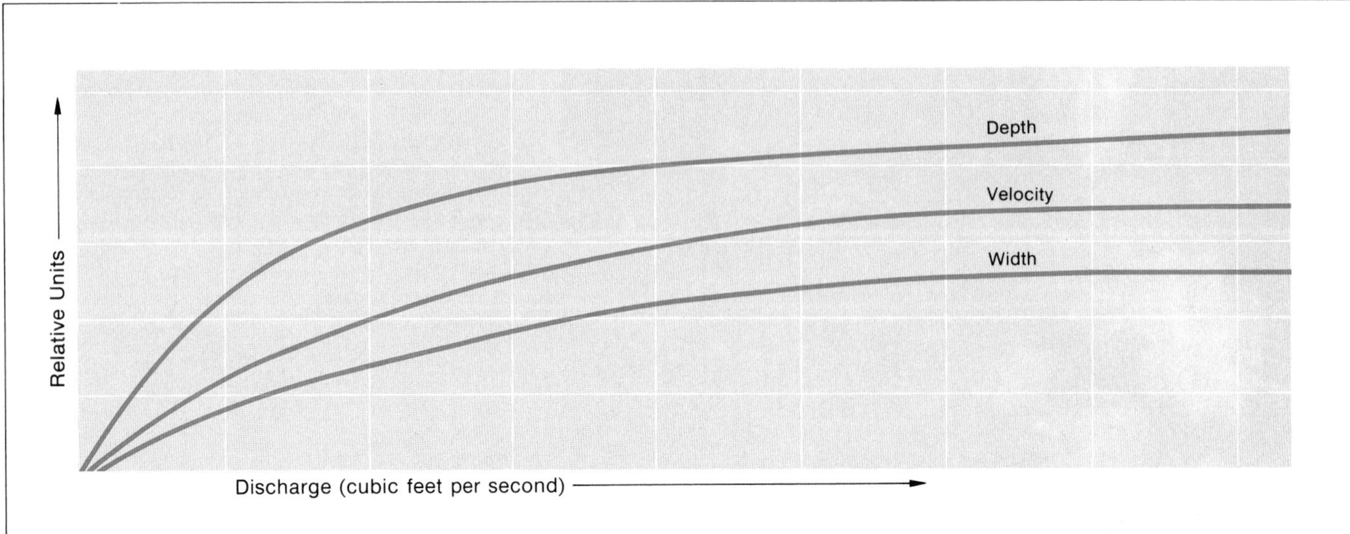
Relative Units
Discharge (cubic feet per second)
Depth
Velocity
Width

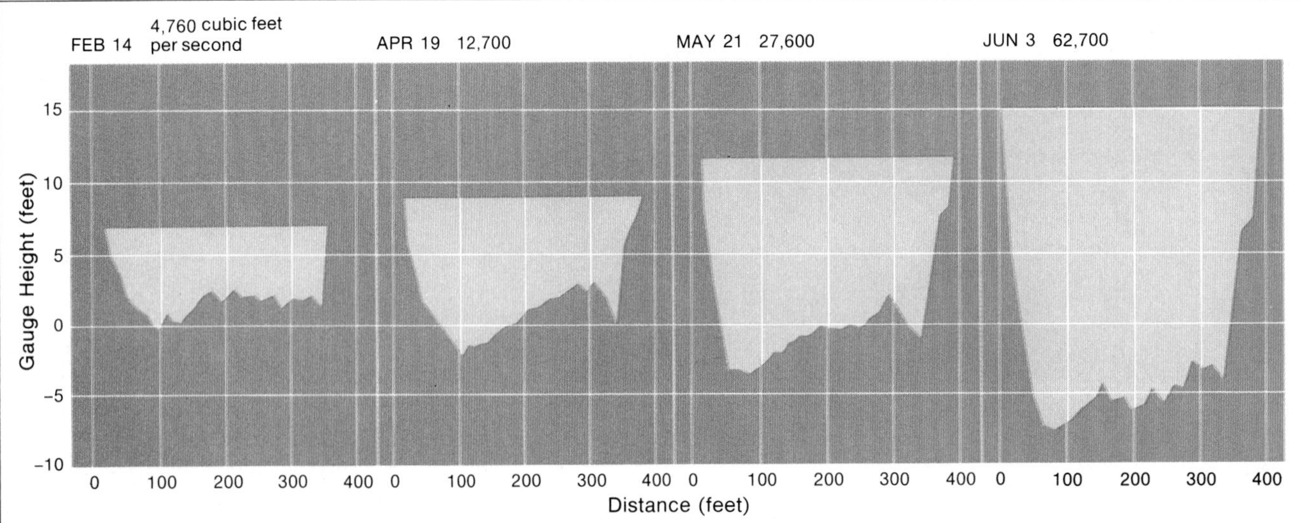
FEB 14 4,760 cubic feet per second
APR 19 12,700
MAY 21 27,600
JUN 3 62,700
Gauge Height (feet)
15
10
5
0
-5
-10
0 100 200 300 400
Distance (feet)

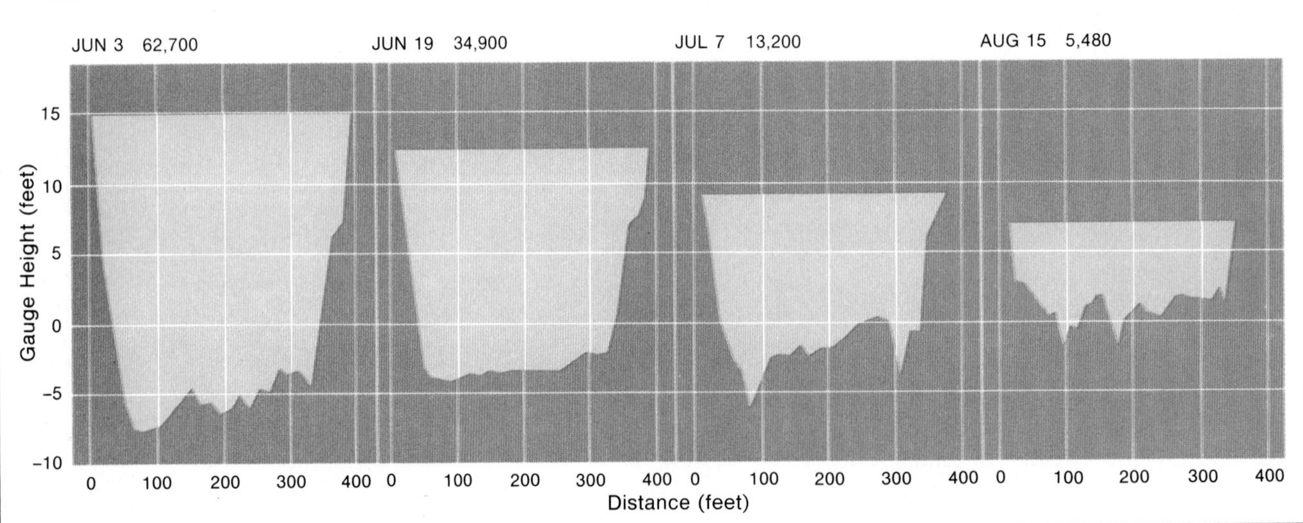
JUN 3 62,700
JUN 19 34,900
JUL 7 13,200
AUG 15 5,480
Gauge Height (feet)
15
10
5
0
-5
-10
0 100 200 300 400
Distance (feet)

charge and load is difficult to predict. The complex interrelationship of a stream's properties can be illustrated by the example of a stream in flood. Because each section of a stream is called upon to carry a greater discharge than it normally does, the velocity of flow increases greatly during flood. According to the schematic diagrams in Figure 12.12, an increased cross-sectional area and a larger velocity of flow characterize the equilibrium state of a stream that has an increased discharge. But at the beginning of a flood, the stream is not in equilibrium with its new discharge: its area is too small and its velocity of flow is too great. The high velocity of flow enables the water to scour the bed and banks of the stream, which enlarges the cross-sectional area. If the flood persists for some time, and if the bed of the stream is easily eroded, the stream will scour its channel until it reaches equilibrium under the conditions of increased discharge.

As the floodwaters subside and the discharge of the stream returns to normal, the stream is no longer in equilibrium: its area is too large for the given discharge, and its flow velocity is too low. The decreased velocity of flow reduces the stream's ability to carry its solid load, and the stream bed begins to fill with sediment. Eventually the stream returns to equilibrium by reducing the area of its channel to accommodate its flow. Figure 12.12 shows the cross section of a stream at different stages of a flood; the channel area increases during conditions of greater discharge and decreases when the flood subsides.

Stream characteristics are interrelated. As the example of a stream in flood indicates, several properties of a stream often readjust over a period of time in response to a change in discharge or load. The adjustments always occur in a direction that accommodates the altered conditions, and the stream tends toward a state of equilibrium. Streams incised in hard rock may require a long period of time to reach equilibrium, whereas streams in erodable material respond far more rapidly.

Vertical Adjustments

Long-continued scouring, or degradation, produces a landscape of canyons or valleys. Over shorter periods of time, such as during floods, a stream also scours or fills the land bordering its normal channel of flow.

Whether a stream takes up additional load or deposits part of its load depends on the size of the sediment particles and the velocity of flow in various parts of the stream. Scouring is likely to occur where the speed of flow increases. Where the flow is slower than normal, the amount of material transported as suspended load and bed load decreases. When a stream overflows its banks, friction with the ground causes the flow velocity of the shallow floodwaters to be significantly lower than the velocity in the main stream. Because their speed is checked, floodwaters deposit most of their load not far from the channel.

Artificial channels have been used in laboratory experiments to study how a stream's ability to transport material is related to the

Figure 12.12 (opposite) (a) This schematic diagram illustrates the tendency of the depth and width of a stream channel and the velocity of flow to increase with increased discharge at a given location along the stream. The curves refer to the equilibrium situation in which the discharge has continued long enough for the stream to complete readjustment by scouring or filling.

(b) (c) These channel cross sections for the Colorado River at Lees Ferry, Arizona, show the scouring (b) and subsequent filling (c) that occurred during a high-water period in 1956. Discharge is in units of cubic feet per second. The marked enlargement of the channel by scouring was accompanied by an increase in flow velocity to accommodate the greatly increased discharge. In August, when the discharge returned to almost its initial February value, the cross section of the channel also returned to almost its original size. (After *Fluvial Processes in Geomorphology* by Luna B. Leopold, M. Gordon Wolman, and John P. Miller. W. H. Freeman and Company. Copyright © 1964)

Figure 12.13 The goosenecks of the San Juan River in southeastern Utah are a striking example of the ability of a stream to continue downcutting on a meandering course as the land undergoes a relative uplift. The stream has cut a deep, narrow canyon in sedimentary rock while following a tightly curved meandering course.

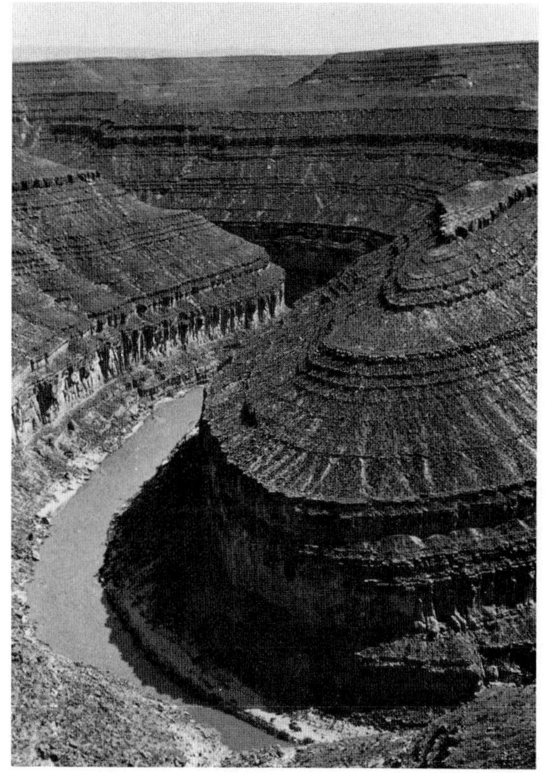

stream's discharge and channel characteristics. In general, the weight of load a stream is able to carry increases with increased discharge, increased cross-sectional area, increased slope, and increased smoothness of bed. In addition, a stream is able to carry a heavier load of fine particles than of coarse particles.

Slopes in Stream Beds

The average longitudinal profiles of several streams and rivers are shown in Figure 12.10. In each case the slope of the stream becomes flatter toward the mouth of the stream, so that the longitudinal profile is concave upward. There is no simple explanation for why a stream's profile tends toward concavity, but several factors are known to be important. One factor has to do with a stream's discharge. When a stream is part of a large drainage net, its discharge increases downstream because its tributaries contribute additional water. The tributaries also add to the load carried by the stream, but as the discharge of a stream increases, its efficiency in carrying a load increases as well. The stream is therefore able to decrease its slope downstream and still carry its load.

Even the Nile River exhibits a concave longitudinal profile, although its discharge remains essentially constant for thousands of kilometers as the Nile flows through the Egyptian desert. The shape of a stream's longitudinal profile seems to depend on the distribution of particle sizes in the stream's load as well as on the discharge. For many rivers, such as the Nile and the Mississippi, the particles in the solid load become finer downstream. Because a stream carries fine particles more efficiently than it carries coarse particles, it is plausible for its slope to decrease downstream; the stream needs less energy to transport a given load if the particles are fine.

Graded Streams

Scouring and filling processes change the cross-sectional area and gradient of a stream and tend to bring a stream into adjustment with the work it does in transporting water and load. In principle, a stream should eventually reach a state of equilibrium in which the gradient, flow, and area of cross section are adjusted to carry the flow of sediment and water delivered to the stream from the drainage basin. A stream in such condition, by definition a condition of equilibrium, has been termed a *graded stream*. However, a stream's discharge varies through the year because of variations in precipitation, and a true state of equilibrium is never maintained long. So a graded stream is in equilibrium in the sense that it adjusts itself to slight changes in conditions by scouring or filling for short periods of time; no net scouring or filling occurs on the average over a long period of time. The stream is adjusted to do just the amount of work required to transport the load presented to it.

Knickpoints

At a waterfall, cascade, or section of rapids, the slope of a stream steepens in an abrupt discontinuity, or *knickpoint*. Such knickpoints

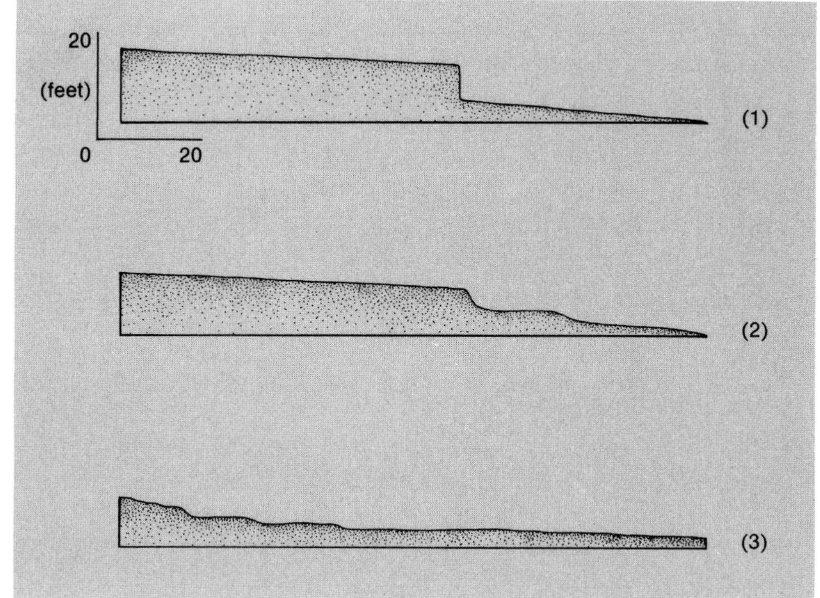
20
(feet)
0
20
(1)
(2)
(3)

Figure 12.14 Streams generally act in such a way as to reduce knickpoints and to attain a smooth longitudinal profile. The diagram shows a knickpoint (1) that was formed on Cabin Creek, Montana, by an earthquake. The stream began to scour its channel above the knickpoint and to fill below the knickpoint. Because the material of the channel was soft and easily cut, the knickpoint was significantly reduced in only a few months (2). Three years later (3) the profile showed no trace of the knickpoint. (After Morisawa, 1968)

are often associated with a change in the underlying geologic structure, such as a vertical displacement of a block by faulting or a change from a more resistant to a less resistant type of bedrock. At the immense knickpoint at Niagara Falls, for example, the bed of the Niagara River above the Falls is underlain by unusually resistant limestone that makes downcutting by the river very slow. Beneath the resistant limestone is less resistant shale, which is easily eroded at the exposed face of the Falls.

Knickpoints can also form in coastal regions where the sea level has lowered over a short period of time. A newly exposed coastal slope is usually steeper than the comparatively flat slope of a river in its lower reaches, and a discontinuity of slope, or a knickpoint, occurs near the original mouth of the river.

Some knickpoints erode upstream while maintaining their profile. Niagara Falls migrates upstream at an average rate of approximately 1 meter per year (3 feet per year) and has retreated 11 kilometers (7 miles) since its formation. In time, however, most knickpoints gradually undergo progressive flattening as the stream approaches a graded condition. The bed above the knickpoint tends to scour, and the bed below the knickpoint tends to fill. If the bed material is readily eroded, a knickpoint may disappear in only a few years (see Figure 12.14).

Lateral Migration of Stream Channels

Streams can adjust themselves laterally as well as vertically to achieve equilibrium. If the surrounding land is easily eroded, the bed of a stream may move a considerable distance as the stream adjusts itself. Figure 12.15 shows a section of the Mississippi River near the state boundary between Arkansas and Mississippi. Drawn in earlier times, the boundary line followed the channel of the river

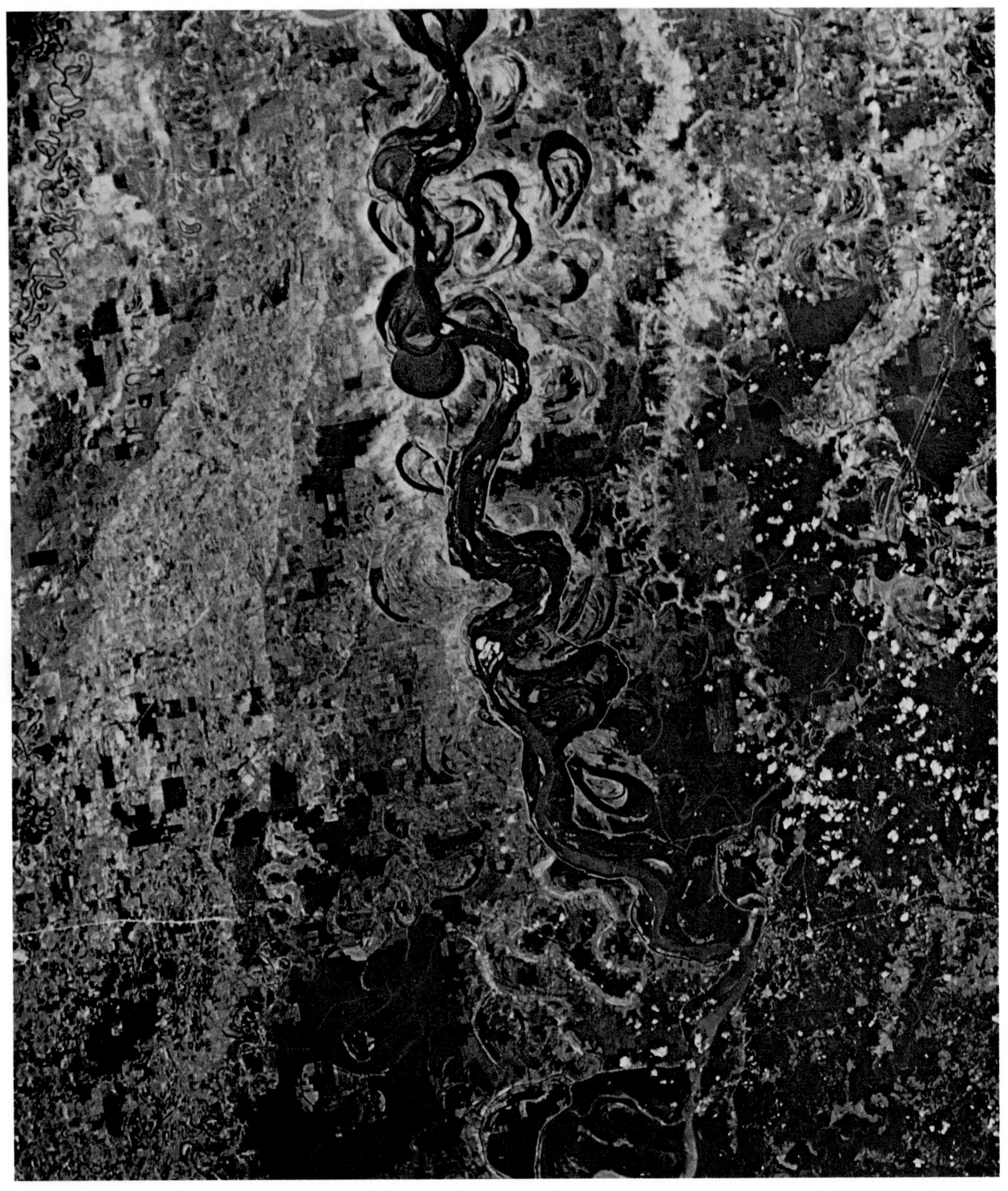

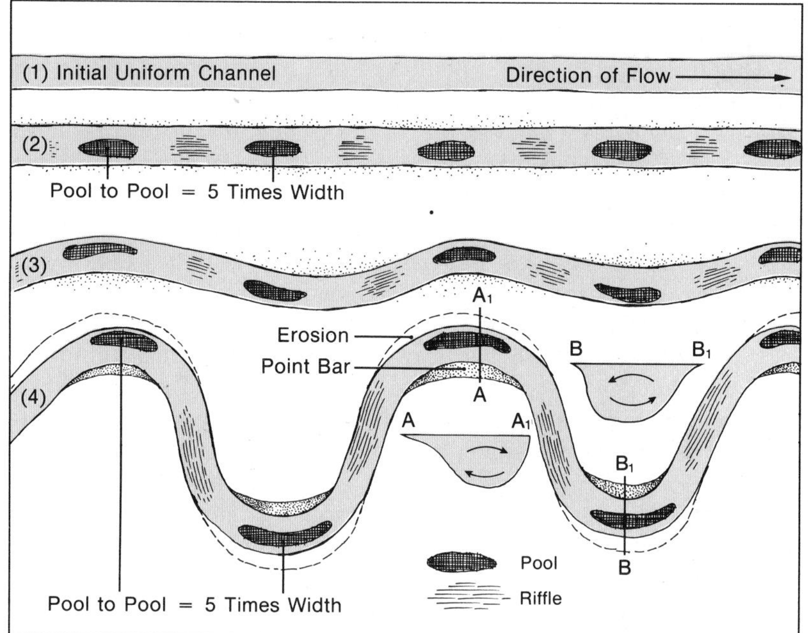
(1) Initial Uniform Channel
Direction of Flow
(2)
Pool to Pool = 5 Times Width
(3)
A_1
Erosion
Point Bar
A
B
B_1
(4)
A
A_1
B_1
Pool
B
Riffle
Pool to Pool = 5 Times Width

Figure 12.15 (opposite) This photograph spans a 100-mile section of the Mississippi River north of Vicksburg, Mississippi. It was taken from the Apollo 9 spacecraft in 1969 using color infrared film, which shows vegetation in shades of red (see Appendix II). The sinuous Mississippi has frequently changed channels; remnants of earlier channels are clearly visible. Boundaries between the state of Mississippi and the neighboring states of Louisiana and Arkansas were laid out in the nineteenth century to coincide with the channel of the Mississippi River. Because of subsequent changes in the channel, portions of each state now lie on both sides of the river.

Figure 12.16 (top left) Streamflow in a straight channel of uniform cross section (1) seems to be unstable, and if the channel is readily erodable, the stream tends to develop a sequence of alternate deep pools and shallow gravel bars, or riffles (2). The straight channel becomes more sinuous, with pools forming toward the outer concave banks (3). In time, a meandering channel may be developed (4). A sinuous or meandering stream tends to scour its outer concave banks and to fill its inner convex banks; the speed of flow is greatest near the concave banks and least near the convex banks. There is also a lateral flow of water, as the cross section diagrams indicate. The lateral flow scours a meandering channel so that the deepest part lies near the concave banks, forming a pattern of pools analogous to the pattern shown in (2). (After Dury, 1969)

as it was then. Major changes in the position of the channel have obviously taken place since that time.

The channel of a stream is often specified by the line, called the *thalweg*, that follows the deepest part of the stream's cross section. Stream channels are seldom straight and uniform for any appreciable distance, and the thalweg usually sweeps back and forth across the stream bed.

As viewed from above, stream channels exhibit three main types of channel patterns: meandering, braided, and "straight," or nonmeandering. Meanders occur in streams of all sizes, from a tiny channel in sand to the kilometer-wide Mississippi River. Except for differences in overall scale, meander curves are strikingly similar on all streams; the length of a meander on a stream of any size is 7 to 10 times the width of the stream. The development of meanders appears to be normal behavior for streams, particularly for those flowing in homogeneous deposits of easily eroded sediment. Meandering is so characteristic of streamflow that water flowing across the surface of packed sand in a laboratory soon excavates a meandering path for itself. Meandering flows are common in nature. They occur even in the Gulf Stream, where there are no stream banks or solid load but only boundaries between warm and cool water; it is not clear that the mechanisms of meandering are the same in all cases, however.

One clue to understanding the mechanism that generates meanders is that even streams without well-developed meanders often show a succession of deep pools separated by shallow bars, or *riffles*. Like meanders, the scale of pool and riffle spacing is proportional to the width of the stream; the distance from one pool to the next is typically 5 to 7 times the width of the stream. Furthermore, successive riffles tend to be located alternately near one bank

Figure 12.17 (below) The diagram shows how the wavelength, amplitude, and radius of a meander are defined. The radius around a meander is not constant, however, because the form of a meander is more complex than a simple circular arc. (After Dury, 1969)

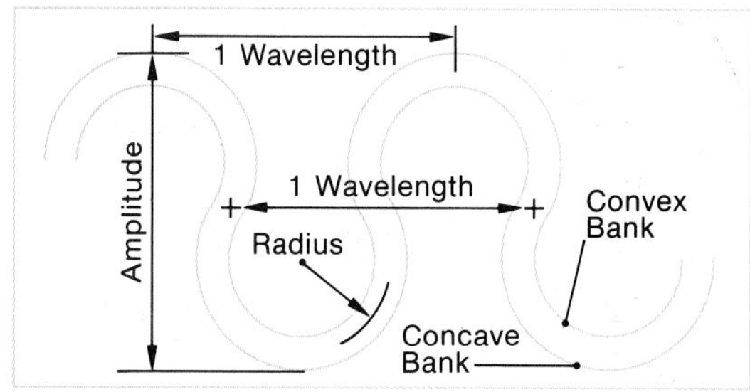
1 Wavelength
Amplitude
1 Wavelength
Radius
Convex
Bank
Concave
Bank

of the stream and the other, in a pattern suggestive of meanders.

The flow velocity in a meander is maximum along the thalweg, which follows the outer bank of each curve. Scouring occurs at the concave banks nearest the thalweg, and filling occurs at the convex banks on the other side of the stream, enabling the shape and position of the meander to change with time.

Another example of lateral migration is the development of braided channels, which are associated with the growth of bars in a broad, shallow stream that is overloaded with gravel or sand. The development of bars tends to produce separate shallow but steeply sloping channels in which the water pushes vigorously along the bed, helping the stream adjust to a large bed load of debris. The bars and channels of a braided stream usually undergo continual change because of sediment transport. Scouring and filling in some of the channels may result in the development of a pattern of meanders. The overall pattern of a braided stream usually persists with only slow modifications, and braiding appears to be a stable way for a stream to adjust laterally to certain conditions of load and bed material.

Landforms Associated With Rivers

Flowing water is the paramount force shaping most of the landscape, but certain landforms, such as floodplains, river terraces, and deltas, are specifically associated with the work of streams.

Floodplains

A *floodplain*, the flat plain that borders a wide river channel, is built up from sediment deposited by the river. The sediment deposits, renewed by periodic floods, usually make fertile farmland. Floodplain deposits originate partly from sediment deposition on the inner curves of meanders. Sometimes one loop of a meander becomes so curved that it closes at the neck. The stream then flows mainly across the neck, abandoning its former channel around the meander. Deposition of sediment in the abandoned loop closes off the old meander from the stream, which forms an isolated *oxbow lake* that gradually fills in with fine sediment and vegetation.

Some rivers, such as the Colorado River in the Grand Canyon, cut primarily vertically and develop a narrow valley with no floodplain. Other rivers migrate laterally and develop moderately wide valley floors. The vast lowland of the Mississippi, Missouri, and Ohio river valleys also received large quantities of debris carried by meltwater from glaciers during the recent ice age. In some places the fill is 50 meters deep (160 feet deep).

The timing and amount of precipitation on a river's drainage basin sometimes result in a discharge too great for a river to carry within its banks, and a flood occurs. The floodwaters spill over the river banks onto the valley floor and deposit sediment over part of the floodplain. Floodwaters tend to deposit more material on floodplains than they remove, particularly where vegetation is present to help hold the soil. When a river floods its banks, floodwaters

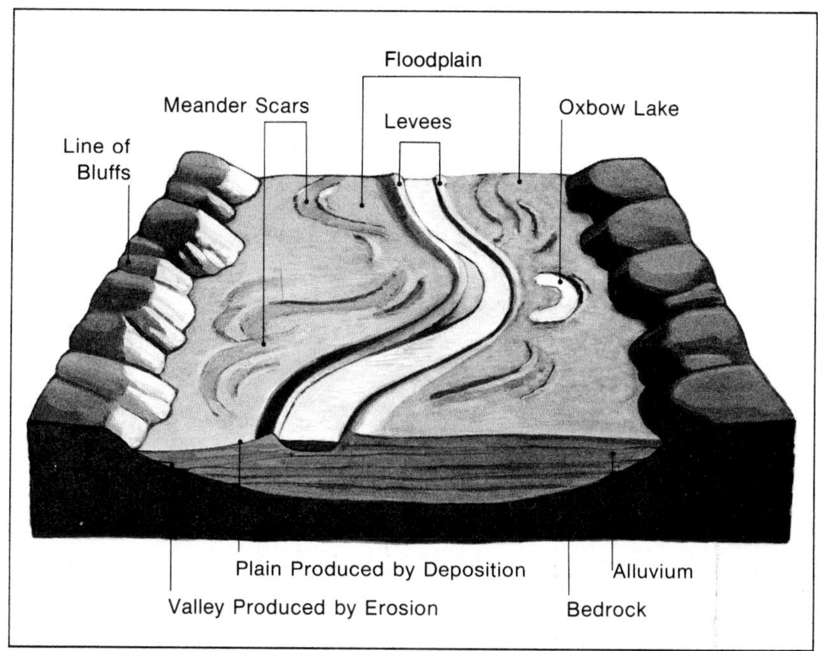
Floodplain
Meander Scars
Levees
Oxbow Lake
Line of Bluffs
Plain Produced by Deposition
Alluvium
Valley Produced by Erosion
Bedrock

Figure 12.18 (top) In maturity, a stream attains a graded condition and cuts laterally rather than downward. Its valley becomes covered with a deep deposit of alluvium, forming a floodplain. The floodplain is widened and planed off as the channel of the meandering stream gradually migrates from side to side across the valley. Bluffs on either side of the valley become cut back. *Oxbow lakes* are the remnants of recently abandoned meanders. When the lake is eventually filled with sediment and vegetation, a *meander scar* remains. In times of flood, the overflowing river carries new sediment onto its floodplain and builds raised banks, or *natural levees* by deposition of sand and silt close to the river channel. (After Bunnett, 1968)

(bottom) The sinuous Animas River near Durango, Colorado, has cut a floodplain much wider than the width of the stream by lateral migration. Meander scars marking earlier channels are clearly visible.

move out of the channel, lose speed, and drop their largest particles of sediment, which builds up the banks into *natural levees*. When a flood overtops its levees, water is trapped between the natural levees and the valley walls.

Terraces

Suppose the Mississippi River suddenly began to cut vertically as vigorously as the Colorado River cuts its banks, perhaps because of an increased discharge from upstream. As the river cut deeper into its floodplain, the remaining floodplain would eventually form *terraces* above the level of the river (see Figure 12.19). The presence of alluvial terraces along the channel of a river indicates that for some geologic, climatic, or hydrologic reason the river was required to seek a new equilibrium profile sometime in the past. The process of filling and cutting may be repeated many times, forming "steps" of terraces on the slopes of a river valley. Part of the geologic history of a river valley can be inferred from a study of its terraces. An episode of downcutting that begins on a level floodplain, for example, produces paired terraces of equal height on both sides of the river channel.

Underfit Streams

Sometimes a small stream flows in a valley that appears to be much too large for a stream of its size and discharge. Several mechanisms may cause such *underfit* streams. Melting glaciers at the end of an ice age provide enough meltwater to produce many large streams, which cut valleys in sizes appropriate to their discharges. But after the glaciers disappear, and when climate changes significantly, precipitation may no longer provide the runoff that had been carried by streams during the glacial periods. So the streams are left with much smaller discharges. Underfit streams may also be formed if,

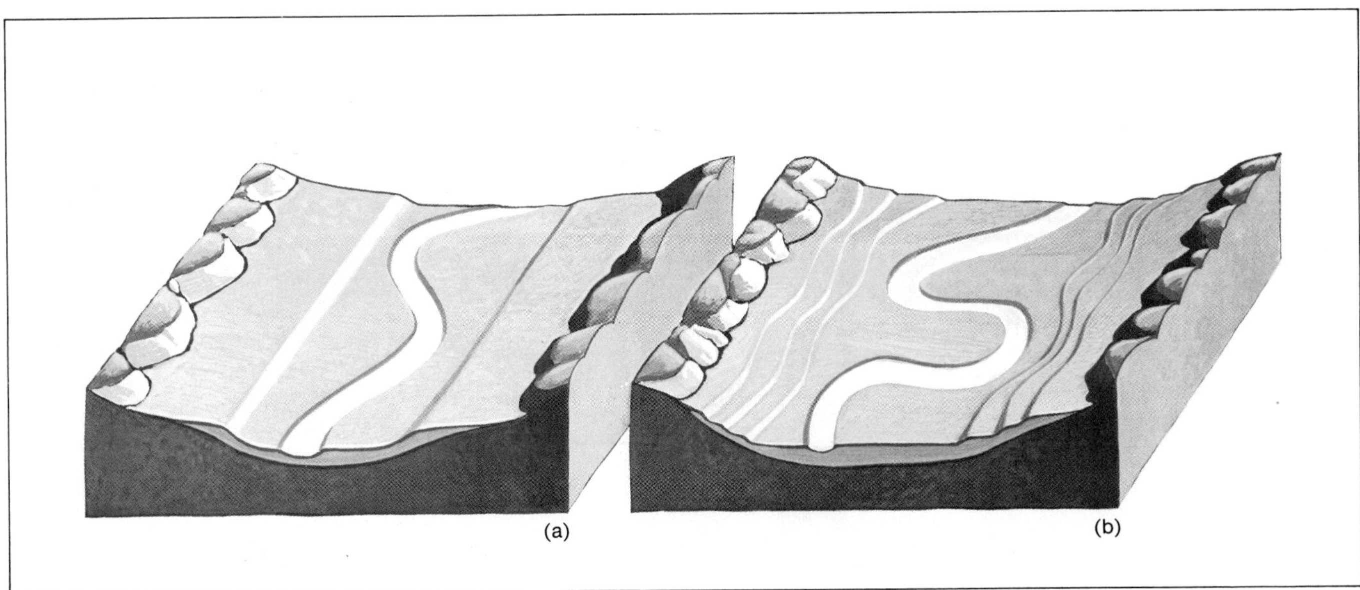

(a)
(b)

Figure 12.19(a) Symmetric, *paired alluvial terraces* are cut in opposite sides of a floodplain when a stream is able to meander back and forth across its valley while cutting slowly downward.

(b) Unsymmetric, *unpaired alluvial terraces* are formed when a stream cuts downward rapidly compared to the rate at which it meanders across its valley floor. Each terrace is cut separately while the stream remains on that side of its valley.

for some reason, a portion of the water flowing into a stream is diverted to another course, which diminishes the stream's discharge.

Deltas

A notable feature of sediment-carrying rivers is the deposit of sediment at their mouths, forming a *delta*. When a river such as the Nile or the Mississippi reaches the sea, the velocity of flow lessens significantly and the river drops most of its load. Whether a delta is built at a river mouth depends on the balance between the river's deposition of sediment and the sea's erosion of that sediment. Also, some deltas, such as that of the Mississippi, slowly sink under the weight of deposited sediment.

Figure 12.20 shows the deltas of two major rivers. Each delta has its own characteristic shape, determined largely by the pattern of offshore currents, which carry material away from the delta and into the ocean. When a river enters its delta, its discharge often becomes divided among several *distributary* stream channels. (Distributaries flow *out* of a stream, whereas tributaries flow *into* the stream.) The channels of distributaries often migrate laterally by local scouring and filling. Like floodplains, the lowland of a delta is fertile, so deltas are often densely populated. Loss of life can be heavy during floods, especially if distributaries change courses.

Alluvial Fans

Alluvial fans are the equivalent on land of deltas. When a stream flows from a steep mountain canyon onto a plain, the gradient often flattens abruptly, the water spreads out, and some of it infiltrates the ground. The decrease in depth and discharge causes the stream to deposit sediment at the base of the mountain, building fan-shaped deposits known as alluvial fans. Adjacent fans sometimes join laterally to form an alluvial apron. Sometimes sediment from an ex-

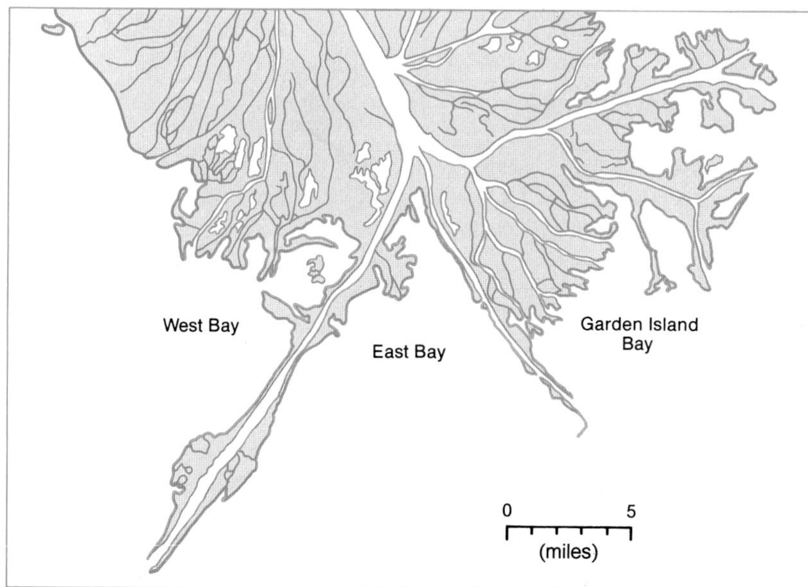
West Bay
East Bay
Garden Island Bay
0
5
(miles)

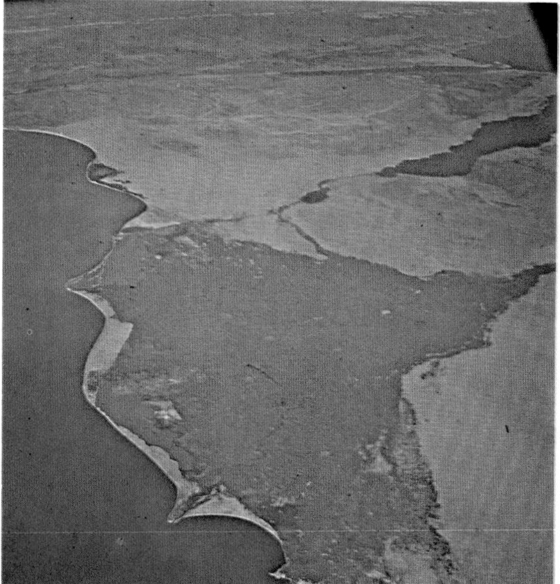

Figure 12.20 (bottom left) The delta of the Mississippi River in Louisiana resembles a bird's foot because of the way the distributary streams build out talons, or digits, of sediment deposits.

(below) The fan-shaped delta of the Nile River in Egypt resembles the triangular shape of the Greek letter "delta," which accounts for the origin of the term. The delta of the Nile has a different shape from the delta of the Mississippi River because of different patterns of sediment transport, stream deposition, and erosion by currents of the Mediterranean Sea. The construction of the Aswan Dam has changed the sedimentation pattern of the Nile, and its delta may begin to be eroded.

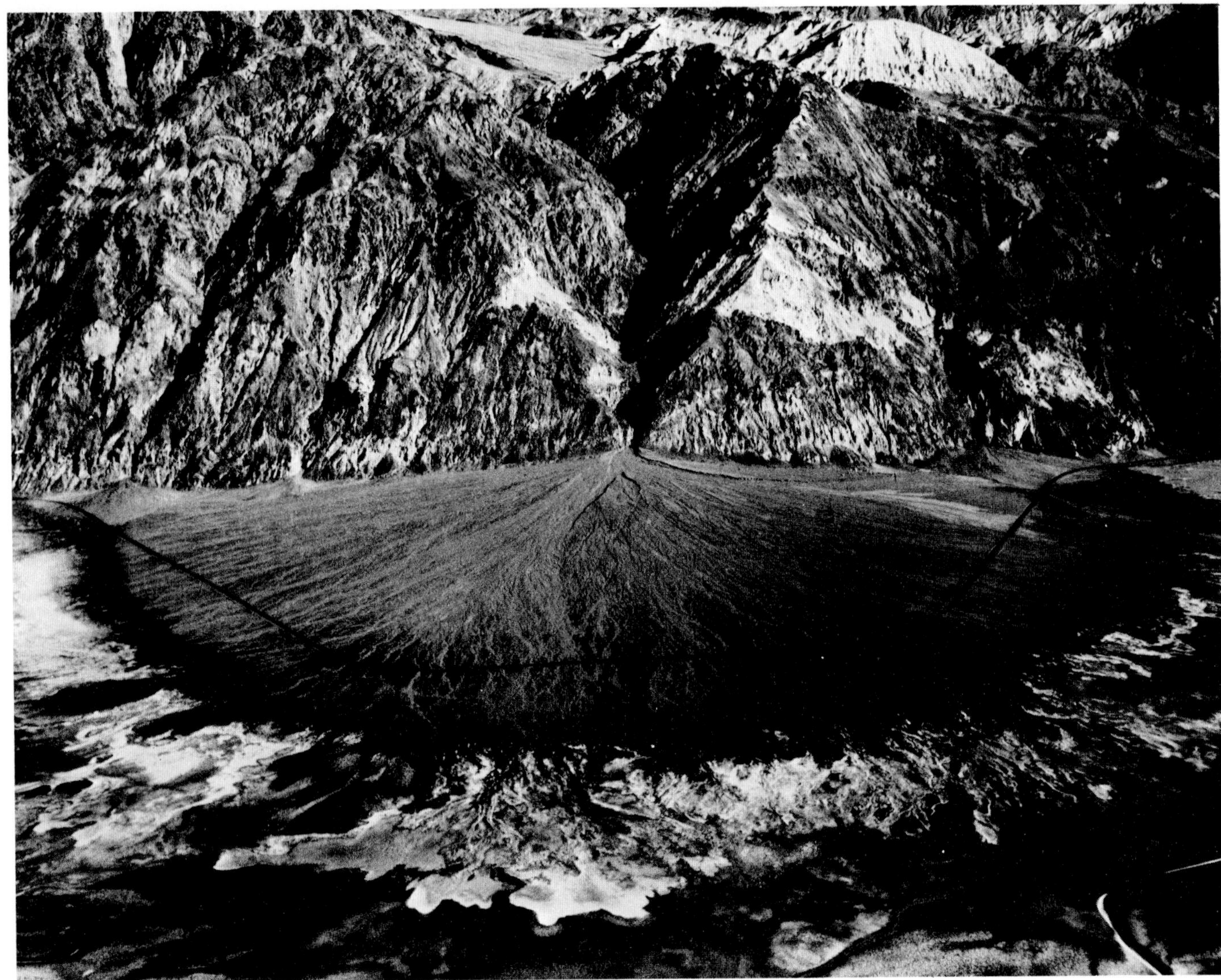

Figure 12.21 This alluvial fan is located on the east side of Death Valley, California; its size can be judged from the road crossing it. Rains are infrequent in this region, but torrential storms occur in the mountains from time to time, carrying down large quantities of coarse sediment to be deposited on the fan.

tensive mountain range covers a plain several kilometers outward from the alluvial fans. Salt Lake City, Utah, is built on such an alluvial plain. Water easily infiltrates the loose, coarse sediment of the alluvial fans, so that the alluvial plain below often has an excellent supply of groundwater.

Summary

Surface runoff and groundwater flow are supplied by surplus water that remains from precipitation after evapotranspiration and soil moisture needs are met. Local flows of surface runoff join progressively larger and fewer streams to form a drainage net.

Flowing water gains the ability to do work as it flows downhill. Some of the energy of flowing water is used to overcome friction with the banks and bed of a stream channel and some is dissipated as heat in turbulent flow. Additional energy can be used to transport material in suspension, in solution, or along the stream's bed. The greater the stream's velocity of flow, the larger the particles it can transport.

Streams can change the form and position of their channels by their ability to scour and to fill. The form of a stream depends on the interaction of many variables, such as the discharge, amount of load, particle-size distribution in the load, width and depth of the channel, velocity of flow, and bed roughness. The longitudinal profile of a stream usually tends toward a concave upward shape, steeper upstream and flatter toward the mouth. The lateral displacement of streams in homogeneous alluvial deposits frequently generates meanders. Streams build characteristic landforms by deposition of sediment. Floodplains, deltas, and alluvial fans are some of the principal landforms created by the work of streams.

Case Study: The Flood of '73

From time to time man has to deal with natural disasters. Earthquakes, volcanoes, floods, droughts, hurricanes, and tornadoes can strike unexpectedly, often wreaking havoc on lives and property. But often people court calamity by building homes, planting crops, and erecting cities in problematic locations; for example, by building on a floodplain.

Floods have brought bounty as well as tragedy to people through the years: the yearly flooding of the Nile River was necessary to replenish the fertile soil along its banks. But generally, precipitation and consequent flooding are unpredictable from year to year, so attempts are made to control the flow and distribution of river waters with dams, levees, and canals. Even with such controls, nature occasionally surprises us.

One of the greatest flood impacts of recent years occurred in 1973 in the Mississippi Valley. Residents along the river knew high waters were coming months before they came. The question was how high the water would rise.

Historically, floods of the Mississippi River occurred every few years and the best way to avoid them was to move to high ground. But as the valley became more settled, people were no longer content to adjust their affairs to the natural workings of the river. So the goal became to eliminate nearly all floods.

The lower Mississippi River system and adjacent river basins are shown on this map. (After Muller, 1974)

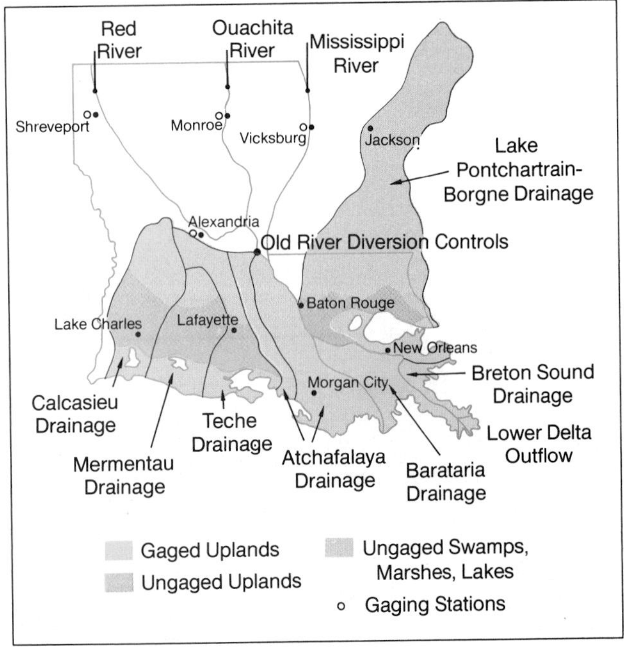

Red River
Ouachita River
Mississippi River
Shreveport
Monroe
Vicksburg
Jackson
Lake Pontchartrain-Borgne Drainage
Alexandria
Old River Diversion Controls
Baton Rouge
Lake Charles
Lafayette
New Orleans
Morgan City
Breton Sound Drainage
Calcasieu Drainage
Teche Drainage
Lower Delta Outflow
Mermentau Drainage
Atchafalaya Drainage
Barataria Drainage
Gaged Uplands
Ungaged Uplands
Ungaged Swamps, Marshes, Lakes
Gaging Stations

Until recently man's principal defense against floods has been the construction of artificial levees to augment the natural levees that the river deposits during high water. However, confining the river within its normal channel, and out of its normal floodplain during peak discharge, raises its level considerably. A break in the levee under those conditions can produce a flood of disastrous proportions. In the 1930s the opening of the Bonnet Carré spillway just north of New Orleans, Louisiana, inaugurated a new approach to flood control on the Mississippi. The spillway is a controllable crevasse 3 kilometers wide and 7 kilometers long leading from the Mississippi above New Orleans to the lower level of Lake Pontchartrain. During periods of great discharge, river water can be diverted to the lake through the spillway, which eases the pressure on the levees downriver at New Orleans and in the lower delta. The Bonnet Carré spillway has been used four times since the 1930s, successfully averting floods each time.

In the 1950s the Old River Diversion was built to prevent the capture of the Mississippi water by the nearby Atchafalaya River. Until it was built, some of the high waters during flood conditions were naturally diverted from the Mississippi into the Atchafalaya. Other spillways, such as the Morganza, were also built to control the flow of the Mississippi during flood conditions. During the flood of 1973, the Old River Diversion was weakened, so the Morganza spillway was opened for the first time.

The flood story of 1973 actually began in late 1972. Rainfall had been particularly high in the central Mississippi Valley during the summer and fall. Farther south rainfall became heavy in the late fall and early winter. By winter the lower Mississippi was unusually high and there was already talk of possible spillway openings. Spectacular deluges occurred during the spring, particularly in the south, and excessive rain continued to fall over the entire basin. The water came dangerously close to the tops of levees. In February the Bonnet Carré spillway was opened, and in mid-April plans were made to open the Morganza spillway to release water into the Atchafalaya basin to reduce pressure on the Old River Diversion upstream.

There is no perfect flood control system: relieving pressure in one area means adding pressure to another. Opening the Morganza spillway appeared to be the way to protect the greatest number of lives and property, but taking the pressure off Old River and New Orleans meant more water downstream in the Atchafalaya River. Several hundred people as well as livestock were evacuated from the Morgan City area in the Atchafalaya basin. Oil and gas wells were flooded and their production was stopped, at least temporarily. Farmland was under water, so crops of cotton and corn could not be planted that spring. However, varieties of late season soybeans were planted instead.

Another side effect of the flooding was the breakage of sewer lines, which resulted in sanitation problems. The rising water also caused docking difficulties for ships in port cities.

There was also concern in the Atchafalaya basin that sediment and silt-laden waters from opening the Morganza spillway might fill in the shallow lakes and swamps. Short-term adverse effects were also felt by the crawfish industry in the Atchafalaya area.

Opening the Bonnet Carré spillway affected the ecology of Lake Pontchartrain, for a time at least, by adding sediment and fresh water that changed the salinity and damaged the oyster population. However, there was some speculation that over time, the ecology of the lake would be improved by the flushing action of the water, because it washed out debris and water-clogging vegetation and brought in a fresh supply of nutrients.

The opening of spillways such as the Morganza and the Bonnet Carré prevented the main river from flooding, so that flooding occurred only on the Mississippi's tributaries. Because water in the mainstream was so high, the other rivers were unable to flow into it. So they overflowed their banks locally, forming "inland seas" of backwater flooding in the basins and valleys.

Backwater flooding occurred mainly in Mississippi and Louisiana. Farther north the flooding was worse; the heaviest flooding occurred around St. Louis, Missouri. In one spot—Kaskaskia Island in Illinois—the main river overtopped the levees. In some places levees broke or washed out and the water poured through them; in other places the levees held, but were not high enough, so the water cascaded over them. Near St. Louis, people fled their homes, moved back to their sodden, muddy houses when the rivers dropped, and then had to leave again, and yet again, as new crests came down the river.

The rains and the flooding finally stopped in early summer. The flooding had lasted months, and four separate flood crests had flowed down the Mississippi. The long-term results of the 1973 flood were difficult to assess. Parts of nine states—Kansas, Iowa, Illinois, Missouri, Kentucky, Tennessee, Arkansas, Mississippi, and Louisiana—had been declared disaster areas. At least two dozen people were killed by the floods and about 35,000 were evacuated from their homes. In Louisiana alone about 3.7 million acres of land were under water, and damages and losses in crops and livestock reached an estimated 36 million dollars.*

Because the waters subsided so slowly, it was too late that season to plant a cotton crop in the fertile delta area of Mississippi. Soybean, corn, and sugar cane crops were also affected because planting was delayed. The heavy losses in food crops and livestock led to higher food prices throughout the country. Thousands of homes had to be rebuilt, roads and levees had to be repaired, and massive clean-up projects were undertaken.

During the flood, emotions ran high as various means to divert and control the overabundant water were instigated. The oyster fishermen of Louisiana didn't want the spillways opened. The people who had homes in the floodplains of backwater areas didn't want their regions inundated with spillover from the main river. Concern was expressed over the deer and other wildlife who were not able to reach higher ground, who returned blindly to their flooded territories, or who became the easy victims of poachers.

But even with the great losses, many engineers believed that the flood control measures proved as effective as could be expected. The levees and flood walls prevented the Mississippi River itself from flooding except in the one place and water was kept out of the major cities. Most of the flood warnings came early enough that people could be evacuated, and loss of life attributed to the flood was relatively small. After the flood, study of the positive and negative effects of flood control systems continued.

*After *U.S. News and World Report* (May 14, 1973)

The Old River Diversion, in the center of the photograph, was threatened by the flood but did not give way. The Mississippi, on the right, is at its maximum water level and is flowing into the Atchafalaya, on the left.

The city of Jonesville, Louisiana, is surrounded by backwater flooding, but the city itself was not flooded. Surrounding farmland was under water for as long as two months in some areas. All along the river valley, rural areas tended to be flooded, whereas cities were protected.

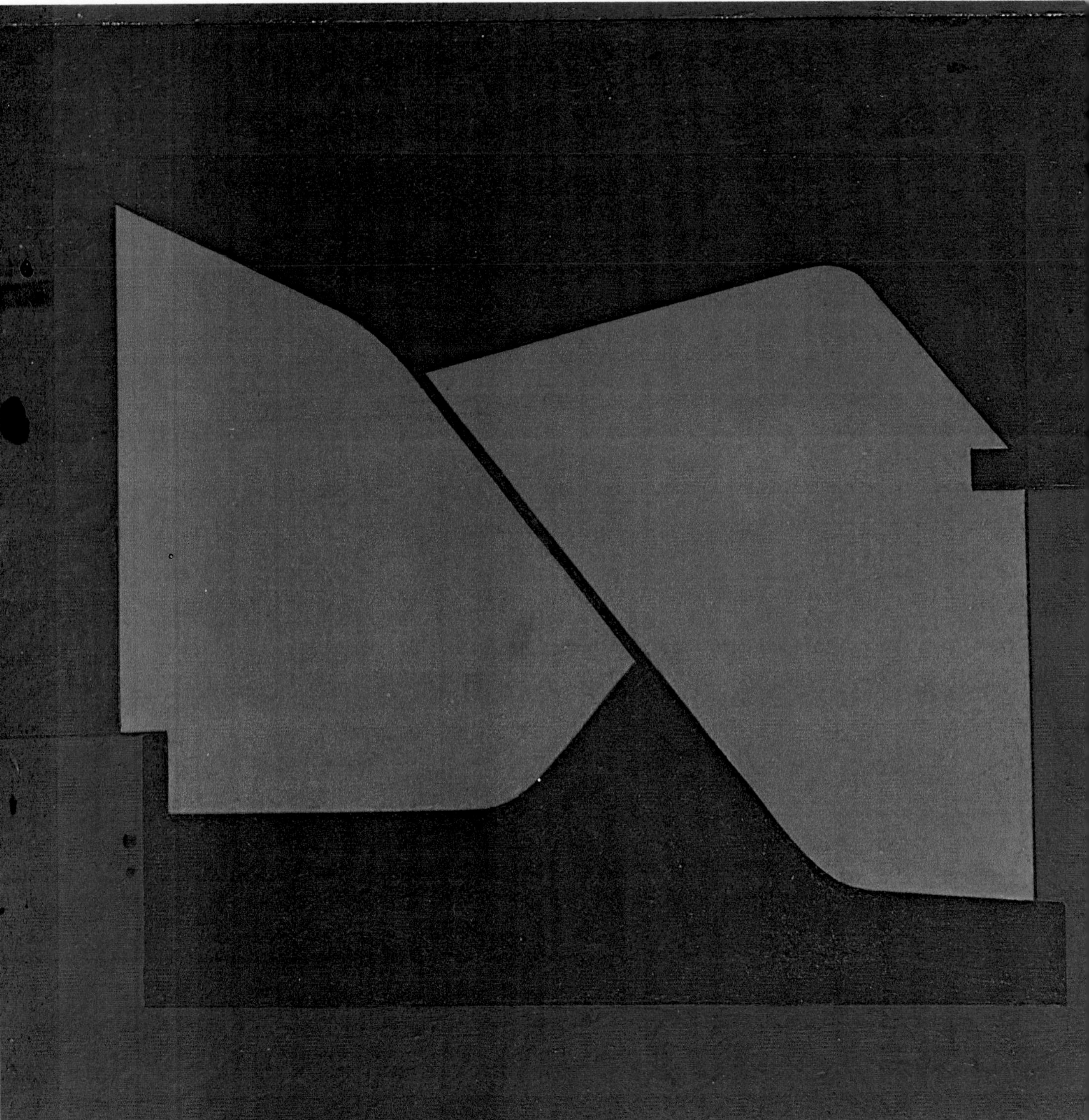

13 Landforms and the Shifting Crust

The Structural Foundation
Igneous Rock
Sedimentary Rock
Metamorphic Rock

Differentially Eroded Rock Structures

Volcanism: The Production of New Igneous Rock
Geothermal Energy
The Source of Volcanic Energy
Ejected Materials and Volcanic Hazards

Landforms Associated With Crustal Motion
Faults and Folds
Landforms and Faulted Structures
Landforms and Folded Sedimentary Rocks

The land may be altered abruptly by volcanic eruption or by a sudden earthquake. Both dramatic processes of change are powered by the earth's internal energy. Landforms produced by movements of the earth's crust are subject to continued change by the processes of weathering and erosion, which are powered by external energy.

Kandahar by Victor Vasarely, 1951.

Landforms and the Shifting Crust

In Italy it is against the law to divert lava flows. The prohibition dates from an incident in the year 1669, when a stream of red-hot lava from Mount Etna in Sicily crept toward the village of Catania. Men from Catania, dressed in wet leather to ward off the heat, used iron bars to open a breach on the flanks of the seething stream and divert the flow. Their efforts were successful; the lava flow stopped its advance. Unfortunately, however, the side flow began to creep toward the village of Paterno. When the outraged villagers of Paterno drove the men of Catania away from the lava, the lava resumed its flow toward Catania and a large part of the village was destroyed. Remnants of the lava can still be seen in the streets—testimony, perhaps, to the vulnerability of man's schemes in the face of the indifferent forces of nature.

The earth's crust continually adjusts itself to the forces acting upon it. With the exception of a sudden volcanic eruption or a jolting earthquake, these adjustments are usually slow and difficult to perceive without the use of scientific instruments. Occasionally, however, slow movements can be detected in unexpected ways. A winery near Hollister, California, is built directly over an active branch of the San Andreas fault. The rocks on either side of the fault are slowly moving horizontally relative to each other, and the movement is gradually fracturing the walls of the winery, which serve to record the displacement. The slippage is not uniform; on the average it amounts to a centimeter or so per year. Walls in the winery have been broken, and the displacement of the ground has offset rows of vineyards. Other parts of the San Andreas fault system, which extends nearly the length of California, do not show steady movement of this magnitude.

Over vast spans of geologic time, such small and uneven motions can bring about large changes in the earth's crust. For example, many small, upward displacements, each on the order of a few meters, eventually raised marine rock masses in Asia to heights of 10 kilometers (6 miles) or more above sea level; today those rock masses are known as the Himalaya Mountains. The crust can twist, rupture, fold, crack, or move vertically as well as horizontally, and all of these distortions are eventually expressed in contours of the land. One of the fundamental ideas of geomorphology is that underlying geologic structures exert a strong controlling influence on the shape of landforms.

Processes of erosion and weathering work on the raw features of such primary deformations. In regions where small crustal motions have occurred over long periods of time, the appearance of deformed structures has gradually been softened by erosional forces. Consider the vertical displacement of a crustal block, which generates a sheer rock face: over time, flowing water may incise it and create numerous gentle slopes. Or consider the massive sandstones, laid down over the ages: differential erosion by wind and rain may

Figure 13.2 (opposite) The layers of sedimentary rock visible in this photograph of the Canadian Rocky Mountains have been subjected to intense folding and deformation by movements of the earth's crust. The once-horizontal sedimentary rock strata have been compressed and distorted by stresses caused by motion of crustal plates.

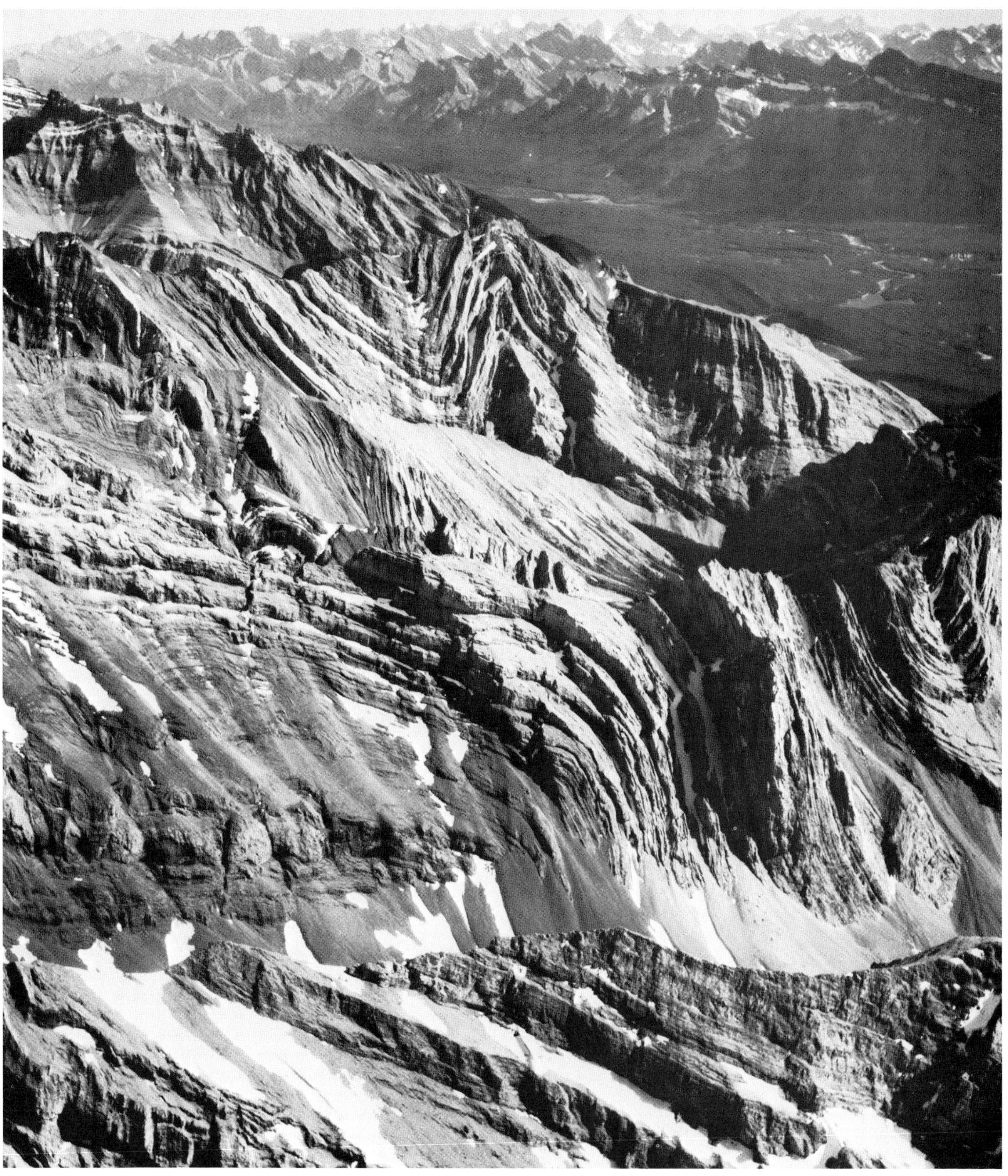

Figure 13.3 (below) in the absence of erosion, an uplifted fault block would form a continuous, straight *fault scarp*, as shown at (a). Processes of erosion result in the dissection of the fault scarp as shown at (b), (c), and (d). (After Cotton, 1960)

(right) The east front of the Sierra Nevada is a fault scarp that has been partly dissected by erosion. Note the straight and regular appearance of the mountain range and its abrupt rise from the neighboring plain. The form of the scarp has been softened by erosion.

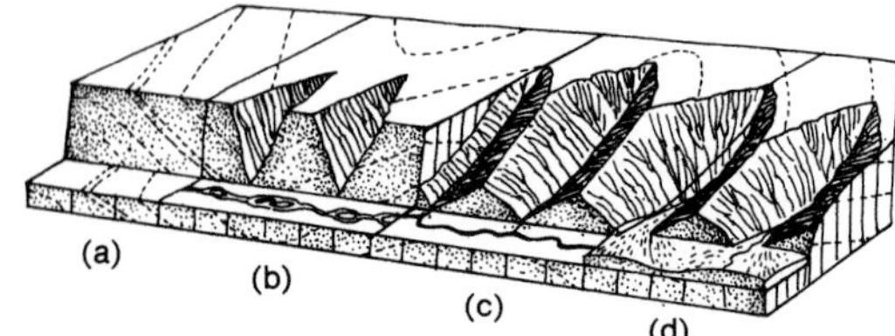
(a)
(b)
(c)
(d)

create the fantastic sandstone pillars and archways prevalent in the American Southwest.

Landforms are a function of *structure*, *process*, and *stage of development*. As applied to physical geography, the term "structure" has broad meaning. It concerns the mechanical strength, hardness, and chemical stability of rocks, their porosity, their crystalline forms, and the nature of their weathered products. The term also applies to the large-scale arrangements of rocks: whether they are arranged in sheets or in homogeneous masses, whether joints are present or absent, whether layers are level or inclined, and so on. Because extensive bodies of rock are seldom structurally uniform, different parts of geologic structures offer different degrees of resistance to erosion. It is the combination of varied structure and varied process—as well as the stage of structural development in time—that underlies the diversity of landforms that can exist even in the same climatic region.

The Structural Foundation

The rocks of the earth's crust are composed of two broad categories of material. The main constituents of rocks making up the continental crust are silica (silicon dioxide) and a smaller proportion of aluminum compounds. The general name for such rock is *sial* (silicon plus aluminum). The oceanic crust and the crustal layer that underlies the continents are on the average composed of a material that consists of silica and magnesium compounds in approximately equal proportions, together with aluminum, calcium, and other constituents in lesser amounts. This rock, which is denser than sial, is called *sima* (silicon plus magnesium). When a mixture of undifferentiated rock-forming material is heated to its melting point in the earth's interior the less dense sial tends to float to the top of the sima. The sialic continents therefore float on a layer of sima.

Volcanic eruptions on continents usually bring up molten rock, or *magma*, that is sialic in composition, reflecting the make-up of the continental crust. Magma from volcanoes on the floor of the Pacific Ocean usually has the composition of sima, like the oceanic crust. The rock of the Atlantic, Indian, and Pacific basins is composed almost entirely of sima, but the floors of the Atlantic and Indian Oceans contain patches of sial.

Igneous Rock

Sial and sima are *igneous* in origin, which means they have resulted from the crystallization of molten magma. Local differences in the chemical composition of sial and sima give rise to a wide variety of igneous rocks. Igneous rocks have a crystalline structure, which is exemplified by a chunk of granite. For a rock of given composition, the size of its crystals depends largely on the rate at which the molten rock cooled. Rock that has cooled rapidly usually has the smallest crystals, because atoms cannot drift together into large crystals unless there is enough time for them to do so.

Common igneous rocks in the continental crust are coarse-grained *granite* and relatively fine-grained *basalt*. Granite and basalt both show large-scale jointed structures: granite is usually split into sheets, whereas basalt often divides into five- or six-sided parallel columns. Granite is formed deep in the earth under great pressure, and when granite is uplifted to the surface the internal pressure often causes the rock to split into sheets. The columnar joints characteristic of basalt are formed by thermal stresses caused when molten basalt solidifies and contracts. Crystal structure and joints are significant in the development of landforms because they affect the response of rock materials to weathering processes. The rates of chemical reaction are different for crystals of different minerals in rocks, and the weathering of a rock may be governed by the rate of disintegration of the least resistant mineral. Joints expose surfaces to chemical action, so that a jointed mass of rock weathers on many surfaces at the same time.

Sedimentary Rock

Weathering and erosion work to disintegrate rocks and transport the debris. Streams carry rock debris to lakes, oceans, and inland seas, where it accumulates in layers. Accumulations of sediment occur on the land as well because of wind action or because of deposition from streams and glaciers. Sometimes layers of sediment appear as *unconsolidated sediment*, as they do on beaches or deserts. Unconsolidated deposits often have a high silica content because of the abundance of this mineral in most rocks and because of its resistance to weathering. Other unconsolidated sediments include clay beds and *alluvial deposits*, which are composed of mineral grains of different sizes. During geologic history, the pressure acting on deeply buried beds of unconsolidated sediment converted

Figure 13.4 *Granite* is the most common intrusive igneous rock on the continents. (top left) The enlargement of a thin section shows that granite is a coarse-grained rock consisting of many different minerals. Silicate minerals are the main constituents of granite; when granite weathers into particles, sand is produced. (top right) Granite commonly forms large masses on the continental crust. This dome of granite exhibits the characteristic way granite exposed at the earth's surface splits into sheets, or *exfoliates.* (bottom) The Sierra Nevada range in California, shown here in the vicinity of Mount Whitney, consists largely of granite. The granite boulders in the foreground are separated from massive parent rock by weathering along the joints of the granite.

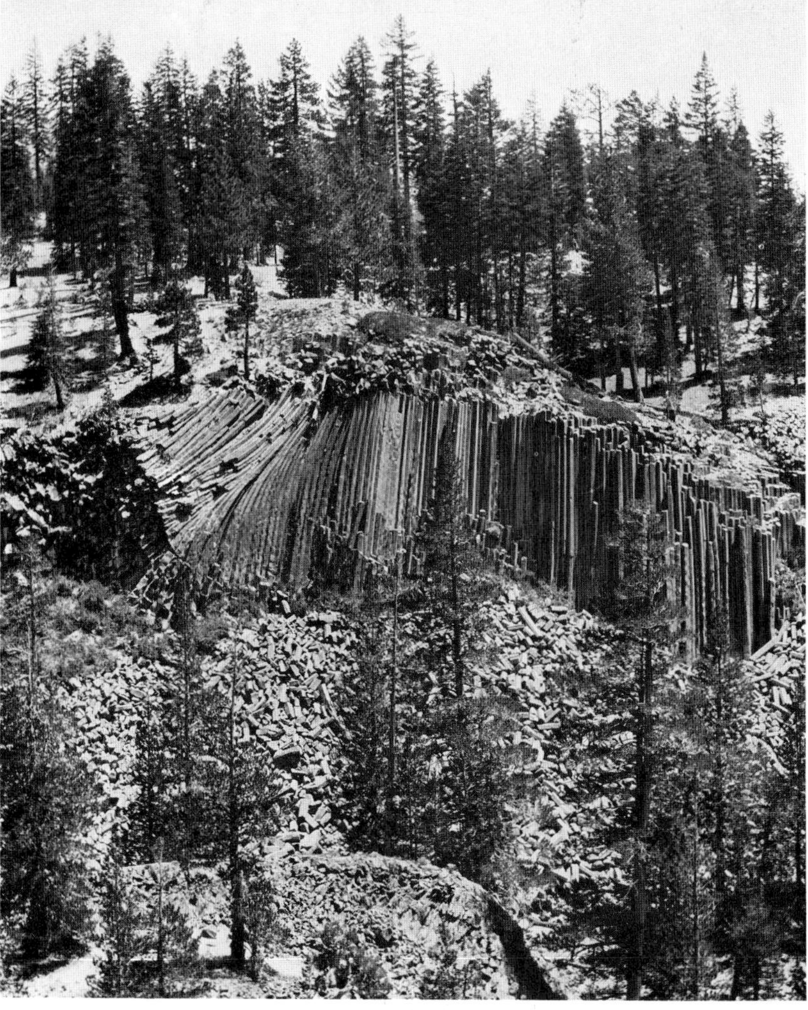

Figure 13.5 *Basalt* is a volcanic rock and is the predominant igneous material of the ocean basins. (top left) The enlargement of a thin section shows basalt to be a fine-grained rock. It consists largely of three different silicate minerals and volcanic glass. (top right) This solidified volcanic lava flow on Kilauea crater, Hawaii, shows the dark color of basalt. (bottom) The Devil's Postpile in the Sierra Nevada of California consists of prismatic columns of andesite, a rock intermediate in composition between granite and basalt. The columns were caused by thermal contraction as the rock solidified.

them from loose aggregates of particles to solid *sedimentary rocks*. Sediment may also be cemented together by chemical action at or near the earth's surface.

Sedimentary rocks compose nearly three-quarters of the surface of the earth's landmasses. The most common sedimentary rock, *shale*, consists largely of consolidated fine particles, many of clay size. *Sandstone*, another common sedimentary rock, is made of silica grains usually cemented together by other minerals. The response of shale and sandstone to weathering reflects their composition and structure. Shale is dense and impervious to water, and its surface weathers slowly. However, shale forms in thin layered sheets, and water can infiltrate the layers where the edges of the sheets are exposed, speeding the process of weathering. Massive sandstone disintegrates into separate silica grains if the cementing material is susceptible to weathering.

Limestone is another common type of sedimentary rock. Limestone deposits are important economically because they supply the raw material for cement—our cities are built mainly from limestone products. Unlike shale and sandstone, which are formed from inorganic clay and sand, most limestone is the product of organisms. Limestone consists primarily of calcium carbonate, one of the principal mineral constituents of sea shells. Limestone beds hundreds of meters thick originally were formed under water by the slow deposition of the calcified shells and skeletons of sea animals.

Metamorphic Rock

Metamorphic rocks result when further heat and pressure recrystallize igneous or sedimentary rocks without actually melting them. Perhaps the best-known metamorphic rock is marble, which arises from the transformation of limestone. Comparatively uncommon, metamorphic rocks compose less than 10 percent of the earth's landmasses. They are often located in areas of intensive crustal deformation because of the high pressures generated when the crust warps.

Differentially Eroded Rock Structures

Each rock type responds in its own particular way to weathering and erosion processes. The underlying rock structure in a given region is seldom homogeneous. Most of the rocks near the earth's surface are sedimentary, for instance, and sedimentary rocks show a characteristic layering resulting from differences in the composition and particle size of the original sediment layers. Each layer in a sedimentary rock may exhibit a different degree of resistance to weathering by virtue of its composition and structure.

Originally, the layers of sedimentary rock were laid down in horizontal sheets, or *strata*, of various thicknesses. Perhaps volcanic eruptions also spread a layer of hard, igneous rock over the sediment from time to time. If the strata remain horizontal, only the exposed uppermost layer is initially subjected to the effects of erosion and weathering. As erosion proceeds, however, the lower layers are exposed, and the rate of erosion varies according to the resistance of the layer encountered. The Grand Canyon is an outstanding

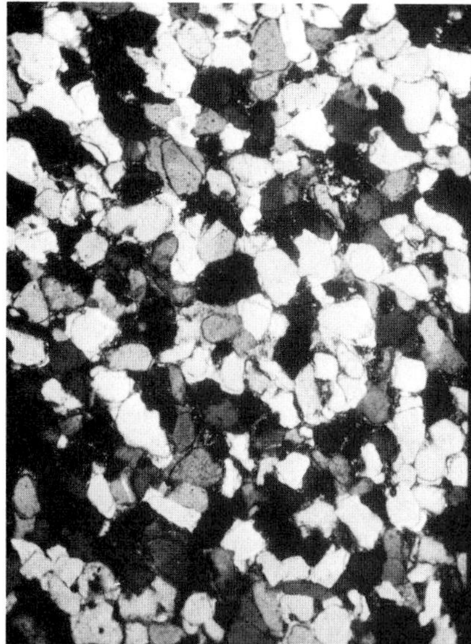

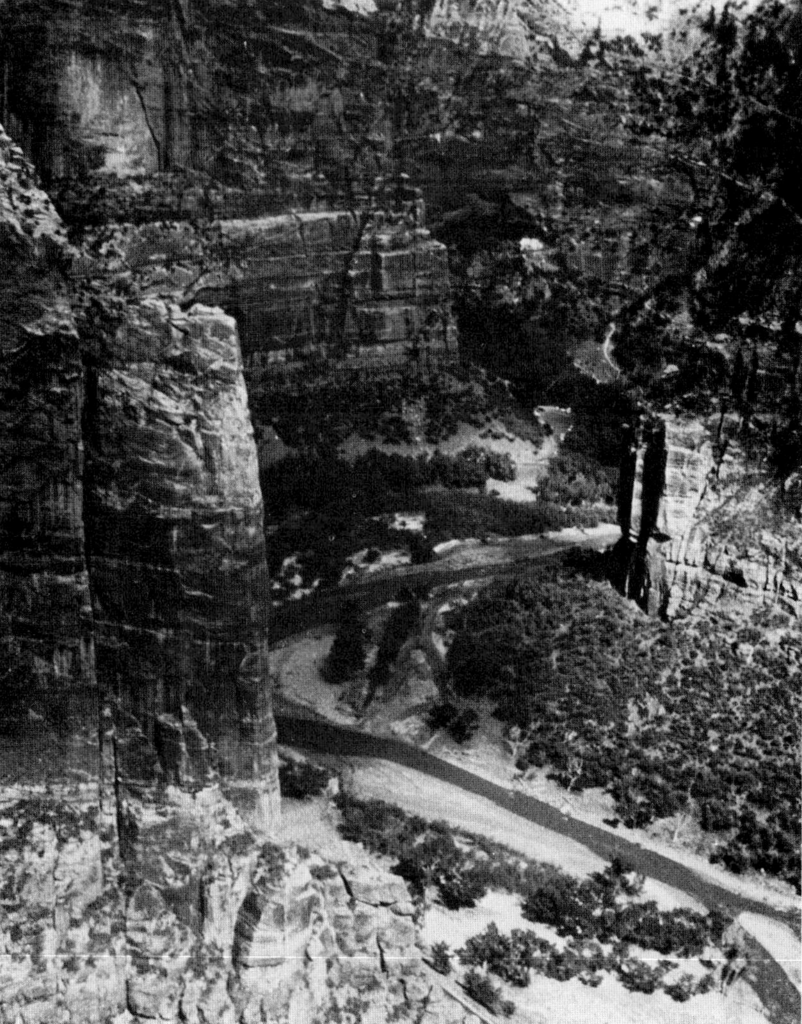

Figure 13.6 *Sandstone* is one of the most common sedimentary rocks on the earth's surface. (top left) Sandstones are usually composed of grains of silicate minerals cemented together by other minerals, as shown in this enlarged thin section. In general, sandstone is any rock composed of fragments that are in the size range of sand grains. When sandstone weathers, the cementing material disintegrates, releasing the individual grains. (top right) Sandstone is formed by the consolidation of beds of sand deposited by wind or water. This outcrop of sandstone in Victoria Land, Antarctica, shows evidence that its layers were laid down by flowing water at a time when Antarctica's climate was warmer than it is today. (bottom) Some of the massive beds of sandstone in Zion National Park, Utah, were deposited by wind action. Erosion by water and the collapse of large masses of rock have produced the spectacular cliff faces.

Figure 13.7 *Limestone* is formed primarily by organic processes. Most limestones are formed by the consolidation of shells of marine animals that have accumulated in shallow seas. (top left) This section of limestone reveals fragments of marine organisms. (top right) Limestone consists largely of compacted and cemented calcium compounds; it is susceptible to erosion by solution, particularly by acidic water. The fluted patterns in this outcrop of limestone on the coast of central Yugoslavia were formed by the action of water when the limestone was buried under moist soil. (bottom) In parts of the world where the bedrock consists of layers of thick, bedded limestone, groundwater has caused a distinctive landform known as karst topography. Groundwater and underground streams dissolve some of the limestone, particularly along its joints, which produces numerous underground caverns and drainage holes, or sinkholes. The photograph shows a portion of the karst region of southern Indiana. The landscape is dotted with sinkholes, which appear as depressions in the soil cover.

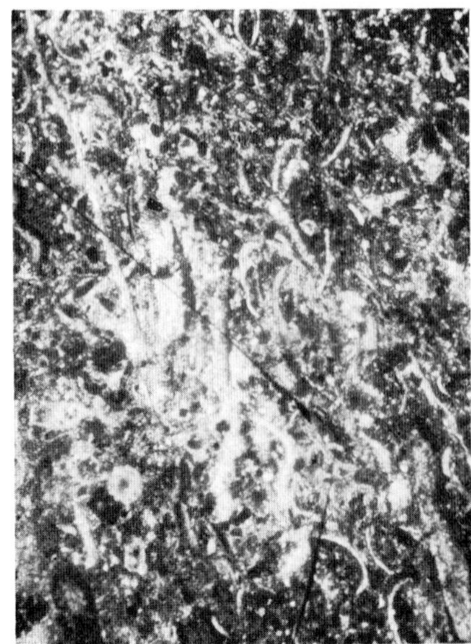

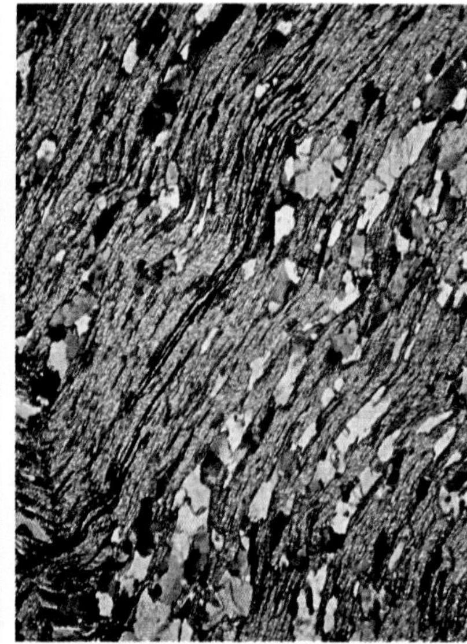

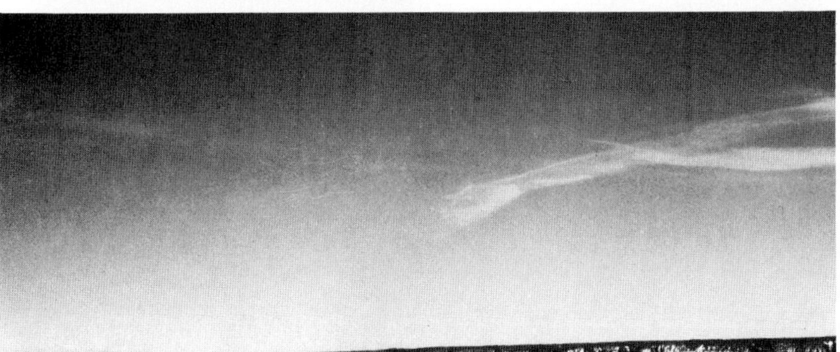

Figure 13.8 *Schist* is a metamorphic rock that forms by the transformation of shale under conditions of heat and high pressure. (top left) This outcrop of schist exhibits the swirled patterns often seen in metamorphic rock. (top right) Schist is a crystalline rock that has an internal structure dominated by a layered arrangement of platy minerals as seen in this enlarged thin section. Because weathering tends to be most effective between the sheets, schists tend to break into thin sheets. (bottom) The inner gorge of the Colorado River in this section of the Grand Canyon consists largely of schist and granite intrusions.

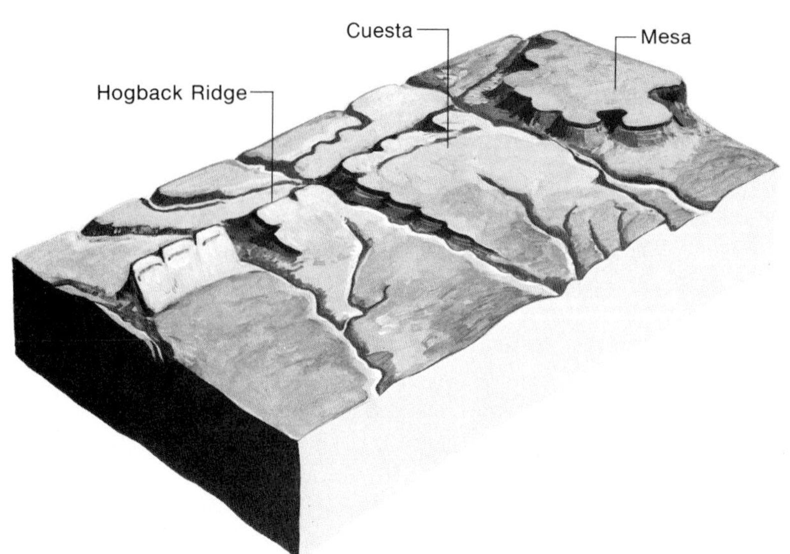
Cuesta
Mesa
Hogback Ridge

Figure 13.9 (right) *Mesas, cuestas,* and *hogback ridges* are typical landforms that can be produced by differential erosion in regions where resistant cap rock overlies less resistant layers. A flat-topped mesa is formed where the rock strata are horizontal. A cuesta, or escarpment, is formed where the strata dip at an angle of a few degrees, so that a cuesta possesses one steep face. A hogback ridge is formed where the strata dip steeply, so that a hogback ridge presents two steep faces of different composition. One face consists of resistant cap rock, and the other is the exposed underlying layers.

(bottom) This photograph of Monument Valley, near the Arizona-Utah border, shows a large butte on the right and mesas in the left background. A butte is a small mesa that has been reduced by erosion so that the summit is no longer flat.

example of a river valley cut into horizontal strata that have different degrees of resistance. Even rock of the same type may have varying degrees of resistance in different parts, so that weathering can result in the formation of fantastically shaped landforms.

Mesas and *buttes* are examples of landforms resulting from the differential erosion of horizontal, stratified rock structures. In a mesa, a resistant upper layer of rock protects softer layers beneath it from erosion. The wearing down of the surrounding landscape leaves an isolated tableland that is characterized by a horizontal, steep-sided cap of resistant rock on top of less resistant rock that has more gentle slopes. Such a tableland is called a mesa if it is extensive in area; it is called a butte if its width is comparable to its height.

Tableland plateaus such as mesas are based on a structure of horizontal layered rock. Suppose, however, that the rock layers are not horizontal but are tilted at an angle of a few degrees. The short side of the structure has a steep face and the long side is gently sloping, eventually merging with the level of the surrounding land. Such a structure is called a *cuesta*. If the angle of dip of a hard rock layer is extreme, exceeding 45° or so, a surface form called a *hogback ridge* results. Hogback ridges are roughly symmetrical, with the same angle of slope on both sides.

The underlying rock structure sometimes controls the pattern of a drainage net. Figure 13.10 shows an arrangement in which bands of highly resistant rock alternate with bands of weakly resistant rock. If the bands are roughly parallel, the drainage pattern takes the shape of a trellis with the major stream beds cut along the bands of softer rock. In locations such as the Adirondack Mountains of

Figure 13.10 The drainage patterns of streams are sometimes strongly influenced by the underlying geologic structure. (a) This diagram shows the stream patterns in the Appalachian Mountains near Harrisburg, Pennsylvania. The shaded areas represent highly resistant sandstone; in the remaining regions, the bedrock is weakly resistant shale. The streams cut downward into the shale more easily than into the sandstone, so that stream courses lie in valleys underlain by weak shale between ridges of hard sandstone. Some small tributary streams have cut into the sandstone ridges by headward erosion. Note that the Susquehanna River, the master stream of the region, cuts directly through several ridges of resistant rock. The reason for this is explained in Figure 13.26. (b) The Zambesi River near Livingstone, Zambia, exhibits a rectangular drainage pattern. Such patterns are developed where the underlying rock is strongly jointed into rectangular blocks. The stream courses tend to develop along the weakly resistant joints.

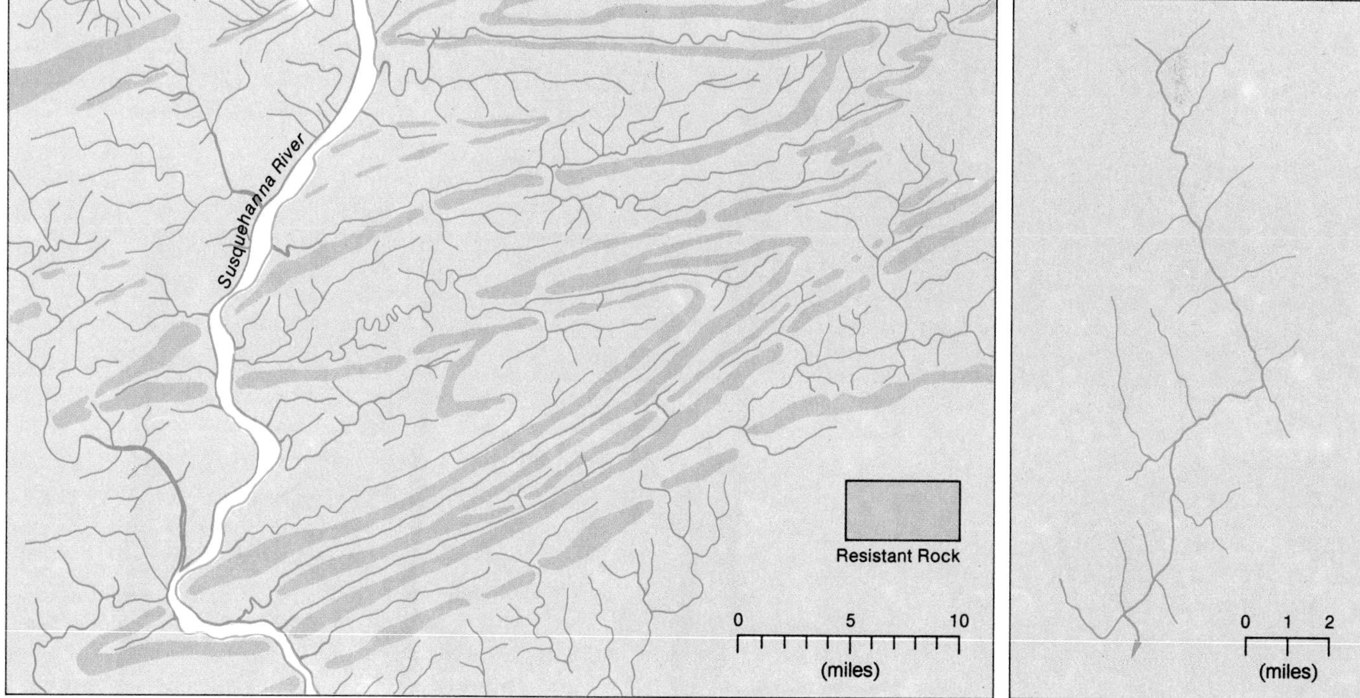
Susquehanna River
Resistant Rock
0
5
10
(miles)
0
1
2
(miles)

New York the granitic bedrock is extensively fractured or jointed. The streams in such regions tend to follow the line of weakness represented by a joint. Drainage patterns over jointed rock usually show a distinct rectangular pattern, with the streams making sharp right-angle bends as they follow the joints.

Limestone presents a special case in the development of landforms because some of its forms are relatively soluble in water and because it is divided into rectangular blocks. In dense limestone, water flows mainly along the joint faces. If the limestone beds are underground, horizontally moving groundwater cuts underground channels and caves that may be further enlarged by underground streams. Landscapes created by the solution of limestone are called *karst* regions, after a well-known limestone area in Yugoslavia. A typical feature of a karst region is the multitude of *sinkholes* dissolved into the surface of the limestone. Some sinkholes are shallow depressions but others drain directly into underground channels. In the karst region of southern Indiana, there are more than 1,000 sinkholes per square mile. Surface streams are scarce in karst regions because few streams are able to cross a limestone plain without disappearing down a sinkhole drain. In some locations limestone beds have been uplifted above the surface of the surrounding land. In warm, humid regions such as southeast China and Puerto Rico, the exposed limestone typically becomes cut into cones or towers. In arid regions, however, limestone is a highly resistant rock. Figure 13.7 shows a portion of the karst region in Indiana.

Structure and process sometimes combine to produce landforms of such rarity that they occur at only a few places on the earth. The *penitent rocks* found in Zambia and Rhodesia in Africa are a good example. Penitent rocks are formed of *schist*, a metamorphic rock that exhibits a low resistance to weathering along one plane.

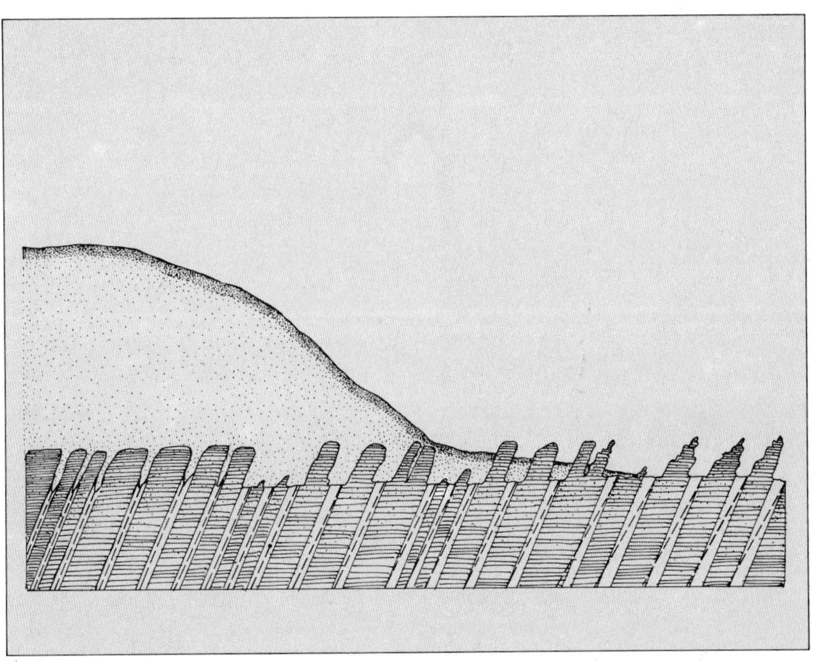

Figure 13.11 The slanting penitent rocks found in Rhodesia consist of schist, which tends to weather along the lines of weakness between the thin layers that constitute the rock. This schematic diagram illustrates how penitent rocks may be formed by weathering under the soil, gradually becoming exposed as the soil cover is eroded. (After Ackerman, 1962)

The diagram in Figure 13.11 illustrates how penitent rocks may be formed by the deep weathering that occurs along weak layers of schist while the schist is still underground. Subsequent erosion of the surface probably then strips away the surrounding material to reveal the rock form beneath.

Dike ridges, another type of landform resulting from differential erosion, are found in Arizona, Colorado, New Zealand, and other regions that have experienced volcanic activity. Molten masses from the interior of the earth are sometimes forced up into vertical fissures of the overlying rock. In a long, narrow fissure, the solidified lava forms a structure in the shape of a vertical wall. Erosion wears the surrounding rock into slopes centered on a backbone of resistant lava, forming a dike ridge. In extreme cases the surrounding rock is worn away completely, leaving the lava wall standing free. If the surrounding rock happens to be more resistant than the lava, the more rapid erosion of the lava leaves a trench between the rock walls.

Volcanism: The Production of New Igneous Rock

The igneous rock that originally formed the whole of the earth's surface about 4.5 billion years ago has since been weathered away or altered by metamorphism. The rocks on the earth's surface are continually being weathered and eroded. Igneous, sedimentary, and metamorphic rocks alike contribute at their own rates to the deposition of sediment layers. Sediment layers then gradually become sedimentary rocks over periods of time ranging from millions to hundreds of millions of years. Some of the sedimentary and igneous rocks are also subjected to heat and pressure and become metamorphic rocks.

The steady destruction of igneous rock by weathering and erosion is balanced by the production of new igneous rock, which completes the cycle of rock destruction and creation illustrated in Figure 13.12. Some igneous and metamorphic rocks return to the molten interior of the earth as lithospheric plates descend into the oceanic trenches. The *rock cycle* is completed when magma wells up during volcanic activity and forms new igneous rock. When molten materials of low density flow upward from the interior to the surface of the earth through fractures in the distorted colliding plates, they produce volcanic eruptions at the surface.

The known active volcanoes are distributed primarily around the boundaries of the Pacific basin, where the oceanic plates of the Pacific descend under the continental plates of Asia and the Americas, or where spreading of the crust occurs along oceanic ridges. There are approximately 500 volcanoes known to have been active in historic times. About 60 of the known active volcanoes are located under the sea, but there are undoubtedly additional active volcanoes on the sea floor that have not yet been discovered.

The other main regions of concentrated volcanic activity are in Italy and Iceland. Located on the rifting Mid-Atlantic Ridge, Iceland is the scene of widespread volcanic action; in fact, the island is

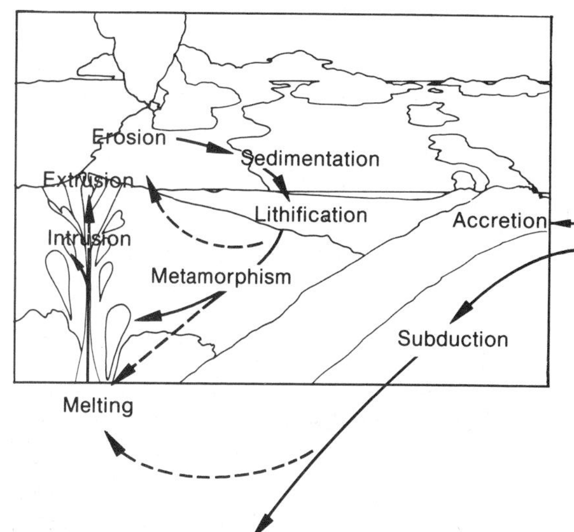
Erosion
Sedimentation
Extrusion
Lithification
Accretion
Intrusion
Metamorphism
Subduction
Melting

Figure 13.12 During the course of geologic time, rock materials pass from one form to another in a process known as the *rock cycle*. This schematic painting, and the inset diagram above, illustrate the principal phases of the rock cycle. At the lower left of the figure, molten rock from the interior of the earth is shown welling up through fissures toward the surface. Solidified molten rock forms igneous rock. Sometimes the rock solidifies beneath the surface, forming *intrusions*, or *batholiths*, as shown in the figure. Erosion of the surface cover may lead to the subsequent exposure of the intrusive rock masses. Some of the molten rock is also directly extruded at the surface.

Rocks exposed at the surface are subject to erosion, principally by flowing water. The fragments and grains of rock are carried by water or wind to the sea or to inland basins, where they collect in beds of sediment. In the presence of heat or pressure, or by the chemical action of a cementing agent, sediment beds become consolidated, or *lithified*, into sedimentary rock.

The motion of crustal plates plays an important role in the rock cycle. The downward thrust of a plate near a continental boundary, depicted in the lower right of the figure, forms a shallow sea trough that slowly subsides, allowing deep sediment beds to accumulate. The downward movement of plates in subduction zones also returns rock material to the interior of the earth, where it can be remelted to continue the rock cycle. Deformation near the boundary of colliding plates generates the pressure required to transform igneous and sedimentary rock into metamorphic rock.

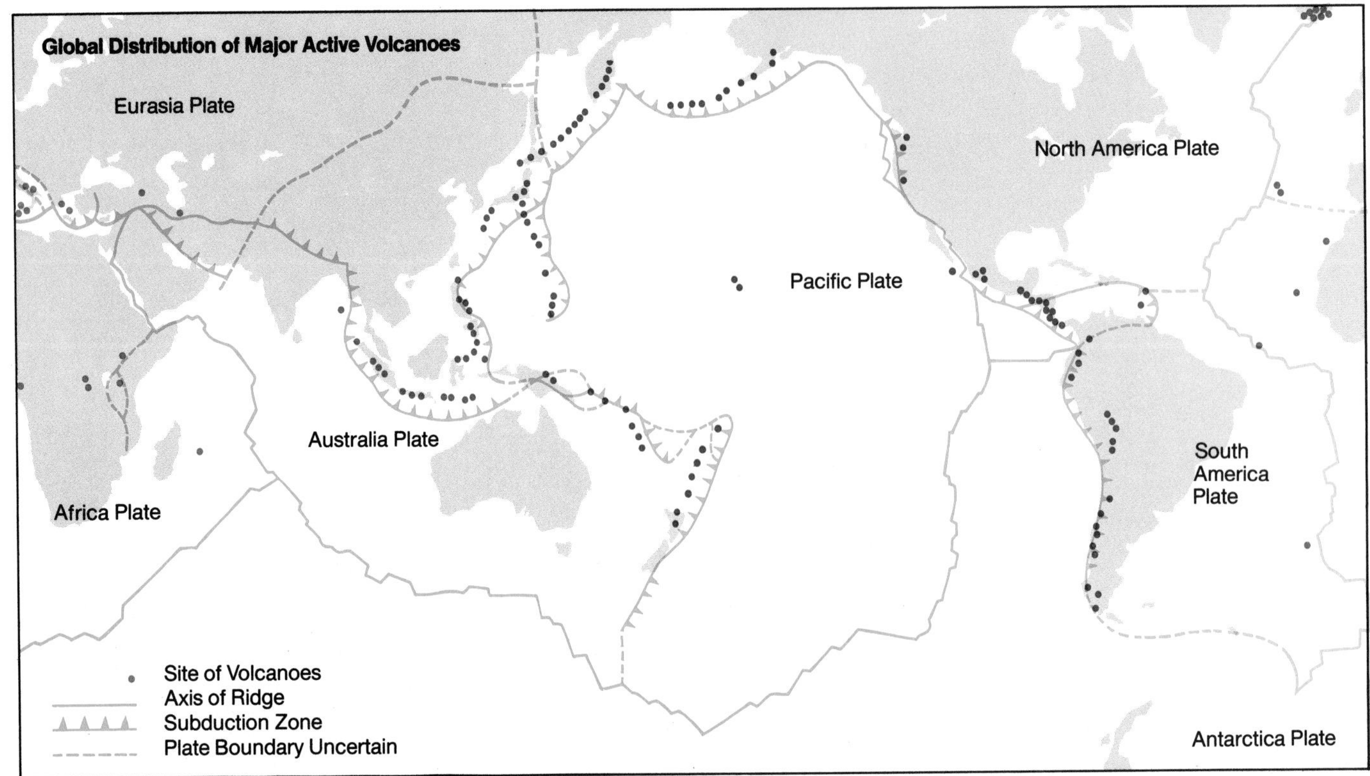
Global Distribution of Major Active Volcanoes
Eurasia Plate
North America Plate
Pacific Plate
Australia Plate
Africa Plate
South America Plate
Antarctica Plate
Site of Volcanoes
Axis of Ridge
Subduction Zone
Plate Boundary Uncertain

Figure 13.13 This map shows many of the known active volcanoes of the world; for clarity, some volcanoes have been omitted from regions where many volcanoes are present. The map also shows the principal plate boundaries, ridges, and trenches associated with the earth's crustal plates. Regions of volcanic activity are frequently located near active plate boundaries, where molten rock from the earth's interior wells up through fissures. Note, for example, the large number of active volcanoes around the Pacific Ocean basin, which is ringed by trenches. The volcanoes in Iceland and some of the volcanoes in the Atlantic Ocean are associated with the spreading center along the Mid-Atlantic Ridge. Volcanoes in the Caribbean, in East Africa, and in the eastern Mediterranean appear to be associated with plate boundaries as well. Conversely, the east coasts of North and South America, where there are no plate boundaries, are devoid of volcanic activity. Isolated volcanic regions far from plate boundaries, such as the Hawaiian Islands, are probably due to fractures in the lithospheric plates, which allow molten rock derived from the earth's interior to reach the surface. (After Bullard, 1962)

formed entirely of volcanic deposits. Early in 1973 a dormant volcano became active on the island of Heimaey off the south coast and buried a fishing port in ashes.

The only volcano currently classified as active in the conterminous United States is Mount Lassen in California. Its last active period was between 1914 and 1917. The states of Alaska and Hawaii are extremely active volcanically; Hawaii is little else but volcanoes, and the islands consist of a complex of eroded volcanic cones built up by the quiet flow of broad sheets of lava. Many of the volcanoes along the west coasts of North and South America are active, although they are quiet at present. The volcanoes are young and the crust's activity continues. The timing of eruptions cannot be predicted, but they will undoubtedly occur in the near geologic future.

Geothermal Energy

The hot magma under the earth's crust is a potential source of bountiful heat energy. In most regions of the earth, magma is covered by a thick layer of crustal rock, so only a negligible amount of heat energy is conducted to the surface.

To tap substantial amounts of heat energy from magma under the crust would require wells many tens of kilometers deep. However, in regions of volcanic activity or in areas where severe faulting has fissured the crust, hot springs or jets of steam can bring heat energy to the surface. Hot springs result when groundwater comes into contact with hot rock or with steam given off by magma. Sometimes steam generated from magma or from heated groundwater is piped through fissures toward the surface, where it may vent directly into the atmosphere or remain at moderate subsurface levels within range of ordinary wells. Regions where hot water and steam are available near the surface include Iceland, Italy, New Zealand, and the southwestern United States.

Inhabitants of Iceland have long used the water from hot springs to heat their homes and even to make their fields warmer for agriculture. Nearly half of the homes in Iceland and almost all of the buildings in the capital of Reykjavik are heated by hot water piped from nearby hot springs. The use of hot springs for heating is restricted to locations within a few kilometers of the springs because of heat losses along the delivery pipes.

Several countries have generating stations powered by steam from the ground. Electric power can be distributed to a wide area to make the best use of energy from localized steam fields. The geothermal power plants in the Larderello area of Italy northwest of Rome make an important contribution to the Italian economy. In 1960, the Larderello stations generated 4 percent of Italy's total electric power output.

The first commercial geothermal station in the United States is located in a faulted region 150 kilometers (90 miles) north of San Francisco. The plant produces electric energy at the rate of several hundred million watts, somewhat less than 1 percent of the total

Figure 13.14 (left) This schematic diagram shows the type of geologic structure in steam fields that can be readily exploited for power generation. The region is covered with dense cap rock over an aquifer permeated with water. Water is heated by contact with hot rock, and steam is generated. The steam comes to the surface through fissures or is led to the surface through wells sunk into fissures or into the aquifer. The high-pressure steam may then be used to operate electric generators. Geothermal energy is a comparatively clean and inexpensive source of power, but at present its use is confined to geologically suitable regions.

(right) The hot water issuing from Castle geyser in Yellowstone National Park, Wyoming, is driven upward by steam generated from groundwater in fissured hot rock deep below the surface.

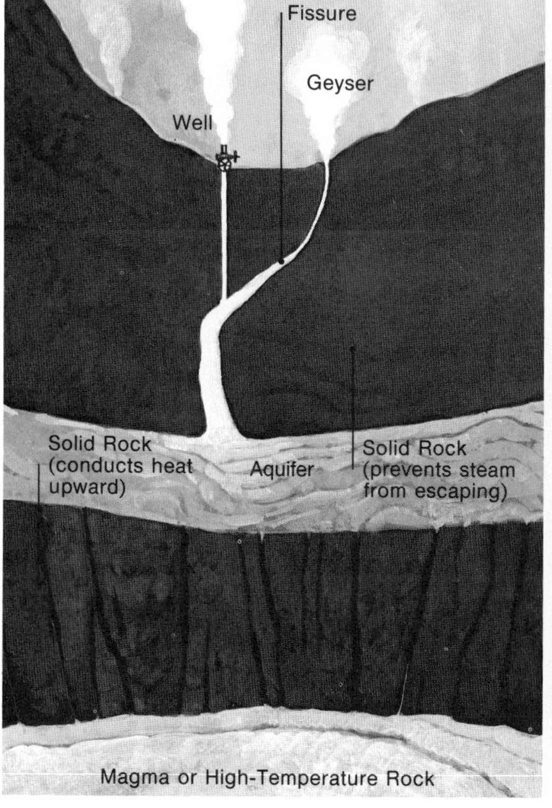
Fissure
Geyser
Well
Solid Rock
(conducts heat
upward)
Aquifer
Solid Rock
(prevents steam
from escaping)
Magma or High-Temperature Rock

Figure 13.15 (opposite top) A *stratovolcano* has a composite cone consisting of layers of ash and lava. Mount Egmont, a stratovolcano in New Zealand, has two cones: the cone on the left is the original cone, and the cone on the right developed because of the shift of volcanic activity to a new vent.

(opposite bottom left) The Mauna Loa volcano in Hawaii is the largest active volcano in the world. Its gently rounded oval dome rises more than 9,500 meters (30,000 feet) above the floor of the Pacific Ocean basin. Such a *shield volcano* is formed by the accumulation of numerous thin flows of highly fluid lava. The shiny cinders in the foreground are glossy fragments from the eruption of nearby Kilauea crater in 1960.

(opposite bottom right) *Cinder cones* are volcanic cones that are built from cinders and ashes ejected from volcanic vents. The unconsolidated material forms slopes at the angle of repose, 35° or so. Numerous cinder cones dot the crater of Haleakala volcano, Maui.

rate of electric power consumption in California. Areas in south central California are believed to have the potential to supply most of California's power needs from geothermal energy.

The Source of Volcanic Energy

Volcanism refers to an upwelling of molten rock from the earth's interior to the earth's surface, but there are many ways in which this upwelling can occur, and it need not be violent. Eruptions of Hawaiian volcanoes, for example, are comparatively quiet affairs, and sightseers watch the flowing lava streams in comparative safety. Nor do symmetric cones necessarily accompany volcanic action. In Iceland the lava issues from extensive fissures and spreads over the land in flat layers, forming great lava plateaus. Similar lava plateaus from past volcanic activity are also found in eastern Washington and in Idaho.

Some volcanic eruptions are truly violent, however. Krakatoa, a volcano off the coast of Java, exploded in 1883 and the noise was heard in Australia, more than 3,000 kilometers (1,900 miles) away. The ash cloud from the eruption was so dense that regions 200 kilometers (120 miles) away were in darkness for an entire day. What is the driving force behind such intense activity?

It appears that steam powers a violent volcanic eruption. The molten rock, or magma, deep in the earth contains dissolved liquids and gases. Even though the magma's temperature may be 1000°C, it can contain water in solution amounting to 5 percent or more of its volume. The reason the water does not turn immediately to steam is that the pressure in the earth's interior is equal to many hundreds of times the normal pressure of the atmosphere. If you have suddenly opened the pressure cap of a hot automobile radiator, you have seen water gush out with great force. Under pressure, water remains liquid well above its normal boiling point, but when the pressure is released some of the water vaporizes and expands with explosive force.

In some regions, such as Hawaii or Iceland, volcanic eruptions are usually nonviolent. The magma in such regions is simatic, which tends to be hot and fluid; therefore, accumulated steam is vented at a moderate rate. The magma composed of sial, however, is heavy and viscous, so steam is trapped and pressures build to the point where huge quantities of material may be suddenly ejected from the volcano. The magma in the volcanoes of the Mediterranean basin is sialic, so eruptions there are often violent and explosive.

Ejected Materials and Volcanic Hazards

Magma that flows from a volcano and loses its dissolved gases is called *lava*. The more violent eruptions may produce a variety of ejected materials in addition to lava. Sometimes magma contains so much dissolved gas that as it rises to the surface it becomes frothy, like foam on beer. The rock foam expelled during an eruption is called *pumice*. When it solidifies, pumice frequently contains so many internal gas bubbles that it can float on water.

Sometimes the violence of an eruption blows magma into small

Figure 13.16 (opposite top) Lava assumes different forms depending on its temperature, composition, and rate of cooling. This advancing *aa* lava flow, photographed on Hawaii, is characterized by chunky, angular blocks mixed with still-molten lava. Temperatures measured in these basaltic composition lavas are about 1200°C (2200°F). When aa lava flow solidifies, an almost impassable field of jagged lava is formed.

(opposite bottom) When basalt lavas pour out onto the sea floor, the rapid cooling develops large globular masses of viscous lava. These masses may accumulate in a heap, and the still-plastic rock may be deformed into pillowlike masses. The rapid cooling forms a glassy margin, and shrinkage during cooling forms radial cracks, or joints, in the pillow. These accumulations of submarine basalt are called *pillow basalts*.

Figure 13.17 (top) *Pahoehoe* lava, shown here at Kilauea crater in Hawaii, is much more fluid than aa lava even though the general chemical composition is similar. Note the thin layer of solidified crust on the molten rock. When the lava solidifies, the surface is left in the form of smooth billows or ropes.

(center) Shiprock, in northwest New Mexico, is a striking example of a volcanic plug. Note the extensive volcanic dikes radiating from Shiprock along fault lines.

(bottom) Flat-topped *table mountains*, such as this one in Iceland, may form when the land surrounding a nearly horizontal lava flow is removed by erosion or where faulting and erosion occur. The distinct layering visible on the hill slope is caused by the differential erosion of layers of volcanic rock of different resistances.

Figure 13.18(a) This sequence of diagrams shows the accepted explanation for the formation of a *caldera*, or collapsed volcanic cone, such as Crater Lake in Oregon and Krakatoa in Indonesia. In (1), magma fills the throat and chamber of the volcano and causes ash and pumice eruptions. (2)(3) Withdrawal of the magma into the chamber is followed by an explosion and expulsion of much of the magma because of a rapid expansion of the gas content of the magma. The violent expulsion causes an *ash flow.* (4) Removal of the contents of the magma chamber deprives the volcanic cone of support, and the summit of the volcano collapses inward, forming the caldera. The solidified ash flow material is called *tuff.* (5) New small cones may subsequently develop inside the caldera. (After Bullard, 1962)

(b) This diagram illustrates several types of rock bodies that are formed by igneous processes. A *batholith* is a huge (hundreds of square kilometers in outcrop area) mass of intrusive material that has cooled slowly beneath the earth's surface. A *stock* is a similar mass, but of smaller size; it may be an offshoot of a batholith, as shown here. Feeders to a volcano solidify to form a volcanic *neck,* or *plug.* Volcanoes are located above magma chambers, which may solidify to form stocks or batholiths, and erosion of dead volcanoes may leave the resistant rock of the volcanic neck standing. Injections of magma may form *sills* or *laccoliths* if they are injected along the joints of stratified rocks. The injections may cause the strata to arch upward, producing a lens-shaped mass of igneous rock; the cooled magma is called a laccolith. *Dikes* are caused by injection of magma along zones of weakness such as a fault. They are distinct from sills in that they cut across stratified rocks, whereas sills are concordant with the strata. Lava flows or sills exposed by erosion may form resistant cap rocks of mesas. The southern portion of the Palisades along the Hudson River is a basalt sill that was injected into sandstone layers, with subsequent removal of the upper layers by erosion. Extensive flat-topped lava flows of the Columbia Plateau in the state of Washington have been eroded to produce numerous mesas. (After Ollier, 1969)

(c) This sequence of diagrams illustrates how a volcanic flow can lead to an *inversion of relief* by transforming a river valley into a locally high ridge. The pink material in the middle diagram represents lava that has filled part of a preexisting river valley. If the volcanic rock is more resistant than the surrounding rocks, differential erosion may leave a ridge capped by volcanic rock along the site of the former river valley. (After Ollier, 1969)

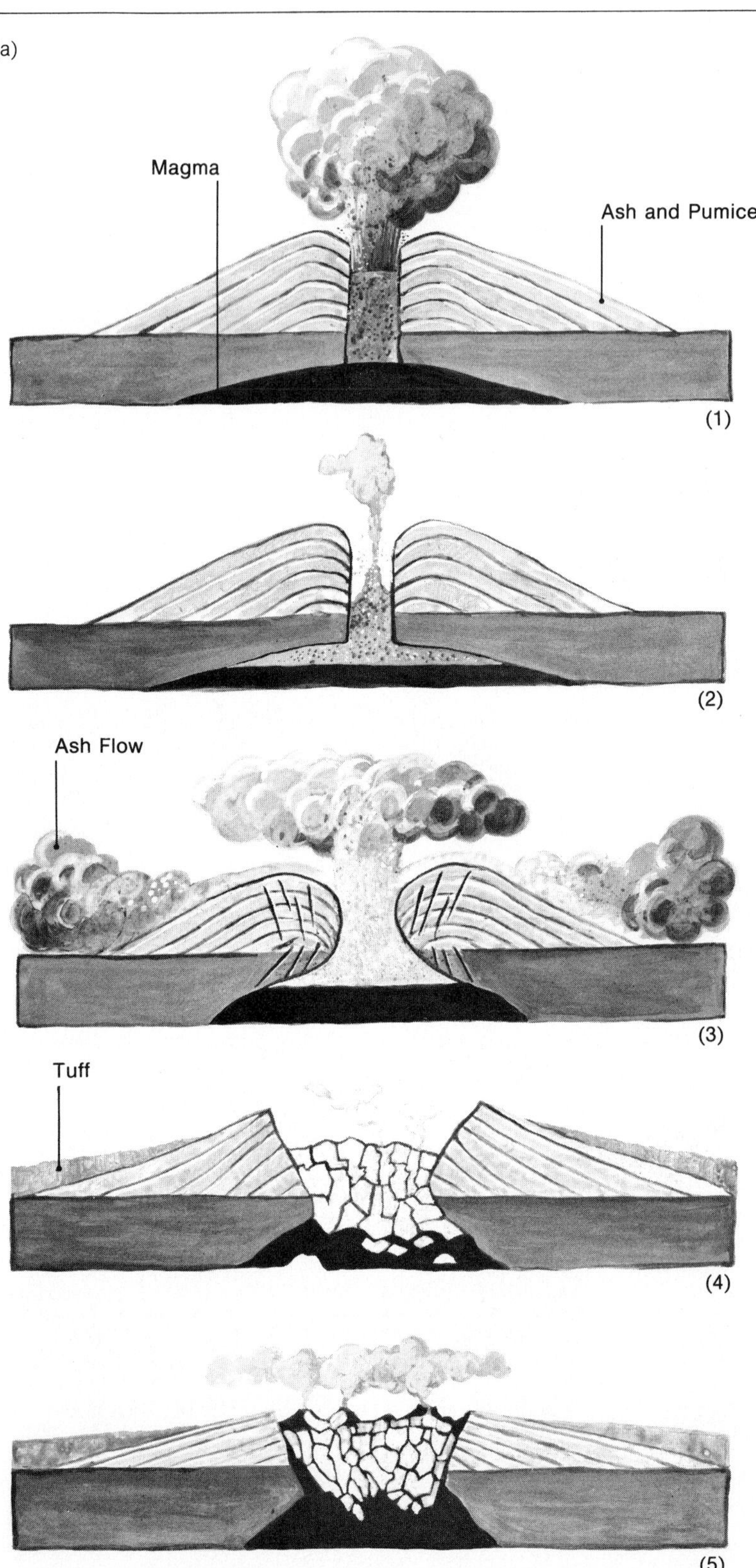
(a)
Magma
Ash and Pumice
(1)
(2)
Ash Flow
(3)
Tuff
(4)
(5)

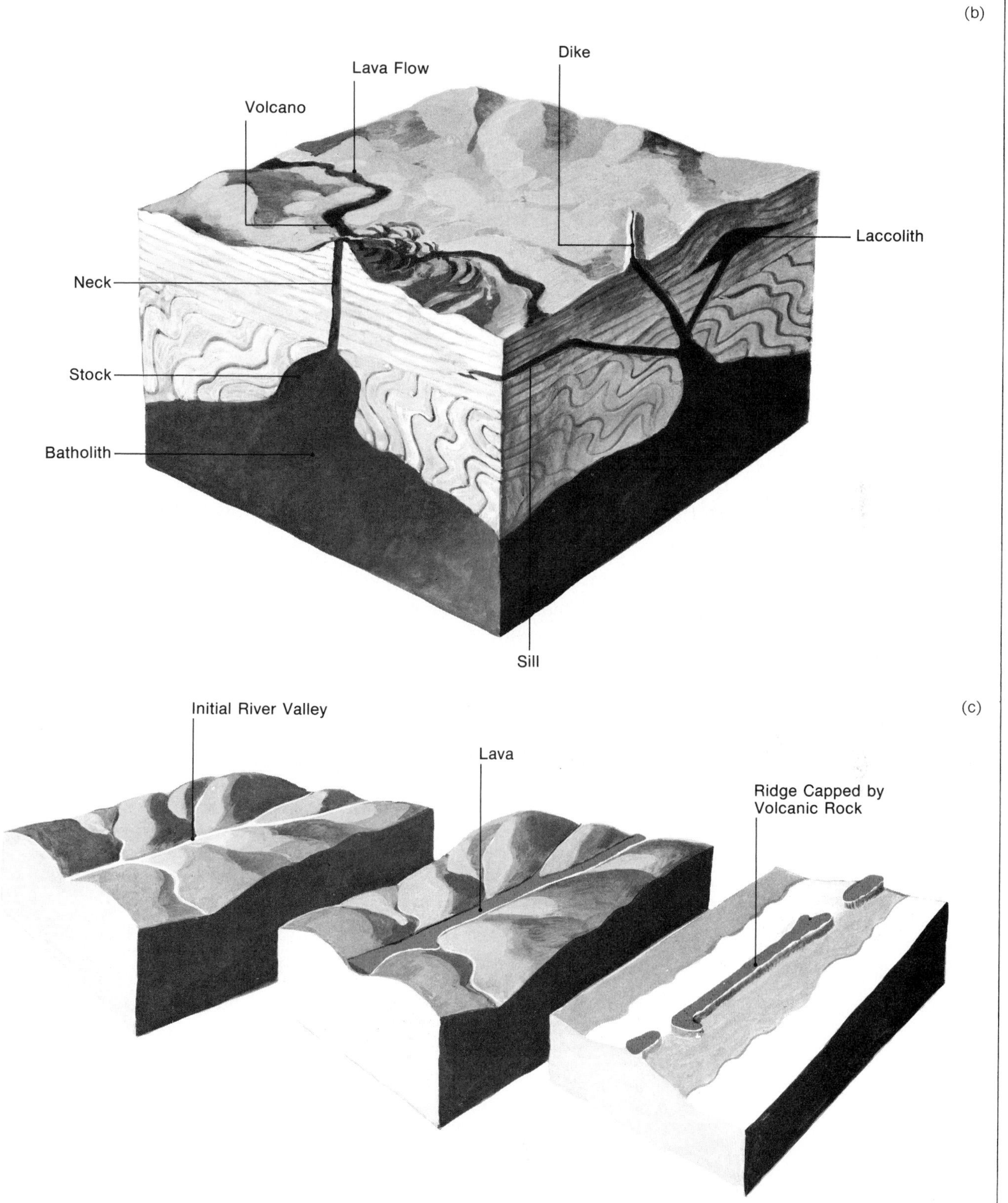

(b)

Dike

Lava Flow

Volcano

Laccolith

Neck

Stock

Batholith

Sill

(c)

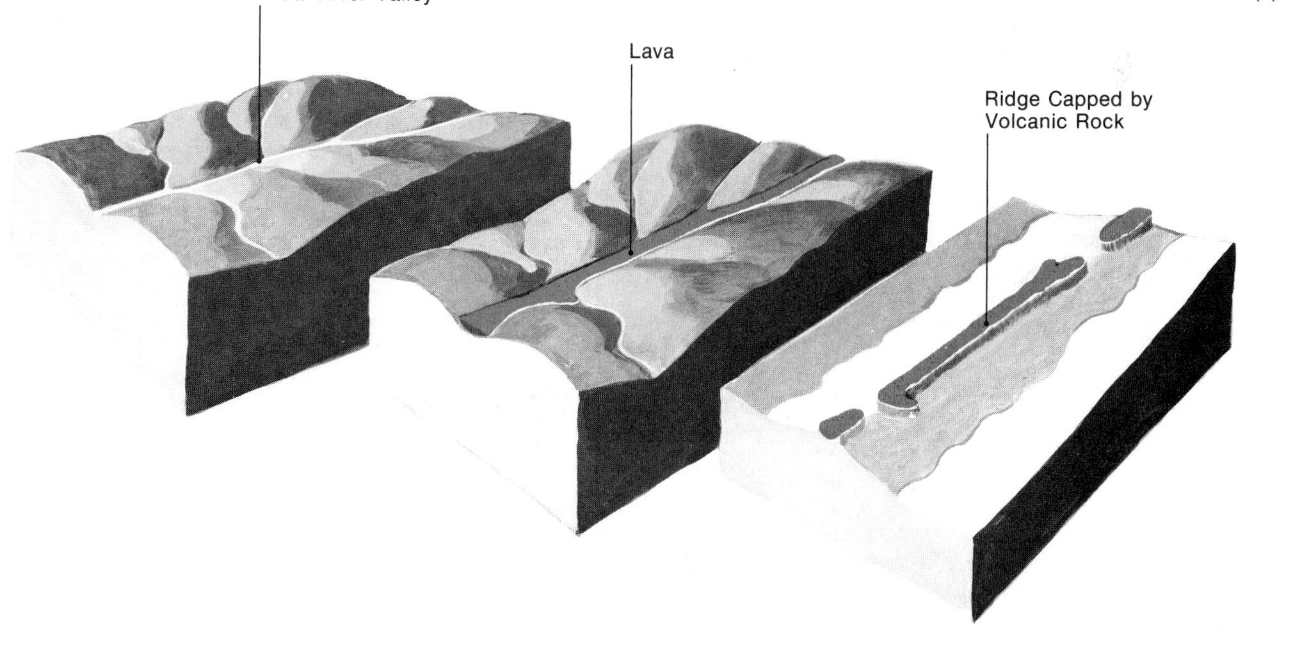

Initial River Valley
Lava
Ridge Capped by
Volcanic Rock

rock fragments a few millimeters in size that fall as *volcanic ash*. The eruption of Mount Vesuvius in Italy in A.D. 79 buried the Roman town of Pompeii in ash and pumice 5 to 10 meters (15 to 30 feet) deep, killing thousands of people and preserving the town until its rediscovery in the eighteenth century. The town of Herculaneum near Pompeii was not in the path of the ash fall, but it suffered a different fate. Under the influence of heavy rains the deposits of ash on the slopes of Vesuvius became a mobile, semiliquid mudflow, or *lahar*. The inhabitants of Herculaneum apparently had warning of the approaching mudflow and most of them escaped, but their town was buried 20 meters (60 feet) deep in volcanic mud.

The rain of falling ash from an eruption is usually cold because the molten rock has had a chance to cool during its descent through the air. Sometimes, however, a body of hot gas venting from a volcano carries glowing ashes in suspension. Part gas and part solid, the cloud, called a *nuée ardente*, acts almost like a liquid and flows with great speed down the flanks of the volcano's cone. A notable nuée ardente occurred during the 1902 eruption of Mount Pelée on the island of Martinique in the West Indies. After a few days of moderate volcanic activity, Mount Pelée suddenly generated a nuée ardente that descended on the town of St. Pierre like a black rolling cloud moving at hurricane speeds. In a few moments the cloud's hot gases had killed all but 2 of the 30,000 inhabitants and had reduced the city to a shell.

Lava flows generally do not constitute a threat to human life because the speed of flow is usually slow—only a few tens of meters per hour. But the richness of volcanic soil has tempted men to build farms and villages near volcanoes, so lava flows often endanger property. Flowing lava forms a crust at its edges and surface, and the hardened sides tend to channel the flow like natural levees. When a lava flow from the Mauna Loa volcano in Hawaii threatened the city of Hilo in 1935, high explosives were used to breach the side of the flow in a modern reenactment of the effort to save Catania. Within a day the forward advance of the flow had stopped entirely, possibly because heat energy was diverted from the main flow of lava.

Landforms Associated With Crustal Motion

Volcanic activity can produce isolated mountains, such as Mount Fujiyama in Japan, but most of the great mountain ranges of the earth are formed by distortions of the earth's crust. The European Alps and the Himalayas, for example, consist primarily of sedimentary rock with a small proportion of igneous and metamorphic rock. The sedimentary rock that constitutes mountain ranges is often more than 10,000 meters (33,000 feet) thick, many times the thickness of sedimentary rock found in regions where mountain building has not been active.

The sedimentary rock that forms mountain ranges appears to have been formed from both shallow and deep water deposits. Much of the sediment for mountain building accumulated in slowly sinking troughs, or *geosynclines,* developed along continental margins. Parts of these geosynclines were situated on *continental shelves,* the shallow regions around the margins of continents. As the troughs sank, enormous thicknesses of sediment were built up while the upper layer remained under a shallow sea near the level of the earth's surface. The deep-water portions of geosynclines are located at oceanic trenches or at *continental rises,* where the sloping edge of a continent meets the ocean floor. The limestone portions of such ranges as the Alps and Himalayas represent former shelf deposits, whereas the interbedded sandstone and shale deposits appear to have accumulated on a continental rise or in a marginal trench.

Most geosynclinal troughs seem to be created by the descent of oceanic plates into the mantle. Recent active regions of mountain building in the Alps, the coastal ranges of the western United States, and the young ranges of Asia coincide with boundaries of actively moving global plates. Melting of the descending plate and the continental interface generates magma that wells up to the surface, so that a chain of active volcanoes is frequently formed in the early stages of mountain building. Igneous rocks such as granite are therefore found among the sedimentary rocks in mountain chains. Mountains begin to form by uplift of rock in the geosyncline when the rate of descent of the trough slackens, which allows the less dense sedimentary rock to rise above the surface. The compression exerted when plates move together can also squeeze sedimentary rock upward. Rock in mountain ranges often exhibits a high degree of wrinkling and rupturing from the pressures to which it has been subjected.

The most recent period of mountain building on the earth's surface began about 10 or 20 million years ago with the uplift of the Alps, Himalayas, and coastal ranges of the western United States. The Appalachian Mountains of the eastern United States were uplifted much earlier, perhaps 200 million years ago when plates were active in that region.

Faults and Folds

At some time in the earth's history, movements of the crust have significantly altered the landscape in most regions. Crustal movements normally are slow. Continental drift, sea floor spreading, and the rising of mountain chains all occur at a rate of approximately 1 centimeter per year (0.4 inch per year), which illustrates the close connection between sea floor spreading and mountain building. The movements are seldom steady. If you push a wet finger across a pane of glass, your finger alternately sticks and slips. Similarly, immense blocks of rock attempting to move past one another as the crust distorts will sometimes stop until enough strain builds up to force a sudden movement. The movement accompanying the sudden release of stored elastic energy is experienced as an earthquake. If the movements are steady, strain is continually released a bit at a time, and earthquakes do not occur. When liquids were pumped into deep storage wells in the area near Denver, Colorado,

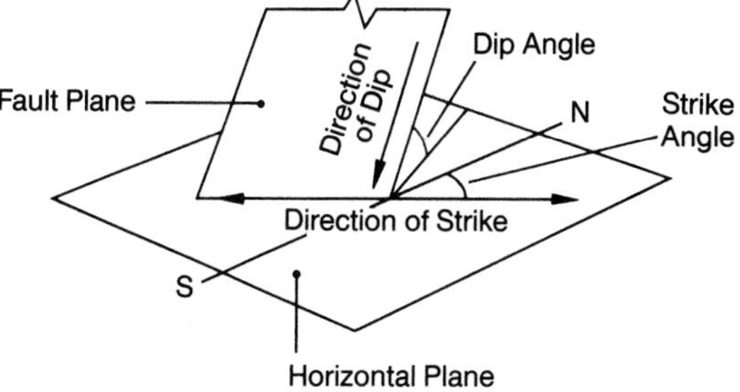
Dip Angle
Direction of Dip
Fault Plane
N
Strike Angle
Direction of Strike
S
Horizontal Plane

Figure 13.19 A break, or fault, in a rock mass often occurs along a plane. As illustrated in this diagram, the orientation of a fault plane can be specified by its *strike* and *dip*. Strike is the compass bearing of a horizontal line on a fault plane; it is measured in degrees from the north-south direction. Dip is the angle of steepest inclination of a fault plane; it is measured in degrees from the horizontal. The direction of dip is at right angles to the direction of strike.

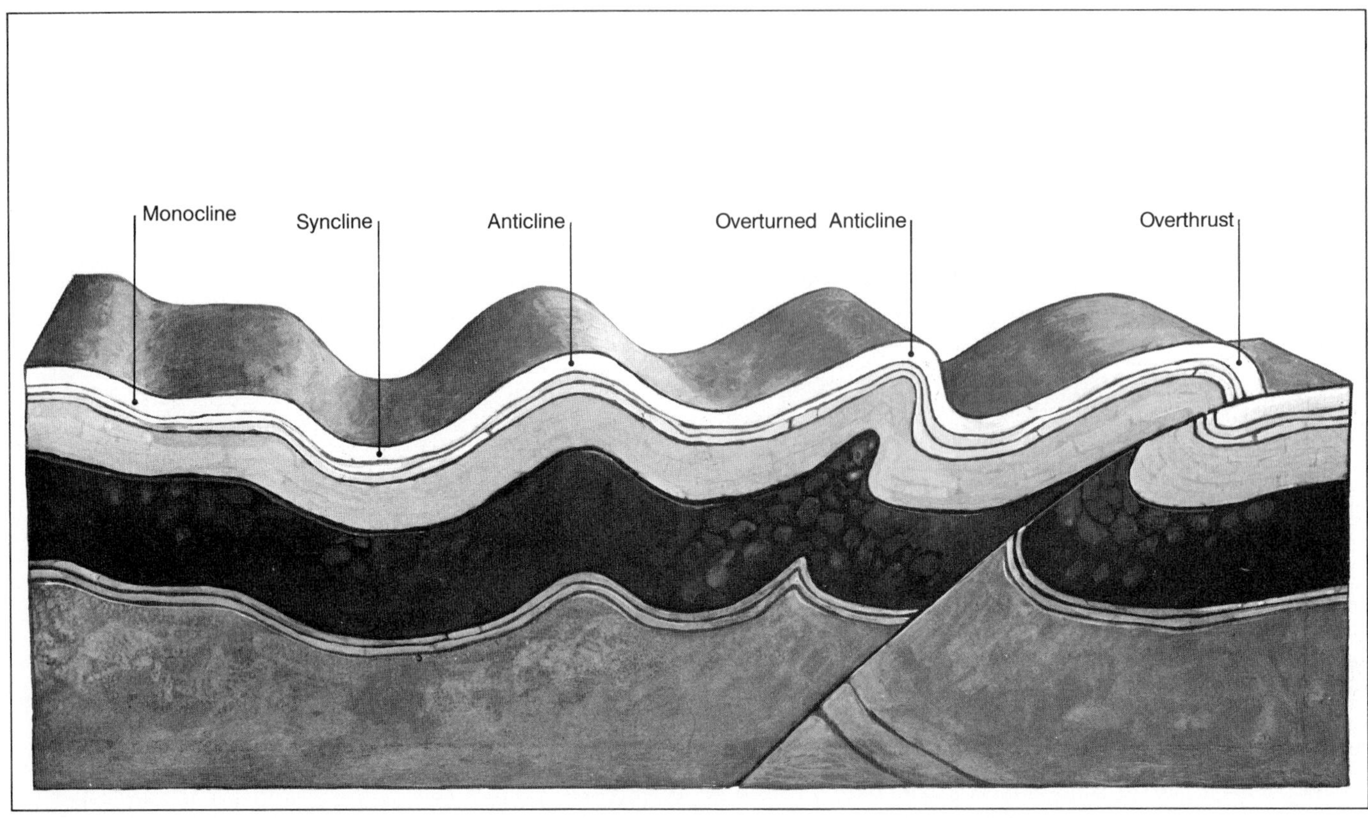
Monocline
Syncline
Anticline
Overturned Anticline
Overthrust

Figure 13.20 This diagram illustrates the geometry of folded rock strata and the terms used to describe it. A sharp flexure in bedded rocks is a *monocline*. A downfold whose sides, or *limbs*, are inclined upward on either side is a *syncline*, and an up-arch whose limbs are inclined downward is an *anticline*. Where folding is extreme, the limbs of the fold may be asymmetrically inclined and the strata on one limb of the fold may be oversteepened or even overturned. If the rock strata are subject to great pressure, an overturned fold may rupture along a zone of weakness, and one limb of the fold may be thrust over the other limb, forming an *overthrust fold*. Fold structures are geomorphically significant because distinctive landforms may develop by erosion of folds exposed at the earth's surface.

a significant increase in small earth tremors occurred. Liquids probably lubricate the contact surfaces between the blocks of rock, facilitating motion and preventing the accumulation of great strains. Pumping liquids into faults might be a way of releasing crustal strain at a controlled low rate, thereby reducing the occurrence of large earthquakes in areas of great strain, such as along the San Andreas fault in California.

The earth's crustal movements, combined with the force of differential erosion on rocks of different resistances, produce landforms of great complexity. Earlier in this chapter you looked at landforms developed in horizontal, undeformed sedimentary rock. Imagine now that the sedimentary rock is broken, twisted, folded, and tilted. Different strata of the rock become exposed at different locations on the earth's surface. The less-resistant rock strata wear away more rapidly than the others, and a highly differentiated landscape results—one that is controlled partly by the underlying geologic structure and partly by the weathering and erosion processes of the particular climatic region.

Many landforms are developed on complex geologic structures, such as *faults* and *folds*, as illustrated in Figures 13.20 and 13.21. Faults arise when separate crustal blocks move relative to one another in a variety of ways. Folds can also take a number of forms. It may seem strange that hard rock can be folded almost double upon itself, but folded structures are clearly visible in many exposed rock faces. Rock is brittle and tends to crack under sudden impact,

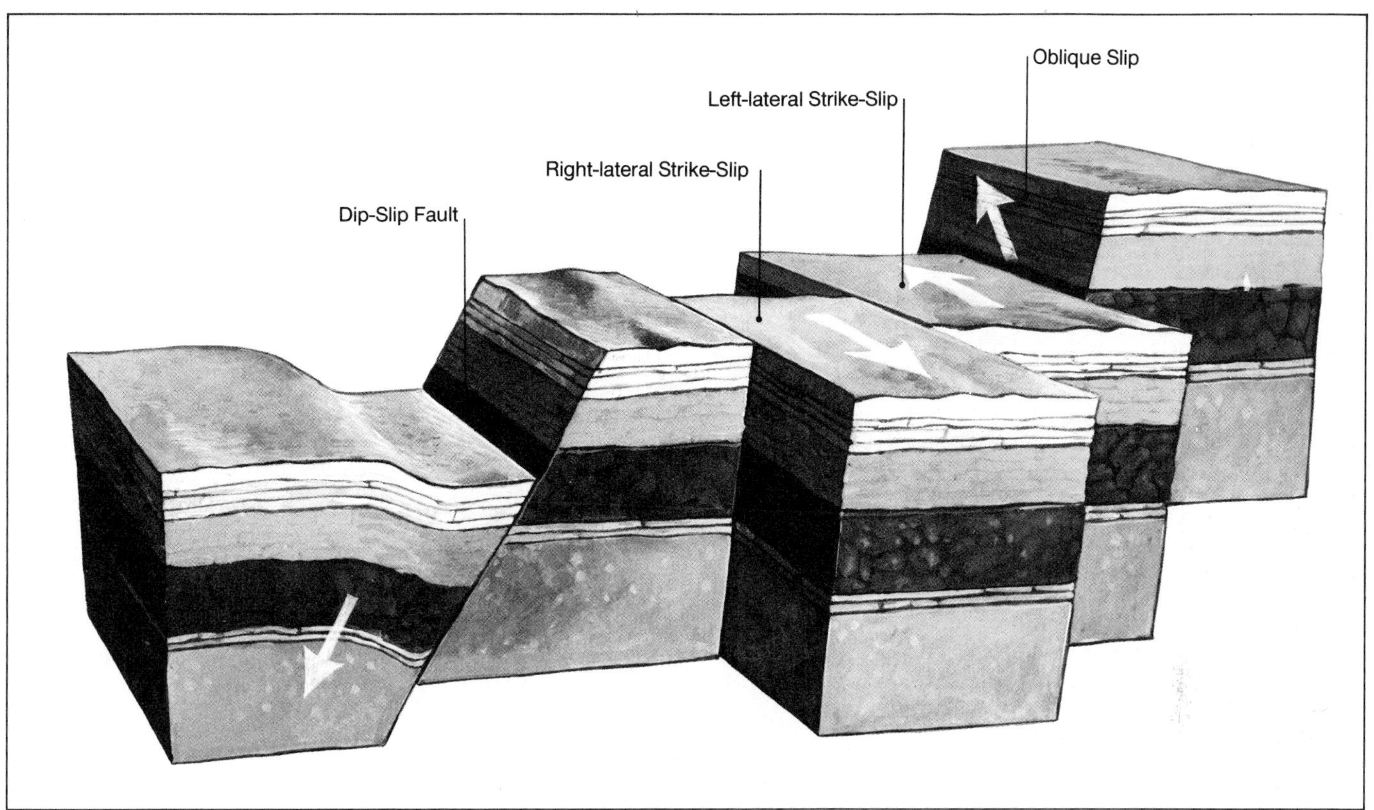
Oblique Slip
Left-lateral Strike-Slip
Right-lateral Strike-Slip
Dip-Slip Fault

Figure 13.21 Faults are the result of the mechanical failure of crustal rocks that occurs because of tension, compression, and tearing. Faults are described by the relative displacement, or *slip,* of rock units adjacent to the fault. The steeply inclined fault on the left of the diagram (a *normal fault*) is the result of tension that has pulled the crust apart; the block over the fault (the *hanging wall*) has moved downward relative to the other block (the *foot wall*). Compression, or crustal shortening, can produce a *reverse fault* in which the hanging wall has moved upward relative to the foot wall. The faults in the center of the diagram have vertical fault planes and exhibit horizontal slip. Such *tear faults* often occur in connection with earthquakes and movement along fault zones.

Faults can also be classified by specifying the direction of slip relative to the fault plane. The fault on the left of the diagram is called a *dip-slip fault* because the slip is along the direction of dip. In a *strike-slip fault*, the slip is along the direction of strike, as the diagram shows. A strike-slip fault is either right or left lateral, depending on the relative direction of slip of the opposite block as seen by someone facing the fault line. An *oblique slip,* shown on the right of the diagram, exhibits motion along both the direction of strike and the direction of dip. Fault structures exposed at the earth's surface develop landforms of characteristic appearance.

such as a hammer blow, but prolonged stress, particularly at high temperatures and high pressures, can slowly distort rock layers without breaking them.

Although the block diagrams that illustrate landforms show cross sections of underlying geologic structures, we usually see only the surface manifestations of the geologic structure. We can interpret the underlying structure by inspecting the surface landforms, studying the nature of the exposed rocks at the surface, and conjecturing about the effect of erosion on the structure. Interpretation may not be clear where complex faulted and folded structures occur, because several different combinations of structures and processes can produce similar landforms. Making the correct interpretation often is important economically, because geologic structure is used as a guide to the location of useful mineral and fuel deposits. In the last few decades improved instrumentation has aided geologists in determining the structure and composition of rock under the surface. Modern geologists use measurements of local magnetism, of the local force of gravity, and of the speed and intensity of seismic waves induced by explosions to clarify the nature of geologic structures and associated ore bodies.

Landforms and Faulted Structures

The Great Basin, centered in Nevada, dominates the map of the western United States. The Great Basin is traversed by numerous north-south ranges of mountains and is bounded on the west by the Sierra Nevadas of California and on the east by the Wasatch Mountains of Utah. Many of the mountain ranges in the Great Basin and on its borders show evidence of block faulting. The Sierra Nevada range is essentially a large tilted block. Its eastern edge stands 3,000 meters (10,000 feet) above the depressed block to the east, Owens Valley, whereas the western edge of the block has sunk below sea level. The Wasatch range is also a tilted fault block, with its western edge uplifted.

Block-faulted mountain ranges such as the Sierra Nevadas display a number of distinguishing features. Particularly striking is the way the steep faces seem to leap up suddenly from the level of the neighboring basin floor. Erosion has deeply dissected and incised the faces of the uptilted blocks forming the numerous mountain ranges in the Great Basin, but in some regions small areas of rock bear polished flat surfaces, or *slickensides*, caused by the abrasion of one fault block slipping past another.

The uplift of a block tends to be along straight lines that mark boundary faults. Many of the mountain ranges in the Great Basin have remarkably straight base lines. Sometimes the base line of a bold mountain range is observed to cut across regions of both strongly resistant and easily eroded rock. Where an escarpment is *discordant* with the rock type, as when a mountain base line cuts across rocks of different resistances, the assumption is strong that faulting rather than erosion created the existing relief.

Faults are fractures that penetrate deep into the earth's crust and act as channels to conduct heat and water from the interior of the earth. Hot and cold springs are therefore frequently found along fault lines; most of the warm springs in the Great Basin are in

Figure 13.22 (bottom) The San Andreas fault, a portion of which forms the dominant straight-line feature in the center of this photograph of the Carrizo Plain in California, is a region of shear at the boundary between the crustal plates of the Pacific Ocean and the crustal plates of North America.

(opposite) This orange grove is located in the zone of the San Andreas fault. Note the offset of the rows due to motion along the fault. The right lateral strike-slip fault has displaced the trees by a meter (3 feet) or more.

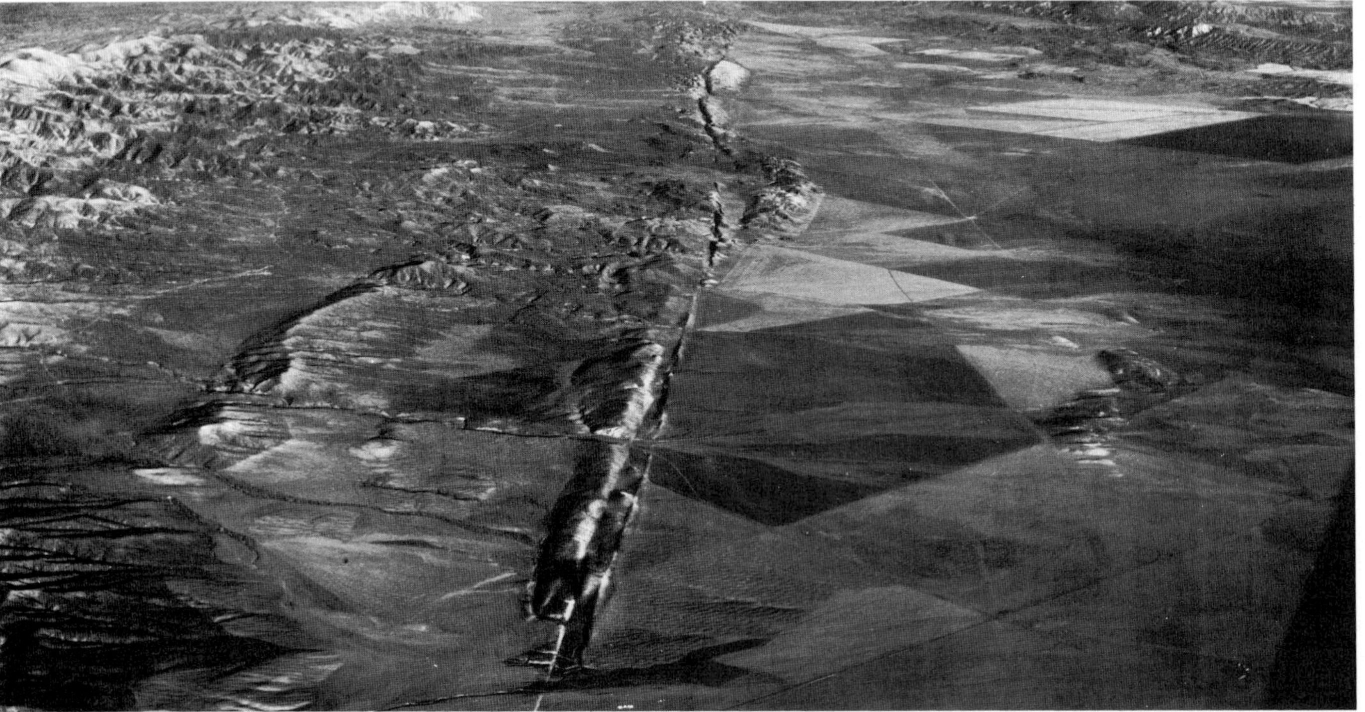

regions that bear other signs of faulting. Magma may also well up along such fractures, resulting in volcanic activity at the surface.

A fault in which the relative movement of blocks is principally horizontal is called a *strike-slip fault* because the slippage occurs in the direction of the strike (see Figure 13.21). One of the most famous strike-slip faults is the San Andreas fault, which stretches north and south for 800 kilometers (500 miles) along western California. The San Andreas fault system is not just a single fracture but consists of numerous essentially parallel faults located in a zone many kilometers wide.

Special landforms are associated with strike-slip faults. The line of the San Andreas fault is marked by straight valleys at the foot of straight mountain fronts. The straight valleys are caused primarily by the rapid erosion of a band of crushed rock, which movements of the fault blocks have fragmented. The differential movements of the earth in a large fault zone sometimes cause a local area to sink below the surrounding land or become hemmed in on all sides by higher ground. If the depression is closed, it acts as a local drainage basin for the higher land around it, forming a *sag pond*. Sag ponds along the San Andreas fault often occur in mountainous areas, and where they occur in alignment, they are often the clearest surface evidence of the fault. Springs are common along the San Andreas fault. Many of the springs are caused by displacement of the local water table, but other springs are hot and mineralized, indicating a connection with regions deep within the earth's crust.

One of the interesting features of a strike-slip fault system is the horizontal offset of features such as stream beds caused by the relative displacement of the fault blocks. Displacement of the fault

Figure 13.23 These two photographs of a volcanic fissure were taken at the same location near Kilauea crater, Hawaii, in 1959 (top) and in 1971 (bottom). The fissure was formed by tensional forces that pulled apart the upper crust. The earlier photograph shows vegetation well established in the developing volcanic soil of the fissure. By 1971, however, eruptions of fresh lava had destroyed the vegetation and discolored the rocks with mineral oxides.

Volcanic activity continues to alter the surface of the earth. Although at present volcanic activity tends to be restricted to isolated local regions, in the geologic past the upwelling of lava from the earth's interior has covered large areas of the earth, including portions of Idaho, Oregon, and Washington.

occurs in small increments, and at each stage streams crossing the fault tend to retain their old channels above and below the fault. In time, the streams crossing an active fault may suffer a total horizontal offset of hundreds of meters.

Landforms and Folded Sedimentary Rocks

Because faults are linear features, they define linear blocks and rifts. Compressive stresses commonly force the crust to shorten by wrinkling, or *folding*, producing a quite different landscape.

In a series of folds the arches are called *anticlines*, and the troughs are known as *synclines*, as Figure 13.20 illustrates. The highest point in any cross section of an anticline is the *crest*. The type of landform that develops on folded sedimentary rock depends on the climate and vegetation cover, the size and shape of the folds, the relative resistances of the rock strata, the depth of erosion, and the vertical and horizontal orientation of the folds.

Begin with a simple example. Suppose that the anticlinal and synclinal folds have level crests and troughs over long distances, as illustrated in Figure 13.20. Suppose further that one of the thick folded layers is highly resistant to erosion compared to the other layers. If the block is uplifted above the surface so that erosion begins, the more rapid erosion of the less resistant rock eventually results in a landscape of regularly spaced parallel ridges separated by valleys. Each ridge is the edge of a resistant layer, and each valley is cut in softer erodable materials. In actuality, the folds are never continuous but die out eventually. This causes the ridges to converge and meet along the fold axes.

Figure 13.25 shows the wide variety of ridge and valley shapes that can be generated in folded rock depending on the orientation of the folds and the differential erosion of the rock layers. The

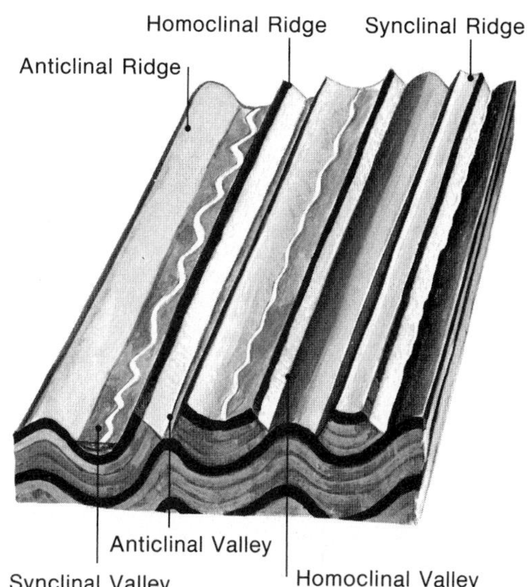
Homoclinal Ridge
Synclinal Ridge
Anticlinal Ridge
Anticlinal Valley
Synclinal Valley
Homoclinal Valley

Figure 13.24 This schematic diagram illustrates the variety of structures that can be formed by erosion of folded rock strata that crop out at the earth's surface. The dark-shaded rock layers are assumed to be more resistant than the other layers. Erosion of the weaker rock overlying hard rock on an anticline forms an *anticlinal ridge*. A *synclinal valley* of hard rock is formed similarly when weaker rock is removed from a syncline. If a portion of the top layer of resistant rock over an anticline has been worn away, *homoclinal* ridges are formed by the exposed edges of rock strata, with an *anticlinal valley* between them. A *homoclinal valley* is formed at the junction of a homoclinal ridge with the steeply sloping limb of an exposed anticline. A concave *synclinal ridge* is formed where a remnant of resistant rock from a syncline acts as cap rock to retard the erosion of the underlying layers. Ridges and valleys developed on structures of folded rock are common in many landscapes, such as the Appalachian Mountain region of the eastern United States. (See also Figure 13.10).

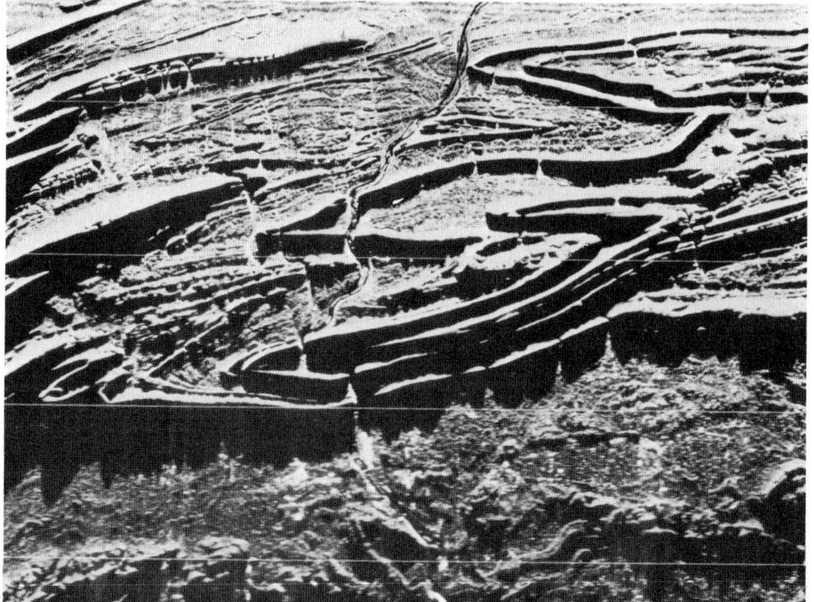

Figure 13.25 Folded strata form a distinctive ridge and valley landscape in southeastern Pennsylvania. This aerial view near Harrisburg was made with side-looking airborne radar (see Appendix II); the image spans a distance of approximately 50 miles. The ridges are formed of highly resistant sandstone, and the valleys are cut into weakly resistant shale or limestone. Note the nested hairpin ridges formed where anticlines dip downward. A diagram of the drainage pattern of this region is presented in Figure 13.10. The gaps in the ridges through which the Susquehanna River flows are clearly visible in the images.

Figure 13.26 These diagrams illustrate how rivers such as the Susquehanna River in the Appalachian Mountains of Pennsylvania are able to cut gaps through ridges of highly resistant rock. (a) A region underlain by folded rock strata becomes worn to (b) a plain of low relief. Rivers establish themselves on the plain, perhaps in a cover of undifferentiated alluvial deposits, and gradually incise the plain (c), cutting across the bands of strongly and weakly resistant rock. (d) Eventually, erosion of the weaker rock exposes a ridge system of resistant rock. However, if some of the streams are able to cut their channels downward rapidly enough to keep pace with the removal of the weaker rock, they may maintain their original courses and cut notches through the resistant rock ridges. Newly established streams in the ridge and valley landscape are unable to cut through the resistant ridges and must flow parallel to them along the bands of weaker rock. (After Lobeck, 1956)

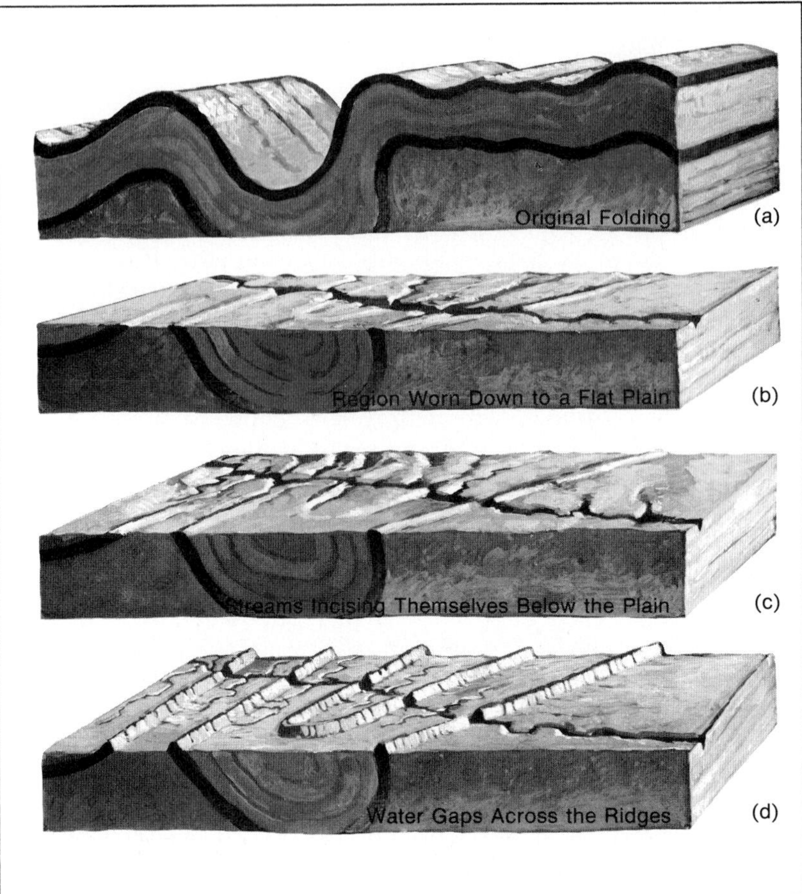

Original Folding
(a)
Region Worn Down to a Flat Plain
(b)
Streams Incising Themselves Below the Plain
(c)
Water Gaps Across the Ridges
(d)

nested and hairpin ridges of central Pennsylvania are a direct expression of underlying folded geologic structures.

The streams in central Pennsylvania form a trellis-shaped drainage pattern. Tributary streams flow along their parallel valleys between the long ridges before they join the master streams. Clearly the geologic structure strongly controls much of the drainage pattern in this region. Yet the larger rivers in Pennsylvania do not detour around ridges but cut through them. The Susquehanna River near Harrisburg cuts through three successive high ridges in the space of 10 kilometers (6 miles) instead of paralleling the ridges as the smaller streams do. It is postulated that in the geologic past, a long period of erosion reduced central Pennsylvania to a landscape of low relief. The Susquehanna became established in a flat depositional cover on this surface, perhaps in deposits overlying the bedrock landscape. Thus, the course of the river was not determined by the underlying folded rocks. Later, when the land was slowly uplifted, the river was able to maintain its preexisting channel by cutting down into the uplifted rocks. In some places the river cut through weaker overlying layers of rock, and on meeting more resistant rocks in the folded structure beneath, it continued to erode these more resistant rocks, forming narrow gorges. The general erosion of the less resistant rocks by streams and their tributaries gradually exposed the pattern of ridges and valleys seen today. Many of the river gorges that cut through mountain ranges may have been formed in a way similar to the way the Susquehanna cut through the uplifted ridges.

Summary

The rocks of the earth's crust consist primarily of silica with mixtures of other constituents such as aluminum or magnesium compounds. Igneous rock is formed by cooling from the molten state, sedimentary rock is formed by the cementing together of sediments, and metamorphic rock is formed by the recrystallization of other rocks. The chemical composition and physical structure of rock determine the relative ease with which it can be weathered and eroded. Differential erosion of rock, particularly on layered or jointed rock, plays an important role in the formation of landforms.

Landforms are a function of structure, process, and stage of development. Some landforms are formed primarily by deformations of the earth's crust. Weathering and differential erosion act on geologic structures to produce a variety of complex landforms and drainage patterns. The comparatively high solubility and jointed structure of limestone give limestone karst regions unique landforms.

Volcanic activity is found at active plate boundaries, particularly around the basin of the Pacific Ocean and along the oceanic ridges. Volcanic activity is usually nonviolent in Iceland and Hawaii because steam easily escapes from the simatic magma. Violent volcanic eruptions, such as those in Italy, occur when steam is trapped in heavy viscous sialic magma; rock fragments are frequently ejected in these eruptions. Volcanism produces characteristic landforms such as stratovolcanoes, lava plateaus, and volcanic plugs and dikes. Hot water and steam for heating and power are often available near the earth's surface in regions of volcanic activity or deep faulting.

Block faulting produces linear structures such as fault valleys and mountain ranges with straight base lines. Folding, particularly the folding of layered sedimentary rock, can produce a variety of ridge and valley landforms depending on the geometry of the folds and the relative resistance of the exposed layers to erosion. Where some of these landforms are distributed on the earth is the subject of Chapter 14.

14 Regional Landforms

Climate and Landforms
Climatic Regions and Weathering Processes
Vegetation and Erosion

Slopes and Their Formation
Forces That Hold Slopes Together
Creep: Slippage of Slope Material
Soil Wash and Rainsplash

Local and Regional Climatic Differentiation
Hoodoo Rocks
Asymmetry of Slopes
Morphogenetic Regions

Slope Processes in Different Climates
Slopes in Periglacial Climates
Slopes in Moist Midlatitude Regions
Slopes in Arid and Semiarid Regions
Slopes in Hot, Moist Climates

A Gallery of Landscapes

The towering pillars of limestone in the karst region of China are an example of a unique landform produced by certain combinations of weathering processes with underlying rock structures. Climatic controls influence many of the processes that act on landforms, giving slopes their characteristic shapes.

Winter Landscape from the Ch'ing Dynasty. (Courtesy of the Smithsonian Institution, Freer Gallery of Art, Washington, D.C.)

Regional Landforms

For anyone accustomed to the more gentle countryside of the eastern United States, seeing the western United States for the first time may be as astonishing as landing on another planet. In the East, the hills hide their rocks under a cover of soil and vegetation. In the West, the steep exposed rock slopes reveal the lessons of geology to anyone who cares to study them. Why are the landscapes of the East and West so different? The contrast is too great to be explained entirely by differences in rock types and geologic structures.

What kind of landform does granite produce? Look again at the examples of slopes in Figure 11.11. The same rock type can underlie a wide variety of landforms. Near Atlanta, Georgia, granite forms the smooth, bald dome of Stone Mountain, whereas in the Mojave desert a boulder slope is composed of rough granite chunks. In the Alps, granite juts out in sawtooth ridges. A resident of New Hampshire may think of granite in terms of the worn slopes of the White Mountains. And a person living in a tropical rainforest in an area of granitic bedrock has probably never seen a landform made of exposed granite because the rocks disintegrate rapidly in a hot, moist climate.

The physical and chemical properties of rock are one of the controls that determine the shapes of landforms. The underlying geologic structure and varying processes of weathering and erosion also control the evolving prominences and contours that give the landscape its shape. The way a particular rock type responds to a given process depends on the nature of both the rock and the operative processes. Granitic rock, for instance, may resist mechanical weathering, but it deteriorates rapidly under chemical weathering; the reverse is true of shale.

Climate and Landforms

Climate controls many of the processes that act on landforms. Weathering processes, both mechanical and chemical, are strongly dependent on the temperature and moisture delivered by climate. Frost weathering, for example, requires low temperatures and a supply of liquid water for effective action, and chemical weathering is enhanced by high temperatures and abundant moisture. Erosion by flowing water, an important agent of geomorphic change, depends on the amount of runoff generated by water surplus from the local water budget. Climate acts on landforms indirectly by the strong influence it exerts on the soil and vegetation in a region. And the soil and vegetation cover affect the development of landforms by modifying the action of weathering and erosion processes.

The importance of a given geomorphic process may be greater in one climatic region than in another. A process that dominates in one part of the world, such as the transport of sediment by the wind in some arid regions, may be relatively insignificant in another area. The differences in process between one climatic region and

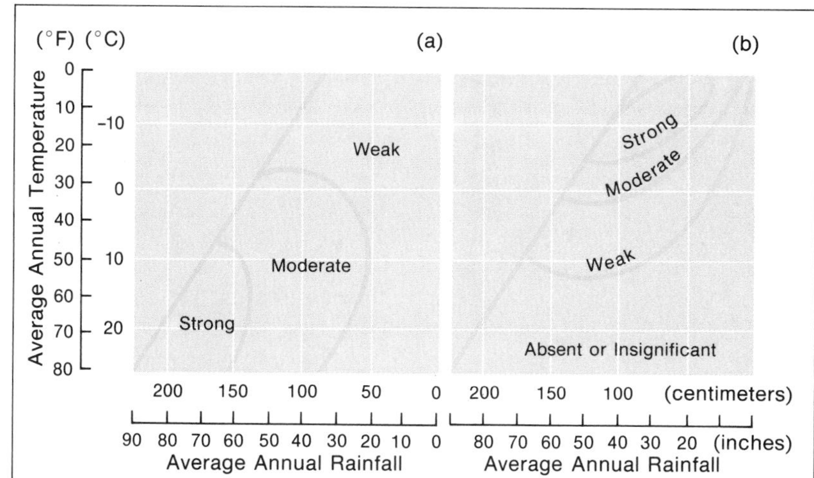

(°F) (°C)
(a)
(b)
Average Annual Temperature
0
10
20
30
40
50
60
70
80
-10
0
10
20
Weak
Moderate
Strong
Strong
Moderate
Weak
Absent or Insignificant
200 150 100 50 0
200 150 100 (centimeters)
90 80 70 60 50 40 30 20 10 0
80 70 60 50 40 30 20 (inches)
Average Annual Rainfall
Average Annual Rainfall

Figure 14.2(a) This schematic diagram suggests the relative importance of chemical weathering in various regimes of temperature and precipitation. (Certain possible regimes, such as low temperature and high precipitation, do not occur on the earth and are not included in the diagram.) Chemical weathering is most active where temperatures are high and moisture is abundant. The solvent action of rainwater and surface runoff is primarily responsible for chemical weathering, which therefore tends to be most intense near the surface of the ground and to decrease with depth. Although chemical weathering is weakest in dry regions, it may still be an important form of weathering. Chemical weathering dominates in many hot, dry regions where other weathering processes may be comparatively inactive. (After Peltier, 1950)

(b) This schematic diagram suggests the relative importance of mechanical frost weathering in various climatic regimes. Frost weathering occurs where free moisture is available to penetrate rock joints and cracks, and where the temperature drops below freezing from time to time. It is unimportant in hot, dry regions or in regions that are perpetually frozen. The diagram does not take into account relatively minor processes of mechanical weathering that may be locally important, such as abrasion by wind-borne particles. (After Peltier, 1950)

another sometimes lead to the development of landforms characteristic of a particular climatic region. When traveling you realize that the familiar landscape of home is by no means the norm over the entire earth. The wooded hills of the eastern United States seem as exotic to a desert nomad as the curving dunes and rippled sands of the desert appear to someone from New England.

Many landforms are not uniquely associated with a climatic region but occur in several regions, and in general, the dependence of landforms on climate is by no means as great as the dependence of soils or vegetation on climate. Processes may differ in intensity from one climatic region to another, but the landforms shaped by the processes may look very similar. Furthermore, the response time of landform systems to climate is usually long, so that many present-day landforms still show the influence of earlier climates. Nevertheless, the processes occurring today are controlled to some extent by the present climate. This chapter discusses the general relation of climatic processes to landforms, and it surveys the distinctive landforms of several regions.

Climatic Regions and Weathering Processes

One of the most significant ways that climate affects the development of landforms is through its control of weathering processes. The same type of rock material may weather at greatly different rates in different climatic regions. Limestone blocks quarried to provide facing stone for the great pyramids in the deserts of Egypt have weathered so little in 4,000 years that tool marks are still visible on them, but limestone grave markers in moist New England have worn to the point of illegibility in only 300 years, because limestone is susceptible to chemical erosion by flowing water and by acidic groundwater.

Figure 14.2(a) illustrates the importance of chemical weathering in various regimes of temperature and rainfall. As the figure indicates, high temperatures and abundant moisture, the conditions found in rainforests, favor active chemical reactions that result in weathering. Most chemical reactions require moisture in order to proceed, because in solution atoms become ionized and interact strongly with one another. A high temperature speeds up chemical reactions because it causes the ions to move more rapidly and thus to collide more frequently, enhancing the probability that they will react with one another. The rate of many chemical reactions roughly doubles for every 10°C (18°F) increase in temperature. The weathered surface layer in hot, moist climates is frequently tens of meters deep because of the intensity of chemical weathering and because many tropical regions have experienced their climates for long periods of geologic time. Chemical weathering is less significant in hot, dry climates, where little water is available to ionize the atoms, or in cold climates, where chemical reactions are slow and water is frozen much of the time. Figure 14.2(b) suggests the relative intensity of frost action in different climatic regions. Frost action, the principal process of mechanical weathering, is unimportant in warm regions and becomes more important in cool regions that have a satisfactory moisture supply.

The relative importance of all weathering processes in different regimes of temperature and moisture is illustrated in Figure 14.3. The diagram indicates that weathering processes are relatively inactive in dry climates, both hot and cold. A diagram that classifies weathering processes with respect to climatic factors should be interpreted only as a broad generalization and should be used only as a framework for thinking about different regions. The figure

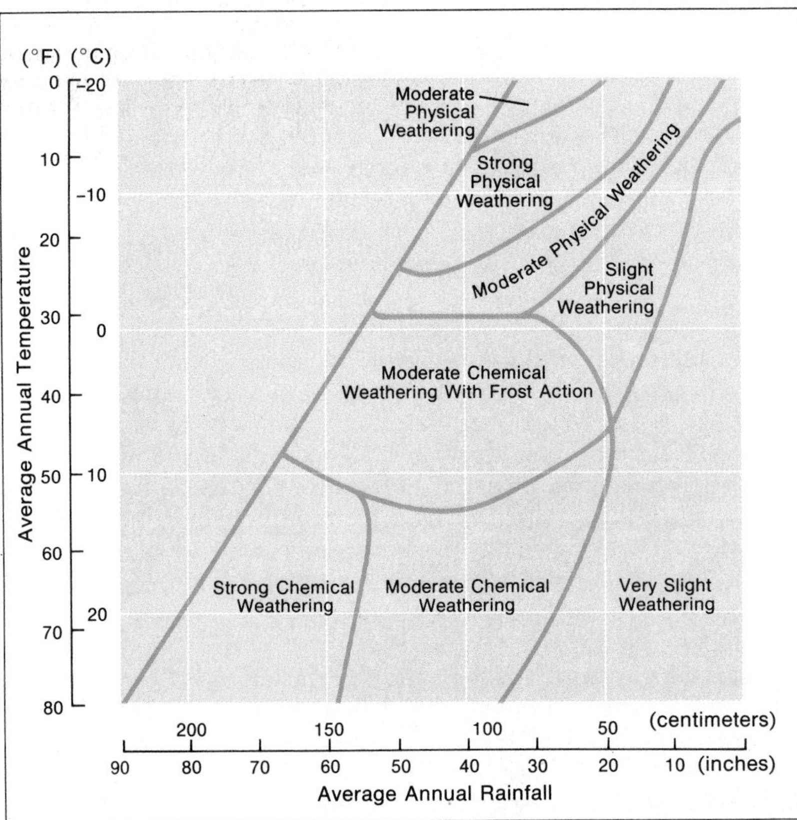
(°F) (°C)
0
-20
10
-10
20
30
0
40
50
10
60
20
70
80
Average Annual Temperature
Moderate Physical Weathering
Strong Physical Weathering
Moderate Physical Weathering
Slight Physical Weathering
Moderate Chemical Weathering With Frost Action
Strong Chemical Weathering
Moderate Chemical Weathering
Very Slight Weathering
200
150
100
50
(centimeters)
90
80
70
60
50
40
30
20
10 (inches)
Average Annual Rainfall

Figure 14.3 This schematic diagram suggests the dominant weathering process in various climatic regimes. Chemical weathering and frost weathering cannot occur unless moisture is available, so that weathering is slight in dry regions where precipitation and surface runoff are meager. However, even in regions where weathering is slight, the continuous action of relatively weak processes can eventually produce characteristic landforms. It should also be noted that many parts of the earth's surface were subjected to different regimes of weathering through time because of changes in climate. The landforms in such regions may have been formed partly by weathering processes that are no longer active there today. (After Peltier, 1950) omits the effect of seasonal variations in the timing of temperature and precipitation, and it does not consider the detailed processes involved in weathering. However, insufficient data are available to allow a more accurate description of the relation between climate and weathering processes.

Vegetation and Erosion

Vegetation frequently contributes to weathering and erosion processes on landforms. Plants play a role in weathering, particularly by the effects of their nutrient cycles, acid secretions, and decay products, which affect the chemistry of the water in contact with soil and subsurface rock.

However, biological weathering is insignificant compared to the important effect vegetation has on landform development through its modification of erosion. A dense vegetation cover helps maintain an open surface structure in the soil, partly by the action of roots in keeping soil grains separated and partly by preventing raindrop splatter from sealing the open spaces between grains at the surface. Because vegetation tends to maintain a high rate of infiltration into a soil, it significantly reduces the amount of moisture available for surface runoff and consequent erosion. To a lesser degree, plants reduce erosion by holding and stabilizing soils.

If land in a moist climate is kept free of vegetation, erosive gulleying will soon begin to carry off soil materials. The sediment yield from open, unplanted land is 50 to 100 times greater than the sediment yield from the same area of grassland under conditions of equal precipitation. In agriculture, land sometimes remains free of plant cover for part of the year, and erosion is greater during such periods. Studies of lake sediments near an ancient Roman town indicate that the average sediment yield from nearby land increased tenfold soon after the town was founded and agricultural activity began. Recent estimates of man's effect on rates of erosion indicate that the average erosion rate over all the earth's landmasses has doubled or tripled because of such activities as clearing land for agriculture, removing trees in logging operations, and sealing the ground with pavement, all of which increase the amount of water available for surface runoff and erosion.

Local water budgets show that the amount of moisture remaining for surface runoff and erosion in a particular location depends on the amount of precipitation and on the amount of moisture entering the soil. In arid regions, precipitation is small, but because plant cover is sparse, little moisture enters the soil and surface runoff can be locally high in sudden rainstorms. Where moisture is more abundant, however, the vegetation cover is thicker, which increases the amount of water entering the soil. So the amount of runoff generated in a moist region actually may be less than the amount generated in a drier region, because of the effect of vegetation.

Figure 14.4 illustrates typical rates of lowering, or *denudation*, on a moderate (10°) slope in a hot climate where the average monthly temperature is about 25°C (77°F). Potential evapotranspiration in such climatic regions is approximately 80 to 100 centimeters (30

Figure 14.4 This diagram illustrates how the rate of lowering by erosion, or denudation, is affected by precipitation and vegetation. The example considered is a 10° slope in a warm region where the average annual temperature is 25°C (77°F). Soil wash by surface runoff is the most important process of denudation. The rate of lowering is least where the annual precipitation is about 100 centimeters (40 inches), because in such regions potential evapotranspiration by the vegetation cover is equal to precipitation, and little surface runoff is generated. The rate of denudation is greater in drier regions, because vegetation cover is sparse and surface runoff readily erodes the exposed bare ground. Creep and rainsplash on bare ground also play a role. In very moist regions, more moisture is available than the vegetation can use, and large moisture surpluses are generated. The abundant moisture carries away material in solution as well as by soil wash.

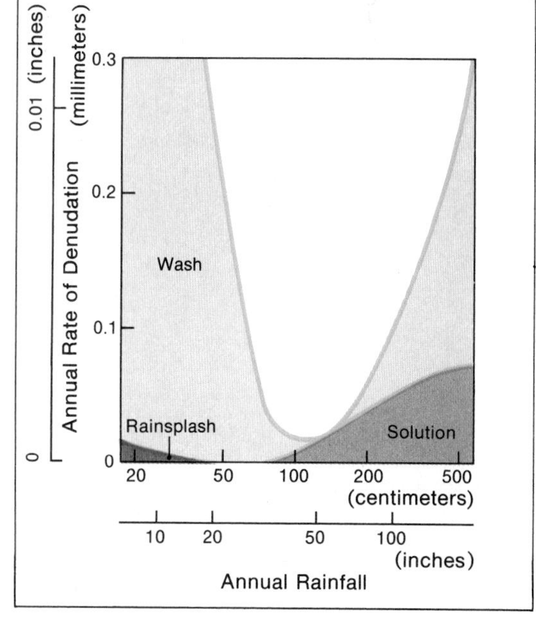
0.01 (inches)
0
(millimeters)
0.3
0.2
0.1
0
Annual Rate of Denudation
Wash
Rainsplash
Solution
20
50
100
200
500
(centimeters)
10
20
50
100
(inches)
Annual Rainfall

to 40 inches), so vegetation tends to be sparse where the annual precipitation is less than 50 centimeters (20 inches) or so. The figure shows that denudation rates are high in hot semiarid and arid regions, where the amount of moisture is insufficient to maintain a dense cover of vegetation. In regions where precipitation greatly exceeds potential evapotranspiration, surplus moisture becomes available to erode material or to dissolve and carry it away in solution. Denudation is therefore also high in warm regions where the annual precipitation is much greater than 100 centimeters (40 inches). There does appear to be a relationship between denudation rates and precipitation. The rate of denudation is usually highest in semiarid and arid regions where vegetation is sparse or absent. Denudation is primarily by erosion in such regions, because removal of material by solution is significant only in exceptionally humid regions of the world.

Slopes and Their Formation

The shape of a slope—whether it is a mammoth mountain range or a ridge of dirt—depends on the relation between the rate at which material is disintegrated on the slope and the rate at which material is transported away. Understanding the main processes that act on slopes is helpful in considering how landforms develop. Furthermore, many processes of slope formation are influenced by climate and sometimes lead to the development of landforms characteristic of particular regions.

Forces That Hold Slopes Together

Slopes of soil, of rock fragments, and even of solid rock are held together by forces between their particles. Without such forces, a hillside would collapse under the ever-present downward pull of gravity. The steeper the slope, the more effective gravity is in urging materials to move down the slope. A slope can be stable only if the force of gravity acting on each volume of rock or soil is balanced by the cohesive force acting through the volume's area of support. Weathering of jointed rocks on a slope can reduce the cohesive force to the point where rocks break loose, or where slabs crash to the ground. The movement of large masses of material down a slope under the direct influence of gravity is called *mass wasting*.

For a given material on a hillside—whether a blanket of soil or a layer of loose debris—there is a maximum slope angle, called the *angle of repose*, that the material can make without slipping downhill. If new material is added to a slope that is already at its angle of repose, a portion of the slope may slide downward. Wet soils are particularly likely to become detached from a slope, because the absorbed moisture increases the weight of the soil and lessens the cohesive force between the soil particles.

Figure 14.5 (opposite) Weathering products from the upper slopes of Postern Peak in the Canadian Rockies have formed talus slopes of coarse, light-colored material near the base of the mountain. Talus slopes do not have an opportunity to weather to soils that could hold vegetation, because new material is constantly transported down the slope to be carried away by valley streams below.

Where cliffs of exposed rock appear in the landscape, the segment of a slope immediately below the bare cliff face is sometimes covered with *talus*, a layer of loose rock fragments that have broken away from the cliff face by weathering. Talus slopes are often seen

in semiarid climates where rock remains exposed and free of vegetation. Talus slopes usually accumulate until the angle of repose is reached, so that a slight downward push can sometimes start the talus sliding downward under the force of gravity.

Creep: Slippage of Slope Material

If a slide of loose material accelerates its speed and accumulates additional debris, it may become a cascading landslide. The scar from a catastrophic landslide down a mountain slope or the pile of broken rocks at the foot of a cliff offers clear evidence that slopes can be unstable. But even the rock and soil on a seemingly stable pasture slope slip downhill at a fraction of a centimeter per year in a form of mass wasting known as *creep*. Rates of creep are measured by setting pins or rods into the soil and, after an interval of time, observing how far the pins or rods are displaced from their original positions. Creep as rapid as a few centimeters per year can tear soil horizontally and cause fence posts to tilt visibly. Creep

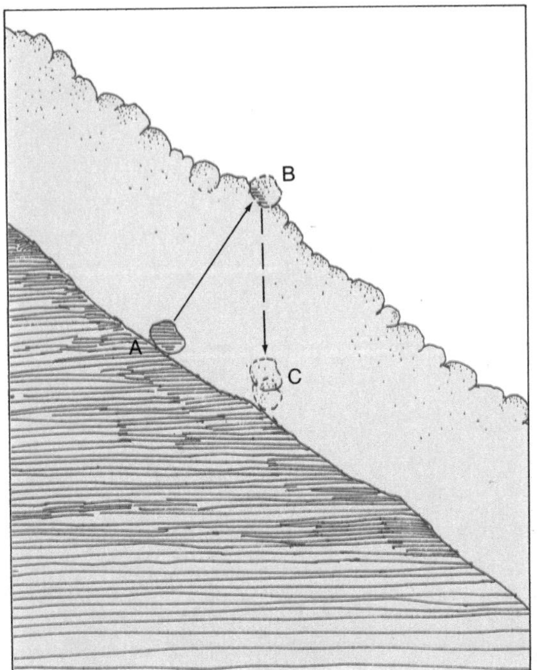

B
A
C

Figure 14.6 Soil tends to creep downslope, on the average, under the influence of gravity. This schematic diagram illustrates one of the mechanisms responsible for soil creep. When soil becomes moist it tends to expand at right angles to itself, but when the soil dries and contracts the soil particles tend to move vertically downward, which induces a slight downslope migration.

Figure 14.7 The ice stems, or *pipkrakes,* shown in this photograph are a minor mechanism for moving soil downslope. When water in the ground freezes and expands outward, it tends to carry soil particles with it at right angles to the surface. Soil particles tend to fall vertically downward when pipkrakes melt, and a downslope movement is induced.

Figure 14.8 (opposite top) Glaciers, such as this small valley glacier in Victoria Land, Antarctica, form in regions where the cold climate allows snow to persist throughout the year. As the glacier ice flows slowly down the valley, rock fragments become incorporated in the ice. Armed with these tools, the glacier forcefully erodes the bedrock and the valley walls. Glaciers are perhaps the most powerful agents for the geomorphic alteration of slopes.

(opposite bottom) Landslides result in the sudden transport of large amounts of material down slopes. The landslide scar shown in the photograph was produced in 1955 by a landslide that moved tons of soil, rock, and vegetation into Emerald Bay on Lake Tahoe at the border of California and Nevada.

Figure 14.9 (left) Soil saturated with water has little cohesive strength, and sometimes large bodies of mud break loose and flow rapidly downslope in a *mudflow*. The mudflow in the photograph occurred in Idaho, near the headwaters of the Pahsimeroi River visible in the foreground. It has a characteristic lobed appearance, and its speed enabled it to spread out into the valley before coming to rest. Mudflows are frequently found in dry regions where sudden heavy rains saturate the soil and where the vegetation cover is too sparse to help bind the soil.

Figure 14.10 (right) Sometimes a portion of a slope slides downward and outward in a *slump*. The photograph shows a slump at Point Fermin on the Palos Verdes Peninsula, California. The forward section of coast became detached from the mainland after wave action at the base of the cliffs undermined its support. The mass slumped downward as a whole, so that its surface is now considerably lower than before. Note the broken edge of the parking lot in the lower right.

probably occurs on all soil-covered slopes that are steeper than a few degrees, but the rates are often slow and difficult to measure.

One explanation for the mechanism of soil creep is that moistened or freezing soil expands at right angles to the slope. When the soil dries or thaws, however, it tends to settle in a vertical direction, which results in a small displacement down the slope. The presence of vegetation on a slope tends to reduce the rate of creep because roots have a binding effect on the soil. Typical rates of creep on a grassy slope are usually of the order of a centimeter or less per year, but the measured creep of a steep grassy slope in England was much slower—equivalent to 1 centimeter in 50 years (0.4 inch in 50 years).

Solifluction is a process of mass wasting that occurs in cold climates where the soil is permanently frozen with only the upper meter or so thawing each summer. Solifluction is movement of soil as a whole when the spaces between the soil particles become saturated with water from thawing ground ice. The rate of solifluction is typically a few centimeters per year. Occasionally soil contains so much moisture that the soil particles become separated and flow as viscous liquid, producing a *mudflow*. This can occur in any climate in which all water is not frozen permanently.

Soil Wash and Rainsplash

Flowing water is usually the most important transporter of material on slopes. *Soil wash* often transports 50 to 100 times as much material as soil creep if water is available for runoff.

Rainsplash can move small particles when raindrops fall forcefully on unprotected soil surfaces. Sometimes rainsplash on a hillside moves particles as large as 1 centimeter (0.4 inch), and rainsplash can move larger stones indirectly by cutting away the finer material that supports the stones. Rainsplash on a slope results in a net downward movement of material; the particles thrown into the air by the impact of a raindrop usually land somewhat downslope on the average. A thick cover of vegetation almost completely prevents transport by rainsplash, so like erosion, rainsplash tends to be more effective in semiarid climates than in moist climates.

The convex summit of a ridge is sometimes nearly horizontal, so that water cannot gain the energy required to detach and carry away soil particles. Thus, convex ridge tops commonly exhibit a narrow belt where little or no erosion occurs.

Figure 14.11 Solifluction lobes, shown here on a slope on the Seward Peninsula, Alaska, exhibit an irregular margin because the downslope movement of the thawed soil is intermittent.

Local and Regional Climatic Differentiation

Processes controlled by climate operate on landforms at all scales, from individual frost-riven boulders to sloping hillsides. Following are examples of climatic influences on local and regional landforms.

Hoodoo Rocks

What processes could have produced a rock like the Goblet shown in Figure 14.12? The Goblet is an unusual sandstone formation located in semiarid country near Blanding, Utah. The sculpting of such *hoodoo rocks* has sometimes mistakenly been attributed to the erosive power of sand grains carried by the wind, but abrasion is unimportant compared to differential weathering in the presence of moisture.

Even in arid regions where the annual rainfall is only 10 or 20 centimeters (4 or 8 inches), water is the principal weathering agent. When infrequent rains fall on a rock pinnacle, some of the rain water runs down the vertical face to the base. The shaded lower parts of the rock dry more slowly than the upper parts, and where the rock is wet, minute physical and chemical changes occur. Water infiltrates some of the rock's crystals, expanding mineral crystals and partially dissolving the cementing substances. With enough repetitions of the process, the grains will loosen from the parent rock and eventually fall out. The base and stem become indented most because they stay wet the longest after a rainfall. The rock surfaces of the pedestal and undersides of the cap of a hoodoo rock give direct evidence that the loosening of grains, and not abrasion by wind-borne sediment, is the major shaping process in rocks such as the Goblet.

Figure 14.12 This hoodoo rock, known locally as the Goblet of Venus, stood near Blanding, Utah, until it was toppled several years ago. Its narrow stem was cut by the slow action of trickling water on the individual grains of sandstone.

Asymmetry of Slopes

The hoodoo rock illustrates the effect of climate in shaping a single landform. Climate also influences the shape of larger features, such as slopes. In many climatic regions the slope angles on the opposite sides of east-west valleys differ, although there is no corresponding difference in rock type. In the middle latitudes north-facing slopes are steeper on the average than south-facing slopes by about 5°. The vegetation and soil on opposing slopes are also different.

Part of the observed asymmetry of slopes is due to differences between the "microclimate" of one slope and the other. In the

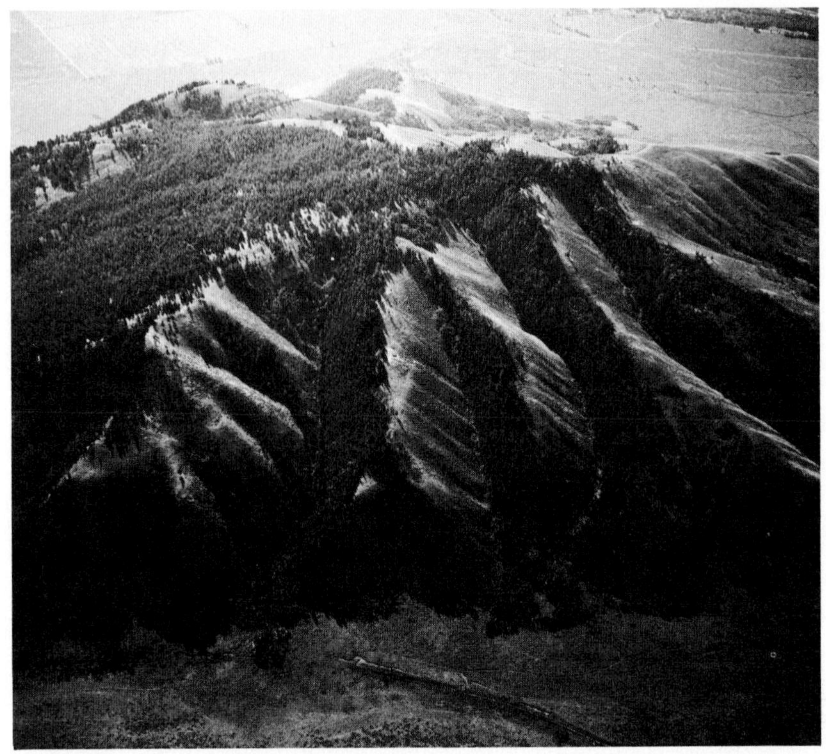

Figure 14.13 The dissected slopes on Black Tail Butte, Wyoming, all show an asymmetry between the north-facing side and the south-facing side. The north-facing slopes are thickly covered with coniferous forest, whereas the drier south-facing slope has only sparse clumps of shrubs and grasses. The nature of the vegetation cover influences the development of the slopes. Flowing water will tend to erode the barren south-facing slope more rapidly than the north-facing slope, which is protected from rainsplash by its vegetation cover and which has less surface runoff because of its more open, permeable soil structure.

Northern Hemisphere, south-facing slopes receive more solar radiation than north-facing slopes, so that south-facing slopes tend to be warmer and drier. In a cold climate, a north-facing slope may be covered only by a layer of rubble while a segment of the warmer south-facing slope supports a layer of soil covered by tundra vegetation. In Virginia, for example, the drier south-facing slopes of some valleys are forested mainly by pine, whereas the moister slopes on the opposite side of the valley carry oak trees as well as pine.

Different processes operate to produce valley asymmetry in different climatic regions. In cooler climates the difference in rates of snow melting on opposite sides of a valley can help induce asymmetry, and in warmer climates, the complicated interplay of available moisture, soil type, and vegetation type affects the form of slopes in a valley.

Part of valley asymmetry is probably caused by the asymmetric erosive action of streams flowing in the valley and is not directly attributable to local differences in climate. The transport of material down a slope, hence the form of the slope, is influenced by the rate of removal of material from the slope's base. If a stream removes material more rapidly from one side of its channel than from the other, perhaps because of differences in underlying rock structure, slopes on opposite sides of the stream will follow different courses of development.

Morphogenetic Regions

The relative importance of various slope-forming processes differs from one region to another. A classification system of *morphoge-*

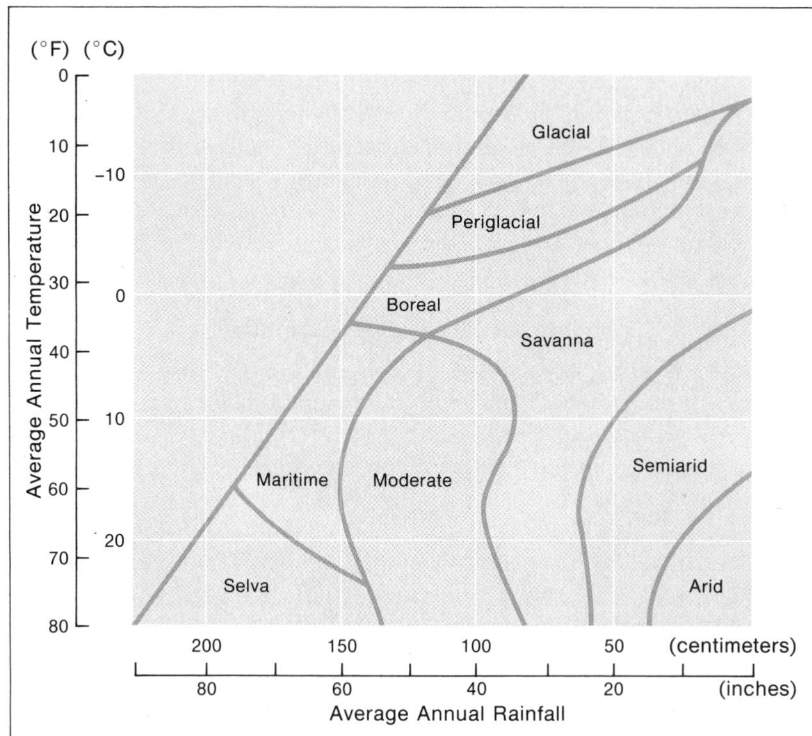
(°F) (°C)
Average Annual Temperature
0
10
-10
20
30
0
40
50
10
60
70
20
80
Glacial
Periglacial
Boreal
Savanna
Maritime
Moderate
Semiarid
Selva
Arid
200
150
100
50
(centimeters)
80
60
40
20
(inches)
Average Annual Rainfall

Figure 14.14 As this schematic diagram suggests, each region on the earth can be characterized as a morphogenetic region possessing a certain combination of weathering and erosion processes that influences the development of landforms in the region. The boundaries in the diagram should be considered only as qualitative guides. The processes considered in classifying the morphogenetic regions include glacier and snow action, mass movement, frost weathering, and transport by water and wind. A detailed description of each region is given in Appendix III. Note that because of climatic change the morphogenetic character of a region may vary with time. The northeastern and north central United States are now included in the moderate morphogenetic region, but 10,000 years ago they were glacial or periglacial. (The term "boreal" refers to climatic regions that are also known as subpolar, or subarctic.) (After Peltier, 1950)

netic regions based on the weathering, transport, and erosion processes that predominate in particular regions is useful for seeing how climate influences landform development. Figure 14.14 illustrates one classification system of morphogenetic regions with their suggested climatic boundaries. The region labeled *moderate* corresponds to the moist midlatitude region Davis considered in his cycle of erosion. Erosion in the moderate region is accomplished primarily by running water with moderate rates of mass wasting; frost action is comparatively unimportant. Detailed definitions of the nine morphogenetic regions according to this scheme are listed in Appendix III.

If climate were the dominant factor controlling the shape of landforms, it would be possible to link a distinctive set of landforms with each of the morphogenetic regions. However, no landform is shaped solely by climate, or by rock type, or by geologic structure; the three factors cannot be completely disentangled. So it is not surprising that similar landforms may arise in different morphogenetic regions. Furthermore, the climate in a given region is not necessarily constant and may change more rapidly than landforms can respond, particularly in regions subject to recent glaciation. For this reason, the landforms in a region may reflect past rather than present climates. The classification into morphogenetic regions specifies only the processes that are presently active in a given region; it does not consider how the landscape may have developed. At the climatic extremes, however, certain landforms are associated with only one morphogenetic region. It is difficult to imagine 1,000 square kilometers of sand dunes being anywhere but in an arid climate where there is active wind transport and no vegetation.

Slope Processes in Different Climates

Different geomorphic processes dominate in different climatic regions, which the concept of morphogenetic regions illustrates. As the following brief descriptions of slopes in the principal climatic regions indicate, there are a wide variety of processes creating slope forms; in many instances the slopes closely reflect the climate in which they are formed.

Slopes in Periglacial Climates

Periglacial regions are found at the margins of Pleistocene or present-day ice sheets. Such regions are marked by sparse vegetation cover and active frost weathering. Freezing and thawing of water in the soil can move portions of soil, and solifluction is common. Typical slopes in periglacial regions consist of an upper exposed face of nearly vertical rock and a talus slope immediately below produced by frost shattering of the cliff. The comparatively steep (30° to 40°), dry talus slope merges into a more gentle, wetter layer, where mass movements and solifluction commonly occur. The solifluction segment of the slope is often concave upward with a slope angle of 10° or so at its lower end, and it merges into an outwash slope covered by material washed down from the slopes above. Because periglacial regions have generally been subject to climatic change during geologic time, they are probably not in equilibrium with present climatic processes.

Slopes in Moist Midlatitude Regions

The softly curving slopes of the eastern United States are examples of slopes in the moderate morphogenetic region. In moist midlatitude climates, soil wash is relatively ineffective because the dense vegetation cover minimizes surface runoff and rainsplash. However, over long periods of time soil wash may be responsible for the lower concave portion of slopes. The summits of slopes in the moderate region are almost always convex unless the slope is being cut back rapidly. The convex shape of soil-covered hilltops is attributed to the process of creep. In semiarid regions where soil wash is more important than soil creep, convex slopes also are frequently developed, but by the nature of overland flow of water rather than by mass wasting. In such climates subtle variations in rock type create conspicuous differences in slope angle and form.

Slopes in Arid and Semiarid Regions

In semiarid regions, such as the southwestern United States, the vegetation cover is limited, and bare rock and rock fragments are frequently exposed. The slopes exhibit greater angularity than other gently inclined erosion surfaces. A typical slope in a dry region has a steep cliff in its upper reaches followed by a comparatively straight talus slope with a slope angle of 25° to 35°. The slopes tend to have only a thin layer of rock debris. The amount of material leaving the lower section of the slope segment equals the amount of new material entering the segment from above, and little material accumulates on the slopes. The sparse vegetation and compacted soil structure cause infiltration into the soil to be slow, so that runoff and erosion are heavy during occasional rainstorms. Wash is more effective than creep in dry regions; rainsplash also is important in dry regions.

The lower concavity of slopes in dry regions often forms an extensive region called a *pediment*. Pediments are surfaces of transport of sediment by water. In terms of surface area, they are the dominant landform in dry regions, and desert plains are largely gently sloping concave pediments cut by gullies and water channels.

Slopes in Hot, Moist Climates

Hot, moist climates can be considered as two regimes, the savanna zone, where precipitation is less than potential evapotranspiration for several months of the year, and the rainforest, where precipitation almost always exceeds potential evapotranspiration. The large amount of runoff characteristic of both regions is caused by sparse vegetation in the savanna and by excess precipitation in the rainforest. The abundant vegetation in the rainforest sometimes covers slopes as steep as 70° and protects them from erosion by running water. Soil wash and removal of material by solution are important processes of denudation in both regions, but the sliding of saturated soil on steep slopes is conspicuous in rainforests. The weathered mantle, or *regolith*, in a hot, moist climate is generally thick, sometimes attaining 100 meters (330 feet) vertically. The concentration of less soluble iron or silica produces soils that tend to harden when deep sections are exposed to the atmosphere. Thus the landscape acquires a crust of reddish laterite.

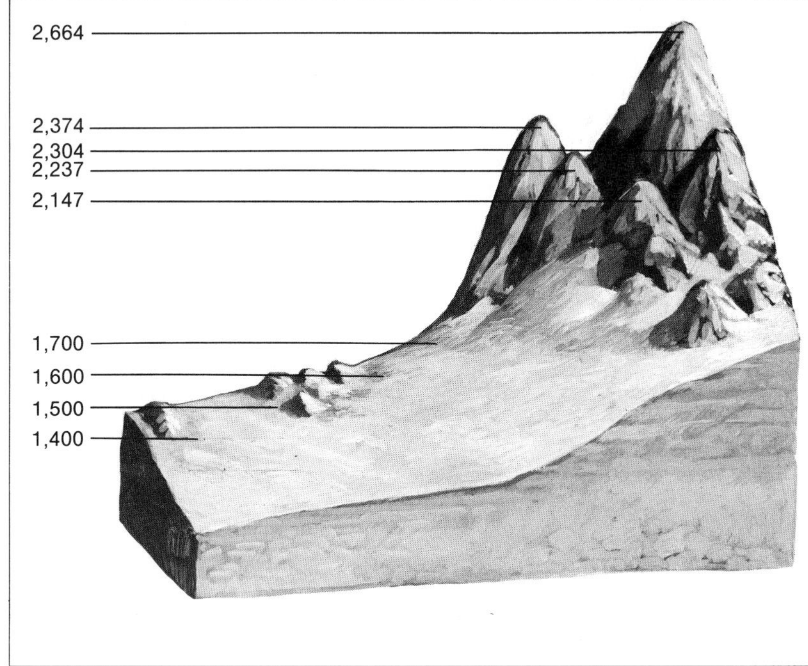

2,664
2,374
2,304
2,237
2,147
1,700
1,600
1,500
1,400

Figure 14.15 This illustration shows a rock pediment near Antelope Peak, Arizona. (Elevations are given in feet.) The pediment is the smooth slope that begins abruptly at the mountain front. The slope of the pediment is steepest near the mountains and more gradual farther away. Pediments are characteristic features in semiarid mountainous regions; they are believed to be formed by flowing water in streams, by unchanneled sheetwash flowing across the surface, and by the weathering retreat of the mountain front. The surface of a pediment is usually covered with a thin layer of alluvial deposit cut into a dense pattern of drainage rills.

A Gallery of Landscapes

Figure 14.16 **Cold Climate Landscape.** Where temperatures remain below freezing the larger part of the year, the landscapes take on a highly distinctive appearance. Below the top meter or so of the land surface, the water in all pore spaces in rock and soil remains frozen solid throughout the year. Permanently frozen rock and soil is known as *permafrost.* Permafrost reaches to depths exceeding 600 meters in northern Alaska and Siberia. Above the permafrost, the surficial blanket of soil thaws each summer, becoming saturated with water that cannot escape downward due to the frozen substrate. This water here and there lubricates the soil to the degree that the soil slips downward as a mass, a few inches in a matter of a week or two. The water-saturated soil moves in lobes that are conspicuous on hillslopes as the inset (bottom right) indicates. This type of movement is known as *solifluction.* Solifluction lobes are a few centimeters to a meter or so in height. Studies show that despite the appearance of great activity produced by these solifluction lobes, most are inactive, movement occurring only in a few places during each thaw. However, considered over a long period of time, the entire soil cover is draining downslope at a far more rapid pace than that produced in warmer regions by the creep process. Thus the portion of the soil that freezes and thaws annually, expanding, contracting and moving downslope, is known as the *active layer*. The long-term effect of the solifluction process is to smooth the landscape, filling preexisting depressions, and peeling down projections. Small valleys are infilled with *colluvium* (slope deposits), forming flat-floored *dells* with poorly developed watercourses. Water leaks through the colluvial valley fills.

Bare rock projecting above the active layer, or exposed by removal in the active layer, is susceptible to intense freezing during the long winter season. The expansion of water in passing from the liquid to the crystal state exerts enormous pressures on confining walls, causing rocks to split and, occasionally, to crumble. Thus, projecting solid rock masses are rapidly reduced to rubble, which becomes incorporated into the solifluction lobes. High areas are therefore lowered effectively. Broad summits are covered with angular rock rubble produced by the intensity of the freezing process in exposed sites. The result is a "sea of rocks," or *felsenmeer.* At the edges of rock exposures, large *talus* accumulations reflect the downslope movement of frost-riven blocks.

The annual freeze and thaw process produces a host of unusual minor landforms in addition to solifluction lobes. Repeated volume changes in the weathered mantle has the effect of sorting out the coarse and fine material it includes. Due to the efficiency of freezing in springing loose slabs of rock, the weathered mantle on hills contains an abundance of coarse debris. This becomes shunted away from the fines which accumulate in masses surrounded by rings of slabs which are more or less "on end." The result is the stone rings shown in the inset (top center). On slopes these rings are drawn out downhill into "garlands" and "stripes" by movements of the active layer. In lowlands composed of silty alluvial deposits another type of *patterned ground* is developed. During the freezing process, water-soaked silt first expands, but at prolonged low temperatures, eventually begins to contract. Wet ground, frozen in the cold season, thus cracks, producing a network of polygonal fissures. Ice begins to form in these and eventually produces clear

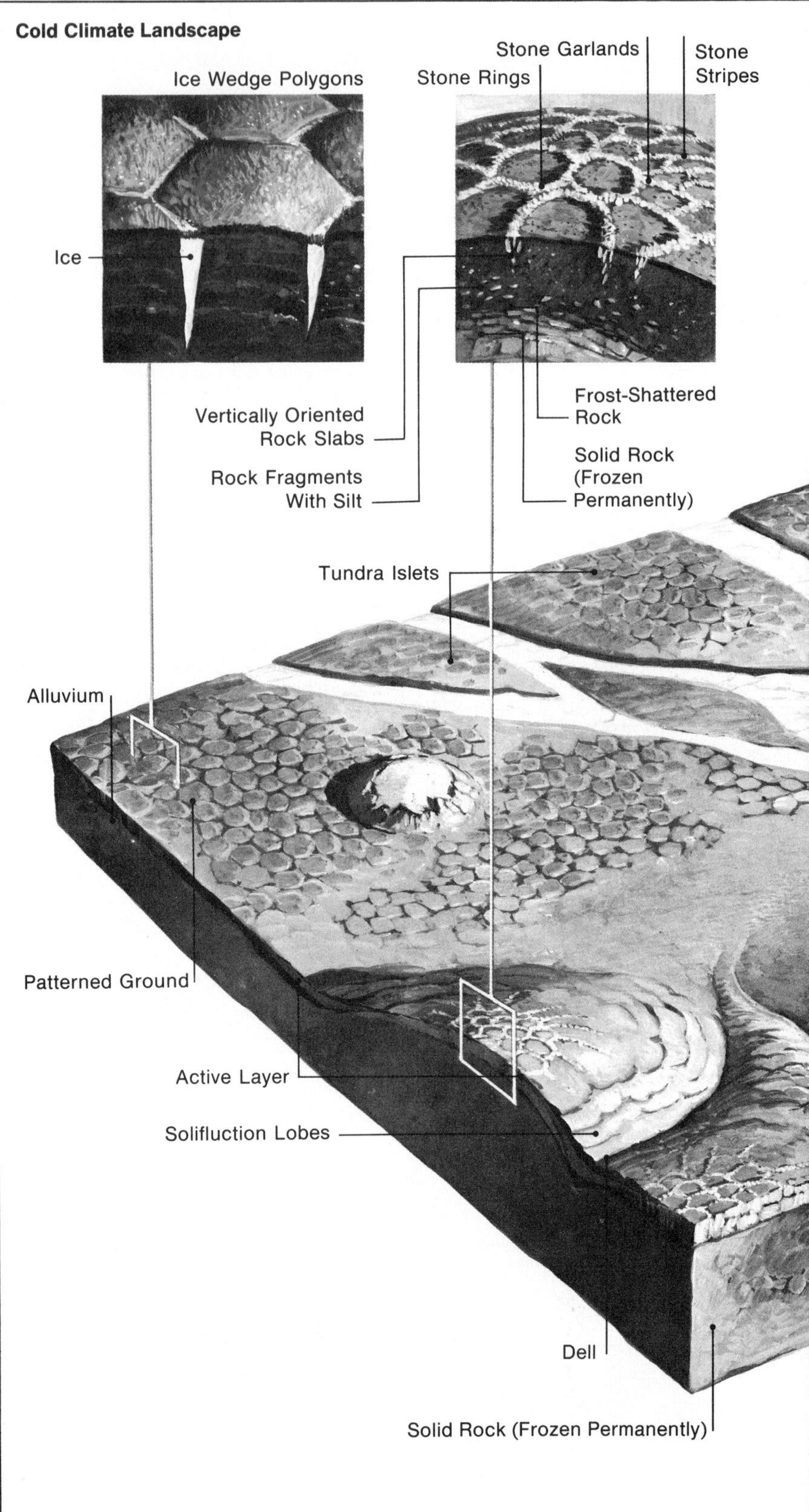
Cold Climate Landscape
Ice Wedge Polygons
Stone Garlands
Stone Rings
Stone Stripes
Ice
Vertically Oriented Rock Slabs
Rock Fragments With Silt
Frost-Shattered Rock
Solid Rock (Frozen Permanently)
Tundra Islets
Alluvium
Patterned Ground
Active Layer
Solifluction Lobes
Dell
Solid Rock (Frozen Permanently)

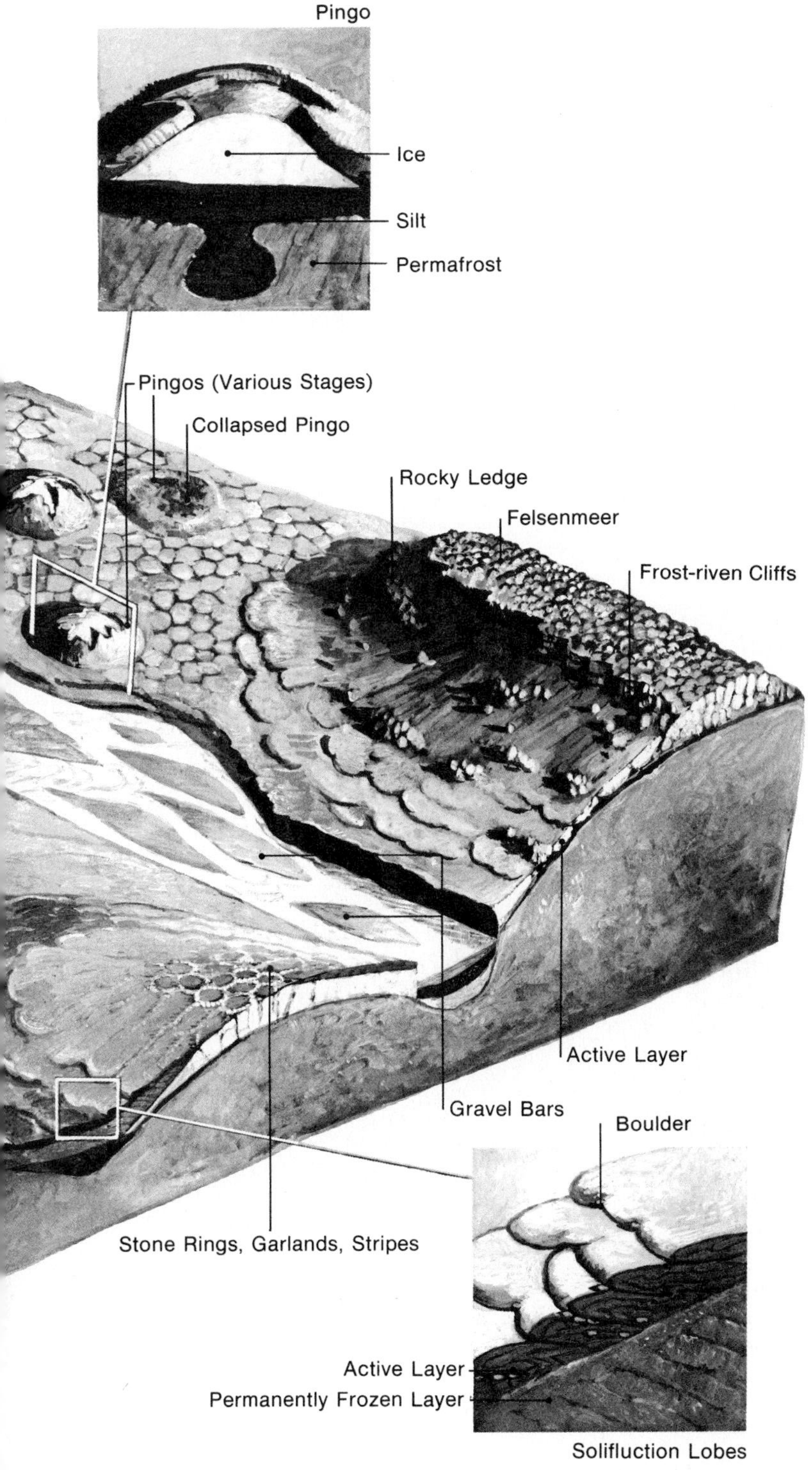

Pingo
Ice
Silt
Permafrost
Pingos (Various Stages)
Collapsed Pingo
Rocky Ledge
Felsenmeer
Frost-riven Cliffs
Active Layer
Gravel Bars
Boulder
Stone Rings, Garlands, Stripes
Active Layer
Permanently Frozen Layer
Solifluction Lobes

wedges, often a meter across and 5 meters deep (see inset at top left). These are *ice wedge polygons.* Once formed they are permanent, continuing to grow very slowly. Clear ice also forms in the ground in lenslike masses that heave up the overlying sod (see inset at top right). This unusual, but not uncommon, phenomenon is known as a *pingo.* Pingoes develop in old lake beds or marshes that are filling with sediment and vegetation. At a certain point the encroachment of permafrost into the fill traps a lens of unfrozen water and fill. Freezing of the water in turn draws moisture from the fill to produce a growing mass of ice that eventually domes up the recently developed sod above it. If a pingo is destroyed by thawing it leaves a depression ringed by an earth rampart. A large pingo may be 100 meters across and 30 meters high.

Other oddities of landscapes formed by *cryergic* (low-temperature) processes are overhanging (frozen) river banks, and unusually smooth stream profiles due to the abundance of fresh rock waste which provides streams with abrasional tools. (After Oberlander, 1974)

Figure 14.17 The small, ice-cored domed hill is a pingo near the delta of the Mackenzie River in the Northwest Territories, Canada. The surface of the ground exhibits the polygonal pattern often found in cold climate regions underlain by perennially frozen ground.

Figure 14.18 **Desert Landscape—Relief by Geologic Movement.** In this landscape the difference in elevation of the highest mountain crests and the basin floors is a consequence of rupture of the land surface. The basin is an area that has dropped with respect to the high land adjacent. The line along which the rupture and displacement occurs is a fault. The vertical displacement may attain thousands of meters. In arid regions, water collects in the basin after rains, producing a temporary lake. This soon evaporates, leaving a dry lake bed, or *playa*. This may be a mud flat, or it may develop a crust of saline minerals precipitated during evaporation of the lake water or brought to the surface by capillary rise from a body of saline groundwater that is near the surface. Where a mud flat is present the water table is very low and very large lakes have not been present recently (in the last several thousand years). The minerals of saline playas are zoned in order of solubility from carbonates at the edge, through sulphates, to chlorides, such as pure rock salt, in the center. These evaporite minerals are extremely valuable, and contribute upwards of 50 million dollars annually to the economy of California alone.

Such geologic structures may occur in any climate. What is distinctive in arid regions are the *playas, alluvial fans,* well-preserved *evidences of fault movement,* and *angularity* of form. In a humid area, such a basin would fill with water, overflowing eventually to produce an outlet channel to the sea. The lake would be eliminated by infilling with sediment to a point where a graded outlet channel would cross it, carrying off immediately all runoff from the surrounding mountains. The alluvial fans reflect rapid torrential runoff from slopes lacking vegetative protection. Loose material is flushed off during cloudbursts, moves into flooding channels and is carried out of the mountains. Where the bedrock channel ends, the floodwater spreads and percolates into the gravelly debris around the basin periphery. The resulting loss of velocity causes the coarse sediment load to settle out, adding to the volume of the fan. Two types of fan are present in this view. On the left, large fans drown the spurs of the down-tilting block, filling the canyons with alluvium. This is because filling of the basin creates a constantly rising base level. Across the basin the fans are small. They lie against the young fault scarp, and the valleys feeding them are actually "hanging" due to rapid uplift along the fault. The outer "toes" of these fans are drowned by alluvium due to subsidence along the right side of the basin block, and this partly accounts for their small size. Ordinarily, these fans are small, because their catchment areas, on the fault scarp, are smaller than catchments on the "backslope" of the block. The inset shows how fault uplift often creates a wineglass valley, by rejuvenating erosion, causing slotlike gorges to be cut below older V-shaped valleys. The bowl of the wineglass is the open upper valley; the stem is the new erosional slot; the base is the alluvial fan below.

Fan surfaces in deserts are interesting due to their microtopography. Frequently, hardened streams of mud studded with sizeable boulders (1 to 3 meters in diameter) are present, indicating that the fan is built largely by *debris flows*—mixtures of mud, gravel, and boulders that rumble out of the canyons during infrequent very heavy rains that fall on barren surfaces of loose material. The inset (top right) shows a recent debris flow covering a slightly older one. Surfaces of different ages are apparent in the *color* they display, because the desert rocks tend to become coated with a patina of iron and manganese oxides with

Desert Landscape in Areas of Block Faulting

Large Alluvial Fans Coalesce to Form a Bajada (Alluvial Apron)
Water Table
Fan Deposits

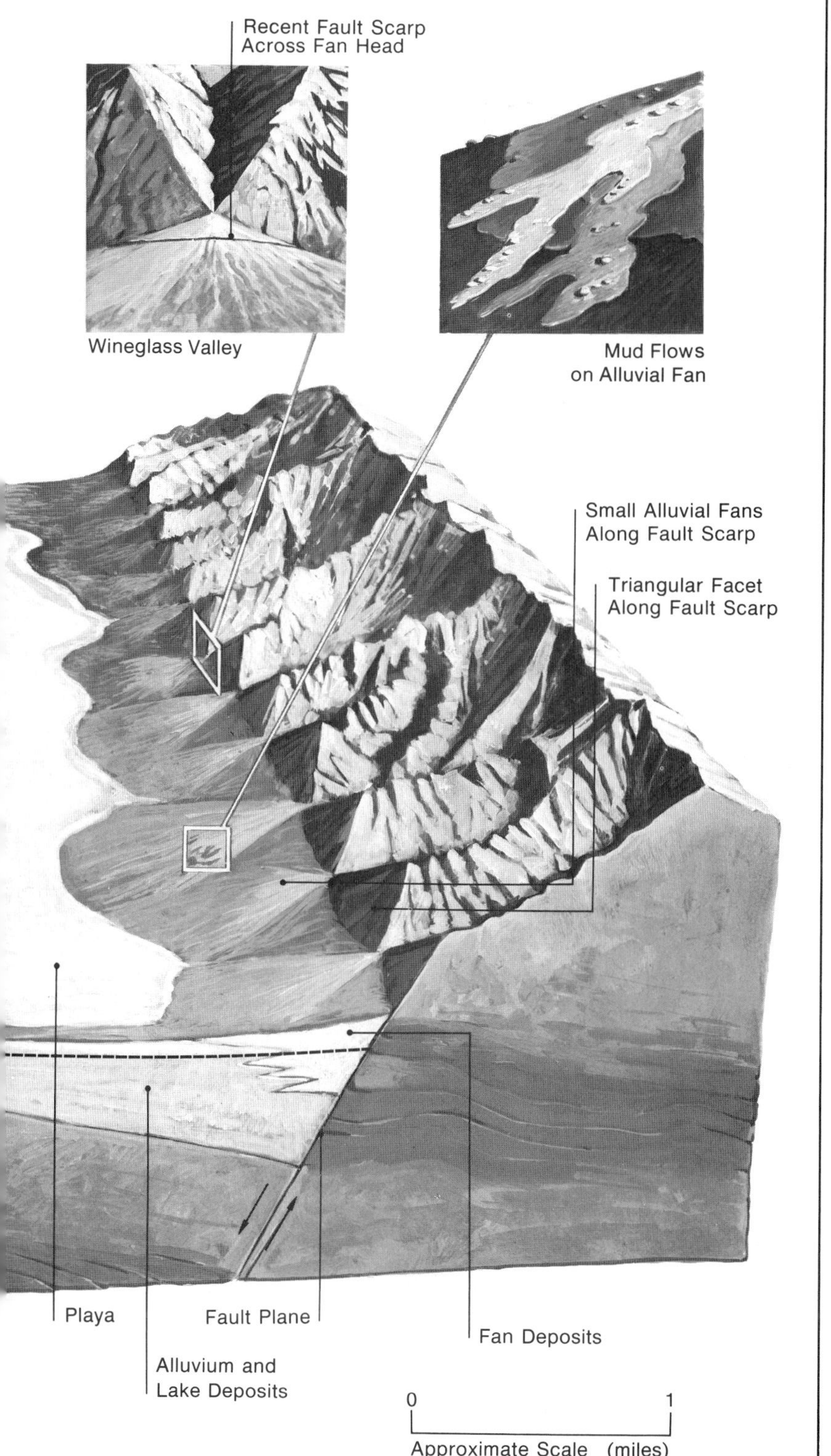

Recent Fault Scarp Across Fan Head

Wineglass Valley

Mud Flows on Alluvial Fan

Small Alluvial Fans Along Fault Scarp

Triangular Facet Along Fault Scarp

Playa

Fault Plane

Fan Deposits

Alluvium and Lake Deposits

0 1

Approximate Scale (miles)

the passage of time—*desert varnish.* Thus, new deposits are light in color, and successively older ones are increasingly darker. All sizeable alluvial fans are a mosaic of tones as a consequence.

The prevailing angularity of form results from the absence of a smoothing cover of soil. Chemical weathering is almost eliminated due to lack of moisture, and the absence of a protective vegetative cover precludes soil accumulation, loose material being flushed away by the infrequent rains faster than it is formed by the weathering processes.

Fault scarps composed of both bedrock and alluvium are clearer than in humid areas, because the limitation of weathering and absence of soil creep allows unusually long preservation of newly formed surfaces, even where they are merely banks of gravel and mud. Scarps commonly cut across the upper parts of alluvial fans, indicating recent uplift of the mountain block above them. (After Oberlander, 1974)

Figure 14.19 Alluvial fans slope down toward a basin from the mountains bordering Death Valley, California. The view shows the west side of Death Valley looking toward the south.

Figure 14.20 **Desert Landscape—Relief by Erosion.** This landscape is created by erosion by running water in an area of very gently dipping sedimentary rocks of varying resistance. The basic element is the *cuesta:* an escarpment produced by the edge of a layer of resistant rock, usually sandstone or limestone. In such a landscape we find escarpments or cuestas wearing backward through time, always in a down-dip direction. In the diagram, the dip is to the right, and the main cliffs are being pushed back by erosion to the right. Cliff retreat commences at structural highs, or domes, which are the first areas to be cut through by erosion, producing "holes," which subsequently enlarge through time. Cliffs are also produced where streams cut across the landscape as in this view. The main streams may flow obliquely to the structural slope, as a consequence of having been present here *before* tilting commenced. Such streams are *antecedent* to the geological movement. If they flow on a structural slope, across the dip, their canyons are asymmetric: Surface water will run down into them on one side, producing tributary gullies, but not on the other, leaving a straight cliff there. The Grand Canyon of the Colorado in Arizona is a notable example. Commonly springs are present on the up-dip side of the canyon as groundwater leaks through the inclined permeable sandstone or limestone layers. Large springs create great cavernous alcoves, which have been used by humans for shelter—whole villages have been constructed within them, as by the Pueblo Indians of the Southwestern United States (see inset at top right).

The cliffs actually retreat by being undermined by the more rapid erosion of less resistant rocks at their base. The *shale* in this view is rapidly removed, causing the more resistant *cap rock* (sandstone) to collapse (see inset at top left). Collapse of rock faces occasionally produces an opening through a narrow wall of rock, forming a *natural arch*. These should not be confused with *natural bridges*, which are created where two meander loops along a canyon stream wear away the rock between them. Both arches and bridges are rare except where the sandstone is very thick and massive.

Mesas result where once-continuous layers of resistant rock have been fragmented by erosion into widely separated flat-topped remnants. *Buttes* are remnants too small to preserve a flat summit. Both mesas and buttes commonly form erosional outliers along the faces of retreating cuestas.

Where the less resistant rocks, usually shale or marl, have been stripped off over a wide area, exposing the top of a little-eroded resistant layer, a *stripped plain* results. Stripped plains commonly expose joint sets, and these may be exploited by weathering and erosion to form very rough rock badlands consisting of humps, fins, and other strange bedrock forms.

Sandstone that collapses due to undermining often shatters into loose sand. Streams in such areas also convey a large quantity of sand, which may blow out of the dry stream beds between the rare water flows. As a consequence, sand dunes are common in these landscapes. In the extreme deserts, such as the Sahara, the flow of sand driven by the prevailing winds creates vast areas of sand accumulation, known as *ergs*. Dry sand moves when subjected to a wind exceeding about 10.5 miles per hour. The smallest relief forms produced are sand *ripples*. These crawl over dunes, causing them to advance by dumping sand over their crests. In the

Desert Landscape Formed by Erosion of Sedimentary Rock
Cap Rock
Oldest (Deepest) Formation
Chimney Rock
Stripped Plain
Cuesta
Mesa
Butte
Footslope
Mesa
Sand Dunes
Outlier
Natural Bridge
0
(miles)
1
Approximate Scale

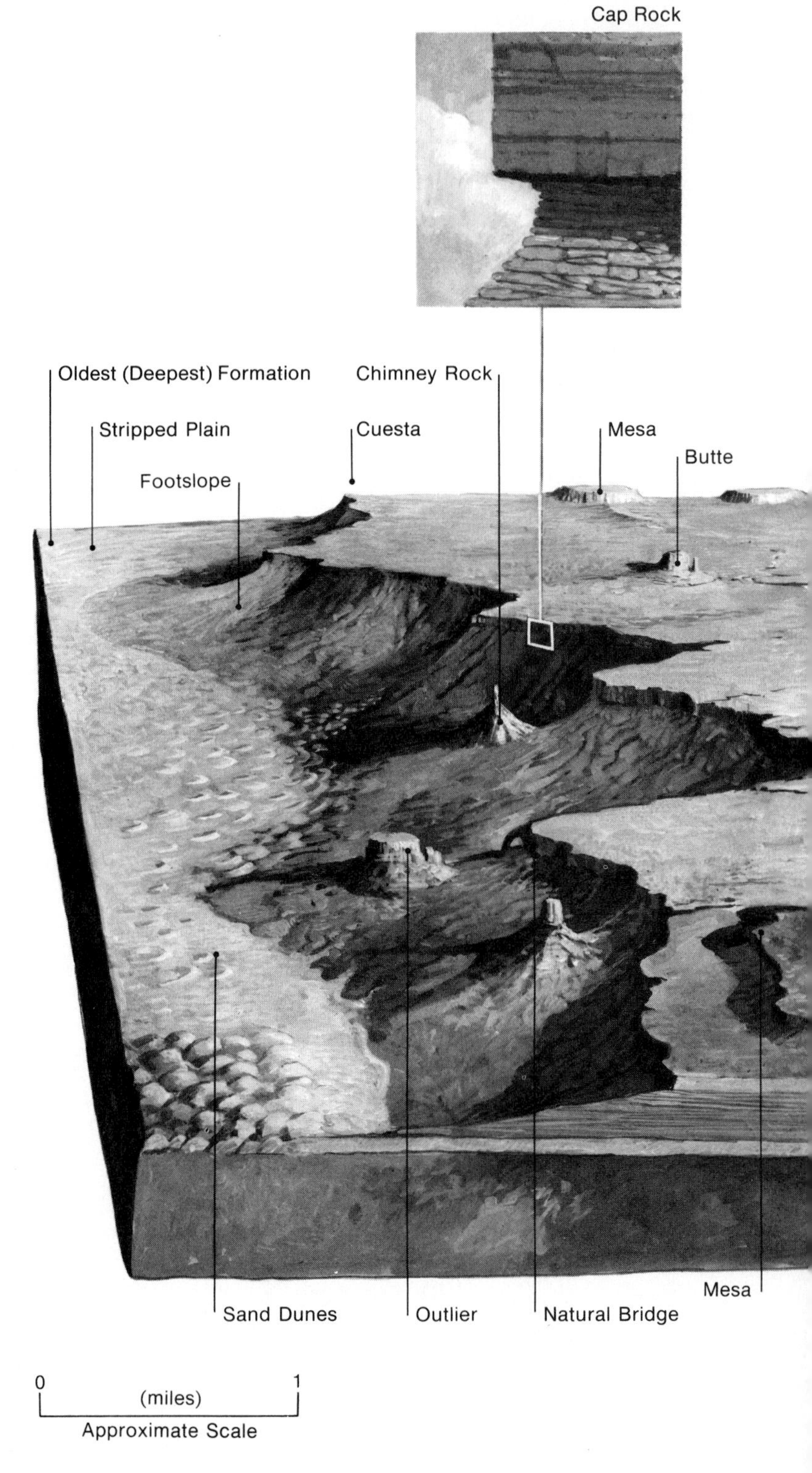

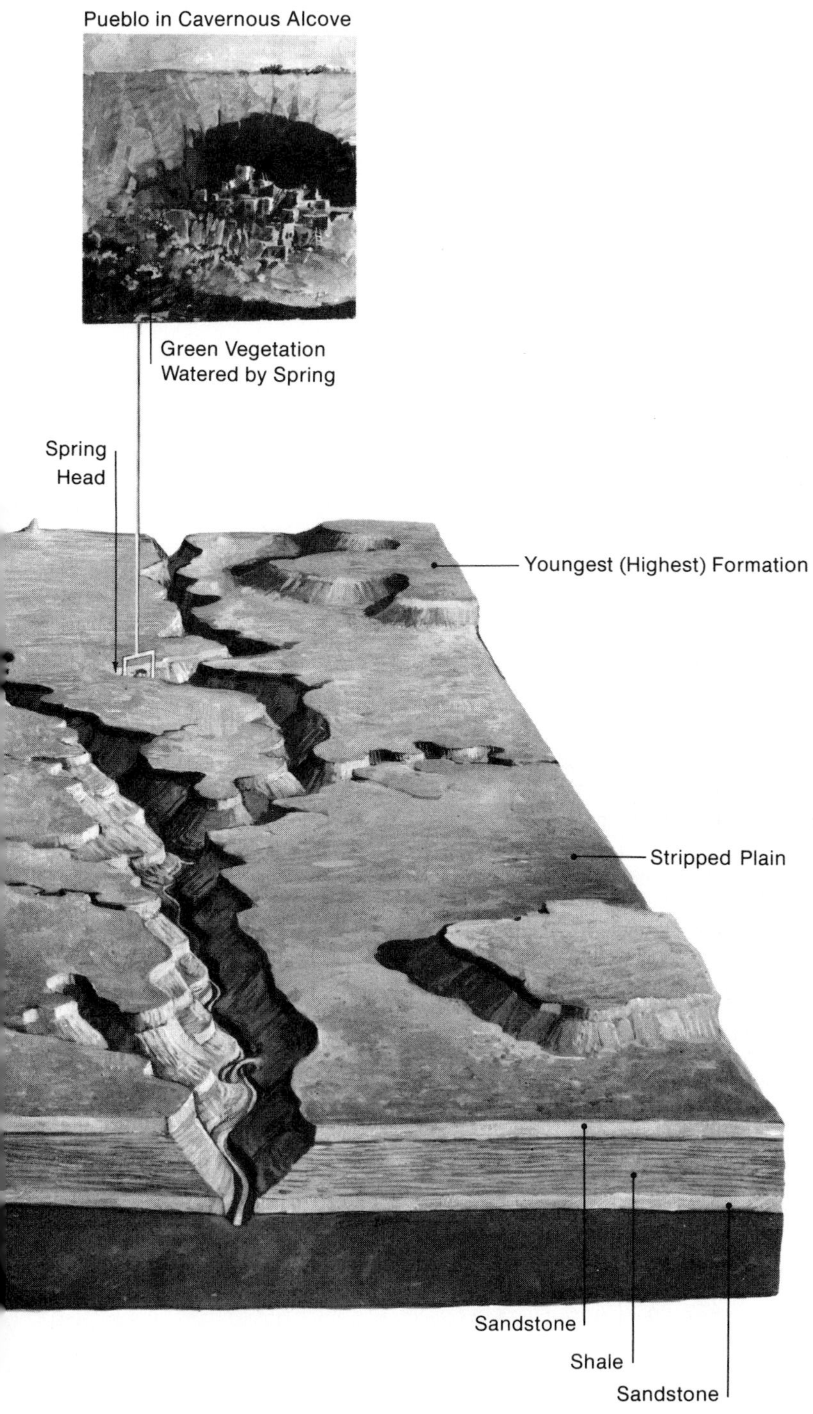

Pueblo in Cavernous Alcove

Green Vegetation Watered by Spring

Spring Head

Youngest (Highest) Formation

Stripped Plain

Sandstone

Shale

Sandstone

world's great sand ergs, forms larger than individual dunes occur, having characteristic sizes and regular spacing. These forms are called *draa;* they are whale-backed mountains of sand, and active dunes crawl over them in the same way that ripples creep over dunes.

Layered structures like those illustrated also are common in humid areas. However, the humid landscape result is different. Cliffs are less distinct due to a blanket of soil that creeps downslope, protected by a vegetative cover. Chemical decay affects nearly all rock types so that lithological variations are less distinct. Natural bridges and arches are rare due to weakness of rock resulting from deep chemical weathering. Dunes are absent, because vegetation colonizes debris accumulations, preventing detachment of particles by wind. The often highly colored bare rock that lends fascination to the desert scene is altogether obscured by a soil cover strongly anchored by vegetation.
(After Oberlander, 1974)

Figure 14.21 Mesas and buttes dominate the landscape in Monument Valley in southeastern Utah. Vertical jointing in horizontally bedded rocks leads to the development of the vertical rock faces characteristic of landforms in arid regions with sparse vegetation. Landforms in regions of sand accumulation are depicted in Figure 11.10.

Figure 14.22 **Desert Landscape—Relief by Erosion in Crystalline Rocks.** Where granitic or gneissic rocks appear in deserts, the landscape is commonly one consisting of long, smooth ramps leading up to extremely rough bouldery slopes. The projecting relief forms rise very abruptly above the ramplike surfaces, resembling islands projecting above the sea; hence these summit relief forms are called *inselbergs* (island mountains). The lower portion of the ramp is an alluvial apron, but the upper part is an erosion surface, known as a pediment, that truncates solid granitic rock. The summit relief clearly is a remnant of a long period of erosion, but unlike a monadnock on a Davisian peneplain, it is very distinct from the erosion surface below it, which it meets at a sharp angle in most cases. These steep-sided residuals (inselbergs) are thought to have been formed by the retreat of slopes at a constant angle, which is unlike the mode of slope development thought to characterize most non-desert landscapes, in which slope angle is imagined as diminishing through time. Thus, in humid regions the landscape is thought to "wear *down*," whereas in arid regions it is believed to "wear *back*." The crucial factor in the two types of development is thought to be the presence or absence of a mantle of soil subject to the gravitational creep process. Soil covers slopes in humid areas, being protected by a vegetative cover; this soil creeps downslope as a consequence of gravitational force, and both on theoretical grounds and by actual measurement, creep causes slopes to decrease in angle as time passes. The absence of a protective vegetative cover in arid regions prevents soil formation, as weathered material is washed away as fast as it is formed. Slopes retreat through the erosive effect of running water rather than the creep process, and this leads to parallel slope retreat, both theoretically and by measurement in the field.

In the landscape shown, the bouldery slopes are a consequence of the tendency of granitic rocks to be well jointed, forming a rigid mass of plane-faced blocks, which weathering tends to loosen and round at edges and corners. The steep slopes have been regarded as the angle of repose of the loose blocks produced by the weathering process.

The lower blocks of the inselberg slopes are commonly perforated by cavities produced in the weathering process. The exteriors of the boulders may be very dark due to patination by desert varnish.

According to recent investigations of such landscapes, their development may have been misunderstood by geomorphologists. Evidence of the presence of a soil formerly covering them suggests that they are relict from a more humid landscape, having been altered to their present form by relatively recent erosion under desert conditions—the erosion being triggered by loss of the former vegetative cover. Nevertheless they evidence parallel slope retreat, even in the former soil-covered landscape, which poses new questions of slope development in general. (After Oberlander, 1974)

Desert Landscape Formed by Erosion of Crystalline Rock
Base of Inselberg
Alluvial Apron
Inselberg
Dry Channel
Pediment
0
(miles)
1
Approximate Scale

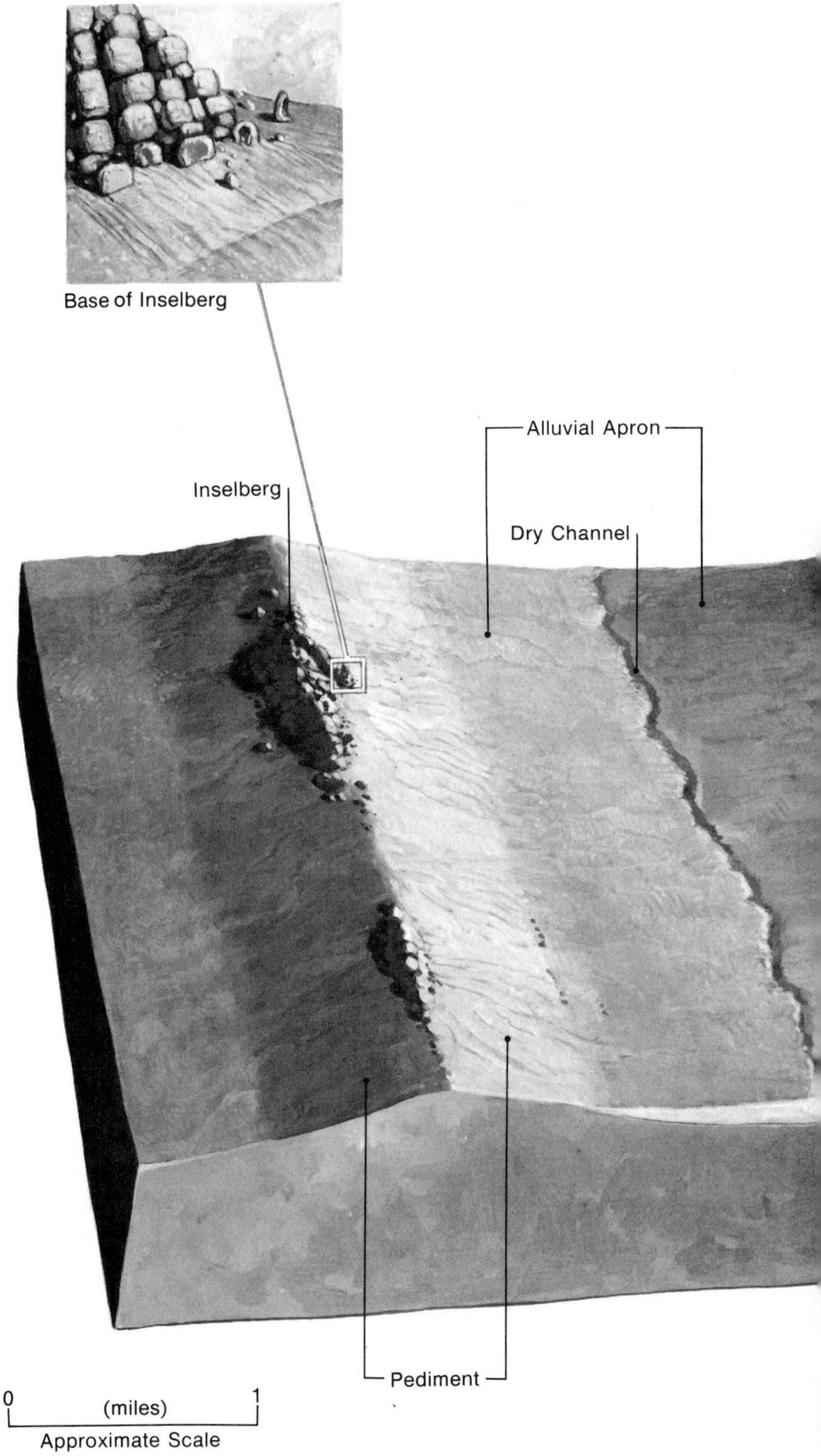

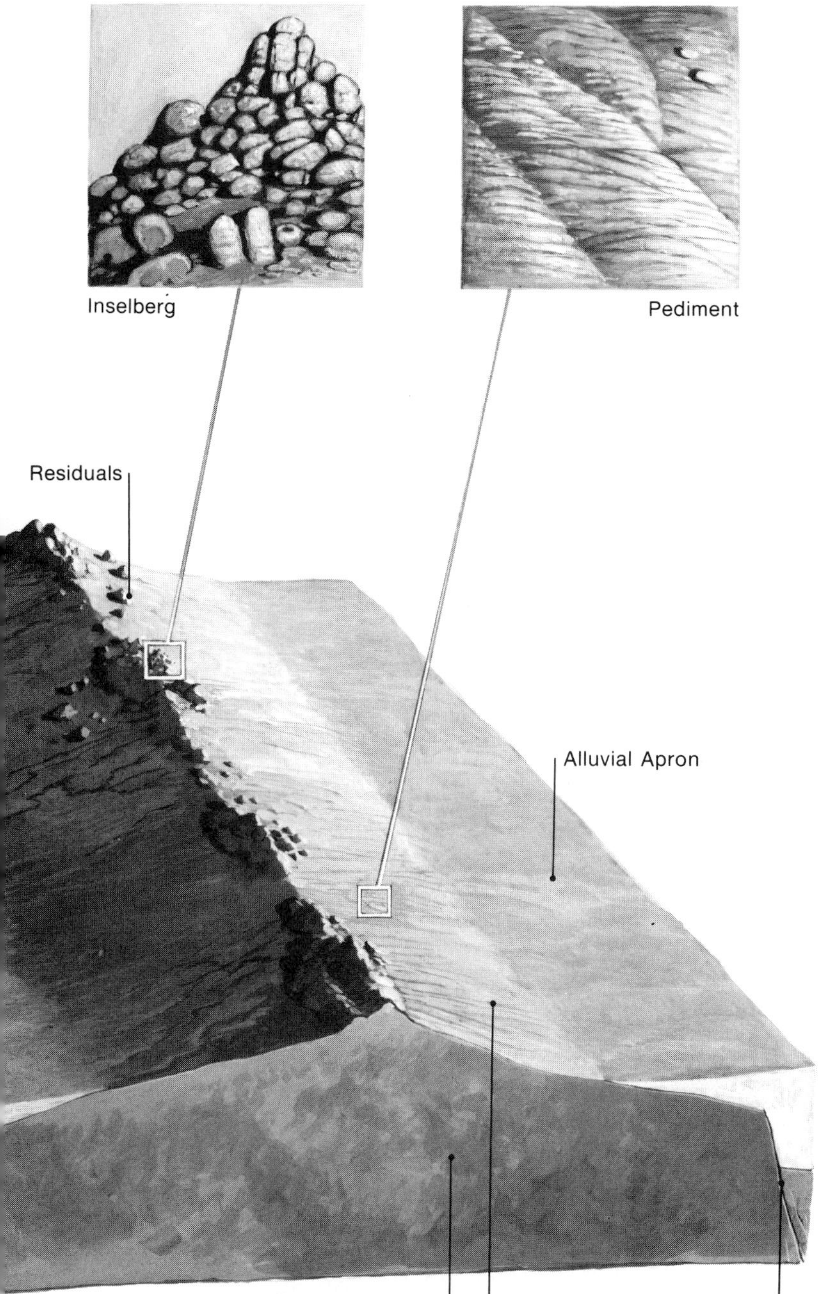

Inselberg
Pediment
Residuals
Alluvial Apron
Granitic Bedrock
Joint-guided Gullies
Fault

Figure 14.23 Most of the area of desert landscapes formed by erosion in crystalline rock consists of gently sloping pediments, such as this pediment located in the Mojave Desert east of Victorville, California.

Figure 14.24 **Tropical Savanna Landscape.** This landscape is widespread in the seasonally wet tropics on stable continental areas wherever crystalline rocks are dominant, as in Brazil, much of Africa south of the Sahara, India, and Australia. The term savanna refers to the vegetation association consisting of drought and fire-resisting tree species rising from a sea of grass. The important geomorphic features are very *flat plains,* into which only shallow saucer-shaped valleys are incised by ephemeral streams; *escarpments,* over which streams cascade spectacularly; and enormous rock domes, termed *bornhardts* (after the German who brought those in Africa to the outside world's attention). This peculiar landscape is the consequence of an aggressive climate and the peculiarities of granitic rock. The mode of landscape development has been characterized as *double planation,* as there are two separate interfaces along which geomorphic activity takes place. One is the *land surface,* subject to rainwash, gravitational transfer, and stream erosion. Erosional removal lowers the land surface and wears back projections upon it. The second interface is that at which decayed rock meets fresh, unaltered rock—the *weathering front*—which lies below the land surface. The unusual aspect of this landscape is the aggressive chemical weathering under warm tropical conditions that decomposes rock in the subsurface as fast or faster than surface erosion can wear down the land. Boreholes in some locations in Africa indicate the weathering front to be hundreds of feet below the surface. They also show it to be highly irregular. Where the granite is closely jointed the rock is deeply decayed, as water can penetrate it effectively. But where jointing is sparse, water is excluded and chemical decay cannot begin. *Bornhardts,* rising above the surface as much as 500 meters, are cores of monolithic granite that resisted chemical decay below the surface. They have been left behind as the land surface wears down and escarpments retreat. *Tors* are piles of boulders, isolated by chemical decay below the land surface, but exposed by erosion before they are completely decomposed.

The flat surrounding plains are armored with a *laterite* crust: a residue of oxides and hydroxides of iron and aluminum resulting from aggressive chemical alteration of the decomposed rock, such that all bases and even the silica are released and leached away. Mobilization of iron and aluminum and their fixation in the crust seems to require a water table that fluctuates in level seasonally, as it does in the savanna regions—the iron and aluminum being derived from the zone in which the fluctuation in level occurs. Because the water table varies in height most under hills and least under valleys, there is more chemical erosion under hills, so that their crowns literally sink as subsurface removal occurs. This is probably the cause of the conspicuous flatness of the laterized plains.

A last peculiarity of the savanna is the presence of ungraded streams and the absence of gorges and canyons of any significant extent. Of course the two are associated. Canyons are cut by streams carrying abundant tools for abrasion, which permit irregularities in stream beds to be filed smooth. The absence of canyons and presence of ungraded profiles including many waterfalls and cascades is a consequence of a paucity of abrasional tools. Streams of many tropical savanna regions are poor in cobbles and boulders because thorough chemical decay destroys coarse granitic rock waste before it reaches stream channels, which, as a consequence, transport sand and silt exclusively.

Tropical Savanna Landscape
Bornhardt
Laterite Crust
Decomposed Rock
Saucer-shaped Valley (Dry)

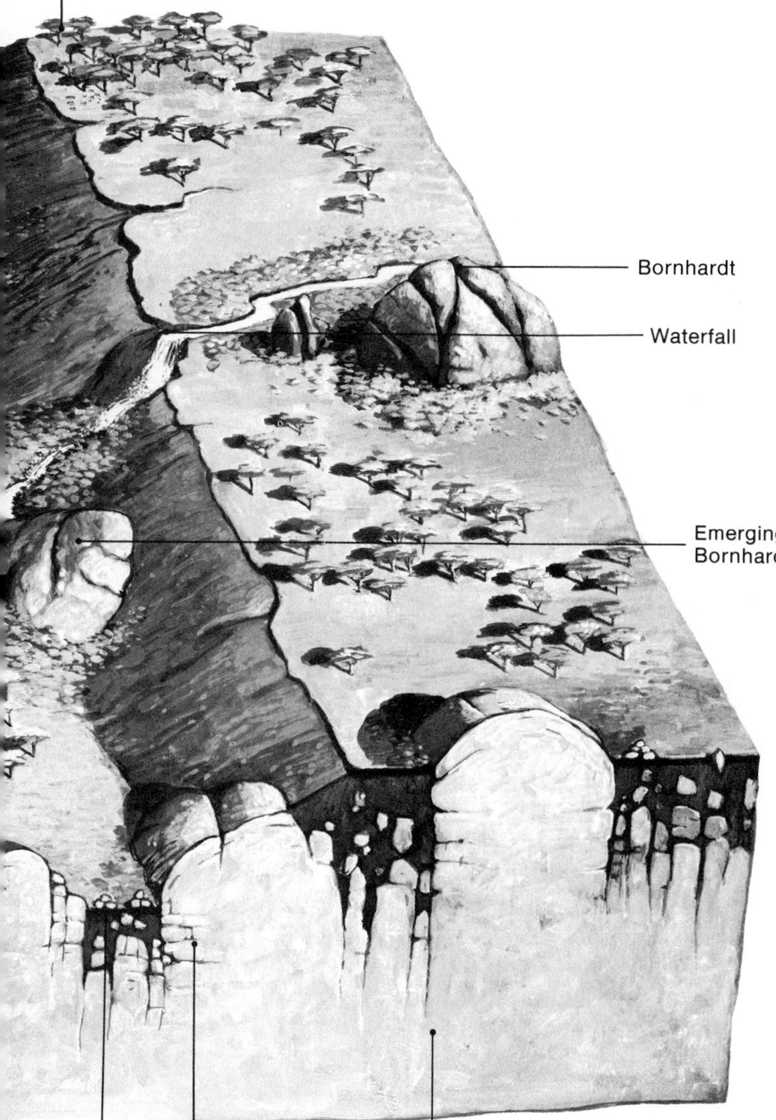

Laterite Crust

Bornhardt

Waterfall

Emerging Bornhardt

Tor

Weathering Front

Granitic Bedrock

Rocks other than granite and related types do not decay as massively in tropical climates, although they may decay equally rapidly. Whereas chemical weathering eventually causes a granitic block to collapse into sand by permeating its structure, it will attack a basaltic block on the surface only, constantly reducing its size, but permitting a durable kernal to persist in its interior. Thus basaltic outcrops are an excellent source of tools whatever the climate (see Figure 14.26). (After Oberlander, 1974)

Figure 14.25 The massive granite dome of this bornhardt in Rhodesia, Africa, rises abruptly from the flat surrounding plain of the savanna.

Figure 14.26 **Tropical High Island Landscape—Volcanic Dome.** The distinctive aspect of humid tropical landscapes is the effect of water in very copious amounts, combined with high temperatures. Chemical weathering is very active, producing soils even on near-vertical slopes. The soil is often saturated with moisture, leading to landsliding, and water runoff attains enormous volumes that produce highly intricate sculpture of steep slopes. Areas of granitic, metamorphic, and sedimentary rocks are generally maturely dissected, having a very high drainage density and slopes steeper than those in higher latitudes.

Certain geological configurations produce truly unusual landscapes in the tropics, such as those illustrated, which occur widely. The figure shows the eroded flank of a volcanic dome, such as those composing the Hawaiian Islands. The landscape is peculiar in several ways relating both to the climate and the geological components present. The valley walls are as steep as 80 degrees, yet covered with vegetation rooted in a shallow soil. Due to saturation with moisture, landslides are frequent and their scars are everywhere, exposing the bare basaltic lava composing the dome. The valley walls are fluted, each flute becoming a waterfall or cascade during the frequent rains. These flutes have been characterized as "vertical valleys." Where adjacent valleys have expanded so that their walls merge at the top, extremely jagged crestlines result. The main streams drop into the troughlike valleys over a series of waterfalls, each one terminating in a small plunge basin. The ephemeral waterfalls along the valley sides also drop from one plunge basin to another. Above the abrupt valley heads there may be an upland swamp; that on the Hawaiian Island of Kauai may be the wettest place on earth (annual precipitation may approach 600 inches). Many of the valleys have flat floors resulting from subsidence of the islands and infilling by alluvium. It appears that the precipitous nature of the valleys dissecting these volcanic domes may be due to the low water table resulting from the permeability of the lava flows composing them—thus they are almost tantamount to gigantic spring alcoves, expanding due to sapping by spring action at their base. (After Oberlander, 1974)

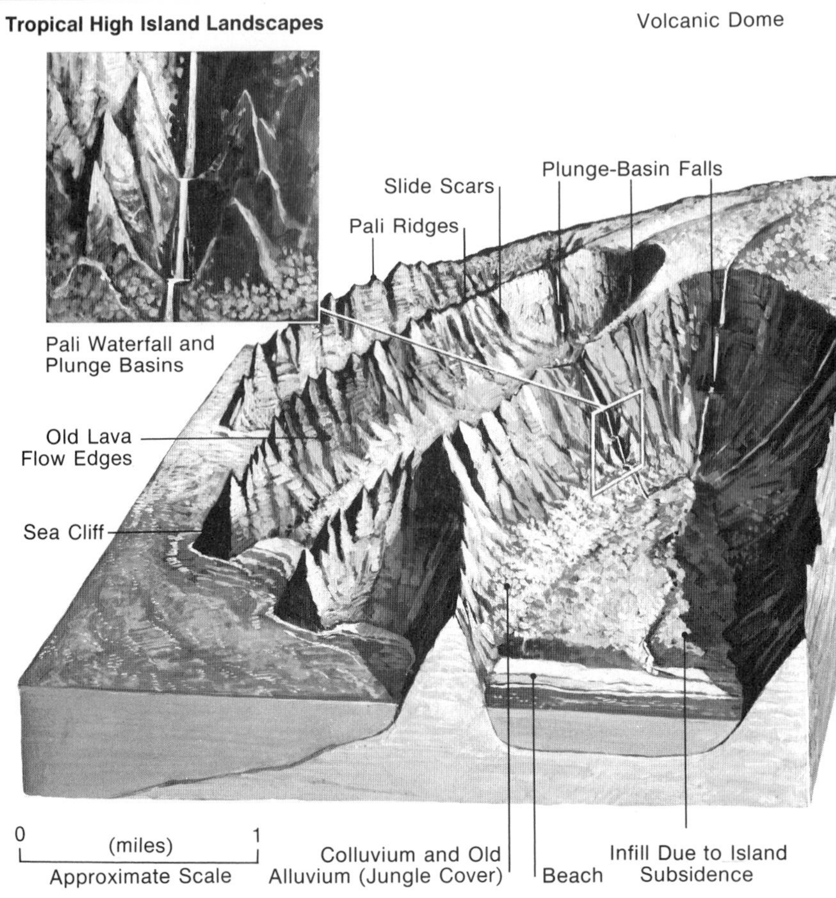

Tropical High Island Landscapes
Volcanic Dome
Slide Scars
Plunge-Basin Falls
Pali Ridges
Pali Waterfall and Plunge Basins
Old Lava Flow Edges
Sea Cliff
0
(miles)
1
Approximate Scale
Colluvium and Old Alluvium (Jungle Cover)
Beach
Infill Due to Island Subsidence

Figure 14.27 Pancake rocks on South Island, New Zealand, are an unusual karst formation in limestone with closely spaced joints.

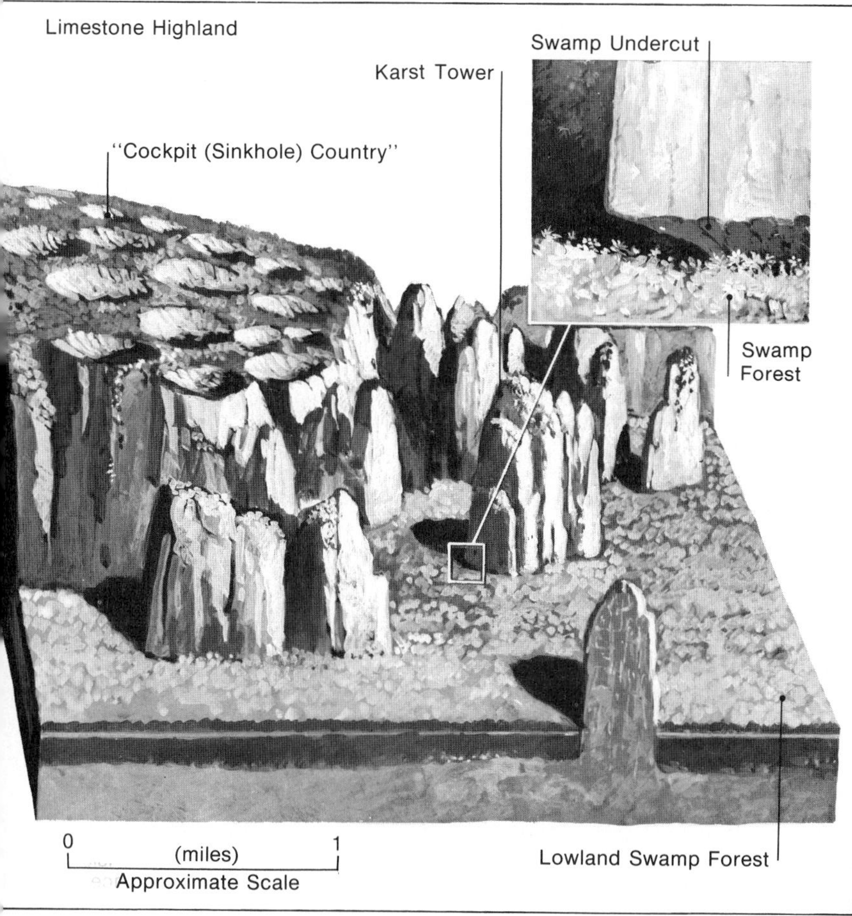
Limestone Highland
Swamp Undercut
Karst Tower
"Cockpit (Sinkhole) Country"
Swamp Forest
0
(miles)
1
Approximate Scale
Lowland Swamp Forest

Figure 14.28 **Tropical High Island Landscape—Limestone Highland.** A second highly distinctive landscape is found where massive limestones appear in moist tropical areas of strong relief. Plateau tops are pock-marked with large sinkholes, produced by the solution of limestone along intersecting joints. Due to its content of organic acids, groundwater in the tropics is particularly aggressive toward limestone, which it corrodes rapidly. Between the sinkholes is a maze of often sharp-crested limestone ridges. This terrain has been termed *cockpit karst.* Solution working deep into the joints at the plateau edge detaches enormous slabs and pinnacles which become isolated as fantastic vertical-sided mountains, hundreds of meters in height. These have been made famous in Oriental art. They are common in Southwest China, North Vietnam, Malaysia, and Indonesia. Once isolated these *karst towers* are kept steep-sided by undercutting and collapse. They rise from swampy lowlands, in which the water has been made acidic by contact with peat and other products of vegetative decay. This water corrodes a notch, sometimes 5 or 6 meters deep, around the karst towers. The overhangs collapse periodically, after which a new notch is formed, keeping the towers permanently "girdled." Tower karst develop only where limestone free of impurities is divided by well-defined but widely spaced joints. Where the limestone contains impurities of clay, marl, or chert, the towers are replaced by less imposing cone-shaped hills, which are common in Southeast Asia, Puerto Rico, and Cuba. Thus most tropical karst topography may be resolved into "cockpit," "tower," or "cone" karst, none of which is present in higher latitudes. (After Oberlander, 1974)

Summary

Climate controls many of the processes of weathering and erosion that shape landforms today. Especially in cold and hot, dry climatic regions, the combination of processes characteristic of a climatic region may develop distinctive landforms. However, landforms of similar appearance occur in several different climatic regions, even when the processes of slope formation differ among the regions. In general, landforms cannot be classified according to climatic regions.

Slopes are the basic elements of landforms. After various processes of weathering produce loose material on a slope, a number of different mechanisms can act to transport the material. Processes of mass wasting include mud, earth, and rock slides; rock and soil creep; wash; solifluction; and rainsplash.

Different mechanisms of erosion may be important in different climatic regions. Soil wash tends to be less important where vegetation cover is dense, because vegetation maintains the infiltration capacity of the soil and reduces runoff. Soil creep is usually more effective than soil wash in moist midlatitude regions. Many of the distinctive landforms of the world can be partially understood in terms of the dominant processes acting to shape them.

15 The Power of Ice

Glaciation: Present and Past
Glaciation of the Pleistocene
Indirect Effects of Pleistocene Glaciation

Anatomy and Dynamics of Glaciers
The Making of Glacial Ice
The Glacier Budget
Movement of Glaciers

The Work of Glaciers on the Landscape
Mechanisms of Glacial Erosion
Glacial Erosion and Landforms
Glacial Deposition and Landforms

Glaciation and Man

Many of the earth's most spectacular landforms were created by the power of glaciers. Flowing ice is one of the most forceful agents of geomorphic change, and several regions of the earth still bear the chiseled marks made by ancient glaciers.

Maligne Lake, Jasper Park by Lawren S. Harris. (The National Gallery of Canada, Ottawa)

The Power of Ice

Slightly more than 15,000 years ago, the world's climate was cooler than it is today, and the areas of northern Europe and North America that are most densely populated today were mantled with more than a kilometer of glacial ice. An exploding volcano may be more spectacular than a slow-moving *glacier*, but the image of a colossal ice sheet many kilometers thick moving inexorably across the land has an undeniable grandeur. Man had not begun writing history at the time of the last great ice advance, so the human side of the drama remains unknown. But abundant documents from the Little Ice Age in Europe in the seventeenth and eighteenth centuries tell of farmers abandoning their fields and villagers being forced to emigrate as alpine glaciers advanced. Will the residents of Minneapolis, Detroit, or Chicago ever have to flee because of approaching glaciers? The sites of these cities were covered by ice little more than 10,000 years ago.

Where are the glaciers today? In most parts of the United States that receive snow, the winter snowfalls usually come in late autumn and melt by early spring. But farther poleward or at higher elevations, snow persists for a longer portion of the year. In Alaska, Greenland, New Zealand, and Antarctica, and in the Canadian Rockies, the European Alps, the Himalayas, and the Andes, snow that falls during the winter does not melt completely during the warm season. Under these conditions, snow accumulates to form glaciers.

Glaciers take two principal forms: *ice sheets,* or *icecaps,* and *valley glaciers.* Ice sheets, or icecaps, such as those in Greenland and Antarctica, are masses of glacier ice, sometimes continental in size. Valley, or alpine, glaciers are streams of ice that form in high mountain valleys and flow gradually downward under gravitational stress. Valley glaciers and ice sheets are dynamic systems, constantly changing, constantly moving. And as they move, they exert powerful forces on the landscape. As agents of geomorphic change, glaciers have left a rich legacy of distinctive landforms especially suited for our recreation and pleasure. The majestic peak of the Matterhorn in Switzerland, the ski slopes in Tuckerman Ravine in New Hampshire, the sandy arm of Cape Cod, the Finger Lakes of central New York State, the fjords of Norway—all can be attributed to the work of moving ice.

Glaciers also function as storehouses of water. Three-fourths of all the world's fresh water, or approximately 25 million cubic kilometers, is stored as ice, primarily in the ice sheets of Antarctica and Greenland. This amount is equal to 50 years of precipitation over the entire earth. If the Antarctic ice sheet melted at a steady rate, it would provide water flow equal to the discharge of the Amazon River for 5,000 years or equal to the discharge of the Mississippi River for 50,000 years.

When water is stored in ice sheets, however, it is effectively removed from the day-to-day workings of the hydrologic cycle. But

Figure 15.2 (opposite top left) Valley glaciers form in mountain regions where snow persists throughout the year. If the annual accumulation of snow exceeds annual losses due to processes such as melting, a glacier eventually forms. The river of ice slowly flows down mountain valleys, as shown in the diagram. In some regions, the glacier ice from one or more valley glaciers spreads out over a plain and forms a *piedmont glacier,* which is analogous to an ice sheet but smaller.

(opposite top right) The surface of this valley glacier in the French Alps is broken into numerous crevasses.

(opposite bottom) The Malaspina Glacier in Alaska forms an extensive piedmont ice sheet.

Valley
Glacier
Piedmont Glacier

ice sheets provide an important mechanism by which climate can control the addition or removal of liquid water from the world's supply. On the global scale, changes in temperature and precipitation patterns influence the advance or retreat of ice sheets, thereby affecting the level of water in the world's oceans. On the local scale, valley glaciers are also sensitive to the effects of climate. Whether a glacier grows or retreats depends on the balance between the amount of snow added to the glacier in the winter and the amount of ice melted away in the summer. Water that the melting valley glaciers provide to streams and rivers may be an important source of water during dry summer months. The city of Tacoma, Washington, depends on meltwater from a valley glacier to generate electric power, for example.

Glaciation: Present and Past

Glaciers grow only in regions where the snow added to the glacier each year more than compensates for the ice and snow lost by melting and other processes. Glaciers therefore persist in the cool climates of the higher latitudes or in high mountains where the low temperatures associated with high altitudes provide a favorable environment and where precipitation is in the form of snow. Great ice sheets exist only in polar regions, but valley glaciers can be found even in the tropics. Mount Kenya, located on the Equator in Africa, has a number of small glaciers at altitudes above 4,500 meters (15,000 feet).

The ice sheet that covers Antarctica extends over an area of 12 million square kilometers, whereas a small valley glacier may have an area of only a fraction of a square kilometer. There are more than 1,000 glaciers in the conterminous United States, most of them located in the Cascade Mountains of Washington. Nearly all of the glaciers in the United States outside of Alaska are small; their combined area is only about 500 square kilometers. Nevertheless, they provide meltwater for thousands of streams each year. The glacial melting occurs primarily in the summer months, just when water demands for irrigation are highest and when streams fed by precipitation are normally lowest. Alaska's glaciers cover more than 40,000 square kilometers. One glacier system alone, the Malaspina, has an area larger than the state of Rhode Island. Meltwater from large glaciers feeds some of Alaska's largest streams; an extensive glacier can provide meltwater from its lower end throughout the year while new snow nourishes its upper reaches.

Glaciation of the Pleistocene

The present geologic age, called the *Recent*, is usually taken to have begun 10,000 years ago, at the close of the last period of glaciation. The preceding age, the *Pleistocene*, comprised four or five major episodes of glaciation interspersed with longer, milder interglacial periods. Ice has been common in the earth's history; previous ice ages occurred hundreds of millions of years ago. The Pleistocene is variously dated as having begun 1 to 2 million years ago. Figure 15.3 shows the extent of glaciation in the Northern Hemisphere

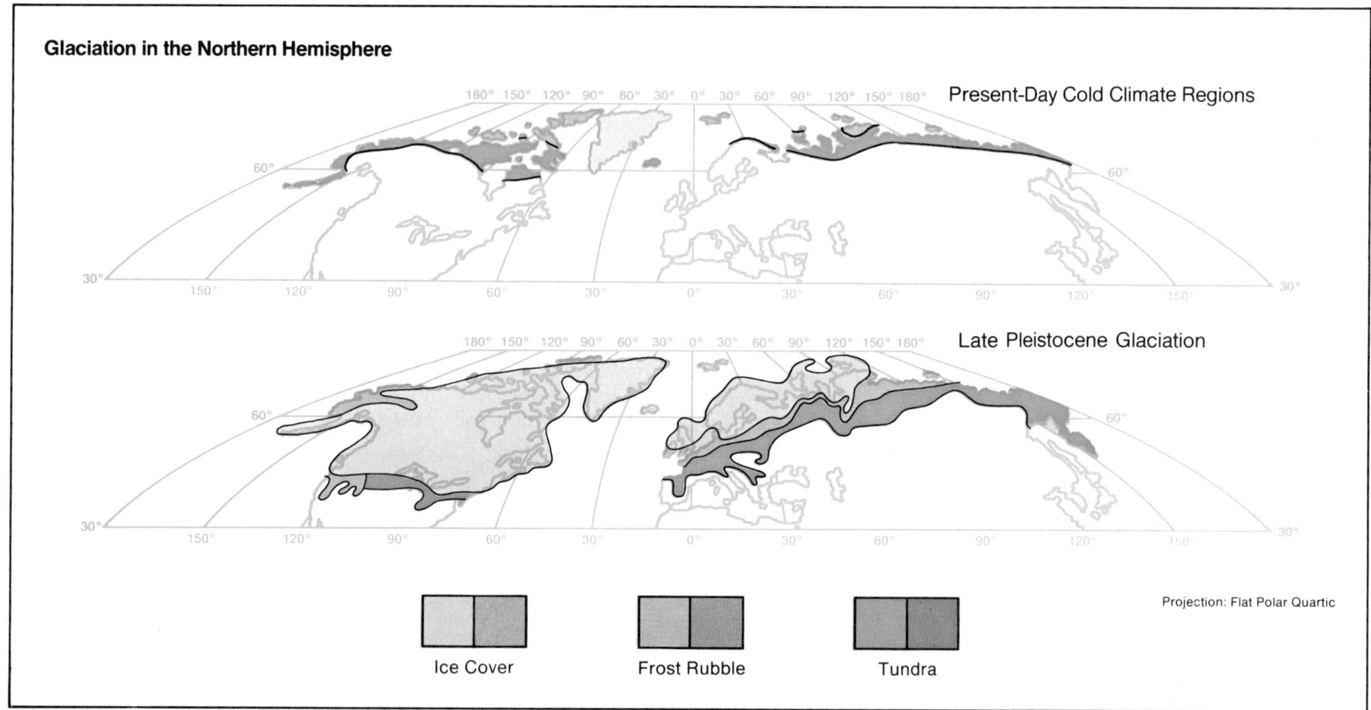
Glaciation in the Northern Hemisphere
Present-Day Cold Climate Regions
Late Pleistocene Glaciation
Projection: Flat Polar Quartic
Ice Cover
Frost Rubble
Tundra

Figure 15.3 (top) At present, regions in the Northern Hemisphere that exhibit glaciation and other features characteristic of cold climates generally are restricted to the high latitudes. Isolated regions of glaciation also occur at high altitudes in mountain ranges such as the Alps, the Rocky Mountains, and the Alaskan mountains. (bottom) During the last major period of glaciation, approximately 15,000 years ago during the late Pleistocene, large parts of northern North America and northern Europe were subjected to the direct or indirect effects of glaciation. Many sections of the northeastern and north central United States exhibit landforms characteristic of glacier action, and periglacial regions not directly covered by ice sheets possess features such as extensive deposits laid down by glacier-fed streams. (After Davies, 1969)

at present compared to the glaciation during the last part of the Pleistocene. Because glaciation exerts such a powerful and characteristic shaping force, the landscape bears the imprint of glaciers long after they have retreated. Although it is not difficult to determine the extent of a glacier's advance, each successive wave of glaciation obliterates much of the work of preceding glaciers, and often the effects of only the most recent glaciation are apparent in the landscape.

Indirect Effects of Pleistocene Glaciation

The effects of glaciation are not limited to the area directly covered by ice. Meltwater from the massive accumulation of ice during the glacial ages of the Pleistocene generated runoff that shaped the landscape far beyond the boundary of the ice. Drainage patterns were altered as the advance of the ice sheets blocked river systems and forced streams to change their courses, which created new patterns of stream erosion and deposition. Ice on North America, for instance, stopped at the drainage basins of the Ohio, Missouri, and Mississippi rivers, which caught the meltwater like funnels. The cold weather associated with the glaciers accelerated the process of frost weathering in regions far beyond the glacier's boundaries.

The average thickness of the Pleistocene ice sheets is estimated to have been 1 or 2 kilometers (0.6 or 1.2 miles). The weight of the ice sheets was enormous—approximately 1,000 tons for each square meter of area—and the earth's semiplastic crust sagged under the burden. The weight of the ice depressed the crust by as much as

500 meters (1,600 feet) under the thickest parts of the ice sheets. The crust has been returning to its original level since the retreat of the ice sheets, but the process is not yet complete. Parts of Scandinavia and of North America are currently rising at the rate of a fraction of a centimeter per year, with a rise of perhaps 100 to 200 meters (330 to 660 feet) yet to be accomplished.

Because the oceans supply water for the growth of ice sheets, a world-wide lowering of sea level accompanies a period of widespread glaciation. The sea level has been estimated to have lowered 100 to 150 meters (330 to 500 feet) during the Pleistocene glaciations, based on the estimated volume of the ice sheets. An independent line of evidence that leads to the same conclusion is the observation that river valleys appear to extend into submerged coastal shelves down to a depth of about 100 meters (330 feet) below the present sea level. The lowering of the sea gave rivers an opportunity to cut and widen their valleys near their outlets. When the sea returned to its present level, submerged valleys were left that now form some of the world's best harbors.

The oceans are shallow near many coastlines. If the sea level were to drop by 150 meters (500 feet), a considerable amount of new land would appear. The Florida peninsula would double in area, for example, and land bridges would join Ireland, England, and France. Coastal shelves exposed during the glacial period undoubtedly played a role in prehistoric times by providing new migration routes for people and animals to pass freely from one land to another. Peat deposits and mammoth remains have been found on the sea floor off New England where they were left when the shallow coastal regions were exposed.

At present, sea level appears to be rising slowly, at an average global rate of 1 millimeter per year (0.04 inch per year). Along the north Atlantic coast of the United States, the average rate of sea level rise relative to the land has been 3 millimeters per year (0.12 inch per year) during the last 40 years. Part of this local rise in sea level is attributed to the melting of ice sheets in the warmer climate of the first half of this century, and part may be due to local sinking of coastlines.

Anatomy and Dynamics of Glaciers

The mechanism that starts and stops periods of glaciation remains a puzzle. Ice sheets are in a delicate balance between growth and retreat, and relatively small changes in climate may cause the balance to shift either way. The hypotheses that have been advanced to account for climatic change, discussed in Chapter 10, include the effect of increased altitude of the landmasses after a period of geologic uplift and the effect of variations in the amount of radiation the earth receives—variations caused by changes in the earth's relationship to the sun, changes in the atmosphere's carbon dioxide and dust content, and changes in the sun's energy output. But no single theory seems able to account for the growth and recession of ice sheets four times in the comparatively short period of 2 million years. Indeed, there have been glaciations in the geologic past that occurred hundreds of millions of years ago. Proposed mechanisms may be effective when acting in concert with one another, but a generally accepted mechanism for the cause of glaciation has not yet been devised.

The Making of Glacial Ice

Glaciers are formed above the annual *snowline*, the boundary at which some winter snow is able to persist throughout the year. In present-day climates, the snowline reaches sea level at approximately 70°N and 70°S latitudes, but at lower latitudes, it is found in the heights of mountains. The snowline in California's Sierra Nevada is about 4,500 meters (15,000 feet) on warm slopes and between 3,600 and 4,000 meters (12,000 and 13,000 feet) on shaded slopes. Glaciers are born above the snowline, where snow is present at all times.

What happens to snow that accumulates above the snowline of a mountain? Snow falls originally in the form of skeletal hexagonal ice crystals, but after a time, pressure, local melting, and vaporization at the fine crystal points convert the lacy crystals to rounded

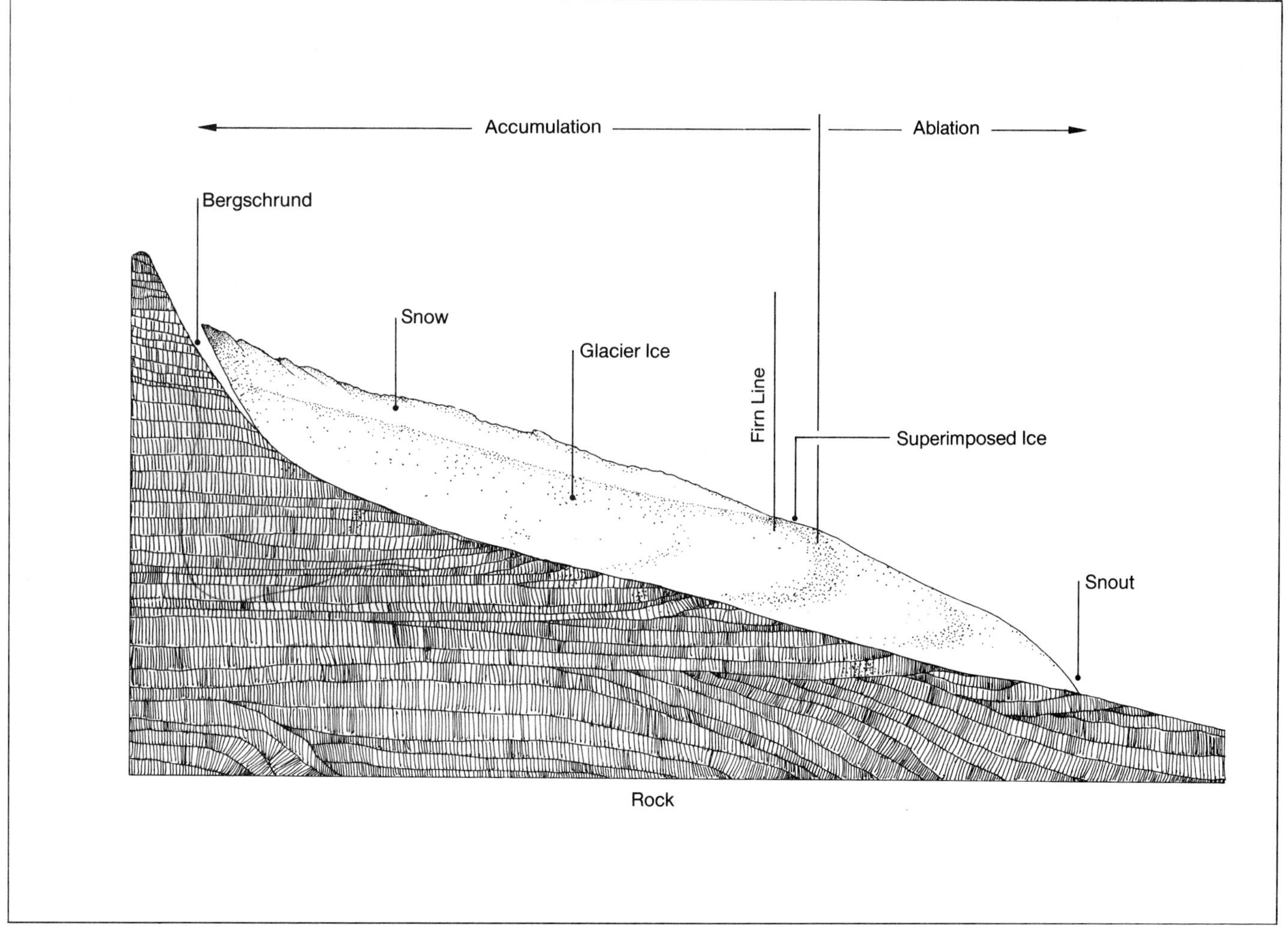

Accumulation
Ablation
Bergschrund
Snow
Glacier Ice
Firn Line
Superimposed Ice
Snout
Rock

Figure 15.4 This schematic diagram illustrates the principal features of a valley glacier. The vertical scale has been exaggerated for clarity. The firn line marks the boundary between the annual accumulation of snow and glacier ice; the glacier's surface is white upslope from the firn line and bluish on the downslope side. The boundary between the region of accumulation and the region of ablation is slightly downslope from the firn line, because melting and refreezing cause the lower edge of the snow to become superimposed ice. Many glaciers exhibit a deep crevasse called a *bergschrund* at their headwalls. It has been suggested that glacier erosion of the rock headwall occurs because water from melting snow and ice seeps into the crevasse and in some manner helps to pry rocks loose when it freezes.

granular crystals of ice. The speed of conversion to granules depends on the climate. In extreme cold, the snow is dry and powdery, and snow in layers remains fluffy for a long time. At temperatures closer to freezing, the snow is wet and heavy, and the transformation of snow crystals to granular crystals requires a few days at most.

The *specific gravity* of a substance is its weight relative to an equal volume of water. The specific gravity of solid, bubble-free ice is 0.92; newly fallen snow has a specific gravity between 0.05 and 0.15, because the snowflakes are loosely packed and have air spaces between them. With time, and under the pressure of overlying snow, ice granules formed from snow gradually pack closer together. After one season, the granular material reaches a specific gravity of about 0.55 and is called *firn*. The appearance of firn signifies that the granules have begun to join with one another but that open interconnecting pore spaces still exist among the granules. In tens or hundreds of years, the pore spaces gradually disappear, until at a specific gravity of 0.85, firn becomes glacier ice, with crystals that are frequently the size of walnuts. Then, as air bubbles gradually disappear from the ice, the specific gravity increases to 0.9 or so and the ice is bubble-free blue glacier ice.

Because a layer of firn develops in one season, glaciers often show a succession of distinct annual layers. Dust, dirt, and pollen accumulate on the snow surface each summer and become locked in the ice to form annual layers. Old glaciers and ice sheets are repositories of climatic history. The layers can be dated either by counting layers or by using carbon 14 dating on carbon dioxide dissolved in the ice. By studying the pollen and dust in each layer, knowledge can be gained of climatic variations through time. Periods of volcanic activity can be traced by ash deposits in the layers. Even man's pollution of the atmosphere is recorded in the layers of firn: analyses of annual layers have revealed an increase in lead, for example.

The Glacier Budget

The budget concept can be applied to a glacier's gain and loss of ice. If a glacier begins to develop well above the snowline, snow will accumulate year after year, and the glacier will steadily increase in size and begin to flow. Where part of the glacier flows below the snowline, however, ice is lost from the glacier during the summer. Loss of ice occurs primarily by melting. The general term for loss of ice is *ablation*.

Mature valley glaciers are divided into two zones, an upper *zone of accumulation*, which has a net annual gain from snowfall and little melting, and a lower *zone of ablation*, which has a net annual loss from less snowfall and more melting (see Figure 15.4). The *firn line* separates the two zones. The surface of the glacier is snowy and white above the firn line; below it, the glacier is blue-gray glacier ice. If a glacier were stationary, its firn line would coincide with the snowline for that locality, but mountain glaciers flow downhill, so the firn line lies somewhat below the local snowline.

During heavy winter snowfalls, *accumulation* is the dominant

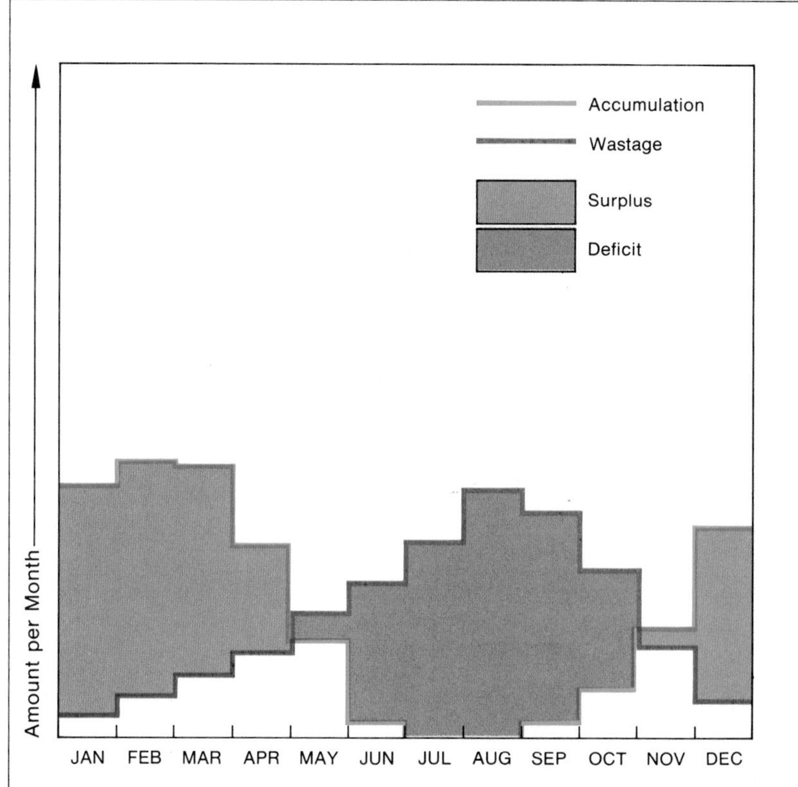
Accumulation
Wastage
Surplus
Deficit
Amount per Month
JAN
FEB
MAR
APR
MAY
JUN
JUL
AUG
SEP
OCT
NOV
DEC

Figure 15.5 This diagram illustrates a typical glacier budget. During the winter months, the accumulation of new snow on the glacier exceeds the loss of ice and snow by melting and other processes of ablation. During the warm summer months, ablation exceeds accumulation, and the glacier experiences a deficit. Note that both accumulation and ablation occur during most months; the glacier may be losing ice near its snout by melting while gaining new snow in its upper, colder reaches. The mass of the glacier increases during periods of surplus and decreases during periods of deficit. If the net annual surplus exceeds the net annual deficit on the average, the glacier grows and advances.

process on a glacier, but in the summer, when snowfalls are less frequent, ablation is greater than accumulation. Glaciers grow in the winter and diminish in the summer. Whether a glacier shows a net growth for the year depends on the balance between annual accumulation and annual ablation.

The rates of accumulation and ablation determine the glacier's degree of activity. As an example, consider two different glaciers, each having no net annual growth. For one glacier, light snowfall and a low rate of summer melting accomplish the balance of accumulation and ablation. Such conditions occur on the ice sheets of central Antarctica and Greenland, where only 5 to 15 centimeters (2 to 6 inches) of precipitation accumulate each year and where ablation rates are also small. The second glacier may be in a region of heavy winter snowfall and warm summer weather, so that the rates of accumulation and ablation are both large. But surplus material gathered in the zone of accumulation cannot be lost by ablation until it is transported into that zone. A glacier with high rates of accumulation and ablation transports a considerable amount of ice from the upper zone to the lower zone, and it is an *active* glacier. If the rates of accumulation and ablation are both low, the glacier transports little ice, and it is relatively *passive*.

The snow that falls on a glacier may not be transported to the zone of ablation for a long period of time. Large, passive glaciers such as those in Antarctica may require thousands of years to respond to climatic change. But small, active valley glaciers show

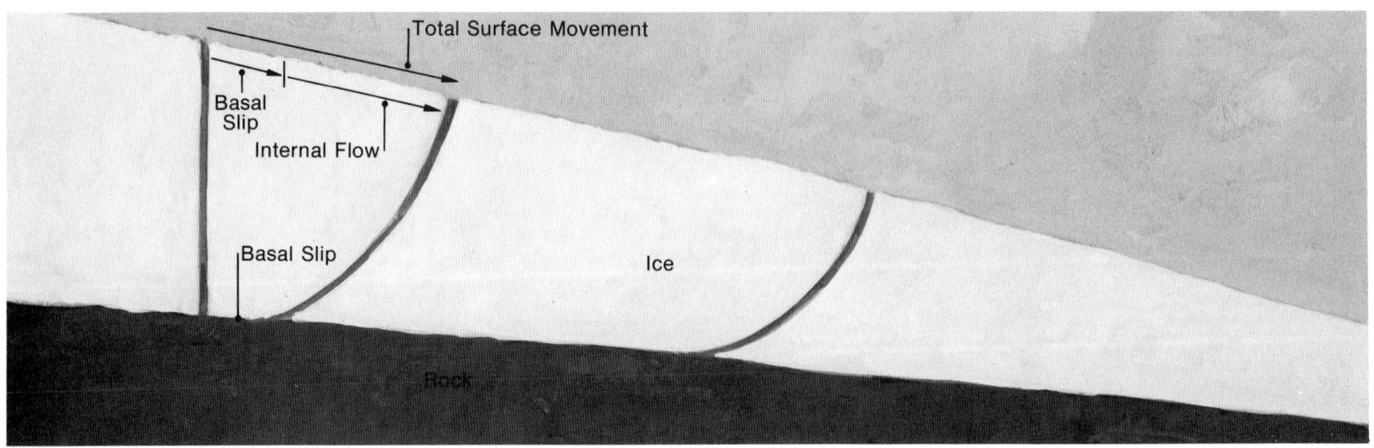
Total Surface Movement
Basal Slip
Internal Flow
Basal Slip
Ice
Rock

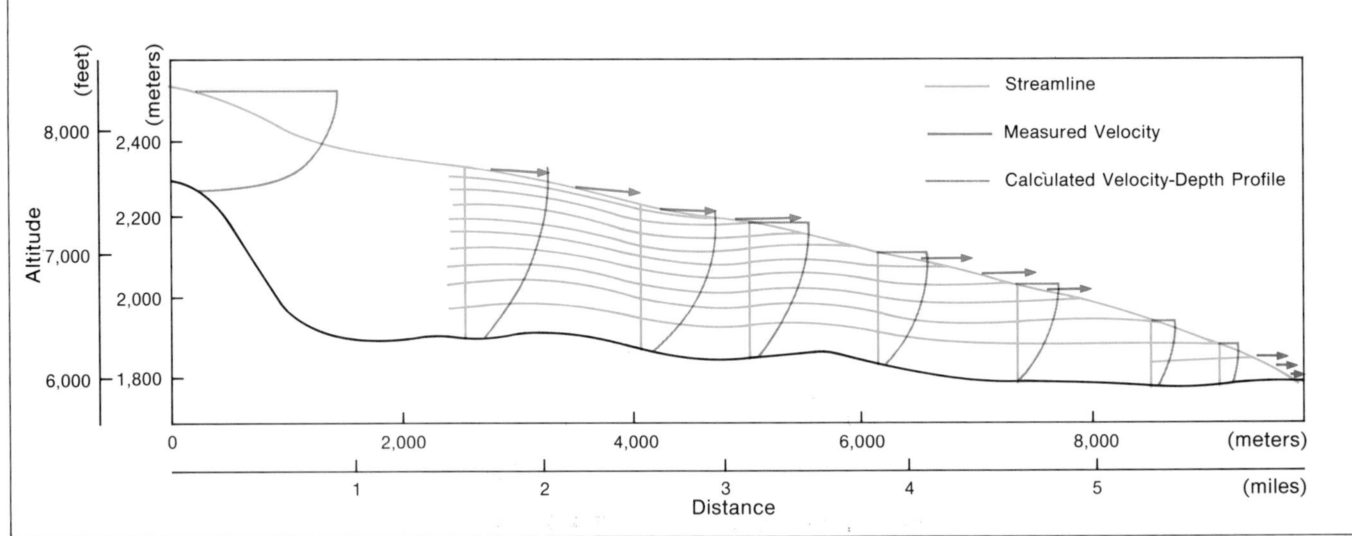
Altitude
(feet)
(meters)
8,000
7,000
6,000
2,400
2,200
2,000
1,800
Streamline
Measured Velocity
Calculated Velocity-Depth Profile
0
2,000
4,000
6,000
8,000
(meters)
1
2
3
4
5
(miles)
Distance

Figure 15.6 (top) The mass movement of a glacier is accomplished partly by basal slip over its rock base and partly by internal flow of the ice. As the diagram indicates, the process of basal slip causes the glacier to move as a whole, such that the distances traveled by the base and by the surface are the same. In the part of the motion due to internal flow, however, the ice deforms in such a way that the surface moves more rapidly than deeper portions of the glacier. (After Sharp, 1960)

(bottom) This diagram, in which the vertical scale has been exaggerated, shows the ice velocity at various points in the Saskatchewan Glacier. Only velocities at the surface of a glacier can be measured directly; the internal velocities indicated in the diagram are inferred. The measured velocity of this glacier's surface in its thicker upper portion is of the order of 400 meters per year (1,300 feet per year). Near the snout, the velocity is approximately 100 meters per year (330 feet per year). Note that a point on the upper portion of the glacier's surface tends to move slightly downward, whereas points on the lower portion of the surface move slightly upward. (After Meier, 1960)

the effects of reduced or increased snowfall and other climatic changes within a few years or decades, because the proportion of ice transported each year is greater in active glaciers than it is in passive glaciers.

Movement of Glaciers

The energy for transporting ice from the zone of accumulation to the zone of ablation comes from the decrease in gravitational energy associated with movement from a higher to a lower altitude. The actual mechanism of glacier movement is complex, however, because of the special properties of ice.

One of the properties of ice is that it melts under pressure. Less pressure is required to melt ice at 0°C than ice that is colder. Ice skating is difficult on an extremely cold day, for example, because the pressure the skate blade applies to the ice is insufficient to produce the thin film of water required for lubrication.

Glaciers, or parts of glaciers, may be classified as *temperate* or *polar*, depending on whether the temperature of the glacial ice is at the pressure melting point or considerably below it. In temperate glaciers, meltwater produced by pressure melting or meltwater from the surface pervades the body of the ice. One way that temperate glaciers can move is by *basal slip*, in which a glacier slides bodily on the lubrication provided by a thin film of water between the glacier and its bed. Cold polar glaciers, such as those in Antarctica, are usually partly frozen to their beds and are relatively incapable of basal slip. The distinction between temperate and polar glaciers is important in geomorphology, because a glacier cannot actively erode its bed unless basal slip occurs.

Ice has the ability to flow slowly when pressure is applied, and the flow of ice is an important mechanism for the motion of both temperate and polar glaciers. *Internal flow* occurs when applied pressure causes crystals to slip along lines of internal weakness. In valley glaciers, the weight of the upper part of the glacier applies pressure on the lower part, and in ice sheets the pressure comes from the ice in the central part of the sheet pushing outward with its weight. Because of the pressures, different parts of the ice move at different speeds. Often the maximum surface speed of a mountain glacier occurs near the center of the flow stream, with the speed decreasing toward the edges. The layer of rocks and dirt on the surface of a mountain glacier often exhibits curved ribbonlike patterns along the ice stream because of the difference in surface speeds between the center and the edges of the ice flow. The speed of ice flow in glaciers varies greatly, but a typical speed is 1 meter per day (3 feet per day) or less. An active glacier flowing down a steep slope may move up to 30 meters per day (100 feet per day) or more.

Glaciers sometimes exhibit wavelike behavior called a *surge*, in which a thicker region of ice moves down the glacier from the zone of accumulation at a rate several times the glacier's normal speed. When the surge reaches the glacier's terminus, or *snout*, the glacier advances rapidly for a time; speeds up to 100 meters per day (330 feet per day) have been reported. Surges occur when accumulation builds above a certain threshold amount, which results in a rapid transfer of ice to the ablation zone. Surges also occur when the ice is literally floated off its bed by an accumulation of meltwater under hydrostatic pressure.

The Work of Glaciers on the Landscape

There are no glaciers in the northeastern United States, because nowhere in this region does the present climate allow snow to persist throughout the year. Nevertheless, the landscape of New England and New York State bears numerous marks of glacial action. The landforms left by the ice sheets of the Pleistocene have not yet come into equilibrium with the shaping forces of the present climate.

Like a stream, glaciers modify the landscape both by erosion and by deposition. Glaciers are far more powerful than streams as agents of geomorphic change, however. A valley glacier can shape the floor and sides of a wide valley simultaneously, whereas a stream is confined to work in its floodplain. Furthermore, the comparative solidity and other special properties of glacial ice enable it to erode the land by mechanisms not possible with flowing liquid water.

Mechanisms of Glacial Erosion

Glaciers have great ability to transport rock debris. Unlike a stream, a glacier's speed of flow has little influence on the size of the debris particles it can carry. A glacier's debris load varies in size from rock dust as fine as flour to boulders weighing thousands of tons. One of the first tasks a glacier accomplishes when it invades a new region is the removal of soil and weathered material. The resulting exposure of the underlying rock enables the glacier to begin removing solid material that would remain relatively untouched by other erosion processes.

As a glacier flows over bedrock, it leaves scratches and grooves parallel to the direction of flow. Glacier ice itself is too soft to cut rock, so the *abrasion* of rock is the work of stone fragments held in the ice and propelled by the glacier. Even with the glacier's tools for eroding rocks, many hard rocks seem resistant to grooving. On a smaller scale, the finer sediments in glacial ice scratch or polish rock surfaces.

Glaciers can also erode rock by *quarrying*, or *plucking*. As a glacier moves over jointed rock, it can tear loose stone blocks from the parent rock and transport them. Glaciers remove far more material by plucking than by abrasion, particularly where well-developed jointing divides the rock into smaller blocks. Plucking also provides many of the large rocks used as tools in abrasion. The mechanism of plucking is not fully understood, and several processes may contribute to it. Ice flowing over an outcrop may simply push rock away from the downstream face of the rock. Me-

Figure 15.7 The ice of a glacier is armed with rock fragments of various sizes, which can abrade the bedrock under the flowing glacier. Glacial abrasion has polished and striated the rock shown in the photograph. The rock also exhibits a prominent gouge caused by the plucking action of the glacier.

chanical weathering from frost action is important in climates that support the growth of glaciers, and some form of mechanical weathering associated with frost action is believed to occur even in rocks that are covered by glaciers. Once a glacier erodes a thickness of rock, possibly by abrasion, the reduced pressure on exposed underlying rocks may cause them to crack from internal stresses. The jointing developed in the unloaded exposed rock enables plucking to begin.

The difference between abrasion and plucking has geomorphic significance. As Figure 15.8 indicates, abrasion works primarily to smooth and polish the upstream face of rock knobs, whereas plucking steepens and roughens rocks on the downstream side. Glaciers tend to shape the landscape into streamlined but rugged features, smoothed in one place and roughened in another. Frequently, the landscape also exhibits long grooves and numerous rock basins that fill with water.

The rate of glacial erosion is not easy to measure because the area of active erosion is covered by ice. One of the more accurate methods is to measure the rate at which sediment is carried away from the glacier by meltwater streams. The few data that are available suggest that glacial erosion may be perhaps 50 times more rapid for a given area than the rate of erosion of an area of similar size in the drainage basin of the Mississippi River, for example.

Glacial Erosion and Landforms

The effects of glacial erosion can be seen on many scales, from an individual rock outcrop to an entire countryside. On the smallest scale, abrasion produces grooving, scratching, rounding, and polishing of rock faces. Crescent-shaped chips and gouges are also charac-

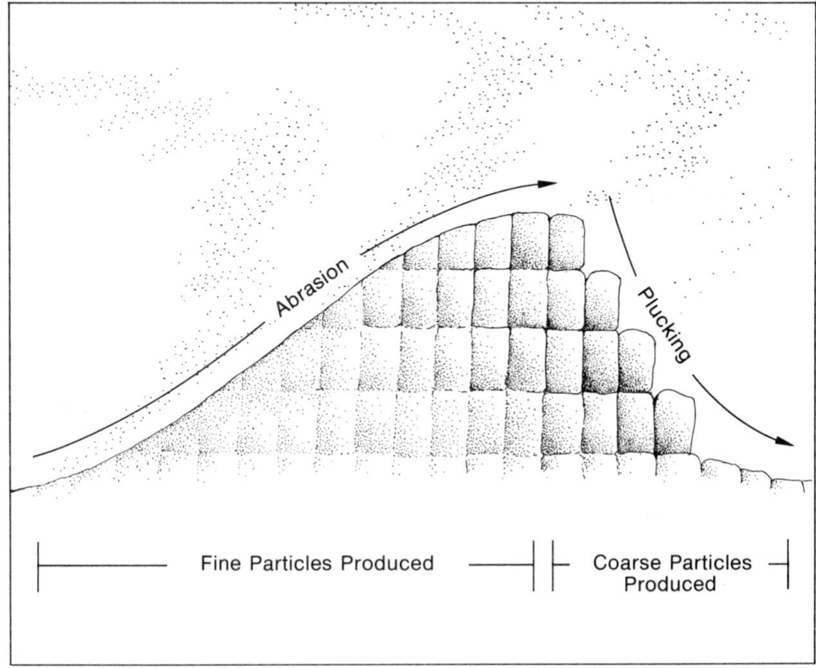
Abrasion
Plucking
Fine Particles Produced
Coarse Particles Produced

Figure 15.8 In this diagram illustrating the characteristic action of glaciers on outcrops of jointed rock, the direction of glacier flow is from left to right. The upslope side of the rock is smoothed and polished by abrasion, and the downslope side is roughened as the glacier plucks large rocks away from the outcrop.

teristic effects of glacial erosion on individual rocks. Sometimes rock surfaces are sculpted into smoothly contoured furrows and bowls, which may have been hollowed out by flowing meltwater under a glacier.

Large-scale landforms caused by glacial erosion include glacial cirques, troughs, lakes, and fjords. *Cirques* are the amphitheaterlike forms developed at the sources of valley glaciers. Their enlargement eats away the preglacial upland topography, which becomes characterized by knife-edged ridges called *arêtes*. These meet in bold pyramidal peaks known as *horns*. The presence of cirques, arêtes, and horn peaks is the diagnostic feature of *alpine topography* produced by mountain glaciation. Valley glaciers usually follow an existing stream valley, and the work of the glacier frequently alters the valley cross section from a V shape to a broad-bottomed, smooth-

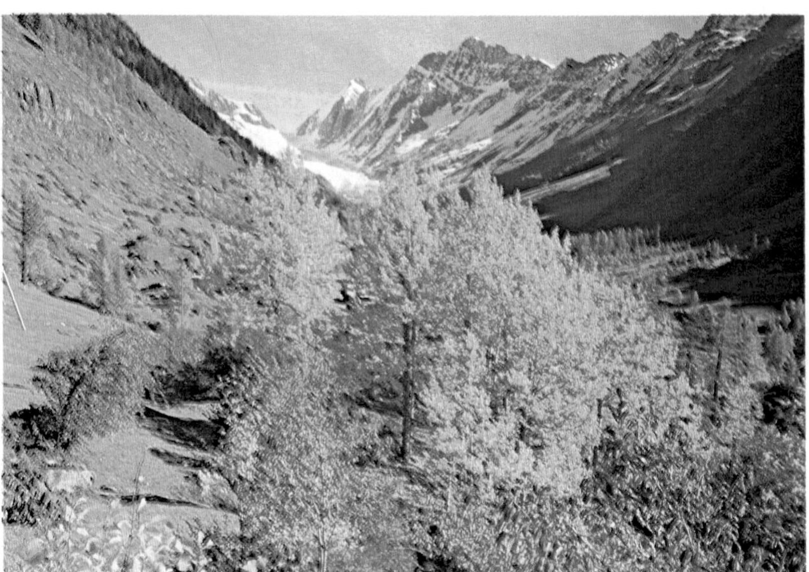

Figure 15.9 Valley glaciers follow pre-existing valleys as they flow toward lower elevations. The glacier scours the valley floor and walls, eroding irregularities and forming a smooth *U-shaped valley,* or glacial trough. The U-shaped valley in the photograph is the Lötschental in Switzerland.

Figure 15.10 When a valley glacier reaches the sea it can scour out its channel. When the glacier melts or retreats it leaves a wide, deep ocean inlet, or *fjord,* such as this fjord at Milford Sound, New Zealand.

sided, U-shaped trough. The longitudinal profile of a glacier valley is sometimes marked by sharp breaks in slope and by steep sections called *valley steps*, because a glacier seems to accentuate discontinuities, rather than smoothing its bed as a stream would. The steep slopes are made steeper by plucking. When continental ice sheets advance into an existing river valley, they act like valley glaciers and deepen the valley by erosion. The long, narrow Finger Lakes of central New York State lie in former stream valleys that were deeply gouged by glaciers during the Pleistocene.

Fjords present the outward appearance of submerged river valleys, but they are actually the deeply eroded valleys of glaciers that once flowed into the sea. Some fjords are eroded to more than 1,000 meters (3,300 feet) below present sea level, well beyond the depth of any known change in sea level. Thick glaciers excavated

Figure 15.11 The circular basins in this photograph of the Wind River Mountains, Wyoming, are *cirques* formed by the quarrying action of ice and snow at the head of valley glaciers. The steep walls of a cirque are believed to be produced by collapse when frost action at the bottom of the bergschrund removes the base support of the cliffs. Small lakes, or *tarns*, fill the cirques.

fjords under the surface of the water, a feat impossible for streams, which are ineffective agents of erosion below sea level. Glaciers cut fjords into the coastline of Alaska in the past, and they continue to do so today.

Glacial Deposition and Landforms

Glaciers can carry great loads of debris, and they can produce landforms by deposition as well as by erosion. The debris deposited directly by glacial ice is known variously as *moraine, drift,* or *till.* Glaciers are quite dirty, especially in their lower reaches, from the load of debris generated by abrasion and plucking. The ice sheets of the Pleistocene gained a heavy load by picking up most of the weathered material from the ground they passed over. Even if a glacier is not growing, the moving ice acts like a conveyor belt to transport debris toward the snout. As the ice melts, debris is deposited in an *end moraine,* a ridge that consists largely of uncompacted and unsorted sand, gravel, and rock. The depositional topography is highly diverse, consisting of ice-streamlined till (*drumlins*), ice tunnel deposits (*eskers*), hills formed from jumbles of debris deposited in masses against wasting ice (*kames*), and depressions left where iceberg-sized masses of glacial ice were buried and subsequently melted (*kettles*).

Ice is solid water, and the work of glaciers cannot be dissociated from the work of flowing water. Except where ablation occurs primarily by the breaking off of ice chunks, active glaciers generate

Figure 15.12 (opposite top) The quarrying action of two adjacent glaciers may erode the intervening walls to the narrow, ragged rock ridges, or *arêtes*, seen here in the French Alps.

(opposite bottom left) The deepening of stream valleys by glaciers is often more rapid in the main valleys than in the smaller side tributary valleys. After the glaciers melt away, the tributary valleys may be left as *hanging valleys,* which join the deep glacial trough high above the valley floor. The cascade in the photograph is a stream falling from a hanging valley in Yosemite National Park, California.

(opposite bottom right) The Matterhorn in Switzerland is the classic example of a *horn* formed by erosion where the headwalls of three or more cirques intersect.

Figure 15.13 (bottom left) This *till plain* in the state of Washington is deeply covered with glacial debris. The *ground moraine* in the foreground is unsorted debris deposited as sheets beneath the glacier.

(bottom right) These hills in the mountains of Wales are *cirque moraines.* They are accumulations of transported debris left at the margin of a former small glacier in an upland ampitheatre or cirque. Similar glacial deposits are found in most cirques throughout the world. They represent the latest Pleistocene depositional activity by relatively small ice masses in the mountains.

large outflows of meltwater in the zone of ablation. The meltwater often contains quantities of silt from glacial erosion; some glaciers produce meltwater that is milky from suspended particles. Meltwater streams from the Pleistocene ice sheets deposited silt on their floodplains; later, wind carried the silt, called *loess*, to new locations. Deposits of loess form the parent material for soil in much of the best farmland in the Mississippi and Missouri river valleys.

Glacial till contains assorted particles with a range of sizes. It often reflects the composition of local rock; where the bedrock is sandstone, till has a high proportion of silica, for example. Pleistocene ice sheets deposited till over much of the midwestern United States, as indicated by the map in Figure 15.3. The depth of the till often exceeds 30 meters (100 feet), which is enough to cover much of the preglacial topography and leave a relatively featureless fertile plain.

Glaciation and Man

The ice sheets of the Pleistocene profoundly altered the glaciated areas of northern North America. Some of the work of the glaciers made the land useful to man for agriculture, industry, or recreation. Glacial deposits of sand and gravel are used today for construction, for example, and in the Midwest, a large preglacial valley filled with glacial till supplies groundwater to cities in Illinois, Indiana, and Ohio.

One of the major features produced by the ice sheets on the North American continent is the Great Lakes. Their basins were formed as ice sheets scoured and excavated the preglacial river valleys in weakly resistant layers of sedimentary rock. The Great Lakes now provide transportation routes, produce hydroelectric power at Niagara Falls, and supply a seemingly abundant supply of fresh water. The principal industrial section of the United States is oriented around the Great Lakes, and the pattern of cities and transportation routes in that region strongly reflects the influence of the Lakes.

Not all of the changes wrought by glaciers have been beneficial to man, however. North of the Great Lakes, the land is a barren, swampy, sparsely populated region. This land, too, resulted from the work of the ice sheets, but there, the ice sheets removed the soil from the Canadian bedrock and left a relatively inhospitable landscape of shallow basins and bare rock. In New Hampshire, the ice sheets left a deep bed of stony soils. Each spring, freezing and thawing ice in the soil pushes a new crop of rocks to the surface of the farmlands, a phenomenon that encouraged many New Hampshire farmers to move to the fertile till plains of Ohio in the nineteenth century.

Summary

Valley glaciers and glacial ice sheets form where snow persists throughout the year. Snow accumulates on the cooler portion of a glacier, and the packed annual layers of snow gradually become dense glacier ice. Ice ablates from the warmer sections of a glacier, and the net growth of a glacier is governed by the relative rates of accumulation and ablation.

At present glaciers are found only in high mountains or in the higher latitudes, but in the cooler climate of the late Pleistocene, glacier ice covered large sections of North America and Europe. The glaciers left a distinctive imprint on the land; landforms such as cirques, drumlins, fjords, and till plains are produced only by glacial action. Glaciers carved out such features as the Great Lakes basins.

Glaciers move by slippage on their beds and by the flow of ice, and they transport large amounts of unsorted rock debris, which they obtain by picking up loose material from weathered surfaces and by abrading and plucking bedrock.

Meltwater from glaciers carries loads of debris far beyond the boundaries of the glaciers. The glacier itself also deposits great quantities of sediment; much of the Midwest is covered with deposits of glacial till tens of meters deep.

16 The Edge of the Land

The Continental Shelves

The Geomorphic Significance of Tides
Cause of the Tides
Estuaries: Gift From the Sea

Wave Motion

Wave Action Against the Land
The Work of Waves
Tsunamis

Beaches
Sediment Transport and Deposition
Beach Erosion and Man

Relief Forms of the Ocean Floor

Pounding waves carve out cliffs, bays, and fantastic shapes along the coastlines. But where is the edge of the land? Landforms also exist under the sea, and some of the mountains and canyons on the ocean floor are more massive than those on the continents.

Early Morning After a Storm at Sea by Winslow Homer. (The Cleveland Museum of Art; gift from J. H. Wade)

The Edge of the Land

Someone who goes to a beach only in the summer may think of beaches as relatively enduring parts of the landscape, like hills or river valleys. But beaches are among the most ephemeral of geomorphic features. They show an annual cycle of growth and retreat, filling with sand during the summer and losing sand during the winter. Year-to-year changes are often all too apparent; some beaches have disappeared completely within a few decades. Figure 16.2 shows the beach on the Pacific Ocean at La Jolla in southern California, sandy in the summer and rocky in the winter. A recent assessment predicts that if present processes continue unchecked, all of the sand on the beach will be washed away within 20 years.

Beaches are the work of moving water. Water transports sediment to and from beaches, and beaches retreat when annual loss of sediment exceeds annual accumulation. Beaches respond rapidly to changes in patterns of erosion and deposition. Because of man's special interest in maintaining beaches, he has sought to promote deposition and retard erosion by constructing breakwaters and jetties to modify sediment-bearing currents. But often when local remedies have been applied, increased deposition along one part of the coast occurs at the expense of increased erosion elsewhere.

Beaches are not the only landform characteristic of coastal areas. Along many coastlines, steep cliffs plunge directly into the sea and there is no beach at all. In other places, the coast is low and is bordered by tidal mud flats and brackish lagoons. In some places sand spits project out from the land or barrier beaches run parallel to the shore. There are landforms under the surface of the ocean as well. You have already read that great undersea mountain ridges

Figure 16.2 Beaches are not fixed landforms, but continually change as material is deposited or swept away. During the summer (left), this beach at La Jolla, California, is covered with sand deposited by ocean currents, but during the winter (right), currents sweep the sand away, leaving a beach of rocks too large for the currents to transport. (After *Geology Illustrated* by John S. Shelton. W. H. Freeman and Company. Copyright © 1966)

follow the oceanic plate boundaries where new sea floor is being generated. Other features on the ocean floor, quite unknown to us before the development of echo-sounding techniques, include major valley systems, fracture zones, and truncated submerged mountains.

The Continental Shelves

Where does the land end? The continental plates do not break off abruptly; rather, the land slopes gradually under the water to form *continental shelves.* Despite their name, many shelves show considerable relief and are far from being smooth topographically. Along glaciated coasts, such as those of New England or Norway, the inner continental shelf exhibits many features produced by glaciers. If the sea level were suddenly lowered, the exposed shelf off New England would appear as hilly as the sand ridges of the present coastal landscape.

The continental shelves along the Atlantic coast of the United States typically descend 1 or 2 meters for every kilometer away from the coast (5 or 10 feet per mile). The continental shelf off Jacksonville, Florida, is shallower than the average depth of shelves; it is only 50 meters (160 feet) below the ocean's surface 100 kilometers (60 miles) off the coast. Along the Pacific coast, the continental shelves tend to be narrower and more steeply sloping than the shelves along the Atlantic. At San Diego, California, the shelf only 10 kilometers (6 miles) offshore is at a depth of 50 meters (160 feet). At a depth of 100 to 200 meters (330 to 660 feet) continental shelves show a rapid change of slope and become the much steeper *continental slopes.* The continental slope off the Atlantic coast of the United States descends at an average rate of 50 meters per kilometer (250 feet per mile) of distance, whereas off the Pacific coast, the slope's average rate of descent is approximately 100 meters per

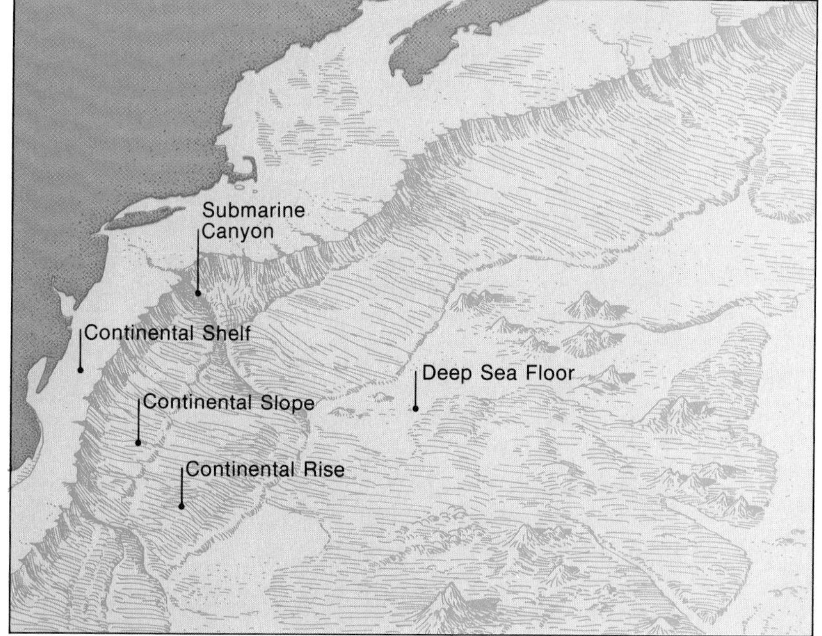
Submarine
Canyon
Continental Shelf
Deep Sea Floor
Continental Slope
Continental Rise

Figure 16.3 The topography of the ocean bottom off the coast of the northeastern United States shown in this figure exhibits several major divisions. The vertical scale is exaggerated approximately 20 times for clarity. Near the margin of the continent, the water is shallow and the ocean bottom forms a gentle slope called the *continental shelf.* At the edge of the shelf, the bottom slopes steeply down the *continental slope* to the deep sea floor. The *continental rise,* where the continental slope meets the sea floor, is a region where sediments accumulate. Note also the smaller features of relief such as the deeply trenched *submarine canyon* associated with the Hudson River and the relief forms on the sea floor. (After Heezen and Tharp) kilometer (500 feet per mile). The *continental rise* is a gentle, sediment-covered slope that borders the steeper continental slope and extends hundreds of kilometers to the deep sea floor.

The wide continental shelves typical of the Atlantic coast apparently are features associated with the changes in sea level that accompanied periods of glaciation. The depth where the continental shelves terminate and the continental slopes begin corresponds approximately to the position of sea level during the Pleistocene ice ages. The shelves appear to be terraces cut by wave action, either during the Pleistocene or earlier. The sea's ability to erode the land is evident on many shorelines; some shores have been cut back several kilometers in historic times. Normally, wave action develops a narrow, gently sloping terrace from the debris produced by the erosion of coastal sea cliffs. But if the sea level is lowered, the cutting extends farther offshore and the terrace becomes wider. Without marked changes in sea level, wide continental shelves probably could not have been formed.

Conversely, if the level of the sea has ever been higher than it is now, features caused by wave erosion should be present at altitudes above the present sea level. Such landforms are exposed to weathering and erosion, however, and they become less easy to identify as time goes on. Nevertheless, emerged marine terraces have been found at many altitudes along various coasts, which indicates that at times in the geologic past the level of the sea was higher relative to the land than it is now. Marine features have been found in England 200 meters (660 feet) above the present sea level, and in California there are marine terraces 400 meters (1,300 feet) above the present sea level. The melting of all polar

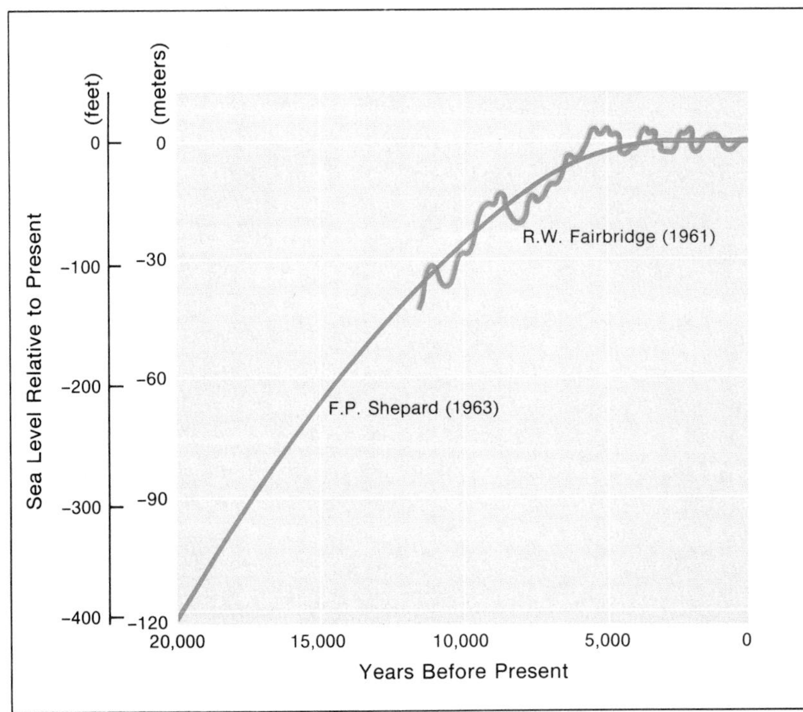
Sea Level Relative to Present
(feet)
(meters)
0
-100
-200
-300
-400
0
-30
-60
-90
-120
R.W. Fairbridge (1961)
F.P. Shepard (1963)
20,000
15,000
10,000
5,000
0
Years Before Present

Figure 16.4 As this graph indicates, the level of the sea has risen by 120 meters (400 feet) since the glacial period of the late Pleistocene. Melting of the ice sheets returned large quantities of water to the oceans. Numerous coastal landform features that were exposed 20,000 years ago are now deep under water; many of the world's major harbors are located in wide river valleys that were submerged by the rising seas. (After Bird) ice would raise the level of the oceans by 30 to 60 meters (100 to 200 feet) or so—less if there were an accompanying deformation of the ocean basins. Some geomorphologists have argued that during Pleistocene interglacial periods the sea level was never more than 10 meters (30 feet) above what it is today and that the marked height of some marine features above the present sea level is a result of subsequent uplift of coastal regions.

As the sea rose during the melting of the Pleistocene ice sheets, shoreline plants and animals were submerged. Radiocarbon dating of their remains, taken from different heights on the slopes of submerged river valleys, gives an indication of sea level through the recent geologic past. Figure 16.4 shows two sets of results from such measurements, which are in reasonably good accord except for the detailed variation of sea level during historic times.

Figure 16.5 shows the change in sea level relative to the land along the coasts of the United States during the past 80 years. Part of the general trend toward higher levels has been attributed to a lowering of the land brought about by the removal of groundwater and oil. But the trend also seems to reflect a genuine increase in the amount of water in the oceans from polar ice melting.

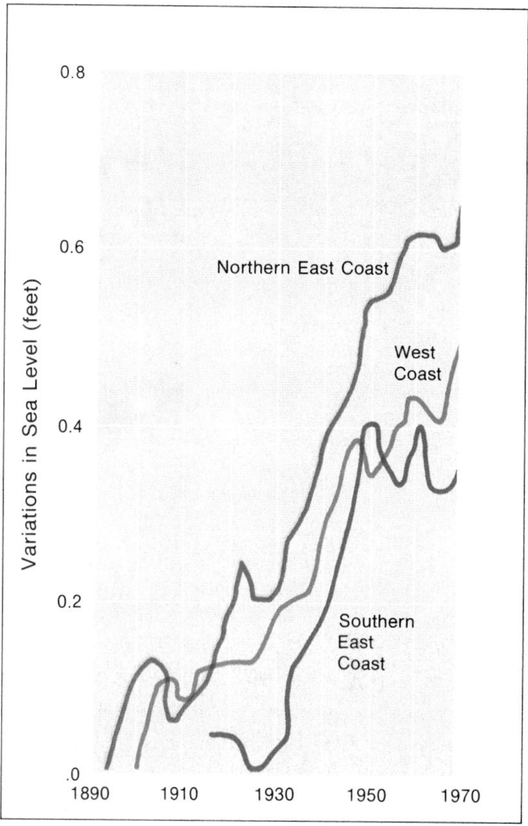

Variations in Sea Level (feet)
0.8
0.6
0.4
0.2
.0
1890
1910
1930
1950
1970
Northern East Coast
West Coast
Southern East Coast

Figure 16.5 The height of sea level relative to the coastlines of the United States has been increasing during this century. Part of the increased height is attributed to local lowering of the land.

The Geomorphic Significance of Tides

The edge of the land, taken to be the boundary between water and land, has changed position during geologic time because the sea level has changed relative to it. Beyond the continuing variation in that boundary brought about by wave action, there are also daily and seasonal variations brought about by the ebb and flow of the tides. Tides normally raise and lower the water level at a coast, and they are significant geomorphically because tidal changes in water level expose different parts of the coast to the erosive action of ocean waves.

Cause of the Tides

The gravitational effects of the sun and the moon on the waters of the earth cause the tides. To understand how these gravitational pulls affect the oceans, begin by considering the effect of the sun. The sun's gravitational pull constantly attracts the earth, but as the earth "falls" toward the sun, its forward motion carries it around the sun in its orbit. The strength of the sun's gravitational force decreases with distance, so water on the side of the earth nearest the sun experiences a greater pull than water on the far side. Consequently, water on the near side falls toward the sun faster than water on the far side, and bulges of water develop on both sides of the earth. The moon has a similar effect. It is much closer to the earth, however, and it is twice as effective as the sun in producing tides. As the earth rotates, the double bulge moves relative to features on the earth's surface. Normally, two high tides occur during every 24-hour, 40-minute period—the time required for a point on the earth to return to nearly the same position relative to the moon, which moves rapidly in its orbit.

Shoreline features of the earth's landmasses strongly influence

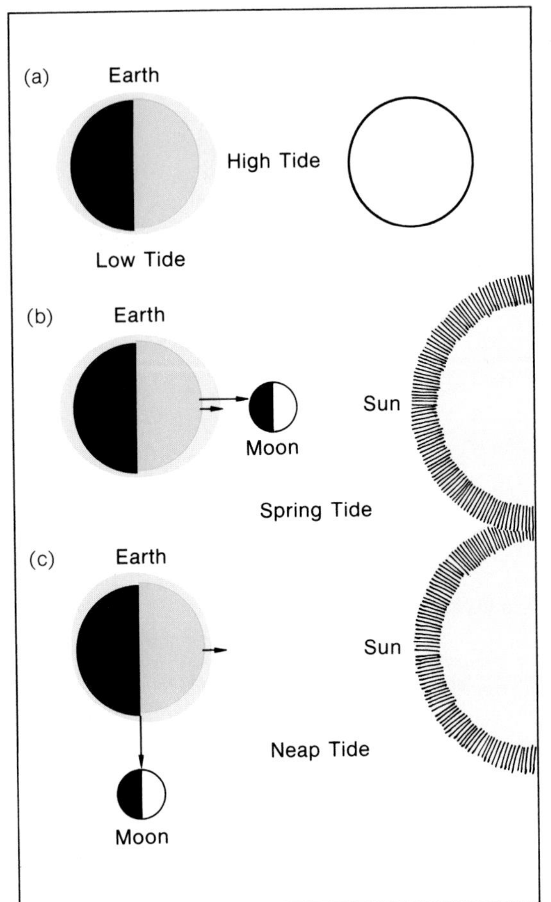
(a)
Earth
High Tide
Low Tide
(b)
Earth
Moon
Sun
Spring Tide
(c)
Earth
Sun
Neap Tide
Moon

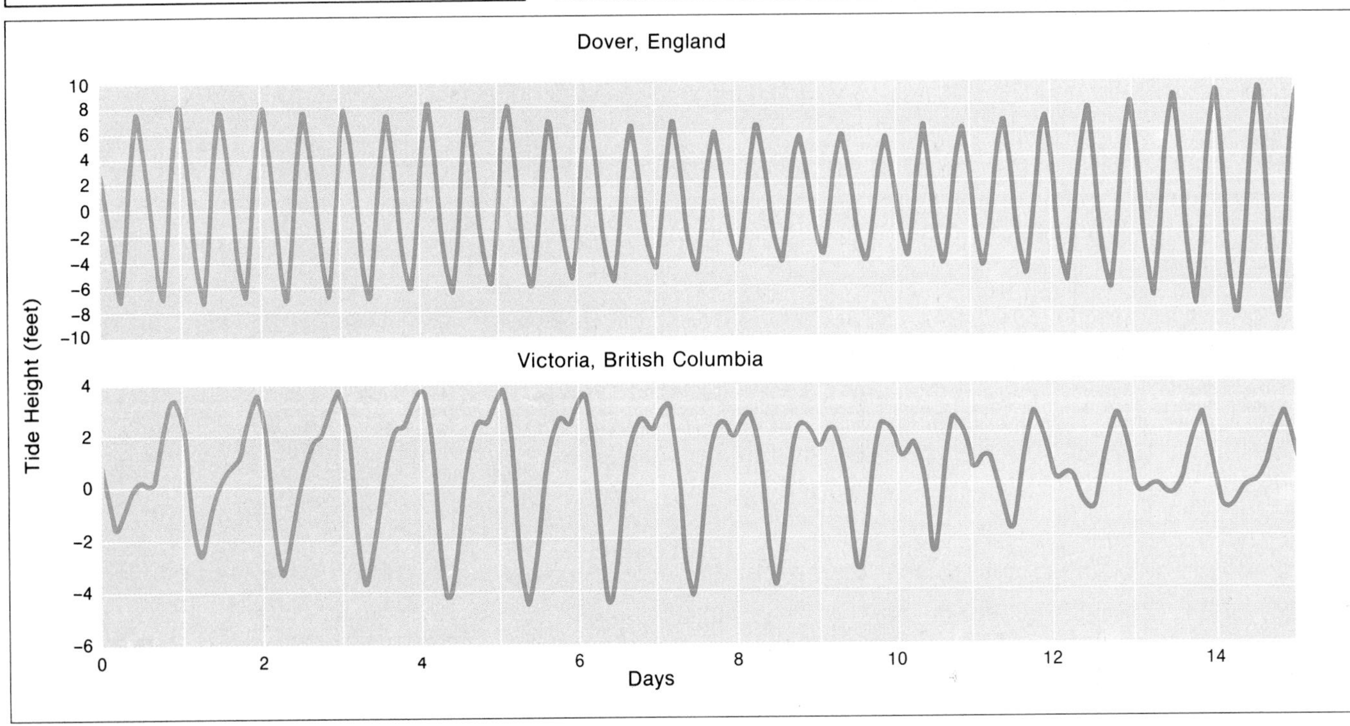
Dover, England
Victoria, British Columbia
Tide Height (feet)
Days
10
8
6
4
2
0
-2
-4
-6
-8
-10
4
2
0
-2
-4
-6
0
2
4
6
8
10
12
14

the pattern of the tides. Some locations have two high tides of unequal height each period, and other locations have only one major high tide. High tides are typically 1 or 2 meters (3 or 6 feet) above average sea level, but they can be many times higher if tidal waters are funneled into a narrow channel. Because both the sun and moon influence the tides, the height of high tide varies from day to day in a complicated fashion that depends on the relative positions of the earth, the moon, and the sun. At new moon or full moon, the earth, moon, and sun are nearly in line and the tidal range is high because the sun and the moon act together. Tides during this period are called *spring tides*. When the moon is in its first or last quarter, the moon and the sun are far out of alignment and the pull on the earth's waters is in two different directions. Low tidal ranges occurring at such times of the month are called *neap tides*. Tidal patterns vary from year to year because the moon requires nearly 19 years to return to exactly the same position relative to the earth and sun positions.

Estuaries: Gift From the Sea

Tidal action is especially significant in estuarine environments. During the general rise in sea level that followed the retreat of the Pleistocene ice sheets, the mouths of many river valleys became flooded. Where a river valley terminated in a wide, flat coastal plain, it became possible for tides to move inland for a distance. The zone in which the tides meet the river water is called an *estuary*. In an estuary, there is appreciable mixing of salt water and fresh water (see Figure 16.7).

Estuaries are one of the most productive biological regions on the earth for a given area. The pulse of the tides carries nutrients from the sea into estuaries, carries out waste, and helps maintain a continual circulation of water. Even creatures that do not live in estuaries during their full life cycles may use these regions as breeding grounds. In fact, estuaries have an ecological significance far beyond their relatively confined boundaries. For example, 10 square kilometers of estuary in Rhode Island are the breeding area for flounder that constitute about 25 percent of the region's flounder catch.

Because estuaries receive large quantities of sediment from upstream, they are characterized by low flats of clay and silt. Vegetation such as salt marsh grass frequently establishes itself on these flats. The marsh grass helps to protect the mud from erosion and also helps to build the mud flat by filtering additional sediment from the tidal waters. Salt marshes therefore tend to encroach upon the open waters of an estuary.

Wave Motion

The surface of the ocean is never still. On a windy day the water is ruffled into intricate patterns of disturbances; even when the air is relatively still, the ocean usually shows a smooth rising and falling *wave motion*. The physical significance of waves—whether they are light waves, sound waves, or water waves—is that they carry

Figure 16.6 (opposite top left) (a) Tidal bulges of water, here greatly exaggerated in height, are produced on the earth because of the gravitational effect of the moon and the sun. Two bulges occur because the earth is falling (technically, accelerating) toward the moon or toward the sun; the water falls fastest on the nearest side, where the gravitational force of the tide-producing body is strongest. The water on the far side falls most slowly and therefore tends to be left behind. As the earth rotates, the tidal bulges sweep across the earth, producing twice-daily high tides in general. Detailed analysis shows that tides on the earth also possess a once-daily component produced because of the positions of the moon and the sun relative to the earth's rotation.

(b) The highest high tides and the lowest low tides, or spring tides, occur during new moon and full moon, when the moon and the sun act together.

(c) The lowest high tides, or neap tides, occur during the moon's first quarter and last quarter, when the moon and the sun are arranged as shown in the diagram.

(opposite top right) The moon, shown here over the coast of southern California, is the principal cause of the tides. The moon has twice the effect of the sun in raising tides. Tides also occur in the atmosphere and, to a small extent, in the solid earth.

(opposite bottom) The graphs show the tide heights relative to average sea level at Dover, England, and Victoria, British Columbia, for a period of 15 days. The tides at Dover exhibit the usual pattern of twice-daily high tides. Note that the heights of the daily tides vary through the month because the earth, moon, and sun change their relative positions. The tides at Victoria show effectively only one high and one low tide per day. Once-daily tides can occur where the twice-daily tides are unusually small because of special land configurations. If tides arrive at a location by way of two channels, with a relative time delay of 6 hours between the separate flows, the twice-daily tides may be canceled. The tides at many locations show a mixture of twice-daily and once-daily tides. (After Warbug)

Figure 16.7 (below) The estuary of the Meon River, England, has low flats of alluvial material that the river has deposited. The flats are stabilized by the reeds growing on them.

energy from one place to another. Waves on the ocean are generated by the action of winds on the water surface. Waves can carry the energy from a storm a distance of thousands of kilometers and use it to erode a distant shore. The low, regular waves visible in the ocean on a still day are called *swell*; they are produced by distant storms over the ocean.

The distance between the highest point, or *crest*, of one wave and the crest of the next wave is called the *wavelength*. The lowest point of a wave is called the *trough*. In deep water, such as in the ocean, the speed of water waves depends only on wavelength; the farther apart the crests are, the greater the speed. Where the depth of the water is less than the length of the wave, speed is independent of wavelength and depends on the depth of the water. Swell has fairly long wavelength, typically 100 meters (330 feet), and in deep water it moves at approximately 15 meters per second (30 miles per hour).

In a propagating wave such as a wave traveling in water, the wave disturbance is passed on from particle to particle, and an individual particle of water travels only a short distance as the wave passes. A given water particle in a deep water wave executes nearly circular motion, moving with the wave near the crest and moving against the wave near the trough. The motion is not perfectly circular, and each time one wavelength passes, the particle advances slightly in the direction of the wave motion. Swell that originates far offshore does not bring distant water with it, although it does bring distant energy.

Although wind can generate waves with a wide range of wavelengths, only waves with long wavelengths are able to travel long distances. The reason is that as water particles move up and down as a wave disturbance propagates through them, a small part of the wave energy is dissipated as heat because of friction between particles. Because waves of short wavelength bob up and down more frequently than waves of long wavelength, they dissipate their energy more rapidly. After a mixture of waves with different wavelengths has traveled a long distance, the waves of shorter wave-

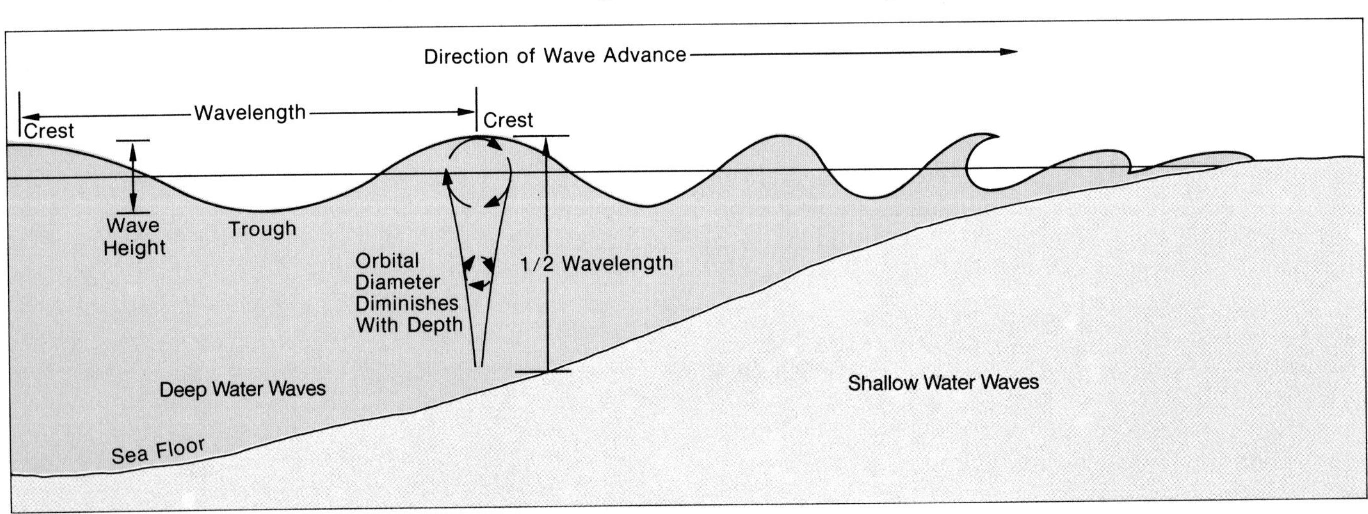
Direction of Wave Advance
Wavelength
Crest
Crest
Wave Height
Trough
Orbital Diameter Diminishes With Depth
1/2 Wavelength
Deep Water Waves
Shallow Water Waves
Sea Floor

Figure 16.8 The nearly circular motion executed by individual water particles as a wave passes is negligibly small at depths greater than half a wavelength. Waves in water deeper than half a wavelength, which are called deep water waves, are not influenced by the presence of the ocean bottom. As the wave enters shallow water near a shore, the ocean bottom begins to interact with the wave motion. The wave becomes steeper and its wavelength becomes shorter. The water particles at the top of the wave eventually reach a speed greater than the speed of the wave, and the wave breaks. The wave dissipates its remaining energy as it washes up on the beach.

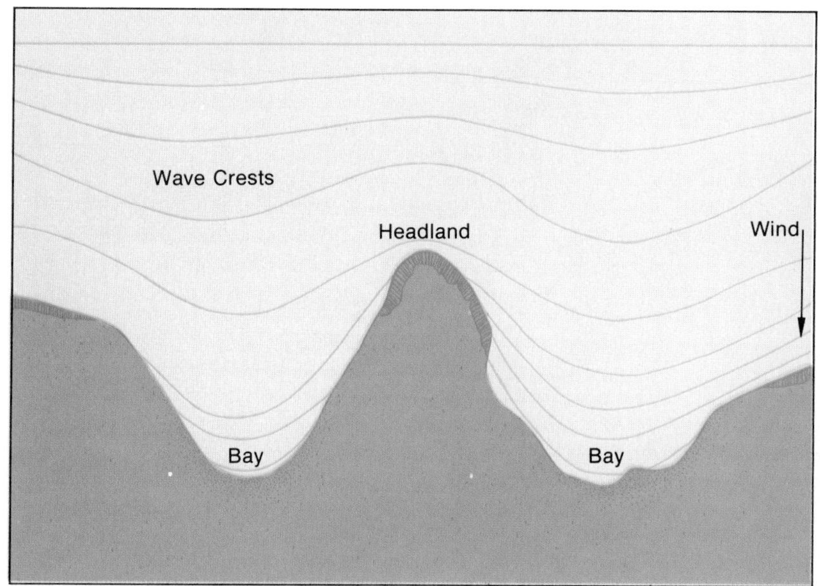
Wave Crests
Headland
Wind
Bay
Bay

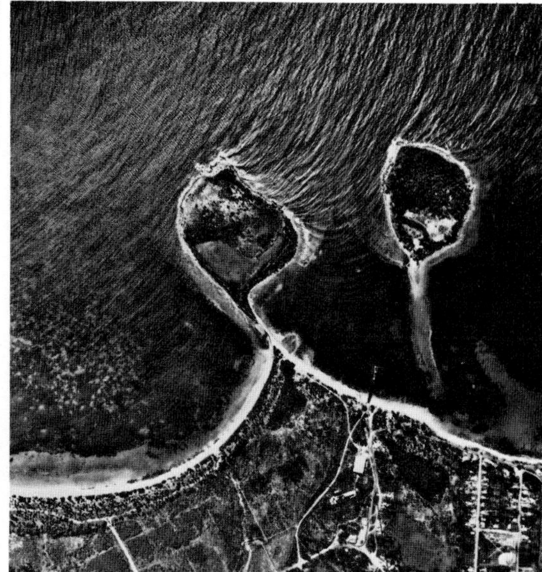

Figure 16.9 (left) Waves may be altered in direction, or refracted, when they interact with shorelines. The diagram shows waves refracted at a headland, or promontory. Refraction may cause waves to concentrate their energy on the sides of a headland, hastening erosion and causing the formation of features such as sea caves hollowed into the side walls of the headland.

(right) The wave patterns in this photograph have been altered by refraction near the coastlines of the islands and the mainland.

length have dissipated most of their energy and have largely died out, leaving only the waves of longest wavelength. The waves with longest wavelengths also travel fastest and arrive at distant shores sooner.

Waves cannot disappear unless there is some way for them to dissipate their energy. When a wave encounters an obstacle such as a sea wall, the wave may be *reflected* and reverse its direction. Waves can change direction by *refraction* as well as by reflection. When ocean waves approach a region where the sea floor gently slopes up to the shore, they change from deep water waves, with a speed that is dependent on wavelength, to shallow water waves, with a speed dependent on water depth. The shallower the water is, the slower the speed of the waves. Consider what happens when waves approach a shore at an oblique angle. The leading parts of the waves begin to slow as they reach the shallow water, and the waves effectively pivot (see Figure 16.9). This process of refraction changes the waves' direction of motion. Refraction causes waves moving at an angle to the shore to reach the shore more nearly head on. Even with refraction, however, the waves still approach the shore at a slight angle. Refraction at an obstacle can cause waves to become focused, much the way a lens focuses light rays. The refraction of waves can cause their energy to be concentrated on a small region of coastline.

Wave Action Against the Land

Deep water waves are confined largely to the ocean surface. The motion caused by the passage of a deep water wave is negligible at a distance of half a wavelength below the surface. For swell, this depth may be 50 to 100 meters (160 to 330 feet), so that as swell approach a shore the sea floor begins to influence the wave motion. In shallow water, water particles can no longer complete their circular motion, so the shape of the wave changes: the crests are

Figure 16.10 (top left) Fungus Rock, on an island off the coast of Malta, has been hollowed by wave action to form sea caves. If the waves can cut entirely through the rock without collapsing it, an open sea arch will be formed.

(top right) Durdle Door, on the southern coast of England, is a sea arch cut into chalk. When the upper part of the arch becomes weakened by erosion and collapses, the free-standing leg of the arch will be a sea stack.

(opposite top left) The chalk cliffs at Étretat on the coast of Normandy, France, were cut by waves to form sea arches and sea stacks. These cliffs are cut in the same geologic formation as Durdle Door, but are on the opposite side of the English Channel. Note the vertical face of the cliffs; wave action at their base undermines the rock and causes the collapse of the upper portions. The cliffs thus maintain a vertical face as they retreat.

steeper and the troughs become flatter. As the speed of the wave decreases, the speed of water particles at the crest eventually exceeds the speed of the wave, and the wave *breaks*. The remnants of the wave rush up onto the beach as a foaming sheet called *swash*. The flow of water that returns to the sea is called *backwash*.

The Work of Waves

Some coasts end in sheer cliffs that are battered by waves. On beaches, waves contact only sand or pebbles; at cliffs, waves erode the land. Apart from the erosion that moving water can always accomplish, there are some erosion processes characteristic of waves. In the impact of a wave upon jointed, cracked rock, the inrush of water suddenly compresses the air in the cracks. The intense pressures developed by the compressed air and the pounding of the wave are momentary, but individual blocks are loosened and eventually fall out of the cliff. Waves can usually erode cliffs only near their bases, but in storms waves can leap higher and can even hurl rock fragments at cliffs. Furthermore, removal of material at the base of a cliff removes support from the upper portions so that they fall. The power of waves, especially during a storm, is enormous. The sea has swept heavily constructed lighthouses completely away, and it has tossed about concrete blocks weighing thousands of tons.

Features on the shoreline or on the ocean bottom offshore can refract waves and concentrate their energy. Because of refraction, waves tend to attack jutting promontories of the coastline especially vigorously. Figure 16.9 shows how refraction concentrates wave energy on a promontory, particularly on its sides, cutting hollow sea caves. In time, wave erosion can smash a hole through a narrow promontory, forming an arch. Continued erosion from above and below eventually separates the tip of the promontory from the main cliff mass, which leaves a free-standing *stack*.

The rate of erosion by waves depends on the resistance the rock

offers as well as on the energy the waves transfer. Sandy cliffs in England have been observed to retreat at an average rate of several meters per year, an unusually rapid rate of erosion. However, an intense storm cut the cliffs back by more than 10 meters in a few hours. By contrast, more resistant sea cliffs have exhibited an immeasurably small rate of erosion, estimated to be less than a few centimeters per year.

The wind drives waves, but the transfer of energy from wind to water is a slow process. Waves reach their maximum height for a given wind only when the wind is able to blow for many hours over a great length of open water, the distance being known as the *fetch*. A coast's vulnerability to the powers of wave erosion depends on the pattern of swell from distant storms, the presence or absence of a long fetch, the direction of the prevailing winds in relation to the fetch and to the coastline, and the possibilities for refraction at the coast.

Tsunamis

Tsunamis are the most powerful waves that travel the ocean. They are generated by underwater earthquakes or volcanic eruptions, and they generally originate in the active trenches around the Pacific Ocean basin. A tsunami is a train of waves, with wavelengths often exceeding 100 kilometers (60 miles) and average speeds in the open ocean of 800 kilometers per hour (500 miles per hour). Because of its great wavelength, a tsunami in the deep ocean behaves like a shallow water wave, and its speed depends on the ocean's depth.

Tsunamis in the mid ocean are difficult to detect; the distance from crest to trough may exceed 50 kilometers (30 miles), and the wave is so spread out that its height is small. When the tsunami reaches the shallow waters near land, however, its speed decreases and it steepens into a massive, enormously destructive wall of water sometimes 5 or 10 meters (15 or 30 feet) high. Tsunamis are practically unknown in the Atlantic basin, but a tsunami did strike

Figure 16.11 The height of waves on open water depends on the speed of the wind and on the length of time the wind has been blowing. These calculations are based on a distance from the windward shore that is great enough to preclude the influence of other factors on wave height. (After Sverdrup and Munk, 1947)

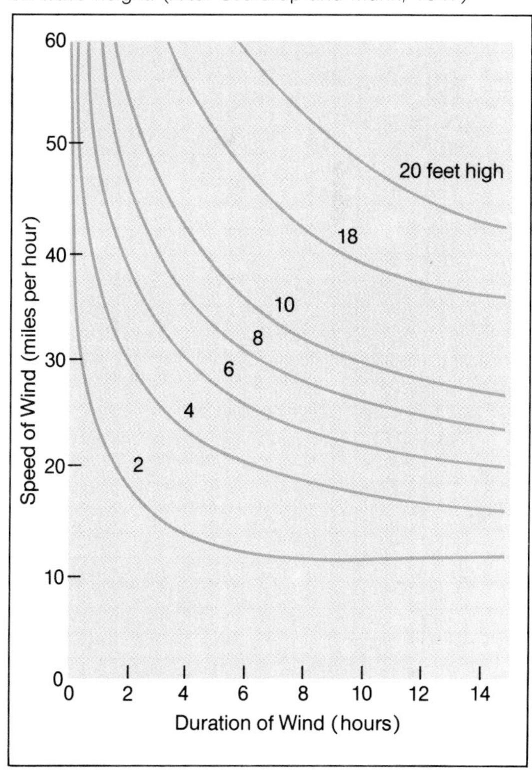
Speed of Wind (miles per hour)
0
10
20
30
40
50
60
Duration of Wind (hours)
0
2
4
6
8
10
12
14
20 feet high
18
10
8
6
4
2

Figure 16.12 A *beach* is the zone of transition between the land and sea. It extends between the level of lowest water to the upper limit reached by the highest storm waves, which is the area subject to alternate erosion and deposition of sand. The actual location and profile of a beach is constantly changing. The diagram shows the principal geomorphic divisions of a beach. The *berm* is the nearly flat portion at the top of a beach; it is covered with material deposited by waves and constitutes the *backshore*. The *foreshore* extends from the edge of the berm to the low tide line. The *offshore*, which is permanently under water, contains *bars* and *troughs*. (After Shepard, 1967)

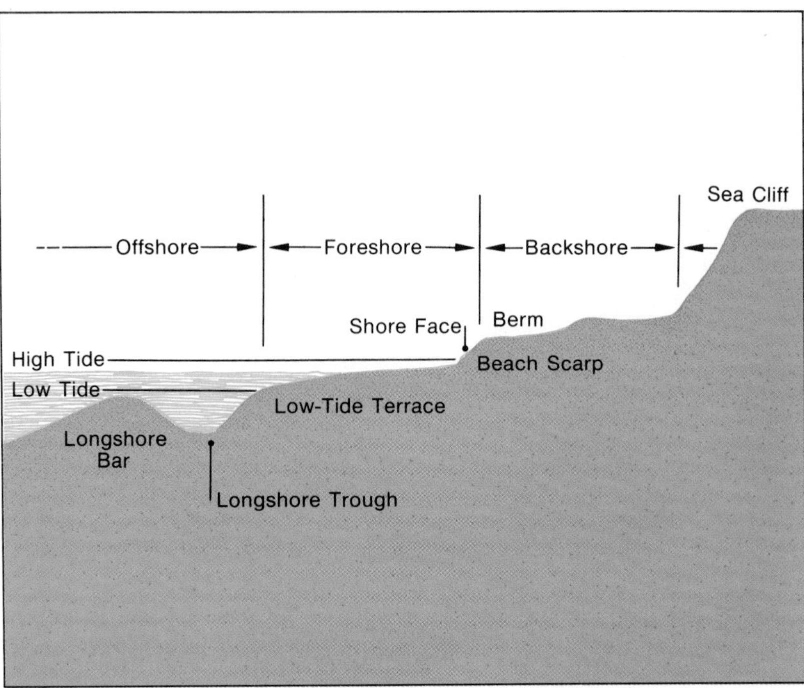
Sea Cliff
Offshore
Foreshore
Backshore
Shore Face
Berm
High Tide
Beach Scarp
Low Tide
Low-Tide Terrace
Longshore Bar
Longshore Trough

Portugal 200 years ago. The frequency of occurrence of tsunamis in the Pacific basin is perhaps one per decade; they have caused considerable damage and loss of life in Japan and in Hawaii.

Beaches

Beaches move. There are beaches made of sand, of stones, and even of tin cans, but in each case the material is free to be carried by moving water. Like glaciers, beaches have a mass budget: water delivers material to beaches and it carries material away. The growth or recession of a beach depends on the balance between accumulation and loss.

Wave action generally can move sediment particles from the ocean bottom if the water is less than 10 meters (30 feet) deep. Sand dumped offshore at greater depths is seldom transported to the beach. If the extent of a beach is defined in terms of sediment transport, beaches extend from the high-water line on the shore to an offshore depth of approximately 10 meters.

Many beaches are stripped of sediment during the winter when waves are high. The gentler summer waves deposit more sediment than they remove, which causes the beach to grow in size. Figure 16.13 shows measured winter and summer profiles for a beach at Carmel, California. Essentially the whole beach is reconstructed every year.

Figure 16.13 (opposite) (a) The shape of a beach changes throughout the year. During the winter, waves remove material from the berm and deposit it in offshore bars. During the summer, the process is reversed, and the berm grows at the expense of the bars. The vertical scale in the diagram is exaggerated 25 times for clarity.

(b) The growth of the berm on the beach at Carmel, California, is shown in this series of measured profiles. By February of the following year, most of the berm had been once more cut back. The vertical scale is exaggerated 10 times. (After "Beaches" by Willard Bascom. Copyright © 1960 by Scientific American, Inc. All rights reserved)

Sediment Transport and Deposition

An important factor in the dynamics of beach growth is the way sediment is transported parallel to a beach. As you read earlier, waves approaching a beach at an oblique angle are bent by refraction but still make a slight angle with the line of the beach. As

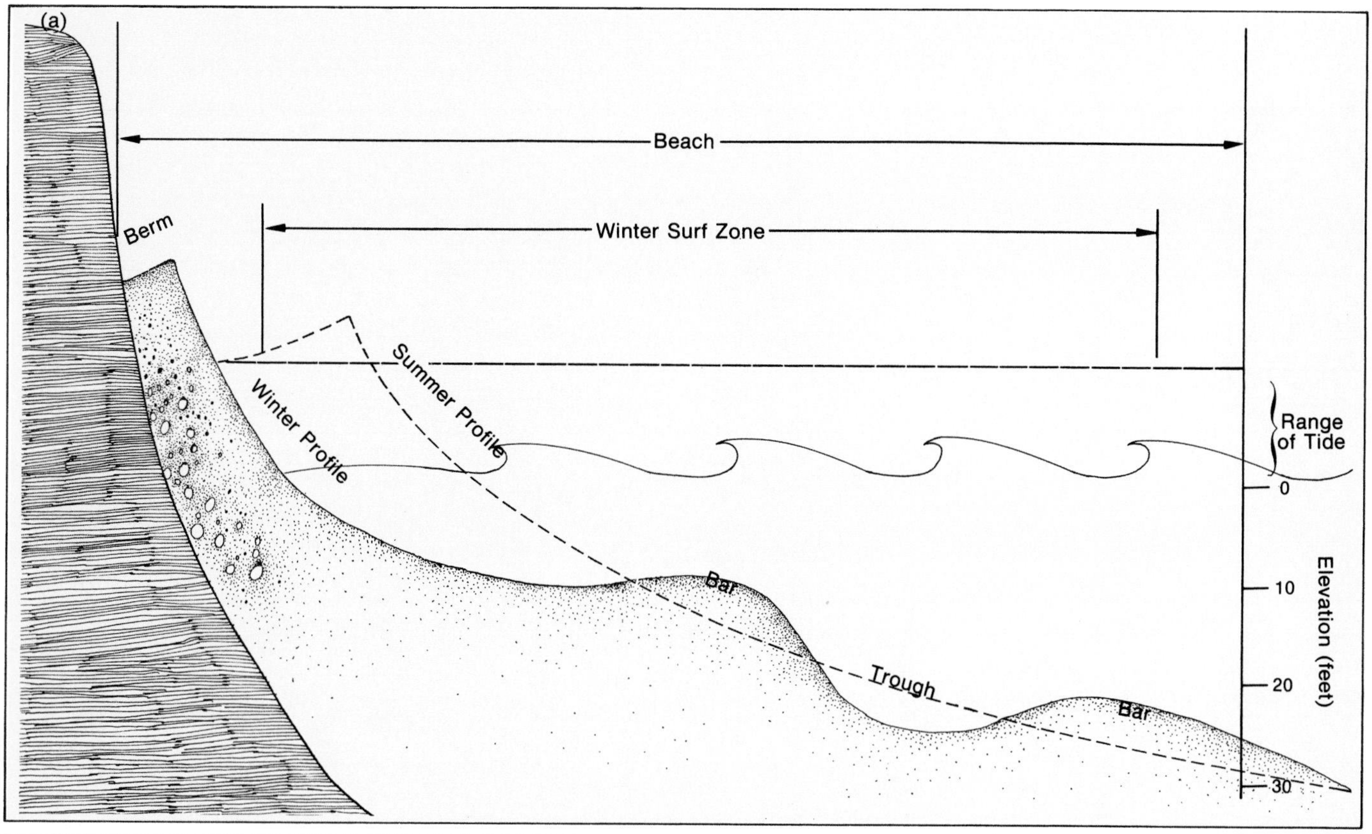

(a)

Beach

Berm

Winter Surf Zone

Summer Profile

Winter Profile

Range of Tide

0

10

20

30

Elevation (feet)

Bar

Trough

Bar

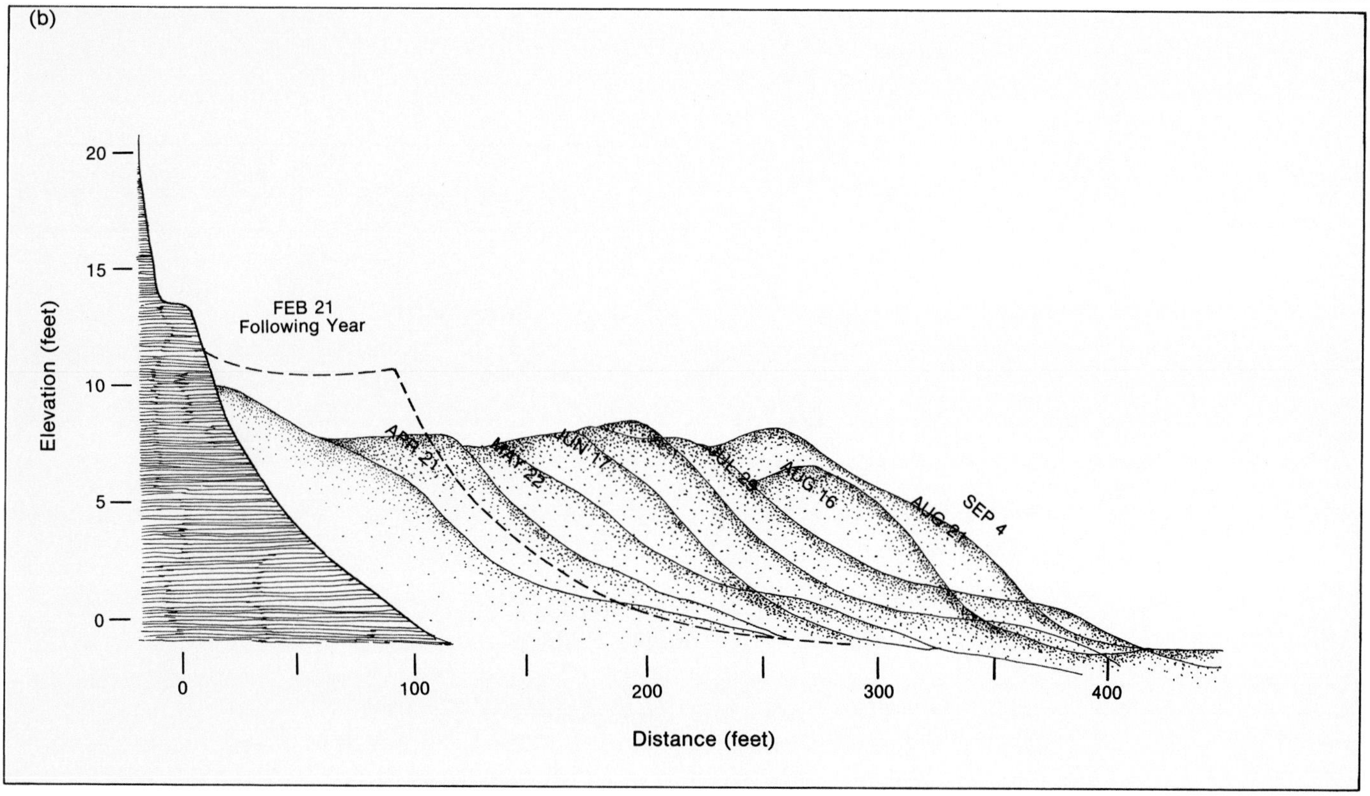

(b)

20

15

10

5

0

Elevation (feet)

FEB 21
Following Year

APR 21

MAY 22

JUN 17

JUL 23

AUG 16

AUG 21

SEP 4

0

100

200

300

400

Distance (feet)

the swash comes up the beach, it moves sediment obliquely up the beach. The backwash tends to retreat straight downslope, so that sediment is shunted along the beach in zigzag fashion. Although waves strike a beach from numerous directions, swell driven by prevailing winds establish an average direction of sediment flow. The oblique approach of the incoming water implies that there is a flow of water parallel to the beach. This flow, called a *longshore drift*, can transport sediment parallel to the beach.

Waves transport sediment by removing old sediment from a beach and replacing it with new sediment. Beaches must therefore be steadily supplied with sediment. Sometimes the sediment comes from wave erosion of the coast, but the most important source of new sediment is rivers carrying products of erosion from the land. Many beaches have been losing sand in recent years because the construction of upstream dams has seriously reduced the amount of sand that rivers normally supply to the coast.

Along a coast that has a prevailing longshore drift, sand eroded from one beach can be passed along to the next beach. Sand moves from east to west along the beaches on the southern coast of Long Island, New York. The new sand appears to come mainly from glacial deposits in eastern Long Island. Loss seems to be outrunning accumulation, however, and some beaches are retreating at a rate of 1 or 2 meters per year (3 or 6 feet per year).

Beach Erosion and Man

One local solution to beach erosion is to interrupt the longshore drift with a jetty that extends into the water at right angles to the beach. The longshore drift loses speed at the jetty and drops part of its sediment, just as a stream drops part of its sediment when its velocity decreases. The enhanced accumulation that the jetty

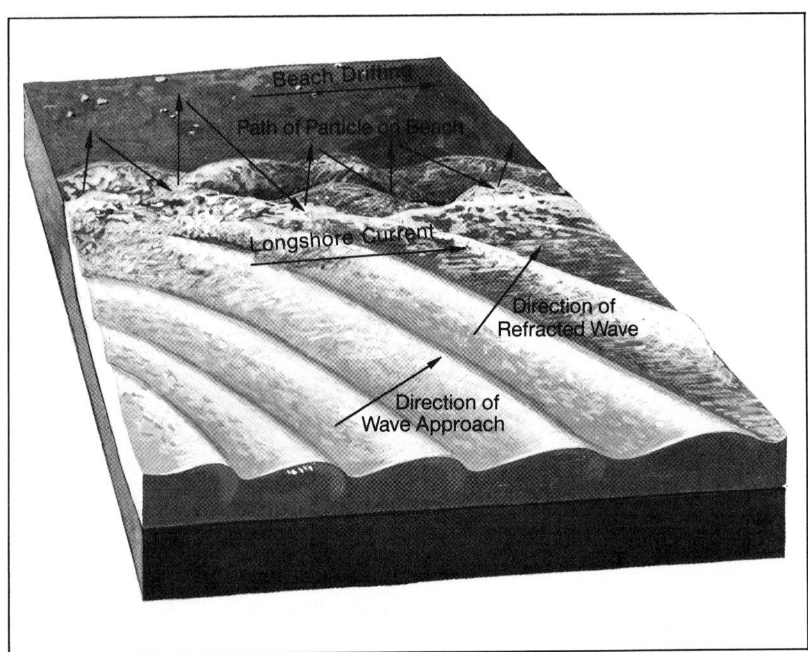

Beach Drifting
Path of Particle on Beach
Longshore Current
Direction of
Refracted Wave
Direction of
Wave Approach

Figure 16.14 Waves that strike a beach obliquely cause transport of material parallel to the beach. Some of the material is transported by waves, which wash up the beach obliquely and return straight downslope. The offshore current established by the oblique waves also transports material parallel to the beach.

Figure 16.15 Longshore currents along this section of coast in Oregon have removed most of the sand from the beach in the foreground. Beyond the point, the current has built up a beach by depositing sand.

promotes inevitably robs beaches farther down the coast of their new sediment. Those beaches erode all the more rapidly, as resort owners know. Miami Beach in Florida has lost much of its sand by erosion despite efforts of individual hotels to maintain the beaches by constructing jetties. The most recently developed method of attack on the problem of beach erosion has been to increase the supply of new sediment by dredging sand from land deposits or from the deep ocean and then dumping it where prevailing longshore drifts can pick it up for transport to beaches.

In the construction and improvement of harbors, the problem is to stop the accumulation of sand that would block the channels. Detailed knowledge of sediment transport paths is needed to increase the size of beaches or to avoid the growth of sand bars in a harbor. Sediment transport has been studied by releasing sand marked with dye or with radioactive material and then checking along the coast to see where the marked sand has been deposited.

Relief Forms of the Ocean Floor

Not long ago the ocean basins were thought to be featureless plains. Although the study of the oceans is still in its infancy, it is clear

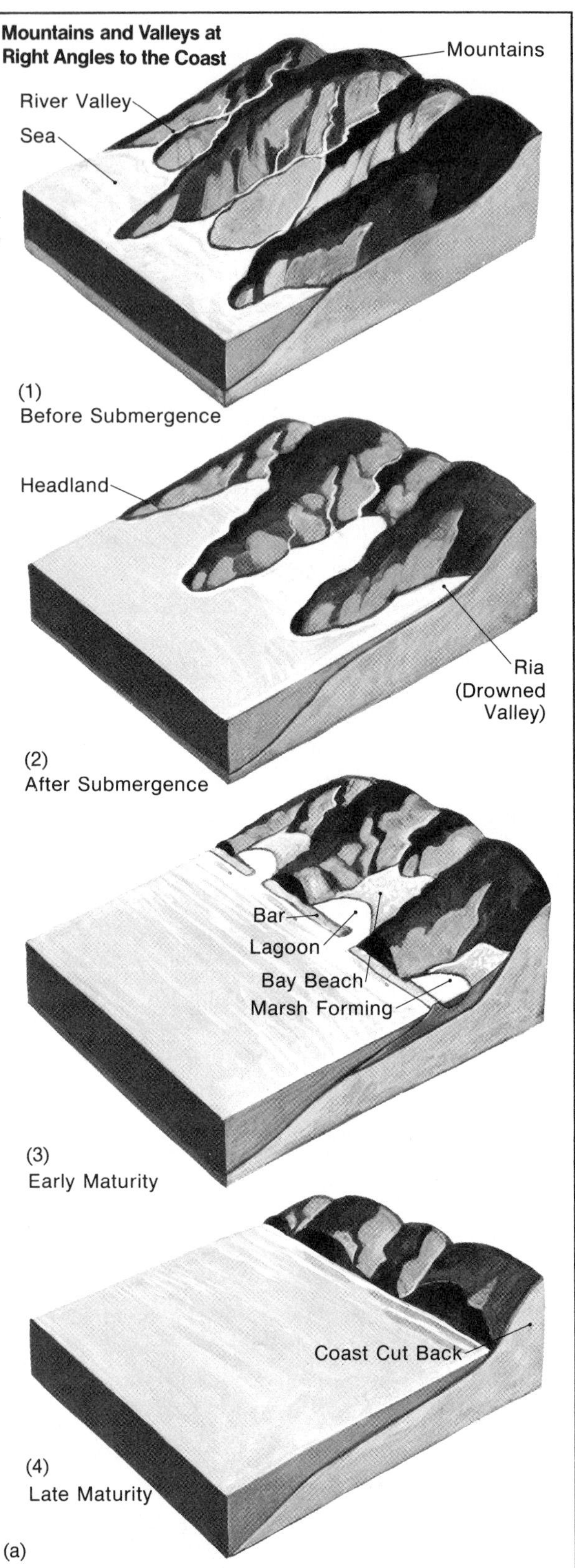

Mountains and Valleys at Right Angles to the Coast
Mountains
River Valley
Sea
(1)
Before Submergence
Headland
Ria
(Drowned
Valley)
(2)
After Submergence
Bar
Lagoon
Bay Beach
Marsh Forming
(3)
Early Maturity
Coast Cut Back
(4)
Late Maturity
(a)

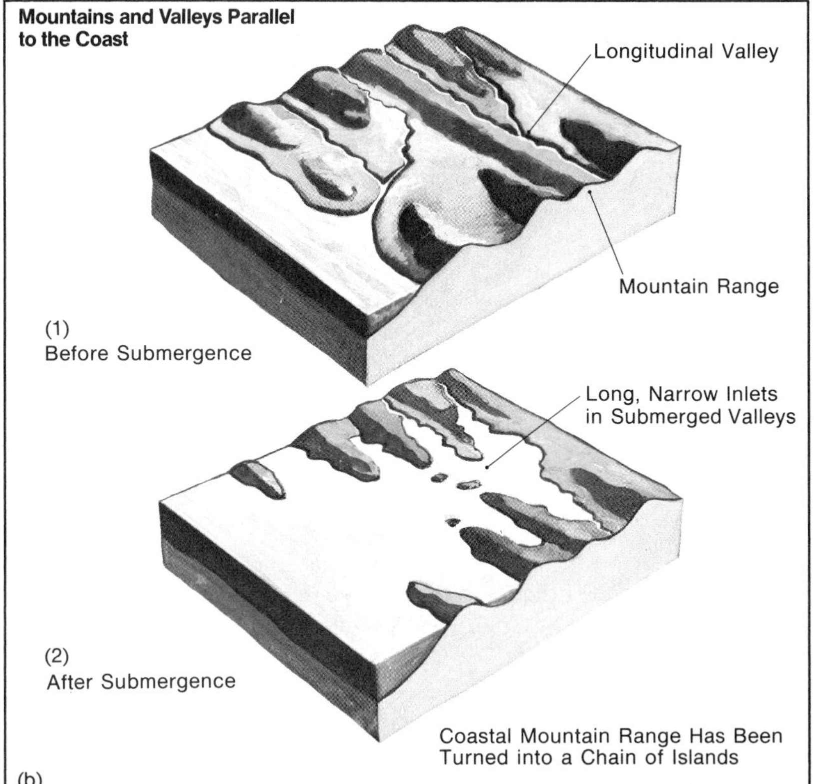
Mountains and Valleys Parallel to the Coast
Longitudinal Valley
Mountain Range
(1)
Before Submergence
Long, Narrow Inlets in Submerged Valleys
(2)
After Submergence
Coastal Mountain Range Has Been Turned into a Chain of Islands
(b)

Figure 16.16 (opposite top left) Wave erosion has cut an extensive *marine terrace* on this coast at Palos Verdes, California. A marine terrace forms when waves erode the base of cliffs, causing them to retreat. The surface of the terrace tends to remain rocky while the terrace is being actively cut, but when equilibrium is reached, deposits accumulate and a beach forms.

(opposite bottom left) These chalk cliffs in England have been cut back to form a comparatively straight coastline bounded by nearly vertical cliffs.

Figure 16.17(a) (opposite right) The sequence of schematic diagrams shows the typical evolution of a coastline formed where valleys and ridges transverse to the coast are submerged by a rise of sea level relative to the land.

(1) The initial coastline consists of valleys and ridges running to the sea.

(2) The relative rise of sea level forms drowned river valleys, or *rias*, and headlands jutting into the ocean.

(3) Wave erosion cuts back the headlands and forms vertical cliff faces. Deposition of sediment by currents builds spits and beach areas; a spit that closes off a bay, forming a lagoon, is called a *bar*.

(4) Continued erosion of headlands wears the coast back to a straight line bordered by vertical cliffs and a narrow beach. Stream erosion will reduce the elevation of the ridges and highlands of the land area.

(b) (top) (1) The diagram shows a coastal region where ridges and valleys are parallel to the coast.

(2) Submergence of the coastline is caused by a relative change of sea level with respect to the land. Wave erosion of the ridges leaves numerous islands oriented parallel to the coast, such as those off the Dalmatian Coast of Yugoslavia. (After Bunnett, 1968)

Figure 16.18 (center) A long period of continuous deposition by longshore currents has caused the gradual extension of this curved *spit* near the mouth of the Colorado River in Mexico. The bay behind the spit may eventually be closed off, forming a lagoon.

(bottom) The long, parallel projections of land and elongated islands that jut into the ocean off the coast of Maine were continuous ridges before being submerged. Submergence of the coast filled the valleys with water and exposed the ridges to wave erosion. Eventually, the islands and ridges will be reduced to sea level.

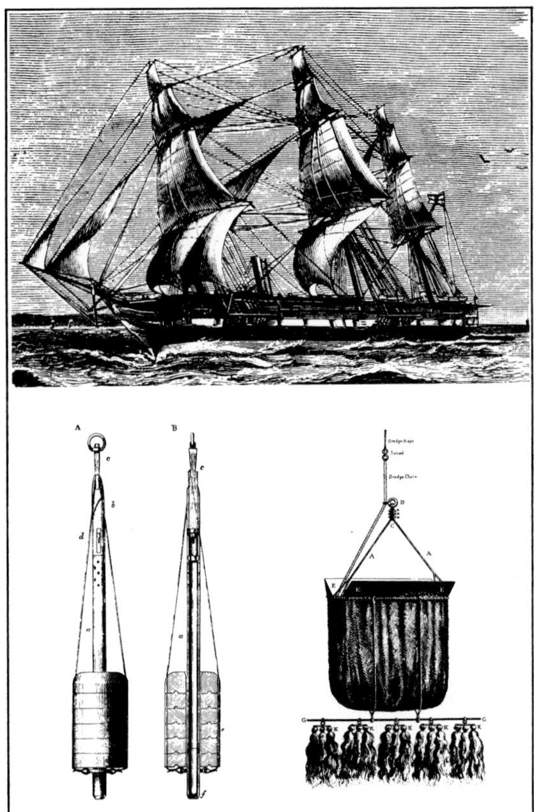

(miles)
(kilometers)
Depth
Diving Saucer
Deepstar
Bathysphere
Alvin
Aluminaut
Echo Sounding
Seismic Reflection Profiling
Deep Sea Drilling
Rock Dredging
Core Tube
Underwater Photography
Rock Dredging
Bottom Sampling
Deepest Diving Bathyscaphe
Mariana Trench (10,000 meters)

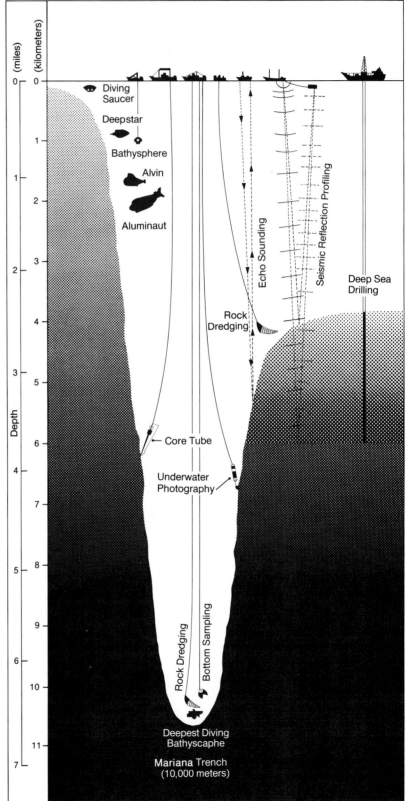

Figure 16.19 Detailed study of the ocean floor began 100 years ago with the voyage of the British research ship *HMS Challenger* (top left) in 1872. The *Challenger* made the first world-wide measurements, or soundings, of ocean depths. Some of its equipment (above left) included weights on lines used to measure ocean depths and dredges to bring up samples from the ocean floor. Only about 500 deep sea soundings were made during the 3-year voyage because of the primitive and laborious methods available.

(right) Modern methods of oceanographic study include echo sounding and seismic profiling for the determination of ocean floor topography; deep sea drilling for the retrieval of sediment samples from the ocean bottom; and the use of a variety of deep sea and surface craft, underwater cameras, and dredging tools.

(Adapted from B. C. Heezen and C. D. Hollister, *The Face of the Deep*. ©1971 by Oxford University Press, Inc.)

that the ocean floor has many intriguing features as well as considerable relief. Geomorphic formations appear beyond the "edge" where the sea meets the land.

Much of the ocean bottom lies at depths of 5,000 meters (16,000 feet) or more, well beyond the range of deep-sea craft, which are presently restricted to depths of 2,000 meters (6,500 feet) or less. Because the landforms of the deep ocean floor cannot be observed at close range, information about them must be gained by indirect means. One of the most widely used methods for charting the relief of the ocean floor is to generate sound waves at the surface and measure the time it takes for the echo to return from the bottom. Sound waves reflected from an undersea mountaintop return to the surface sooner than sound waves that descend into a trench, so that a profile of the ocean floor can be constructed from echo data. Special coring and digging tools lowered from surface ships have brought back samples of the ocean bottom and its sediments.

Off the immediate coast of southern California, where the continental shelf is especially narrow, the ocean floor contains a number of basins that are somewhat similar in size and shape to such basins on land as Death Valley. The basins in the ocean floor tend to have flat bottoms, however, and their side slopes show few of the canyons that are common to eroded basins on land.

Two striking third-order features of the ocean floor are *submarine canyons* and *guyots*. Submarine canyons are found throughout the world on continental slopes; their existence has been known for nearly a century. According to soundings and inspection using scuba diving gear and submersible vehicles, submarine canyons are topographically quite similar to ordinary river valleys. One of the most famous submarine canyons appears to be an extension of the lower Hudson River. Submarine canyons are puzzling because they often cut through the entire continental slope down to the deep sea floor. The Hudson canyon extends to a depth of nearly 5,000 meters (16,000 feet), which is much too deep to be a drowned river valley.

The most plausible explanation for the origin of submarine canyons is that they are products of erosion in the deep sea caused by *turbidity currents*. A turbidity current is a fluidlike suspension of silt that is cohesive enough to remain intact and dense enough to flow down a continental slope to the sea floor. Many submarine canyons are located near the mouths of major rivers, which is a reflection of the great amount of sediment there. Clean-washed sand and the remains of shallow-water animals have been found at the bottoms of submarine canyons, which provides evidence that sediment transport from long distances has been active. The ability of turbidity currents to carry on erosion deep under water has not been fully demonstrated. However, the occasional rupture of submarine cables on continental slopes and rises may indicate the presence of such forceful movements of material as turbidity currents.

Another puzzling feature of the sea floor is the *guyot*, or tablemount. A guyot is an isolated flat-topped mountain peak on the ocean floor. Volcanic activity often forms mountains on the sea floor, but

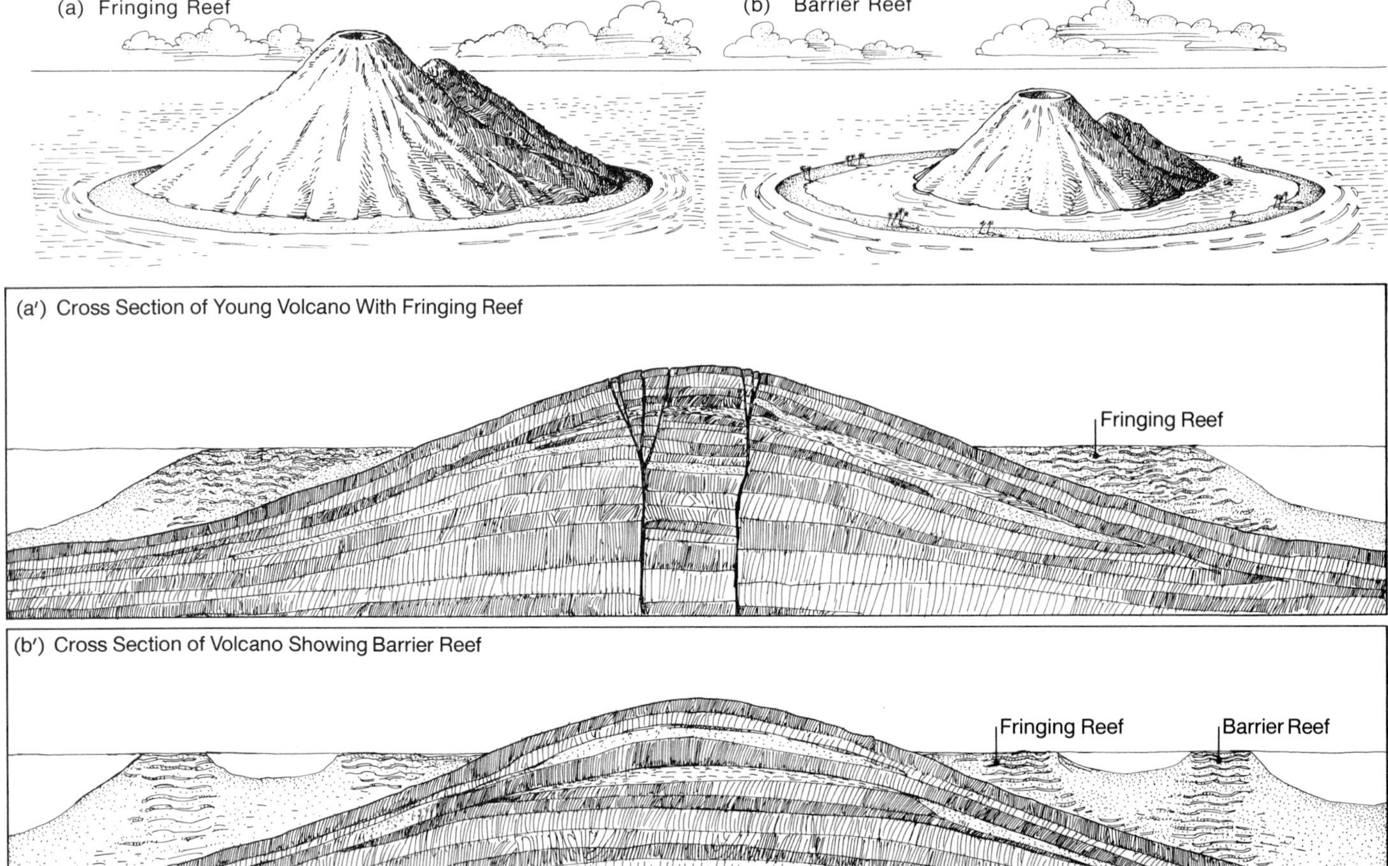
(a) Fringing Reef
(b) Barrier Reef
(a′) Cross Section of Young Volcano With Fringing Reef
Fringing Reef
(b′) Cross Section of Volcano Showing Barrier Reef
Fringing Reef
Barrier Reef

Figure 16.20 These perspective and cross section diagrams detail the probable explanation of atoll formation that was first advanced by Charles Darwin. Atolls are ring-shaped reefs of accumulated coral built around volcanic islands. The base of a coral reef may rest on bedrock a mile under the water, but because coral can grow only in the top few hundred feet of water, Darwin proposed that coral reefs grew upward as the volcanic island slowly subsided.

(above) (a) (a′) A fringing reef of coral establishes itself around a volcanic island.

(b) (b′) If the island begins to subside slowly, upward growth of the coral may be able to keep pace with subsidence. The upper living layer of coral thus remains in the shallow waters where growth is possible.

(opposite) (c) (c′) Continued subsidence of the volcanic island, plus the effects of erosion, may leave only a ring of coral reef. A coral island inside the reef will form when corals are established on the subsiding roots of the old volcano.

(d) (d′) A slight drop in sea level, or rise of the sea floor, causes the emergence of the coral island and its surrounding lagoon and coral reef.

the guyot's flat top is difficult to explain. Although guyots are often 1,000 meters (3,300 feet) or more below the surface of the ocean, their flat tops appear to have been planed off by erosion at the ocean's surface. Shallow-water fossils found on the tops of guyots also indicate that guyots have not always been submerged so deeply.

It is generally agreed that the top of a guyot once projected above sea level and that its top was worn away by erosion. According to one view, the guyot was submerged when the ocean floor subsided, perhaps as a result of the development of oceanic trenches. Alternatively, guyots may be formed as volcanic islands at oceanic ridges and then carried under water down the flanks of the ridge as sea floor spreading continues. In either case, the clustering of guyots in certain areas of the Pacific basin seems to indicate that the guyots are in some way connected with the dynamics of the oceanic crust.

Summary

Coastlines and undersea areas possess distinctive landforms. The continental plates descend underwater along gently sloping continental shelves that were cut by wave action during periods of glacia-

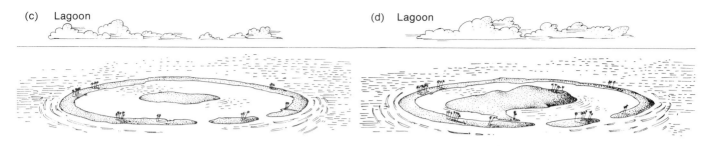

(c) Lagoon

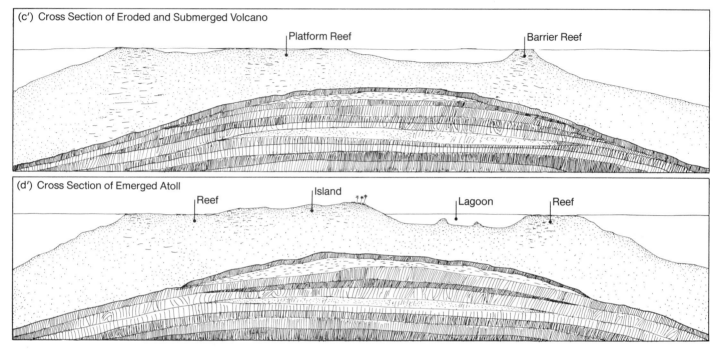

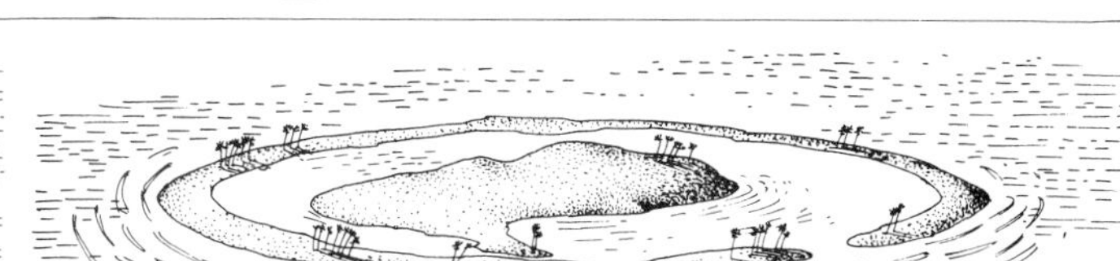
(d) Lagoon
(c′) Cross Section of Eroded and Submerged Volcano
Platform Reef
Barrier Reef
(d′) Cross Section of Emerged Atoll
Reef
Island
Lagoon
Reef

tion, when the sea level was lower than today. At a depth of 100 meters or so the continental slope descends more steeply until it meets the continental rise and the ocean floor.

Tides are not important agents of geomorphic change, but they allow wave action to reach higher areas on coastlines. The gravitational attraction of the moon and the sun on the earth's waters governs the periodic rise and fall of the tides. The nutrients and wastes that tides transport in and out of estuaries help maintain the high productivity of estuarine ecosystems. The wind is the major driving force of waves on the ocean's surface. Waves carry energy, and their ability to concentrate their energy by refraction enhances their erosive power on coastal features.

Sand is continually transported to and from beaches, so that whether a beach grows or recedes depends on its mass budget. Because longshore drifts transport much of this sediment, jetties projecting into the water interrupt the transport of sand.

Third-order forms of relief under the sea include submarine canyons on the continental slopes and flat-topped guyots. The origins of such landforms continue to be a matter of speculation.

Graphic Essay

III

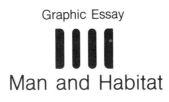

Man and Habitat

Man's distant evolutionary ancestors were semierect, apelike herbivores that made their homes in the high canopies of tropical and subtropical forests. But millions of years ago some of the forest dwellers began to leave their tree homes for a time on foraging expeditions into the grasslands at the forest margins, returning to the trees at night for protection against predators. Eventually some groups developed patterns of social behavior appropriate to life in the grasslands which, together with the immense reproductive advantages of dwelling on the ground, enabled them to move into—and survive in—the new environment. They would evolve, over time, into modern-day apes—and into man.

The story of man's effective use of the land begins at least that long ago. The record between then and now becomes more puzzling and incomplete the further back we go in time, but the glimpses we get of man's progression in securing his place in the world are fascinating.

Consider, for example, the excavations at Lazaret, a cave in France on the shores of the Mediterranean. Here the archaeologist has given us an insight into man's use of natural landforms 130,000 years ago. Excavations have uncovered a rough semicircle of rocks enclosing an area along the cave wall; clearly

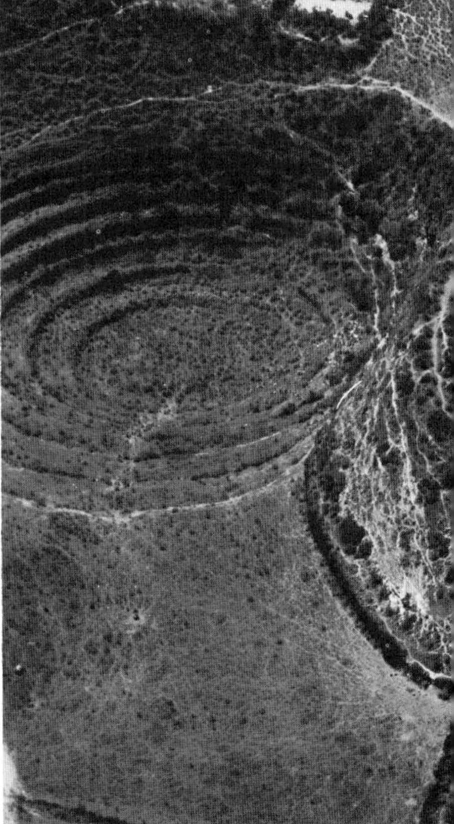

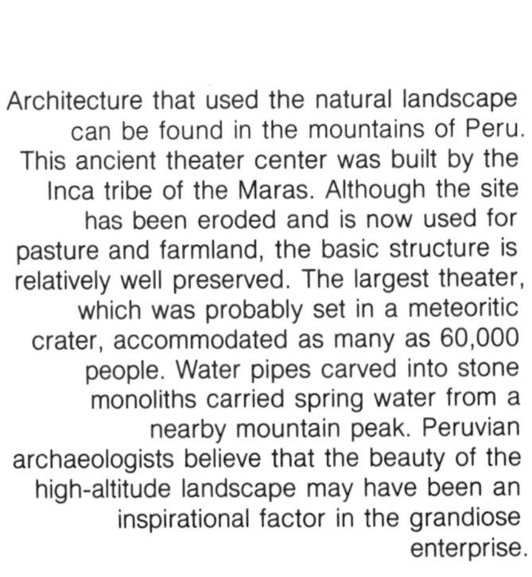

Architecture that used the natural landscape can be found in the mountains of Peru. This ancient theater center was built by the Inca tribe of the Maras. Although the site has been eroded and is now used for pasture and farmland, the basic structure is relatively well preserved. The largest theater, which was probably set in a meteoritic crater, accommodated as many as 60,000 people. Water pipes carved into stone monoliths carried spring water from a nearby mountain peak. Peruvian archaeologists believe that the beauty of the high-altitude landscape may have been an inspirational factor in the grandiose enterprise.

a shelter had been erected there. The excavations also exposed the remains of sea shells that often are attached to seaweed that is washed ashore during storms, and a pattern of animal claws—all that would remain of furs—concentrated with the shells. And suddenly we realize the cave dwellers had assembled within the protection of the natural cave a mattress of seaweed and furs from the animals they hunted.

Many additional examples can be cited of the successive natural environments that provided the raw materials for early man's existence and that channeled the development of his cultural expressions. Through the high Zagros Valley in Iranian Kurdistan, which typifies the natural habitat where domestication of plants and animals began and where dwellings were built not only for shelter but for food storage; through the monumental cities of bricks made of mud from the vast plains of the Tigris and Euphrates rivers; through the enigmatic citadel of Machu Picchu built out of stone on a nearly inaccessible plateau in the Peruvian Andes; through the medieval fortress-castles built high on hilltops for defense—through all of this tangible history we see unique expressions of man's ingenuity in working with natural landforms—how he has shaped them for his own purposes, and how his culture has been shaped by the material available to him.

His use of the land has often represented celebrations of his imagination. High in the mountains of Peru is an ancient amphitheater which has no counterpart in the history of that time. Today we find another use for natural amphitheaters—Arecibo, an immense radio telescope used for scanning the far reaches of the universe, was constructed in the bowl of a natural limestone sinkhole in the hills of Puerto Rico.

When man is in tune with the habitats the earth provides we can point with pride to his accomplishments. But today we must reexamine the thrust of man's interaction with the land and ask ourselves, Is it indeed a harmonious relationship?

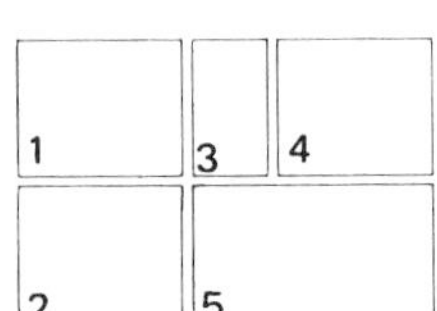
1
3
4
2
5

1. Not far from the amphitheater in Peru are the remains of an Inca city, Machu Picchu, whose site took advantage of a nearly inaccessible plateau, most likely for defense. The elaborate city was never discovered by the Spaniards who swept through the land during the sixteenth century A.D. At the beginning of this century it was cleared of vegetation and partly restored. At the top of the high peak are agricultural terraces and a few rooms that may have served as lookout posts.

2. In Arecibo, Puerto Rico, man has turned a natural bowl to the sophisticated use of twentieth-century radio astronomy. The bowl, a large sinkhole cut out by water in a limestone karst region (see also Chapter 14), required little excavation to attain the spherical shape needed for the telescope. The lack of temperature extremes in the tropical climate and freedom from sources of electrical interference were other considerations in favor of the site. The surface of the bowl was originally covered with open chicken-wire mesh to reflect weak interstellar radio waves to the sensitive receiver suspended over the bowl. To improve the sensitivity, the bowl is to be covered with aluminum plates. However, the plates will block sunlight to the natural vegetation in the bowl, and without vegetation, the exposed limestone surface may erode rapidly.

3.4. In Afghanistan men carved entire towns out of rock. This close-up shows one of the Goreme cones, sculpted by nature on the outside and by man on the inside. These dwellings, which are still inhabited, range in size from a tent to that of a "skyscraper" with as many as sixteen floors.

5. These cave homes are built into eroded mountains at Purullens near Gaudix in the Province of Granada, Spain.

MANITOBA
SASKATCHEWAN
WASH
OREGON
NEB
IOWA
MO
TENN
TEXAS
SONORA
FLA

Epilogue: Physical Geography USA

The Eastern Seaboard: Boston, Massachusetts

Land's Edge: Cape Cod

First Stop: Niagara Falls

Rolling Farmland: The Midwest

The Land of the Big Sky: The Great Plains

Mudstone Mountains: The Badlands of South Dakota

Devil's Tower: A Wyoming Monument

The Continental Divide: The Rocky Mountains

Arches, Pillars, Balancing Rocks: The Southwest

A Man-made Garden: The Imperial Valley

Journey's End: California

A patchwork of different climatic and geomorphic regions, the United States is a natural laboratory for exploration and field study. Wherever you live, wherever you travel, you can apply the principles and processes of physical geography to understanding your natural environment.

Map 1962 by Jasper Johns. (Frederick R. Weisman Collection)

Epilogue: Physical Geography USA

The study of physical geography is a rewarding career in itself, and each year many students decide to pursue this particularly fascinating inquiry into the world we live in. Many others, however, turn to other disciplines that are closer to their individual interests in the intellectual pursuit of knowledge. What, then, is the value of this brief encounter with physical geography for those readers who move on?

Simply put, the knowledge we gather—whether it be in physics, or biology, or geography—can enrich our vision of the world around us in many unexpected ways. Where before we would drive rapidly through a lowland valley, dismissing the countryside rolling past us without much thought, perhaps now we will recall the incredible processes that have worked over vast spans of time to shape that valley into what it is today; perhaps we will contemplate what the valley will be thousands of years hence. In this sense the study of physical geography has the potential to free us from the brief moments in time and space in which we find ourselves. It is a passport to horizons beyond horizons that have existed in times past, and to ones that have not yet emerged.

Is this an exaggeration? Perhaps you may think so, perhaps not. But consider, through the final pages of this book, an imaginary journey you conceivably could take to test the validity of all that you have learned about the earth, this home of ours for all time.

The United States is a huge country, spanning some 4,400 kilometers, or one-eighth the circumference of the earth, and a cross-country trip could take any number of routes. The itinerary for this "trip" was planned to cover several major regions in order to apply the physical principles and processes to a number of landscapes and climates. The trip begins in New England, goes west through New York State, crosses the farmland of the Midwest and the prairies of the Great Plains, passes through the southwestern deserts, and ends at the Pacific coast.

The Eastern Seaboard: Boston, Massachusetts

Suppose you begin your journey in the city of Boston—steeped as it is in tradition and in the colorful history of early America. If you are acquainted with Boston you realize that it is equally well known for its unpredictable weather, which runs the gamut from spring days alive with forsythia, to torrential downpours in the fall, to freezing cold in the winter months.

But if you find yourself in Boston in the summer, your clothing will stick to your skin because of the heat and high humidity. The sticky summers in Boston are dominated by the flow of air from the south, which brings sultry conditions of heat and high humidity from the subtropical Atlantic Ocean off the coast of Florida. The air flows northward over the East Coast of the United States as part of the clockwise circulation around the high-pressure cell stationed over Bermuda. The cell weakens in winter, leaving Boston open to the chill winds from the interior of the continent and from the North Atlantic.

The temperature of the Atlantic Ocean off the coast at Boston is frequently less than 13°C (55°F) during the summer, which is almost too cold for swimming. A sea breeze from the ocean waters often cools the coastal areas to comfortable temperatures 5 to 8°C lower than a few kilometers inland. Fog is common along the coast when warm, humid air from the southwest flow is chilled over the cool ocean and then drifts back to the shore.

Bunker Hill, the site of the first major battle of the American Revolution, is, like many of the hills in the Boston area, a drumlin of sand and gravel formed by the ice sheets that once spread over the Northeast. Much of the sand and gravel used for fill in Boston's marshland during the nineteenth century came from another drumlin, Beacon Hill, which was reduced in height considerably. Boston Harbor is dotted not only with sailboats but with numerous small islands that have unusual elongated elliptic shapes. These islands, like the hills on the nearby coast, are drumlins. Some consist of several neighboring drumlins joined by sand the harbor currents have deposited.

Erosion of the coast and the harbor islands supplies quantities of sand for deposition elsewhere. Just north of Boston lies Nahant, an island deposited by the coastal currents that is tied to the mainland by a narrow strip of sandy beach 3 or 4 kilometers (2 miles) long. A bar that joins an island to the mainland is called a *tombolo*. When you began the study of physical geography, the number of special terms may have seemed overwhelming—drumlin, caldera, podzols, peneplain. But the terms can be useful for brevity and accuracy in the same way that scientific terms for bones and muscles are useful.

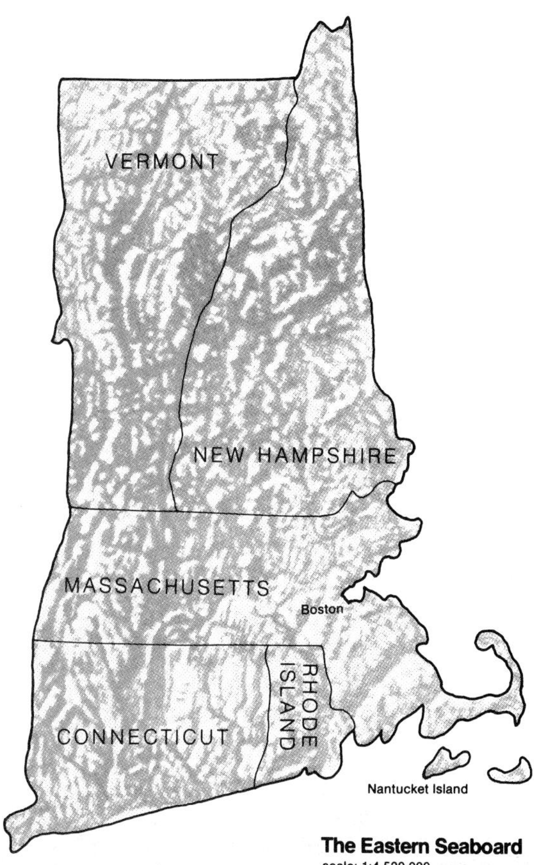
VERMONT
NEW HAMPSHIRE
MASSACHUSETTS
Boston
RHODE ISLAND
CONNECTICUT
Nantucket Island

The Eastern Seaboard
scale: 1:4,500,000

Land's Edge: Cape Cod

A cooler place to spend the summer is Cape Cod, a short distance south of Boston. As you drive to the end of Cape Cod, which takes just a few hours, you may be surprised at how markedly the scenery changes. The sandy, hilly landscape of the Cape contrasts with the farmland, woods, and well-developed soils on the mainland not far away. Along the unprotected outer coast of the Cape the vegetation consists of low shrubs such as bayberry, with scrub pine growing somewhat farther back from the coast. Large, shady deciduous trees, sustained by the 100 centimeters (40 inches) of annual precipitation, line the streets of the towns.

Like many features in the Northeast, Cape Cod is a product of the ice sheets that once covered so much of North America. It was formed only 50,000 to 70,000 years ago when two lobes of retreating glacier ice became stationary for a time at the Cape's present location. The melting face of one lobe was oriented along an east-west line, and it deposited a moraine that now forms the hilly backbone of the east-west section of the Cape. The second ice lobe lay to the east of the first lobe and was oriented with its

Many of the low hills and depressions that cover large portions of Cape Cod are formed from moraines deposited by glaciers.

melting face in the north-south direction. This lobe deposited a moraine that initially formed the outer coast of the upper arm of the Cape. Each lobe also deposited quantities of outwash, so today most of the Cape's land is a glacial outwash plain. Wave action has eroded the moraine deposited on the outer coast by the second lobe, so that the upper arm of the Cape now consists only of outwash plain.

The outwash plain terminates at the outer coast in a sheer sea cliff of unconsolidated sediment at least 30 meters (100 feet) high. The exposed face of the cliff shows little or no layering, evidence that the cliff consists of one massive deposit of sediment. The cliff was formed by waves cutting at its base; without a base of support, material from above shears off the cliff, and the entire cliff face retreats. A wide, sandy beach runs along the shoreline between the water and the cliffs.

Parts of the Cape are undergoing erosion from wave action at a noticeable rate. Not long ago the foundations of the now-abandoned Coast Guard station at Nauset hung on the edge of the cliffs. A new station has been built farther inland—but the sea comes closer each year.

The Cape is building up in areas where longshore currents are depositing sand from eroding sea cliffs elsewhere on the peninsula. Not far from the rapidly eroding shore at the Nauset Coast Guard station a great spit of sand extends parallel to the shore for many kilometers. If left unchecked, the spit will, in time, completely close off an inlet from the sea, converting the inlet to a pond. Many examples of ocean inlets being closed off are found along the southern shore of Nantucket Island, south of Cape Cod.

Round kettle ponds on Cape Cod were formed when glaciers deposited sediments around iceberg-sized chunks of ice that subsequently melted to leave round depressions. The open areas marked with rectangular grid patterns are cranberry bogs.

If the Atlantic is too cool for comfortable swimming at Nauset, the warm, shallow waters of Salt Pond in nearby Eastham may be more inviting. The numerous round ponds were formed when great isolated blocks of glacier ice did not completely melt until after the outwash plains had been laid down around them. Scattered halfway between the coast and Salt Pond are large isolated boulders, called *erratics*. These rocks, which make fine picnic perches, are further evidence of glacial action. Because glaciers are able to carry loads without much regard to the size or weight of the particles, the ice sheets brought with them a few boulders from the granite of northern New England.

First Stop: Niagara Falls

In less than a day you can drive west of Boston and Cape Cod to see another example of glacial action—the spectacular Niagara Falls. The thundering Falls are an awesome sight in themselves, but from your reading in geomorphology, you can also appreciate the underlying structure of the Falls. The immense Niagara escarpment is the striking geologic feature of the area. About 10 kilometers (6 miles) north of the city of Niagara Falls, halfway between the Falls and Lake Ontario, a ridge drops suddenly to a low, wide plain bordering the southern coast of Lake Ontario. The Niagara escarpment, which is about 50 meters (150 feet) high—the same height as the Falls—is still evident along Lake Ontario 100 kilometers to the east, and it also continues into Canada on the west. The cap rock is a distinctive, highly resistant limestone called *Niagara limestone*. The state of Michigan rests on a saucer of Niagara limestone, and the edges of the saucer crop to the surface to form the Saugeen

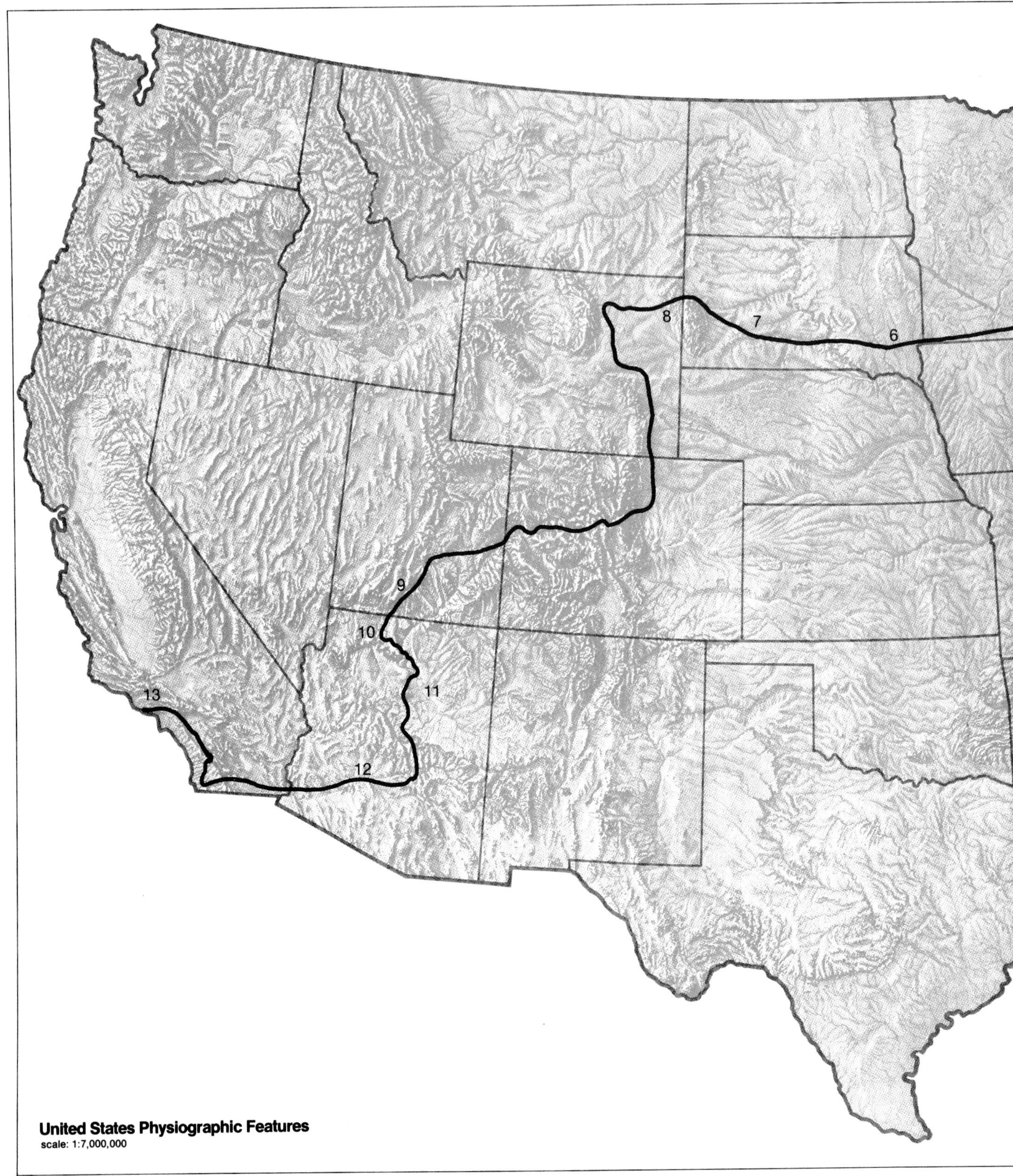
8
7
6
9
10
11
13
12
United States Physiographic Features
scale: 1:7,000,000

2
1
3

Photo Journey

1. Cape Cod, 424
2. Kettle Ponds, Cape Cod, 425
3. Niagara Falls, 429
4. Sandstone Castle, Driftless Area, 430
5. Mississippi River at Wisconsin, 431
6. Great Plains, 432
7. Badlands, 433
8. Devil's Tower, 434
9. Arches National Monument, 438
10. Grand Canyon, 438
11. Canyon of the Little Colorado, 440
12. Arizona Desert, 440
13. Beach at Oxnard, 442

Peninsula and the Door Peninsula that arch northward around Michigan. The bed of Niagara limestone has been traced as far west as Iowa.

The retreat of the ice sheets from the Great Lakes 11,000 years ago exposed the Niagara escarpment and opened drainage from Lake Erie into Lake Ontario. Initially the Falls flowed over the edge of the escarpment, but gradually they cut a gorge headward toward Lake Erie. Although the Niagara limestone cap rock is highly resistant, the underlying rock is weak, and as the lower layers are eroded, the lack of support causes the jointed hard limestone above to fall in blocks. In perhaps 25,000 years the gorge will reach the outlet of Lake Erie and the lake will rapidly be drained if the Falls are left unattended. The accumulation of blocks at the base of the Falls tends to reduce the slope and convert the knickpoint at the Falls from a sharp plunge to a sloping pile of rubble. The blocks are removed by heavy equipment from time to time to maintain the value of the Falls as a tourist attraction.

The low plain between the Niagara escarpment and Lake Ontario is about 10 kilometers wide. Beach ridges have been found inland on the plain, which indicates that waters of the lake once covered

Lake Superior
WISCONSIN
Lake Michigan
MICHIGAN
Lake Huron
Lake Ontario
Niagara Falls
NEW YORK
Lake Erie
PENNSYLVANIA
OHIO
INDIANA
ILLINOIS

The Great Lakes and the Midwest
scale: 1:10,000,000

The Niagara River has not been able to cut downward significantly into the cap rock of hard dolomite since the river began flowing during the Pleistocene era. Instead, the river has cut a long gorge headward by undermining weak shale layers at the Falls.

part of the plain. This area abounds with orchards of apples, peaches, and plums, because the soil and low relief of the plain are suited to growing fruit. The nearby lake extends the growing season by moderating the cold air from the north.

Rolling Farmland: The Midwest

As you drive west from New York into the farmlands of Ohio and Indiana, the high temperatures will remind you that summer is not an ideal time for traveling. But the hot summer days and nights of the continental climates of the Midwest are ideal for growing crops such as corn. Because air from the Gulf of Mexico regularly delivers moisture, some rainfall is usually available throughout the summer growing season when it is most needed by crops. The region from Ohio west to Iowa is eminently suited to agriculture. The Pleistocene ice sheets covered the area with fine till, giving the land gentle relief and leaving a rock-free soil. In the eastern section of the region, soil developed under forests and became gray-brown forest podzolic soils, whereas in western Illinois and Iowa, the soil developed under tall grass prairie and became prairie soil.

For economic reasons the Midwest is given largely to the growing of corn, a crop of such high productivity that even relatively small farms of one or two 160-acre parcels can be efficient. Much of the corn is used locally to support the production of hogs, because raising hogs on corn is one of the most efficient ways to obtain animal protein.

The Midwest landscape appears to be endless flat cornfields, but in east central Illinois you suddenly come upon an isolated section

The Castle, in the Driftless Area of southwestern Wisconsin, is formed by erosion of sandstone. It would have been destroyed if the ice sheets of the late Pleistocene had invaded this region.

of steep, hummocky hills. The hills are part of a glacial moraine left from one of the four or five major glaciations that advanced over the Midwest and deposited extensive moraines from Michigan and Ohio westward to Minnesota and Iowa. The terminal moraines that were left at stationary edges of the ice sheets are often several kilometers wide and 30 to 50 meters in relief. Major belts of moraine ridges curve around the lower ends of Lake Michigan and Lake Erie and extend for hundreds of kilometers south into Illinois, Indiana, and Ohio.

The southwest section of Wisconsin, known as the Driftless Area, was never covered by major glaciation: apparently, the troughs of Lake Superior and Lake Michigan diverted the advancing lobes of ice. The Driftless Area contains features such as masses of sandstone eroded into fantastic turrets and pillars that would never have withstood the press of massive ice sheets. The glaciers nevertheless left some mark on the Driftless Area: parts of the Area have deposits of alluvium from glacial outwash streams, and lakes of glacial meltwater covered some sections. There is also evidence that mechanical weathering was especially active in the Area during the colder climates associated with episodes of glaciation.

From Boston to Wisconsin and Minnesota, the countryside in summer is a succession of green farmlands and heavily wooded regions. The rainfall in Madison, Wisconsin, is 80 centimeters (30 inches) annually, which is somewhat less than at Boston but which is sufficient to support moist deciduous woodlands. The wooded hillsides of southern Wisconsin are reminiscent of parts of New York State 2,000 kilometers farther east. But an easy day's

The Mississippi River, shown here in its upper reaches near La Crosse, Wisconsin, nearly corresponds to the boundary between the moist climates of the eastern United States and the drier regions of the West.

drive across Minnesota into South Dakota produces a striking change: across the 98th meridian in South Dakota, the countryside suddenly changes to a treeless landscape of golden grassland.

The Land of the Big Sky: The Great Plains

The climate in the Great Plains is much drier than the climate in the regions to the east. Moist air from the Gulf of Mexico is the principal source of precipitation for the Great Plains, but the Gulf air tends to flow northeastward, skirting the Great Plains. Annual precipitation at the eastern boundary of South Dakota is 60 centimeters (25 inches), and 300 kilometers (200 miles) farther west it is only 40 centimeters (15 inches), half that of Madison, Wisconsin. The rapid decrease in precipitation westward of the 98th meridian accounts for the sudden change from the forests of Wisconsin to the grasslands of South Dakota.

Although the amount of precipitation in the Great Plains is less than in the Midwest, the timing is favorable to agriculture because most of the precipitation occurs during the growing season. At Pierre, in the center of South Dakota, 80 percent of the annual precipitation falls in the period from April through October. However, water demands are especially high in the summer when the high temperatures and dry winds of the extreme continental climate make potential evapotranspiration high.

The Great Plains, shown here in central South Dakota, are grasslands of low relief with too little precipitation to support forests.

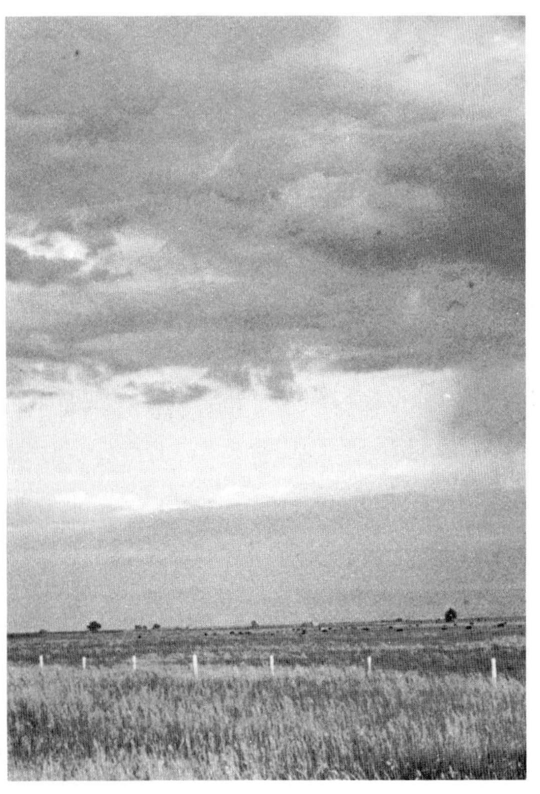

Because the most desirable varieties of wheat grow best where annual precipitation is between 40 and 75 centimeters, much of the wheat production in the United States is concentrated in the Great Plains. Wheat is relatively less productive than corn, so wheat farms are usually several times the size of farms in the Corn Belt.

The history of wheat farming in the Great Plains is closely tied to the marginal nature of precipitation there. During several years of this century, annual precipitation in North Dakota has been as low as 50 percent of the normal precipitation. Years of below-average precipitation in the Great Plains during the past 100 years have shown a tendency to come in groups, which compounds the difficulty of maintaining farm operations. The first settlers arrived after the Civil War, when the Great Plains were experiencing a series of years with above-average moisture, but the agricultural practices appropriate to the humid Midwest proved ineffective on the Great Plains during the drought years from 1880 to 1905, and many farms were abandoned. Twelve years of abundant moisture from 1905 to 1916, however, stimulated expansion of wheat farming, and the grass cover of great sections of the plains was plowed under and replaced by wheat fields. During the inevitable period of drought that followed, exposed dry soil blew away in dark clouds of dust. Much of the Plains area is better suited for cattle grazing than for agriculture. Precipitation during any winter month is seldom more than 1 or 2 centimeters, and except during occasional blizzards, cattle can be safely left on the open range.

The low relief, the lack of trees, and the thinly scattered population give the landscape in South Dakota a strong impression of openness and emptiness. In many regions the only sign of human activity you may see is the roadway and a few fences strung across the fields. Yet if it had double the rainfall and a more moderate winter, the Great Plains would be a settled corn belt like Illinois.

Mudstone Mountains: The Badlands of South Dakota

The morning light on the jagged peaks of the Badlands miles away across the low plains makes them appear as rugged as the rock walls and horns of the European Alps. If you expect to see towering peaks, you may feel out of proportion when you arrive because scale is deceiving on the plains: the features of the Badlands actually are only 100 to 200 meters high. The Badlands are made of compacted mud and silt laid down by rivers 30 million years ago following earlier periods of deposition and erosion. The soft, unconsolidated mudstone is highly susceptible to erosion by flowing water, and the surfaces become soft and slippery after a rainstorm.

Since the sediment layers were deposited, erosion has deeply etched the Badlands, and the tempo of change is comparatively rapid The rate of erosion of exposed peaks is estimated to be several centimeters per year, and many features have experienced noticea-

The Badlands of western South Dakota consist of exposed sediment deposits of soft mudstone that are readily eroded by rainfall. Horizontal bands of sediment can be distinguished by their color in many places.

ble alteration over the span of only a few decades. In some places a cover of sod has retarded the erosion of portions of flat grassland, which has left sod-covered tables elevated above the present level of the plain. The Badlands can be seen as a scale model of a mountain system in which geologic time is running much faster than it normally does.

Erosion in the Badlands has exposed a profile of horizontal sediment layers brightly colored by minerals. Each mound carries its own set of contour lines made by the bands of colored sediment, and patterns can be traced from one formation to the next. Some of the layers are white volcanic ash, which were probably carried there by the wind from ancient volcanic eruptions in Wyoming.

The Black Hills, which lie on the border of South Dakota and Wyoming, rise up like an oval island in the surrounding plains. The geologic structure of the Black Hills is an uplifted dome approximately 150 kilometers (100 miles) long and 80 kilometers (50 miles) wide. The rocky hills rise more than 1,000 meters (3,000 feet) above the plain. The uplift of the dome probably occurred simultaneously

The Devil's Tower in northeastern Wyoming consists of the remnants from a small mass of volcanic rock; the columns were formed during cooling of the igneous material. Note the fractured columns in the talus pile at the base of the tower.

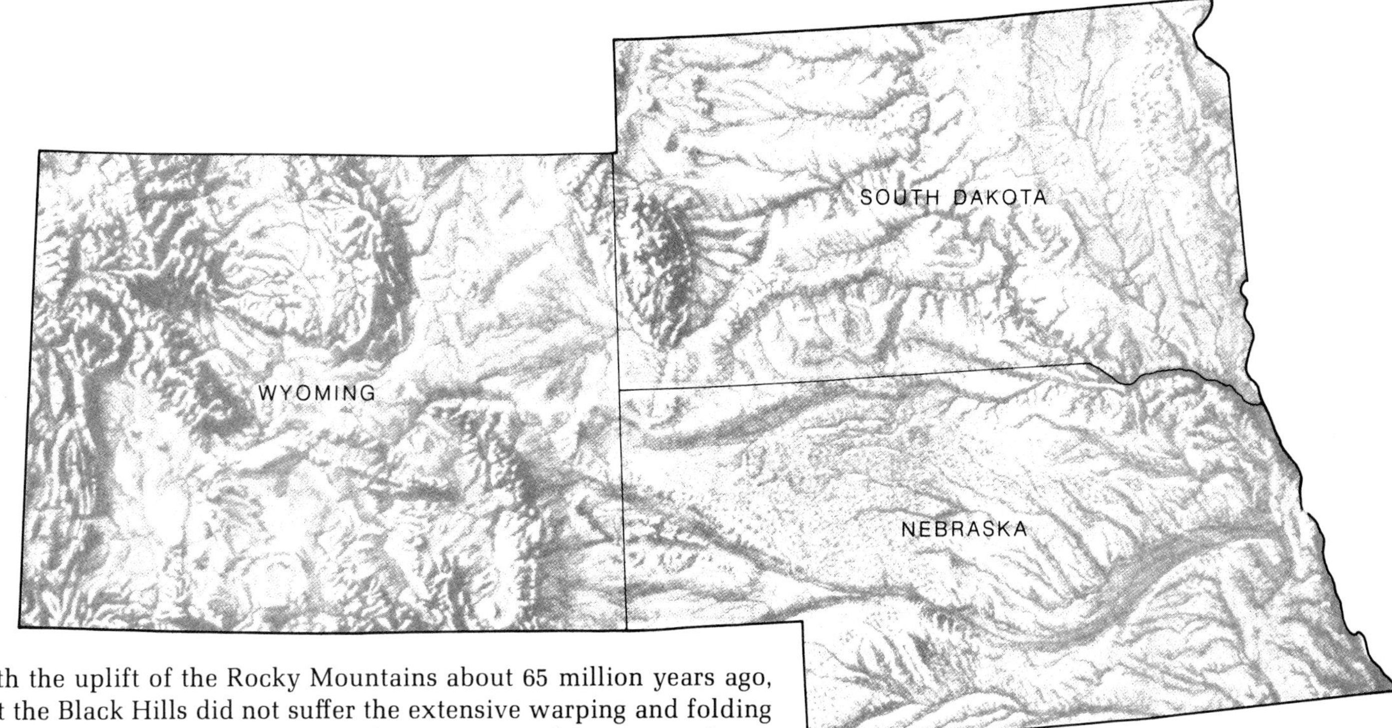
SOUTH DAKOTA
WYOMING
NEBRASKA

The Great Plains
scale: 1:7,000,000

with the uplift of the Rocky Mountains about 65 million years ago, but the Black Hills did not suffer the extensive warping and folding exhibited by rock strata in the Rockies. In many places Precambrian basement rock is exposed because the Black Hills have been worn by erosion. Precambrian rock is normally covered by several thousand meters of sedimentary rock, so at one time the total uplift of the Black Hills must have been greater than the present degree of relief indicates.

The Black Hills are named for the dark mantle of coniferous trees that blankets their heavily forested slopes. The mountain valleys are densely covered with grass, and shrubs and low trees are common. The plains surrounding the Black Hills are essentially treeless, for lack of sufficient moisture, and the green vegetation of the Black Hills in late summer contrasts with the yellowed prairie grass of the plains. The Black Hills are likely to be cooler than the lowlands around them; they may be deeply shrouded in fog, even when skies to the east and west are clear. Orographic condensation and precipitation evidently give the Black Hills more moisture than the surrounding plains receive, so on the local scale, the Black Hills are in a different climatic region from areas only 100 kilometers away.

Devil's Tower: A Wyoming Monument

From the Black Hills you may want to take an interesting side trip into northeastern Wyoming to Devil's Tower. A bare mass of igneous rock 250 meters (800 feet) in diameter, Devil's Tower sweeps steeply upward to a height of over 200 meters (600 feet) above its base. It looms in impressive isolation above the surrounding land and is visible for many kilometers. The sides of the flat-topped Tower are divided into vertical fluted columns, which usually have five sides and are often several meters in diameter. Four-, five-, or six-sided columns characteristically form when a mass of molten rock cools because of the stresses of thermal contraction set up during cooling. The devil is often credited with creating such volcanic landforms—the Devil's Postpile near Yosemite National Park in California is a similar, but smaller, formation.

The Devil's Tower was formed at the same time the Black Hills were uplifted. The Tower is believed to have been originally an intrusion of molten rock into overlying sedimentary rock layers. Erosion of the sedimentary rock left the highly resistant igneous rock as a freestanding tower. The Tower's columns are crossed by horizontal fractures, especially in the upper quarter, and in many places columns have broken away. The base of the Tower is surrounded by a layer of talus that includes great fragments of many-sided columns. The relatively small amount of talus suggests that the Tower has not been extensively eroded.

Most of Wyoming is a mile or more above sea level, and the air on a fair day is remarkably transparent. On a typical day the morning sky may be free of clouds. By afternoon, a few patches of cumulus clouds may form, and as the day wears on, the clouds cover more and more of the sky. Toward evening, as the land cools and rising thermal columns of air weaken, the cumulus will evaporate and the sky will become clear once more.

The plains of Wyoming form gently rolling billows that may remind you of a cresting, dipping sea. The towns are spaced an hour's drive apart and your only company on the road may be flocks of birds that swoop up from the roadside as you pass. In the southeastern section of Wyoming the featureless plains give way to an extension of the Front Range of the Rocky Mountain system. Rock slopes and buttes protrude from the high plateau and the vegetation is sparse. In the northern part of the state, short grass gives the ground almost continuous cover, but in the south the grass and brush grow in clumps, exposing large sections of hardened earth.

The Continental Divide: The Rocky Mountains

About 20 or 30 kilometers (15 miles) west of Fort Collins, Colorado, the Front Range of the Rockies leaps upward from the plain. In a distance of only a few kilometers, the open land changes into a rugged mountain landscape. The highest peak in Rocky Mountain National Park is 4,300 meters (14,000 feet) above sea level, but the local relative relief is less than that because the nearby plains are at an altitude of 1,600 meters (5,000 feet). The road through the Park ascends to an altitude of over 3,500 meters—a stop to photograph or gaze at the view may leave you short of breath. The lower slopes of the mountains are densely covered by tall conifers, but in the colder climate at higher altitudes the trees are stunted. At 3,200 meters (10,500 feet), the last gnarled trees disappear from the slopes, leaving the patches of soil to herbs and grasses and exposing a panoramic view of the mountains.

The geologic history of the Rockies is complex. The main period of uplift began 65 million years ago, and subsequent erosion, periods of uplift, sediment deposition, volcanic activity, and even glaciation through the ages have shaped the mountains. Many peaks show unmistakable evidence of glacier action from valley glaciers

UTAH
COLORADO
ARIZONA

The Rocky Mountains and the Southwest
scale: 1:6,000,000

Turret Rock, in Arches National Monument, Utah, is cut from sandstone, primarily by the action of water. This region receives an average precipitation of 9 inches annually, and sagebrush is one of the dominant species of vegetation. Cactus is seldom found here; it is common only in drier deserts, where the annual precipitation is only a few inches.

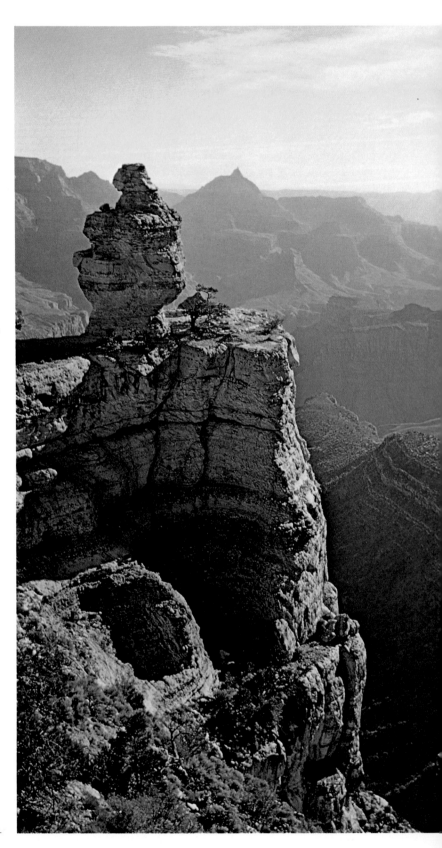

The Grand Canyon of the Colorado River in northern Arizona, seen here from the South Rim, is an incomparable example of the erosive power of flowing water. The Colorado River has cut primarily downward; widening of the canyon has been accomplished largely by water flowing down the side slopes into the gorge.

that were formed in the Rockies during the colder climate of the ice ages. Glaciers have cut bowl-shaped cirques into the rock—Hayden Prong is a fine example of a horn left by two intersecting cirques. Glaciers also rounded the bottoms of V-shaped preglacial valleys and deposited moraines now covered by forest.

Arches, Pillars, Balancing Rocks: The Southwest

The Rocky Mountains mark the Continental Divide of North America; on the western slope of the Rockies rivers flow toward the west, and on the eastern slope, they flow east. The lower lands west of the Rockies are drier than the Great Plains. Moist air from the Pacific is blocked by the Sierra Nevada range, and moist air from the Gulf of Mexico rarely penetrates that far west. At Moab, in eastern Utah near Arches National Monument, the annual precipitation averages only 23 centimeters (9 inches). The vegetation in eastern Utah consists of low clumps of brush that are well separated from one another and that expose the dry, cracked earth beneath. There are few succulent plants such as cactus. Plants in dry regions respond quickly to additional moisture, so vegetation along the side of the road is often greener and denser than the vegetation farther away because it collects the runoff from the impermeable road surface.

Arches National Monument is a collection of fantastically shaped columns, arches, and blocks cut from sandstone by erosion. The sandstone layer is 100 meters (300 feet) thick in places, and it is characteristically red from iron in the cementing matrix that holds the sand grains. The sand is believed to have been laid down by wind 150 million years ago. Episodes of uplift and other crustal movements have left the sandstone severely cracked and twisted, particularly in the lower layers. Some of the vertical sandstone cliffs show slick sides where slabs of rock have literally peeled off. Differential erosion of the heavily jointed rock eventually produced pillars, balancing rocks, and other shapes, many of them rounded, grooved, and pitted from erosion. The Monument also contains dozens of spectacular natural stone arches. Because the lower layers of sandstone tend to exhibit less resistance than the upper layers, erosion usually cuts through the lower section of a thin rock wall first, forming an archway.

The immensity of the Grand Canyon in northwestern Arizona overwhelms the other desert formations. Landforms such as the Black Hills or Devil's Tower are on a scale easy to comprehend, but the vastness of the Grand Canyon—more than a mile deep, 200 miles long and 18 miles wide—is impossible to comprehend in a single view. The Colorado River is a narrow ribbon less than 100 meters wide at the bottom of the Canyon. The average slope of the Colorado River is a drop of more than 1 meter per kilometer (10 times the slope of the Mississippi River at St. Louis, Missouri), and it is able to carry great amounts of sand and silt in suspension. In fact, the Colorado carries half as much suspended load as does the Mississippi, even though the discharge of the Colorado is 100 times less.

The Colorado cuts progressively into deeper, older rocks. Its narrow V-shaped gorge at the base of the canyon is cut into Precambrian rock, so that the exposed layers of strata in the canyon set out more than a billion years of geologic history. As the Colorado cuts downward, processes of weathering and slope erosion work at the side walls and widen the canyon. Resistant layers of rock form steep cliffs and softer rock forms gentler slopes, giving the side walls a steplike appearance.

The nearby canyon of the Little Colorado River makes an interesting contrast with the Grand Canyon. The Little Colorado is carving a canyon with almost vertical walls into a rock plain, so that the width of the canyon is little more than the width of the river. From the road, only a narrow cut can be seen wandering across the plain.

Flagstaff, in the northern part of Arizona, perches on the southern end of the Colorado Plateau 2,100 meters above sea level. Annual precipitation at Flagstaff averages 50 centimeters (20 inches), and the hills are covered with conifers. Because of its elevation Flagstaff has cooler weather than the hot desert to the south, which is only a few hundred meters above sea level. On the road from Flagstaff to Phoenix, a distance of about 200 kilometers (125 miles), the vegetation changes from the conifers of the high country to giant cactus. The average annual precipitation in Phoenix is 20 centimeters compared to 50 at Flagstaff, and the average monthly temperatures at Phoenix are about 13°C (23°F) higher than at Flagstaff. Phoenix is surrounded by green irrigated fields, but the normal landscape of the region is sandy desert.

(opposite top) The Little Colorado River, near the Grand Canyon in northern Arizona, has cut downward to form a narrow, steep-walled canyon. Erosion of the side walls will eventually cause the canyon to become wider.

(opposite bottom) The desert of southwestern Arizona receives only a few inches of precipitation annually, and only plants specially adapted to dry conditions can survive there. The giant Saguaro cactus is a distinctive species in this region, but low shrubs such as creosote bush are more abundant.

A Man-made Garden: The Imperial Valley

The Salton Sea, a new feature of the landscape in the Imperial Valley of Southern California, is not even 100 years old. The Salton Basin is a faulted basin nearly 70 meters (200 feet) below sea level north of the upper end of the Gulf of California. The Colorado River, which has its delta at the head of the Gulf, once drained into the Basin where it deposited a rich layer of sediment. Only water is needed to make the Basin a fertile agricultural region.

In the last half of the nineteenth century, the attention of land developers was drawn to the unique situation of the Salton Basin. They saw the possibility of irrigating it at little expense by constructing a short canal following natural channels and diverting some of the abundant water from the Colorado River. The Colorado River is tens of meters above sea level where it enters its delta, so leading water to the deep Basin presented no problems. Renamed the Imperial Valley for promotional reasons, the Salton Basin was producing profitable crops by 1904, a few years after irrigation was introduced.

Engineers cut through the bank of the Colorado to supply the main irrigation canal for Imperial Valley, but they failed to supply control devices at the cut, and in the spring of 1905 a series of unusually high floods finally tore the bank open at the cut. The Colorado plunged into the Basin and began filling it with water.

The cut was not completely plugged until early in 1907, and by that time a large body of water named the Salton Sea had been formed. Today dams regulate the discharge of the Colorado River and the Imperial Valley flourishes under their protection.

Because the Imperial Valley is below sea level, irrigation water that percolates through the soil has nowhere to drain except into the Salton Sea, so the Sea is becoming increasingly salty from the constant input of saline drainage water.

Journey's End: California

West of the Salton Sea, the dry, unirrigated desert lands stretch to the eastern slopes of the Coast Range mountains, 1,300 meters (4,500 feet) high. To the west of this divide, the belt of land between the Pacific Ocean and the mountains is green and moist, particularly during the winter and spring rainy season. Vegetation varies from deciduous trees along the moist coastal lands in the north to chaparral and eucalyptus in the drier regions south of Los Angeles.

California is a geologically active region. Its sharply rising coastal mountains began their uplift only a few million years ago, and fault movements give a reminder that the process continues today. Hot springs and fault valleys, such as those along the San Andreas fault, are common features throughout the state. The coast, too, shows signs of uplift in its elevated marine terraces and steep cliffs, so different from the low submerged coastline along much of the eastern United States.

The trip could go on and on, to every land of the earth. Each new vista in each new direction would reveal the innumerable physical processes at work on the surface of the earth, shaping the materials at hand into unique combinations of climate, vegetation, soil, and landforms. Confronted by such variety and beauty, our understanding may, at the end, give way to a sense of marvel and wonder.

(opposite) The California coastline shown here at Oxnard is characterized by steep beach cliffs and newly uplifted mountains. By contrast, much of the coast of the eastern United States is low, submerged land with rounded, worn-down hills.

California
scale: 1:12,000,000
San Andreas Fault
CALIFORNIA
Los Angeles
Salton Sea
Imperial Valley

Appendix I: Expressing Scientific Measurements

Scientific Notation: "Powers of Ten"

Very large and very small numbers are cumbersome to manipulate. Expressing such numbers in terms of *scientific notation,* often called "powers of ten" notation, facilitates computations involving these numbers.

Scientific notation for powers of ten is as follows:

$$10 = 10 = 10^1$$
$$100 = 10 \times 10 = 10^2$$
$$1{,}000 = 10 \times 10 \times 10 = 10^3$$
$$10{,}000 = 10 \times 10 \times 10 \times 10 = 10^4$$
$$100{,}000 = 10 \times 10 \times 10 \times 10 \times 10 = 10^5$$
$$1{,}000{,}000 = 10 \times 10 \times 10 \times 10 \times 10 \times 10 = 10^6$$

Note that the "superscript" number, or the *exponent,* at the upper right-hand corner of the 10 indicates the number of times 10 is multiplied by itself. It is also equal to the number of zeros in the corresponding value in the left-hand column. One million is 1,000,000 and has 6 zeros; it may be written as 10^6.

To multiply two or more such numbers, simply add the exponents of each:

$$10^2 \times 10^4 = 10^6$$
$$10^3 \times 10^2 \times 10^4 = 10^9$$

To divide numbers expressed in terms of scientific notation, simply subtract the exponents:

$$10^6/10^2 = 10^4$$
$$10^9/10^3 = 10^6$$

Scientific notation may be used to express any large number, including numbers with decimals, as some small number multiplied by a simple large number:

$$40{,}000{,}000 = 4 \times 10{,}000{,}000 = 4 \times 10^7$$
$$46{,}500{,}000 = 4.65 \times 10{,}000{,}000 = 4.65 \times 10^7$$

In each case, the exponent indicates the number of places the decimal point has been "moved" to the left. To convert the number 4.65×10^7 back to its unabbreviated form, simply move the decimal seven places to the right.

Extremely small numbers are also conveniently written in terms of scientific notation by using negative exponents, as follows:

$$0.1 = 1/10 = 1/10^1 = 10^{-1}$$
$$0.01 = 1/100 = 1/10^2 = 10^{-2}$$
$$0.001 = 1/1{,}000 = 1/10^3 = 10^{-3}$$

The multiplication rule given for large numbers still holds:

$$10^{-1} \times 10^{-4} = 10^{-5}$$
$$10^3 \times 10^{-2} = 10^1$$
$$10^4 \times 10^{-4} = 10^0 = 1$$

A negative exponent signifies the number of places the decimal has been moved to the right:

$$.0075 = 7.5 \times 10^{-3}$$
$$.00000037 = 3.7 \times 10^{-7}$$

Similarly, conversion of these numbers back to the unabbreviated form entails moving the decimal the appropriate number of places to the left.

Significant Figures

No quantity in experimental science can be measured to absolute accuracy. The result of a scientific measurement is therefore usually accompanied by an estimate that indicates the degree of accuracy that can be assigned to the measurement. A measurement of the age of a rock might be reported as 1.2 ±0.2 billion years, for example. The value 1.2 billion years represents the best, or average, value from a series of determinations. The quantity ±0.2 billion years establishes a measure of the accuracy of the measurement. The error limits imply that according to the best estimate of the experimenters, the true age of the rock probably lies between 1.0 and 1.4 billion years. Because of the inherent uncertainty in estimating error limits, there is a small probability that the true age of the rock might lie somewhat outside the assigned error limits. This probability decreases substantially if the limits are extended to ±0.4 billion years.

The value of a quantity should be stated in a way that reflects its estimated accuracy. It is misleading, for instance, to report the age of a rock as 1.233 billion years if the error limits are ±0.2 billion years. The additional figures *33* are not *significant* because they are swamped by the inherent error of the measurement. Only the first two figures, *1.2,* are significant, so giving more than two figures is not meaningful in this case. Similarly, if repeated measurement of the depth of the ocean in a certain location gives an average value of 2,312 feet with an estimated error of ±200 feet, the result should be reported to two significant figures, or as 2,300 ±200 feet. The trailing zeros are not considered to be significant unless they are terminated by a decimal point; the number 500. has three significant figures, for example, and 500.0 has four significant figures.

Conversion factors are often expressed to a limited number of significant figures when only moderate accuracy is required. To two significant figures, 1 centimeter can be said to be equal to 0.39 inch, for example. Many of the conversions stated in the text between the metric system of units and the English system have been rounded off to a similar number of significant figures to convey the same degree of accuracy. A length of 300 meters is approximately 984.3 feet, but expressed to one significant figure it is 1,000 feet.

Metric to English Conversions

Length
1 kilometer = 1,000 meters = 0.6214 mile = 3,281 feet
1 meter = 100 centimeters = 1.0936 yards = 3.281 feet = 39.37 inches
1 centimeter = 10 millimeters = 0.3937 inch
1 micron = 10^{-6} meter = 10^{-4} centimeter = 3.937×10^{-5} inch

Area
1 square kilometer = 10^6 square meters = 0.3861 square mile = 247.1 acres
1 square meter = 10^4 square centimeters = 1.196 square yards = 10.764 square feet = 1,550.0 square inches

Volume
1 cubic kilometer = 10^9 cubic meters = 0.2399 cubic mile
1 cubic meter = 10^6 cubic centimeters = 1.308 cubic yards = 35.31 cubic feet = 61,024 cubic inches

Mass
1 metric ton = 1,000 kilograms = 2,204.6 avoirdupois pounds
1 kilogram = 1,000 grams = 2.2046 avoirdupois pounds

Time
1 day = 86,400 seconds
1 year = 3.156×10^7 seconds

Speed
1 meter per second = 3.281 feet per second
1 meter per second = 3.6 kilometers per hour = 2.237 miles per hour
1 kilometer per hour = 0.62 mile per hour
1 knot = 1 nautical mile per hour = 1.151 miles per hour

Pressure
1 atmosphere = 1,013.2 millibars = 760 millimeters of mercury = 29.92 inches of mercury

Temperature
Temperature in °C = 5/9 (temperature in °F − 32°)
Temperature in °F = 9/5 (temperature in °C) + 32°
Temperature in °K = temperature in °C − 273.15°

Energy
1 calorie = 4.186 joules = 3.968×10^{-3} British Thermal Unit
1 langley = 1 calorie per square centimeter

Power
1 calorie per second = 4.186 joules per second = 4.186 watts
1 calorie per minute = 251 watts

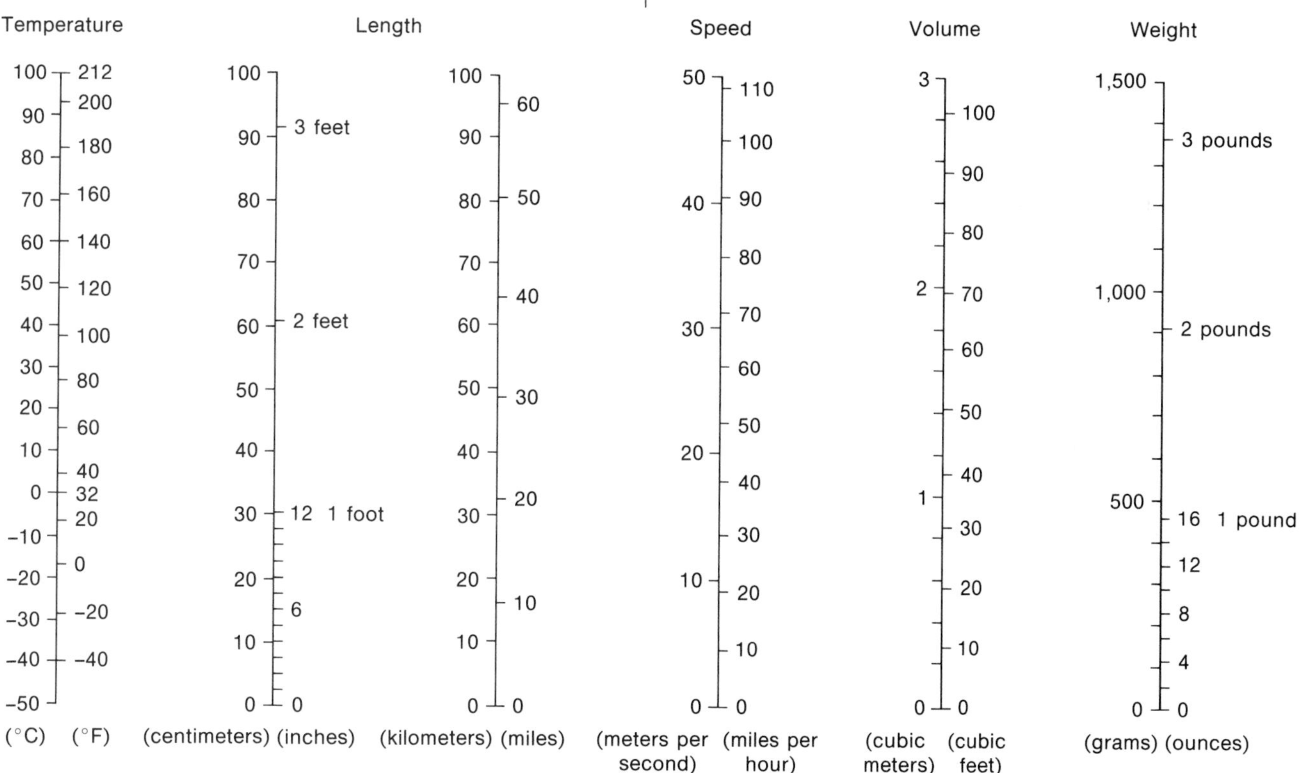

Temperature
Length
Speed
Volume
Weight
(°C) (°F)
(centimeters) (inches)
3 feet
2 feet
1 foot
(kilometers) (miles)
(meters per second) (miles per hour)
(cubic meters) (cubic feet)
(grams) (ounces)
3 pounds
2 pounds
1 pound

Appendix II: Tools of the Physical Geographer

Maps: A Representation of the Earth's Surface

Physical geographers are concerned with the surface of the earth—with the circulation of the atmosphere and the oceans, with the distribution of climates, soils, and vegetation, and with the shape and location of landforms. They use a number of special tools and techniques in their study of the earth. Maps, which represent the earth's surface, are one of the primary tools used in geography for recording, interpretation, and analysis.

Modern maps exist in great variety, from simple street maps to complex navigational charts for jet aircraft. A map can convey a large amount of information in a way that is easily assimilated; a well-made map of climatic regions, for example, can make important climatic relationships much clearer than can lengthy written descriptions or tables of data. In addition to displaying global systems of classification, maps have many other uses in geography: geographers may use maps on a local scale to study the particular features of landforms or small drainage basins, for example, and on field trips they draw their own maps to record interesting observations. This section describes how maps display information, and it discusses the principles of their construction in order to show how to make the best use of maps.

A *map* can be defined formally as a two-dimensional graphic representation of the spatial distribution of selected phenomena. (A three-dimensional representation is more properly called a *model.*) A map is *planimetric;* it shows *horizontal* spatial relationships on the earth. A true map is planimetrically accurate in that it shows the correct position of every location. In addition to specifying position on the earth's surface, maps represent distance and direction from one location to another and indicate the sizes or shapes of various regions. With appropriate symbolization, maps can also represent the vertical relief of landscapes, the steepness of slopes, and the location of rivers, or of buildings, roads, and other man-made objects.

Figure II.1 (top) This Babylonian map from 500 B.C. is one of the earliest attempts to portray the world.

(bottom) Claudius Ptolemy's map of the world, reconstructed from his descriptions written in the second century, is a comparatively accurate representation that takes into account the spherical shape of the earth.

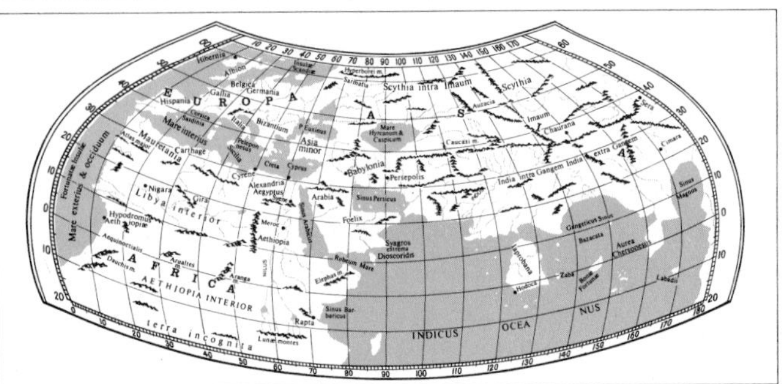
Hibernia
Albion
Belgica
Germania
Gallia
Hispania
E U R O P A
Sarmatia
Scythia intra Imaum
Scythia
A S I A
Imaum
Chaurana
Sera
Cimara
Bizantium
Italia
Mare interius
Mauretania
Carthage
Sicilia
Asia minor
Creta
Cyprus
Cyrene
Alexandria
Aegyptus
Babylonia
Persepolis
India intra Gangem
India extra Gangem
Mare Hyrcanum & Caspicum
Arabia
Sinus Persicus
Sinus Arabicus
Foelix
Nigara
Libya interior
Hypodromus
Mare exterius & occiduum
Fortunatæ Insulæ
Meroe
Aethiopia
Syagros
Dioscoridis
Rubeum Mare
Taprobana
Gangeticus Sinus
Aurea Chersonesus
Sinus Magnus
A F R I C A
AETHIOPIA INTERIOR
Rapta
Lunæ montes
terra incognita
INDICUS
OCEA
NUS

Scale and Distance. Because the physical size of a map is obviously smaller than the area of the earth's surface that the map represents, it is necessary to translate length on the map into actual distance. A map's *scale* gives the relation between length measured on the map sheet and the corresponding distance on the earth's surface.

There are several ways to express the scale of a map. It may be given in a simple *verbal statement*, such as "1 inch equals 3 miles." Other lengths will bear the same proportion to distances on the earth, so that 2 inches on the map will correspond to 6 miles, and so on. The scale of a map also may be indicated by a *graphic scale* marked off in units of distance on the earth, as shown in Figure II.2. One advantage of a graphic scale is that the scale remains correct if the map is copied in a larger or smaller size, whereas a verbal statement of scale becomes incorrect if the size of the map is altered.

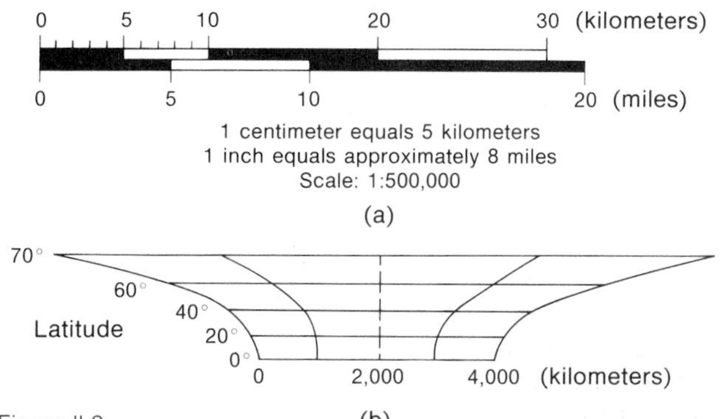

0 5 10 20 30 (kilometers)
0 5 10 20 (miles)
1 centimeter equals 5 kilometers
1 inch equals approximately 8 miles
Scale: 1:500,000
(a)
70°
60°
40°
20°
0°
Latitude
0 2,000 4,000 (kilometers)
(b)

Figure II.2

Scale is often expressed as a fraction, called the *representative fraction.* A representative fraction of 1/5,000 (commonly written 1:5,000) means that a length of 1 unit on the map represents 5,000 units of distance on the earth. A representative fraction makes no reference to any particular system of units, because it represents simply the ratio between length on the map and a corresponding distance on the earth. If the representative fraction is 1:5,000, for instance, 1 inch on the map represents 5,000 inches on the earth. Alternatively, 1 centimeter on the map represents 5,000 centimeters on the earth.

A verbal statement of scale can be restated as a representative fraction by converting both members of the statement to the same units and dividing. So a scale of 1 inch to the mile is equivalent to a scale of 1:63,360 because there are 63,360 inches in 1 mile. Conversely, a representative fraction can be expressed as a verbal statement of scale by assigning units and applying conversion factors as required: a scale of 1:100,000 can be stated as 1 centimeter to 1 kilometer, for example. Table II.1 lists equivalents for selected scales. Note the ease of conversion between representative fractions and verbal statements of scale when the metric system is used.

Table II.1. Map Scale Equivalents

Scale	Inches to 1 Mile	1 Inch Represents (feet)	1 Inch Represents (miles)	1 Inch Represents (kilometers)	1 Centimeter Represents (kilometers)
1:2,500	25.344	208	0.039	0.064	0.025
1:10,000	6.336	833	0.158	0.254	0.10
1:20,000	3.168	1,667	0.316	0.508	0.20
1:24,000	2.640	2,000	0.379	0.610	0.24
1:31,680	2.000	2,640	0.500	0.805	0.3168
1:50,000	1.267	4,167	0.789	1.270	0.50
1:62,500	1.014	5,208	0.986	1.588	0.625
1:63,360	1.000	5,280	1.000	1.609	0.6336
1:100,000	0.634	8,333	1.578	2.540	1.00
1:125,000	0.507	10,417	1.973	3.175	1.25
1:250,000	0.253	20,833	3.946	6.350	2.50
1:1,000,000	63×10^{-3}	83,333	15.783	25.400	10
1:5,000,000	12.7×10^{-3}	0.417×10^{6}	79	127	50
1:10,000,000	6.3×10^{-3}	0.833×10^{6}	158	254	100
1:50,000,000	1.3×10^{-3}	4.17×10^{6}	790	1,270	500
1:100,000,000	630×10^{-6}	8.33×10^{6}	1,580	2,540	1,000
1:250,000,000	250×10^{-6}	20.83×10^{6}	3,950	6,350	2,500
1:500,000,000	130×10^{-6}	41.7×10^{6}	7,900	12,700	5,000
1:1,000,000,000	63×10^{-6}	83.3×10^{6}	15,800	25,400	10,000

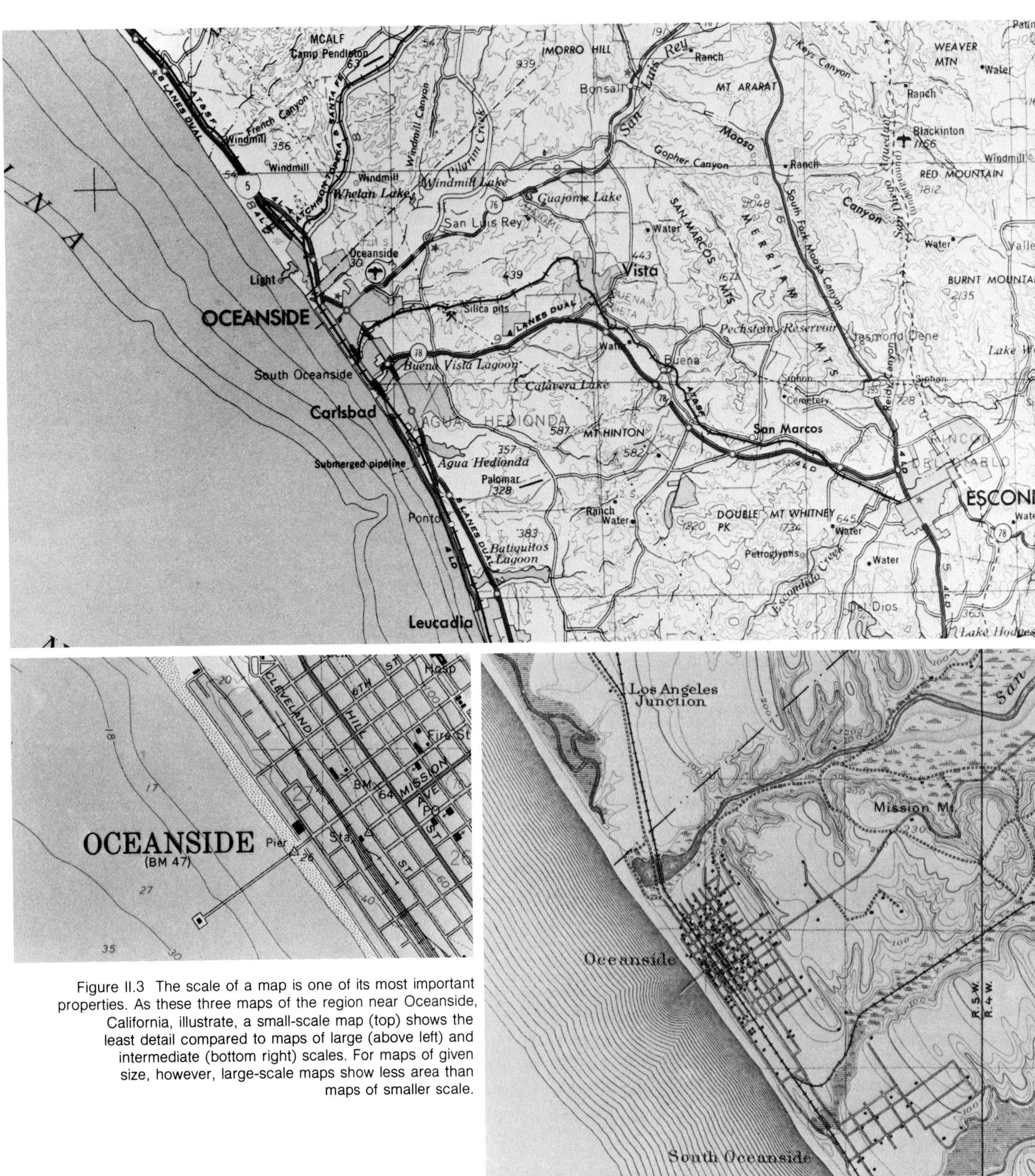

Figure II.3 The scale of a map is one of its most important properties. As these three maps of the region near Oceanside, California, illustrate, a small-scale map (top) shows the least detail compared to maps of large (above left) and intermediate (bottom right) scales. For maps of given size, however, large-scale maps show less area than maps of smaller scale.

When a portion of a globe is transferred to a flat map—which is known as *projecting* a spherical surface onto a plane—distortions inevitably occur. One consequence of the distortion is that the scale of a map cannot be constant for every portion of the map. However, scale does not vary greatly on a flat map of a small region, and even the conterminous United States can be mapped in such a way that the scale does not vary by more than a few percent. Significant variations of scale may occur on global maps, however. For this reason, the map user must keep in mind that the stated scale for a given map is correct only for a limited portion of the map. The scale on a Mercator map of the world, for instance, is several times larger at higher latitudes than at the Equator (Figure II.2).

When the representative fraction is a small number, less than 1/1,000,000 (1:1,000,000), a map is called a *small-scale* map. One unit of length on a small-scale map represents a large distance on the earth; small-scale maps are used when a large fraction of the earth's surface, such as a continent or an ocean, must be represented on a map of limited size. If a map has a representative fraction larger than 1/250,000 (1:250,000), it usually is called a *large-scale* map; it is capable of showing greater detail than a small-scale map. Maps are made at different scales to suit different purposes. A geographer interested in studying a local landform on maps or in planning a day's field trip would need a map on a scale such as 1 inch to the mile (1:63,360) or larger. But to represent world climate distributions, a small-scale map with a representative fraction of 1:100,000,000 would be more suitable.

Location. The principal method for specifying location on the earth's surface is the system of latitude and longitude described in Chapter 3. *Latitude,* the position of a place north or south of the Equator, is expressed in angular measure relative to the center of the earth. The angle of latitude varies from 0° at the Equator to 90° at the poles. *Longitude,* the position of a place east or west of a selected prime meridian, is expressed in angular measure that varies from 0° to 180° east or west of the prime meridian. The most commonly accepted prime meridian is the one on which the Greenwich observatory in England is located, but reference to other prime meridians is sometimes found on older maps or on maps published by some countries. The framework of lines representing parallels of latitude and meridians of longitude on a map is called the *graticule* of the map. Depending on the method chosen to construct a map, the lines of the graticule may be straight or curved, and they may or may not intersect at right angles to one another, although they do intersect at right angles on a globe.

The division of 1 degree of angular measure into 60 minutes and the subdivision of 1 minute of angular measure into 60 seconds makes the system of latitude and longitude cumbersome for accurately specifying the position of objects. Therefore, alternative systems have been devised for giving location, based on the use of rectangular *grid* systems analogous to the grid system of x and y coordinates used for plotting data on graphs.

The first step in constructing a grid system is to choose a standard form of map that meets the needs of the user. (The advantages and drawbacks of different types of maps are discussed later in this appendix.) Once the map is chosen, a square grid is overlaid on the map and numerical coordinates are assigned to the reference lines of the grid. The coordinates are usually expressed in units of distance from a selected origin. The grid coordinates of any location can then be read from the map as illustrated in Figure II.4. By convention, the coordinate to the east, or the *easting,* is specified first. Then the coordinate to the north, or the *northing,* is specified. The rule is to read toward the *right* and *up,* following the same order used for giving the x and y coordinates of a point on a graph. The grid coordinates of a location are often given as one number consisting of an even number of digits; the first half of the number gives the easting, and the second half, the northing.

One of the grid systems most frequently encountered consists of the *Universal Transverse Mercator* (*UTM*) and the *Universal Polar Stereographic* (*UPS*) grids

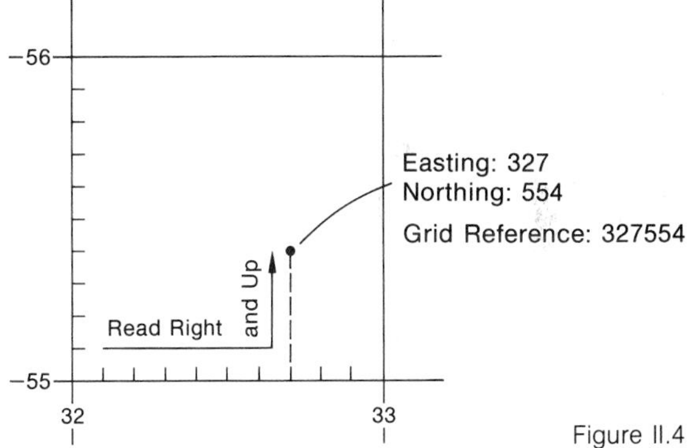

–56
–55
32
33
Easting: 327
Northing: 554
Grid Reference: 327554
Read Right
and Up

Figure II.4

developed for military and civilian use. The UTM grids extend between latitudes 80°N and 80°S, and the UPS grids cover the regions poleward from latitude 80°.

For the purpose of setting up the UTM grids, the globe between latitudes 80°N and 80°S is divided into 60 sections, each 6° of longitude wide. A square grid with 100,000-meter spacing is then superimposed on a transverse Mercator map of each section. The central meridian of each section is assigned an easting of 500,000 meters; the Equator is assigned a northing of 0 meters for locations in the Northern Hemisphere and a northing of 10,000,000 meters for locations in the Southern Hemisphere. Each 6° section is divided into

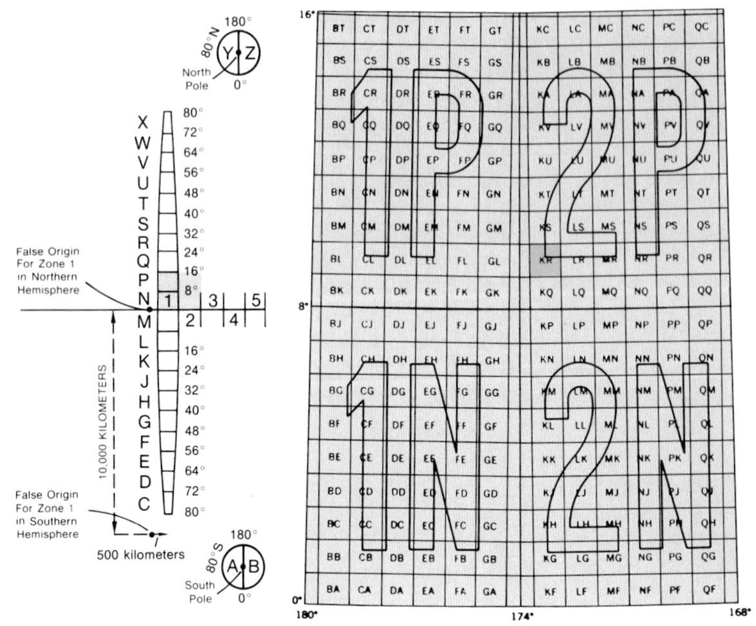
180°
80° N
Y Z
North Pole
0°
X W V U T S R Q P N M L K J H G F E D C
80° 72° 64° 56° 48° 40° 32° 24° 16° 8°
16° 24° 32° 40° 48° 56° 64° 72° 80°
False Origin For Zone 1 in Northern Hemisphere
False Origin For Zone 1 in Southern Hemisphere
10,000 KILOMETERS
500 kilometers
1 2 3 4 5
180°
80° S
A B
South Pole
0°
16°
8°
0°
180°
174°
168°
1P 2P
1N 2N

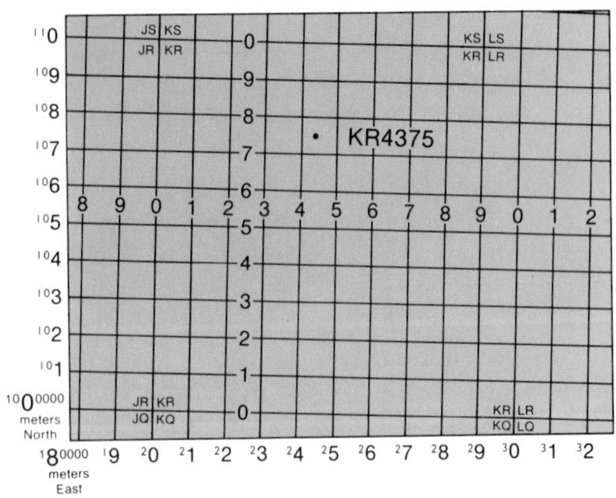
KR4375
JS KS
JR KR
KS LS
KR LR
JR KR
JQ KQ
KR LR
KQ LQ
$^{1}0^{0}0000$ meters North
$^{1}8^{0}0000$ meters East

Figure II.5 (above) The UTM grid is subdivided successively into zones, quadrangles, and 10,000-meter grid squares.

Figure II.6 (below) The 15° square of the GEOREF system is subdivided into 1° squares.

twenty 6° by 8° quadrilaterals, and for purposes of zone identification, each quadrilateral is assigned an index consisting of a number from 1 through 60 and a letter from C through X (with *I* and *O* omitted). Each of the 100,000-meter squares is further assigned a two-letter identification index. The UPS grid zones are designated in a similar way.

The United States National Ocean Survey (formerly the United States Coast and Geodetic Survey) has designed a grid system for each of the states, called the *State Plane Coordinate* system. The basic grid square of the state coordinates is 10,000 feet on a side; eastings and northings for the grid are listed in units of feet.

Some maps contain references to one or more grid coordinate systems, as well as to the conventional system of latitude and longitude. The topographic maps prepared by the United States Geological Survey, for example, bear tick marks around the margins that indicate UTM coordinates, state plane coordinates, and latitude or longitude. On the topographic maps published by the United States Army Topographic Command (TOPOCOM), formerly known as the Army Map Service, the UTM grid is drawn on the map, with latitude or longitude given as tick marks on the margins.

Aeronautical charts employ a system of coordinates that is equivalent to latitude and longitude but that uses a different method of notation, known as the World Geographic Reference System (GEOREF). According to the GEOREF system, longitude is divided into twenty-four 15°zones and latitude is divided into twelve 15°zones. A two-letter index specifies each 15° by 15° quadrangle, and each quadrangle is divided into

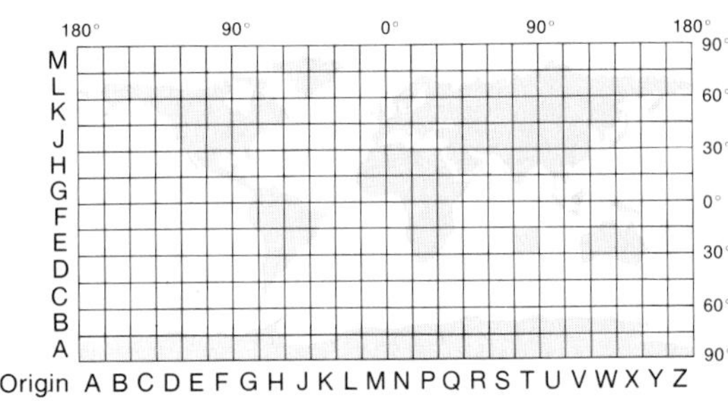
180° 90° 0° 90° 180°
M L K J H G F E D C B A
90° 60° 30° 0° 30° 60° 90°
Origin A B C D E F G H J K L M N P Q R S T U V W X Y Z

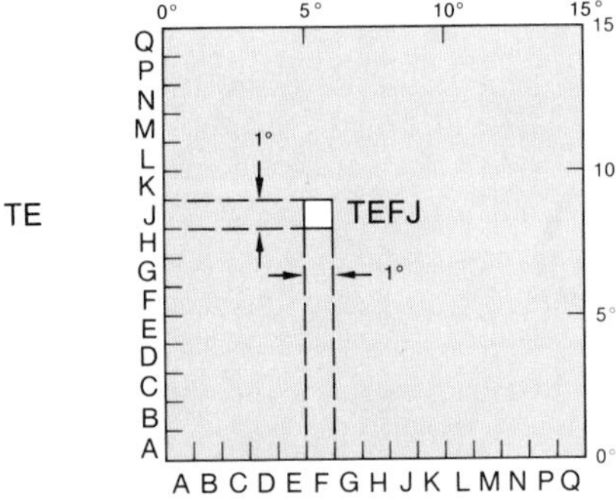
0° 5° 10° 15°
Q P N M L K J H G F E D C B A
15° 10° 5° 0°
TE
TEFJ
1°
1°
A B C D E F G H J K L M N P Q

1° by 1° sections. Four letters are necessary to specify a given 1° by 1° section: the first two provide the designation of the 15° quadrangle, and the last two, the designation of the 1° quadrangle. For accurate specification of location, each 1° interval is divided into 60 minutes of angular measure and subdivided into decimal fractions of a minute. Most aeronautical charts employ the GEOREF system rather than the UTM grid.

A modified grid system has been in use for many

years in connection with the survey of public lands conducted by the Bureau of Land Management. The basic land unit of the survey, which was begun in the eighteenth century, is the *township*, a square plot 6 miles on a side. Townships are laid out with two sides along meridians and the other two sides along parallels of latitude. Because meridians converge toward the north, the north-south sides of the townships usually take a jog eastward or westward every 24 miles to maintain the size of the 6-mile square.

Townships are laid out with respect to a north-south *principal meridian* and an east-west *base line*. Different land surveys established 31 sets of principal meridians and base lines for the conterminous United States and 5 sets for Alaska. The location of each township in a survey region is given with respect to the point at which the principal meridian and the base line intersect. The coordinates that specify a particular township are read off as the number of townships north or south of the base line; the number of townships east or west of the principal meridian is called the *range*. The system for locating townships is an exception to the "*right* and *up*" rule of reading because northings are read before eastings. Townships are further subdivided into 36 squares, 1 mile on each side, which are called *sections;* sections are numbered 1 through 36 in a serpentine fashion, beginning in the upper right corner of the township (see Figure II.7).

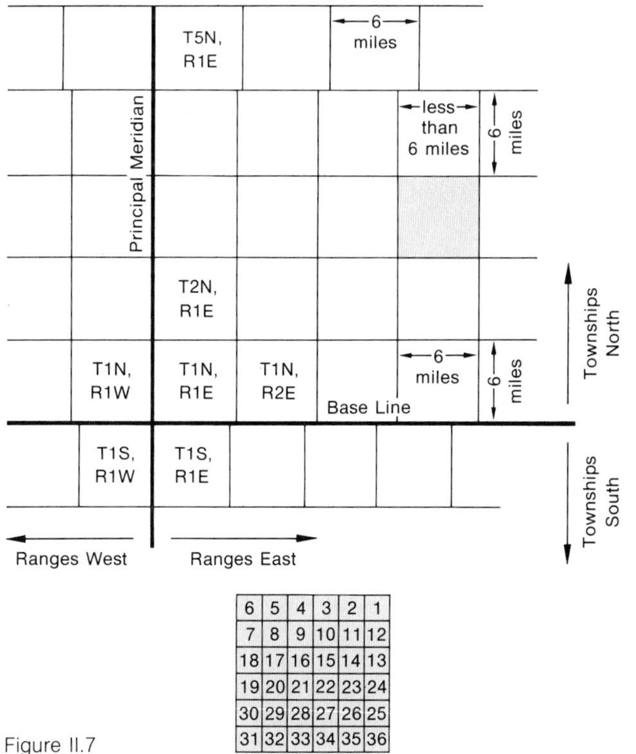
T5N,
R1E
6
miles
Principal Meridian
less
than
6 miles
6
miles
T2N,
R1E
T1N,
R1W
T1N,
R1E
T1N,
R2E
6
miles
Base Line
6
miles
Townships
North
T1S,
R1W
T1S,
R1E
Townships
South
Ranges West
Ranges East
6 5 4 3 2 1
7 8 9 10 11 12
18 17 16 15 14 13
19 20 21 22 23 24
30 29 28 27 26 25
31 32 33 34 35 36
Figure II.7

Direction. By definition, meridians of longitude lie along a true north-south direction and parallels of latitude lie along a true east-west direction. Because of the distortions inherent in mapping the surface of a sphere onto a flat sheet of paper, meridians or parallels often vary in direction across a map sheet. There is sometimes no uniform direction relative to the map sheet that represents north, although the north-south line at any given location is always coincident with the local meridians. But for large-scale maps that cover only a local region, the distortions are small and the map sheet can be aligned with respect to a single standard direction to establish a sense of orientation.

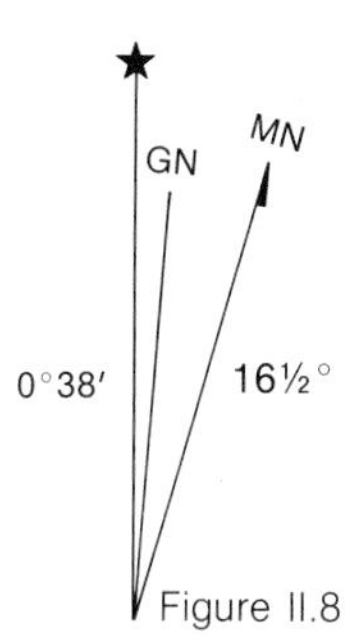
GN
MN
0°38′
16½°
Figure II.8

Many large-scale maps indicate the direction of *true north* by means of a star-tipped arrow or the symbol *TN*. However, this direction is usually not the same as *magnetic north*, the direction in which a magnetic compass needle points. Large-scale maps usually indicate the direction of magnetic north by means of a half-headed arrow and the symbol *MN*. The earth's magnetic field is not uniform, and the magnetic poles do not coincide spatially with the geographic poles, so the relation between magnetic north and true north must be specified separately for each local region. The difference between magnetic north and true north is known as *magnetic declination* and is expressed in degrees east or west of the true meridian of a given location. Across the conterminous United States, for example, the magnetic declination varies from 0° to as much as 25° east or west, and in polar regions 90° or more is possible. Furthermore, the direction of magnetic north at a given location varies with time, often by as much as 1° in 20 years. For precision map work, therefore, reference should be made to recent compilations of magnetic declinations, such as those prepared by the National Ocean Survey.

Direction on a map may also be specified as *grid north*, the northerly direction arbitrarily determined by a particular grid system, and symbolized on the map by *GN*. Grid north generally does not coincide with either magnetic north or true north. The grid north directions specified by two different grid coordinate systems usually differ from one another as well.

Directions other than north can be expressed in terms of *azimuth*, which is the angle of the desired direction measured clockwise from a chosen reference direction and expressed in degrees between 0° and 360°. According to the choice of true north, magnetic north, or grid north as a reference, the corresponding azimuths are termed *true azimuth*, *magnetic azimuth*, or *grid azimuth*.

The Shape of the Earth. An exact determination of the earth's shape is required for the highest precision in map making and in position finding on the earth's surface. The earth bulges slightly at the Equator, so that the diameter of the earth measured in a plane through the Equator is 43 kilometers (27 miles) longer than the polar diameter.

In the United States, Canada, and Mexico, the reference surface for map making is the *Clarke spheroid of 1866*, a smooth mathematical surface that closely approximates the shape of the earth. The Clarke spheroid of 1866 assumes an equatorial diameter for the earth of approximately 12,756 kilometers (7,926 miles) and a polar diameter of approximately 12,713 kilometers (7,899 miles). These values are in agreement with a 1968 determination of the earth's shape made by an orbiting satellite. The difference between the equatorial and polar diameters is so small that the earth is considered a sphere in most map work.

Table II.2 lists the lengths corresponding to 1° of longitude and 1° of latitude according to the Clarke spheroid of 1866. The length of 1° on a meridian is nearly constant from the Equator to the poles, but the length of 1° on a parallel tends toward 0 at the poles. At latitude 60°, the length of 1° on a parallel is approximately half its value at the Equator.

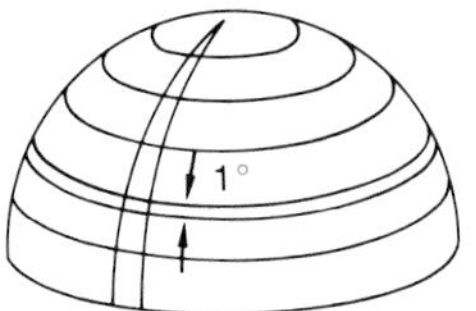
1°

Table II.2(a). Length of 1° on a Meridian (Clarke Spheroid of 1866)

Latitude	Kilometers	Statute Miles	Nautical Miles*
0°-1°	110.57	68.70	59.70
10°-11°	110.60	68.73	59.72
20°-21°	110.71	68.79	59.78
30°-31°	110.86	68.88	59.86
40°-41°	111.04	69.00	59.96
50°-51°	111.24	69.12	60.06
60°-61°	111.42	69.24	60.16
70°-71°	111.57	69.33	60.24
80°-81°	111.67	69.39	60.30
89°-90°	111.70	69.41	60.31

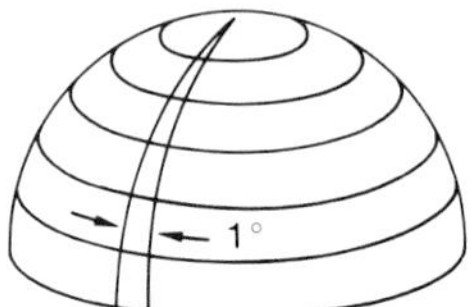
1°

Table II.2(b). Length of 1° on a Parallel (Clarke Spheroid of 1866)

Latitude	Kilometers	Statute Miles	Nautical Miles*
0°	111.32	69.17	60.11
10°	109.64	68.13	59.20
20°	104.65	65.03	56.51
30°	96.49	59.96	52.10
40°	85.40	53.06	46.11
50°	71.70	44.55	38.71
60°	55.80	34.67	30.13
70°	38.19	23.73	20.62
80°	19.39	12.05	10.47
89°	1.95	1.21	1.05

*1 nautical mile is approximately equal to 6,076.1 feet, or 1.15 statute miles.
Source: adapted from: List, Robert (ed.). 1971. *Smithsonian Meteorological Tables*. 6th rev. ed. Washington, D. C.: Smithsonian Institution Press.

Map Projections

A model globe is the only way to represent large portions of the earth's surface with accuracy, because only a globe correctly takes into account the spherical shape of the earth. A flat piece of paper cannot be fitted closely to a sphere without wrinkling or tearing, so small-scale maps that represent regions hundreds or thousands of miles in extent inevitably introduce noticeable distortions. Maps printed on flat sheets are far more portable and convenient to use than globes, however, and it is important for the geographer to be able to employ flat maps with an understanding of their properties.

The fundamental problem of map making is to find a method of transferring a spherical surface onto a flat sheet in a way that minimizes undesirable distortions. Any method of relating position on a globe to position on a flat map is called a *projection*, or *transformation*. A projection is a correspondence between a globe and a flat sheet such that every point on the globe or selected portion of the globe can be assigned a corresponding point on the sheet. Numerous projections have been devised for mapping, each with its characteristic advantages and each with its characteristic distortions. Because no projection is free of distortion, the choice of a projection should be made with regard to its proposed application. Several different projections are used for the maps in this book: the flat polar quartic projection for global maps, the stereographic projection for maps of polar regions, and

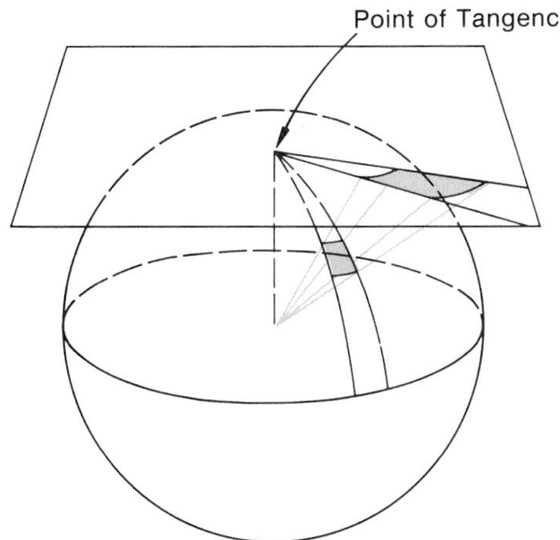
Point of Tangency
Figure II.9 Construction of the Gnomonic Projection

the Albers' conic equal area projection for maps of the United States. Each projection has advantages for its particular application, and no single projection is best for all uses.

The principles of projection are illustrated in Figure II.9 using the example of the gnomonic projection. This projection can be constructed by tracing the rays of light from a light source at the center of a transparent globe onto a plane that touches the globe at one point, called the *point of tangency*. Each point on the portion of the surface of the globe that is projected onto the plane is assigned a corresponding point on the plane, which constitutes a map. However, only a few projections, such as the gnomonic, can be visualized geometrically. Many projections can be expressed only as a mathematical rule that relates points on a globe to points on a flat sheet.

Projections and Distortions. The principles of projection ensure that every properly drawn map shows correct location on the earth's surface, regardless of the projection used. However, maps are often relied upon to show other properties, such as direction and distance from one location to another and the shape and size of areas. A single flat map cannot depict all of these properties without distortion, so every projection represents a compromise. The map user should realize where and to what extent inaccuracies are present in the projection he is using. The gnomonic projection illustrated in Figure II.9, for example, exaggerates distances far from the point of tangency compared to distances near the point of tangency. Distortion of distance (scale distortion) is therefore greatest near the outer borders of a gnomonic projection and least near the point of tangency.

Scale distortion may lead to distortion of direction, shape, and size, depending on the projection. On some projections, the scale in a small region is not the same in all directions, which necessarily leads to distortion of direction. An azimuth of 45° with respect to a meridian on a globe, for example, will not be mapped as a 45° angle if the scale on the map is different in the east-west and north-south directions. Furthermore, distortion of direction implies that the outline of a small region will not be mapped accurately, and so distortion of shape will be present on such a map.

A number of projections, called *conformal* projections, have been devised so that scale is the same from every direction in a small region of the map. Shapes of small regions are therefore portrayed accurately also. However, the scale on a conformal map necessarily changes from one region to another. Regions that span a large portion of the globe tend to exhibit overall distortion when mapped on a conformal projection; each small section is accurately depicted, but the relative sizes of separate sections may be incorrect. The well-known standard Mercator projection is conformal, but it represents areas in the higher latitudes several times too large compared to areas near the Equator.

Another important class of projections is the *equal area*, or *equivalent*, projections. On such projections, the scale in a small region is different in different directions, and shapes are distorted. However, the scale is designed to vary over the map in such a manner that the relative sizes of any two areas are correct. Figure II.10 illustrates how shapes can be distorted

without altering their areas. Equal area projections are especially useful for displaying distributions in physical geography, because areas are represented in their correct proportions and misleading relative comparisons are avoided. Many projections are neither conformal nor equal area but represent compromises to obtain adequate representation of shape without badly distorting size.

An impression of a projection's major properties can sometimes be obtained by seeing how the lines of latitude and longitude are portrayed. On a globe, parallels and meridians intersect at right angles. If a certain map shows them crossing at right angles, the projection may or may not be conformal, but if they are shown crossing at a different angle, distortion of direction is certainly present and the projection is not conformal. Distortions of shape or size can be seen by comparing the small quadrangles bounded by parallels and meridians on a projection to those on a globe.

Map projections can be made from a globe onto a plane, cylinder, or cone, which are the only surfaces that can be spread on a flat sheet without distortion. Projections are conventionally classified into families: *azimuthal*, or *zenithal*, projections of the globe onto a plane, *cylindric* projections onto a cylinder, and *conic* projections onto a cone. A fourth family, sometimes called *pseudocylindric* projections, is reserved for certain projections usually used to portray the entire globe. The projections in a given family tend to have similar properties and similar distortion characteristics.

Azimuthal Projections. Azimuthal projections are projections of a globe onto a plane tangent to the globe at some point. The point of tangency is often taken to be the North or South Poles or a location on the Equator, but in principle any point on the globe may be used. Most azimuthal projections can depict only one hemisphere of the earth or less at one time.

In azimuthal projections, azimuths measured around the point of tangency are mapped with no error. The distortion of any azimuthal projection is least nearest the point of tangency and increases with increased distance away from the point of tangency. Figure II.11 exhibits some of the characteristics of selected azimuthal projections. Distortion patterns are depicted by degrees of yellow shading, with the deeper tints corresponding to regions of greater distortion.

Cylindric Projections. A cylinder closely fitted to a globe makes contact with the globe along a great circle, called the *circle of tangency*. Cylindric projections are usually designed so that the circle of tangency is the Equator or a meridian. If the Equator is chosen as the circle of tangency, the whole surface of the earth, except for the polar regions, can be shown on one map.

Most cylindric projections with the Equator as the circle of tangency show parallels of latitude and meridians of longitude as sets of straight parallel lines intersecting at right angles. The best known such projection, the standard Mercator, is conformal; others are not. The standard Mercator projection is frequently used for navigation charts because a straight line on this projection represents a line of constant azimuth, which simplifies navigation.

Distortions on a cylindric projection are least nearest the circle of tangency and increase with increased distance from the circle of tangency. On a standard Mercator projection, for example, areas in the higher latitudes are grossly exaggerated. Conversely, on a cylindric equal area projection based on the Equator, shapes at higher latitudes are badly distorted.

The base maps for the UTM grid system are prepared using a Mercator projection fitted to two arcs of tangency in each of the 60 separate sections between 80°N and 80°S latitude. No point on one of the 60 sectional maps is more than 150 miles or so from a meridian of tangency, and distortion is correspondingly small. Figure II.12 shows the properties of several cylindric projections.

Conic Projections. A cone placed upon a globe contacts the globe along a circle of tangency. If the apex of the cone is above a pole, the circle of tangency coincides with a parallel of latitude, known as the *standard parallel* of the conic projection. Parallels of latitude are shown as curved arcs in such conic projections, and meridians of longitude, as converging straight lines. The distortions of a conic projection are least nearest the standard parallel and increase away from the standard parallel. Conic projections are therefore useful in mapping countries, such as the United States, that extend primarily along the east-west direction. Conic projections are usually restricted to portions of the earth's surface that cover a limited range of latitude. Such a map usually does not extend more than 100° in longitude to avoid excessively curved parallels.

Conic projections of greater accuracy are produced by allowing the cone to cut through the globe, so that contact is made at two standard parallels. If the standard parallels are not too far apart, the whole region bounded by the parallels can be mapped to excellent accuracy. Albers' equal area conic projection, which was chosen for the maps in *The National Atlas of the United States*, can show the conterminous United States with a linear scale distortion that does not exceed 2 percent. For the conterminous United States, the standard parallels of the Albers' conic projection are chosen to be 29½°N and 45½°N. For Alaska, the standard parallels are 55°N and 65°N, and for Hawaii, 8°N and 18°N. The properties of several conic projections are shown in Figure II.13.

Pseudocylindric Global Projections. A projection that shows the entire earth is useful for displaying global distributions of climate, soils, vegetation, and other quantities important in physical geography. Such a map should be an equal area projection so that relative areas can be compared without error. If the projection displays parallels of latitude as parallel straight lines, regions with the same latitude are placed along the same straight band on the map, a useful property because many quantities of interest in physical geography depend on latitude.

Although significant distortion always accompanies any attempt to portray the entire earth on a single flat map, several projections have been developed that give creditable representations. On most of the commonly used global projections the Equator and the prime meridian are shown as straight lines that intersect at

Table of Azimuthal Projections

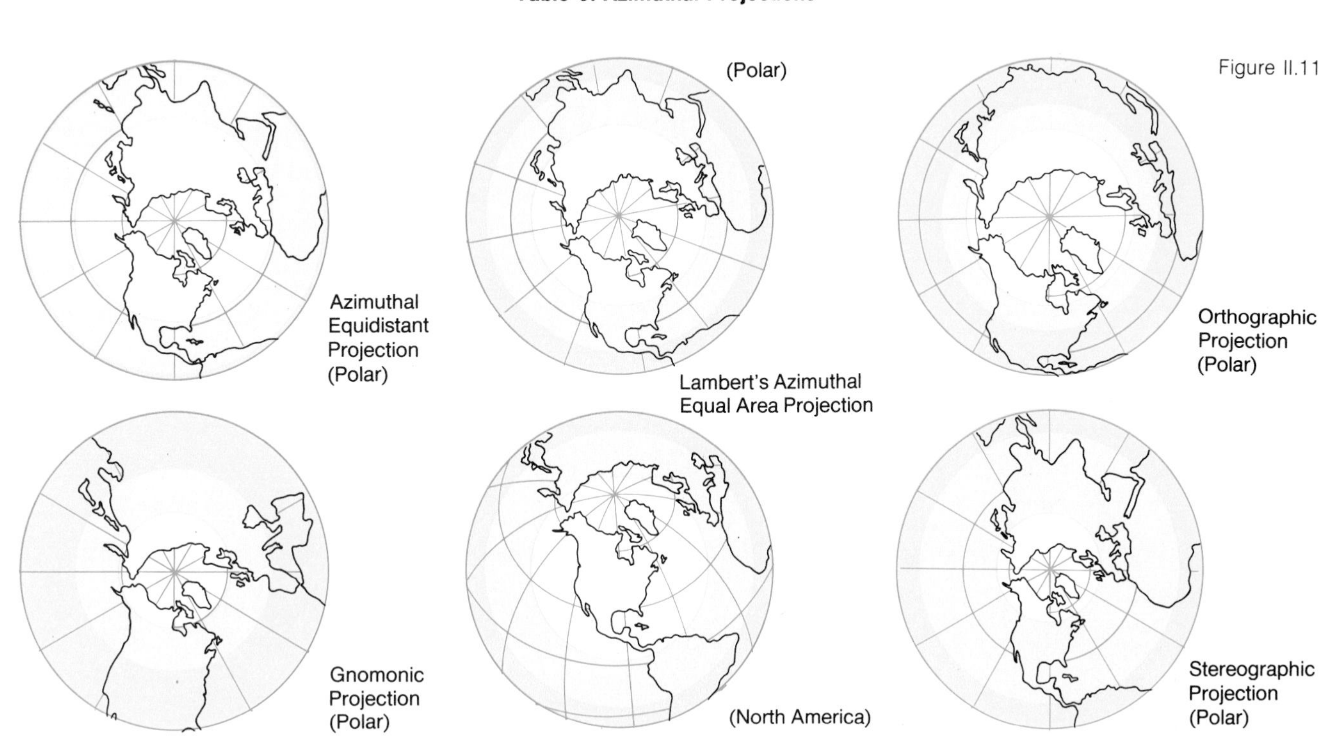

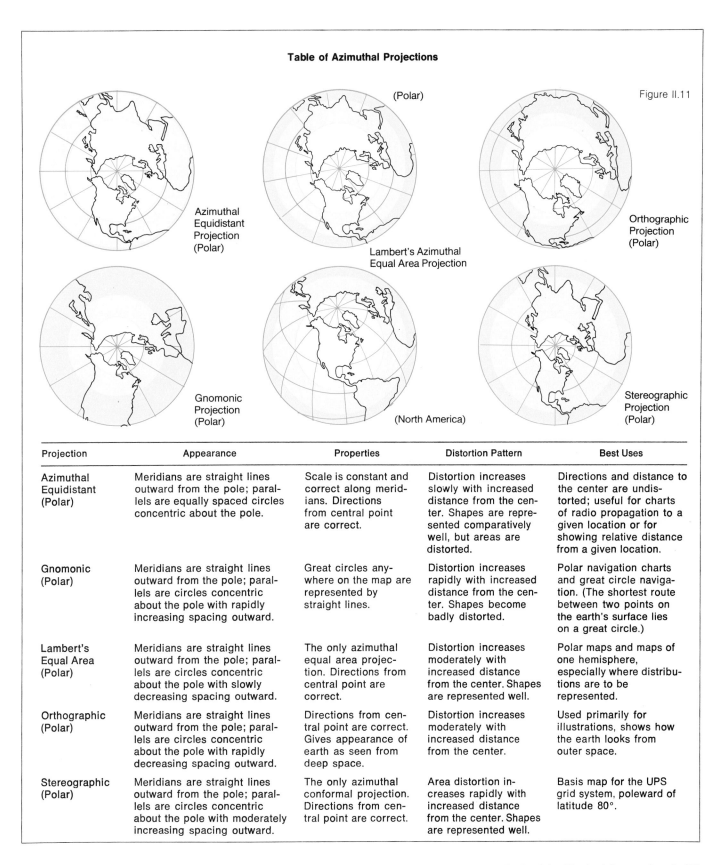

(Polar)
Figure II.11
Azimuthal Equidistant Projection (Polar)
Lambert's Azimuthal Equal Area Projection
Orthographic Projection (Polar)
Gnomonic Projection (Polar)
(North America)
Stereographic Projection (Polar)

Projection	Appearance	Properties	Distortion Pattern	Best Uses
Azimuthal Equidistant (Polar)	Meridians are straight lines outward from the pole; parallels are equally spaced circles concentric about the pole.	Scale is constant and correct along meridians. Directions from central point are correct.	Distortion increases slowly with increased distance from the center. Shapes are represented comparatively well, but areas are distorted.	Directions and distance to the center are undistorted; useful for charts of radio propagation to a given location or for showing relative distance from a given location.
Gnomonic (Polar)	Meridians are straight lines outward from the pole; parallels are circles concentric about the pole with rapidly increasing spacing outward.	Great circles anywhere on the map are represented by straight lines.	Distortion increases rapidly with increased distance from the center. Shapes become badly distorted.	Polar navigation charts and great circle navigation. (The shortest route between two points on the earth's surface lies on a great circle.)
Lambert's Equal Area (Polar)	Meridians are straight lines outward from the pole; parallels are circles concentric about the pole with slowly decreasing spacing outward.	The only azimuthal equal area projection. Directions from central point are correct.	Distortion increases moderately with increased distance from the center. Shapes are represented well.	Polar maps and maps of one hemisphere, especially where distributions are to be represented.
Orthographic (Polar)	Meridians are straight lines outward from the pole; parallels are circles concentric about the pole with rapidly decreasing spacing outward.	Directions from central point are correct. Gives appearance of earth as seen from deep space.	Distortion increases moderately with increased distance from the center.	Used primarily for illustrations, shows how the earth looks from outer space.
Stereographic (Polar)	Meridians are straight lines outward from the pole; parallels are circles concentric about the pole with moderately increasing spacing outward.	The only azimuthal conformal projection. Directions from central point are correct.	Area distortion increases rapidly with increased distance from the center. Shapes are represented well.	Basis map for the UPS grid system, poleward of latitude 80°.

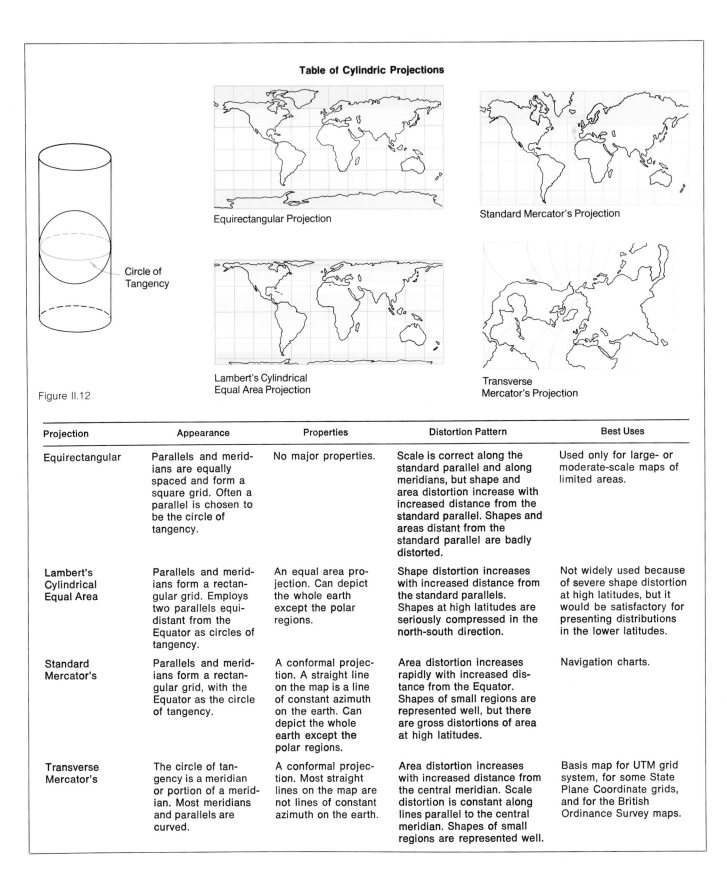

Table of Cylindric Projections

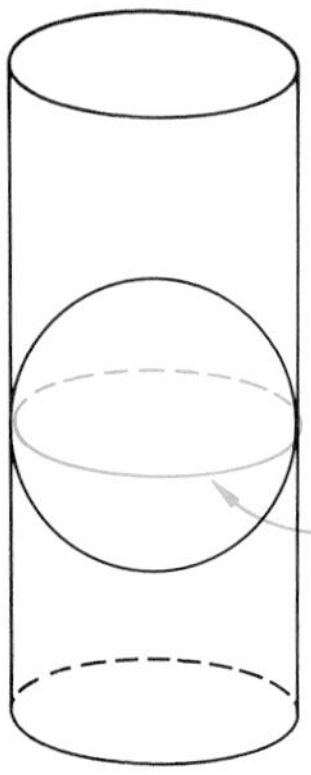

Circle of Tangency

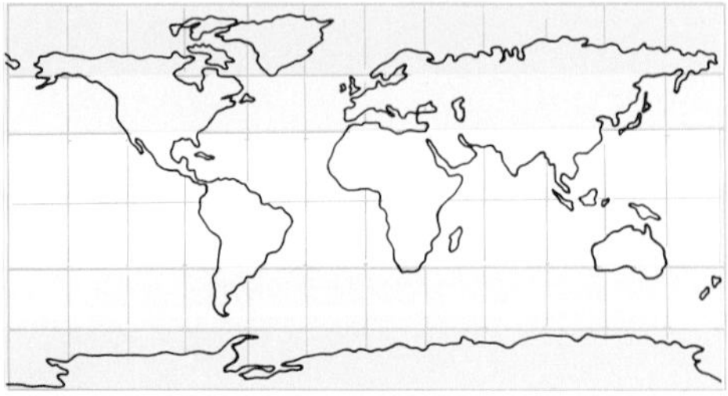

Equirectangular Projection

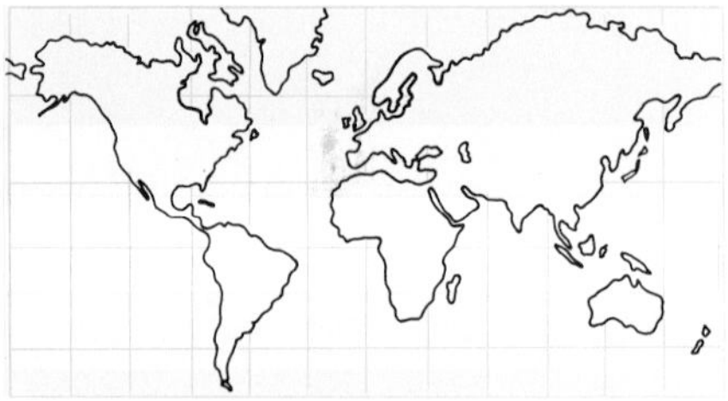

Standard Mercator's Projection

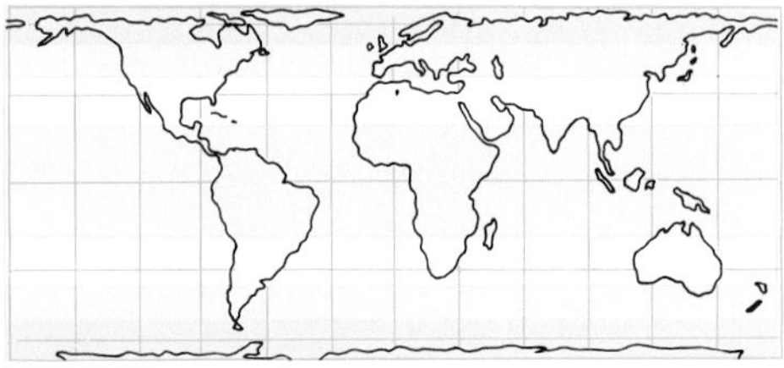

Lambert's Cylindrical Equal Area Projection

Transverse Mercator's Projection

Figure II.12

Projection	Appearance	Properties	Distortion Pattern	Best Uses
Equirectangular	Parallels and meridians are equally spaced and form a square grid. Often a parallel is chosen to be the circle of tangency.	No major properties.	Scale is correct along the standard parallel and along meridians, but shape and area distortion increase with increased distance from the standard parallel. Shapes and areas distant from the standard parallel are badly distorted.	Used only for large- or moderate-scale maps of limited areas.
Lambert's Cylindrical Equal Area	Parallels and meridians form a rectangular grid. Employs two parallels equidistant from the Equator as circles of tangency.	An equal area projection. Can depict the whole earth except the polar regions.	Shape distortion increases with increased distance from the standard parallels. Shapes at high latitudes are seriously compressed in the north-south direction.	Not widely used because of severe shape distortion at high latitudes, but it would be satisfactory for presenting distributions in the lower latitudes.
Standard Mercator's	Parallels and meridians form a rectangular grid, with the Equator as the circle of tangency.	A conformal projection. A straight line on the map is a line of constant azimuth on the earth. Can depict the whole earth except the polar regions.	Area distortion increases rapidly with increased distance from the Equator. Shapes of small regions are represented well, but there are gross distortions of area at high latitudes.	Navigation charts.
Transverse Mercator's	The circle of tangency is a meridian or portion of a meridian. Most meridians and parallels are curved.	A conformal projection. Most straight lines on the map are not lines of constant azimuth on the earth.	Area distortion increases with increased distance from the central meridian. Scale distortion is constant along lines parallel to the central meridian. Shapes of small regions are represented well.	Basis map for UTM grid system, for some State Plane Coordinate grids, and for the British Ordinance Survey maps.

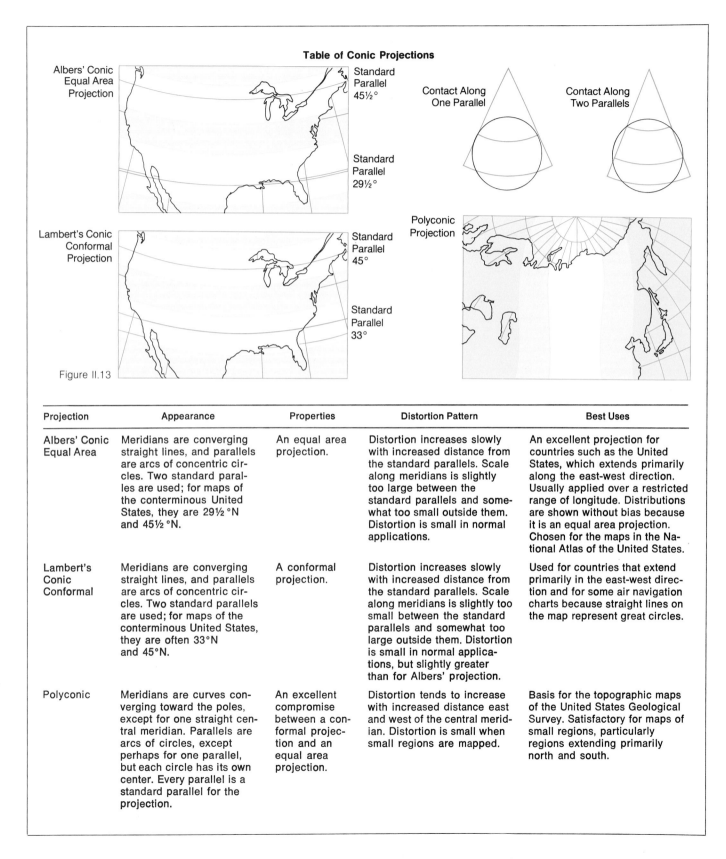

Table of Conic Projections

Albers' Conic Equal Area Projection

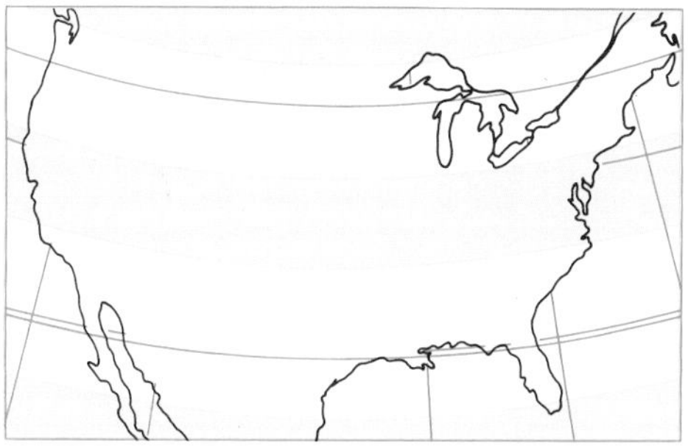

Standard Parallel 45½°

Standard Parallel 29½°

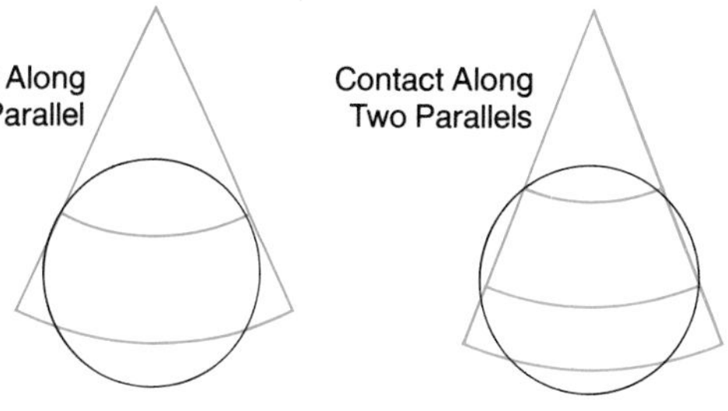

Lambert's Conic Conformal Projection

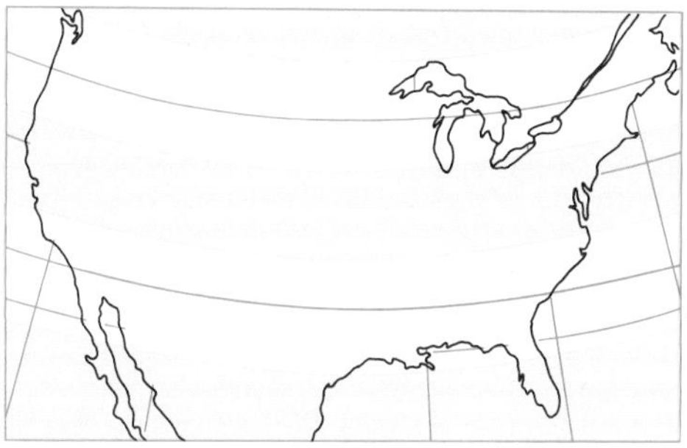

Standard Parallel 45°

Standard Parallel 33°

Contact Along One Parallel

Contact Along Two Parallels

Polyconic Projection

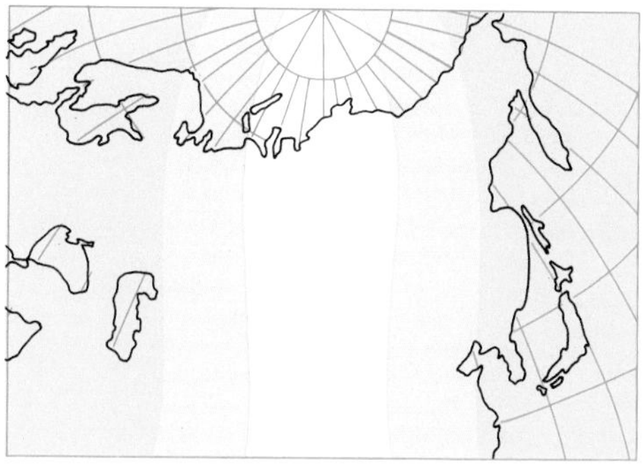

Figure II.13

Projection	Appearance	Properties	Distortion Pattern	Best Uses
Albers' Conic Equal Area	Meridians are converging straight lines, and parallels are arcs of concentric circles. Two standard paralles are used; for maps of the conterminous United States, they are 29½°N and 45½°N.	An equal area projection.	Distortion increases slowly with increased distance from the standard parallels. Scale along meridians is slightly too large between the standard parallels and somewhat too small outside them. Distortion is small in normal applications.	An excellent projection for countries such as the United States, which extends primarily along the east-west direction. Usually applied over a restricted range of longitude. Distributions are shown without bias because it is an equal area projection. Chosen for the maps in the National Atlas of the United States.
Lambert's Conic Conformal	Meridians are converging straight lines, and parallels are arcs of concentric circles. Two standard parallels are used; for maps of the conterminous United States, they are often 33°N and 45°N.	A conformal projection.	Distortion increases slowly with increased distance from the standard parallels. Scale along meridians is slightly too small between the standard parallels and somewhat too large outside them. Distortion is small in normal applications, but slightly greater than for Albers' projection.	Used for countries that extend primarily in the east-west direction and for some air navigation charts because straight lines on the map represent great circles.
Polyconic	Meridians are curves converging toward the poles, except for one straight central meridian. Parallels are arcs of circles, except perhaps for one parallel, but each circle has its own center. Every parallel is a standard parallel for the projection.	An excellent compromise between a conformal projection and an equal area projection.	Distortion tends to increase with increased distance east and west of the central meridian. Distortion is small when small regions are mapped.	Basis for the topographic maps of the United States Geological Survey. Satisfactory for maps of small regions, particularly regions extending primarily north and south.

Table of Pseudocylindric Projections

Flat Polar Quartic Equal Area Projection

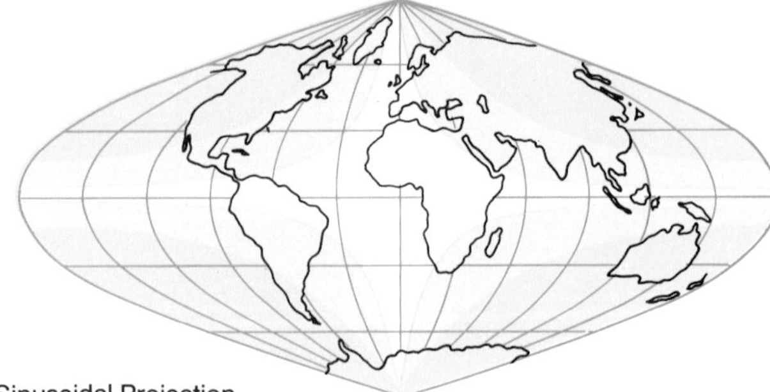

Sinusoidal Projection

Mollweide's Projection

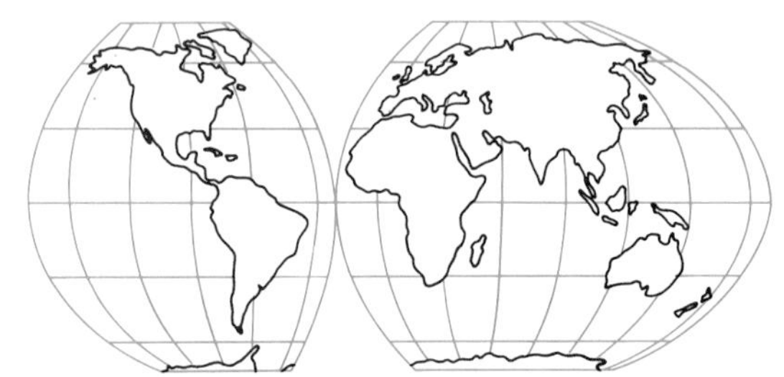

Interrupted Flat Polar Quartic Projection

Figure II.14

Projection	Appearance	Properties	Distortion Pattern	Best Uses
Flat Polar Quartic Equal Area	Parallels are straight parallel lines. Meridians are curves in general but the central meridian is straight. The poles are represented by straight lines one-third the length of the Equator. Meridians converge toward the poles and are equally spaced along each parallel. The spacing between parallels decreases slightly with increased latitude. The boundary of the map is a complex curve.	An equal area projection.	Distortion is least nearest the Equator and central meridian and greatest at high latitudes near the boundaries.	Generally useful as a world map and for depicting global distributions.
Mollweide's	Parallels are straight parallel lines. Meridians are elliptical in general, but the central meridian is a straight line half the length of the Equator. The meridians converge to a point at each pole and are equally spaced along each parallel. The spacing between parallels decreases slightly with increased latitude. The boundary of the map is an ellipse.	An equal area projection.	Distortion is least in midlatitude regions near the central meridian and greatest at high latitudes near the boundaries.	Generally useful as a world map and for depicting global distributions.
Sinusoidal (Sanson-Flamsteed)	Parallels are straight parallel, equally spaced lines. Meridians and the boundary are sinusoidal curves. The central meridian is straight and half the length of the Equator. Meridians converge to points at the poles. The length of each parallel is equal to its length on a globe of corresponding size.	An equal area projection.	Distortion is least nearest the Equator and the central meridian and greatest at high latitudes near the boundaries.	Somewhat inferior to other projections as a global map, but useful for maps of individual continents.

right angles at the center of the projection. In general, the regions of greatest distortion on such projections lie near the outer margins. Regions of least distortion are at the center of the projection, often at the intersection of the Equator and the prime meridian or near the prime meridian in the midlatitudes. Figure II.14 shows the properties of several pseudocylindric projections.

The *flat polar quartic* projection, an equal area projection, is the basis for many of the global distribution maps in this text (see page 69). In this projection, the poles are represented by lines one-third the length of the Equator, and the boundary of the map is a complex curve. Distortion on the flat polar quartic projection tends to be smallest near the intersection of the Equator and the prime meridian and greatest at high latitudes near the margins.

Interruption and Condensation of Projections. On a map showing global distributions, the landmasses of the earth are often of greater interest than the oceans. In such a case, the projection can be *interrupted* in the Atlantic and Pacific Oceans. The projection may then be fitted, or reprojected, to standard meridians in each of the subdivisions, so that every landmass is in a region of small distortion near a standard meridian, which improves the overall accuracy of the map. If ocean areas are of no interest in a particular application, portions of the Atlantic and Pacific Oceans can be omitted entirely, producing a *condensed interrupted map.* Condensation does not reduce the distortion of the map, but condensed maps make good use of available space in displaying landmasses.

Simplification in Maps

No map can be sufficiently detailed to show every minor aspect of a landscape. The smaller the scale of the map, the less the detail that can be presented on a map of given size. The map maker must therefore simplify the information that his map is to present.

Simplification, sometimes referred to as generalization, entails omitting detailed information that would fill a map with unreadable clutter, but not at the expense of obscuring the general character of the represented landscape. A large-scale map is not merely an enlarged portion of a small-scale map; in general, a large-scale map shows a greater amount of detail than does a small-scale map of the same region.

Generalization is a process that depends on the skill of the map maker and his ability to retain essential features on a map while eliminating detail. A map maker cannot, for example, show the individual peaks and slopes of the Andes Mountains on a global map. However, by artful choice of symbols he can communicate impressions of their height and ruggedness.

Symbolization: The Representation of Relief

A variety of graphic symbols represent natural and man-made objects on maps. Some symbols have a pictorial quality: railroad tracks, for example, are often represented by a line with regularly spaced cross ticks. Other symbols are purely conventional choices, and because there is no standard system of symbolization for maps, the map legend or symbol table should always be consulted.

The graphic methods used to denote objects such as rivers, roads, and buildings on a map must often be symbolic generalizations because of the difficulty of showing the object in true scale. A stream 80 feet wide would have to be rendered as a nearly invisible line 1/1,000 inch wide if scale were followed exactly on a map with a scale of 1:1,000,000. Instead, the map maker must employ a narrow yet easily visible line to portray the river. Similarly, the individual turns and bends of a river are averaged on a small-scale map, leaving only the larger curves to denote the stream's course.

The physical geographer interested in landforms is especially concerned with the symbols that depict vertical relief on maps. *Contour lines,* special kinds of *shading,* and *color tints* are some of the methods for indicating relief on maps. For quantitative work, particularly with large-scale maps, relief is represented most accurately by contour lines.

A contour line on a map represents a line of constant altitude above a chosen reference level, called a *datum plane,* such as mean sea level, the average height of the surface of the sea. Consider the hilly island in Figure II.15. The figure shows horizontal planes at a regular vertical spacing, or *contour interval,* of 200 feet. Each horizontal plane cuts the surface of the ground in a

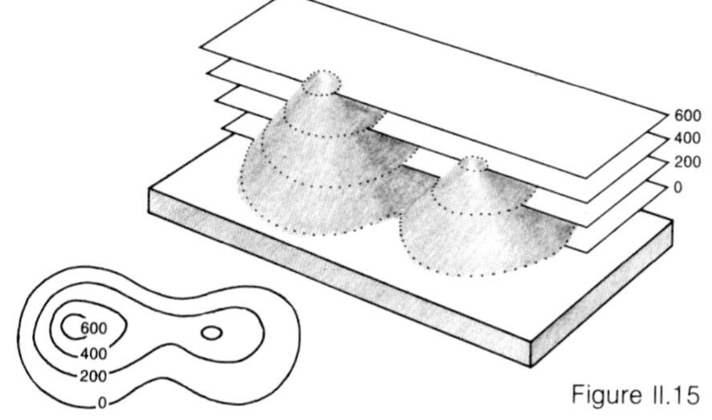

600
400
200
0
600
400
200
0

Figure II.15

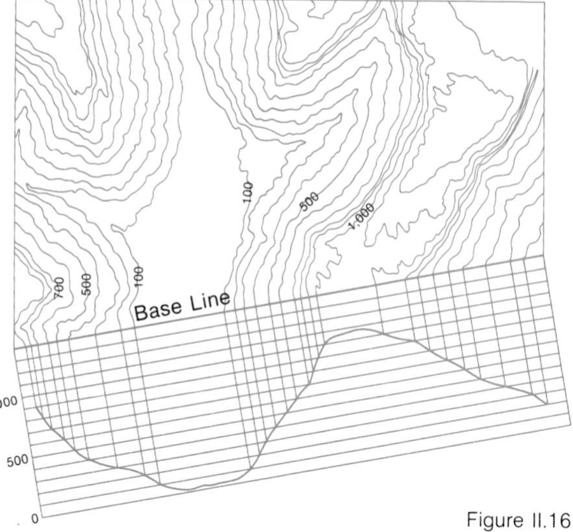
Base Line
100
500
1,000
700
500
100
Altitude (feet)
1,000
500
0

Figure II.16

curve, which is the contour line for that altitude; every point on a given contour line is the same altitude above the datum plane.

As shown in Figure II.16, contour lines can be transferred to a flat map to represent vertical relief. For accuracy without unnecessary clutter, the contour interval for a map should be chosen to be commensurate with the nature of the landscape being depicted; the contour interval will normally be larger for a mountainous region than for a gently sloping plain. The choice of contour interval also depends on the scale of the map. On a large-scale map, for example, the contour interval may be only a few feet, and the contour lines will show comparatively small features of slopes. A small-scale map of the same region will employ a larger contour interval, and in addition the small kinks and bends of the contours will be smoothed and averaged, so that less detail will be represented. To make contour lines easier to read, every fifth contour line is thickened and labeled with its altitude. The user of a contour map should bear in mind the value of the contour interval between adjacent contour lines.

As Figure II.16 indicates, the horizontal spacing of the contour lines on a contour map can be interpreted in terms of the local slope of the landform. Contour lines that are relatively close together represent a steeper slope than do contour lines that are relatively far apart. The convexity or concavity of a slope can also be inferred by inspecting the horizontal spacing of the contour lines. For quantitative purposes, a *topographic profile* of the landscape along a given direction can be prepared from a contour map according to the method illustrated in Figure II.16. The vertical scale of a topographic profile is usually exaggerated with respect to the horizontal scale in order to portray relief features more clearly. The altitudes of hilltops and the

Figure II.17 (above) The relation between contour lines and topography is illustrated in this landscape model. The model was cut from a block of plastic by a cutter set to trace contour lines from a map; the depth of cut was adjusted according to the elevation of each contour line. Although individual layers are distinguishable on the model, the closely spaced contours give a good impression of the topographic relief.

Figure II.18 (opposite) Relief can be depicted on maps by hachuring (left) or by shaded relief (right).

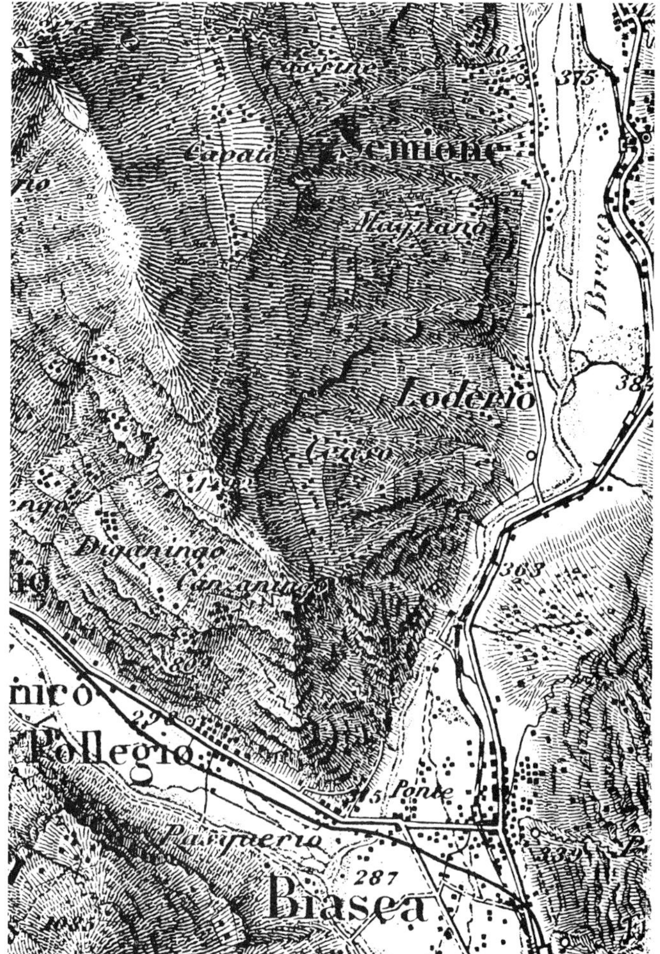
375
Loderio
363
Pollegio
Ponte
287
Biasca

FLATTOP MTN
TRAIL
Bear Lake Ranger Station
Nymph Lake
Flattop Mtn
12324
Tyndall
Emerald Lake
Dream Lake
Tyndall Gorge
Tyndall Cr
Tyndall Glacier
Hallett Peak
12713
Lake Haiyaha
Creek
Chaos Canyon
Chaos Cascades
Chaos
Chaos
Canyon
Glacier
TUNNEL
11932
Otis Peak
12486
Glacier Falls
The Loch
Andrews Tarn
Andrews Cr
Brook
Mills Lake
Andrews Pass
Andrews Glacier
Creek
Embryo Lake
The Gash
Vale
ADAMS
Timberline Falls
Loch
The Sharkstooth
Glass Lake
Thatchtop
12668
Sky Pond
Icy
Taylo Peak
13153
Shelf Lake
Shelf
Solitude Lake
Taylor Glacier
Arrowhead
Powell Peak
McHenrys Notch
McHenrys Peak
GRAND CO
Stone Man Pass
Lake Powell
The Spearh

bottoms of depressions are sometimes stated explicitly as *spot heights* on a map.

In addition to contour lines and spot heights, a variety of qualitative or partly quantitative artistic techniques can be used on a map to indicate relief. In *hachuring*, a method popular among cartographers in the nineteenth century, a slope is depicted by straight line segments drawn along the direction of maximum steepness of slope, so that hachure marks are at right angles to contours. The marks are drawn so that their density, either in width or spacing, increases with increased steepness; steep slopes appear darker than gentle slopes in hachuring. Hachured maps present a pleasing appearance when well drawn, but they are seldom used now because of the difficulty of reading slope angle from them and because of the labor involved in their construction.

A modern method of symbolizing relief on a map is to add shading to give an impression of height and depth. Such *shaded relief* is commonly drawn as though the area were illuminated from the upper left, or the "northwest," corner of the map.

Altitude tinting, the use of color tints for successive ranges of altitudes, is often employed, particularly on small-scale maps where contour lines cannot be read easily. Green is usually used for altitudes near sea level, and in a typical system, colors for higher altitudes progress through yellow, orange, brown, and blue. The green tint used for low altitudes should not be taken as indicative of vegetation cover. The methods of hachure, shaded relief, and altitude tinting are sometimes used in combination on a contour map to depict the general character of a landscape while retaining the quantitative accuracy of contours.

Three-dimensional models with exaggerated vertical relief are mass-produced by molding thin sheets of plastic; on large-scale models, the quality of the molding is sufficiently good to represent the landscape accurately. Global relief maps sometimes are prepared from photographs of three-dimensional models, which are illuminated from the upper left to emphasize relief by lights and shadows.

A *block diagram,* such as the one in Figure II.19, is a means of representing landforms and their underlying

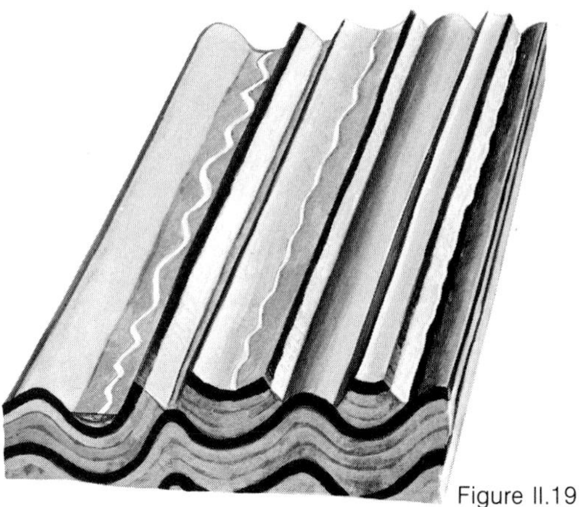
Figure II.19

geologic structure in one diagram. A number of block diagrams are used in this text.

Strictly speaking, a block diagram is not a map, because it does not show position directly, but the landforms depicted on block diagrams are usually drawn with reference to contour maps of the region. The vertical scale on a block diagram is usually designed to be the same as the horizontal scale, in order to portray the geologic structures in proper relation to the landforms. The landform drawings in a block diagram often omit features such as vegetation to give the structures of interest greater emphasis.

Mapping in the United States

Regional and city planning and the development of a country's resources are closely tied to the availability of specialized maps. The study of geographic features such as soils, landforms, and water usage is also largely dependent on information compiled in the form of maps. In addition to maps of topography and location required for navigation and planning, government agencies prepare maps of water, soil, geologic features, timber, and other features. Maps are also employed to report statistical, economic, or demographic information; a map devoted to a single topic, such as the distribution of population or rainfall, is known as a *thematic* map.

The responsibility for mapping the United States is divided among many civil and military agencies, depending on the purpose of the map. Only a few examples are given here to indicate the extent of the mapping services carried on by the government.

The United States Geological Survey (USGS) publishes several series of topographic maps in different scales that cover the United States, Puerto Rico, and other territories. The USGS Topographic Maps are discussed in detail in the following section of this appendix because of their particular importance to physical geographers.

Maps related to navigation purposes are known as *charts*. The preparation of nautical and aeronautical charts is carried out by several agencies. Nautical charts of foreign waters are prepared by the Naval Oceanographic Office (formerly the Hydrographic Office of the Navy). This Office also issues small-scale bathymetric charts that portray the topography of much of the ocean floor. The National Ocean Survey (formerly the United States Coast and Geodetic Survey) prepares nautical navigation charts of United States coastal and offshore waters, and some inland waterways, using Mercator's projection and a range of scales. The harbor charts prepared by the Ocean Survey may have scales as large as 1:18,000 to show accurate locations for important features such as main channels, marker buoys, and underwater cables. The Army Corps of Engineers issues nautical charts of the Great Lakes and navigation charts for numerous rivers and inland waterways.

Aeronautical charts are produced by several agencies, including the National Ocean Survey and the United States Air Force Aeronautical Chart and Information Center. The scales employed vary from 1:250,000 on charts for aircraft of moderate speed to 1:2,000,000 and smaller for the needs of global jet transport. Special features of aeronautical charts include identification of airports and air navigation radio beacons, and a simplified representation that emphasizes major relief features and landmarks easily visible from the air. Aeronautical charts are revised frequently from aerial photographs and can be helpful to geographers.

The Geological Survey publishes several types of maps useful in physical geography. In addition to topographic maps, the Geological Survey publishes maps of national forests and grasslands, geologic maps, and hydrologic maps for water-resource planning. The geologic maps, which indicate the distribution of mineral deposits, fault zones, and landforms, are available for approximately half the United States at a scale of 1:250,000; a project to produce geologic maps for the whole country at a scale of 1:24,000 is under way. The hydrologic atlases prepared by the Geological Survey detail the expected flow rate from wells sunk into groundwater supplies. Other maps in hydrologic atlases show the extent of floods for help in land-use planning. The Map Information Office of the Geological Survey is a useful source of information concerning the availability of maps and aerial photographs.

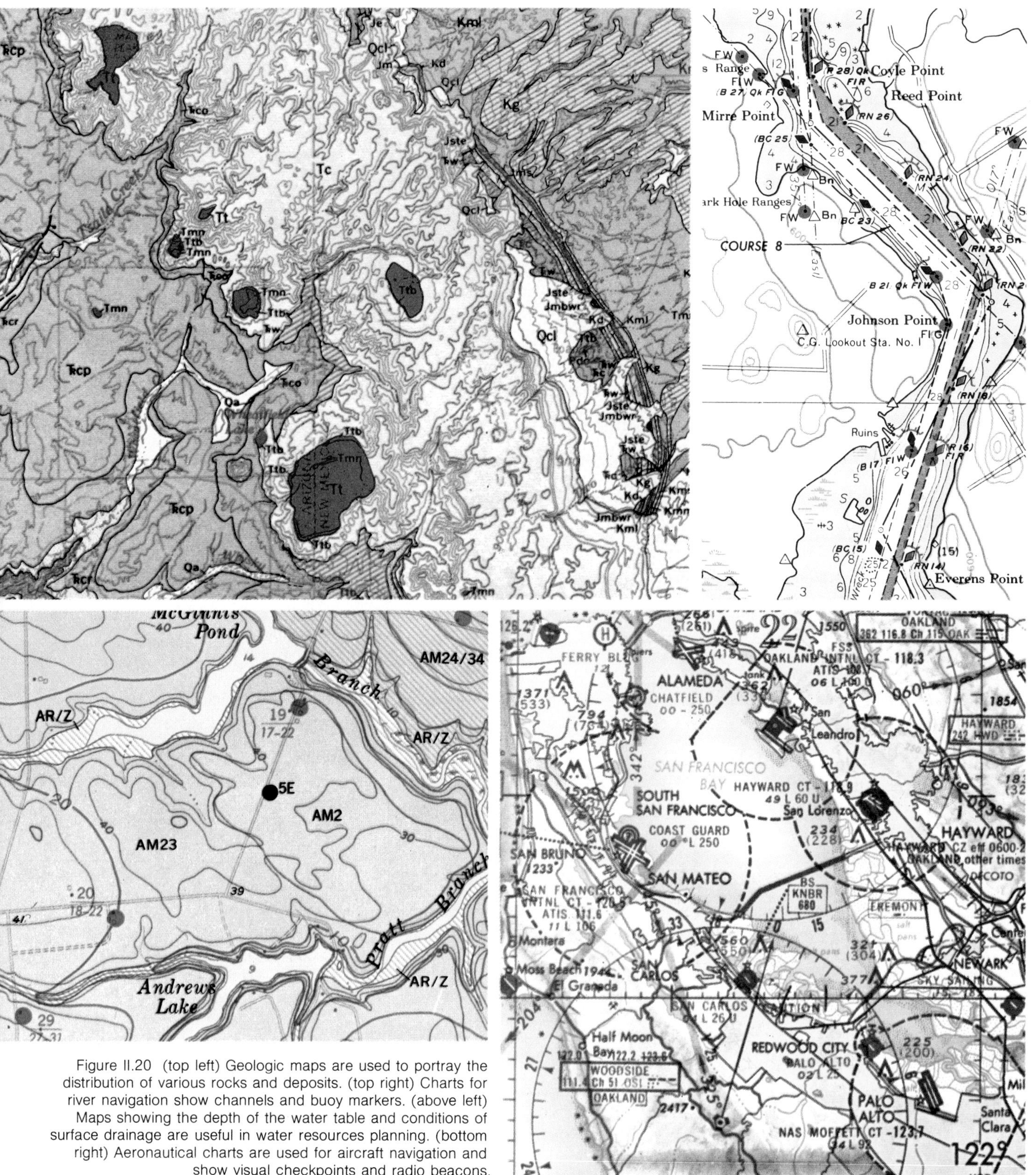
Tc
Kg
Qcl
Ttb
Tt
Tmn
Qa
Coyle Point
Reed Point
Mirre Point
COURSE 8
Johnson Point
C.G. Lookout Sta. No. 1
Ruins
Everens Point
McGinnis Pond
Branch
AM24/34
AR/Z
5E
AM2
AM23
Pratt Branch
Andrews Lake
ALAMEDA
CHATFIELD
San Leandro
SAN FRANCISCO BAY
SOUTH SAN FRANCISCO
COAST GUARD
HAYWARD
San Lorenzo
SAN MATEO
SAN BRUNO
Montara
Moss Beach
El Granada
SAN CARLOS
REDWOOD CITY
PALO ALTO
Half Moon Bay
WOODSIDE
OAKLAND
NEWARK
Santa Clara

Figure II.20 (top left) Geologic maps are used to portray the distribution of various rocks and deposits. (top right) Charts for river navigation show channels and buoy markers. (above left) Maps showing the depth of the water table and conditions of surface drainage are useful in water resources planning. (bottom right) Aeronautical charts are used for aircraft navigation and show visual checkpoints and radio beacons.

USGS Topographic Maps

A *topographic map* is a graphic representation of selected man-made and natural features of a portion of the earth's surface, plotted to a definite scale. In particular, topographic maps show vertical relief. The United States has maintained a topographic map program since 1879 under the direction of the Geological Survey. The Geological Survey publishes the National Topographic Map Series, an invaluable source of information for physical geographers in the United States who are concerned with landforms or who must plan and execute field studies. The maps are compiled primarily from aerial photographs, but field surveys are sometimes used to verify details of photo interpretation.

A *map series* is a family of maps produced according to the same general specifications or with some features, such as scale, in common. The Topographic Series is itself made up of several series of maps. Each of the maps is drawn so that its sides coincide with parallels of latitude or meridians of longitude, and a similar symbolism is used in all series. Table II.3 lists the principal series comprising the Topographic Series, together with some of their properties. In addition to the series listed, there is a set covering national parks, monuments, and historic sites, a set for certain metropolitan areas, a set for rivers and floodplains, and sets for Puerto Rico, Antarctica, and other areas.

Table II.3. Principal Map Series in the National Topographic Map Series

Series and Scale	1 inch Represents	Projection	Angular Size (latitude-longitude)	Area Covered* (square miles)	Remarks
7½-Minute Quadrangle 1:24,000	2,000 feet	Polyconic	7½ minutes by 7½ minutes (⅛° by ⅛°)	49 to 70	Available for most of Hawaii, much of 48 states, and only a small part of Alaska. To meet national standards of map accuracy, at least 90 percent of mapped objects must be shown within 40 feet of true location. Also available without contour lines and woodland color tint.
15-Minute Quadrangle 1:62,500	0.986 mile	Polyconic	15 minutes by 15 minutes (¼° by ¼°)	197 to 282	Available for most of the United States except Alaska. Compiled from same data as 7½-minute series, but show less detail because of smaller scale. To meet standards of accuracy, at least 90 percent of mapped objects must be shown within 100 feet of true location. Also available without contour lines and woodland color tint.
Alaska 1:63,360	1 mile	Polyconic	15 minutes by 20 to 30 minutes	207 to 281	Covers most of Alaska. Comparable in scale and content to 15-minute series. More detailed 1:24,000 data are not available. Also available without contour lines and woodland color tint.
United States 1:250,000	3.95 miles	Transverse Mercator	Usually 1° by 2°	4,580 to 8,669	The conterminous United States and Hawaii are completely mapped in 473 separate maps. Alaska is completely mapped in 153 separate maps.
State Maps 1:500,000	7.89 miles	Lambert conic conformal		Individual states of the conterminous United States	Also available for most states as base maps without relief, as topographic maps, and as maps with shaded relief. Base maps for the conterminous United States are also available with a scale of 1:1,000,000.
United States 1:1,000,000	15.78 miles	Lambert conic conformal	4° by 6° (4° by 12° for Alaska)	73,734 to 122,066	Maps in this series conform to specifications for the International Map of the World. Prior to 1962 the chosen projection was a modified polyconic, but the Lambert conic conformal has since been adopted. Available for Alaska and most of the conterminous United States.

*Area covered depends on latitude.

Source: United States Department of the Interior Geological Survey. 1970. *The National Atlas of the United States of America.* Washington, D.C.: United States Department of the Interior.

Using USGS Topographic Maps. The 7½-minute quadrangle series and 15-minute quadrangle series are large-scale maps suited to field survey work and landform studies on a local scale. Such maps contain a large amount of information, often in abbreviated or symbolic form.

Topographic maps employ an extensive set of symbols to depict natural and man-made features. The USGS topographic map symbol table, reproduced in Figure II.21, is not printed on the map sheets but is available separately. The colors on a topographic map are an integral part of the symbolization. As Table II.4 shows, each color is restricted to a particular class of symbols; blue, for example, indicates water features, and a red tint indicates urban areas where only landmark buildings are mapped individually.

Table II.4. Color Code for USGS Topographic Maps

Color	Uses
Black	Man-made objects, names, notes, most labels
Blue	Water features, including streams, lakes, canals, and marshes
Brown	Contour lines and relief features
Green	Vegetation of various types: woodlands, orchards, vineyards, brush, and thick ground cover
Purple	On interim maps, areas that have been revised since the previous edition
Red	Urban areas, land boundaries, and some roads

Source: Adapted from USGS Symbol Table.

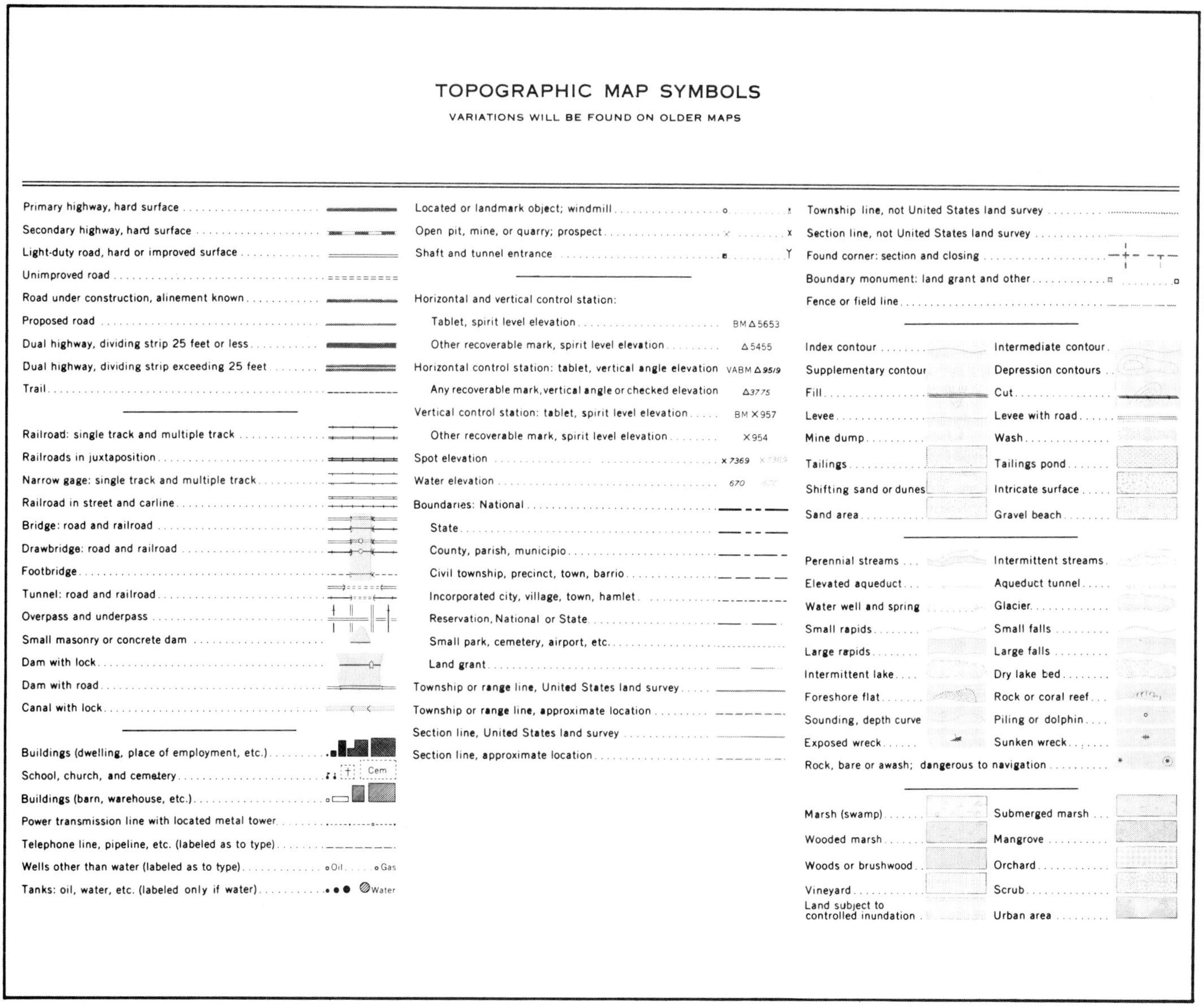

TOPOGRAPHIC MAP SYMBOLS

VARIATIONS WILL BE FOUND ON OLDER MAPS

Primary highway, hard surface
Secondary highway, hard surface
Light-duty road, hard or improved surface
Unimproved road
Road under construction, alinement known
Proposed road
Dual highway, dividing strip 25 feet or less
Dual highway, dividing strip exceeding 25 feet
Trail

Railroad: single track and multiple track
Railroads in juxtaposition
Narrow gage: single track and multiple track
Railroad in street and carline
Bridge: road and railroad
Drawbridge: road and railroad
Footbridge
Tunnel: road and railroad
Overpass and underpass
Small masonry or concrete dam
Dam with lock
Dam with road
Canal with lock

Buildings (dwelling, place of employment, etc.)
School, church, and cemetery — Cem
Buildings (barn, warehouse, etc.)
Power transmission line with located metal tower
Telephone line, pipeline, etc. (labeled as to type)
Wells other than water (labeled as to type) — Oil, Gas
Tanks: oil, water, etc. (labeled only if water) — Water

Located or landmark object; windmill
Open pit, mine, or quarry; prospect
Shaft and tunnel entrance

Horizontal and vertical control station:
Tablet, spirit level elevation — BM Δ5653
Other recoverable mark, spirit level elevation — Δ5455
Horizontal control station: tablet, vertical angle elevation — VABM Δ9519
Any recoverable mark, vertical angle or checked elevation — Δ3775
Vertical control station: tablet, spirit level elevation — BM × 957
Other recoverable mark, spirit level elevation — × 954
Spot elevation — × 7369
Water elevation — 670
Boundaries: National
State
County, parish, municipio
Civil township, precinct, town, barrio
Incorporated city, village, town, hamlet
Reservation, National or State
Small park, cemetery, airport, etc.
Land grant
Township or range line, United States land survey
Township or range line, approximate location
Section line, United States land survey
Section line, approximate location

Township line, not United States land survey
Section line, not United States land survey
Found corner: section and closing
Boundary monument: land grant and other
Fence or field line

Index contour — Intermediate contour
Supplementary contour — Depression contours
Fill — Cut
Levee — Levee with road
Mine dump — Wash
Tailings — Tailings pond
Shifting sand or dunes — Intricate surface
Sand area — Gravel beach

Perennial streams — Intermittent streams
Elevated aqueduct — Aqueduct tunnel
Water well and spring — Glacier
Small rapids — Small falls
Large rapids — Large falls
Intermittent lake — Dry lake bed
Foreshore flat — Rock or coral reef
Sounding, depth curve — Piling or dolphin
Exposed wreck — Sunken wreck
Rock, bare or awash; dangerous to navigation

Marsh (swamp) — Submerged marsh
Wooded marsh — Mangrove
Woods or brushwood — Orchard
Vineyard — Scrub
Land subject to controlled inundation — Urban area

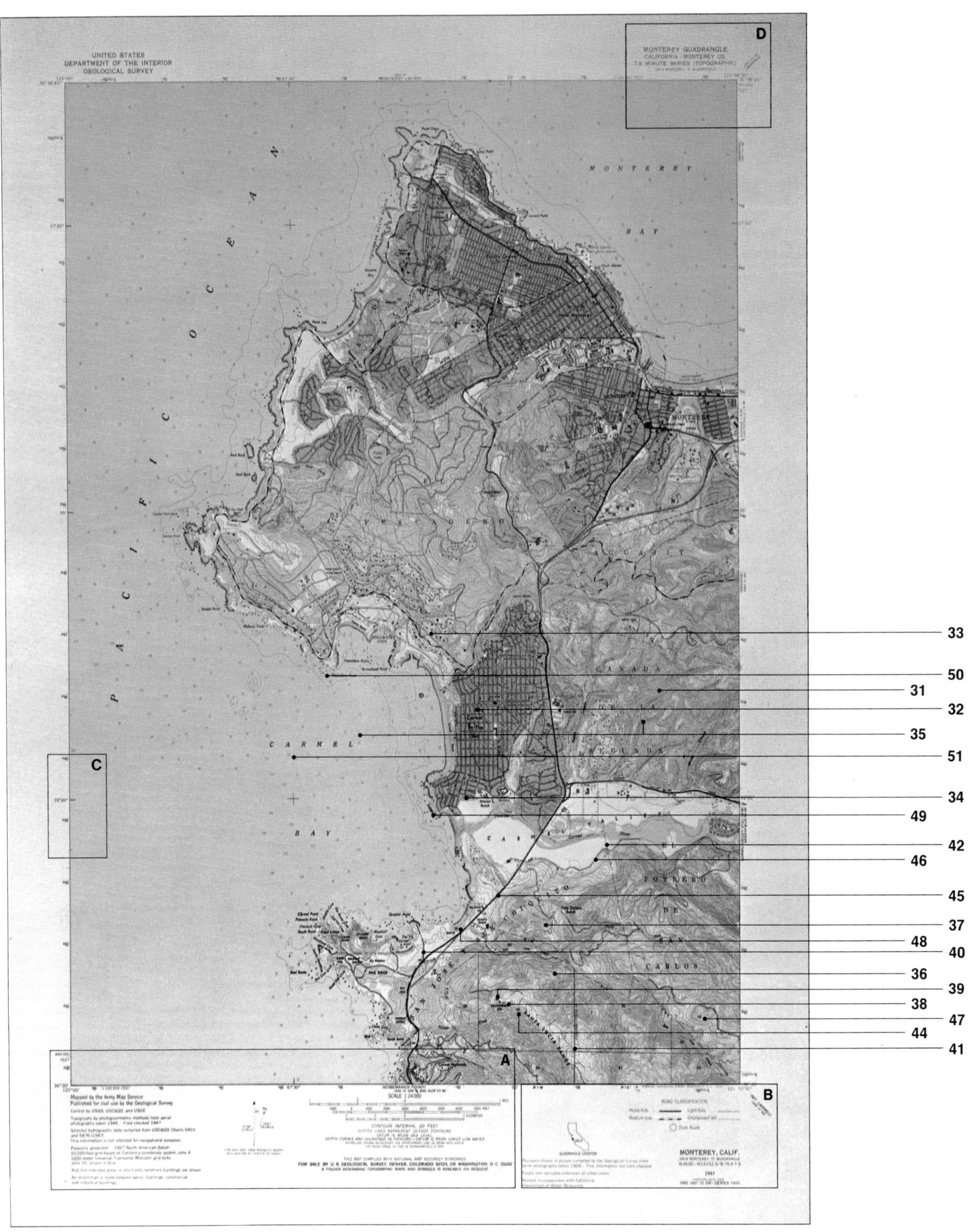
UNITED STATES
DEPARTMENT OF THE INTERIOR
GEOLOGICAL SURVEY
D
MONTEREY QUADRANGLE
CALIFORNIA - MONTEREY CO.
7.5 MINUTE SERIES (TOPOGRAPHIC)
P A C I F I C O C E A N
M O N T E R E Y
B A Y
C A R M E L
B A Y
C
A
B
33
50
31
32
35
51
34
49
42
46
45
37
48
40
36
39
38
47
44
41
SCALE 1:24000
CONTOUR INTERVAL 20 FEET
MONTEREY, CALIF.
1947

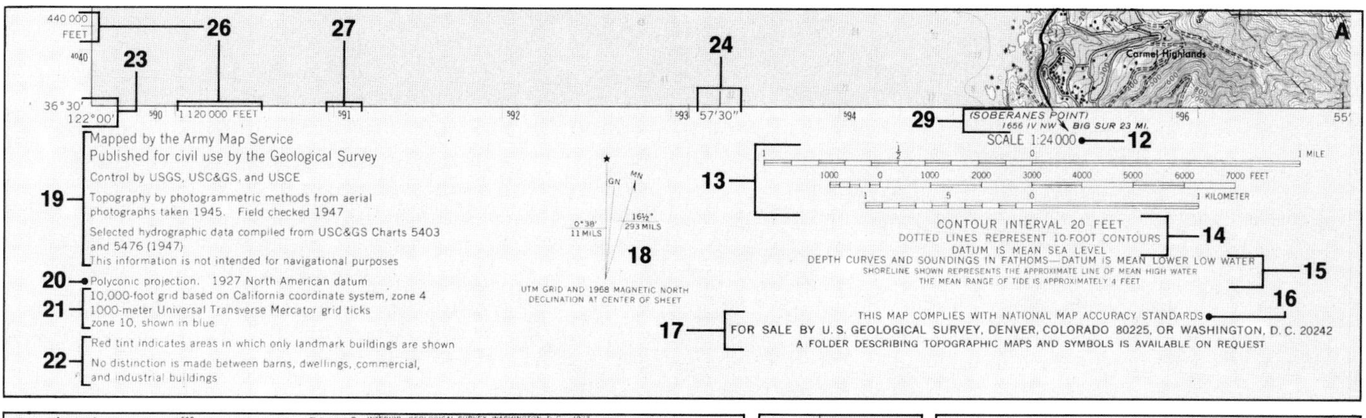
A
440 000
FEET
4040
36°30′
122°00′
590
1 120 000 FEET
591
592
593
57′30″
594
595
596
55′
26
27
23
24
Carmel Highlands
29
(SOBERANES POINT)
1656 IV NW
BIG SUR 23 MI.
SCALE 1:24 000
12
19
Mapped by the Army Map Service
Published for civil use by the Geological Survey
Control by USGS, USC&GS, and USCE
Topography by photogrammetric methods from aerial photographs taken 1945. Field checked 1947
Selected hydrographic data compiled from USC&GS Charts 5403 and 5476 (1947)
This information is not intended for navigational purposes
20
Polyconic projection. 1927 North American datum
21
10,000-foot grid based on California coordinate system, zone 4
1000-meter Universal Transverse Mercator grid ticks zone 10, shown in blue
22
Red tint indicates areas in which only landmark buildings are shown
No distinction is made between barns, dwellings, commercial, and industrial buildings
GN
MN
0°38′
11 MILS
16½°
293 MILS
18
UTM GRID AND 1968 MAGNETIC NORTH DECLINATION AT CENTER OF SHEET
13
1 MILE
7000 FEET
1 KILOMETER
CONTOUR INTERVAL 20 FEET
DOTTED LINES REPRESENT 10-FOOT CONTOURS
DATUM IS MEAN SEA LEVEL
14
DEPTH CURVES AND SOUNDINGS IN FATHOMS—DATUM IS MEAN LOWER LOW WATER
SHORELINE SHOWN REPRESENTS THE APPROXIMATE LINE OF MEAN HIGH WATER
THE MEAN RANGE OF TIDE IS APPROXIMATELY 4 FEET
15
16
THIS MAP COMPLIES WITH NATIONAL MAP ACCURACY STANDARDS
17
FOR SALE BY U. S. GEOLOGICAL SURVEY, DENVER, COLORADO 80225, OR WASHINGTON, D. C. 20242
A FOLDER DESCRIBING TOPOGRAPHIC MAPS AND SYMBOLS IS AVAILABLE ON REQUEST

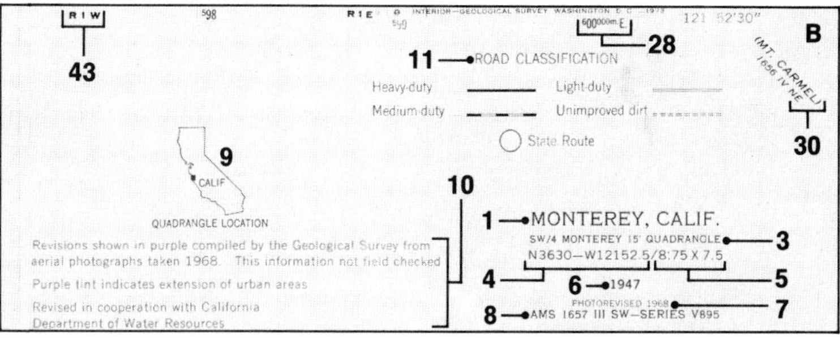
R 1 W
43
598
R 1 E
121°52′30″
B
28
11
ROAD CLASSIFICATION
Heavy-duty
Light-duty
Medium-duty
Unimproved dirt
State Route
(MT. CARMEL) 1656 IV NE
30
9
CALIF.
QUADRANGLE LOCATION
10
Revisions shown in purple compiled by the Geological Survey from aerial photographs taken 1968. This information not field checked
Purple tint indicates extension of urban areas
Revised in cooperation with California Department of Water Resources
1
MONTEREY, CALIF.
SW/4 MONTEREY 15′ QUADRANGLE
3
N3630—W12152.5/8:75 X 7.5
4
5
6
1947
PHOTOREVISED 1968
7
8
AMS 1657 III SW—SERIES V895

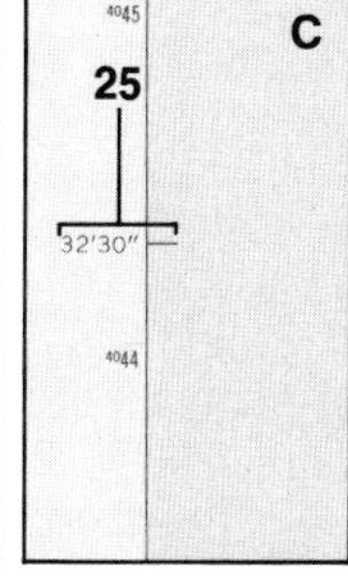
C
4045
25
32′30″
4044

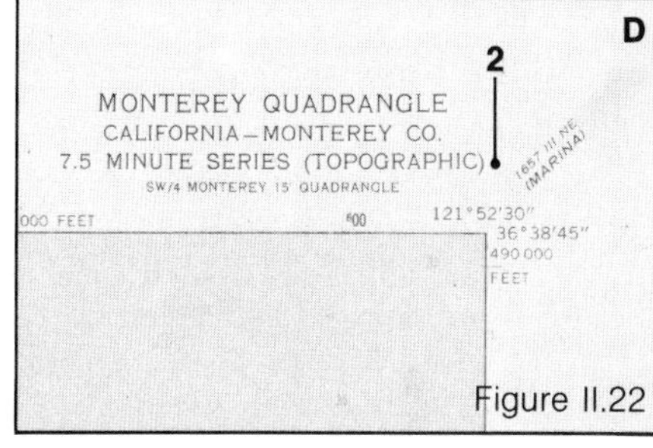
D
2
MONTEREY QUADRANGLE
CALIFORNIA—MONTEREY CO.
7.5 MINUTE SERIES (TOPOGRAPHIC)
SW/4 MONTEREY 15′ QUADRANGLE
(MARINA) 1657 III NE
000 FEET
600
121°52′30″
36°38′45″
490 000
FEET
Figure II.22

The margins of topographic maps contain latitude and longitude markings, references to grid coordinate systems, scales, and much other helpful data. Figure II.22 takes the example of the Monterey, California, 7½-minute quadrangle topographic map and shows the location and meaning of several dozen pieces of information that are typically found on maps in the 7½-minute and 15-minute series. Items numbered on the map are explained in the following list:

1. Name of map sheet
2. Name of map series
3. This map is the southwest portion of the Monterey 15-minute quadrangle
4. Latitude and longitude of southeast corner of map
5. Extension of map in minutes of latitude and longitude
6. Date of edition
7. Date of revision
8. Army Map Service sheet number and series number
9. Location of quadrangle in the state
10. Note on revision
11. Road symbol table
12. Scale of map as a representative fraction
13. Graphic scales in several units of length
14. Note on contour interval and contours
15. Note on depth curves, soundings, and shorelines (1 fathom equals 6 feet)
16. Statement of map accuracy
17. Central offices from which map is available
18. Declination diagram showing true north, UTM grid north, and 1968 magnetic north (1 mil equals 3.375 minutes)
19. Credits, sources, and dates of preparation
20. Projection
21. Grids referred to on this map
22. Note on symbolization
23. Latitude and longitude reference for southwest corner
24. Longitude tick mark based on reference at left margin
25. Latitude tick mark based on reference at bottom margin
26. California state coordinate grid reference points in feet
27. UTM grid reference (blue tick marks)
28. UTM grid reference 600,000 meters east of origin
29. Name and Army Map Service sheet number of neighboring map sheet in same series to the south
30. Name and Army Map Service sheet number of neighboring map sheet in same series to the southeast
31. Green tint indicates woodland
32. Red tint indicates urban areas; only major buildings or landmarks such as churches and schools are shown
33. Isolated buildings are indicated in sparsely settled areas
34. Purple tint indicates areas that have been revised since the previous edition
35. Blue is used for water, such as intermittent streams and the ocean
36. Contour lines are brown; every fifth contour line is thickened and marked with its altitude
37. Unchecked spot height
38. Checked spot height
39. Bench mark, a point with carefully measured location and (usually) altitude that is used in surveying
40. Boundary of state reservation
41. Principal meridian for laying out townships for United States land survey
42. Spanish land-grant survey boundary (meets and bounds system)
43. Township coordinate, Range 1 West
44. Section number
45. Heavy-duty road
46. Unimproved dirt road
47. Quarry
48. Sand
49. Shoreline rocks
50. Bare rocks indicated by star symbol
51. Depth sounding in fathoms

Field Work and Mapping

The physical geographer engaged in the study of streams, slopes, land use, or soils may find that the information he requires for the analysis of an area is not available from published sources. The smallest slope details may not be indicated on maps of the area of interest, for example. Or the geographer may be concerned with phenomena occurring during a short span of time, such as the erosion of stream banks during spring flood conditions, the study of which may require daily or weekly observation. In such cases, the geographer must make his own survey of the area. Because maps are often the simplest and best way to record and display the data of interest, geographers need to be familiar with the field study and mapping techniques suitable for their purposes. This section discusses some of the observational methods employed to study regions of small size using limited manpower and instrumentation.

An Approach to Field Work. After the geographer has decided that field study in a small region is necessary for the solution of a particular research problem, some preliminary planning must be done. Circumstances in the field may dictate later changes, but it is essential to begin a field study with a definite plan in order to avoid wasteful, undirected effort. The first step in planning field work is to choose a region for study suited to the problem at hand. Maps and aerial photographs can be helpful in locating a suitable region, but first-hand inspection of possible sites may also be needed. The size of the site chosen depends on the nature of the research problem and on the available resources. A study of stream erosion processes, for example, might involve an intensive study of only a few hundred feet of stream bank, whereas a study of land use in farming country might entail a broad survey covering several square miles.

Usually the geographer is interested in studying the land use of an area, in the spatial distribution of a quantity, such as a tree species, or in some aspect of the physical landscape. Data from such field studies are often best compiled in the form of a map, and proper execution of the map is therefore central to the success of the study. The scale of the map is determined by the needs of the problem and the degree of detail that must be represented. A study of erosion processes on a slope may, for example, require the mapping and description of every few square feet of slope, whereas the basic unit of area in a local soil survey might be several acres in extent. If the area to be studied is a few square miles or so, the geographer may be able to record data directly on a map such as a USGS topographic map or on an aerial photograph of the region. Many of the map series in the USGS Topographic Map Series are available without contour lines and woodland tint, making them suitable as base maps for some studies. However, the scale of USGS topographic maps is not large enough to represent in detail areas of a few thousand square meters or less, and the geographer may have to prepare his own maps.

Before going into the field, the geographer must give some consideration as to how the data are to be selected and classified. Every map is the result of selection and generalization, and the geographer must ensure that the required data are obtained without making serious omissions and without collecting quantities of irrelevant information. If an extensive field study is planned, the data-gathering procedures should be tested beforehand on a small part of the region. The data from the test run should be analyzed as completely as possible to determine whether the proposed field techniques are adequate for gathering the required data.

Mapping Techniques. Simple mapping techniques suitable for field use include *field sketching,* the *compass traverse,* and use of the *plane table.* A field sketch is a panoramic sketch of a local region; it can be used to show landforms and the relation of landforms to their surroundings and can be labeled to map out spatial distributions of vegetation, soils, and so forth. For many purposes a field sketch is superior to a photograph of the region, because the geographer can emphasize important features in a sketch.

A field sketch need not be executed artistically to be of value in a field study; often a simple uncluttered drawing is all that is required. The rule in sketching is to begin with the main outlines of a landscape, such as the skyline, and work toward smaller details. To show the correct relative locations of features, it is helpful to draw a horizontal line on the sketch pad to indicate the horizon reference line, and a vertical line to indicate the central axis of the sketch. A ruler held at arm's length can then be used to estimate the relative distances of selected landmarks, as illustrated in Figure II.23. Once the major topographic features and a few landmarks have been sketched, the relative positions of other details can be located by eye.

A compass traverse is a simple map-making technique, which requires only a compass, to measure directions. A *traverse* is a series of straight lines running from one point to the next. A *closed* traverse ends at the original starting point, and an *open* traverse does not. A closed traverse is used when features within a particular area are to be mapped. An open traverse is more suitable when features on either side of a central line are to be recorded and mapped.

Once the starting point and the approximate route of the traverse have been selected, a landmark along the first leg of the route is sighted from the starting point. The azimuth of the landmark is then measured using a compass, taking precautions to keep the compass away from magnetic materials that might influence its reading. When the azimuth has been measured and recorded in the data book, the investigator then walks toward the landmark while pacing off the distance. If the investigator makes observations along the traverse, the data and the location should be recorded. After reaching the end of the first leg, a sighting can be taken to a new landmark and the traverse can be extended.

The field record of a compass traverse consists of a list of azimuths, distances, and observations. To convert these data to a map, the legs of the traverse can be laid out on drawing paper to their correct scale and with their correct azimuths (see Figure II.23). Observational data can then be added.

Sometimes the position of an object away from the line of the traverse must be measured. One method of finding its location, known as *radiation*, is to take a sighting on the object from a known point on the traverse, measure the azimuth by compass, and pace off the distance to the object. Steps can be saved by employing the technique of *intersection* instead. As Figure II.23 illustrates, the position of the object relative to the traverse is completely specified by measuring the azimuth of the object from two known points along the traverse. For best accuracy, the two sighting points should be spaced so that the lines of intersection meet at the object with an angle of between 60° and 120°.

The methods of radiation and intersection are the basis for map making with a plane table, which is a horizontal drawing board supported by a stand or tripod. A map can be drawn while in the field using a plane table, allowing the work to be checked until satisfactory accuracy is obtained.

When using a plane table, the first step is to establish a *base line* of known distance and azimuth between two sighting points that command good views of the area to be mapped. After the base line has been drawn to an appropriate scale on a piece of drawing paper fastened to the plane table, the table is placed over one of the sighting points and oriented so that the base line on the map sheet is exactly parallel to the base line on the ground. Once this has been accomplished, a sighting device such as a triangular rule can be laid on the map sheet and aligned from the sighting point toward the object to be mapped (see Figure II.23). In this way the line to the object can be

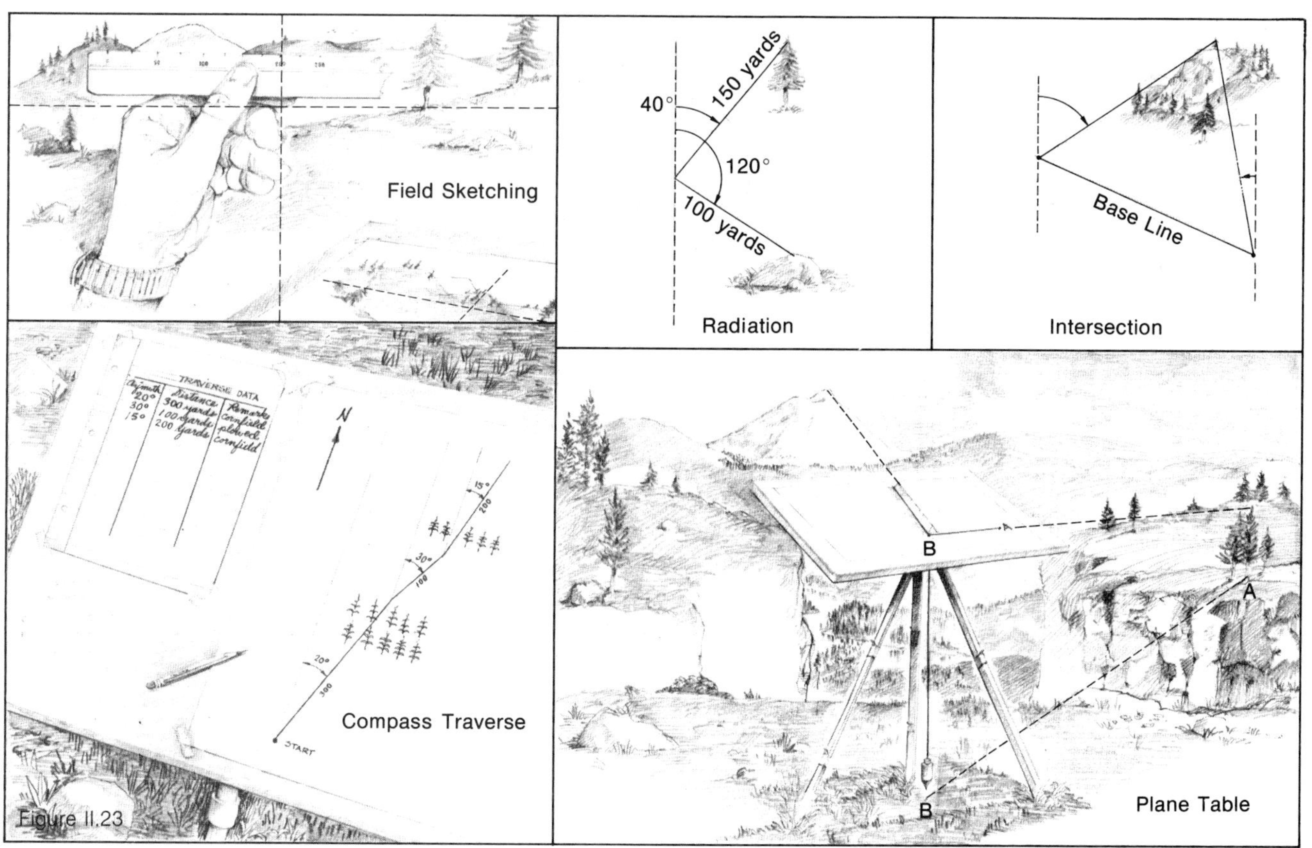
Field Sketching
40°
150 yards
120°
100 yards
Radiation
Base Line
Intersection
TRAVERSE DATA
N
15°
30°
20°
START
Compass Traverse
B
A
B
Plane Table
Figure II.23

drawn on the map at the correct azimuth without the need for compass measurements. The location can be fixed by employing the method of radiation, pacing off the distance from the sighting point to the object. Alternatively, the determination of azimuths from both sighting points fixes the location of the object by intersection. To maintain good accuracy, the orientation of the plane table should be checked from time to time by sighting along the base line toward the other sight point. Additional sighting points and base lines can be set up if parts of the area to be mapped are not visible from the original points.

Remote Sensing

The study and mapping of the earth's surface through direct field work is unsatisfactory for many purposes because of the limited coverage possible, particularly in relatively inaccessible areas, and because of the great expense involved. New capabilities, generally known as *remote sensing,* have been developed in recent years for studying the surface of the earth. Remote sensing methods employ the interaction of electromagnetic energy with the surface of the earth. Special sensing devices can acquire data at a distance from the objects or phenomena being studied. In some methods, the electromagnetic radiation of sunlight, laser light, or microwave radar that is reflected from objects on the earth is sensed and recorded, and in other methods of remote sensing the infrared thermal radiation directly emitted by objects is detected.

Remote sensing equipment is usually mounted in aircraft or space satellites. The advantages of remote sensing include the ability to cover large regions in a very short time, the ability to provide data from very large to very small scale, the possibility of repetitive measurements to follow the progress of selected events, and the ability to penetrate areas inaccessible to ground survey. The techniques of remote sensing have been greatly extended recently in an effort to obtain the information needed to manage the earth's resources more efficiently, and much of the information being gathered is of direct interest to physical geographers. The capabilities of remote sensing technology include the detection of vegetation types and their seasonal changes, the measurement of surface water distribution and soil moisture, and the depiction of landforms and surface geologic structures.

Aerial photography, one of the first remote sensing methods developed, is now widely used for most original mapping. Aerial mapping surveys generally employ the visible light reflected from the earth and detected using cameras with black and white film. Optical techniques allow the distortions inherent in a photograph taken at low or moderate altitudes to be rectified so that distant areas in the photograph have the same scale as areas directly below the camera. Two exposures can be used to record a stereoscopic (three-dimensional) image from which contour lines can be drawn. A single photograph may cover an area of a few square miles to hundreds of square miles, depending on the cameras and the altitude of the aircraft. Ground features as small as 1 centimeter in size can be detected on photographs made from aircraft flying at an altitude of a few kilometers, and satellite imagery can provide resolution of objects only a few meters in size from an altitude of several hundred kilometers. These advanced capabilities are gradually becoming available for civilian uses, such as the detection of local sources of air pollution.

Aircraft fitted with detectors of long-wave infrared thermal radiation can provide images of objects on the earth's surface with intensities proportional to their temperatures. Infrared imagery methods can be employed to detect forest fires and the mixing of warm water from industrial effluents with cooler river water.

Application of remote sensing methods to the study of the earth has been revolutionized by using remote

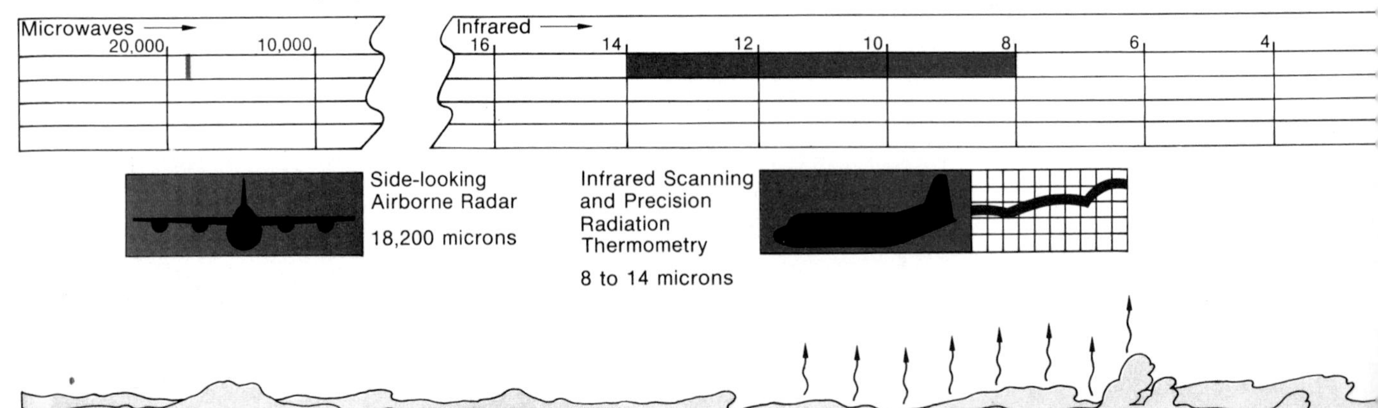

Microwaves
20,000
10,000
Infrared
16
14
12
10
8
6
4
Side-looking
Airborne Radar
18,200 microns
Infrared Scanning
and Precision
Radiation
Thermometry
8 to 14 microns

sensors in conjunction with space satellites. Surveillance of the earth's cloud cover from weather satellites has greatly improved the accuracy of weather predictions and the ability to track potentially damaging hurricanes and typhoons. Weather satellites usually orbit at altitudes of several hundred kilometers.

Sunlight falling upon the earth consists of electromagnetic radiation in the visible region and in the near-visible infrared region of the electromagnetic spectrum. Different objects may reflect a different relative proportion of each wavelength, so that the light reflected from an object, when analyzed in terms of wavelength, forms a means of identification known as the *spectral signature* of the object. Vegetation, for example, usually reflects proportionately more short wavelength infrared radiation than does bare soil. Furthermore, different types of vegetation can usually be distinguished by their signatures. Diseased plants or the lack of soil moisture can also be detected.

Several remote sensing techniques have been developed for using characteristic signatures to distinguish and identify objects on the earth's surface. *Color infrared film* is similar to ordinary color film, except that it depicts the previously invisible infrared radiation as red, the red colors as green, and the green colors as blue. Vegetation usually appears red on a color infrared photograph of a region; the intensity of the color is directly proportional to the amount of chlorophyll in the vegetation. Color infrared photographs of test fields have proven that different crops and their stages of growth can be distinguished by such means. Another way to make use of spectral signatures is to employ several cameras, each fitted with a filter that allows only wavelengths in a certain band to enter. The Earth Resources Technology Satellite ERTS-1, which views the earth from an altitude of nearly 1,000 kilometers, employs four multispectral scanners and three television cameras, each sensing the same scene on a different band of wavelengths. These images can be artificially "combined" to produce a "color picture" of the area covered. From its orbit between the poles, ERTS-1 can scan a given portion of the earth once every 18 days to allow the progress of crops to be followed through the growing season; to monitor streamflow, snow depth, and soil moisture; and to detect other short-term changes in the environment.

Many remote areas in the tropics are almost continuously covered by clouds, hiding the ground from view. *Side-looking airborne radar,* which utilizes radio waves of short wavelength, can penetrate clouds and yield a high resolution picture of the surface. The radar apparatus is usually fitted to the side of the aircraft to give an unrestricted lateral view; radar waves emitted from the aircraft are partly reflected from the ground and detected by a receiving apparatus on the aircraft. Measurement of the time required for the radar waves to make the round trip is electronically translated into a measure of distance, producing a map of the region. Airborne radar has been used to map the Amazon Basin, Panama, and portions of Africa and New Guinea. Although radar methods do not differentiate between types of vegetation as clearly as the signature method, forest, range, and agricultural land can be separately identified.

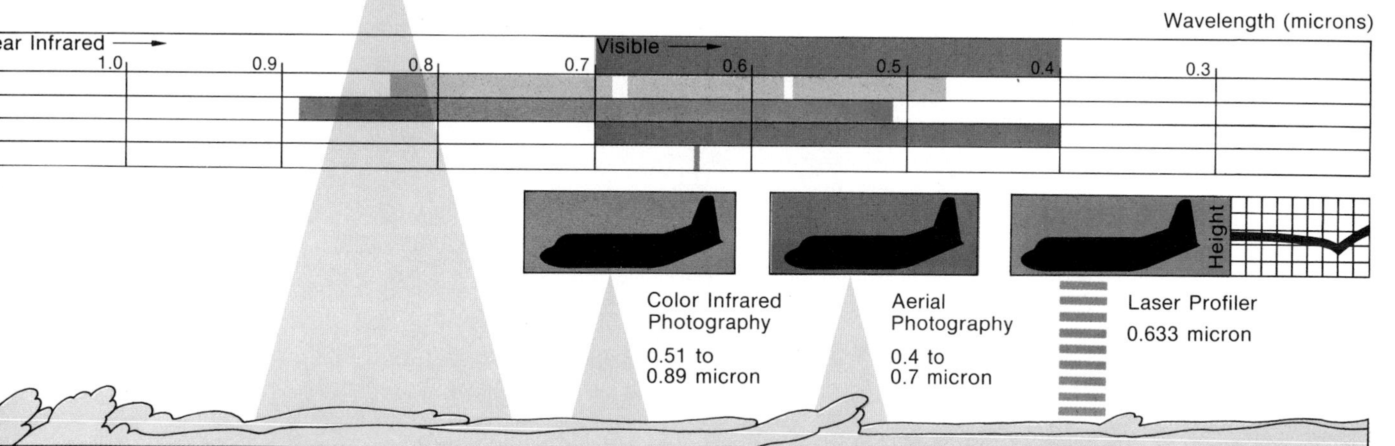
NASA
ERTS Television Cameras
0.475 to 0.575 micron
0.580 to 0.680 micron
0.690 to 0.830 micron
Wavelength (microns)
ear Infrared
Visible
1.0
0.9
0.8
0.7
0.6
0.5
0.4
0.3
Height
Color Infrared Photography
0.51 to 0.89 micron
Aerial Photography
0.4 to 0.7 micron
Laser Profiler
0.633 micron

Figure II.24 Remote sensing techniques utilize many portions of the electromagnetic spectrum, as shown in this chart. The visible and near infrared regions are used for photography and television, and the infrared is used for scanning surface temperatures. Relief features are mapped by laser profiling and radar.

Figure II.25 (top) The aerial view on the right was photographed with ordinary color film. The same fields photographed with color infrared film are shown on the left; healthy vegetation appears bright red.

(above left) Stereo pairs are used to make topographic relief apparent in aerial photographs. When viewed properly, the hills in this stereo view of a Japanese coast stand out clearly above the level of the sea. To see the stereo effect, the left-hand photograph should be viewed with the left eye, and the right-hand photograph with the right eye. A card held between the photographs helps to separate the images.

(bottom right) Vegetation on irrigated land is clearly differentiated from dry desert in this color infrared photograph of the Imperial Valley of California taken from earth orbit. The Salton Sea appears as a large dark region at the upper left. The pattern of agricultural fields is clearly distinguishable; note in particular the border between the United States and Mexico that is evident toward the bottom of the photograph. The well-irrigated crops on the United States side show as bright red, compared to the bluish hues representing vegetation on the Mexican side of the border.

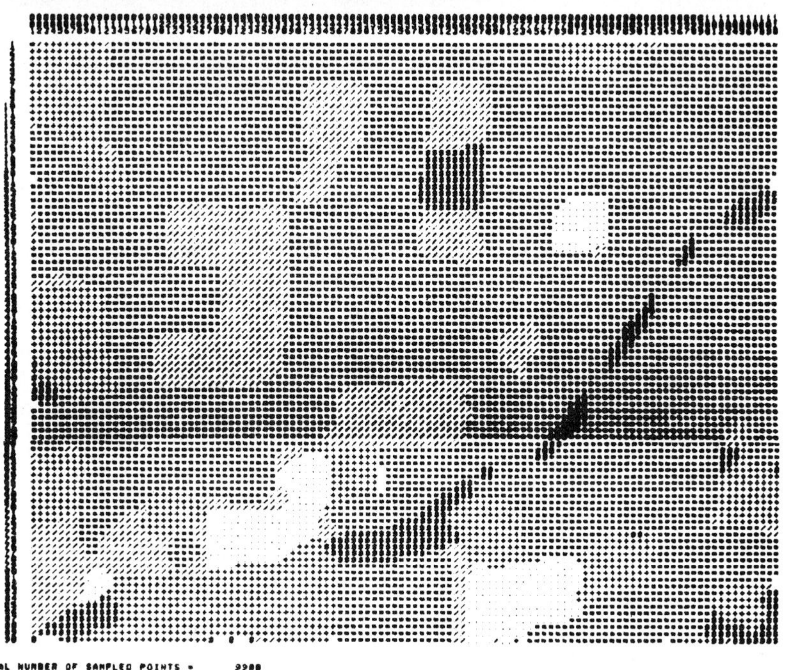
TOTAL NUMBER OF SAMPLED POINTS = 9988

Figure II.26 (top left) Monterey Bay, California, appears at the left of this photograph taken by ERTS-1 from an altitude of 900 kilometers (560 miles). The color print is made from a composite of three black and white photographs, each exposed to a different band of wavelengths; the image is similar to that produced by color infrared film. Vegetation areas stand out in red, and linear features associated with the San Andreas Fault zone run from the upper left to the lower right. A patch of fog hides part of the coast south of the Monterey Peninsula. A topographic map of the Monterey Peninsula is shown in Figure II.22.

(top right) Relief features, including linear features in the San Andreas Fault zone, show crisply in this side-looking radar image of part of the San Francisco peninsula. The thin linear feature at the lower right is the 2-mile-long linear accelerator at Stanford University.

(bottom left) The areas of different shading on this computer print-out correspond to different vegetation types or to water. The computer map was prepared directly by computer analysis of multiband images from ERTS-1. Accuracy of identification exceeded 95 percent.

Appendix III: Classification Systems

Köppen System of Climate Classification

The Köppen system of climate classification recognizes five principal climate regions, which are symbolized by *A, B, C, D,* and *E*. The *B* regions, which are the dry realms, are further divided into two major subregions, *BS* (steppe) and *BW* (desert), and the *E* region is divided into the regions *ET* (tundra) and *EF* (perpetual frost). Figure III.1 shows Köppen's criteria for establishing the boundaries between the regions *A, C,* and *D*; the region *BS*; and the region *BW*. Each of the three sets of criteria are based on average annual temperature and on annual precipitation. Which criterion is used depends on whether precipitation occurs evenly throughout the year or whether precipitation occurs primarily during the summer or during the winter. Köppen considered a region to have a dry winter if at least 70 percent of the precipitation occurs during the 6 summer months, and to have a dry summer if at least 70 percent of the precipitation occurs during the 6 winter months. Regions not fitting either category are considered to have an even distribution of precipitation. Summer is interpreted as the season when the sun is high over the horizon at midday in the particular hemisphere being considered, and winter is interpreted as the low-sun season.

For a given average annual temperature, the regions *A, C,* and *D* receive more precipitation annually than do *BS* regions; *BS* regions in turn receive more precipitation than *BW* regions. The regions *A, C, D, ET,* and *EF* are distinguished from one another according to various criteria based on temperature, with the warmest being *A* and the coldest being *EF*. The criteria are as follows:

A:	Average temperature of the coldest month exceeds 18°C (64.4°F).
C:	Average temperature of the warmest month exceeds 10°C (50°F). Average temperature of the coldest month lies between 18°C (64.4°F) and -3°C (26.6°F).
D:	Average temperature of the warmest month exceeds 10°C (50°F). Average temperature of the coldest month is below -3°C (26.6°F).
ET:	Average temperature of the warmest month lies between 10°C (50°F) and 0°C (32°F).
EF:	Average temperature of the warmest month is below 0°C (32°F).
H:	Unclassified highland climates.

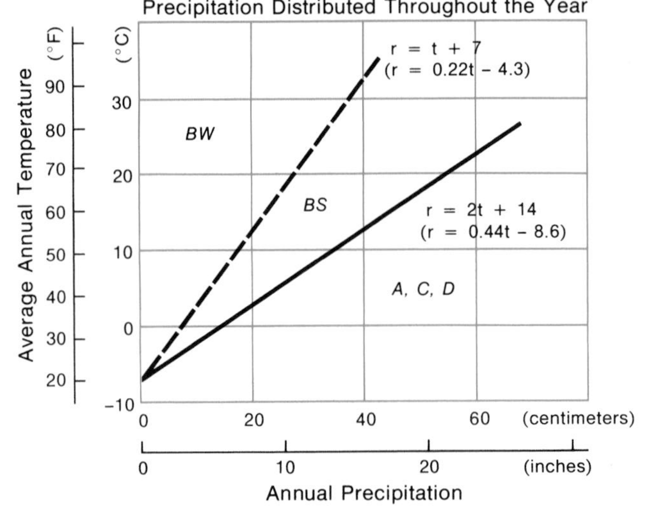

Precipitation Distributed Throughout the Year
Average Annual Temperature (°F) (°C)
90
80
70
60
50
40
30
20
30
20
10
0
-10
BW
BS
A, C, D
r = t + 7
(r = 0.22t - 4.3)
r = 2t + 14
(r = 0.44t - 8.6)
0
20
40
60
(centimeters)
0
10
20
(inches)
Annual Precipitation

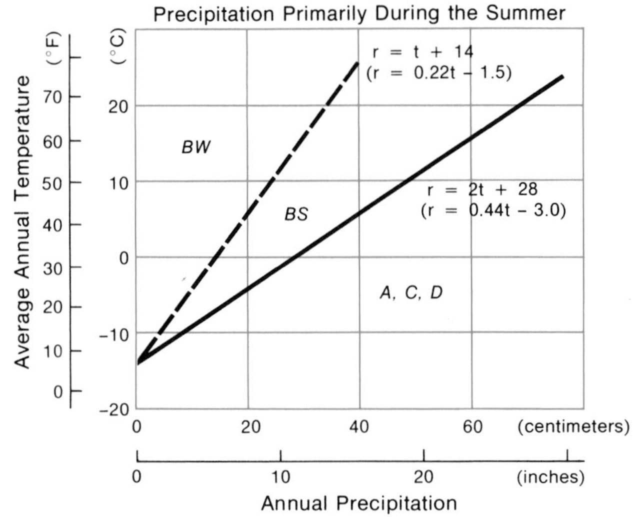

Precipitation Primarily During the Summer
Average Annual Temperature (°F) (°C)
70
60
50
40
30
20
10
0
20
10
0
-10
-20
BW
BS
A, C, D
r = t + 14
(r = 0.22t - 1.5)
r = 2t + 28
(r = 0.44t - 3.0)
0
20
40
60
(centimeters)
0
10
20
(inches)
Annual Precipitation

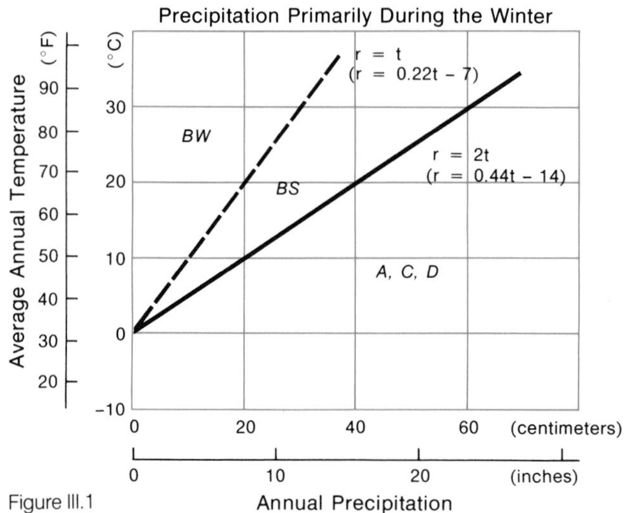

Precipitation Primarily During the Winter
Average Annual Temperature (°F) (°C)
90
80
70
60
50
40
30
20
30
20
10
0
-10
BW
BS
A, C, D
r = t
(r = 0.22t - 7)
r = 2t
(r = 0.44t - 14)
0
20
40
60
(centimeters)
0
10
20
(inches)
Annual Precipitation

Figure III.1

Principal Subdivisions of A Regions. The *A* climate regions are subdivided into *Af*, *Am*, and *Aw* regions on the basis of the amount and the seasonality of precipitation. If precipitation in the driest month exceeds 6 centimeters (2.4 inches), the region is classified as *Af*, as indicated in Figure III.2. *Af* regions, such as equatorial lowland rainforests, receive abundant moisture for plant growth throughout the year. If a region has a winter dry period during which precipitation for the driest month is less than 6 centimeters, the region is classified as *Aw* or as *Am*. The distinction between *Aw* and *Am* regions depends upon the relation between the amount of precipitation during the driest month and the annual precipitation, as Figure III.2 shows. Tropical wet and dry regions are classified as *Aw*. The *Am*, or monsoon, climate regions have enough annual precipitation so that the moderate winter dry season does not exhaust supplies of soil moisture, and plant growth is not seriously affected.

A fourth subdivision, the *As* region, would in principle be characterized by a summer dry season. However, *As* regions only occur locally near the Equator, and the *As* classification is not significant on the global scale.

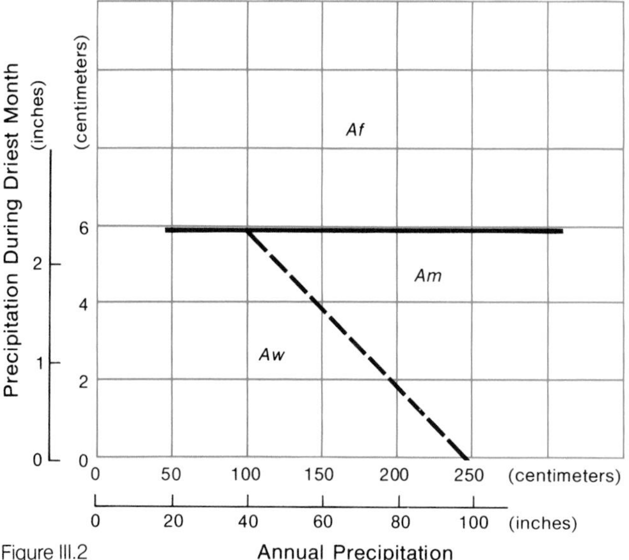
Precipitation During Driest Month
(inches)
(centimeters)
Af
Am
Aw
0
1
2
0
2
4
6
0
50
100
150
200
250
(centimeters)
0
20
40
60
80
100
(inches)
Annual Precipitation

Figure III.2

Principal Subdivisions of BS and BW Regions. The *BS* and *BW* regions are divided on a thermal basis, as specified by the additional notation *h*, *k*, or *k'*:

h (hot):	Average annual temperature exceeds 18°C (64.4°F).
k (cold winter):	Average annual temperature is less than 18°C (64.4°F) and average temperature of the warmest month exceeds 18°C.
k' (cold):	Average annual temperature is less than 18°C (64.4°F) and average temperature of the warmest month is less than 18°C.

The deserts of North Africa and central Australia are examples of *BWh* regions. *BWk* regions occur, for example, in the high plateaus of central Asia.

Principal Subdivisions of C and D Regions. The Köppen *C* or *D* climate regions are further differentiated according to the seasonality of precipitation. A second letter of notation *s* denotes regions with a dry summer (*Ds* seldom occurs), *w* denotes regions with a dry winter, and *f* denotes regions with no marked dry season.

Cs, Ds:	The driest month occurs during the warmest 6 months, and the amount of precipitation received during the driest month is less than one-third the amount received during the wettest month of the coldest 6 months. Also, precipitation during the driest month must be less than 4 centimeters (1.6 inches).
Cw, Dw:	The driest month occurs during the coldest 6 months, and the amount of precipitation received during the driest month is less than one-tenth the amount received during the wettest month of the warmest 6 months.
Cf, Df:	No marked dry season occurs, so that the *s* or *w* categories are not applicable.

The subdivisions *Cs*, *Ds*, *Cw*, *Dw*, *Cf*, and *Df* may be further differentiated according to the seasonality of temperature by adding a third letter of notation *a*, *b*, *c*, or *d*.

a:	Average temperature of the warmest month exceeds 22°C (71.6°F).
b:	Average temperature of the warmest month is less than 22°C (71.6°F), but at least 4 months have an average temperature greater than 10°C (50°F).
c:	Average temperature of the warmest month is less than 22°C (71.6°F), fewer than 4 months have an average temperature greater than 10°C (50°F), and the average temperature of the coldest month is greater than −38°C (−36.4°F).
d:	Same as *c*, except that the temperature of the coldest month is less than −38°C (−36.4°F).

The Köppen system of climate classification possesses additional symbols to denote special features such as frequent fog or seasonal high humidity, but they are seldom used on the global scale. A global map of the Köppen climate classification is shown in Figure 7.10 on page 176.

Thornthwaite's Formula for Potential Evapotranspiration

The tables and graphs presented in this section make it possible to calculate potential evapotranspiration for a given month at a given location from Thornthwaite's formula, if the average monthly temperatures are known. Thornthwaite's formula is designed to be used with temperatures expressed in degrees Celsius, and temperature data on the Fahrenheit scale should be first converted to Celsius before beginning the calculation of potential evapotranspiration.

The first step in applying Thornthwaite's method is to calculate a monthly heat index *i* for each of the 12 months. The montly heat index is defined according to the formula

$$i = \left(\frac{T}{5}\right)^{1.514}$$

where *T* is the long-term average temperature of the month in Celsius degrees. Approximate values of the monthly heat index can be read from the graph in Figure III.3.

The sum of the twelve monthly heat indexes is the annual heat index, *I*. The annual heat index is

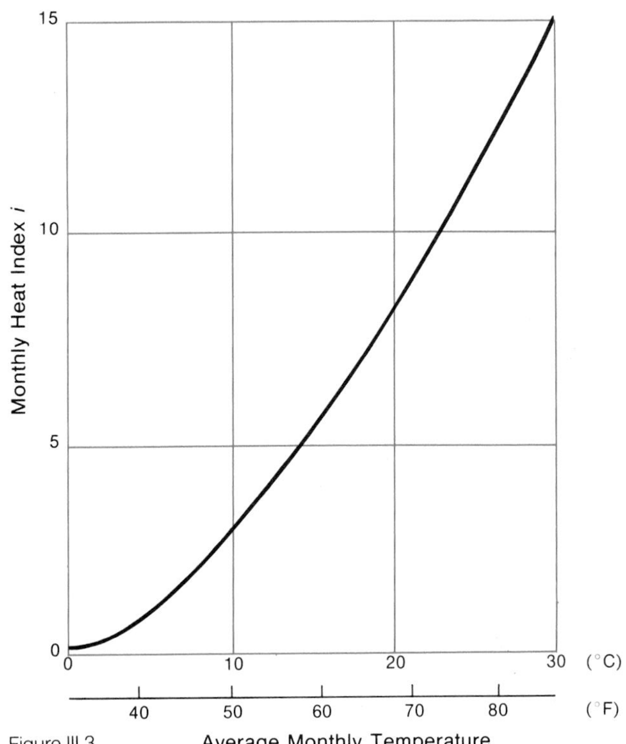

Monthly Heat Index i
15
10
5
0
0
10
20
30
(°C)
40
50
60
70
80
(°F)
Average Monthly Temperature

Figure III.3

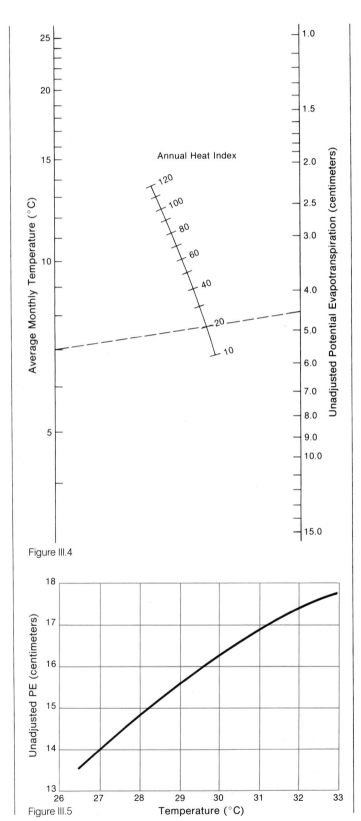

Average Monthly Temperature (°C)

Annual Heat Index

Unadjusted Potential Evapotranspiration (centimeters)

Figure III.4

Unadjusted PE (centimeters)

Temperature (°C)

Figure III.5

Table III.1. Daylength Correction Factors for Potential Evapotranspiration

Latitude	JAN	FEB	MAR	APR	MAY	JUN	JUL	AUG	SEP	OCT	NOV	DEC
0°	1.04	0.94	1.04	1.01	1.04	1.01	1.04	1.04	1.01	1.04	1.01	1.04
10°N	1.00	0.91	1.03	1.03	1.08	1.06	1.08	1.07	1.02	1.02	0.98	0.99
20°N	0.95	0.90	1.03	1.05	1.13	1.11	1.14	1.11	1.02	1.00	0.93	0.94
30°N	0.90	0.87	1.03	1.08	1.18	1.17	1.20	1.14	1.03	0.98	0.89	0.88
40°N	0.84	0.83	1.03	1.11	1.24	1.25	1.27	1.18	1.04	0.96	0.83	0.81
50°N	0.74	0.78	1.02	1.15	1.33	1.36	1.37	1.25	1.06	0.92	0.76	0.70
10°S	1.08	0.97	1.05	0.99	1.01	0.96	1.00	1.01	1.00	1.06	1.05	1.10
20°S	1.14	1.00	1.05	0.97	0.96	0.91	0.95	0.99	1.00	1.08	1.09	1.15
30°S	1.20	1.03	1.06	0.95	0.92	0.85	0.90	0.96	1.00	1.12	1.14	1.21
40°S	1.27	1.06	1.07	0.93	0.86	0.78	0.84	0.92	1.00	1.15	1.20	1.29
50°S	1.37	1.12	1.08	0.89	0.77	0.67	0.74	0.88	0.99	1.19	1.29	1.41

Source: Thornthwaite, C. Warren. 1948. "An Approach Toward a Rational Classification of Climate," *Geographical Review*, 38: 55-94.

representative of climatic factors at a given location because it is based on long-term averages. Thornthwaite found an empirical formula that gives potential evapotranspiration, *PE*, for a given month of a particular year in terms of *I*. His formula for *PE*, unadjusted for duration of sunlight, is

$$\text{Unadjusted } PE \text{ (centimeters)} = 1.6\left(\frac{10\ T}{I}\right)^m$$

where *T* is the average temperature in °C for the specific month being considered, and *m* is a number that depends on the annual heat index *I*. To a good approximation, *m* is given by the formula

$$m = (6.75 \times 10^{-7})I^3 - (7.71 \times 10^{-5})I^2 + (1.79 \times 10^{-2})I + 0.492$$

With these formulas, unadjusted potential evapotranspiration can be calculated using tables of logarithms or a computer. Alternatively, approximate values of unadjusted potential evapotranspiration can be read from the nomogram in Figure III.4 with enough accuracy for most purposes if the values for the average monthly temperature and for the annual heat index are known.

If the average temperature of the month is below 0°C, the potential evapotranspiration is taken to be 0. If the average monthly temperature exceeds 26.5°C, unadjusted *PE* is given directly in terms of temperature according to the graph in Figure III.5.

Values of unadjusted potential evapotranspiration must be corrected for the duration of daylight in order to obtain the desired final values. Unadjusted potential evapotranspiration for a given month at a given location should be multiplied by the correction factor listed in Table III.1. The correction factors for latitude 50°N are used for all latitudes farther to the north, and the factors for latitude 50°S are used for all latitudes farther to the south.

Soil Classification Systems

The comprehensive 7th Approximation of soil classification has superseded the earlier 1938 United States Department of Agriculture classification system for use in the United States. However, many of the terms from the 1938 classification remain in common use.

The 1938 Soil Classification System. The 1938 classification divides soils into three orders: *zonal*, *intrazonal*, and *azonal*. The orders are divided into suborders, which are in turn divided into the great soil groups. The listing given in this section is a modification of the 1938 classification prepared in 1949 by James Thorp and Guy Smith, soil correlators for the Division of Soil Survey of the United States Department of Agriculture. The great soil groups are listed under their appropriate suborder.

Zonal Soils

1. Soils of the cold zone
   - Tundra soils
2. Light-colored soils of arid regions
   - Desert soils
   - Red desert soils
   - Sierozem
   - Brown soils
   - Reddish brown soils
3. Dark-colored soils of semiarid, subhumid, and humid grasslands
   - Chestnut soils
   - Reddish chestnut soils
   - Chernozem soils
   - Prairie soils
   - Reddish prairie soils
4. Soils of the forest-grassland transition
   - Degraded chernozem
   - Noncalcic brown soils

5. Light-colored podzolic soils of the timbered regions

   Podzols
   Gray podzolic soils
   Brown podzolic soils
   Gray-brown podzolic soils
   Red-yellow podzolic soils

6. Lateritic soils of forested warm midlatitude and tropical regions

   Reddish brown latosols
   Yellowish brown latosols
   Latosols

Intrazonal Soils

1. Halomorphic (saline and alkali) soils of imperfectly drained arid regions

   Saline soils (solonchak)
   Solonetz soils
   Soloth soils

2. Hydromorphic soils of marshes, swamps, seep areas, and flats

   Humic-gley soils
   Alpine meadow soils
   Bog soils
   Half-bog soils
   Low-humic gley soils
   Planosols
   Groundwater podzolic soils
   Groundwater lateritic soils

3. Calcimorphic soils

   Brown forest soils
   Rendzina soils

Azonal Soils (no suborders)

   Lithosols
   Regosols (includes dry sands)
   Alluvial soils

The 7th Approximation. The following list gives the orders and suborders of the 7th Approximation. The suborders are further divided into a large number of great groups, as tabulated in 1960 and 1967 publications of the United States Department of Agriculture Soil Conservation Service. The descriptions presented in this section are adapted from *The National Atlas of the United States of America.* Some indication of the use of the soil is also listed.

1. **Alfisols:** Moderate to high alkalinity, gray surface horizon, subsurface clay; usually moist, perhaps dry for part of warm season.

   Aqualfs: Seasonally wet; for general crops (drained) or woodland (undrained).
   Boralfs: In cool to cold regions; woodland and pasture.
   Udalfs: In midlatitude to tropical regions; usually moist, dry for short periods; row crops, small grain, and pasture.
   Ustalfs: In midlatitude to tropical regions; reddish brown; dry for long periods; range, small grain, and irrigated crops.
   Xeralfs: In regions with rainy winters and dry summers; dry for a long period; range, small grain, and irrigated crops.

2. **Aridisols:** Semidesert and desert soils with definite horizons; low in organic matter; never moist for more than 3 consecutive months.

   Argids: Accumulation of clay in a horizon; range and some irrigated crops.
   Orthids: Accumulation of such salts as calcium carbonate or gypsum; range and some irrigated crops.

3. **Entisols:** No definite horizons (generally azonal).

   Aquents: Permanently or seasonally wet; limited pasturage.
   Fluvents (Alluvial Soils): Organic content decreases irregularly with depth, formed in loam or clay alluvial deposits; range, irrigated crops (dry regions), and general farming (humid regions).
   Orthents: Loam or clay; organic content decreases regularly with depth; range, irrigated crops (dry regions), and general farming (humid regions).
   Psamments: Texture of loamy fine sand or coarser; range and wild hay (Alaska), woodland, small grains (warm, moist regions), pasture and citrus (Florida), and range and irrigated crops (warm, dry regions).

4. **Histosols:** Wet organic peat and muck soils, formed in swamps and marshes; truck crops where drained.

5. **Inceptisols:** Weakly developed horizons; materials have been altered or removed but have not accumulated; usually moist except during part of warm season.

   Andepts: Formed in volcanic ash; woodland, range, and pasture.
   Aquepts: Seasonally wet, organic surface horizon; pasture and hay (Alaska) or woodland and row crops if drained (southeastern United States).
   Ochrepts: In crystalline clay minerals, light-colored surface horizons; woodland and range (Alaska, Northwest), pasture, wheat, and hay (Oklahoma and Kansas), and pasture, corn, and small grain (Northeast).
   Tropepts (Latosols): In tropical regions; pineapple and irrigated sugar cane (Hawaii).
   Umbrepts: In crystalline clay minerals, thick dark surface horizon; low in alkalis; woodland and range.

6. **Mollisols:** In subhumid and semiarid warm to cold climates; dark surface horizon rich in organic matter, highly alkaline.

   Albolls: Seasonal perched water table, bleached subsurface horizon; small grain, peas, hay, pasture, and range.
   Aquolls: Seasonally wet; gray subsurface horizon; where drained, small grains, corn, and potatoes (north central United States), and rice and sugar cane (Texas).
   Borolls: In cool and cold regions; small grain, hay, and pasture (north central United States), and range and woodland (western United States).
   Rendolls (Rendzinas): Subsurface horizon with large amount of calcium carbonate; cotton, corn, small grains, and pasture.
   Udolls: In midlatitude climates; usually moist, no accumulation of calcium carbonate or gypsum; corn, small grains, and soybeans.
   Ustolls: Mostly in semiarid regions; intermittently dry for a long period; wheat and some irrigated crops.
   Xerolls: In regions with rainy winters and dry summers; dry for a long period; wheat, range, and irrigated crops.

7. **Oxisols:** In tropical or subtropical lowland regions; mixtures of kaolin (aluminum clay) and silicon dioxide, low in weatherable minerals.

   Humox: Usually moist; high in organic content, low in alkalinity; sugar cane, pineapple, and pasture (Hawaii).
   Orthox (Latosols): Usually moist: moderate to low in organic content, relatively low in alkalinity; sugar cane, pineapple, and pasture (Hawaii).
   Ustox (Latosols): Partly dry for long periods; pineapple, irrigated sugar cane, and pasture (Hawaii).

8. **Spodosols (Podzols):** In humid, mostly cool, midlatitude regions; acid, low supply of alkalis, with compounds of aluminum or iron.

   Aquods: In seasonally wet regions; pasture, range, or woodland, and citrus and truck crops (Florida).
   Orthods: Accumulation of organic matter and compounds of aluminum and iron in a horizon; woodland, hay, pasture, and fruit.

9. **Ultisols:** Usually moist but dry part of the year; low supply of alkalis, subsurface horizon of clay accumulation.

   Aquults: Seasonally wet; woodland, limited pasture; hay, cotton, corn, and truck crops where drained.
   Humults: In warm humid regions; high in organic content; small grain, truck, and seed crops (Oregon, Washington), and pineapple and irrigated sugar cane (Hawaii).

Udults: In continuously moist regions; relatively low in organic content; general farming, woodland, and pasture, and cotton and tobacco in some regions.
Xerults: In regions with rainy winters and dry summers; dry for long periods; low in organic content; range and woodland.

10. Vertisols: Clay soils; wide, deep cracks when dry.

Pelluderts (Grumusols): Black or dark gray surface horizon.
Torrerts (Grumusols): Usually dry; wide, deep cracks open most of the time; range and some irrigated crops.
Uderts: Usually moist; cracks not open for more than 2 or 3 months at a time; cotton, corn, small grains, pasture, and some rice.
Usterts: Wide, deep cracks open and close more than once each year, usually open for 3 months or more; general crops and range, and irrigated crops (Rio Grande valley).
Xererts: Wide, deep cracks open and close once each year, remain open for more than 2 months at a time; irrigated small grains, hay, and pasture.

Morphogenetic Regions

The listing in Table III.2 gives the general characteristics of morphogenetic regions according to Louis Peltier. Peltier based his classification of regions upon temperature, precipitation, and characteristic processes of erosion.

Table III.2. Morphogenetic Regions

Morphogenetic Region	Annual Temperature	Annual Precipitation (centimeters*)	Morphologic Characteristics
Glacial	−18 to −7°C (0 to 20°F)	0 to 115 (0 to 45)	Glacial erosion Nivation (snow movement or snowmelt runoff) Wind action
Periglacial	−15 to −1°C (5 to 30°F)	13 to 140 (5 to 55)	Strong mass movement Moderate to strong wind action Weak effect of running water
Boreal	−9 to 3°C (15 to 38°F)	25 to 150 (10 to 60)	Moderate frost action Moderate to slight wind action Moderate effect of running water
Maritime	2 to 21°C (35 to 70°F)	125 to 190 (50 to 75)	Strong mass action Moderate to strong action of running water
Selva	16 to 29°C (60 to 85°F)	140 to 230 (55 to 90)	Strong mass action Slight effect of slope wash No wind action
Moderate	3 to 29°C (38 to 85°F)	90 to 150 (35 to 60)	Maximum effect of running water Moderate mass movement Frost action slight in colder part of region No significant wind action except on coasts
Savanna	−12 to 29°C (10 to 85°F)	65 to 125 (25 to 50)	Strong to weak action of running water Moderate wind action
Semiarid	2 to 29°C (35 to 85°F)	25 to 65 (10 to 25)	Strong wind action Moderate to strong action of running water
Arid	13 to 29°C (55 to 85°F)	0 to 40 (0 to 15)	Strong wind action Slight action of running water and mass movement

*Numbers in parentheses indicate inches.

Source: Peltier, Louis. 1950. "The Geographic Cycle in Periglacial Regions As It Is Related to Climatic Geomorphology," *Association of American Geography Annals*, 40: 214–236.

Contributors

Physical Geography Today: A Portrait of a Planet was developed by a group of scholars and educators who contributed their ideas and expertise in an effort to present a unified, dynamic view of physical geography. The book evolved from seminars during which the group first developed a detailed blueprint and later reviewed and reworked manuscripts. The intention was to present the underlying physical principles and processes that explain why the earth is the way it is. Robert J. Kolenkow, who has shown a special ability to use physical principles to explain the environment, was asked to write the text, based on the outline of Robert A. Muller and the other advisors. The contributors also recommended a graphics program that amplified and clarified the textual material.

The CRM Books staff worked as a book team to bring all the parts of the book together. Special features, such as the case studies and graphic essays, were developed by the editors and designers to provide another dimension—that of the human perspective—to physical geography.

Robert A. Muller, principal advisor for *Physical Geography Today,* earned his doctorate at Syracuse University after undergraduate studies at Rutgers University. As a student he was elected to both Phi Beta Kappa and Sigma Xi. Prior to his current appointment as professor of geography at Louisiana State University in Baton Rouge, he held a research appointment in forest climatology with the United States Forest Service in California, and he taught physical geography, meteorology, and climatology at the University of California, Berkeley, at the University of Delaware, and at Rutgers. Dr. Muller's special interests include aspects of the hydrologic cycle at the earth's surface and the interrelationships of weather and climate with man and his works. Dr. Muller's recent research has been directed primarily toward water-balance evaluations of the effects of land management, reforestation, and urbanization on streamflow and floods. In addition to advising and reviewing on the text, Dr. Muller provided technical data for a number of graphic pieces.

Robert J. Kolenkow was a member of the Research Staff of the Radioactivity Center at the Massachusetts Institute of Technology and, later, an associate professor in the Department of Physics there. He has also been a lecturer in physics at the University of California, San Diego. Before receiving his doctorate in physics from Harvard University in 1959, he studied at MIT and at the University of Göttingen, Germany, as a Fulbright exchange student. At Harvard, he was the recipient of a Danforth Fellowship and a National Science Foundation Fellowship. In 1968 he was awarded the Everett Moore Baker award for outstanding undergraduate teaching at MIT. Dr. Kolenkow's research interests are focused on the interaction of nuclear radiation with matter, medical physics, atomic and molecular beam studies, and atomic and molecular interactions. He has recently coauthored a freshman textbook in mechanics. Dr. Kolenkow, the principal writer and graphics consultant for this project, developed the text based on ideas and outlines from the other advisors.

Reid A. Bryson, director of the Institute for Environmental Studies at the University of Wisconsin, Madison, established the Department of Meteorology there in 1948 and in 1962 began the University's Center for Climatic Research. He is using his educational background and interests in geology, geography, and meteorology to investigate "total systems" of regional and global climatic change. In his special field of world climatology, he has conducted research in Puerto Rico, South America, and south Asia, and on several islands in the Pacific. Dr. Bryson has also carried out extensive field work in Alaska, the Canadian Arctic, and India. One of his major studies involves man-made climatic modifications occurring in the Rajasthan Desert of India. He has written three books and more than one hundred articles in journals of meteorology, geography, oceanography, hydrology, and archaeology. Dr. Bryson advised primarily on the weather and climate chapters of this text.

Douglas B. Carter, chairman and professor in the Department of Geography at Southern Illinois University in Carbondale, received his doctorate from the University of Washington in 1957. He has served on a number of committees, including the Committee on Climatology Advisory to the Weather Bureau of the National Academy of Sciences and the Panel on Physical Geography for the Commission on College Geography, National Science Foundation–Association of American Geographers. Dr. Carter has been a contributing editor to *Publications in Climatology* since 1964, and he has published several professional articles on climate and its hydrologic relationships. As a general writing consultant for this textbook, Dr. Carter contributed primarily in the areas of ecology, weather, and climate.

R. Keith Julian, who prepared the instructors guide and served as a general reviewer and cartographic consultant for this text, is teaching and completing work on his doctorate in geography at the University of California, Los Angeles. Originally trained as an economist at Harvard University, he is now specializing in research in energy and resource utilization, as well as in work in theoretical cartography, geographic applications of remote sensing technology, and regional development. In 1972 he founded a teaching preparation program for graduate students at UCLA and is presently serving as director of that program as well as doing research and writing in the field of geographic education.

Theodore M. Oberlander, Associate Professor of Geography at the University of California, Berkeley, received his doctorate from Syracuse University in 1963. Professor Oberlander's interest is in natural landscapes, expressed professionally in the analysis of landforms and avocationally in landscape painting and photography. For the past several years his research has focused on the role of pre-Pleistocene climates and climatic changes in the evolution of California desert scenery. His recent work has shown that certain landforms long regarded as classic features of arid morphogenesis have been inherited from nondesert landscapes some 10 million years in age. Dr. Oberlander advised in the area of geomorphology and contributed graphic material on regional landforms for this text.

Robert P. Sharp, professor of geomorphology at the California Institute of Technology in Pasadena, received his doctorate in geology from Harvard University. Prior to joining the staff at Caltech in 1947, Dr. Sharp taught at the University of Minnesota and at the University of Illinois. He was chairman of the Division of Geological Sciences at Caltech from 1952 until 1968. Dr. Sharp is a Fellow of the American Academy of Arts and Sciences and the Geological Society of America, and a member of the National Academy of Science, the Glaciological Society, the American Geophysical Union, and other professional societies. His current research interests include the geomorphology of arid regions, the Pleistocene glaciation of the Sierra Nevadas, the geology of Mars, and the implications of oxygen isotope data from Antarctica. He has published approximately eighty geological papers and has authored two books, *Glaciers* and *Geology: Field Guide to Southern California.* Dr. Sharp advised on the geomorphology portions of this book.

M. Gordon Wolman, a general advisor for this text, was awarded his doctorate from Harvard University in 1953 and is currently professor of geography at The Johns Hopkins University in Baltimore, Maryland. His principal research has centered on the study of natural river systems and the effect of man's activities on the natural environment. His recent research has been concerned with measures of water quality in rivers, problems of sedimentation associated with urbanization, and the environmental impact of urban development. Dr. Wolman's publications include *Fluvial Processes in Geomorphology,* which he coauthored with Luna B. Leopold and J. P. Miller.

A number of other people also contributed to the development of this book. In particular, we gratefully acknowledge the contributions of **George Cox**, California State University, San Diego, for his consultation on the Ecologic Energetics chapter; **James Hawkins**, Scripps Institution of Oceanography, La Jolla, California, for his advice on the illustrations and captions in the geomorphology portions of the book; **Dale Rice**, National Fisheries Services, La Jolla, California, for his review of the case study on whales; and **Payson R. Stevens**, CRM Books, Del Mar, California, for his conceptualization of the graphic essays.

Bibliography and Suggested Readings

1 In the Beginning

Bates, David Robert (ed.). 1960. *The Earth and Its Atmosphere*. rev. ed. New York: Basic Books.

Dietz, Robert S., and John C. Holden. 1970. "The Breakup of Pangaea," *Scientific American*, 223 (October): 30–41. Also Offprint No. 892.

Eicher, Don L. 1968. *Geologic Time*. Foundation of Earth Science Series. Englewood Cliffs, N.J.: Prentice-Hall.

Heirtzler, J. R. 1968. "Sea-Floor Spreading," *Scientific American*, 219 (December): 60–70. Also Offprint No. 875.

Hurley, Patrick M. 1959. *How Old is the Earth?* Garden City, N.Y.: Anchor Books.

Kurten, Bjorn. 1969. "Continental Drift and Evolution," *Scientific American*, 220 (March): 54–64. Also Offprint No. 877.

Landsberg, Helmut E. 1953. "The Origin of the Atmosphere," *Scientific American*, 189 (August): 82–86. Also Offprint No. 824.

McAlester, A. Lee. 1968. *The History of Life*. Englewood Cliffs, N.J.: Prentice-Hall.

Tarling, D. H., and M. P. Tarling. 1971. *Continental Drift*. London: G. Bell & Sons.

Urey, Harold C. 1952. "The Origin of the Earth," *Scientific American*, 187 (October): 53–60. Also Offprint No. 833.

2 The Dynamic Planet

Bolin, Bert. 1970. "The Carbon Cycle," *Scientific American*, 223 (September): 124–132. Also Offprint No. 1193.

Chorley, Richard J., and Barbara A. Kennedy. 1971. *Physical Geography: A Systems Approach*. London: Prentice-Hall International.

CRM Books. 1972. *Physical Science Today*. Del Mar, Cal.: CRM Books. pp. 135–147.

Penman, H. L. 1970. "The Water Cycle," *Scientific American*, 223 (September): 98–108. Also Offprint No. 1191.

Revelle, Roger. 1963. "Water," *Scientific American*, 209 (September): 92–108. Also Offprint No. 878.

3 Energy and Temperature

Barry, R. G., and Richard J. Chorley. 1970. *Atmosphere, Weather, and Climate*. New York: Holt, Rinehart, and Winston.

Budyko, M. I. 1958. *The Heat Balance of the Earth's Surface*. Washington, D.C.: United States Department of Commerce.*

Geiger, Rudolf. 1965. *The Climate Near the Ground*. Scripta Technica, Inc. (tr.). Cambridge, Mass.: Harvard University Press.

Lowry, William P. 1967. "The Climate of Cities," *Scientific American*, 217 (August): 15–23. Also Offprint No. 1215.

Riehl, Herbert. 1972. *Introduction to the Atmosphere*. New York: McGraw-Hill.

Sellers, William D. 1965. *Physical Climatology*. Chicago: University of Chicago Press.

4 Moisture and Precipitation

Battan, Louis J. 1962. *Cloud Physics and Cloud Seeding*. Science Study Series. Garden City, N.Y.: Doubleday.

Geiger, Rudolf. 1965. *The Climate Near the Ground*. Scripta Technica, Inc. (tr.). Cambridge, Mass.: Harvard University Press.

Lehr, Paul E., and R. Will Burnett. 1965. *Weather*. New York: Golden Press.

Lowry, William P. 1969. *Weather and Life*. New York: Academic Press.

Mason, Basil John. 1962. *Clouds, Rain, and Rainmaking*. London: Cambridge University Press.

Myers, Joel N. 1968. "Fog," *Scientific American*, 219 (December): 74–82. Also Offprint No. 876.

Penman, H. L. 1955. *Humidity*. New York: Barnes & Noble.

Riehl, Herbert. 1972. *Introduction to the Atmosphere*. New York: McGraw-Hill.

Sewell, W. R. Derrick (ed.). 1966. *Human Dimensions of Weather Modification*. Department of Geography Research Paper No. 105. Chicago: University of Chicago Press.

Woodcock, A. H. 1957. "Salt and Rain," *Scientific American*, 197 (October): 42–47. Also Offprint No. 850.

*Available from United States Government Printing Office, Washington, D.C.

5 Atmospheric and Oceanic Circulation

Battan, Louis J. 1961. *The Nature of Violent Storms*. Science Study Series. Garden City, N.Y.: Anchor Books.

Cole, Franklyn W. 1970. *Introduction to Meteorology*. New York: Wiley.

Das, P. K. 1968. *The Monsoons*. New Delhi, India: National Book Trust, India.

Flohn, H. 1969. *Climate and Weather*. New York: McGraw-Hill.

Gross, M. Grant. 1967. *Oceanography*. Robert J. Foster and Walter A. Gong (eds.). Merrill Physical Science Series. Columbus, O.: Charles E. Merrill.

Hare, F. K. 1966. *The Restless Atmosphere*. 4th ed. W. G. East (ed.), Geography. London: Hutchinson.

Lehr, Paul E., and R. Will Burnett. 1965. *Weather*. New York: Golden Press.

Reiter, Elmar R. 1967. *Jet Streams: How Do They Affect Our Weather?* Garden City, N.Y.: Doubleday.

Riehl, Herbert. 1972. *Introduction to the Atmosphere*. New York: McGraw-Hill.

Scientific American. *Readings in the Earth Sciences*. 1969. Vol. I, Scientific American Resource Library. Offprints 801–843. San Francisco: W. H. Freeman.

Stewart, R. W. 1969. "The Atmosphere and the Ocean," *Scientific American*, 221 (September): 76–86. Also Offprint No. 881.

Sutton, Oliver G. 1961. *The Challenge of the Atmosphere*. Harper Modern Science Series. New York: Harper & Row.

Trewartha, Glenn T. 1968. *An Introduction to Climate*. 4th ed. John C. Weaver (ed.), McGraw-Hill Series in Geography. New York: McGraw-Hill.

Weyl, Peter K. 1970. *Oceanography: An Introduction to the Marine Environment*. New York: Wiley.

6 The Hydrologic Cycle and Local Water Budget

Kazmann, Raphael G. 1972. *Modern Hydrology*. 2nd ed. New York: Harper & Row.

Overman, Michael. 1969. *Water: Solutions to a Problem of Supply and Demand*. Doubleday Science Series. Garden City, N.Y.: Doubleday.

Penman, H. L. 1970. "The Water Cycle," *Scientific American*, 223 (September): 98–108. Also Offprint No. 1191.

Revelle, Roger. 1963. "Water," *Scientific American*, 209 (September): 92–108. Also Offprint No. 878.

Sayre, A. N. 1950. "Ground Water," *Scientific American*, 183 (November): 14–19. Also Offprint No. 818.

Thornthwaite, C. Warren, and J. R. Mather. 1955. *The Water Balance*. Vol. VIII, No. 1, Publications in Climatology. Centerton, N.J.: Drexel Institute of Technology.

Ward, R. C. 1967. *Principles of Hydrology*. London: McGraw-Hill.

7 Global Systems of Climate

Blüthgen, Joachim. 1966. *Allgemeine Klimageographie*. 2nd ed. Berlin: Walter de Gruyter & Co.

Carter, Douglas B., and John R. Mather. 1966. *Climate Classification for Environmental Biology*. Vol. XIX, No. 4, Publications in Climatology. Elmer, N.J.: Drexel Institute of Technology.

Felton, Ernest L. 1965. *California's Many Climates*. Palo Alto, Cal.: Pacific Books.

Hare, F. K. 1966. *The Restless Atmosphere*. 4th ed. W. G. East (ed.), Geography. London: Hutchinson.

Haurwitz, Bernard, and James S. Austin. 1941. *Climatology*. New York: McGraw-Hill.

Kendrew, W. G. 1953. *The Climates of the Continents*. 4th ed. Oxford, England: Clarendon Press.

Rumney, George R. 1968. *Climatology and the World's Climates*. New York: Macmillan.

Trewartha, Glenn T. 1968. *An Introduction to Climate*. 4th ed. John C. Weaver (ed.), McGraw-Hill Series in Geography. New York: McGraw-Hill.

8 Global Systems of Soils

Archer, Sellers G. 1956. *Soil Conservation*. Norman, Okla.: University of Oklahoma Press.

Bridges, E. M. 1970. *World Soils*. New York: Cambridge University Press.

Bunting, Brian T. 1965. *The Geography of Soil*. Chicago: Aldine.

Edwards, Clive A. 1969. "Soil Pollutants and Soil Animals," *Scientific American*, 220 (April): 88–89. Also Offprint No. 1138.

Eyre, S. R. 1968. *Vegetation and Soils*. 2nd ed. Chicago: Aldine.

Hodge, Carle (ed.). 1963. *Aridity and Man: The Challenge of the Arid Lands in the United States*. Publication No. 74, American Association for the Advancement of Science. Washington, D.C.: AAAS.

Kellogg, Charles E. 1950. "Soil," *Scientific American*, 183 (July): 30–39. Also Offprint No. 821.

Money, D. C. 1965. *Climate, Soils and Vegetation*. London: University Tutorial Press.

Ollier, Clifford D. 1969. *Weathering*. New York: American Elsevier.

Russell, E. Walter. 1961. *Soil Conditions and Plant Growth*. 9th ed. London: Longmans, Green & Co.

Thorp, James, and Guy D. Smith. 1949. "Higher Categories of Soil Classification: Order, Suborder, and Great Soil Groups," *Soil Science*, 67: 117–126.

United States Department of Agriculture. 1938. *Soils and Men*. Yearbook of Agriculture. Washington, D.C.: United States Department of Agriculture.* pp. 970–978.

*Available from United States Government Printing Office, Washington, D.C.

9 Ecologic Energetics

Boerma, Addeke H. 1970. "A World Agricultural Plan," *Scientific American*, 223 (August): 54–69. Also Offprint No. 1186.

Boughey, Arthur S. 1969. *Contemporary Readings in Ecology*. Belmont, Cal.: Dickenson.

Brown, Lester R. 1967. "The World Outlook for Conventional Agriculture," *Science*, 158 (November): 604–611.

Chang, Jen-Hu. 1970. "Potential Photosynthesis and Crop Productivity," *Annals of the Association of American Geographers*, 60 (March): 92–101.

Flanagan, Dennis (ed.). 1970. "The Biosphere," *Scientific American*, 223 (September).

Kormondy, Edward J. 1969. *Concepts of Ecology*. William D. McElroy and Carl P. Swanson (eds.). Englewood Cliffs, N.J.: Prentice-Hall.

———. 1965. *Readings in Ecology*. William D. McElroy and Carl P. Swanson (eds.). Englewood CLiffs, N.J.: Prentice-Hall.

Major, Jack. 1963. "A Climatic Index to Vascular Plant Activity," *Ecology*, 44 (Summer): 485–497.

McLusky, Donald S. 1971. *Ecology of Estuaries*. London: Heinemann Educational Books.

Odum, Eugene P. 1971. *Fundamentals of Ecology*. 3rd ed. Philadelphia: W. B. Saunders.

———. 1969. "The Strategy of Ecosystem Development," *Science*, 164 (April): 262–269.

Phillipson, John. 1966. *Ecological Energetics*. No. 1, Studies in Biology. London: Edward Arnold.

Slobodkin, L. Basil. 1959. "Energetics in *Daphnia Pulex* Populations," *Ecology*, 40 (April): 232–243.

Ward, Barbara, and René Dubos. 1972. *Only One Earth: The Care and Maintenance of a Small Planet*. New York: W. W. Norton.

Whittaker, Robert H. 1970. *Communities and Ecosystems*. Current Concepts in Biology Series. New York: Macmillan.

10 Global Systems of Vegetation and Climate

Brooks, C. E. P. 1949. *Climate Through the Ages: A Study of the Climatic Factors and Their Variations.* rev. ed. London: Ernest Benn.

Carter, Douglas B., and John R. Mather. 1966. *Climate Classification for Environmental Biology.* Vol. XIX, No. 4, Publications in Climatology. Elmer, N.J.: Drexel Institute of Technology.

Claiborne, Robert. 1970. *Climate, Man, and History.* New York: W. W. Norton.

Eyre, S. R. 1968. *Vegetation and Soils.* 2nd ed. Chicago: Aldine.

Ladurie, Emmanuel Le Roy. 1971. *Times of Feast, Times of Famine.* Garden City, N.Y.: Doubleday.

Lamb, H. H. 1966. *The Changing Climate: Selected Papers.* London: Methuen.

Landsberg, H. E. *et al.* 1965. *World Maps of Climatology.* 2nd ed. New York: Springer-Verlag.

Money, D. C. 1965. *Climate, Soils, and Vegetation.* London: University Tutorial Press.

Nelson, H. L. 1968. *Climatic Data for Representative Stations of the World.* Lincoln, Neb.: University of Nebraska Press.

Rumney, George R. 1968. *Climatology and the World's Climates.* New York: Macmillan.

Trewartha, Glenn T. 1968. *An Introduction to Climate.* 4th ed. John C. Weaver (ed.), McGraw-Hill Series in Geography. New York: McGraw-Hill.

General Geomorphology References

Bloom, Arthur L. 1969. *The Surface of the Earth.* Englewood Cliffs, N.J.: Prentice-Hall.

Cotton, C. A. 1960. *Geomorphology.* 7th ed. Wellington, New Zealand: Whitcombe and Tombs.

Easterbrook, Don J. 1969. *Principles of Geomorphology.* New York: McGraw-Hill.

Lobeck, Armin K. 1956. *Things Maps Don't Tell Us.* New York: Macmillan.

Machatschek, Fritz. 1969. *Geomorphology.* 9th ed. New York: American Elsevier.

Pitty, Alistair F. 1971. *Introduction to Geomorphology.* London: Methuen.

Selby, M. J. 1967. *The Surface of the Earth.* London: Cassell.

Small, R. J. 1970. *The Study of Landforms.* New York: Cambridge University Press.

Thornbury, William D. 1966. *Principles of Geomorphology.* New York: Wiley.

Upton, William B., Jr. 1970. *Landforms and Topographic Maps.* New York: Wiley.

11 Sculpture of the Landscape

Clayton, Keith. 1966. *Earth's Crust.* London: Aldus Books.

Davis, William Morris. 1954. *Geographical Essays.* New York: Dover.

Leopold, Luna B., M. Gordon Wolman, and John P. Miller. 1964. *Fluvial Processes in Geomorphology.* San Francisco: W. H. Freeman.

Penck, Walther. 1953. *Morphological Analysis of Landforms.* London: Macmillan.

Tarling, D. H., and M. P. Tarling. 1971. *Continental Drift.* London: G. Bell & Sons.

12 Flowing Water and Its Work

IBM Research Reports. 1972. *Hydrology by Computer.* Yorktown Heights, N.Y.: IBM Research.

Judson, Sheldon. 1968. "Erosion of the Land, or What's Happening to Our Continents?" *American Scientist,* 56 (Winter): 356–374.

Leopold, Luna B., and W. B. Langbein. 1966. "River Meanders," *Scientific American,* 214 (June): 60–70. Also Offprint No. 869.

———, M. Gordon Wolman, and John P. Miller. 1964. *Fluvial Processes in Geomorphology.* San Francisco: W. H. Freeman.

Matthes, Gerard H. 1951. "Paradoxes of the Mississippi," *Scientific American,* 184 (April): 18–23. Also Offprint No. 836.

Morisawa, Marie. 1968. *Streams: Their Dynamics and Morphology.* Patrick M. Hurley (ed.). Earth and Planetary Science Series. New York: McGraw-Hill.

13 Landforms and the Shifting Crust

Bullard, Fred M. 1962. *Volcanoes: In History, in Theory, in Eruption.* Austin, Tex.: University of Texas Press.

Iacopi, Robert. 1964. *Earthquake Country.* Menlo Park, Cal.: Lane Book Company.

Ollier, Clifford D. 1969. *Volcanoes.* Cambridge, Mass.: The M.I.T. Press.

———. 1969. *Weathering.* New York: American Elsevier.

Shelton, John S. 1966. *Geology Illustrated.* San Francisco: W. H. Freeman.

14 Regional Landforms

Birot, Pierre. 1968. *The Cycle of Erosion in Different Climates.* Berkeley, Cal.: University of California Press.

Carson, M. A., and M. J. Kirby. 1972. *Hillslope Form and Process.* London: Cambridge University Press.

Leopold, Luna B., M. Gordon Wolman, and John P. Miller. 1964. *Fluvial Processes in Geomorphology.* San Francisco: W. H. Freeman.

Ollier, Clifford D. 1969. *Weathering.* New York: American Elsevier.

Penck, Walther. 1953. *Morphological Analysis of Landforms.* London: Macmillan.

15 The Power of Ice

Davies, J. L. 1969. *Landforms of Cold Climates*. Cambridge, Mass.: The M.I.T. Press.

Embleton, Clifford, and A. M. King. 1968. *Glacial and Periglacial Geomorphology*. New York: St. Martin's Press.

Hough, Jack L. 1958. *Geology of the Great Lakes*. Urbana, Ill.: University of Illinois Press.

Sharp, Robert P. 1960. *Glaciers*. Eugene, Ore.: University of Oregon Press.

16 The Edge of the Land

Bird, E. C. F. 1969. *Coasts*. Cambridge, Mass.: The M.I.T. Press.

Gross, M. Grant. 1967. *Oceanography*. Columbus, O.: Charles E. Merrill.

McLusky, Donald S. 1971. *Ecology of Estuaries*. London: Heinemann Educational Books.

Shepard, Francis P. 1967. *The Earth Beneath the Sea*. Baltimore, Md.: The Johns Hopkins Press.

Steers, J. A. (ed.). 1971. *Introduction to Coastline Development*. London: Macmillan.

Epilogue: Physical Geography USA

Atwood, Wallace W. 1940. *The Physiographic Provinces of North America*. Boston, Mass.: Ginn & Co.

Hunt, Charles B. 1967. *Physiography of the United States*. San Francisco: W. H. Freeman.

Lee, W. Storrs. 1963. *The Great California Deserts*. New York: G. P. Putnam's Sons.

Paterson, J. H. 1962. *North America*. 2nd ed. London: Oxford University Press.

Strahler, Arthur N. 1966. *A Geologist's View of Cape Cod*. Garden City, N.Y.: The Natural History Press.

Thornbury, William D. 1965. *Regional Geomorphology of the United States*. New York: Wiley.

Watson, J. Wreford. 1967. *North America*. New York: Praeger.

Appendix I: Expressing Scientific Measurements

No references.

Appendix II: Tools of the Physical Geographer

Birch, T. W. 1964. *Maps: Topographical and Statistical*. 2nd ed. London: Oxford University Press.

Colwell, Robert N. 1973. "Remote Sensing as an Aid to the Management of Earth Resources," *American Scientist*, 61 (March–April): 175–183.

Dickinson, G. C. 1969. *Maps and Air Photographs*. London: Edward Arnold.

Garland, G. D. 1965. *The Earth's Shape and Gravity*. New York: Pergamon Press.

Greenhood, David. 1964. *Mapping*. Chicago: The University of Chicago Press.

National Aeronautics and Space Administration. 1970 (March). *Earth Resources Program Synopsis of Activity*. Washington, D.C.: NASA.*

———. 1970. *Ecological Surveys From Space*. Washington, D.C.: NASA.*

———. 1971 (July). *Exploring Earth Resources From Space*. Washington, D.C.: NASA.*

Raisz, Erwin. 1962. *Principles of Cartography*. New York: McGraw-Hill.

Robinson, Arthur H., and Randall D. Sale. 1969. *Elements of Cartography*. 3rd ed. New York: Wiley.

Steers, J. A. 1965. *An Introduction to the Study of Map Projections*. 14th ed. London: University of London Press.

United States Department of the Army. 1960 (October). *Map Reading*. Department of the Army Field Manual FM 21-26. Washington, D.C.: United States Department of the Army.*

United States Department of the Interior Geological Survey. 1970. *The National Atlas of the United States of America*. Washington, D.C.: United States Department of the Interior.*

*Available from United States Government Printing Office, Washington, D.C.

Appendix III: Classification Systems

Haurwitz, Bernard, and James S. Austin. 1941. *Climatology*. New York: McGraw-Hill.

Köppen, W. 1936. "Das geographische System der Klimate," *Allgemeine Klimalehre*. Part C, Vol. I. W. Köppen and R. Geiger (eds.), Handbuch der Klimatologie. Berlin: Gebrüder Borntraeger.

Thornthwaite, C. W. 1948. "An Approach Toward a Rational Classification of Climate," *Geographical Review*, 38: 55–94.

Thorp, James, and Guy D. Smith. 1949. "Higher Categories of Soil Classification: Order, Suborder, and Great Soil Groups," *Soil Science*, 67: 117–126.

United States Department of Agriculture. 1938. *Soils and Men*. Yearbook of Agriculture. Washington, D.C.: United States Department of Agriculture.* pp. 970–978.

United States Department of the Interior Geological Survey. 1970. *The National Atlas of the United States of America*. Washington, D.C.: United States Department of the Interior.*

*Available from United States Government Printing Office, Washington, D.C.

Glossary–Index

Index entries for the technical terms used in the text are accompanied by glossary definitions, and names of locations and geographic features include summary descriptions. Italicized page numbers refer to figure captions.

a

aa lava (pronounced "AH-ah"), *333:* a rough, angular form of solidified lava.

Aberdeen, Scotland, summer and winter temperatures, 162

ablation, *381,* 382, *383:* any of the processes, such as melting, that result in the loss of ice or snow from a glacier.

abrasion: the wearing of rock by mechanical friction with other particles.
  by glaciers, 386, *386,* 387

absolute humidity, 83–84: the weight of water vapor present in a given volume of moist air, usually expressed in grams of water per cubic meter of air.
  versus temperature, *83*

absolute scale of temperature, 30: the Kelvin scale of temperature.

absorption, 57: (electromagnetism) the process in which the energy of electromagnetic radiation is taken up by a molecule and changed to another form of energy.
  of electromagnetic radiation by the atmosphere, *57*

Acacia thorn trees, *370*

accumulation, *381,* 382–383, *383:* any of the processes that result in an increase of snow or ice in a glacier.

acid: any substance containing hydrogen ions that are available to take part in chemical reactions.

acidity, 184–185: the availability of hydrogen ions to take part in chemical reactions.

active glacier, 383–384: a glacier that transports material relatively rapidly.

actual evapotranspiration, 149: the actual amount of water evaporated and transpired by a large stand of plants.
  as an index of plant growth, 212–213
  measurement of, 149

adiabatic cooling, 96: the drop in temperature occurring in a parcel of gas that is forced to expand without receiving or losing energy.

adiabatic heating, 96: the rise in temperature occurring in a parcel of gas that is forced to contract without receiving or losing energy.

adiabatic lapse rate: the rate of change of temperature with altitude occurring in a parcel of air rising freely through the atmosphere without significant gain or loss of energy.
  dry, 96
  moist, 99

adiabatic processes, 96, *96, 97,* 99
  See also adiabatic cooling and adiabatic heating

advection, 93: (meteorology) the horizontal movement of parcels of air or air masses.

advection fog, 92, *93:* fog formed when warm, moist air moves over a cool surface and is chilled to its dew point.

AE: *See* actual evapotranspiration.

aerial photography, 470, *471,* 472

aggradation by streams, 291: prolonged filling.

air mass, 115–117: a parcel of air of relatively uniform temperature and humidity, which can move as a unit into regions with different temperatures and humidities.
  classification of, 116
  over United States, *117*

air temperature and ground temperature, 66–67, *67*
  effect of bodies of water on, 68, *68,* 70

albedo, 59, 62: the fraction of incident electromagnetic radiation reflected by a surface.

alkaline, 185: having fewer hydrogen ions than does neutral water.

alluvial apron, 305–306, *364:* an extensive, continuous alluvial deposit formed by the coalescence of adjacent alluvial fans; a bajada.

alluvial fan, 305, *306, 364, 365:* a fan-shaped deposit laid down by a stream where it leaves a mountainous region.
  and groundwater supply, 306

alluvial plain, 306: a plain covered with alluvium.

alluvial soils, *200:* soils developed from alluvial deposits; they occur in river valleys and in mountain valleys.

alluvial terrace, 304, *304:* steplike levels in the alluvium along the edges of a stream's floodplain, formed when the central part of the floodplain is cut down by the stream.
  and geologic history, 304
  paired, unpaired, *304*

alluvium, *368*: sediment deposited by streams.
  and sedimentary rocks, 315, 318

alpine topography, 388

altitude, 52: the elevation above a given reference level, such as average sea level.

altitude tinting, 461, *461:* the use of color tints on maps to indicate ranges of altitudes; also called hypsometric tinting.

altocumulus clouds, *94:* clouds of medium height that form an extensive area of fluffy white patches called mackerel sky; commonly associated with good weather.

and cloud streets, *94*

altostratus clouds, 118: clouds that blanket the sky in a gray layer; commonly associated with bad weather.

and weather fronts, *118*

Amboseli Park, Kenya, savanna vegetation, *238*

Angkor Wat, Cambodia, laterite, 198

angle of repose, 352, 354, *368:* the steepest angle that can be assumed by loose fragments on a slope without downslope movement.

Animas River, Colorado, floodplain and meanders, *303*

anion (pronounced "AN-eye-on"), 184: a negatively charged ion; an atom with a surplus of electrons.

Antarctic Circle, *51*

Antelope Peak, Arizona, pediment, *361*

anticlinal ridge, *343:* a ridge formed along the exposed crest of an anticline.

anticlinal valley, *343:* a valley that follows the axis of an anticline.

anticline, *338,* 343: an up-arched fold in rock strata, in which the limbs, or flanks, of the fold dip downward.

overturned, *338*

anticyclones, or highs, 115–116: a large system of air circulation centered on a region of high atmospheric pressure where air descends; air in anticyclones rotates clockwise in the Northern Hemisphere but counterclockwise in the Southern Hemisphere.

divergence out of, 115, *116*
speed of movement of, 116
and weather, 116–117

Appalachian Mountains, Pennsylvania

folded structure of, 344, *344*
radar view of, *343*
ridge and valley topography of, *323*

7th Approximation classification of soils, 194, 478–479

aqueduct, *146*

aquifer, *137,* 141–144, *142:* an underground layer of permeable material that can store and supply water.

confined, *142,* 143
speed of water flow in, 143
structure of, 142–143, *142*
unconfined, *142*
water quality and temperature of, 142

arch, eroded sandstone, *271, 438,* 439

Arches National Monument, Utah, sandstone formations, *438,* 439

Arctic Circle, *51*

Arecibo, Puerto Rico, radio telescope, 417, *418*

arête, 388, *391:* a sharp ridge formed by erosion where two neighboring cirque walls intersect.

arid climate: *See* dry climate region.

Arroyo de los Frijoles, New Mexico, longitudinal profile, *294*

asteroids, *6:* numerous heavenly bodies, satellites of the sun; smaller than planets in size, they orbit primarily between the orbits of Mars and Jupiter.

asthenosphere, *9:* a layer of rock in the upper mantle, 75 to 175 kilometers beneath the surface; this rock is hot and flows readily, and the lithospheric plates drift on it.

Astoria, Oregon, monthly radiant energy balance, *72*

asymmetry of slopes, 357–358, *358*

in different climatic regions, *358*
effect of streams on, 358

Atchafalaya River basin, Louisiana (pronounced "a-chaff-uh-LIE-uh")

and flood of 1973, 308–308, *309*
model of, *45*

Atlantic Ocean Basin, 2

formation of, 264

atmosphere, 10: the layer of gases that envelopes the surface of a body such as a planet.

composition of, 57
evolution of, 10–11, 18
interaction with radiation, 57–59
pressure variation with altitude, *37, 58*
temperature variation with altitude, *58*
United States standard of 1962, *58*

atmospheric pressure, 37–38: the force per unit area exerted by the gases in the atmosphere.

versus altitude, *37, 58*

atmospheric research, 128–129

atmospheric window, *57,* 63: the wavelength range in which the atmosphere does not strongly absorb electromagnetic radiation; principally from 8.5 to 11 microns.

atoll, *414:* a coral reef in the form of a ring or partial ring, which encloses a lagoon.

formation of, *414*

atoms, *10*

Australia, thorn and scrub forest, *239*

autotrophs, 215: organisms capable of subsisting on inorganic materials and energy; primary producers.

average solar time, *53*

azimuth, 52, 451: angular distance from north, usually measured clockwise and expressed in degrees as an angle between 0° and 360°.

See also grid azimuth, magnetic azimuth, true azimuth

azimuthal projection, 454, *455:* a projection of a portion of the globe onto a tangent plane.

azonal soils, 194: soils that do not have a developed profile, such as those formed of volcanic ash.

b

Babylonian world map, *446*

backshore, *406:* the upper part of a shore, which is beyond the reach of ordinary waves and tides.

backwash, 404: the water that flows back to the sea after the advance of a wave.

badlands, *366*

Badlands, South Dakota, 432–434, *433*

bajada (pronounced "bah-HAH-dah"): *See* alluvial apron.

Baldwin Creek, Wyoming, velocity of flow, *286*

bar, *406, 411:* (geomorphology) a ridge of sand or gravel deposited offshore or in a river, but not meeting the shore; (meteorology) a unit of pressure.

Barahona, Dominican Republic, actual evapotranspiration, 213, *213*

barchan (pronounced "bahr-KAHN"), *366:* a sand dune shaped like a crescent, with its convex side facing the wind and its horns pointed downwind.

basal slip, *384,* 385: a form of glacial motion in which the glacier as a whole moves by slipping over its bed.

basalt, 315, *317, 372*

  Devil's Tower, 434, *435,* 436

base flow, 145, *145:* the portion of a stream's discharge that is maintained by the comparatively steady supply of groundwater.

base level of erosion, 275: the lowest point to which a stream can wear its bed.

base line: an east-west reference line for the United States Bureau of Land Management land survey, which is laid out along a parallel; in surveying, a line of known length and position used as a reference line for map making.

  and field surveying, 469, *469*
  in land survey, 451, *451*
  and topographic profile, *460*

basic: *See* alkaline.

batholith, *327, 334:* an extensive intrusion of magma, often having a surface area of hundreds of square miles.

Baton Rouge, Louisiana

  local water budget, 150–155, *151, 152, 162*
  microclimate, *67*

beach, 396, 406–409, *406,* 425: the part of a shore where wave erosion and deposition is active.

  ephemeral nature of, 396, 406
  mass budget of, *406*
  seasonal variation in, 406, *406*
  and sediment transport, 406–409

bed load, 288: the amount of material transported by motion along the bed of a stream.

bedrock: the rock underlying a region.

Belém, Brazil, thermoisopleth, 235, *236*

Bergeron, Tor, 90

Bergeron ice crystal model, *89,* 90–91: a model of precipitation appropriate where a cloud is below freezing, in which an early stage is the growth of ice crystals from water vapor in an atmosphere containing supercooled water droplets.

bergschrund (pronounced "BERK-shrōont"), *381:* a crevasse between the head of a glacier and a rock wall; the scene of active erosion of rock by snow and ice.

berm, *406:* the relatively flat area at the backshore of a beach, formed of material deposited by wave action.

big bang theory, 4–5, *4:* the hypothesis that the energy and matter in the universe were gathered together at a remote time in the past, rose to a high temperature, and then expanded rapidly outward.

biomass (standing crop), 212: the amount of energy contained in a group of organisms at a given time.

  contrasted with productivity, 212

biomass pyramid, *219:* a schematic diagram that represents the biomass in each of the stages of a food chain; usually, but not always, biomass decreases along a food chain.

black body, 56: the most efficient emitter of thermal radiation at a given temperature.

Black Hills, South Dakota

  climate of, 434
  domed geologic structure of, 269, 434
  precipitation in, 78

Black Tail Butte, Wyoming, asymmetric slopes, *358*

Blanding, Utah, hoodoo rock, *357*

block diagram, 462, *462:* a perspective diagram for representing landforms and their underlying geologic structures.

blow-out dune: *See* parabolic sand dune.

Blue Canyon, California, water budget surplus and AE/PE ratio, 170, *171*

blue-green algae, *7, 17*

Blythe, California, actual evapotranspiration, 213, *213*

bog and meadow soils, *193,* 202, *202*

  agricultural potential of, 202
  distribution of, 202
  profile, *202*
  and vegetation, *193,* 202

Bonnet Carré spillway, Louisiana, 308–309

boreal: the subpolar, or subarctic, climate region.

bornhardt (tropical inselberg), *370, 371:* an isolated summit, usually a granite dome, rising above a savanna plain; a feature of exhumation.

Boston, Massachusetts

  climate of, 422
  drumlins, 423

braided stream channel, 294, *295,* 301, *301,* 302: a channel made up of numerous intertwining streams, which are separated by sand and gravel deposits.

breakwater, 396: a barrier built to reduce the force of the waves against a shore.

Brookhaven forest, Long Island, New York, energy flow, 220, *220*

Brownsville, Texas, water budget surplus and AE/PE ratio, *171*

brunizems: *See* prairie soils.

Bryce Canyon National Park, Utah, erosion spires, *271*

budget, 39–40: an accounting of the energy and material that enters, leaves, or is stored in a system.

buoyant force, *36, 38*

Bureau of Land Management, land survey, 451, *451*

butte, 322, *323, 366, 367:* a small, eroded mesa that often has a flat top.

Buys Ballott's law, 108: states that an observer standing with the geostrophic wind at his back will have high pressure to his right and low pressure to his left in the Northern Hemisphere, but high pressure to his left and low pressure to his right in the Southern Hemisphere.

C

Cabin Creek, Montana, knickpoint, 299

calcified soil, 190–191, *193*

caldera, *334:* a large basin formed by the collapse of a volcanic cone.

California State Water Project, 147–148, *147*

calorie, 34: a unit of energy equal to the energy required to heat 1 gram of water from 14.5°C to 15.5°C.

calving, *175:* the breaking off of icebergs from the edge of ice sheets.

Canterbury Plains, New Zealand, tussocked grassland, *244*

Cape Cod, Massachusetts
- beach, *424*, 425
- erratics, 425
- and glaciation, 423–425
- kettles, *424*, 425
- moraines, 423, *424*, 425
- sea cliffs, *424*, 425
- topography, 423

capillary action, 140: the upward movement of liquids in confined spaces, caused by attractive forces between the liquid and the walls.

Caracas, Venezuela, local water budget, 238, *239*

carbon cycle, 41, *42:* on the earth, the passage of carbon into the atmosphere as carbon dioxide, and its return to the surface to be stored in vegetation and other systems.

carbon 14 dating, 24
- of glacier ice, 382
- of sea level, 399
- of sea water, 130

carbon dioxide
- and the carbon cycle, 41, *42*
- and the cycle of life, *208*
- in the early atmosphere, 11

Carmel, California, seasonal variation of beach, *406*

carnivore, 216: a mammal that subsists chiefly on flesh.

Carrizo Plain, California, and San Andreas fault, *340*

Castle, in southwestern Wisconsin, sandstone formation, *430*

Castle Geyser, Yellowstone National Park, Wyoming, *329*

Catania, Italy, lava flow, 312

cation (pronounced "CAT-eye-on"), 184: a positively charged ion; an atom deficient in electrons.

cation exchange, 185, *185*, 187–188: the replacement of one type of cation bound to the clay-humus complex by another type that is more strongly bound.
- and leaching, 190, *190*
- and soil acidity, 187–188

Celsius scale of temperature, 29–30, *30*

Centigrade scale of temperature, 29: now called the Celsius scale.

Challenger, H.M.S., *412*

chemical reaction rate, and climate, 350

chemical weathering, and climate, 349–350, *349*, *350*

chernozems, *189*, *192*, 198, *198*, *200:* soils of the tall-grass prairie; they are dark and fertile soils.
- agricultural potential of, 198
- and climate, *189*
- global distribution of, *200*
- profile, *198*
- United States distribution of, 198, *200*
- and vegetation, *192*, 198

chestnut and brown soils, *189*, 198–199, *199*, *200:* soils of the semiarid shortgrass prairie; they are rich in calcium compounds.
- agricultural potential of, 199
- and climate, *189*
- global distribution of, *200*
- profile, *199*
- United States distribution of, 198–199, *200*
- and vegetation, 198–199

Chicago, Illinois, summer and winter temperatures, 162

Chlamydomonas algae, 217

chloroplast, *82*

cinder cone, *330:* a steep-sided peak formed from volcanic ash.

circle of illumination, 48, *51:* the great circle that theoretically separates day from night on the surface of the earth.

circle of tangency, 454, *456*

circulation: the movement of air in the atmosphere or of water in the oceans; usually large in scale.
- *See also* general circulation, local circulations, monsoon circulation, secondary circulation

cirque, 388, *389*, 439: a steep-walled, horseshoe-shaped bowl formed by the erosive action of snow and ice at the head of a valley glacier.

cirrus clouds, *94:* feathery, high-altitude clouds that are composed of ice crystals.
- and weather fronts, *118*

Clarke spheroid of 1866, 452

classification
- artificial nature of boundaries in, 161
- of climate, 160–162, 474–475
- of landforms, 267–268
- of morphogenetic regions, 479
- of soils, 191, 194, 477–479
- of systems, 41

1938 classification of soils, 194, 477–478

clay, 183, *183*, *184:* inorganic particles with diameters of less than 2 microns, the most minute of all sediments.

clay-humus complex, 183, *185:* a combination of humus gel and fine rock particles, which can act much like a large ion.

Cleveland, Ohio, hourly temperature distribution, 68, *68*

climate, 160: the average long-term meteorologic conditions of a region, based on indexes such as temperature and precipitation, or on the interaction of energy and moisture with systems.
- comparison between Aberdeen and Chicago, 162–163
- contrast between Cloverdale, California, and Manhattan, Kansas, *167*, 170
- distribution of, on hypothetical continent, 163–166, *165*
- and the general circulation, 160
- global distribution of, *168*
- and interactions at the surface, 166–167, 170–172
- and landforms, 348–352
- and the local water budget, 167, *167*, 170–172, *171*
- man's adaptations to, 254–257
- variations in, *171*
- and weathering, 349–351

climate region, 107, *164*, 165–166, *165*, *168*, 245, 247, *247:* a climatic classification, based on indexes such as rainfall, temperature, humidity, wind, and so on.
- *See also* dry climate region, humid continental climate region, humid subtropical climate region, marine climate region, polar climate region, subpolar climate region, subtropical dry summer climate region, tropical wet climate region, tropical wet and dry climate region

climatic change, 247, 250–253
- and carbon dioxide, 253
- and dust, 252–253
- history of, 250–252
- man's effect on, 253
- present trend in, 252
- theories of, 252–253
- time scale of, 250
- and yearly variations, 247, 250, *251*

climatic index, 161: a measure of some aspect of climate.
ratio of actual to potential evapotranspiration as, *171*
water surplus as, 170–172, *171*
climatic optimum, 251: a time about 5,000 years ago, when the climate was the warmest it has been during the Holocene (Recent).
secondary climatic optimum, 251
climax vegetation, 220, 232–233: the final stabilized pattern of vegetation to establish itself in a region.
Clinton, Mississippi, model of Mississippi River, *45*
clouds
effect of on radiant energy balance, 59, 62–63, *63*
formation of, 93–96, *97*
types of, *94*
cloud seeding, 100–101, *100:* the dispersal of material into clouds to foster precipitation.
effectiveness of, 101
cloud streets, *94:* parallel bands of clouds extending over a wide area.
Cloverdale, California, local water budget, 167, *167,* 170
Clutha River, New Zealand, sediment transport, *289*
coalescence, 88: the merging of water droplets to form larger drops, or the collection of water droplets by a large drop.
coalescence model, 88–90, *89:* a model of precipitation appropriate to clouds warmer than freezing, in which raindrops start from an initial large droplet and grow by coalescence during collisions with smaller droplets.
and thickness of clouds, 88–90
coastline development, *411*
submerged parallel coast, *411*
submerged transverse coast, *411*
coastline features, *411*
cockpit karst, *373*
cold front, *118,* 119, *119:* a boundary between air masses, formed when cold air invades a region of warm air and pushes the warm air upward.
and occluded front, *120*
colluvium, *362:* unconsolidated debris carried by wash from slopes, consisting of material such as soil and talus.
Colorado Plateau of Utah and Arizona, semidesert, 239
Colorado River
downcutting of, *17*
and Grand Canyon, *17, 438,* 439–440
irrigation use at Imperial Valley, 440, 443
scouring and filling of, *297*
spit near mouth of, *411*
color infrared photography, 471, *471,* 472
and identification of vegetation, 471, *472*
community: *See* plant community.
compass traverse, 468–469, *469*
condensation, 80, *80:* the change of a substance's physical state from gaseous vapor to liquid or solid.
heat energy in, 85–86
and snowmelt, 86
condensation nuclei, 87–88, *89:* the fine particles in the atmosphere that act as centers to promote the condensation of water from water vapor.
giant, 90
number of, 87
salt particles as, 90, 94–95
sizes of, 87, *87*
sources of, 87
conduction, 64: the flow of heat from one part of a solid body to the next, by means of collisions between atomic particles.
cone karst, *372, 373*
confined aquifer, *142,* 143: a layer of permeable, water-bearing material that is enclosed between layers of impermeable rock.
conformal projection, 453: a map projection in which angular relationships are shown correctly; conformal projections depict the shapes of small areas accurately.
conic projection, 454, *457:* a projection of a portion of the globe onto a cone.
conservation of energy, 31
conservation of matter in a stream, *286,* 287
conterminous United States: the adjoining 48 states; all the United States except Alaska and Hawaii.
Continental Divide
in Rocky Mountains, 436
and stream drainage, *291*
continental drift, 2, 19–22, *20,* 261–264: the gradual movement of crustal plates across the earth's surface.
continental plate, 264: a section of crust that underlies a continent.
continental rise, *397,* 398: the submarine region where a continental slope meets the ocean floor; sediment often accumulates there.
continental shelf, 397, *397:* the shallow, gently sloping underwater regions at the margins of continents.
formation of, 398
and geosynclines, 337
rate of slope of, 397
continental slope, 397–398, *397:* the steeply sloping underwater region from a continental shelf to a continental rise.
rate of slope of, 397–398
contour interval, 459, *459,* 460, *460:* the difference in elevation between successive contour lines.
contour line, *433,* 459–460, *459, 460:* a line on a map joining points that have the same elevation above a reference plane.
and topography, 460, *460*
convection: (meteorology) the vertical movement of parcels of air or air masses.
convective cell, 109–110, *109, 123:* a current of vertical circulation characterized by ascending warmer air and descending cooler air; a Hadley cell, especially if on a large scale.
convergence of air, 115, *116*
conversion factors, metric to English, 445, *445*
Convict Lake, Sierra Nevada, California, glacial lake and ground moraine, *278*
coral atoll, *414*
core, *9,* 10: the central part of the earth's structure, which begins at a depth of about 2,900 kilometers.
Coriolis, Gaspard, 108
Coriolis force, 108, *108:* a force apparent to an observer on the rotating earth, which tends to deflect moving objects to the right in the Northern Hemisphere but to the left in the Southern Hemisphere.
and the general circulation, 109
variation of with latitude, *109*
corrasion, 289: mechanical erosion caused by particles being transported by wind, flowing water, or glacial ice.
Crater Lake, Oregon
landslide, *334*
pillars, *271*
crest, 402, *402:* the highest point of a wave or wavelike structure.

crevasse, *376:* a deep crack in a glacier, which usually cuts across the direction of glacier flow.

crust, *9,* 10: the thin outermost layer of the earth's structure, separated from the mantle by a boundary of seismic wave discontinuity; the lithosphere.

crustal plate: a section of the lithosphere underlying an ocean or a continent.

cuesta (pronounced "KWESS-tah"), *322,* 323, *366, 367:* an extensive, gently sloping ridge with one steep face.

Cuiabá, Brazil, temperature and precipitation of, 237, *237*

cumulonimbus clouds ("thunderheads"), *94:* towering, anvil-headed clouds associated with thunderstorms, hail, and, occasionally, tornadoes.

and hailstorms, *91*
and weather fronts, *118*

cumulonimbus mammatus, *94:* an uncommon cloud type of extensive area, with downward bulges on its lower surface; it is often associated with severe storms and tornadoes.

cumulus clouds, *94,* 95–96, *175,* 436: puffy white clouds that form locally on rising columns of air.

cycle, 40–41, *41, 42:* the passage of energy or material through successive systems, with eventual return to the starting point.

man's input to, 41

cyclogenesis, *120:* the formation of cyclones in the secondary circulation.

cyclones (lows), 115–116: a large system of air circulation centered on a region of low atmospheric pressure, where air ascends; air in cyclones rotates counterclockwise in the Northern Hemisphere but clockwise in the Southern Hemisphere.

convergence of air into, 115, *116*
development of, 120, *120, 122*

cylindric projection, 454, *456:* a projection of part of the globe onto a cylinder.

d

Dallas, Texas, water budget surplus and AE/PE ratio, *171*

Daphnia, water flea, 217, *217*

Darwin, Charles, 14, 23, *414*

Davis, William Morris, 275–277, *277*

Davisian cycle of erosion, 275–278, *276, 277:* an evolutionary sequence of landforms devised by William Morris Davis to describe midlatitude landscapes shaped by fluvial processes.

criticisms of, 277, 278
dependence on structure, process, and stage, 277
importance of, 277
present view of, 277–278

Death Valley, California

alluvial fan in, *306, 365*
basin, playa, alluvial fans in, *365*

deciduous: shedding its leaves during a definite season of the year; applied to plants, particularly trees.

deep circulation of the oceans, 130

effect of temperature and salinity, 130

deep water waves, 402, *402:* waves in water that is deeper than half their wavelength; their speed is independent of the depth of the water.

deficit, *151, 152,* 154: in the local water budget, the amount by which available moisture supplies fall below potential evapotranspiration.

during wet, dry, and average years, *152,* 154–155

degradation by streams, 291: prolonged scouring.

dell, *362:* a flat-floored low region covered with unconsolidated debris washed down from slopes.

delta, 305, *305:* the accumulated alluvial deposits laid down by a stream at its mouth.

processes that form deltas, 305

dendritic: branching like a tree; applied particularly to irregular drainage patterns in which tributaries enter the main stream at varying angles.

dendritic drainage pattern, *270*

denudation: the removal of material from a location by erosion.

influence of precipitation and vegetation on rate of, 351–352, *351*

deposition, 269: the laying down of material by an agent of erosion or transport, such as water or wind.

desert, 440, *441*

desert soils, *189, 193,* 199, *200,* 202:

agricultural potential of, 202
calcification in, 202
and climate, *189*
global distribution of, *200*
profile, *199*
United States distribution of, 199, *200*
and vegetation, *193*

desert varnish, *364:* a thin coating of colored mineral oxides that forms on exposed rock surfaces in desert regions; usually manganese or iron oxides.

Devil's Postpile, California, basalt columns, *317,* 436

Devil's Tower, Wyoming, 434, *435,* 436

Devon, England

cow pasture, *272*
morning mist, *175*

dew, 85–86, *85:* the moisture that condenses from moist air onto surfaces at the dew point.

and vertical transport of moisture, *64, 86*

dew point, 85–86: the temperature at which a sample of moist air becomes saturated.

dike, volcanic, 325, *333, 334:* a solidified intrusion of igneous rock that forms a vertical wall-like structure.

and differential erosion, 325
and Shiprock, *333*

dip, *337, 339:* the acute angle between the bedding layers of sedimentary rocks and the horizontal plane, or between the steepest line on a fault plane and the horizontal plane.

dip-slip fault, *339:* a fault in which the relative displacement of the blocks is along the direction of dip.

discharge, 286–287, *286:* the volume of water transported by a stream past a given point during a standard time interval; often measured in cubic meters per second or cubic feet per second.

dissolved load, chemical load, 288: the amount of material carried in solution by a stream.

factors affecting, 288

distributaries, 305: channels into which a stream's discharge divides; they do not rejoin; characteristic of delta regions.

divergence of air, 115, *116*

Dokuchaiev, Vasilii, *186,* 187, 194

dominant vegetation, 231–232

Dorset, England, sea cliffs, *271*

Dove, Heinrich Wilhelm, 117

Dover, England, tides, *401*

downcutting, 440, *441:* vertical adjustment by a stream in which material is removed primarily from the bed of the stream; occurs when the amount of material transported is less than a stream's carrying capacity.

drainage basin, 291–292: the area that supplies water for the discharge of a given stream.
- graphic model of, *44*
- hierarchy of streams in, 291–292, *293*
- mathematical model of, *44*
- of Mississippi River, *44, 291*
- physical model of, *45*
- and runoff, 144–145
- stream ordering in, 292, *293*

drainage net, 291: the pattern of streams in a drainage basin.

drainage pattern
- on jointed rock, 323–324, *323*
- in ridge and valley landscapes, *323*

drift, 391: any deposited material that has been transported by wind, water, or ice, but especially by glacier ice.

Driftless Area, Wisconsin, effects of glaciation, *430,* 431

drizzle, 90: precipitation in which drops are much smaller than normal raindrops.

drought
- on Great Plains, 432
- Oklahoma, *79*
- United States East Coast, 247, 250, *250*

drumlin, 391, *391,* 423: a streamlined elongated hill formed from glacial deposits; the long axis parallels the direction of the glacier's flow.

dry climate region, 165, *165,* 166, *168*

duration of daylight, 49, *51, 52*

Durdle Door, England, sea arch, *404*

duricrust, *370:* a lateritic crust that forms a hard, rocklike surface on the soil, in semiarid climates.

e

earth
- age of, 23–24
- formation of, 6-7, *7*

earthquake, *9,* 10, *21, 260, 262, 264, 266:* a sudden movement of the earth's crust, caused by displacement of material below the earth's surface.

earth-sun relationship, 48–53, *51*

easterly wave, 124, *124:* an extended weak, low-pressure trough that drifts westward in the region of the trade winds; it is often accompanied by thunderstorms.

easting, 449, *449:* in a map grid system, the coordinate of a location to the east of the grid origin.

ecosystem, 206–207: a system consisting of a plant and animal community and its environment, often with particular regard to the transfer and circulation of energy and material.
- complexity of, 206
- efficiency of, 207, 217–218
- energy flow in, 206–207
- laboratory, 217, *217*
- and man, 223, 226–227

ecosystems, agricultural, 220–221
- corn and rice yields of, 221, *221*
- energy cost of, 221–222
- instability of, 221–222
- new plant strains in, 221
- and protein in diets, 221

ecosystems, energy transfer in, 217–220, *218, 219, 220*
- in Brookhaven forest ecosystem, 220, *220*
- in natural ecosystem, schematic, 218–219, *218*
- in Silver Springs ecosystem, 219–220, *219*

Ecuadorian rainforest, *235*

eddies, 6, 287: currents of local circulation, which cause turbulent motion of gases or liquids.

electromagnetic radiation: the means by which electric and magnetic fields are able to transport energy.

electromagnetic spectrum, 55–57, *56:* the various kinds of electromagnetic radiation, ordered according to wavelength or frequency; the electromagnetic spectrum includes radio waves, infrared radiation, visible light, ultraviolet light, x-rays, and gamma rays.
- and remote sensing, *471*

Elko, Nevada, hourly temperature distribution, 68, *68*

eluviation, 185: the loss of material downward from a soil layer, through its movement in suspension or in solution.

end moraine, 391: an accumulation of material deposited at the terminal position of a glacier; a terminal moraine.

energy, 30–38, *33:* the ability to do work.
- energy budget of a forest, *237*
- sources of, 31, *33*

energy, chemical, 30: energy stored in the molecules of chemical compounds.

energy, elastic, 30: energy associated with a condition of mechanical strain in a body.
- in the earth's crust, 337–338

energy, geothermal, 28, *33,* 328–330, *328:* energy derived from the earth's interior heat.

energy, gravitational, 30–31: energy associated with the height of an object; also called potential energy.

energy, heat, 31–36: energy associated with the internal random motion of molecules in matter.

energy, kinetic, 30, 31: energy associated with the motion of objects; also called energy of motion.

energy, nuclear, 31: energy bound in the nucleus of an atom.

energy, solar radiant, 28, 30, *33,* 34: energy transferred from the sun by electromagnetic radiation.

energy conversion efficiency, 31–34: the amount of energy converted to a specific form relative to the total energy supplied.

energy flow diagram, *220:* a schematic diagram that emphasizes the transfer of energy from one part of a system to another or from one system to another.

energy of flowing water, 284–285

energy flow pyramid, *219:* a schematic diagram that represents the amount of stored energy in each stage of a food chain; stored energy decreases along a food chain.

England, chalk sea cliffs, *411*

equal area projection, 453, *453:* a map projection in which each region depicted has the same area as it would have on a globe of the same scale; directions are distorted, especially near the margins.

equation of time, *53:* the relation between mean solar time and true solar time.

Equator, 49, *51*

equatorial rainforest, 235–237, *235*
adaptations of vegetation in, 235
global distribution of, *232*
temperature and precipitation of, 235, *235*
temperature variations of, 235
thermoisopleth, 235, *236*
vegetation characteristics of, 235
water budget, *236*, *237*

equatorial trough, 110, *110*, 111

equilibrium, 38–39, *39*: a condition of stability in which a system tends to resist change.

equinox, vernal and autumnal, 49, 51: at the autumnal equinox (about September 21) and the vernal equinox (about March 21) the sun is directly overhead at the Equator.

equivalent projection: *See* equal area projection.

erg, 366: (geomorphology) a vast desert region covered with an accumulation of sand.

erosion, 209, *270*, *271*, *272*: the various processes by which rock debris is obtained and transported.
in the Badlands, 432–434, *433*
on beaches, 406–409
by glaciers, 385–387
and landform modification, 273
and milpa cultivation, 248–249, *249*
by streams, 289–290, 297–298, 304
and stream valleys, 273
by wave action, 404–405

erosion and deposition
on beaches, 406–409
and dams, 408
and man, 408–409
study of, 409

erratic, 425: a large boulder that differs in composition from the local bedrock; presumably transported by glaciers or floating ice from another region.

ERTS-1, 73–74, 128–129, 471, *471*, *473*: Earth Resources Technology Satellite.

escarpment, 425, 428: a steep face at the edge of a region of high local relief.

esker, 391: a ridge of sand and gravel, sometimes many miles long, deposited by streams beneath or in a glacier.

estuary, 401, *401*: the mouth of a river where fresh water and sea water mix under the pulse of the tides.
mud flats in, 401
productivity of, *212*, 401

Etretat cliffs, Normandy, France, sea cliffs, arch, stack, *404*

evaporation, 80–82, *80*, 135, *137*, 138: the change of a substance's physical state from liquid to gaseous vapor.
heat energy and, 80–81
meteorologic conditions affecting, 81–82
molecular picture of, 80–81

evapotranspiration, 65, 82–83: the transfer of water vapor to the atmosphere by the combined action of evaporation and transpiration.
See also actual evapotranspiration, potential evapotranspiration

evapotranspirometer, 149: a device used to measure potential evapotranspiration; it consists of a tank containing well-watered vegetation in which water is carefully measured.

evolution, 14–15, *17*, 18–19: (biology) the development of new species from preexisting species by natural selection and genetic change.
of man, 19

exfoliation, *270*, *370*, *371*: weathering in which layers peel off curved surfaces of rock in sheets; common in granite.

exponential notation, 5, 444: a method for expressing very large or very small numbers in terms of powers of a number; in scientific work, powers of ten are employed.

f

Fahrenheit scale of temperature, 30, *30*

fault, 268, 337–343: a fracture or fracture zone in the earth's crust where relative displacement of rocks has occurred.
See also dip-slip fault, normal fault, reverse fault, strike-slip fault, tear fault.

fault block, 336–343, *337*, *339*: a rock mass that moves as a whole during fault movements; it is bounded by faults on at least two opposing sides.

fault scarp, *314*: a cliff formed by faulting; usually modified by subsequent erosion.

fault zone, features of, 340–341, 343

felsenmeer, *362*: a "sea of rocks"; an extensive area covered by large, irregular, frost-shattered rocks; characteristic of high mountain slopes in cold climates.

fertilizer, 182, 203
limits to usefulness of, 222

fetch, 405: a continuous expanse of open ocean, across which the wind blows in a generally constant direction.

field capacity, 141, *150*: the maximum amount of soil moisture that can be retained by a piece of ground.

field sketching, 468, *469*

field work and mapping techniques, 468–470, *469*

Fiji, cumulus clouds, *175*

filling, 289–291, *297*: the deposition of material from the waters of a stream into its bed.

Finger Lakes, and glaciation, 389

firn, 382: an intermediate form of glacier ice, formed from snow during the first year after its accumulation.

firn line, *381*, 382: the line on the surface of a glacier between accumulated snow (white) and glacier ice (blue).

first order of relief, 261–264, *262*: the continents and ocean basins.

fissure, *342*: an extended crack or fracture in rock, especially in the earth's crust.

Fitzroy, Robert, 117

fjord, 388, *388*, 389, 391: a long, deep inlet exclusive to coasts of high relief that have been severely glaciated.

Flagstaff, Arizona, climate, 440

flood of 1973, 308–309, *309*

floodplain, 302, *303*, 304: a flat plain bordering a river, built up of sediments deposited by the river.

floodwater, 284, 302, 304: stream discharge that is many times larger than normal; the waters that overflow a stream's normal channel.

fluted valley walls, *372*

fog, *92:* a cloud of water droplets formed at or near the ground.

dispersal of, 101
formation of, 93
frequency of in United States, *92*
off northeast coast of United States, 423
See also advection fog, orographic fog, radiation fog

fold, 338–339, *338,* 343–345: a bend in rock strata, caused by movements of the earth's crust.

overthrust, *338*

food chain, 215–220: a series of organisms having interrelated feeding habits, each in turn serving as food for the next.

energy transfer along, 217–218, *218*
limitation on length of, 218

food web, 215–217, *216:* the interconnected feeding relationships between interacting species; a series of interwoven food chains.

of a salt marsh in winter, *216*

foot wall, *339:* the block of rock below a fault plane.

forces in the atmosphere, 107–109, *107*

foreshore, *406:* the lower shore zone of a beach, between the normal levels of high tide and low tide.

fossil, 13–14, *17:* remains or traces of plants or animals that have been preserved by natural processes, often by incorporation into sedimentary rock.

fossil record, 13–14, *17:* the history of life forms recorded in rock strata, with earliest forms in the deepest layers.

French Alps

arête, *391*
valley glacier, *376*

friction, 107, *107,* 108–109: a force that acts to oppose or retard motion, usually accompanied by the transformation of kinetic energy into heat.

and stream flow, 285–286

front, 117–120, *118, 119, 120:* the boundary between two air masses that have different temperatures or humidities.

See also cold front, occluded front, warm front

frost weathering, 187, 360, *362:* the breaking of rock into fragments by the freezing and thawing of water.

Fungus Rock, near Malta, sea cave, *404*

g

gases, *10, 36*

laws of, *36*
properties of, *36*

gauge height, *286:* a measure of the discharge of a stream; the height of a stream calibrated in terms of discharge.

gauging stations, *286, 287*

general circulation of the atmosphere, 106: the average global features of the atmosphere's movements and pressure systems.

driving force of, 109
in the low latitudes, 109, *109*
in the Northern Hemisphere, *111*
pressure cells of, 111, *113*
seasonal shift of, 111, *113,* 114
on a uniform earth, 109–111, *110*

general circulation of the ocean, 126–127, 130

and climate, 129, 160
deep circulation, 130
global, *131*
idealized, *127*
and winds, 127

geodesy, *2,* 452: the study of the shape of the earth.

geologic ages, *17*

geologic movement and landform development, 268–269, *269*

geologic structure and landforms, 318, 322–325, 336–345

geology, *2:* the study of the materials, structure, and history of the earth.

geomorphic processes

work done by, 279–280, *280*

geomorphology, 261: the study of the earth's landforms, including their relation to underlying geologic structure.

GEOREF, 450, *450:* World Geographic Reference System.

geostrophic wind, *107,* 108–109: a wind blowing with steady speed parallel or nearly parallel to lines of constant pressure (isobars).

and the general circulation, *110, 113*

geosyncline, *327,* 336–337: an extended trough in rocks that fills with sediment as the trough subsides; an early stage in the development of some kinds of mountains.

geothermal energy, 28, *33,* 328–330, *328*

and geologic structure, *329*
in Iceland, 329
in Italy, 329
source of, 328–329
in the United States, 329–330

glacial till, *198:* unstratified and unsorted sedimentary material that is transported or deposited by glaciers.

glaciation: the alteration of a region's surface features by glacier ice.

on Cape Cod, 423–425
and crustal rebound, 379–380
distribution, past and present, *379*
indirect effects of, 379–380
and landforms, 387–392
of the Pleistocene, 378–379, *379*
proposed causes of, 380–381
in Rocky Mountains, 436, 439
and sea level, 380

glaciation and man, 392

formation of lakes, 389, 392
removal of topsoil, 392

glacier: a flowing mass of ice formed by an accumulation of snow that is compacted by pressure.

as an agent of geomorphic change, 354, 376
anatomy of, 380–385, *381*
and climatic change, 376
debris carried by, 391
and depositional landforms, 391–392
movement of, 384–385
rate of erosion caused by, 387
rate of erosion compared to streams, 385–386
sizes of, 378
speed of flow of, *384,* 385
as storehouses of water, 376, 378
and the work of flowing water, 391–392
See also active glacier, ice sheet, passive glacier, piedmont glacier, polar glacier, temperate glacier, valley glacier

glacier budget, 382–384, *383:* an accounting of the accumulation and ablation of snow and ice on a glacier.

glacier ice, 381–382: dense, pore-free ice formed in glaciers from accumulated snow with the passage of time; its specific gravity is 0.85 or greater.

formation of from snow, 381–382

glacier valley, *270*

gley soil (pronounced "glee" or "gly"), 191, 195: a waterlogged soil horizon, colored a bluish gray by iron compounds.

Glomar Challenger, 21

GN, 451, *451,* 467: grid north.

gnomonic projection, construction of, 453, *453*

Goblet, 357, *357*

Gondwanaland, *20:* a hypothetical supercontinent formed in the Southern Hemisphere by the breakup of Pangaea.

gouges, glacier, *386,* 387–388

graded stream, 298: a stream that has reached a condition of equilibrium in which neither filling nor scouring occurs consistently.

gradient: the degree to which a quantity varies with distance.

Granada, Spain, cave homes, *418*

Grand Canyon of the Colorado River
- age of, *17*
- erosion in, *438*, 440
- schist formations of, *321*
- size of, 439

granite, 315, *316*
- arid landscape and, *368*
- and bornhardts, *370*, *371*
- boulder slope of, *272*
- dome, *272*
- exfoliation of, *316*
- jointed nature of, *186*, *274*, *274*
- weathering of, *316*

graphic scale, 447, *447*

graticule, 449: the network of lines of latitude and longitude on a map.

gravel: in the study of soils, particles of moderate size, which are larger in diameter than 0.2 centimeter (0.08 inch).

gravitational (potential) energy, 30–31
- and streamflow, 285

gravity, 30, 31: a force exerted by an object by reason of its mass.
- and slope processes, 272
- and tides, 399, *400*, 401

gray-brown podzolic soils, *189*, *193*, 196, *196*, *200*: soils of the midlatitude deciduous forests.
- agricultural potential of, 196
- and climate, 189
- global distribution of, *200*
- profile, *196*
- United States distribution of, 196, *200*
- and vegetation, *193*, 196

Great Basin, western United States, basin and range topography, 339–341

great circle, *51*: any circle on the surface of the earth whose center coincides with the center of the earth; the shortest distance between two points on the earth's surface is an arc of a great circle.

Great Lakes
- and glaciation, 392, 428
- map of, *428*

Great Plains
- agriculture of, 431–432
- history of drought, 432
- map of, *435*
- precipitation in, 431–432
- in South Dakota, *433*

Great Sand Dunes National Monument, Colorado, barchan dunes and ripples, *272*

great soil groups, 194–202, 477–478: subdivisions of the 1938 soil classification system.
- global distribution of, *200*
- *See also* bog and meadow soils, chernozems, chestnut and brown soils, desert soils, gray-brown podzolic soils, latosols, podzols, prairie soils, red-yellow podzolic soils, tundra soils

greenhouse effect, 64

grid azimuth, 451: azimuth measured with respect to grid north.

grid coordinates, 449, *449*
- rule for reading, 449, *449*

grid north, 451, *451*: the reference direction determined by a grid system.

grid systems, 449–451
- GEOREF, 450, *450*
- land survey, 451, *451*
- State Plane Coordinates, 450
- UPS, 449–450, *450*
- UTM, 449–450, *450*, 454

gross productivity, 210: the total amount of food energy produced by plants during a specified interval.

ground moraine, *391*: debris laid down in a sheet beneath a glacier.

groundwater, *137*, 141–145: the water in the fully saturated pores and crevices of underground rock.
- and aquifers, 141–144
- artificial recharge of, 144, *144*
- and the hydrologic cycle, *137*
- storage of, *138*
- and stream base flow, 145, *145*
- use for irrigation, 143–144

grumusols, *200*: dark, dry, cracked soils found in the tropics; vertisols.

guard cells of stomata, *83*

Gulf of California opening, 263, *266*

Gulf Stream, 127, 130, *131*: a fast-moving, warm, surficial ocean current, which moves northward along the western edge of the Atlantic basin.

gully, 292: a small surface drainage channel, too deep to be crossed in a wheeled vehicle.

guyot, 413–414: a flat-topped undersea mountain; a tablemount.

gyre, 127, *127*, *131*: a large-scale, looping current of circulation, particularly in the oceans.

h

hachuring, 461, *461*: a method of indicating relief on maps by shading with parallel lines; the steeper the slope, the denser the lines; shading is made darker on the south and east sides of slopes.

Hadley, John, 109

Hadley cell, 109–110, *109*: a convective cell of large size in the atmosphere.
- subtropical, 165
- *See also* convective cell

hail
- formation of, 91, *91*, 93
- frequency in the United States, *102*
- layering in, 91
- modification of hailstorms, 101–102
- and violent storms, 92, *102*

Hail Alley, *102*

hailstones, 91: a form of precipitation consisting of layered ice balls formed by repeated vertical cycling in clouds.

hailstorm modification, 102, *102*

Haleakala crater, Maui
- cinder cone, *330*
- volcanic landscape, *11*

Half Dome, Yosemite National Park, California, glaciated granite dome, *270*

half-life, 23: the time required for half the atoms in an initial radioactive sample to decay.

hamada, *366*: a desert plateau or high plain that wind has denuded of fine-grained materials, exposing bedrock; a hamada may have a veneer of pebbles or boulders.

Hancock, Wisconsin, hourly radiation balance, *64*

hanging valley, *391*: a valley that has a floor noticeably higher than the level of the valley or shore it intersects.

hanging wall, *339*: the block of rock above a fault plane.

hardpan, 196: a relatively impermeable layer in a soil, formed chiefly of clay that has been cemented by iron compounds or other materials.

Hawaiian Islands
- aa lava, *333*
- formation of, 21
- map and age of, *22*

headland, *403*: a cape or promontory that juts into the sea; usually high and prominent.

heating and cooling of the ground, 66–67

heat island, 72–73, *73*: a local region where the temperature is generally higher than that of surrounding areas; large cities are often heat islands.

Hebgen Lake, Montana, effects of earthquake, *260*

herbivores, 215: organisms that subsist on plants for food.

heterotrophs, 215: organisms that rely on food produced by other organisms; secondary producers.

Hilo, Hawaii, lava flow, 336

hoar frost, 86: the feathery crystals of ice that form on cold surfaces when the dew point is below the freezing point of water.

hogback ridge, *322,* 323: a long, narrow, sharp-edged ridge with two steep slopes, formed by rock strata tilted at a steep angle.

Hollister, California, winery and fault zone, 312

Holocene, *17:* the present (or Recent) geologic epoch; it began approximately 10,000 years ago.

homoclinal ridge, *343:* a ridge developed on the dipping limb of an anticline or a syncline.

homoclinal valley, *343:* a valley formed in weaker rock along the flank of a fold.

Honduras, milpa cultivation, *249*

hoodoo rock, 357, *357:* a rock pillar or pedestal of unusual shape, formed by erosion in regions where rainfall is concentrated in a brief season.

horizon, 185–186, *186:* in soil studies, the distinctive horizontal layers that are apparent in vertical cross sections of many soil types.

horizontal energy transfer on the earth's surface, 66, *66*

horizontal mixing, *111*

horn, 388, *391,* 439: a faceted pyramidal peak formed at the intersection of three or more cirques.

Horton, Robert, 292, *293*

Horton analysis of streams, 292, *293*

  importance in geomorphology, 292

Hudson River, submarine canyon, *397,* 413

humid continental climate region, *165,* 166, *168*

humidity, 83–85: the amount of water vapor in the air.

  See also absolute humidity, mixing ratio, relative humidity, specific humidity

humid subtropical climate region, 165, *165*

humus, 183: decomposed and partly decomposed organic matter in soil.

Huron, South Dakota, local water budget, 244, *245*

hurricane, 124–126, *125, 126:* a term used to describe a tropical cyclonic storm that forms on the east or west coast of North America.

  modification of, 102
  structure of, *125*
  tracks of, *126*

hurricane Camille, 128–129, *128*

Hutton, James, 22–23

hydrogen ion, 184, 185, *185:* a hydrogen atom that has lost its electron; a proton.

  and acidity, 184–185, *185*
  and leaching, 190, *190*

hydrograph, *145:* a graph plotting the discharge of a river at different times.

hydrologic cycle, 40–41, *41,* 134, *137:* the processes by which water passes from the surface of the earth to the atmosphere and back to the earth.

  atmospheric part of, 78, 80, *80*

hydrology, *137,* 138–145: the study of water over, on, and under the earth's surface.

hydrostatic pressure: the pressure exerted by stationary fluids.

hypothetical continent, 163–166, *165:* a model used to explain how the delivery systems for energy and moisture determine regional climates.

  and atmospheric controls on climate, 165–166
  basis for division into climatic regions, 163
  comparison with global climates, 166
  and seasonal change, 165, 166

i

ice age, 251, 376: a period when glaciers generally advance and cover extensive areas; a glacial period.

ice cap: *See* ice sheet.

ice-forming nuclei, *89,* 90: fine particles in the atmosphere, which act as centers to promote the formation of ice crystals from water vapor.

  Dry Ice as, 100, *100*
  silver iodide as, 101

Iceland

  Heimaey, volcano, 328
  icebergs, 175
  table mountain, *333*

ice sheet, 376: an extensive and nearly flat layer of ice that spreads in all directions from the center of a glacier.

  and crustal rebound, 379–380
  thickness of, in Pleistocene, 379

ice wedge, *192, 362:* a wedge of clear ice that forms and persists under the surface; occurs in regions of moist, cold climate.

Idaho, mudflow, *356*

igneous rock, 12, 315: rock formed by the direct solidification of molten minerals; it usually exhibits a microscopic crystalline structure.

  basalt, 315, *317*
  crystallization in, 315, *316, 317*
  granite, 315, *316*
  intrusions of, *334*
  and mountain ranges, 337
  and the rock cycle, 325, *327*

illuviation, 185–186: the deposition of material in an underlying soil layer from above, by leaching; occurs in humid climates.

Imperial Valley, California, irrigation of, 440, 443

income, 150: in budgets, the amount of energy or material available to a system.

Indiana, karst region, *320*

Indus River, longitudinal profile, *294*

inert gas, 10: any member of a family of essentially unreactive chemical elements, including helium, neon, argon, krypton, and xenon; a noble gas.

infiltration, *137,* 140–141: the movement of water from the surface into the ground.

  effect of compaction on, *140,* 141, 203
  and the local water budget, *150*
  and runoff, 144
  into various soils, *140*

infiltration rate or capacity, *140,* 141: the rate at which water enters the ground from the surface.

infrared radiation, 56, *56:* the part of the electromagnetic spectrum lying between radiowaves and visible light, spanning the range of wavelengths from 1,000 microns to 0.7 micron; long-wave infrared radiation has wavelengths exceeding 4 microns.

interaction with molecules in the atmosphere, 58–59

infrared scanning and thermometry, *471*

input, 38–39, *39:* energy or material delivered to a system.

inselberg, *368:* an "island mountain"; an isolated, steep-sided summit rising abruptly from a plain of low relief; formed by erosion.

interception, *137,* 138–140, *139:* the capture of precipitation by plant cover, which prevents it from reaching the soil.

internal flow of glaciers, *384,* 385: a form of glacial motion in which the ice crystals slip past one another; the speed of internal flow decreases with depth.

intersection, 469, *469:* in surveying, a technique for specifying locations by measuring azimuths from both ends of a known base line.

intertropical convergence zone, 110, 111, *113:* a weak belt of low pressure near the Equator where the trade winds meet.

seasonal shift of, *113,* 165

intrazonal soils, 194: soils that develop primarily according to local conditions; they sometimes have characteristics of associated zonal soils.

intrusion, *327:* igneous rock that has invaded older rock and solidified below the surface.

inversion of relief, 334: the development of a locally high structure on a region of low relief, by any of several processes.

and anticlinal valley, *343*
and synclinal ridge, *343*
and table mountain, *333*

ion, 184, *185:* an electrically charged atom; an atom with an excess or deficiency of electrons.

movement in soils of, 190–191, *190*

ionically bound molecule, 184: a molecule that consists essentially of ions bound together by direct electrical attraction; an ionic molecule.

Iowa, snowy cornfield, *175*

Iowa River, hydrograph, *145*

irrigation, 148, 440, 443

island arc, *266, 327:* a line of islands formed by volcanic activity on the continental side of an oceanic trench.

isobar, *107, 108:* a line connecting points of equal atmospheric pressure.

and easterly wave, *124*
and weather map, *115*

isotherm, 235: a line connecting points of equal temperature.

isotopes, 23: atoms of the same chemical element that have different masses.

j

James River, South Dakota, floodplain and meanders, *295*

jet stream, *111,* 114, *114,* 120, *122:* a narrow stream or ribbon of rapidly moving air, occurring at an altitude of about 10 miles.

and monsoon circulation, 123–124

jetty, 408–409: a long man-made barrier extended outward from the coast to modify tidal currents or stream flow; used to stabilize or deepen a channel and to prevent deposition.

joint, 274, *274:* a fracture in rock that has not been accompanied by appreciable displacement; often occurs along planes of weakness.

jointed rock, 274, *274,* 387: rock masses that have numerous extensive fractures.

k

Kabir Kuh Mountains, Iran, crustal deformation, *269*

kame, 391: a mound of sand and gravel deposited by a stream associated with a glacier.

karst region, *320,* 324, *347:* a region characterized by caves, sinkholes, and underground streams; formed where thick beds of jointed limestone are dissolved by solution in water.

- locations of, 324
- scarcity of streams in, 324
- tropical, *373*

Kawera River, New Zealand, sediment transport, *289*

Kelvin scale of temperature, 30, *30:* a scale of temperature fundamental for physical processes; its zero point, called absolute zero, is approximately -273°C (-459°F).

kettle, 391, *424,* 425: a depression formed by the melting of a massive ice block embedded in a glacial till plain.

Key West, Florida, water budget surplus and AE/PE ratio, *171*

Kilauea crater, Hawaii

- ferns in lava, *191*
- pahoehoe lava, *333*
- solidified basalt, *317*
- volcanic fissure, *342*

kinetic energy, 30, 31

Klagenfurt, Austria, thermoisopleth, 242, *243*

knickpoint, 298–299, *299:* a sharp vertical break in the longitudinal profile of a stream valley.

- disappearance of, 299, *299*
- at Niagara Falls, 428

Köppen, Wladimir, 172, 474–475

Köppen system of climate classification, 172–173, 176–177, 474–475

- criticisms of, 176
- detailed description of, 474–475
- global map of, *176*
- terminology of, 173
- and vegetation regions, 173, 176

Krakatoa volcano, Indonesia

- caldera of, *334*
- dust in atmosphere, 87
- effect on climate, 252
- explosive eruption of, 330

Kribi, Cameroon, local water budget, *236, 237*

Kurushio current, 127: a fast-moving, warm, surficial ocean current that moves northward along the western edge of the Pacific basin.

l

laccolith, *334:* a roughly lens-shaped intrusion of magma that pushes overlying rock upward to form a dome at the surface.

lagoon, *411, 414:* a shallow body of water entirely or nearly cut off from the sea by a narrow strip of land.

lahar, 336: a mass flow of volcanic ash deposits.

La Jolla, California, beach, 396, *396*

Lake Ponchartrain, Louisiana, use in flood, 308–309

Lake Tahoe, landslide, *354*

laminar flow, 287: streamline flow in which the fluid moves smoothly; the speed and direction of flow at a given point are steady.

- contrasted with turbulent flow, 287

land breeze, *123:* a local circulation that blows from the land toward the sea, usually at night.

landforms: the natural features of the earth's surface, such as hills, valleys, beaches, and so on.

- associated with faults and folds, 337–345
- differentiation of, 268–274
- early, 11–12, *11*
- and energy, 261, 268
- and erosion and deposition, 269, 272
- evolution of, 268–274
- and geologic movement, 268–269
- and glaciation, 387–392
- and habitat, 416–419
- and life, 18–19
- and man, 261
- relict, 278–279, *278*
- response time of, 349
- structure, process, and stage in development of, 314
- and underlying geologic structure, 312

landforms and climate, 348–352, 359–360

- and vegetation, 351–352

landforms on faulted structures, 339–343

- features of, 340

landforms on folded structures, 343–345, *343*

landforms produced by erosion or differential erosion, 318, 322–325

landscape, arid, on granite, *368, 369*

landscape, arid, on sedimentary rock, *366, 367*

landscape, arid, on uplifted fault block, *364, 365*

landscape, cold climate, *362, 363*

landscape, tropical limestone highland, *372, 373*

landscape, tropical savanna, *370, 371*

landscape, tropical volcanic dome, *372*

landslide, 354, *354*

land survey township system, 451, *451*

langley, 54: a unit of energy flux defined as 1 calorie per square centimeter.

Langmuir, Irving, 100, *100*

lapse rate, 96–99: the rate of change of temperature with altitude in the atmosphere.

Larderello, Italy, geothermal power station, 329

large-scale map, *448,* 449: a map on which a representative area corresponds to a comparatively small area on the earth; thus, a large map of a small area.

laser profiler, *471:* a method of scanning topography remotely by timing the passage of laser light to and from the surface.

latent heat of condensation, *86:* the heat that can be obtained when 1 gram of gaseous vapor condenses to a liquid at the same temperature; numerically equal to the latent heat of vaporization at that temperature.

latent heat of fusion, 35, *35:* the amount of energy required to change the state of 1 gram of substance from a solid to a liquid at the same temperature.

latent heat of sublimation, *35:* the amount of energy required to change the state of 1 gram of substance from a solid to a gaseous vapor at the same temperature.

latent heat of vaporization, 35, *35:* the amount of energy required to change the state of 1 gram of substance from a liquid to a gaseous vapor at the same temperature.

lateral adjustment of streams, 299–302, *301, 303*

lateral moraine, 391–392: elongated deposits of material formed along the valley slopes at the margins of a valley glacier.

laterite, 198, 361: a lateritic soil; or, the material resulting from the hardening of a lateritic soil.

lateritic crust, *370:* a hard surface crust that forms on latosols or lateritic soils; a duricrust.

latitude, *51,* 449: distance north or south of the Equator on the earth's surface measured in degrees; lines of constant latitude are called parallels.

length equivalent to 1° of, 452

latosols, 189, 191, *193,* 197–198, *197, 200:* soils of the warm, humid rainforest; they are deeply weathered and are colored red by iron compounds.

and climate, *189*
global distribution of, *200*
and nutrients, 197–198
profile, *197*
and vegetation, *193,* 197

Laurasia, *20:* a hypothetical supercontinent in the Northern Hemisphere, formed by the breakup of Pangaea.

lava, 330: molten rock from which the gases have escaped.

aa, *333*
pahoehoe, *333*
pillow, *333*

lava flow, diversion of

in Hawaii, 336
in Italy, 312

leaching, 190, *190:* the downward movement of cations in a soil, caused by the influx of hydrogen ions from above.

of podzols, *192*

levee: *See* natural levee.

Lever House, New York City, glass architecture, *256*

lichens and soil formation, 189, *190*

limb of a fold, *338:* the sides, or arms, of a fold in rock strata.

liquid, *10*

lithification, *327:* the process by which unconsolidated material is converted to solid rock.

lithosphere, *9,* 21: the outer layer of rigid rock on the earth, consisting of the crust and upper mantle, and including the crustal plates.

lithospheric plates, 21, 261–262, *264:* comparatively rigid plates, which consist of the earth's crust and the upper portion of the mantle; the plates that drift across the surface of the earth.

converging, *20, 33, 266, 327*
and interacting boundaries, *265*

Little Colorado River, downcutting, 440, *441*

Little Ice Age, 251–252: a period extending from 1550 to 1850, in which average temperatures in much of Europe were low enough to allow the advance of glaciers.

load: the total material carried by a stream by any of various means.

local circulations, *123*

See also land breeze, mountain wind, sea breeze, valley wind

local standard time, *53*

local water budget, 145, 148–156: the local expression of the hydrologic cycle.

applications of, 145, *155*
of Baton Rouge, Louisiana, 150–155, *151, 153*
of Cloverdale and Manhattan, 167, *167,* 170
components of, 145, 148, *150*
components of, as climatic indexes, 167, *167,* 170–172, *171*
improvements in, 155–156
in wet, dry, and average years, *152,* 154–155

loess (pronounced "less," "luss," or "LOW-es"), *199, 272,* 392: fine silt deposited by the wind; often forms deep, partly consolidated beds.

logarithmic scale, *37:* a graphic scale that is laid out proportionally to the logarithms of numbers; convenient for displaying wide ranges of quantities, because each decade (1 to 10, 10 to 100, and so on) occupies the same length interval on the scale.

longitude, *51,* 449: distance east or west on the earth's surface, measured in degrees from a specified prime meridian. Lines of constant longitude are called meridians.

length equivalent to 1° of, 452

longitudinal profile, 293, *294,* 298: the vertical elevation of a stream bed versus horizontal distance along the channel.

of a glacier valley, 389
knickpoints in, 298–299, *299*
and size of particles in load, 298
and work done by stream, 298

longshore drift, 408, *408:* a current that parallels a shore.

long-wave infrared radiation, *56:* infrared radiation with wavelength longer than 4 microns; thermal radiation from the earth is long-wave infrared radiation.

Los Angeles, California

brush fire, *175*
smog, *74*

Lötschental, Switzerland, U-shaped valley, *388*

lysimeter, 149: a device for measuring actual evapotranspiration; consists of tanks containing a stand of vegetation in which water use by plants is carefully measured.

m

Machu Picchu, Peru, Inca city, 417, *418*

Mackenzie River, Northwest Territories, Canada, pingo, *363*

Madison, Wisconsin, monthly radiant energy balance, *72*

magma, 315, 325, 330: the molten rock in the earth's interior; it contains dissolved gases and can reach the surface during volcanic activity.

and geothermal energy, 328–329
and the rock cycle, *327*
sialic, 330
simatic, 330

magnetic azimuth, 451: azimuth measured from magnetic north.

magnetic declination, 451: the angle of difference in direction between magnetic north and true north; magnetic variation.

magnetic north, 451, *451:* the northerly direction determined by a magnetic compass.

magnetic stripes, *266:* a pattern of varying rock magnetism on the ocean floor; caused when a record of the earth's magnetism is "frozen" into solidifying crust.

Maine, coastal islands, *411*

Malaspina glacier, Alaska, piedmont glacier, *376*

Maligne Lake, Jasper National Park, Alberta, Canada, *375*

Manhattan, Kansas, local water budget, 167, *167,* 170

mantle, *9,* 10: the major division of the earth's structure, which lies between the crust and the core.

map: a two-dimensional representation of all or part of the earth's surface, showing selected natural or man-made features, or data; preferably constructed on a definite projection, with a specified scale.

symbolization and generalization in, 459–462
use in geography, 446
See also projection, USGS topographic map

Map Information Office, 462

mapping in the United States, 462, *463,* 464–467

map series, 464: a set of maps that share a common property, such as scale.

maquis (pronounced "mah-KEE"), *240:* vegetation that consists of drought-adapted scrub, occurring in certain regions near the Mediterranean Sea.

marble, 318

marine climate region, 165–166, *165, 168*

marine terrace, 398, *411:* a plain that has been cut by wave action along a coast; a shore platform.

mass wasting, 272, 352: the movement of large masses of rock or soil down slopes under the direct influence of gravity.

Matterhorn, Switzerland, horn, *391*

maturity, 275, *276:* a descriptive term applied to a landscape with rounded hills and well-developed stream valleys; in the Davis cycle, an intermediate stage of landform evolution.

Maui, Hawaiian Islands, slopes, *272*

Mauna Loa volcano, Hawaii, shield volcano, *330*

Mayan cities, Copan, Honduras, tropical seasonal forest, *237*

meander, 294, *295,* 301, *301, 303:* one of a series of looping, sinuous curves in the course of a stream.

 and Mississippi River, 294, 301, *301*
 spacing of, 301, *301*

meander geometry, *301*

meandering channel, 294, *295,* 301, *301, 303:* a stream channel whose course exhibits sinuous curves several times broader than the width of the stream.

 and lateral adjustment of streams, 301–302
 lateral flow of water in, *301*
 scouring and filling in, 301, *301*

meander scar, *303:* a meander that has been abandoned by a stream and that has become filled with soil and vegetation.

mean solar time, *53:* time measured according to the average position of the sun in the sky; differs from standard time at the rate of 4 minutes of time for every degree of longitude from the standard time meridian in a time zone.

mechanical weathering, 269, *349,* 350, *350:* any physical process of weathering; commonly, frost action.

Mediterranean climate: *See* subtropical dry summer climate region.

meltwater: the liquid water resulting from the melting of snow or ice.

Meon River, England, estuary, *401*

meridians of longitude, *51:* lines of constant longitude on the earth; a meridian forms half a great circle through the poles.

mesa, *322,* 323, *366, 367:* a flat-topped area of high local relief, bounded by a steep cliff on at least one side.

metamorphic rock, 12, 318: rock formed by the recrystallization of igneous or sedimentary rock, as a result of heat, pressure, or chemical action.

 and the rock cycle, 325, *327*
 schist, *321*
 weathering of, *321*

meteorites: rock masses from outer space that have reached the earth's surface.

 age of, 23–24

meteorology: the study of weather, especially of physical processes in the atmosphere.

metric system, *29*

 conversion factors for, 445, *445*

Miami, Florida, water budget surplus and AE/PE ratio, *171,* 172

microclimate, *162,* 163: the climate of a highly localized region, such as a forest floor.

 of slopes, 357–358

micron, 55: a metric unit of length defined as 10^{-6} meter; equivalent to 3.937×10^{-5} inch.

Mid-Atlantic Ridge, 20–21, *21,* 262–263, *264:* the spreading center in the central Atlantic Ocean basin where new crust is being formed.

 in Iceland, 262–263
 and volcanism, 325, 328

Middlesex County, New Jersey, water budget and drought, 247, *250*

midlatitude deciduous forests, 241–242, *243*

 global distribution of, *232*
 temperature and precipitation, 242, *243*
 thermoisopleth, 242, *243*
 vegetation characteristics, 242

midlatitude grasslands, 242, 244, *244*

 global distribution of, 232
 vegetation characteristics, 242, 244
 water budget, *245*

midlatitudes: the regions lying approximately between latitude 30° and 50°, on both sides of the Equator.

midlatitude shrublands, 240–241, *240*

 global distribution of, *232*
 and Mediterranean climate, 240–241
 temperature and precipitation, 241, *241, 242*
 vegetation characteristics, 240–241
 water budget, 241, *241*

Midwest

 agriculture in, 428
 climate of, 428
 and glaciation, 428, 431
 map of, *428*
 moraines, 431

Milford Sound, New Zealand, fjord, *388*

Miller, Stanley, 12

millibar, 38: a metric unit of pressure used in meteorology, equal to approximately 0.75 millimeters on a mercury barometer; standard atmospheric pressure is 1,013.25 millibars.

milpa cultivation, 223, 226–227, 248–249, *249:* the term used to describe slash-and-burn agriculture in Latin America.

Minneapolis, Minnesota, solar altitude and azimuth, 52–53

minute: in angular measure, 1/60 of a degree.

Mississippi River

 delta, *305*
 description of, 293–294
 drainage basin, *44, 291, 309*
 during flood of 1973, 308–309, *309*
 at La Crosse, Wisconsin, *430*
 lateral migration of, 299, 301, *301*
 model of, *45*

Missouri River, 293

mixing ratio, 84: the weight of water vapor contained in a given sample of moist air compared to the weight of the air alone; usually expressed in grams of water per kilogram of dry air.

MN, 451, *451, 467:* magnetic north.

model: a representation of a system, used to explain or to predict the effects of changing conditions.

 graphic, *44*
 in mapping, 446, 461
 mathematical, *44*
 of nature, 44–45
 physical, *45*
 of river basin, 44–45, *44, 45*

moderate morphogenetic region, 359, *359,* 479

Moho discontinuity, *9:* Mohorovičić discontinuity or M discontinuity; a boundary, 10 to 35 kilometers (6 to 20 miles) below the earth's surface, between regions in which seismic waves move with different speeds; taken to be the boundary between the crust and the mantle.

moisture index, *173*, 177–178, 195, *195*, 431: a measure of the deficiency or surplus of precipitation compared to potential evapotranspiration.
classification of United States according to, 178, *179*
values of, 178
and vegetation, 178, *179*

moisture realm, 172, 173, 176: a division of a climate system; regions are classified in one of three realms, according to whether potential evapotranspiration is greater than, equal to, or less than precipitation.

Mojave Desert, California
pediment, *369*
tor, *274*

molecule, *10*: the smallest unit of a substance; an atom or a bound combination of several atoms.

monadnock, *276*: an isolated summit in a landscape of low relief; in the Davis cycle of erosion, a landform of old age.

monocline, *338*: a steplike fold that dips in one direction and then levels off to near horizontal.

monsoon circulation, *76*, *113*, 114, 122–124: a wind system that reverses direction seasonally.
and climate, 166
and jet streams, 123–124
timing, in India, 123

Monterey, California
ERTS photograph of, *473*
USGS topographic map quadrangle, *466*, *467*

Monument Valley, Arizona and Utah, mesas and buttes, *322*, *367*

moon
a dead world, 28, 261
formation of, 6
and tides, 399, *400*, 401

moraine, 391, *391*: debris deposited directly by a glacier; usually undifferentiated in particle size.
on Cape Cod, 423, *424*, 425
in the Midwest, 431
in the state of Washington, *391*
See also end moraine, ground moraine, lateral moraine

Morgan City, Louisiana, water budget surplus and AE/PE ratio, 170, *171*, 172

Morganza spillway, Louisiana, 308–309

morphogenetic region, 358–360, *359*, 479: a region characterized by the processes of erosion and transport that are active there.
change with time, *359*
detailed description of, 479

Motueka River Valley, New Zealand, glaciated valley, *270*

mountain building
and geosynclines, 336–337
and plate tectonics, *264*, *266*
rate of, 337
See also uplift

mountain ranges
as features of second order of relief, 264–265

mountains
effect of, on climate, 166, 266–267
effect of, on precipitation, *93*

mountain wind, *123*: a wind that blows downslope out of mountain valleys, usually at night.

Mount Diablo, California, in fog, *92*

Mount Egmont, New Zealand, stratovolcano, *330*

Mount Kilimanjaro, Tanzania, orders of relief, *267*

Mount Pelée, Martinique, eruption, 336

Mount Vesuvius, Italy, eruption, 336

Mount Waialeale, Kauai, precipitation, 94

Mount Whitney, California, granite slopes, *272*, *316*

mudflow, 356, *356*: the downslope flow of a thick, fluid mixture of soil and water.
volcanic, 336

Murray River, Australia, longitudinal profile, *294*

n

Nahant, Massachusetts, tombolo, 423

natural bridge, *366*: a bridge of rock, usually cut by flowing water.

natural levee, *303*, *304*: a ridge of material along a stream's banks, deposited by floodwaters.

natural selection, 14: the proliferation of a species that has qualities enabling it to survive the stress of an environment.

natural vegetation, 233: the climax vegetation in a region, especially where man's influence is minimal.
and climate regions, *233*
effects of man on, 233
global distribution of, *232*

neap tides, 401, *401*: the tides having the smallest daily range during the month, usually 10 to 30 percent below mean tidal range.

net productivity, 210: the total amount of stored food energy produced by plants less the amount used by the plants.

New Orleans, Louisiana, solar altitude and azimuth, 52–53

New York, New York, heat island, thermogram, *73*

Niagara Falls
escarpment, 425, 428, *429*
future development of, 428
geologic structure of, 429
and glaciation, 428
glaciers uncover, *17*
knickpoint, 299
lake terrace, 428

niche, 231: in ecology, the position of an organism in an ecosystem.

Nile River, 284
delta, *305*
longitudinal profile, *294*
sediment deposit, 284

nimbostratus clouds and weather fronts, *118*

normal fault, *339*: a fault in which the hanging wall, or the block above the fault plane, moves downward in relation to the footwall.

northing, 449, *449*: in a map grid system, the coordinate of a location to the north of the grid origin.

Norway
northern coniferous forest, *246*
tundra vegetation, *247*

nucleus, 23: (physics) the dense, massive, positively charged central part of an atom.

nuée ardente (pronounced "NOO-AY ar-DAHNT"), 336: a mixture of hot gases and solid particles, which can flow down the slopes of an erupting volcano like a fluid.

o

Oaxaca, Mexico, effects of milpa cultivation, *249*

oblique slip fault, *339:* the slip of fault blocks along the directions of both strike and dip.

occluded front, *118,* 119, *120:* the contact area at the boundary between cold and cool air, with warmer air aloft.

ocean: the body of salt water that covers about two-thirds of the earth's surface, or one of its major divisions.

general circulation of, 126–127, *127,* 129, 130, *131,* 160
interaction with the atmosphere, 129

ocean basin, 261: vast depressions in the earth's surface, which hold the waters of the oceans.

as features of first order of relief, 261
formation of, 264
rate of widening of, 264

ocean floor relief, forms of, 409, 413–414

oceanic plate, 264: a section of crust that underlies an ocean basin.

oceanic ridge, 20–21, *21,* 262–264, *264:* a boundary between crustal plates where new crust is being formed.

as features of second order of relief, 264

oceanic trench, 21, *262,* 264, *264:* a deep trench in the ocean floor, formed where an oceanic crustal plate is forced beneath a continental crustal plate.

at converging plate boundaries, *266*
as features of second order of relief, 264

oceanographic research, 21, 128–129, *412,* 413

and Glomar Challenger, 21
and H.M.S. Challenger, *412*
techniques of, *412,* 413

Oceanside, California, maps at three scales, *448*

Odum, Howard, 219

offshore, *406:* the part of a beach from the low tide line seaward to where erosion and deposition by waves does not occur.

old age, *276, 277*: a term used to describe a landscape worn to a plain of low relief with broad river valleys and dotted with isolated hills, or monadnocks; in the Davis cycle, an advanced stage of landform evolution.

Old River Diversion, Louisiana, 308–309, *309*

orbit, 2: (astronomy) the path of a satellite such as a planet or moon.

of the earth, 48, *51*
of the planets, *6*

orders of soils, 194: the principal divisions of the 1938 soil classification system.

See also azonal soils, intrazonal soils, zonal soils

Oregon, beach, *409*

origin of life, 12–13

origin of stars, 5

origin of the universe, 4–5

orographic effect: (meteorology) effects such as rain or fog induced by the presence of mountain relief.

orographic fog, 92: fog formed when moist air is chilled to its dew point by being lifted up hills or mountain slopes.

outcrop: rock locally exposed at the surface.

outgo, 150: in budgets, the amount of energy or material transferred out of a system.

outputs, 38, *39:* the energy or material transferred out of a system.

outwash plain: a plain made up of material washed out of a glacier.

overthrust fold, *338:* folded strata in which one limb of an overturned fold is thrust over the other limb because of rupture along a zone of weakness.

overturned anticline, *338:* a fold in which the strata on one limb have been folded more than 90°.

oxbow lake, 302, *303:* a crescent-shaped lake formed from an abandoned stream meander that has become separated from the main stream.

Oxford, England, climatic variation, *251*

Oxnard, California, coastline, *442*

ozone layer, 18, 57, *58:* a layer of ozone (O_3) in the earth's atmosphere located at an altitude of approximately 25 kilometers (15 miles); the ozone layer strongly absorbs ultraviolet light from the sun.

interaction of electromagnetic radiation with, 59

p

Pacific coast

moon and tides, *401*

Pacific Ocean basin

relief features, *262*

pahoehoe lava (prounced "pa-ho-ay-ho-ay"), *333:* a billowy or ropy form of solidified lava.

Palermo, Italy, temperature and precipitation, 241, *241*

pali ridge, *372:* a sharp-edged ridge at the intersection of adjacent valleys on a volcanic dome.

Palos Verdes, California

marine terrace, *411*
slump, *356*

pancake karst, *372*

Pangaea, *17,* 20, *20,* 22, 264: a hypothetical supercontinent that is supposed to have contained all the earth's landmasses approximately 200 million years ago.

parabolic sand dune, *366:* a sand dune shaped like a parabola, with its concave side facing the wind and its horns pointed into the wind; a blow-out dune.

parallel retreat of slopes, 277, *277*

parallels of latitude, *51:* lines of constant latitude on the earth, forming circles that are parallel to the Equator.

parent rock of a soil, 187–188, *188:* the source of minerals for the development of a soil.

passive giacier, 383–384: a glacier that transports material slowly compared to other glaciers.

Paterno, Italy, lava flow, 312

patterned ground, *362, 363*: a division of the ground's surface into a pattern of polygons, caused by frost action; characteristic of landscapes in cold climates.

PE: potential evapotranspiration.

peat, 191, *193,* 202, *202:* partly decomposed plant matter, formed in waterlogged soils.

pedalfers, *192,* 195, *195:* a general class of soils in which aluminum and iron compounds are abundant; they occur in moist regions.

pediment, *277,* 361, *361, 368, 369:* an extensive plain, often faintly concave, that slopes gently from mountains to a basin in arid regions.

pedocals, *192,* 195, *195:* a general class of soils in which calcium compounds are abundant; they occur in arid regions.

pedology, *2,* 187: the scientific study of soils.

peds, 183–184: the large-scale structure of a soil; commonly called clods.

Penck, Albrecht, 172

Penck, Walther, 277, *277*

peneplain, 276, *277*: "almost a plain"; a lowland plain of low relief; in the Davis cycle, the end result of erosion.

penitent rocks, 324–325, *324*: unusual formations of separated, parallel leaning schist rocks.

peppered moths, 14–15, *15*

percolation, 143: the passage of water from the soil into groundwater storage.

perhumid climate region, *178, 179*

periglacial climate: a climate characterized by freeze-thaw phenomena; it usually occurs in a region adjacent to a glacier.

slopes in, 360

permafrost, *192*, 195–196, *195, 362*: a layer of permanently frozen soil, underlying the surface at a depth of a few feet in moist, cold climates.

depth of, *362*

permeable, 140: containing crossconnected channels, usually of capillary size, through which water can move.

Perth Amboy stream, longitudinal profile, *294*

Peru

Inca amphitheater, *416*, 417
Machu Picchu, 417, *418*

pH, 184–185, *185*: a measure of the concentration of hydrogen ions in a solution on a logarithmic scale.

desirable values of, for soils, 185
and hydrogen ion concentration, 184–185
scale, 185, *185*
of water, 185

phase: the physical state of a material, which may be solid, liquid, or gas.

Phoenix, Arizona

climate, compared to Flagstaff, Arizona, 440
water budget surplus and AE/PE ratio, 170, *171*

photorespiration, 210: the increased use of food energy by a plant when it is exposed to light.

photosynthesis, 15–16, 30, 207–210, *208*: the process by which green plants produce stored chemical energy from water and carbon dioxide, with the aid of radiant energy.

and the cycle of life, *208*
efficiency of, *207*, 208
and nutrients, 209–210
variation with light intensity and temperature, 208–209, *210*
variation with solar radiant energy, 208, *209*

physical geography

contribution of, *2*, 421, 422
relationship to other sciences, *2*
use of terms in, 423

piedmont glacier, *376*: a large expanse of glacial ice formed by the coalescence of two or more valley glaciers; it has some of the characteristics of ice sheets.

piezometric surface, *142*: the surface to which water rises in a well sunk into a confined aquifer.

Pike's Peak, Colorado, weathered granite, *186*

pillar, *271*: a column of rock remaining after the erosion of surrounding material.

See also hoodoo rock

pillow basalts, *333*: solidified lava that resembles pillow-like masses.

pingo, *362, 363*: a domed hill formed when the soil cover is pushed up by a lens-shaped mass of ice; occurs in moist, cold climates.

pipkrake, *354*: a stem of ice projecting above the soil, formed during the freezing of subsurface water; usually carries pieces of soil downslope.

Pittsburgh, Pennsylvania, temperature and precipitation, 242, *243*

plane table, 469–470, *469*: a simple surveying instrument.

planetesimals, 5–6: bodies smaller than planets in size, which coalesced to form the planets according to the theory proposed by Harold Urey.

plankton, 224

plant community, 220, 231: an association of vegetation types that are in stable interaction with one another and with the environment.

effect on local environment, 231
lack of competition in, 231

plant succession, 220, *231*, 232–233: the sequence of plant communities in a region, leading to climax vegetation.

and diversity, 232
and stability, 232

plateau, 264: an extensive highland area with low surface relief.

plate tectonics, 21, 261–266, *262*: the theory that attributes crustal deformations, such as mountain uplift and volcanoes, to the movement and interaction of crustal plates.

and the distribution of volcanic activity, 325, 328, *328*
and earthquakes, *262, 264*
and interacting plate boundaries, *266*
and mountain uplift, *264*, 265–266, *265*
and the rock cycle, 325, *327*
and volcanic activity, *262, 264*

Platte River, 293, 294

playa, *364, 365*: a lake bed that is dry for much of the year; a basin in which rainwater accumulates and soon evaporates.

Pleistocene, *17*: the epoch of geologic time, 1 or 2 million years in duration, that ended 10,000 years ago.

plucking, 386, *386, 387*: a process of glacial erosion in which large pieces of rock are torn from outcrops or bedrock.

plunge basins, *372*: a series of pools from which a waterfall successively cascades in its descent to a valley floor.

podzolic soils, *189*, 191

and climate, *189*

podzols, 189, *192*, 196, *196, 200*: soils of the coniferous forest; they are acid and have an ash-gray layer near the surface.

agricultural potential of, 196
and climate, *189*
global distribution of, *200*
profile, *196*
United States distribution of, 196, *200*
and vegetation, *192*, 196

point of tangency, 453, *453*, 454

polar climate region, 165, *165*, 166, *168*

polar front, *111*: the zone where cold air from the polar regions meets warm air from the tropics.

polar glacier, 384–385: a glacier in which the ice is considerably below the temperature of the pressure melting point; a polar glacier is relatively incapable of basal slip.

polishing, glacial, 386, *386, 387*

pollen analysis of climate, 250–251: the study of ancient climates by analyzing deposits of fossil pollen.

pollution: the presence of unwanted materials in quantities so great that natural processes cannot clear them away quickly.

pool, 301–302, *301*: a deep section of a stream, from which material has been eroded.

pool and riffle pattern, 301–302, *301*

spacing in, 301, *301*

Postern Peak, Rocky Mountains, Canada, talus slope, *352*

potential evapotranspiration, 148–149: the amount of water evaporated and transpired from a large stand of plants supplied with all the moisture they are capable of using.

and actual evapotranspiration, 148–149, 154
annual, United States, *148*
factors determining, 149–150
Köppen's estimate of, 173–174
measurement of, 149–150
Thornthwaite's calculation of, 148, 150, 476–477
units of, 148–149

potential photosynthesis, 213–215, *215*: the maximum value of a plant's productivity when all its needs are supplied.

and climate regions, 215, *215*
factors determining, 213, 215
global distribution of, 213, 215, *215*

powers of ten notation, 5, 444

prairie soils, *189*, *192*, 198, *199*: soils of the comparatively moist, tall-grass prairie near the transition zone between forest and steppe; brunizems.

agricultural potential of, 198
and climate, *189*
global distribution of, *200*
profile, *199*
United States distribution of, 198
and vegetation, *192*, 198

precipitation, 80, *80*, *137*, 138: any form of water transferred to the earth's surface from the atmosphere.

global distribution of, *99*
measurement of, 156
United States distribution of, *99*

predator: an animal that preys on other animals.

pressure, 107: the average force exerted on a unit area by molecular collisions.

of a gas, 37
global distribution of, 111, *113*
in the upper atmosphere, *114*

pressure gradient force, 107–108, *107*: a force caused by pressure differences, which tends to push a parcel of air from higher to lower pressure.

pressure melting of ice, 384–385

primary producers, 215: organisms that can subsist on inorganic materials and energy; autotrophs.

primary waves, *9*: seismic waves that are longitudinal waves of compression, analogous to sound waves in air; they are able to propagate through molten material; P waves.

prime meridian, *51*, 449: a reference meridian for specifying longitude; the meridian passing through the observatory in Greenwich, England, is now used as the prime meridian by nearly all countries.

primordial fireball, 4, 5: the extremely hot concentration of matter and energy at the beginning of the universe's expansion.

principal meridian, 451, *451*: a north-south reference line for the United States Bureau of Land Management land survey, laid out along a meridian; a guide meridian.

problems of world agriculture, 221–223

energy cost, 221–222
grain stock depletion, *222*
nutritional problems in the United States, *223*
rate of cereal production, 222–223
social adjustments, 222
United Nations world agricultural plan, 222

productivity, 210–215

of estuaries, *212*
harvest method for measuring, 211
of major crops, 211–212
measurement of, 210–211
measures of, 210–212
net, of major ecosystems, 211, *212*
of the ocean, 211, *212*, 226
units of, 211
See also gross productivity, net productivity

projection, 449, 452–459: any method for relating position on the earth's surface to corresponding position on a map; a transformation.

Albers' conic equal area, 454, *457*
azimuthal, 454, *455*
azimuthal equidistant, 455
condensed, 459
condensed interrupted, 459
conformal, 453
conic, 454, *457*
construction of gnomonic, 453, *453*
cylindric, 454, *456*
and distortions, 453–459
equal area, 453, *453*
equirectangular, 456
flat polar quartic, *458*, 459
gnomonic, 453, *453*, 455
interrupted, *458*, 459
Lambert's conic conformal, *457*
Lambert's cylindrical equal area, 456
Lambert's equal area, 455
Mollweide's, *458*
orthographic, 455
polyconic, *457*
pseudocylindric, 454, *458*, 459
sinusoidal, *458*
standard Mercator's, 454, *456*
stereographic, 455
transverse Mercator's, 449, *456*

pseudocylindric projections, 454, *458*, 459

Ptolemy map, *446*

pumice, 330: a frothy, bubble-filled form of solidified lava.

r

radar, 471, *471, 473:* a method of sensing and distance ranging that involves the detection of short-wavelength radio waves (microwaves) reflected from objects.

radiant energy balance, 59–74, *63:* the difference between incoming energy and outgoing radiant energy at the earth's surface; when the balance is positive, the surface is gaining energy.

- and carbon dioxide, 73
- at the earth's surface, *64*, 65–74, *72*
- and jet condensation trails, 73
- and man, 72–74
- modification of, 70–74
- and turbidity, 72–73
- versus latitude, *65*

radiation, 469, *469:* (physics) the transmission of energy through space by particles or waves; (surveying) a technique for specifying location by measuring azimuth and distance from an origin.

radiation fog, 93: fog that forms when moist air near the surface is cooled to its dew point by contact with ground that has been cooled by radiation heat loss.

radioactive dating, 22, 23–24: the use of the known rate of radioactive decay for the dating of minerals and rocks.

- radioactive clocks, 23

radioactive decay, 6: the spontaneous transformation of one atomic nucleus into another, accompanied by the emission of subatomic particles or gamma rays.

- heat supplied to the earth from, 10

radioactive element: a chemical element that has an unstable nucleus that undergoes radioactive decay.

rain, formation of, 88–91, *89*

rainmaking, 99–101, *100*

rainsplash, *285, 351,* 356–357: the movement of soil particles by the impact of raindrops.

range, 451, *451:* in the township system, the position of a township numbered east or west from a principal meridian.

Rapid City, South Dakota, flooding, 78

rate of geomorphic change, 279–280

- rate of erosion, 279
- rate of uplift, 279

realms of climate, 172

- Köppen's principal divisions, 173, *174*
- Thornthwaite's principal divisions, *178*

Recent, *17:* the present epoch of geologic time, which began 10,000 years ago; the Holocene.

recharge area, 143: the surface region over which water for a particular aquifer is collected.

Red Rock Canyon, California, eroded rock, *270*

Red Sea

- opening of, 17, 263, *266*
- satellite view of, *264*

red-yellow podzolic soils, *189*, 196–197, *197, 200:* soils of the subtropical forests.

- agricultural potential of, 197
- and climate, *189*
- global distribution of, *200*
- profile, *197*
- United States distribution of, 196, *200*
- and vegetation, 197

reflection, 403: the turning back of a wave, or part of a wave, at a surface.

- of radiant energy, 59, 62
- of water waves, 403

refraction, 403, *403:* the change in direction of a wave, caused by interaction with a boundary such as the ocean bottom or a shoreline.

- concentration of energy because of, 403, *403*
- at a headland, *403*

regolith, 361: the layer of loose rock fragments and soil on the earth's surface.

relative humidity, 83–85, *84:* the amount of water vapor present in a given sample of air, compared to the amount the air could contain if saturated, usually expressed as a percentage; it is equal to the ratio of actual vapor pressure to saturation vapor pressure.

- dependence on temperature, *84*
- through the day, *84*

relict landform, 278–279, *278:* a landform developed under climates or conditions different from those prevailing at present.

relief: the range between the highest and lowest elevations of a region.

- in maps, 459–462
- *See also* first order of relief, second order of relief, third order of relief.

relief map of the United States, *426*

remote sensing, 470–473: the study of the earth's surface from a distance by using various forms of electromagnetic radiation.

- advantages of, 470
- and mapping, 470, 471
- and resource management, 470, 471

representative fraction, 447, *447:* a method of specifying the scale of a map by giving the ratio between distance on the map and the corresponding distance on the earth, expressed in the same units of measure; for example, 1:5,000 means that 1 inch on the map represents 5,000 inches on the earth's surface.

respiration, 210, *210:* the use of oxygen and food by organisms to produce energy, along with waste products, such as carbon dioxide.

- energy used for, 210

response time, 39, *39:* a measure of the time required for a system to come to a given state of equilibrium.

- of landforms, 349
- of soils, 189–190

reverse fault, *339:* a fault in which the hanging wall, or the block lying above the fault plane, moves upward in relation to the footwall.

Rhode Island estuary, 401

Rhodesia

- bornhardt, *371*
- penitent rocks, 324–325, *324*

ria, *411:* a valley, usually lying transverse to a coast, that is drowned by submergence of the coast, forming a long inlet.

ridge-ridge transform faults, *21:* crustal fractures at right angles to oceanic ridges, where relative movement can take place.

riffle, 301–302, *301:* a shallow section of a stream where material has been deposited.

rift, *342*

rift valley, 263: a region on land where the earth's crust is separating, in some cases allowing molten rock to well up; analogous to a spreading center.

rill, 292: a very small surface drainage channel, only a few centimeters in size.

Rio Grande, longitudinal profile, *294*

rivers: *See* streams.

river valley: a valley cut by a stream.

rock cycle, 325, *327:* the cycling of minerals from igneous rock to sedimentary and metamorphic rock, and their return to igneous rock.

rocks, composition of, 314–318, *316, 317, 319, 320, 321*

Rocky Mountains

- geologic history of, 436, 439
- and glaciation, 436, 439
- map of, *437*
- topography of, 436, *437*
- uplift of, *17*

Rocky Mountains, Canada, folded rock, *312*

Rossby waves, *122:* extensive undulations in the upper atmosphere above the polar regions.

runoff, 41, *137:* the precipitation that eventually flows into streams, either as surface runoff or as runoff from groundwater.

and streamflow, 144–145, *145*
variation with time, *145*
and water budget surplus, 154

S

Sagastyr, U.S.S.R., thermoisopleth, 247, *247*

sag pond, 341: a pond in a local depression that is caused by subsidence of the ground by faulting.

Sahara desert

man's adaptation to, *256*
oasis, *175*
vegetation, Algiers, *240*

Sakharov, Andrei, 226

saltation, 289: a hopping motion executed by particles driven by wind or flowing water; the particle is repeatedly lifted from the solid surface into the air or water as it moves forward.

Salton Sea, California

formation of, 440, 443
and irrigation, 443

salts: soluble compounds in the soil, usually of calcium or magnesium.

San Andreas fault, California

Carrizo Plain, *340*
earthquakes, 338
movement of, 312
and orange grove, *340*
radar image of, *473*
as strike-slip fault, 341

San Antonio, Texas, monthly radiant energy balance, *72*

sand, 183, *183, 184:* inorganic particles with diameters between 50 and 2,000 microns.

sand dune, *272, 366:* a mound or hill of windblown sand.

See also barchan, parabolic sand dune

San Diego, California, hourly temperature distribution, 68, *68*

sand ripples, *272, 366:* a pattern of small parallel ridges that develop on the surface of sand in regions where sand is not accumulating; the ripples are oriented across the direction of the prevailing wind.

sandstone, 318, *319, 430*

and natural arches, *271, 438,* 439
weathering of, 318, *319*

San Francisco, California

daily temperature and precipitation, *242*
eucalyptus trees, *175*
local water budget, 241, *241*

San Francisco Bay, California, fog, *92*

San Joaquin Valley, California, aqueduct, *146*

San Juan River, Utah, goosenecks, *297*

Saskatchewan glacier, *384*

saturation: (meteorology) a condition in which air at a given temperature contains as much water vapor as normally possible.

saturation vapor pressure, 81–82: the pressure exerted by the water molecules in air that contains the maximum amount of water vapor normally possible.

versus temperature, *81*

Saudi Arabia, satellite view, *264*

savanna, *370, 371:* a seasonally dry grassland, dotted with trees, lying between the equatorial forests and the hot deserts.

scale, 447–449, *447, 448:* in cartography, the relationship between distance on a map and distance on the earth; a small-scale map is a map of a large area, showing limited detail; comparatively, a large-scale map is a map of a small area, with relatively fine detail.

conversion of, 447
equivalents to, 447
graphic scale, 447, *447*
not constant on a map, 449
verbal statement of, 447, *447*
See also representative fraction

scales of climate, 161–163

scattering, 57–58: (electromagnetism) the process in which the energy of electromagnetic radiation is distributed in various directions by interaction with particles.

Schaefer, Vincent, 100, *100*

schist, *321,* 324–325

and penitent rocks, 324–325, *324*

scouring, 289–291, *297:* the erosion of material from a channel by a stream.

sea arch, 404, *404:* an arch cut through a headland or offshore rock by wave action.

sea breeze, *123:* a wind that blows from the sea toward the land, usually during the day.

sea cave, 404, *404:* a hollow cut in a shoreline rock by wave action.

sea cliff, *424,* 425: a shoreline cliff shaped by wave action.

on Cape Cod, *424,* 425

sea floor spreading, 19–22, *21,* 262–264, *266:* the production of new crust by the upwelling of molten material at oceanic ridges.

sea level: the elevation of the surface of the sea, based on the average heights of tides and waves, usually for a given reference period.

along United States coasts, 399, *399*
and glaciation, 380
relative rate of rise at present, 380
since the Pleistocene, 398–399, *398*

sea mount, *262:* a guyot.

sea stack, 404, *404:* a small rocky island or pillar that has been separated from coastal rocks by erosion.

second, *51:* in angular measure, 1/60 of a minute; 1/3600 of a degree.

secondary circulation, 106: short-lived, comparatively localized features of the atmospheric circulation, such as hurricanes and rapidly moving high- and low-pressure cells.

midlatitude, 114-120
tropical, 120, 122–126

secondary producers, 215: organisms that rely on food produced by other organisms; heterotrophs.

secondary waves: seismic waves in which the vibration is transverse to the direction of propagation; they are unable to move through molten material; S waves.

second law of thermodynamics, 33–34

second order of relief, 264, 266–267: surface features that are comparable in size to continents; for example, mountain ranges.

section, 451: one of the 36 subdivisions of a township; approximately 1 mile square, containing 640 acres.

sediment: debris deposited in beds by water, wind, or ice.

sedimentary rock, 12, *13,* 315, 318: rock formed by the consolidation of beds of sediment.

limestone, 318, *320*
and mountain uplift, 336–337
mudstone, in the Badlands, 432–434, *433*
and the rock cycle, 325, *327*
sandstone, 318, *319*
shale, 318

sedimentation

and dating, 23, 24
and fossils, 13–14

sediment movement on beaches, 408, *408*

seismic waves, 9, 10: waves in the earth generated by an earthquake or, artificially, by explosions.

semidesert, *438,* 439

Seneca Creek, Maryland, gauge height, *286*

settling velocity, 87–88, *290:* the steady speed eventually attained by an object falling through a gas or liquid.

Seward Peninsula, Alaska, solifluction lobes, *357*

shale, 318

weathering of, 318

shallow water waves, 402, *402:* waves in water shallower than half their wavelength; their speed depends on the depth of the water.

shape of the earth, 452

sheetwash, 292: material eroded from the ground by runoff in the form of a sheet of water.

shield volcano, *330:* a volcanic cone built up from successive layers of lava.

Shiprock, New Mexico, volcanic neck and dikes, *333*

short-wave infrared radiation, *56:* infrared radiation with wavelengths shorter than 4 microns, such as the infrared radiation emitted by the sun.

sial, *9,* 314–315: the rocks constituting continental crust, which consist primarily of compounds of silicon and aluminum; not as dense as sima.

Siberian high, 111, *113*

side-looking airborne radar, *343,* 471, *471, 473*

sierozem, *189:* gray desert soils, often found in midlatitude continental deserts.

and climate, *189*

Sierra Nevada, California

basin-and-range province, 339–340
fault scarp, *314*
granite boulder slope, *272*
granite slopes, *316*
uplift, *17*

significant figures, 444–445: the number of meaningful digits employed to express the value of a quantity, taking into account the limits of error.

sill, volcanic, *334:* a predominantly horizontal intrusion of magma between horizontal rock strata.

silt, 183, *183, 184:* inorganic particles with diameters between 2 and 50 microns.

silver iodide, 101, 102, *102*

Silver Springs, Florida, energy flow in its ecosystem, 219-220, *219*

sima, *9,* 314–315: the rocks constituting the oceanic crust, which consist primarily of compounds of silicon and magnesium; denser than sial.

simplification in maps, 459

Singapore, temperature and precipitation, 235, *235*

sinkhole, *320,* 324, *373:* a depression in the soil cover over a subterranean drainage hole; common in karst regions.

sinuous stream channel, 294, *295,* 301, *301, 302*

Sisters' Rill, near Washington, D.C., longitudinal profile, *294*

Skellig Michael, Ireland, imported soil, 182

slash-and-burn agriculture, 223, 226–227, 248–249, *249:* a method of agriculture practiced in tropical forests, in which the land is cleared by cutting and burning, then planted and harvested, and later allowed to return to forest, after it loses its productivity; the term "milpa cultivation" is used to describe its practice in Latin America.

sleet, *89:* the ice pellets formed when rain passes through a layer of subfreezing air near the earth's surface.

slickensides, 340: local areas, usually vertical, where rock faces have been smoothed and polished by the abrasion of one rock mass slipping past another.

slip, *339:* a joint or crack where displacement of rocks has occurred; a fault.
See also dip-slip fault, oblique slip fault, strike-slip fault

Slobodkin, Lawrence B., 217, *217,* 226

slope development processes, 274, 352–357

balance between disintegration and transport, 352
effect of moisture on, 352, *354*
forces acting in, 352, 354
and process, resistance, and structure, 274
See also mudflow, rainsplash, slump, soil creep, soil wash, solifluction, solution

slope formation, 277

according to Davis, *277*
according to Penck, *277*

slope processes, and climate, 274, 357–358, 360–361

in arid and semiarid regions, 360–361, *361*
in hot, moist climates, 361
in moist midlatitude regions, 360
in periglacial climates, 360

slopes, study of, 268

diversity of, 272–274, *272*
and landforms, 268, 272–274, *272*

slump, *356:* the downward movement of a large portion of a slope as a whole without rotation.

small-scale map, *448,* 449: a map on which a representative area corresponds to a comparatively large area on the earth.

smog, *74*

smudge pots, 72

snout, of glacier, *383, 384,* 385: the terminus of a glacier.

snow, formation of, *89,* 90

snowline, 381: the boundary line above which some winter snow can persist throughout the year.

snowmelt, 145: the water from melting snow.

soil: a complex of fine rock particles and organic material at the earth's surface, partly bound together at the molecular level.

inorganic matter in, 182–183, 187–188, *188*
need for fertile soil, 182
organic matter in, 183
uses, 182

soil creep, 354, *354,* 356: the slow downslope movement of rock and soil caused by gravity.

rate of, 354, 356

soil development

and climate, 189, *189,* 190–191
factors influencing, 187
and organisms, 189
parent rock, 187–188, *188*
rate of formation, 189–190
and topography, 188–189
and vegetation, 189, *192*

soil management, 202–203

and maintenance of soil productivity, 202–203

soil moisture, *137, 138,* 140–141: the water in the soil that is bound loosely enough to be available for absorption by plant roots.

and the local water budget, 145, 150–156, *150, 151*
measured compared to predicted, *155*
models for availability of, 238

soil pores, 140, 183: the spaces between particles of soil.

and infiltration, 141

soil profile, 186, *186:* a vertical section through a soil; it usually shows various horizons.

soil structure, 183–184: the shape and size of the peds of a soil.

and plowing, 184

soil texture, 183–184, *184:* a characterization of a soil according to the distribution of particle sizes in its mineral content.

soil texture triangle, *184:* a means of specifying the textural classification of a soil in terms of the proportions of clay, silt, and sand.

soil wash, 351, *351,* 356–357: the transport of soil down a slope by flowing water.

solar altitude and azimuth, 52–53, *52*

solar constant, 54–55: the rate at which solar radiant energy is received at the top of the earth's atmosphere on a unit area oriented perpendicularly to the sun's rays; approximately 2 langleys per minute.

solar radiant energy, 28, 30, *33,* 34

daily input, computation of, 55
dependence on angle, *51,* 53
input to top of the atmosphere versus month and latitude, *54, 55*
intercepted by the earth, 53

solar radiation, *57:* the spectrum of electromagnetic radiation emitted by the sun.

solar system, *6*

formation of, 5–6

Soleri, Paoli, *256*

solid, *10*

solifluction, 356, *357,* 360, *362:* a form of mass wasting in which thawed, wet soil moves downslope relatively rapidly; occurs in cold climates.

rate of, 356

solifluction lobe, *357:* a tongue of soil that extends downslope because of solifluction movement.

solstice: *See* summer solstice, winter solstice.

solution, erosion by, *351,* 352: removal of material by solution in water.

South Dakota

Great Plains, *433*
Huron, water budget, 244, *245*

South Island, New Zealand, pancake rocks, *372*

Southwest

map of, *437*

Spain

Granada, cave homes, *418*
maquis scrub, *240*

species: subdivisions of the plant and animal kingdoms; a species is a group of similar organisms that are able to breed with one another.

specific gravity, 382: the ratio of the weight of a given volume of a substance to the weight of the same volume of water.

of glacial ice and snow, 382

specific heat, 36: the amount of energy required to raise the temperature of 1 gram of a substance by 1°C.

specific humidity, 84: the weight of water vapor contained in a given sample of moist air compared to the weight of the moist air. Usually expressed in grams of water per kilogram of moist air.

spectral signature, 471: the characteristic distribution of wavelengths reflected by a substance; can be used to distinguish different types of vegetation.

spit, *411,* 425: a narrow strip of land that extends from the shore into the ocean.

spot height, 461: the altitude of a location, indicated on a map.

spreading center, *266:* a boundary between crustal plates where new crust is being formed by the upwelling of molten rock.

springs and fault zones, 340–341

spring tides, 401, *401:* the tides having the largest daily range in height during the month; usually occur at or near the time of new and full moon.

stage, 274: a phase of landform development or evolution, especially in the Davis cycle.

standard atmosphere, *58:* a description of the average temperature, pressure, or other properties of the atmosphere versus altitude, adopted for use in scientific and engineering calculations.

standard atmospheric pressure, 37–38: 1,013.25 millibars or 760 millimeters of mercury.

standing crop, 212: biomass.

State Plane Coordinate system, 450

stemflow, *137,* 139: the passage of moisture along the branches and stems of plants to the ground.

steppe, *233:* semiarid midlatitude grasslands, extensive in central and northern Asia and in central North America.

stock, *334:* an intrusion of magma, smaller in extent than a batholith.

stomata, 82–83, *82:* microscopic slits, principally in the leaves of green plants, through which carbon dioxide enters the plant and water vapor leaves it.

stone garlands, rings, stripes, *362:* patterns formed by the sorting of stones because of frost action; characteristic of cold climates.

Stone Mountain, Georgia, granite dome, *272*

storage, 39, *39:* the containment of energy or material within a system.

strata, 318–323, *322, 338:* layers of rock, especially sedimentary rock.

stratosphere, 57, *58:* the portion of the earth's atmosphere above the troposphere, extending from an altitude of 10 kilometers (6 miles) to 50 kilometers (30 miles); it is a region of low humidity.

stratovolcano, *330:* a volcano that has a cone consisting of layers of volcanic ash and lava.

stratus clouds, *94:* low clouds that form in a uniform gray layer, often extending from horizon to horizon; common in winter.

and weather fronts, *118*

stream: a body of flowing water.

adjustment of channel by, 295–302
characteristics of selected, 288
as a geomorphic agent, 284
importance to man, 284
power of transport, 289
principles of flow, 285–287
transport of material by, 287–291, *288, 289, 290*

stream bed: the bottom of a stream's channel.

displacement by faults, *340,* 341, 343

stream channel pattern

braided, 294, *295,* 301, 302
meandering or sinuous, 294, *295,* 301, *301, 303*
straight, 301, *301*

stream order, 292, *293:* a labeling of streams according to their position in the hierarchy of a drainage net.

stream valley: a valley cut by a stream.

and the work of flowing water, 273

strike, *337, 339,* 341, 343: the direction of the horizontal line of a fault plane, measured by compass bearing; strike is perpendicular to dip.

strike-slip fault, *339,* 341, *341,* 343: a fault in which the relative displacement is parallel to the direction of the strike.

lateral, *339*

stripped plain, *366:* a plain of hard rock, exposed after less resistant cover rock has been eroded away.

structure of the earth, 6, *9,* 10

subarctic coniferous forests, 244–245, *246*

adaptations of vegetation, 244–245
global distribution of, *232,* 244
limited number of species in, 245

subduction zone, *262,* 264, *264, 266, 327:* a boundary between converging crustal plates where one plate slides or sinks beneath another; often marked by the presence of a deep sea trench, an island arc, and volcanic activity.

subhumid climate region, 178, *178, 179*

submarine canyon, *397,* 413: a deep canyon cut into the continental slope; occurs frequently off the mouths of major rivers.

submerged coastline

parallel, *411*
transverse, *411*

subpolar climate region, 164, *165,* 166, *168, 247*

global distribution of, *168*
vegetation of, 245, 247

subtropical deserts, 239–240, *240*

adaptation of vegetation, 240
global distribution of, *232*
vegetation characteristics, 240, *322*

subtropical dry summer climate region, 165–166, *165, 168, 175*

subtropical highs, 110, *110*

subtropical semidesert vegetation, 239

subtropics: the regions bordering on the tropics; the regions lying approximately between latitude 35° and the tropics.

Sumatra, Indonesia, rice paddies, *256*

summer solstice, 49, *51:* the time when the sun is directly overhead at the Tropic of Cancer, latitude 23½°N; occurs about June 21.

supercontinent, 264: a large hypothetical landmass which is believed to have rifted into the present-day continents.

supercooling, *89,* 90: lowering the temperature of liquid water droplets below the normal freezing point without the formation of ice.

saturation vapor pressure of supercooled droplets, *81*

superposed stream, 344, *344:* a stream whose course was cut in young rocks or sediment overlying older geologic structures, which has maintained its course and cut into the older structures as uplift subsequently occurred.

supersaturated air, 87: air that contains more water vapor than does normally saturated air at the same temperature.

surface runoff, 144–145: the water that flows on the surface as sheets or, more commonly, as streams, rather than as groundwater.

surface tension, 141: an internal force of molecular attraction at the surface of liquids that acts much like a thin membrane on the liquid.

surge, of a glacier, 385: an unusual form of glacial motion, characterized by sudden and rapid advance as a "wave" of ice travels down the glacier.

surplus, *151, 152,* 154: in the local water budget, the amount of available moisture above the amount required by evapotranspiration and soil moisture recharge.

during wet, dry, and average years, *152,* 154–155
and runoff, 154

survey of the earth's resources, 73–74

suspended load, 288, 439: the amount of material carried in suspension by a stream.

deposition in reservoirs, 290

Susquehanna River

radar view, *343*
superposition, 344, *344*
water gaps, *323*

sustainable yield, 217, *217,* 225: the maximum yield that can be obtained from an ecosystem without depleting the average population.

swash, 404: the water that rushes up a beach after a wave strikes the shore.

swell, 402: the regular waves on the ocean, caused by distant disturbances and prevailing winds.

typical speed of, 402

symbolization in maps, 459–462

synclinal ridge, *343:* a ridge that has its axis along a syncline.

synclinal valley, *343:* a valley formed along the trough of a syncline.

syncline, *338,* 343: a troughlike fold in rock strata, in which the limbs of the fold dip upward.

system, 38–39, *39:* the part of the universe selected for study; open systems receive energy or material from outside; closed systems are self-contained.

t

table mountain, *333:* a flat-topped mountain.

taiga: subarctic coniferous-forest regions of the Northern Hemisphere.

talus, 352, *352,* 353, 360, *362:* a layer of loose rock fragments that have accumulated in a sloping pile at the foot of a cliff.

tarn, *389:* a pond in the ice-gouged basin of a cirque.

Taunton, England, mixed agriculture, *256*

tear fault, *339:* a fault that has a vertical fault plane but exhibits horizontal slip; tear faults usually occur in connection with earthquakes and movement along fault zones.

Tempe, Arizona, hourly radiation balance, *64*

temperate glacier, 384–385: a glacier in which the ice is near the pressure melting point, so that motion by basal slip occurs easily.

temperature distributions
- annual global, *69*
- hourly, 68, *68,* 70
- for January, April, July, and October in the United States, *70*

temperature inversion, 96, *97,* 116: a condition in which air at lower altitudes is cooler than air immediately above, so that vertical motion of the atmosphere is impeded.
- and easterly waves, 124

tempo of geomorphic change, 279–280
- rate of erosion, 279
- rate of uplift, 279

Texas, grain and cotton field, *254*

thalweg, 301, 302: the line connecting the deepest points of a stream channel.

thematic map, 462: a map devoted to a single topic such as temperature or precipitation.

thermal radiation, 56, *57:* the characteristic electromagnetic radiation emitted from a body as a result of its temperature.
- from the earth, *57,* 62–63, *63*
- from the sun, *57*

thermogram: a representation, usually by different colors, of the temperature distribution of an object.
- of hurricane Camille, 128, *129*
- of New York City, *73*

thermoisopleth diagram, 235, *236:* a diagram that shows the temperature at a station throughout the day for every day of the year.
- Belém, Brazil, 235, *236*
- Klagenfurt, Austria, 242, *243*
- Sagastyr, U.S.S.R., 247, *247*

third order of relief, 267–268, *267:* local areas of relief, such as hills, valleys, or isolated mountains.

Thornthwaite, C. Warren, 148, 177, 476–477

Thornthwaite system of climate classification, 177–178, *177*
- divisions of, *178, 179*
- indexes for, 177–178

threads, 292: surface drainage that forms filament-like channels.

throughfall, *137,* 139: the moisture that reaches the soil after initial interception by plant cover.

tidal mud flat, 401, *401:* a flat muddy area at a coast, which is covered by water at high tide and exposed at low tide.

tides, 399–401: the rhythmic rise and fall of oceans and the bodies of water extending from them, caused by gravitational effects of the moon and the sun.
- cause of, 399, 401, *401*
- daily rhythm of, 401, *401*
- and estuaries, 401
- geomorphic significance of, 399
- neap, 401, *401*
- spring, 401, *401*
- variation through month, 401, *401*

till, 391: unsorted deposits laid down by glacial ice.
- and groundwater supply, 392

till plain, *391,* 392, 428: a level region covered by till.
- depth of, in the United States, 392

tilt of the earth's axis, 6, 48–52, *51*

time zones, *53*

tombolo, 423: a bar of deposited material that ties an island to the mainland or to another island.

top carnivore, 215, *216:* the animal species at the end of a food chain.

topographic map, 464: a map that shows physical features of the earth's surface, especially relief and contour.

topographic profile, 460–461, *460:* a drawing representing elevation through a cross section of a landscape.

topography: the relief, or pattern of relative elevations, of the surface of the land.

tor, 274, *274, 370:* a mass of rock, often granite, that has weathered into a "pile of boulders."

tornado, 120: a highly localized column of rapidly rotating air, associated with intense thunderstorm activity.

tower karst, *345, 373:* tall, vertical-sided towers of limestone.

township, 451, *451:* the basic land unit of the United States Bureau of Land Management land survey; approximately 6 miles square.

trace elements, 188: plant or animal nutrients that are required only in small amounts.

trade winds, 109, *110:* prevailing winds of the tropics, which blow toward the Equator from the northeast in the Northern Hemisphere but from the southeast in the Southern Hemisphere.

Transeau, Edgar, 208

transpiration, 64–65, *80:* the transfer of water vapor to the atmosphere through the stomata of leaves.
- energy required for, 83
- rate, compared to evaporation from soil, 83

transport by streams, 287–291, *288, 289, 290*
- deposition in reservoirs, 290
- processes, 288–289
- and turbulence, 288–289
- and velocity, 290, *290*

tree ring dating, 24

tributaries, 291–292, 293: streams that feed another stream.

tropical cyclone, 124–126: an intense cyclonic storm in which the winds exceed 75 miles per hour; called a hurricane if it forms on the east or west coasts of North America, and a typhoon if it forms in the western Pacific.
- damaging effects of, 126
- energy supply for, 125, 126

tropical savanna, 237–238, *238*
- annual precipitation in, 238
- global distribution of, *232,* 237
- vegetation characteristics of, 238
- and the water budget, 238, *239*

tropical seasonal forest, 237, *237*
- global distribution of, *232*
- temperature and precipitation in, 237, *237*
- vegetation characteristics of, 237

tropical thorn and scrub forests, 238–239, *239*
adaptations of vegetation in, 239
vegetation characteristics of, 238–239

tropical wet climate region, 165, *165*, 166, *168*

tropical wet and dry climate region, 165, *165*, 166, *168*

Tropic of Cancer, 49, *51*: the northernmost parallel of latitude (approximately 23½°N) at which the sun reaches zenith at some time (about June 21) during the year.

Tropic of Capricorn, 49, *51*: the southernmost parallel of latitude (approximately 23½°S) at which the sun reaches zenith at some time (about December 21) during the year.

tropics: the regions lying between (or near) the Tropics of Cancer and Capricorn.

troposphere, 57, *58*: the lower portion of the earth's atmosphere, extending to an altitude of approximately 10 kilometers (6 miles); the layer in which most clouds form.

trough, 402, *402*, *406*: the lowest part of a wave or wavelike structure, lying between successive crests.

true azimuth, 451: azimuth measured from true north.

true north, 451, *451*: the northerly direction, as measured along the local meridian.

true solar time, *52*, *53*: time measured according to the true position of the sun in the sky.

tsunami, 405–406: a powerful ocean wave, caused by a submarine disturbance, that reaches great heights at a coastline.
height of, 405
wavelength and speed of, 405

tuff, *334*: a rock made up of compacted volcanic ash.

tundra soils, *189*, *192*, 195–196, *195*: black muck soil, usually waterlogged and with a shallow profile because of poor drainage through the permafrost.
agricultural potential of, 195, 196
and climate, *189*
global distribution of, *200*
profile, *195*
and vegetation, *192*, 195

tundra vegetation, 245, 247, *247*
global distribution of, *232*
limited number of species of, 245
and moist soil, 247
thermoisopleth, 247, *247*

turbidity, 72–73: the opacity of the atmosphere to electromagnetic radiation because of the presence of dust, smoke particles, and so forth.

turbidity current, 413: a suspension of silt that can flow much like a fluid through sea water; a possible agent of undersea erosion.

turbulent flow, 287: irregular flow marked by eddies; the speed and direction of flow at a given point are constantly changing.
contrasted with laminar flow, 287

turbulent transfer, 64: the transfer of heat and moisture between the ground and the atmosphere by conduction and air movement.

Turkey, Goreme cone habitat, *418*

Turret Rock, Arches National Monument, Utah, sandstone formation, *438*

typhoon, 124: a term used to describe a tropical cyclonic storm that forms in the western Pacific.
in East Pakistan, 126

U

ultraviolet radiation, 15, 18, 56–57, *56*: the part of the electromagnetic spectrum that spans the range of wavelengths from 0.4 to 0.01 micron, with wavelengths shorter than violet light; ultraviolet light.
interaction with molecules in the atmosphere, 59

unconsolidated sediment, 315: sediment in which the individual particles are not bound strongly enough together to form rocklike material.

underfit stream, 304–305: a stream that is abnormally small in relation to its valley.
as a relict landform, 304

United Nations World Plan for food production, 222

units of energy, 34

uplift, 261, 273, 279: the elevation of part of the earth's surface, relative to other parts.

upper air waves, *122*

UPS, 449–450, *450*: Universal Polar Stereographic grid system, for regions poleward of latitude 80°.

upwelling, 130: the upward movement of deep ocean waters, usually cool and rich in nutrients, along a coast.
off Peru, 130

Urey, Harold, 5–6, 10, 12

USGS, 464: United States Geological Survey of the Department of the Interior.

USGS topographic maps, 464–467
color code for, 465
example of use, *466*, *467*
principal map series in, 464
symbol table, *465*
sing, 465–467
and UTM grid, 450

U-shaped valley, 388–389, *388*: a valley shaped by glaciers, having a gently rounded floor and rounded sides.

Utah, vegetation and climate, *438*, 439

UTM, 449–450, *450*, 454: Universal Transverse Mercator grid system.

V

valley glacier, *354*, 376: a river of ice flowing in a preexisting valley; an alpine glacier.

valley steps, 389: benchlike discontinuities in the longitudinal profile of a valley glacier's bed.

valley wind, 123: a wind that blows upward through a mountain valley, usually during the day.

vaporization, 35–36, *35*: the change in physical state from liquid to gaseous vapor.

vapor pressure of water: the pressure exerted by the water molecules in a given sample of moist air.

variation through the day, *84*

varves, 24: banded layers of sediment laid down in pairs each year; used in estimating dates.

vegetation

adaptation of, to climate, 230–231
and climate, 230–233, *233*
and control of erosion, 351
global distribution of, *232*
and landforms, 351–352

vegetation regions, 233, *233*, 235–242, *238*, *239*, *240*, 244–245, *246*, 247, *247*

See also equatorial rainforest, midlatitude deciduous forests, midlatitude grasslands, midlatitude shrublands, subarctic coniferous forests, subtropical deserts, subtropical semideserts, tropical savanna, tropical seasonal forest, tropical thorn and scrub forests, tundra vegetation

velocity of flow in streams, 285–287, *286*

and cross-sectional area, *286*, *287*

vent: an opening in the earth's crust from which lava or gases issue.

verbal statement of scale, 447, *447*: a statement in words of the scale of a map; for example, 1 inch to the mile.

vertical adjustment of streams, 297–299, *297*, *299*, 440, *441*

factors affecting, 297–298

Victoria, British Columbia, tides, *401*

Victoria Land, Antarctica

eroded sandstone, *319*
glacier, *354*

visible radiation, 55–56, *56*: the part of the electromagnetic spectrum corresponding to visible light, spanning the range of wavelengths from 0.7 to 0.4 micron; visible light.

interaction with molecules in the atmosphere, 59

volcanic activity, 11, *11*, 22, *22*, *33*, *264*, *266*

volcanic ash, 330, 336: fine particles of solidified lava that have been ejected by a volcano.

volcanic energy, source of, 330

volcanic eruptions, 330

energy for, 330
of Mount Pelée, 336
of Mount Vesuvius, 336
quiet, 330
violent, 330

volcanic neck: a volcanic plug.

volcanic plug, *333*, *334*: solidified material in the vent of a dead volcano, revealed as a standing feature when surrounding weak rock is eroded; a volcanic neck.

volcanism, 335–336: the extrusion of molten rock from the earth's interior onto the surface.

distribution of, 325, 328, *328*
in Hawaii, 328
in Iceland, 325, 328

volcanoes, 312, 325–328, *328*, 330–336, *330*, *333*, *334*: vents in the earth's crust where lava is released.

location of, 325, 328, *328*
material ejected from, 330, 336

W

Waimakariri River, New Zealand, braided stream, *295*

Wales, drumlin, *391*

Wallace, Alfred Russel, 14

warm front, 117, *118*, *119*: a boundary formed when warm air invades a region of cool air and rides upward over the cool air.

and occluded front, 119, *120*

Washington State

till plain, ground moraine, *391*
topography, 134

water

appearance on the earth's surface, 11
cost, 135
as energy converter, 34–36, *35*
properties of, 36–37
specific heat of, 36–37
use and management of, 134–135

water balance equation, 150: the relation that expresses the distribution of precipitation between runoff, soil moisture storage, and evapotranspiration.

water budget

as a climatic index, 237
Baton Rouge, Louisiana, 150–155, *151*, *152*, *162*
Caracas, Venezuela, *239*
Cloverdale, California, 167, *167*, 170
global, *137*
Huron, South Dakota, *245*
Kribi, Cameroon, *236*
Manhattan, Kansas, 167, *167*, 170
Middlesex County, New Jersey, *250*
San Francisco, California, *241*

water droplets

formation of, 87
settling velocity of, 88
size of, *87*, 88

water gap, *323*, *343*, 344, *344*: a pass in a resistant ridge of rock, through which a stream flows.

waterlogged soil, 191: a soil that is saturated with liquid water for most of the year.

water storage, 135, *137*, 138, *138*

water table, *142*, 143: the level of water in the saturated zone of an aquifer.

water vapor, vertical transport of, *64*, 86–87, *86*

Watts Branch, Maryland, longitudinal profile, *294*

wavelength, 402, *402*: the spacing between successive crests or successive troughs in a wave.

of light, 55

wave motion, 401–403, *402, 403*
breaking wave, 403–404
and friction, 402–403
generated by wind, 402, *405*
and geomorphic change, 404–405
orbital motion of droplet, 402, *402*
and transport of energy, 401–402

weather: the state of the atmosphere at a given time.

weathering, 37, 186–187, 274: the disintegration of rock in place.
chemical, 187
and chemical composition, 315
contrasted, for different rocks, 348
of granite, *368*
at joints and surfaces, 186, *186*
physical, 186–187
by plants, 186–187

weathering and climate, 349–351, *349*
in arid regions, *364, 366*
in hot, moist climates, *372*
in savannas, *370*

weathering front, *370:* the surface, below the ground, between weathered and unweathered rock.

weather map, *115*
symbols, *115*

Wegener, Alfred, 19–20

West Palm Beach, Florida, monthly radiant energy balance, *72*

whales, 224–225
and the food chain, 224
and significant extinction, 225
and sustainable yield, 225
and the whaling industry, 224

White Cliffs of Dover, England, formation of, 19

Wind River Mountains, Wyoming, tarns and cirques, *389*

winds
January and July global, *113*
on a uniform earth, 110, *110*
of upper atmosphere, 114, *114*

wind speed and wave height, *405*

wineglass valley, *364:* a feature of mountainous semiarid regions of uplift; the bowl of the wineglass is formed by an open mountain valley, the stem is a narrow stream gorge, and the base is an alluvial fan.

winter solstice, 49, *51:* the instant when the sun is directly overhead at the Tropic of Capricorn, latitude 23½°S; occurs about December 21.

Wyoming
cloud formation, 436
topography, 436
vegetation, 436

y

yield, 212: the amount of energy stored in the harvested portion of a crop.
of corn and rice over time, 221, *221*
and level of predation, 217, *217,* 225, 226
versus solar radiation, *209*

Yosemite National Park, California
Half Dome, *270*
hanging valley, *391*

youth, 275, *276:* a descriptive term applied to a landscape with a plain of low relief cut deeply by streams; in the Davis cycle, an early stage of landform evolution.

Yugoslavia
fluted limestone, *320*

z

Zambesi River, Zambia, drainage pattern, *323*

zenith: the point in the sky directly overhead.

Zion National Park, Utah, sandstone formation, *319*

zonal soils, 194: soils with clearly distinguishable horizons, which occur in definite regions of climate and vegetation.
global distribution of, *200*

zone of ablation, 382: the region on a glacier where the local annual loss of ice or snow exceeds the annual gain.

zone of accumulation, 382: the region on a glacier where the local annual gain of snow exceeds the annual loss.

Credits and Acknowledgments

We extend our thanks to all those persons and agencies who assisted us in preparing and assembling the illustrations for this book. In particular, we gratefully acknowledge the cooperation of Doug Armstrong, Laurie Curran, John Dawson, Warren Hamilton, Andy Lucas, the *National Atlas*, G. R. Roberts, John Shelton, and M. Ware/NASA.

Chapter 1

3—Tom Lewis; 4—Phil Stotts; 6—Doug Armstrong; 7—Bill Ralph; 8-9—Robert Kinyon/Millsap & Kinyon after R. Phinney; 9—(right) John Dawson; 10—John Dawson; 11—Butch Higgins; 13—(left) Warren Hamilton, (right) G. R. Roberts, Nelson, New Zealand; 15—From the experiments of Dr. H.B.D. Kettlewell, University of Oxford; 16-17—John Dawson; 20—*Atlas of the Earth*, p. 36, © Mitchell Beazley, Ltd., 1971; 21—John Dawson; 22—Andy Lucas after data by Ian McDougall, *Nature Physical Science*, vol. 231, 1971.

Chapter 2

29-30—Tom Lewis; 32-33—John Dawson; 35-36—Tom Lewis; 37—Doug Armstrong adapted from *Handbook of Geophysics and Space Environment*, edited by Shea L. Valley, Air Force Cambridge Research Laboratories, U.S. Air Force, 1965; 39-41—Tom Lewis; 42-43—John Dawson.

Graphic Essay I

44—Doug Armstrong; 45—(left) Doug Armstrong after A. Armstrong and John Thornes, The *Geographical* Magazine, © IPC Magazines, Ltd., 1972, by permission of New Science Publications, (right) Wide World Photos.

Chapter 3

46—The Museum of Modern Art; 50-51—John Dawson; 52-54—Doug Armstrong after *Smithsonian Meteorological Tables*, edited by Robert J. List, 6th ed., 1971, by permission of the Smithsonian Institution; 55—Doug Armstrong after N. Robinson, *Solar Radiation*, © 1966, Elsevier Publishing Co.; 56—Doug Armstrong; 57—Doug Armstrong after G.M.B. Dobson, *Exploring the Atmosphere*, 1963, by permission of the Clarendon Press, Oxford; 58—Doug Armstrong after U.S. Air Force; 60-61—Andy Lucas and Laurie Curran after Löf, Duffie, and Smith, *World Distribution of Solar Radiation*, Report #21, University of Wisconsin, 1966; 63—Tom Lewis after J. London, 1957; 65—(top) Doug Armstrong after William D. Sellers, *Physical Climatology*, © 1956, University of Chicago Press, (bottom) Doug Armstrong after Herbert Riehl, *Introduction to the Atmosphere*, © 1972 by McGraw-Hill Book Company; 66—(top) Doug Armstrong after F. K. Hare, *The Restless Atmosphere*, 1966, Hutchinson Publishing Group, Ltd.; 66-67—Doug Armstrong, data from Dr. Robert A. Muller; 68—Doug Armstrong after Arnold Court, *Journal of Meteorology*, vol. 8, 1951, American Meteorological Society; 69—Andy Lucas and Laurie Curran after G. Trewartha, *Introduction to Climate*, 1968, McGraw-Hill Book Company, and *Goode's World Atlas*, © 1970, Rand McNally and Company; 70-71—Andy Lucas; 72—Doug Armstrong after William D. Sellers, *Physical Climatology*, © 1965, University of Chicago Press; 73—Howard Sochurek; 74—UPI Compix photos by Bob Flora.

Chapter 4

76—Photo by Derrick E. Witty; 79—(top) Brian Brake/Rapho-Guillumette, (bottom) The Bettmann Archive, Inc.; 80—Joe Garcia; 81—Doug Armstrong after *Smithsonian Meteorological Tables*, edited by Robert J. List, 6th ed., 1971, by permission of the Smithsonian Institution; 82—John Dawson; 83 and 84 (top)—Doug Armstrong after *Smithsonian Meteorological Tables*, edited by Robert J. List, 6th ed., 1971, by permission of the Smithsonian Institution; 84—(bottom) Doug Armstrong adapted from *Hydrology Handbook*, 1949, published by the American Society of Civil Engineers; 85—David Cavagnaro; 86—Doug Armstrong after R. Geiger, *The Climate Near the Ground*, © 1965, Harvard University Press; 87-89—John Dawson; 91—Doug Armstrong after Hermann Flohn, *Climate and Weather*, © 1969 by H. Flohn, used with permission of McGraw-Hill Book Company; 92—(top) David Cavagnaro, (bottom) Andy Lucas and Laurie Curran adapted from Arnold Court and Richard Gerston, *Geographical Review*, vol. 56, 1966, © American Geographical Society of New York; 93—John Dawson; 94—(top) Robert J. Kolenkow, (center) NOAA, (bottom) G. R. Roberts, Nelson, New Zealand; 95—(left) Grant Heilman Photography, (top right) William Burkhardt, (bottom right) NASA; 96-97—Doug Armstrong; 98—Andy Lucas and Laurie Curran; 99—Andy Lucas and Laurie Curran after *Goode's World Atlas*, © 1970, Rand McNally and Company; 100—(left) General Electric Company, (right) The Bettmann Archive, Inc.; 102—Andy Lucas after *Yearbook of Agriculture*, 1941.

Chapter 5

107—Calvin Woo; 108—Doug Armstrong; 109—(top) Doug Armstrong, (bottom) Calvin Woo; 110—(left) Calvin Woo, (right) NASA; 111—Calvin Woo after R. J. Chorley, and P. Haggett, editors, *Models in Geography*, Methuen and Company;

112—Andy Lucas and Laurie Curran after U.S. Weather Bureau, (bottom right) after H. Flohn, *Climate and Weather*, © 1969, McGraw-Hill Book Company; 113—Andy Lucas and Laurie Curran after W. Schwerdtfeger, "The Climate of the Antarctic," *World Survey of Climatology*, vol. 14, edited by S. Orvig, © 1970 by Elsevier Publishing Company; 114—Andy Lucas after H. Riehl, *Introduction to the Atmosphere*, 1972, McGraw-Hill Book Company; 115—Doug Armstrong after Dr. Robert A. Muller; 116—Calvin Woo; 117—Calvin Woo after Dieter H. Brunnschweiler, *Geographica Helvetica*, vol. 12, 1957; 118—Calvin Woo adapted from H. Flohn, *Climate and Weather*, © 1969, McGraw-Hill Book Company; 119—(left) NASA, (right) Doug Armstrong; 121—Calvin Woo; 122—Doug Armstrong; 123—Calvin Woo; 124—Andy Lucas after H. Flohn, *Climate and Weather*, © 1969, McGraw-Hill Book Company; 125—Tom O'Mary after *The Atlas of the Earth*, p. 30, © 1971, Mitchell-Beazley, Ltd.; 126—(top) Andy Lucas, (bottom) NOAA; 127—Doug Armstrong after P. Weyl, *Oceanography*, John Wiley & Sons, Inc., 1970; 129—NASA; 131—Andy Lucas after L. Don Leet and Sheldon Judson, *Physical Geology*, 3rd ed., © 1965, by permission of Prentice-Hall, Inc., Englewood Cliffs, New Jersey.

Chapter 6

136-137—John Dawson; 138—Tom Lewis after R. L. Nace, *Water, Earth, and Man*, edited by R. J. Chorley, 1969, Methuen & Co., Ltd., Publishers; 139— David Cavagnaro; 140—Doug Armstrong after *Yearbook of Agriculture*, 1955, and E. E. Foster, *Rainfall and Runoff*, © 1949, Macmillan Company; 142—John Dawson after Raphael Kazmann, *Modern Hydrology*, 2nd ed., © 1972, by R. G. Kazmann; by permission of Harper & Row, publishers; 144—John Dawson; 145—Doug Armstrong after R. C. Ward, *Principles of Hydrology*, © 1967, McGraw-Hill Book Company (UK) Limited, used with permission, and Bertram S. Barnes, *Transactions*, American Geophysical Union, vol. 20, pp. 721-725, 1939, copyright by American Geophysical Union; 146—(left) Steve Harrison and Louie Neiheisel; 146-147—California State Department of Water Resources; 148—Andy Lucas and Laurie Curran adapted from *Geographical Review*, vol. 38, 1948, © American Geographical Society of New York; 150—Tom Lewis; 151-153—Doug Armstrong; 155—Doug Armstrong after C. W. Thornthwaite and J. W. Mather, *Publications in Climatology*, vol. 8, 1955.

Chapter 7

162—Ron Wiseman; 164—Doug Armstrong; 167—Doug Armstrong adapted from Carter, Schmudde, and Sharpe, by permission from the Commission on College Geography, Technical Report #7, Association of American Geographers; 168-169—Andy Lucas and Laurie Curran after Vernon C. Finch and Glenn Trewartha, *Physical Elements of Geography*, 1949, McGraw-Hill Book Company; 171—Doug Armstrong; 174—(top) G. R. Roberts, Nelson, New Zealand, (bottom) Fred Bauman; 174-175—(top) Robert J. Kolenkow, (bottom) E. Puigdengolas/Carl Östman Agency; 175—(top right) G. R. Roberts, Nelson, New Zealand, (center) Clinton S. Bond/BBM Associates, (bottom) Earl Dibble/Photo Researchers, Inc.; 176—(top left) Doug Armstrong, 176-177—Andy Lucas and Laurie Curran after Köppen-Geiger-Pohl map (1953), Justes Perthes; and Köppen-Geiger in *Erdkunde*, volume 8; and Glenn T. Trewartha, *An Introduction to Climate*, fourth edition, New York; McGraw-Hill, 1968; 178—Doug Armstrong after D. Carter and J. Mather, *Publications in Climatology*, vol. 19, no. 4, 1966; 179—Andy Lucas and Laurie Curran after C. W. Thornthwaite and J. Mather, "The Water Balance," *Publications in Climatology*, vol. 8, 1955.

Chapter 8

180—Photograph by Ronald C. Carriker; 183—John Dawson; 184—Doug Armstrong after E. M. Bridges, *World Soils*, © 1970, Cambridge University Press; 185—(top) John Dawson, (bottom) Doug Armstrong after Lyon and Buckman, *The Nature and Properties of Soils*, 4th ed., © 1943 by Macmillan Publishing Company, Inc.; 186—(top) Brian Bunting, *Geography of Soil*, Hutchinson University Library, 1965, (bottom) Ward's Natural Science Establishment; 188—John Dawson after Arthur N. Strahler, *Physical Geography*, 3rd ed., © 1960 by John Wiley & Sons, Inc., by permission; 189—Doug Armstrong adapted from Blumenstock and Thornthwaite, "Climate and the World Pattern," *Yearbook for Agriculture*, 1941, USDA, Government Printing Office; 190—(top) Warren B. Hamilton, (bottom) Doug Armstrong after E. Bridges, *World Soils*, Cambridge University Press, 1970; 191—Warren B. Hamilton; 192-193—John Dawson, profile information adapted from S. R. Eyre, *Vegetation and Soils*, Edward Arnold Publishers; 195—(top) Andy Lucas after C. F. Marbut; 195 (bottom) to 199—Reproduced from Soil Science Society of America, C. F. Marbut Memorial Slide Collection; 200-201—John Odam and Andy Lucas after Glenn T. Trewartha, Arthur H. Robinson, and Edwin H. Hammond, *Elements of Geography*, © 1967, McGraw-Hill Book Company; 202—Reproduced from Soil Science Society of America, C. F. Marbut Memorial Slide Collection.

Chapter 9

207—Tom O'Mary; 208—Tom Lewis; 209—Doug Armstrong after Jen-Hu Chang, *Climate and Agriculture*, © 1968 by Aldine Publishing Company, reprinted by permission of the author and Aldine; 210—Doug Armstrong after D. M. Gates, *Advances in Ecological Research*, edited by J. Cragg, 1968, Academic Press, London; 212—John Dawson after E. P. Odum, *Fundamentals of Ecology*, 3rd ed., © 1971, W. B. Saunders Company; 213—Doug Armstrong after Jack Major, "Climatic Index to Plant Activity," *Ecology*, 1963, by permission of Duke University Press; 214—Andy Lucas and Laurie Curran after Jen-Hu Chang, reproduced by permission from the *Annals of the Association of American Geographers*, vol. 60, 1970; 216—John Dawson after Robert L. Smith, *Ecology and Field Biology*, 1966, Harper & Row, and R. F. Johnston, *Wilson Bulletin*, 1956, published by the Wilson Ornithological Society; 217—Doug Armstrong after L. B. Slobodkin, *Ecology*, vol. 40, 1959, by permission of Duke University Press; 218—Tom O'Mary after R. Whittaker, *Communities and Ecosystems*, 1970, Macmillan and Company; 219—Tom O'Mary after (top) H. Odum, *Ecological Monographs*, 1957, (bottom) E. P. Odum, *Fundamentals of Ecology*, 3rd ed., © 1971, W. B. Saunders Company; 220—Tom O'Mary after data from G. Woodwell and R. Whittaker, *American Zoologist*, vol. 8, 1968; 221—Doug Armstrong after U.S. Department of Agriculture; 222—Doug Armstrong; 223—Illustration by Tom Lewis, photo by Werner Kalber/PPS, adapted from *Hunger U.S.A.*, a Report by the Citizens' Board of Inquiry into Hunger and Malnutrition in the United States. © 1968. The New Community Press.

Chapter 10

231—John Dawson after E. P. Odum, *Fundamentals of Ecology*, 3rd ed., © 1971, W. B. Saunders Company; 232—John Odam and Andy Lucas after Vernon C. Finch and Glenn T. Trewartha, *Physical*

Elements of Geography, © 1949, McGraw-Hill Book Company; 233—(top) Doug Armstrong adapted from Blumenstock and Thornthwaite, "Climate and the World Pattern," *Yearbook for Agriculture*, 1941, Government Printing Office, (bottom) Doug Armstrong after L. R. Holdridge, *Science*, vol. 105, pp. 367-368, 1947; 234—Sergio Larrain/Magnum Photos; 235—Doug Armstrong after *Climatic Data for Representative Stations of the World* by H. Nelson, by permission of University of Nebraska Press, © 1968; 236—(top) Doug Armstrong after C. Troll, *Oriental Geographer*, vol. 2, 1958, by permission, (bottom) Doug Armstrong after D. Carter and J. Mather, *Publications in Climatology*, vol. 19, 1966; 237—(top) Doug Armstrong after *Climatic Data for Representative Stations of the World* by H. Nelson, by permission of University of Nebraska Press, © 1968, (bottom) photo by James S. Packer; 238—John Lewis Stage/Photo Researchers; 239—(top) Doug Armstrong after D. Carter and J. Mather, *Publications in Climatology*, vol. 19, 1966, (bottom) G. R. Roberts, Nelson, New Zealand; 240—(top) G. Glase/Östman Agency, (bottom) Brian Hawkes/Östman Agency; 241—(top) Doug Armstrong after H. Nelson, *Climatic Data for Representative Stations of the World*, © 1968 by University of Nebraska Press, (bottom) Doug Armstrong after D. Carter and J. Mather, *Publications in Climatology*, vol. 19, 1966; 242—Steve Harrison and Louie Neiheisel after M. Hendl, *Einführung in Die Physikalische Klimatologie*, Band II, 1963; 243—(top left) I. Holmäsen/Östman Agency, (top right) Doug Armstrong after *Climatic Data for Representative Stations of the World* by H. Nelson, by permission of University of Nebraska Press, © 1968, (bottom) Doug Armstrong after C. Troll, *World Maps of Climatology*, © 1965, Springer-Verlag Publishing; 244—G. R. Roberts, Nelson, New Zealand; 245—Doug Armstrong after D. Carter and J. Mather, *Publications in Climatology*, vol. 19, 1966; 246—Eric Lessing/Magnum Photos; 247—(top) Brian Hawkes/Östman Agency, (bottom) Doug Armstrong after C. Troll, *World Maps of Climatology*, © 1965, Springer-Verlag Publishing; 249—Photo by James S. Packer; 250—Doug Armstrong; 251—Doug Armstrong after H. H. Lamb, *The Changing Climate*, 1966, Methuen and Company, Ltd.

Graphic Essay II

254-255—U.S.D.A. Photograph; 256—(top left) G. R. Roberts, Nelson, New Zealand, (bottom left) J. Alex Langley; 256-257—(top) Henri Cartier-Bresson/Magnum Photos, (bottom) Reprinted from *Arcology, The City in The Image of Man* by Paoli Soleri, by permission of the M.I.T. Press, Cambridge, Massachusetts, © 1969; 257—(top right) Courtesy of VISTA, (bottom right) E. Puigdengolas/Östman Agency.

Chapter 11

260—Grant Heilman Photography; 264—NASA; 265-266—Calvin Woo; 267—Picturepoint Ltd.; 269—Aerofilms Ltd., London; 270—(top and bottom left) G. R. Roberts, Nelson, New Zealand, (center left) Florence Fujimoto, (bottom right) David Miller; 271—(top) David Miller, (center left) G. R. Roberts, Nelson, New Zealand, (bottom) Florence Fujimoto; 272—(top) Grant Heilman Photography, (center) Warren Hamilton/U.S.G.S., (bottom) Theodore M. Oberlander; 273—(left) Butch Higgins, (right) G. R. Roberts, Nelson, New Zealand; 274—Theodore M. Oberlander; 276—John Dawson after Longwell, Knopf, and Flint, *Outlines of Physical Geology*, 2nd ed., © 1941, John Wiley & Sons; 277—John Dawson after C. D. Holmes, *American Journal of Science*, vol. 253, 1955; 278—John S. Shelton; 280—Doug Armstrong.

Chapter 12

285—U.S. Navy, Office of Information; 286—(top and bottom) Doug Armstrong, (center) John Dawson; 289—G. R. Roberts/Östman Agency; 290—Doug Armstrong adapted from Marie Morisawa, *Streams: Their Dynamics and Morphology*, edited by P. Hurley, © 1968, McGraw-Hill Book Company; 291-294—Doug Armstrong; 295—(top) John S. Shelton, (bottom) G. R. Roberts, Nelson, New Zealand; 296—Doug Armstrong; 297—Robert J. Kolenkow; 299—John Dawson after Marie Morisawa, *Streams: Their Dynamics and Morphology*, edited by P. Hurley, © 1968, McGraw-Hill Book Company; 300—NASA; 301—Doug Armstrong after G. H. Dury from *Water, Earth and Man*, edited by R. J. Chorley, © 1969, Methuen and Company; 303—(top) John Dawson after R. B. Bunnett, *Physical Geography in Diagrams*, 1968, © Longmans Group, Ltd., (bottom) John S. Shelton; 304—Tom O'Mary; 305—(left) Doug Armstrong after S. M. Gagliano *et. al.*, "Hydrologic and Geologic Studies of Coastal Louisiana," Department of the Army, 1970, (right) NASA; 306—John S. Shelton; 308—Andrea Lindberg after Dr. Robert A. Muller; 309—Dr. Robert A. Muller.

Chapter 13

313—John S. Shelton; 314—(left) John Dawson after C. A. Cotton, *Geomorphology*, 7th ed., © 1960, Whitcombe & Tombs, Ltd., (right) Theodore M. Oberlander; 316—(top left) M. E. Bickford, University of Kansas, (top right) G. K. Gilbert/U.S.G.S., (bottom) U.S. Forest Service; 317—(top left) M. E. Bickford, University of Kansas, (top right) Warren Hamilton/ U.S.G.S., (bottom) U.S. Forest Service; 319—(top left) M.E. Bickford, University of Kansas, (top right) Warren B. Hamilton, (bottom) J. Whitfield Craven; 320—(top left) M. E. Bickford, University of Kansas, (top right) Warren B. Hamilton, (bottom) John S. Shelton; 321—(top left) M. E. Bickford, University of Kansas, (top right) Warren Hamilton/U.S.G.S., (bottom) John S. Shelton; 322—(top) John Dawson, (bottom) Theodore M. Oberlander; 323—Doug Armstrong; 324—John Dawson after E. Ackerman, "Butserstein Zeugen vorzeitlicher Grundwasserschwank-ungen," *Zeitschrift für Geomorphologie*, vol. 6, 1962; 326-327—John Dawson; 328—Andy Lucas after F. M. Bullard, *Volcanoes*, 1962, University of Texas Press; 329—(left) John Dawson, (right) David Miller; 331—(top) G. R. Roberts, Nelson, New Zealand, (left) J. W. Hawkins, (right) Butch Higgins; 332—(top) Charles A. Wood, (bottom) Warren Hamilton/U.S.G.S.; 333—(top) Ward's Natural Science Establishment, (center) John S. Shelton, (bottom) Picturepoint, Ltd.; 334—John Dawson after Fred M. Bullard, *Volcanoes, in History, in Theory, in Eruption*, University of Texas Press, 1962; 335—John Dawson (top) after C. D. Ollier, *Weathering*, Oliver & Boyd, Edinburgh, 1969, (bottom) after C. O. Ollier, *Volcanoes*, An Introduction to Systematic Geomorphology, vol. 6 (A.N.U. Press, Canberra, 1969); 337—Andrea Lindberg; 338-339—John Dawson; 340-341—John S. Shelton; 342—Warren B. Hamilton; 343—(top) John Dawson after Thornbury, *Principles of Geomorphology*, 1966, John Wiley & Sons, (bottom) SLAR Imagery, Grumman Ecosystems Corporation; 344—John Dawson after Armin K. Lobeck, *Things Maps Don't Tell Us*, © 1956, Macmillan Publishing Company.

Chapter 14

349-350—Doug Armstrong after Louis Peltier, reproduced by permission from the *Annals* of the Association of American Geographers, vol. 40, 1950; 351—Doug Armstrong; 353—Canadian National Railways; 354—(top) John Dawson, (bottom) David Cavagnaro; 355—Warren Hamilton/U.S.G.S.; 356—John S. Shelton; 357—(top) P. S.

Smith/U.S.G.S., (bottom) Tom O'Mary; 358–Grant Heilman Photography; 359–Doug Armstrong after Louis Peltier, reproduced by permission from the *Annals* of the Association of American Geographers, vol. 40, 1950; 361–John Dawson; 362-373–drawings by John Dawson after Theodore M. Oberlander, 1974; 363–(top right) R. Belanger/Bedford Institute of Oceanography, Environment Canada; 365–(top right) John S. Shelton; 367–(top right) John S. Shelton; 369–(top right) John S. Shelton; 371–(top right) Ministry of Information, Salisbury, Rhodesia; 372–(bottom) G. R. Roberts, Nelson, New Zealand.

Chapter 15

377–(top left) John Dawson, (top right) Anitra Kolenkow, (bottom) Austin Post/U.S.G.S.; 379–Andy Lucas and Laurie Curran after J. L. Davies, *Landforms of Cold Climates,* An Introduction to Systematic Geomorphology, vol. 3 (A.N.U. Press, Canberra, 1969); 381–John Dawson; 383–Doug Armstrong; 384–(top) John Dawson after Robert P. Sharp, *Glaciers*, © 1960, University of Oregon Press, (bottom) Doug Armstrong after M. F. Meier, 1960; 386–G. K. Gilbert/U.S.G.S.; 387–John Dawson after J. L. Davies, *Landforms of Cold Climates*, An Introduction to Systematic Geomorphology, vol. 3 (A.N.U. Press, Canberra, 1969); 388–(top) William Burkhardt, (bottom) Alvin Lynch; 389–Austin Post/U.S.G.S.; 390–(top) Theodore M. Oberlander, (bottom left) David Miller, (bottom right) William Burkhardt; 391–(left) John S. Shelton, (right) G. R. Roberts, Nelson, New Zealand.

Chapter 16

397–John Dawson after Bruce C. Heezen and Marie Tharp, Lamont Geological Observatory; 398–Doug Armstrong after E.C.F. Bird, *Coasts,* An Introduction to Systematic Geomorphology, vol. 4, by permission of the M.I.T. Press, Cambridge, Mass.; 399–Doug Armstrong after NOAA; 400–(top left) Tom O'Mary, (top right) Jon Brenneis, (bottom) Doug Armstrong after H. D. Warbug, *Tides and Tidal Streams*, with the permission of Cambridge University Press; 401–Eric Kay/Östman Agency; 402–Doug Armstrong; 403–(left) John Dawson after M. J. Selby, *The Surface of the Earth*, © 1967, Cassell & Co., London, (right) Department of the Navy; 404–C. M. Dixon/Östman Agency; 405–(top) Eric Kay/Östman Agency, (bottom) Doug Armstrong after Sverdrup and Munk; 406–John Dawson after F. P. Shepard, *The Earth Beneath the Sea*, © 1967, Johns Hopkins Press; 407–John Dawson; 408–Tom O'Mary; 409–Theodore M. Oberlander; 410–(top left) John S. Shelton, (bottom left) Eric Kay/Östman Agency, (right) John Dawson after R. B. Bunnett, *Physical Geography in Diagrams*, © 1968, Longmans Group, London; 411–(top and center) John S. Shelton, (bottom) John Dawson after R. B. Bunnett, *Physical Geography in Diagrams*, © 1968, Longmans Group, London; 412–(right) Tom Lewis adapted from Bruce C. Heezen and Charles D. Hollister, *The Face of the Deep*, © 1971 by Oxford University Press, Inc.; 414-415–John Dawson.

Graphic Essay III

416-417–Courtesy of the American Museum of Natural History; 418–(top) Picturepoint, Ltd., (bottom) Commonwealth of Puerto Rico; 419–(top) Dr. Naomi Miller Coval, (bottom) George Holton/Photo Researchers.

Epilogue

420–Photograph by F. J. Thomas; 423–Steve Harrison and Louie Neiheisel from *The National Atlas of the United States of America*; 424-425–John S. Shelton; 426-428–Steve Harrison and Louie Neiheisel from *The National Atlas of the United States of America*; 429–Photo by Fritz Goro; 430-434–Robert J. Kolenkow; 435-437–Steve Harrison and Louis Neiheisel from *The National Atlas of the United States of America*; 438-440–Robert J. Kolenkow; 442–John S. Shelton; 443–Steve Harrison and Louie Neiheisel from *The National Atlas of the United States of America.*

Appendix I

445–Andrea Lindberg.

Appendix II

446–(top) The British Museum, London, (bottom) *Atlas of the Universe*, p. 13, © Mitchell-Beazley, Ltd., 1971; 447–Andrea Lindberg; 448–U.S. Geological Survey; 449–Andrea Lindberg; 450–(top left and bottom) Andrea Lindberg after Erwin Raisz, *Principles of Cartography*, © 1962, McGraw-Hill Book Company, (top center) Andrea Lindberg after *Army Field Manual*, FM 21-26, October, 1960; 451-453–Andrea Lindberg; 455-459–Andrea Lindberg after Robinson and Sale, *Elements of Cartography*, 3rd ed. © 1953, 1960, 1969, John Wiley, reprinted by permission of John Wiley & Sons, Inc.; 460–(left) Andrea Lindberg adapted from Whitwell quadrangle, Tennessee, U.S. Geological Survey, (right) Gerald Ratto Photography; 461–(left) Reprinted from Arthur Robinson and Randall Sale, *Elements of Cartography*, © 1953, 1960, 1969 by John Wiley, by permission of John Wiley & Sons, Inc., (right) U.S. Geological Survey; 462–John Dawson; 463–*The National Atlas of the United States of America*; 466–Steve Harrison/Louie Neiheisel, U.S. Geological Survey; 467–U.S. Geological Survey; 469–Tom O'Mary; 470-471–Calvin Woo; 472–(top) Photograph courtesy of the Laboratory for Applications of Remote Sensing (LARS), (bottom left) U.S. Navy, (bottom right) NASA; 473–(top left) NASA, (top right) R.J.P. Lyon, Stanford University, (bottom) Courtesy of Forestry Remote Sensing Laboratory, University of California, Berkeley.

Appendix III

474-476–Andrea Lindberg.

Contributors

480-481–John Dawson.

Physical Geography Today:
A Portrait of a Planet

Book Team

John H. Painter, Jr. • *Publisher*
Bobbye J. Hammond • *Publishing Coordinator*
JoAn Lynch Rice • *Editor*
Susan Harter • *Associate Editor*
John Odam • *Designer*
Linda Higgins • *Associate Designer*
Steve Harrison, Ray Piper • *Design Assistants*
Shelagh Dalton • *Graphics Research*
Donald Umnus • *Production Manager*
Phyllis Barton • *Production Supervisor*
Nancy Hutchison Sjöberg • *Rights and Permissions Supervisor*
Bill Bryden • *Science Marketing Manager*

CRM Books

Richard Holme • *President and Publisher*
Russ Calkins • *Marketing Manager*
Roger G. Emblen • *Publishing Director*
Arlyne Lazerson • *Editorial Director*
William G. Mastous • *Director of Finance and Administration*
Trygve E. Myhren • *Vice-President, Marketing*
John Ochse • *Sales Manager*
Henry Ratz • *Director of Production*
Tom Suzuki • *Director of Design*